Fundamental Developments of Computer-Aided Geometric Modeling

Fundamental Developments of Computer-Aided Geometric Modeling

Edited by
Les A. Piegl
University of South Florida

ACADEMIC PRESS

Harcourt Brace & Company, Publishers
London San Diego New York
Boston Sydney Tokyo Toronto

ACADEMIC PRESS LIMITED
24–28 Oval Road
London NW1 7DX

United States Edition published by
ACADEMIC PRESS INC.
San Diego, CA 92101

This book is printed on acid-free paper.

A CIP record for this book is available from the British Library
ISBN 0-12-554765-X

Typeset by P&R Typesetters Limited,
Salisbury, Wiltshire, UK
Printed and bound in Great Britain by the Bath Press, Avon

Contributors

Andrew P. Armit May Cottage, 21 Greenside, Waterbeach, Cambridge CB5 9HW, UK

Robert E. Barnhill Department of Computer Science and Engineering, Arizona State University, Tempe, AZ 85287-5406, USA

Pierre E. Bézier 12 Avenue Gourgaud, 75017 Paris, France

Carl de Boor Mathematics Research Center, University of Wisconsin-Madison, 610 Walnut Street, Madison, WI 53705, USA

Ian C. Braid Three-Space Limited, 70 Castle Street, Cambridge CB3 0AJ, UK

Paul de Casteljau 4 Avenue du Commerce, 7800 Versailles, France

Maurice G. Cox DITC, Building 93, National Physical Laboratory, Teddington, Middlesex TW11 0LW, UK

Charles M. Eastman Graduate School of Architecture and Urban Planning, University of California, Los Angeles, 405 Hilgard Avenue, Los Angeles, CA 90024-1467, USA

James C. Ferguson BBN Systems and Technologies Corporation, Advanced Simulation Division, 14100 South East 36th Street, Bellevue, WA 98006, USA

William J. Gordon Department of Mathematics and Computer Science, Drexel University, Philadelphia, PA 19104, USA

Robert W. Mann Department of Mechanical Engineering, Massachusetts Institute of Technology, Room 3-144, Cambridge, MA 02139, USA

Michael J. Pratt Design Research Center, Rensselaer Polytechnic Institute, 110 8th Street, Troy, NY 12180, USA

Aristides A. G. Requicha Computer Science Department, University of Southern California, Los Angeles, CA 90089-0782, USA

Richard F. Riesenfeld Department of Computer Science, University of Utah, Salt Lake City, UT 84112, USA

David F. Rogers Aerospace Engineering Department, Department of the Navy, United States Naval Academy, Annapolis, MD 21402-5042, USA

Douglas T. Ross SofTech, Inc., 460 Totten Pond Road, Waltham, MA 02154, USA

Malcolm A. Sabin FEGS Limited, 5 Coles Lane, Oakington, Cambridge CB4 5BA, UK

Herbert B. Voelcker Sibley School of Mechanical & Aerospace Engineering, Cornell University, Grumman & Upson Halls, Ithaca, NY 14853-7501, USA

. . ., we show modern methods for geometry,
so constructed and organized that geometric manipulations
can be performed in a way natural to the computer,
and can yield results that are natural to man.

Steven Anson Coons

Preface

Les Piegl

Welcome to *Fundamental Developments of Computer-Aided Geometric Modeling:* a collection of papers written by pioneers of Computer-Aided Design. This book is for every computer professional who is interested in the scientific basics as well as in the historical/evolutionary aspects of computer geometry.

This book is neither a text book nor a history book. It combines science and history. The reader will find enough material to understand computational geometric methods as used in CAD/CAM, and at the same time the book gives an exposure to how things were discovered from the late 1950s. Computer-Aided Geometric Modeling has helped shape the future of computer supported design. The contributors of this volume had vision, knowledge and dedication to promote progress. Please accept this book as a reflection of their dedication to the betterment of the contemporary society.

HOW THIS BOOK CAME TO BE

Five years ago, Jenny Hayes, former Editor of the journal *Computer-Aided Design*, was soliciting contributions from Advisory Editors to the upcoming 20th anniversary issue. *Computer-Aided Design* was founded in 1968, and it was time to celebrate its 20 years of success. As a fresh member of the board, I was pleased to offer a paper, and to suggest that it would be interesting to pay tribute to pioneering developments in CAD. The field was so young that almost all the pioneers were still alive, and it sounded like a good idea to edit a special issue with contributions from the pioneers of Computer-Aided Design. Jenny liked the idea, however, she argued that a special issue would be too small a forum, and suggested putting a book together instead. After a couple of quick letter exchanges, I was given an offer to edit an anniversary volume on the pioneering works in CAD. Being completely green in book publishing, and having no idea how difficult it was to edit such a volume, I accepted the job. What followed after that was a series of refinements of the initial project. It was clear that one book was insufficient to cover the entire spectrum of CAD, so we restricted our attention to the geometric aspects of Computer-Aided Design, commonly referred to as *Computer-Aided Geometric Modeling* (other names used are Geometric Modeling, Computer-Aided Geometric Design (CAGD), Computational Geometry, etc. – you pick your favorite name). A series of discussions with well-known people followed my initial contact, and lots of modifications were made on the initial draft. By mid-1988, the scope of the project was more or less finalized, and the first chapter was completed (although in French). It took us four more years to bring the book to completion.

Looking back at the five years I spent on this book, I have all kinds of memories ranging from sweet ones to not so sweet ones. I have learned a lot about people, about publishers, and about managing a project in which different people are involved from different countries on different continents. I hand this book over to the reader with great sentiments, and in the hope that he/she will enjoy reading it at least as much as we enjoyed doing it. This book is unique in the sense that everything was written by the inventors themselves weaving their personalities into the subject. It also provides descriptions on all the methods used widely in Computer-Aided Design and Manufacturing.

WHO AND WHAT IS IN THIS BOOK?

There are numerous techniques used to represent and process geometric information using a computer. Some of them are useful, and many of them are considered as mathematical or computational plums only. Most of them have become parts of standard university curriculums in CAD and/or in Graphics, and have initiated substantial research activities in many fields including mathematics, computer science, engineering, art, business, geography; just to name a few.

We start the book with two French pioneers, Paul de Casteljau and Pierre Bézier. Working independently for two French car companies, Citroën and Renault, they discovered an approximation method for free-form curve and surface design. Although company policies and language barriers did not allow de Casteljau to publicize his results, his contribution is equally as important as those who published extensively. De Casteljau writes about polar forms, whereas Bézier gives an overview of the first years of CAD/CAM, and describes the mathematical basis of Renault's UNISURF system.

B-splines were investigated thoroughly by mathematicians way before the advent of computers in engineering design. Two remarkable mathematicians, Carl de Boor and Maurice Cox, have developed algorithms and mathematical tools for their efficient evaluation and processing. The basics of B-splines are presented by de Boor, and algorithms for spline curves and surfaces are given by Cox.

The application of B-splines to Computer-Aided Design was investigated by Richard Riesenfeld. Riesenfeld gives an overview of B-spline curves and surfaces as used in design. He also shares his experience of building a modeler called *Alpha_1* as a research testbed.

Interpolation is another way to represent free-form entities. The chapters by James Ferguson, William Gordon and Robert Barnhill discuss curve and surface techniques based on blending function methods. Ferguson describes the use of Hermite interpolation for curve, surface and hypersurface definition. Gordon's chapter is the first open account of the so-called 'Gordon' surface used at General Motors. Coons' patches and convex combinations are covered by Barnhill, who researched the field over the past 20 years.

The above techniques are used to represent one single entity, i.e. a curve, a surface or a volume. To define more complicated (realistic) objects, the collection of many simple geometric entities is required. The new entity is termed as *solid,* and is dealt with by Ian Braid, Charles Eastman, Herbert Voelcker and Aristides Requicha. There are a number of ways to define a solid, and there are different aspects of solid modeling. Braid overviews boundary modeling where solids are described by their boundary surfaces. Eastman's chapter describes architectural design and research testbeds developed at Carnegie Mellon University. Voelcker and Requicha are well known for their research in Constructive Solid Geometry (CSG), and for the solid modelers developed at the University of Rochester. The history of their research accomplishments at the University of Rochester is described in their chapter.

Malcolm Sabin has made contributions to many branches of geometric design. In his chapter he gives a tour of his 20 years of investigations into many aspects of computer-aided geometry.

Computer-Aided Design, at least its geometric modeling aspects, relies heavily on geometric methods. All kinds of geometric tools, including those of classical analytic geometry, projective geometry, differential geometry and descriptive geometry, are used to define and process shape information. Michael Pratt surveys geometric tools, including those of classical geometry as well as modern computer geometry, used in the CAD, as well as in the CAM process.

Inventing a design tool is not the end of the line in the engineering design process. A computer system has to be built in such a way that it is user friendly enough to be useful for non-computer hackers. No matter how sophisticated the mathematical scheme is, it is completely useless if nobody can make it part of a user-understandable CAD system. Andrew Armit takes us on an excursion into the world of several systems he has built over the last 25 years. He describes two of his better known systems, Multipatch and Multiobject, and follows by describing more recent and, of course, more sophisticated systems.

Graphical communication between the computer and the user, and post-processing of the results for numerical control milling, are two very important ingredients in the complete design process. The information should be brought into the database, and the results should be interpreted in such a way that they are natural to the computer as well as to the user. After the design has been completed and accepted by the designer, the non-existent object has to be turned into a real one by, say, carving it out from some kind of material. The chapter by David Rogers is about interactive computer graphics and numerical control as used at the US Naval Academy's CAD and Graphics Laboratory.

The last two chapters are by two MIT

pioneers, Robert Mann and Douglas Ross. Mann, a close friend and colleague of Steven Coons, describes research activities in Computer-Aided Design at MIT from the late 1950s through to the mid 1960s. Ross concludes the book with a philosophy of Computer-Aided Design as seen from a perspective resulting from 30 years' experience in the field.

REGRETS

I regret that I was not able to convince Bertram Herzog and Robin Forrest to contribute to this volume. I feel their names should be mentioned, and they should be credited for the significant work they have accomplished.

APOLOGIES

I offer my apology to those whom I have left out, or whose contributions just could not fit into this volume. I am sure that the reviewer or the reader will find someone in the computer geometry area whose contribution he/she feels is important to warrant inclusion in this book. I think to give credit to everybody, one really needs to edit an encyclopedia. Well, I was not prepared to edit an encyclopedia.

ACKNOWLEDGEMENTS

It is my great pleasure to acknowledge the help of many people. First I thank Jenny Hayes for the idea of editing this book. If it was not for Jenny, this book would not have been written. Her encouragement at the beginning of the project kept me going even if I had only a slim chance to complete what I had in mind. Jenny, wherever you are, may God bless your heart!

I thank my contributors for their support and co-operation throughout the life of this project. It was a great pleasure working with top notch people who were not only professionals in their scientific fields at the highest degree, but who treated me as a professional colleague despite the fact that I could not even come close to them. Thank you all for understanding the importance of this project, for the excellent chapters, and for the personal encouragement and support that I needed at times.

Special thanks are due to Jack Gregory and Michel Menard for translating de Casteljau's chapter from French to English. Jeff Stevens provided illustrative material for Gordon's chapter, for which I am particularly grateful.

Les Piegl
June, 1992
Tampa, Florida

Contents

1 Polar Forms for Curve and Surface Modeling as used at Citroën

Paul de Casteljau

Originating in 1958 at Citroën, the methodology of mathematical definition of shapes (curves and surfaces) has been taught at the drawing school of Citroën since 1960 [1]. The shapes are defined with the aid of pilot points called poles, which give a good idea of the shape of the curve. From an algebraic point of view, the poles are equivalent to segments or patches of parametric polynomial forms, and possess all of their properties. They allow great flexibility by a progressive deformation of the shape under the influence of the corresponding pole. At the beginning, the limited performance of the computing power available at that time forced us to use simple poles, without considering the continuity between segments.

The word 'pole' comes from the triple repetition of 'pol' in the expression 'Inter*pol*ation of *pol*ynomials with *pol*ar forms'. The construction of shapes from the poles is completely independent of the computation of the poles from sampled data. The theory is pretty much algebraic, and is related to integral calculus, quadratic minimization and orthogonal function theory. Those notions can create more difficulties than they solve, and therefore we do not enter into their details here.

I started within an exceptional team, spirited by M. de la Boixiere, who, in the pure tradition of the Pilgrims (in the sense of caring only for efficiency) was investigating industrialization by numerical control. To the great amazement of the engineers, he constructed, *ex nihilo*, high performance machines previously unavailable at that time. For example, he produced a pneumatic paper tape reader, a stepper motor, a variable speed tape recorder (its impulses were produced by a car distributor!), an electrical power cabinet, a five axis milling machine, etc. Nothing could discourage him. Only one detail remained: the calculation of the pieces to be machined. The objective was a forged or cast shape, which was believed to be mathematically well defined.

Later, my interest turned to car bodies; the designers were astonished and scandalized. Was it some kind of joke? It was considered nonsense to represent a car body mathematically. It was enough to please the eye, the word accuracy had no meaning, and wind tunnel testing was done after the car was already in production. Between orthogonal cross-sections of a piece, no one would admit a 0.8 mm discrepancy. Worse, the difference between the right and left fender of a car reached the centimeter mark! Far from being discouraged, I persisted in my desire to make a test. From then on, there was no doubt, I was considered insane . . .

The first surface computed with an electro-mechanical machine was the size of a postage stamp. It was defined with 350 points of the cutter location offset surface, and the noise of the printer annoyed the people nearby! When M. de la Boixière saw the mound of paper tapes, he became doubtful and claimed: "That will generate a million useless cubics!" Prudently, Citroën designated a draftsman, working currently with the author, who digitized the tool path on a 20:1 drawing so as to achieve acceptable results.

In the end, it was the mathematician who comforted the draftsman. The idea of numerically representing the car body was taken seriously. Citroën bought a Burrough's E101 computer (128 program steps, a memory of about 220, about seven multiplications a second, and it required 5 kW of power!). This allowed more ambitious computations, such as the hood, the dash board, and also 'the flying saucer', a dream of an extinct aviation firm (see Figure 1.1).

Editor's note: the terminology used by the author is different from that used in several well-known textbooks such as I. D. Faux and M. J. Pratt, *Computational Geometry for Design and Manufacture* (Ellis Horwood, Chichester, 1989). The 'parameter sequence' is the *knot vector*, 'Poles' or 'Generalized Poles' are the *control points*, sometimes called *de Boor points*, and 'Simple Poles' are the *Bézier control points*.

Fundamental Developments of Computer-Aided Geometric Modeling
ISBN 012-554765-X

Figure 1.1 A premiere model from 1962: the 'Flying Saucer'. Notice the spiral machining marks and the rectangular and triangular patches.

Being convinced, the management decided to use this technique inside the company in 1963, not without some resentment and jealousy, and this allowed the full drawing of a car, the Citroën GS, for the first time in history. This generated a system at Citroën to compute the shape (SPAC-CAR-G), and another to produce the machining information (SPAC-CAR-U).

DEFINITION OF POLAR FORMS

Parametric curves

Assume that we are given the points $\mathbf{P}_i$, called poles, and a set of polynomial functions $\phi_i(t)$, $i = 0, \ldots, n$. The form

$$\mathbf{P} = \sum_{i=0}^{n} \phi_i(t)\mathbf{P}_i \qquad (1)$$

defines a parametric curve. To make this definition useful practically, we require the polynomials to sum to 1 ($\sum_i \phi_i(t) = 1$), i.e. for a fixed t, $\phi_i(t)$ are the barycentric coordinates of $\mathbf{P}$ with respect to the poles $\mathbf{P}_i$.

The vector $\mathbf{OP}$ can be obtained as the sum of vectors proportional to the vectors $\mathbf{OP}_i$, for any given origin $\mathbf{O}$, and therefore $\mathbf{O}$ is omitted in later discussion. Expression (1) is valid for each coordinate, or projection on any line or axis, and can be extended to several parameters to generate a surface patch.

Example

$$\phi_i(t, u) = C_{n,i}\, t^i u^{n-i} = \binom{n}{i} t^i u^{n-i}$$

$$= (t + u)^n, \quad \text{where } t + u = 1.$$

Two types of surfaces are of practical importance:

- rectangular with $T + U = 1$. Each polynomial is obtained from $(t + u)^n (T + U)^N$.
- triangular with $t + u + v = 1$. The polynomials are of the form $t^\alpha u^\beta v^\gamma$ (with $\alpha + \beta + \gamma = n$) generated by $(t + u + v)^n$.

Without loss of generality, we may assume that t, u, v, T and U vary between 0 and 1.

Parameter sequences

We consider the non-decreasing series of parameters

$$t_0 \leqslant t_1 \leqslant \ldots \leqslant t_i \leqslant t_{i+1} \leqslant \ldots \leqslant t_f \qquad (2)$$

with the assumption that the number of equal parameters is less than or equal to the degree of the curve. The insertion of one (or more) new t value(s) into the interval $[t_i, t_{i+1}]$ $(t_i \neq t_{i+1})$ will generate one (or more) new sequence(s). This operation is called *parameter insertion*. In the discussions that follow, we assume that t denotes either t_i, a constant t_α, or the variable t.

Symmetric function of the parameters

Any function $f(t_1, t_2, \ldots, t_n)$ which is invariant when two variables t_i and t_j are exchanged is called a *symmetric function*. Newton demonstrated that we can express this function with the *elementary symmetric functions* (see [6] for details):

$$f(t_1, t_2, \ldots, t_n) = \sum_{i=0}^{n} g_i(x_1, x_2, \ldots, x_n)$$

where

$$\begin{aligned} x_1 &= t_1 + t_2 + \ldots + t_n \\ x_2 &= t_1 t_2 + t_1 t_3 + \ldots + t_{n-1} t_n \\ &\vdots \\ x_n &= t_1 t_2 \ldots t_n \end{aligned} \qquad (3)$$

These functions can be obtained by expressing

$$\begin{aligned}(1 + t_1)(1 + t_2) \ldots (1 + t_n) \\ = S_{n,0} + S_{n,1} + S_{n,2} + \ldots + S_{n,n},\ S_{n,0} = 1\end{aligned} \qquad (4)$$

One can easily verify, by introducing the new factor $(1 + t)$, the following recurrence formula:

$$S_{n+1,p} = S_{n,p} + tS_{n,p-1} \qquad (5)$$

Example

Introduce the fourth variable t (which can be t_0 or t_4) to the sequence t_1, t_2, t_3 to get

$$\begin{aligned} S_{4,0} &= 1 \\ S_{4,1} &= t_1 + t_2 + t_3 + \mathbf{t}(1) \\ S_{4,2} &= t_1 t_2 + t_1 t_3 + t_2 t_3 + \mathbf{t}(t_1 + t_2 + t_3) \\ S_{4,3} &= t_1 t_2 t_3 + \mathbf{t}(t_1 t_2 + t_1 t_3 + t_2 t_3) \\ S_{4,4} &= \mathbf{t}(t_1 t_2 t_3) \end{aligned} \qquad (6)$$

Operations on poles

Let us assume that $\phi_i(t)$ in equation (1) are symmetric functions expressed in terms of elementary symmetric functions (see (3)) with variables extracted from (2) (the number of variables is equal to the degree). We can consider two types of operation:

1. *Insertion:* obtained by inserting a parameter into a given sequence. This creates two sequences skewed by one unit from one another

$$\begin{array}{l} |\longleftarrow n \text{ values} \longrightarrow| \\ t_i \leqslant t_{i+1} \leqslant \ldots \leqslant t_{j-1} \leqslant t_j \\ \quad\ |\longleftarrow n \text{ values} \longrightarrow| \end{array} \qquad (7)$$

   These two sequences will be abbreviated by s' and s''.

2. *Difference:* the barycentric form

$$M = uA + tB \quad \text{with}$$

$$t + u = 1 \quad \text{or } u = 1 - t,$$

can be expressed as $M = A + \mathbf{t}AB$, where $AB = B - A$. We can do the same on the barycentric form (1) using the sequences s' and s'' (see below).

Multilinear function of degree n

A function E, which is the function of elementary symmetric functions, can be expressed as follows (see (5)):

$$E(t_{i+1}, \ldots, t_{j-1}, \mathbf{t}) = K_{i+1,\ldots,j-1} + \mathbf{t}Q_{i+1,\ldots,j-1} \quad (8)$$

In case t has the values t_i and t_j, and E denotes a polar form of degree n, we have

$$P_{s'} = K + t_iQ$$
$$P_{s''} = K + t_jQ$$

where K and Q stand for $K_{i+1,\ldots,j-1}$, and $Q_{i+1,\ldots,j-1}$, respectively, and s' and s'' are the sequences obtained from (7). K and Q can be expressed as

$$K = \frac{t_jP_{s'} - t_iP_{s''}}{t_j - t_i} = \Delta P_{i+1,\ldots,j-1}$$
$$(9)$$
$$Q = \frac{P_{s''} - P_{s'}}{t_j - t_i} = \delta P_{i+1,\ldots,j-1}$$

Δ and δ are two true operators applied on the polar forms giving multilinear forms of degree $n - 1$ with indices as the common indices. We can reiterate these operators to discover their commutative property, i.e. $\Delta\delta = \delta\Delta$. Consequently, we demonstrated the formula

$$E(t_{i+1}, \ldots, t_{j-1}, \mathbf{t}) = \Delta E_{i+1,\ldots,j-1} + \mathbf{t}\delta E_{i+1,\ldots,j-1}, \quad t_i \leqslant t \leqslant t_j \quad (10)$$

This algorithm is known as the Cox–de Boor algorithm used for index insertion. A similar formula can be obtained using the second differences $\Delta\Delta E$, $\delta\Delta E = \Delta\delta E$ and $\delta\delta E$, and step-by-step we can compute the multilinear form E.

Remark

If we repeat the operators Δ and δ n times, we get the algebraic coefficients of the polynomial (see example below). Under a change of the parameter $t = k\theta + h$, the poles, which have a geometrical nature, will be invariant, and therefore we can restrict ourselves to the interval $[0, 1]$, or sometimes to $[-1, 1]$, to simplify calculations.

Example

Consider a degree 3 curve. We use the following notations: $\mathbf{P}$ and $\mathbf{A}$ denote the vectors $\mathbf{OP}$ and $\mathbf{OA}$ with $\mathbf{O}$ as the origin, and $\mathbf{B}$, $\mathbf{C}$ and $\mathbf{D}$ are difference vectors, e.g. $\mathbf{B}$ points from $\mathbf{A}$ (the startpoint) to the next pole:

$$\mathbf{P}(t_1, t_2, \mathbf{t}) = \mathbf{A} + [(t_1 + t_2) + \mathbf{t}]\mathbf{B} + [t_1t_2 + \mathbf{t}(t_1 + t_2)]\mathbf{C} + \mathbf{t}(t_1t_2)\mathbf{D} \quad (11)$$

$$= [\mathbf{A} + (t_1 + t_2)\mathbf{B} + t_1t_2\mathbf{C}] + \mathbf{t}[\mathbf{B} + (t_1 + t_2)\mathbf{C} + t_1t_2\mathbf{D}] \quad (12)$$

From equation (12) we get the poles P_{012} (for $t = t_0$), and P_{123} (for $t = t_3$), called Riesenfeld points (see Figure 1.2). Similarly, for $t = t_4$ and $t = t_5$ we obtain P_{234} and P_{345}. Using formula (9), we have

$$\delta P_{12} = \frac{t_3P_{012} - t_0P_{123}}{t_3 - t_0} = \mathbf{A} + (t_1 + t_2)\mathbf{B} + t_1t_2\mathbf{C}$$
$$(13)$$
$$\Delta P_{12} = \frac{P_{123} - P_{012}}{t_3 - t_0} = \mathbf{B} + (t_1 + t_2)\mathbf{C} + t_1t_2\mathbf{D}$$

which gives $P(t_1, t_2, \mathbf{t}) = \Delta P_{12} + \mathbf{t}\delta P_{12}$ in harmony with (10). Now, computing ΔP_{23} and δP_{23}, we have

$$\Delta\Delta P_2 = \frac{t_3\Delta P_{12} - t_1\Delta P_{23}}{t_3 - t_1} = \mathbf{A} + t_2\mathbf{B}$$

$$\delta\Delta P_2 = \frac{\Delta P_{23} - \Delta P_{12}}{t_3 - t_1} = \mathbf{B} + t_2\mathbf{C} \quad (14\,\mathrm{a})$$

$$\delta\delta P_2 = \frac{\delta P_{23} - \delta P_{12}}{t_3 - t_1} = \mathbf{C} + t_2\mathbf{D}$$

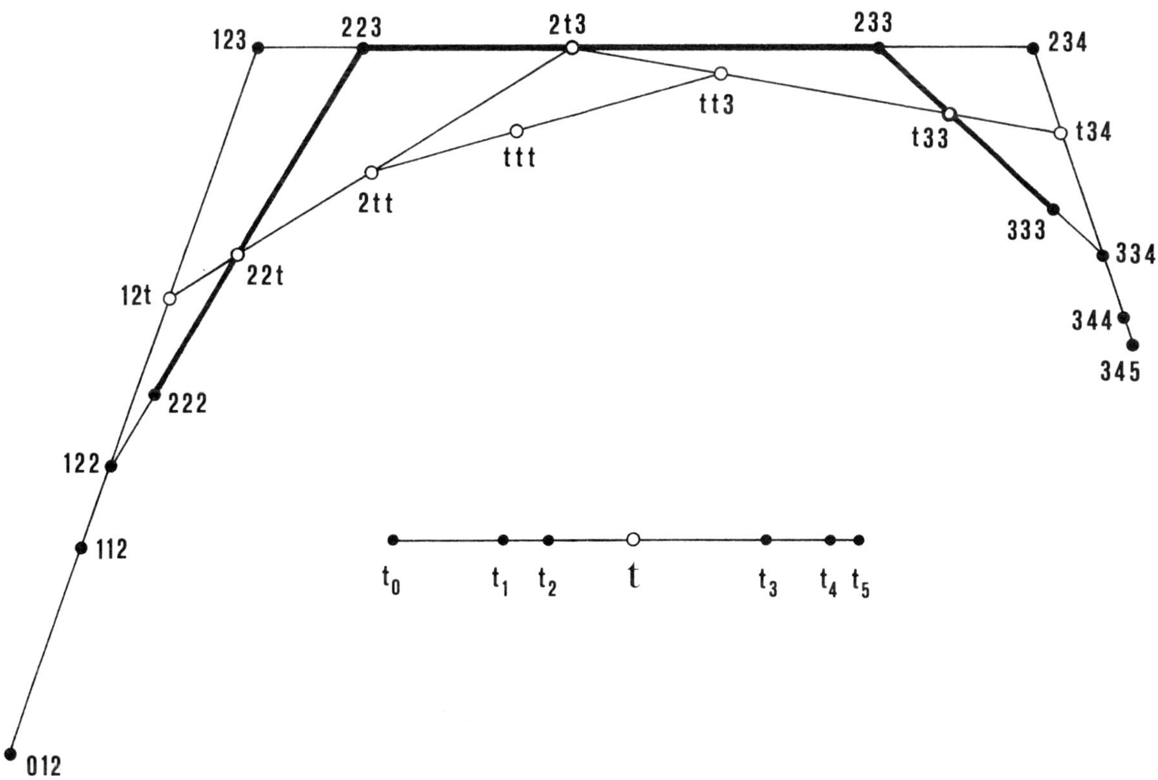

Figure 1.2 Geometric construction of a point for a fixed parameter t. On each segment, only the varying index corresponds to the ratio. The lines joining the poles are drawn in bold.

With P_{345}, we finally get the coefficients of the polynomial

$$\begin{aligned} \Delta\Delta\Delta P &= \mathrm{A} \\ \delta\Delta\Delta P &= \mathrm{B} \\ \delta\delta\Delta P &= \mathrm{C} \\ \delta\delta\delta P &= \mathrm{D} \end{aligned} \tag{14 b}$$

Simple poles

This is a very important particular case. The parameters t_{i-1}, t_i and t_{i+1} can be repeated up to a maximum equal to the degree. For simplicity, we assume that $t_i - t_{i-1} = t_{i+1} - t_i$. Without loss of generality, we can take $t_{i-1} = -1$, $t_i = 0$ and $t_{i+1} = 1$. This will help us to compare two consecutive segments at the point $t = 0$: the first is defined by **A**, **B′**, **C′**, **D′** for $t < 0$, and the second is given by **A**, **B″**, **C″**, **D″** for $t > 0$. Let us compute the difference table for degree $n = 3$:

	Difference 1	Difference 2	Difference 3
$P_{-1-1-1} = \mathrm{A} - 3\mathrm{B}' + 3\mathrm{C}' - \mathrm{D}'$			
	$\mathrm{B}' - 2\mathrm{C}' + \mathrm{D}'$		
$P_{-1-10} = \mathrm{A} - 2\mathrm{B}' + \mathrm{C}'$		$\mathrm{C}' - \mathrm{D}'$	
	$\mathrm{B}' - \mathrm{C}'$		D'
$P_{-100} = \mathrm{A} - \mathrm{B}'$		C'	
	B'		
$P_{000} = \mathrm{A}$			(15)
	B''		
$P_{001} = \mathrm{A} + \mathrm{B}''$		C''	
	$\mathrm{B}'' + \mathrm{C}''$		D''
$P_{011} = \mathrm{A} + 2\mathrm{B}'' + \mathrm{C}''$		$\mathrm{C}'' + \mathrm{D}''$	
	$\mathrm{B}'' + 2\mathrm{C}'' + \mathrm{D}''$		
$P_{111} = \mathrm{A} + 3\mathrm{B}'' + 3\mathrm{C}'' + \mathrm{D}''$			

where, for example, P_{001} denotes $P(0, 0, 1)$ obtained from equation (11).

Now the study of continuity is straightforward, and is a function of the number of equal coefficients **B**, **C** and **D** corresponding to the terms, t, t^2, t^3 (with $C_{3,p}$ omitted):

- common point **A** in all cases
- common tangent if in addition $\mathbf{B}' = \mathbf{B}'' = \mathbf{B}$
- continuous curvature if in addition $\mathbf{C}' = \mathbf{C}'' = \mathbf{C}$
- algebraic identity if in addition $\mathbf{D}' = \mathbf{D}'' = \mathbf{D}$.

Although this verification was done for $n = 3$, it can be extended to any degree. If we set all the variables in (11) to t, the multiplication of the difference table by $\mathbf{C}_{n,p}t^p$ will give the desired polynomial.

Example

Let us take the case $P_{ttt} = t^3$:

	8				
		4			
	12		2		
		6		1	
arc 2,3	18		3		
		9			
	27				
		9			
	36		3		
arc 3,4		12		1	(16)
	48		4		
		16			
	64				
	...				
	n^3				
		n^2			
	mn^2		n		
arc m,n		mn		1	
$(m = n + 1)$	m^2n		m		
		m^2			
	m^3				

Despite the fact that these tables are longer than the well-known Newton differences, they do not have useless coefficients. The successive steps are the poles of the arc. This kind of table presentation is very useful to verify the continuity or to compute the coefficients. For more details, consult [2] or [3].

The simple poles of an arc are unique, and define the arc independently of the parameter (generalized poles are dependent on the parameter sequence). They are also closer to the curve. For any algebraic arc AB, the unique and non-ambiguous references are the simple poles of the arc. It is recognized that Riesenfeld points provide a more condensed form when the degree and the number of points are large.

Algebraic calculus using the poles

The algebraic calculations are based on parameter insertion. For any sequence of parameters $t_0, \ldots, t_i, t_j, \ldots, t_f$, in order to get the simple poles of the arc $[t_i t_j]$ we insert the parameter t_i $n - p$ times and the parameter t_j $n - q$ times, where n is the order of the curve and p and q are the original repetitions of the indices of i and j, respectively.

At the beginning of the arc (defined by t_i with initial repetition p), because we have inserted the parameter t_i $n - p$ times, this will generate $n - p$ algebraic common conditions in the difference table, and therefore set the continuity c to $n - p$. We can see that, not including the point at the beginning, we have, on both consecutive arcs, $n - p$ simple poles which are derived from the same sequence of poles.

Example

Given degree 5, and the parameter sequence 012223345566, what is the continuity c at the point P_3 of parameter $t = 3$? To define arcs $[t_2, t_3]$ and $[t_3, t_4]$, we need at least five indices before 3 and after 5. It is necessary to insert index 3 three times ($n = 5$, $p = 2$), and then inserting index 4

$\mathbf{P}_{22233}$			
	$\mathbf{P}_{22333}$		
P_{22334}		$\mathbf{P}_{23333}$	
	P_{23334}		$\mathbf{P}_{33333}$
P_{23345}		P_{33334}	
	P_{33345}		
P_{33455}			

then: (17)

$$\begin{array}{lll}
\mathbf{P}_{33333} & & \\
 & & \\
\mathbf{P}_{33334} & & \\
 & \mathbf{P}_{33344} & \\
P_{33345} & & \mathbf{P}_{33444} \\
 & P_{33445} & \\
P_{33455} & &
\end{array}$$

We see a continuity $c = n - p = 3$. The reason is that the common algebraic elements are up to P_{22233} or P_{33444} on the simple poles (set in bold). The other simple poles will use different elements containing other indices, and are of a different algebraic nature.

If we want to obtain the endpoints of a curve, it is necessary to repeat the first and the last indices n times each. On the other hand, it is possible to close a curve by taking the parameter sequence as a circular list: $t_0 \leqslant t_1 \leqslant \ldots \leqslant t_f = t'_0 \leqslant t'_1 \ldots \leqslant t'_f$, etc.

Construction of a point on the curve

One has to insert the index t q times where, as above, q is the original repetition of the index. As an example, for $n = 3$ and $q = 2$ we have (see Figure 1.2):

Riesenfeld points: *Simple poles:*

$$\begin{array}{ccccccc}
\mathbf{P}_{012} & & & \mathbf{P}_{222} & & & \\
 & \mathbf{P}_{12t} & & & \mathbf{P}_{22t} & & \\
P_{123} & & \mathbf{P}_{2tt} & P_{223} & & \mathbf{P}_{2tt} & \\
 & P_{2t3} & & P_{ttt} & P_{2t3} & & P_{ttt} \\
P_{234} & & \mathbf{P}_{tt3} & P_{233} & & \mathbf{P}_{tt3} & \\
 & \mathbf{P}_{t34} & & & \mathbf{P}_{t33} & & \\
\mathbf{P}_{345} & & & \mathbf{P}_{333} & & &
\end{array} \qquad (18)$$

We have also solved the subdivision of the arc, also called *sub-poles* (bold).

Subdivision of an arc

This is derived from the polar definition. The $q = 2$ insertions of the parameter t (see above) generate a new sequence where the poles (bold) replace the old ones. This problem has also been solved by Boehm for B-splines. The advantage of the poles resides in their simplicity. All the problems have their solution in the multilinear definition. Furthermore, no distinction is made between 'knots' and 'nodes', because the interpolation is completely disconnected from the definition of the arc.

Geometric construction

Geometric construction reflects exactly what has already been said so far. On any line, the poles are different by only one index, and the corresponding points divide the legs in the same ratio as do the parameters t_i the parameter line (see Figure 1.2). Geometry is independent of the origin of the indices and their absolute length, provided that the proportions are presented. Therefore, the application of poles is a wonderful example of affine geometry. In Figure 1.2, the geometric construction is illustrated using generalized as well as simple poles (which are indicated by bold lines).

Homogeneous form

The homogeneous form $u + t = 1$ is well adapted to compute the simple poles and to preserve the proportion u/t (see Figure 1.2):

$$\begin{array}{llll}
P_{222} & & & \\
 & uP_{222} + tP_{223} \rightarrow P_{22t} & & \\
P_{223} & & uP_{22t} + tP_{2t3} \rightarrow P_{2tt} & \\
 & uP_{223} + tP_{233} \rightarrow P_{2t3} & & uP_{2tt} + tP_{tt3} \rightarrow P_{ttt} \\
P_{233} & & uP_{2t3} + tP_{t33} \rightarrow P_{tt3} & \\
 & uP_{233} + tP_{333} \rightarrow P_{t33} & & \\
P_{333} & & &
\end{array}$$

(19)

and

$$P(t) = P(ttt) = u^3 P_{222} + 3u^2 t P_{223} + 3ut^2 P_{233} + t^3 P_{333}$$

The terms of the expansion of $(u + t)^3$ are the coefficients of the poles. The sub-poles are P_{222}, P_{22t}, P_{2tt}, P_{ttt} and P_{ttt}, P_{tt3}, P_{t33}, P_{333}.

Special applications

Turbine blade: a closed profile such as the turbine blade is defined by the poles A, B, C, D and E. Call the point M the midpoint at parameter $u = t = 0.5$. Subdividing the curve at this parameter value we have sub-poles $A1$, $B1$, $C1$, $D1$, $E1$ and M between 0.0 and 0.5, and $A2$, $B2$, $C2$, $D2$, $E2$ and M between 1.0 and 0.5. The curve having the simple poles M (= midpoint of $E1E2$), midpoint of $D1D2$, midpoint of $C1C2$, midpoint of $B1B2$ and A is the so-called mean curve. The odd degree coefficients are null. When $t^2 = \Theta$, the poles in $\phi(\Theta)$ give the mean curve, including the non-vanishing part for $\Theta < 0$. The differences of the sub-poles give the thickness of the blade.

Polynomials: two polynomials $X = f(t)$ and $Y = g(t)$ generating a planar parametric curve can be expressed in complex form as $Z = X + iY = f(t) + ig(t)$. The roots of this equation can be used to solve difficult problems such as the approximation of the circle. For the circle, $X^2 + Y^2 = R_2 + kt^{2n}$ ($t = 0$ if the circle is the osculating circle). The roots are on a regular polygon, and we keep those (to represent $X + iY$) which are on the upper plane (the other roots represent $X - iY$). One can also use the Tschebyscheff polynomial $T_{2n}(t)$ to average the error, and we have $X^2 + Y^2 = R^2 + kT_{2n}(t)$. An affine transformation of the previous roots, transforming the circle into an ellipse, gives the roots yielding a good approximation of the full circle.

Degree elevation: the choice of the degree n and the number of indices are arbitrary. One can always express a polynomial in terms of a higher degree one by setting the coefficients of the higher degree terms equal to zero. Consequently, we can express a curve with n indices in terms of $n + q$ indices, provided that we do not forget to replace the coefficient $C_{n,i}$ by $C_{n+q,i}$ and setting the minimum repetition to $p + q$ to preserve the continuity c. The polar form will give the new poles.

Surfaces defined by poles

There are two ways to extend the curve definition to surfaces: using either independent or dependent parameters. In the first case there is no difficulty in principle: one can consider two independent and non-decreasing sequences of parameters t_i and T_j. The poles will be doubly indexed such as $P_{tuv,UV}$ for degree 3 by 2. The disadvantage is we need a complete rectangular grid of poles defined along the rows for the parameters t_i and along columns for the parameters T_j. The continuity, or 'spline' effect, is created by holding t_i constant while T_j varies, or by holding T_j constant while t_i varies. These lines are the images of lines parallel to the axis of the parameter space. In fact, any system of lines, each of them dividing the parameter space into two half planes, could be convenient.

The importance of the generalized pole is its flexibility. With parameter insertion one can isolate a zone, with its imposed continuities, and make a vectorial translation of the poles in one direction to make clear the following statement: *The poles of the deformation are the deformations of the poles.* An example is illustrated in Figure 1.3.

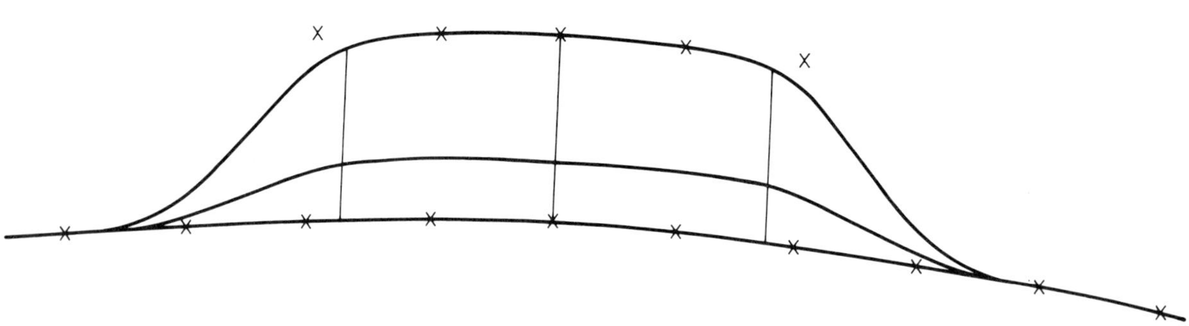

Figure 1.3 Curve deformation. The deformation of the poles gives the poles of the deformation.

Simple poles

We have already seen two cases:

- Independent variables – rectangular patches. The coefficients are obtained by expanding $(u+t)^n(U+T)^N$, where $u+t=1$ and $U+T=1$, and we assume that $0 \leqslant t, T \leqslant 1$.
- Dependent variables – triangular patches. The coefficient of the pole P_{ijk} is given by expanding $(t+u+v)^n$

$$\frac{n!}{i!j!k!} t^i u^j v^k$$

where $t+u+v=1$ and $i+j+k=n$.

INTERPOLATION WITH POLAR FORMS

A good interpolation should satisfy the following requirements:

- Interpolation through a table, with or without smoothing, should be digital, i.e. the interpolation uses a summation of terms relative to each point. Any function $u=f(t)$, defined by a pair of values $u=u_i$ and $t=t_i$, will be expressed as

$$u = \sum_i K_i(t) u_i \tag{20}$$

where u is either a co-ordinate or a function of t, and $K_i(t)$ satisfies the conditions $K_i(t_i)=1$ and $K_i(t_j)=0$ for $i \neq j$.
- Any interpolation has a piecewise form, and therefore $K_i(t)$ will be a polynomial defined between two consecutive parameters t_i and t_j. This polynomial, using the polar form, will have the polar form continuity.
- Any interpolation must recompute the algebraic coefficients of a degree r polynomial if there are sufficient number of points ($>r+1$). This degree is called the *degree of restitution*. It allows recomputation, without any divergence, of the inverse operation of point construction. One cannot tolerate a program that diverges while, inside a loop, it first evaluates some points of a polynomial and then computes the points of a polynomial by interpolating the points. In fact, for the degree $r+1$, the complete curve is translated at each iteration.

Lagrange interpolation

This interpolation satisfies the above requirement using the unique algebraic curve

$$K_i(t) = \prod_{k \neq i} \frac{(t-t_k)}{(t_i-t_k)} \tag{21}$$

Its disadvantage is to lead to too high degrees and, if locally taken, it does not satisfy the continuity of the second requirement.

Given the degree of continuity c, the value of the repetition p is well determined. The polar forms, taken in number p on each segment defined by a Lagrange interpolation of degree r and satisfying the above requirement, are obtained from the following process (for details consult [2] and [3]):

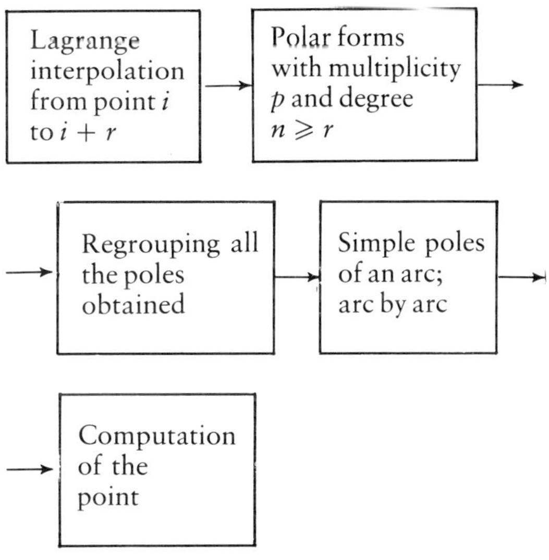

The interpolation is defined by its characteristic (n, c, r), where n is the degree of the polynomial arcs, c $(=n-p)$ is the desired continuity between the arcs (which could be different at each point, and determines the number of continuous derivatives and the continuity of the function), and r is the degree of restitution (if r is variable, then take the smallest one).

If all the points initially belong to an algebraic

curve of degree $\leqslant r$, the Lagrange interpolation strictly preserves the algebraic nature of the curve, and therefore the poles are also the poles of the initial polynomial. Below we consider an example to illustrate this (see the second example of [3]). Let $n = 8$, $c = 5$ and $r = 7$, and let us take a uniform parameter sequence 0123456789, with the points denoted by $P_0, \ldots, P_9$. The nine poles defining the arc t_4t_5 will be chosen on the parameter sequence 2233344455566677 (repetition is $n - c = 3$). We take three sequences of parameters for three Lagrange interpolations:

	(I)	(II)	(III)
Points:	0123 4567	1234 5678	2345 6789
Sequence:	2233344455	3344455566	4455566677
Poles:	$P_{22333444}$	$P_{33444555}$	$P_{44555666}$
	$P_{23334445}$	$P_{34445556}$	$P_{45556667}$
	$P_{33344455}$	$P_{44455566}$	$P_{55566677}$

We use 10 points, 3 Lagrange interpolations on 8 consecutive points, giving 3 poles and the continuity $c = n - p = 5$. This reconstructs the arc t_4t_5. The degree of restitution is obviously 7, because for a polynomial of degree $\leqslant 7$ all the operations are algebraically rigorous.

The discussion in [3] proves that there are 66 conditions of continuity satisfied ($66 = 11 * 6$, where $11 =$ number of points $+ 1$, $6 = c + 1$), and there are 72 conditions of restitution ($72 = 8 * 9$, where $8 = r + 1$, $9 = n + 1$). In total, 138 conditions were satisfied, although 90 coefficients were available in the matrix (see Figure 1.4). These favorable circumstances explain the extraordinary quality of the result. It is by far better than we could expect.

Beyond degree 7, a smoothing is automatically done. With a value $\Delta_8/1047$, it is four times smoother than the quadratic smoothing. Nothing in the definition of this smoothing refers to a minimum, and furthermore, it simultaneously smoothes the function and its first five derivatives (see Figure 1.5).

Choice of parameter

What we have said above may seem to be arbitrary. For stamping, one can take the length

	−77	616	−2156	4312	75250	4312	−2156	616	−77	0
	−41	232	−140	−3752	75250	12376	−4172	1000	−113	0
	−13	−72	1436	−9208	70770	23048	−6628	1464	−157	0
1	0	−224	2234	−11568	61810	35840	−9186	1936	−202	0
$\frac{1}{80640}$	0	−232	2264	−11160	49448	49448	−11160	2264	−232	0
80640	0	−202	1936	−9186	35840	61810	−11568	2234	−224	0
	0	−157	1464	−6628	23048	70770	−9208	1436	−72	−13
	0	−113	1000	−4172	12376	75250	−3752	−140	232	−41
	0	−77	616	−2156	4312	75250	4312	−2156	616	−77

(a)

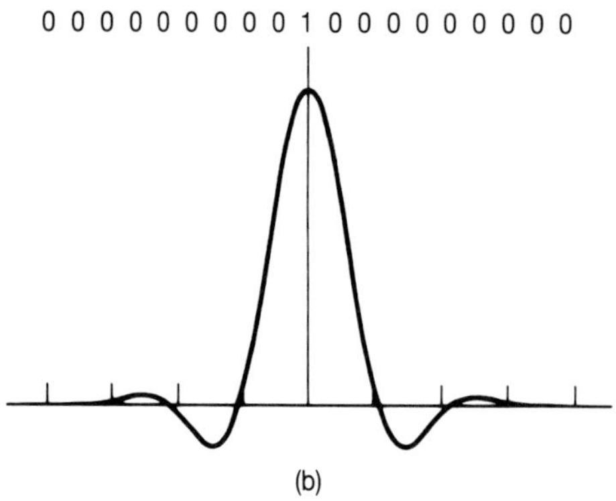

(b)

Figure 1.4 Impulse response: (a) 9×10 matrix with $n = 8$ $c = 5$ $r = 7$; (b) response to an impulse with the above matrix. (Figure courtesy of Hermes Press, Paris [2].)

Figure 1.5 The result of smoothing. The points of this propeller blade were digitized with a stylus!

on the horizontal projection (it is not logical to use the true length which is too closely related to the affine ratio). In fact, in the preceding example, the deviation is represented by the vector Δ_8, which we want to minimize. If the tangential component of Δ_8 is too large, we can reduce it by choosing a better parametrization. For a planar curve, it is preferable to use the affine normal, which is the locus of mid-chords parallel to the initial tangent. For a conic, the affine normal passes through the center of the conic.

Question: Do we have to condemn the use of the orthogonal functions? The security provided by the quadratic minimization $I = \int f''^2(t)\,\mathrm{d}t$ or $\int y''^{1/3}(x)\,\mathrm{d}x$ is illusory. They are not invariant under a non-linear parameter change.

Any quadratic minimization of a function f with a function F such as $I = \int [F(t) - f(t)]^2\,\mathrm{d}t$ should be expressed as $I = \int p^2(t)[F(t) - f(t)]^2\,\mathrm{d}t$, $p(t)^2$ being a positive 'weight' function which has to be determined. Facing this problem, mathematics is confused. One will never finish enumerating the 'best' solutions: Legendre, Laguerre, Hermite, Tschebyscheff, Euler, Bernstein, Frobenius, etc. The problem is therefore not well defined. In a confused way, one hopes for a solution which simultaneously smoothes both the function and its derivative, but it cannot be expressed. The minimum is useless. If these arguments did not convince the reader, take the trigonometric example:

$$u = \frac{\cos\phi}{|\cos\phi|}$$

$$= \cos\phi - \frac{\cos 3\phi}{3} + \frac{\cos 5\phi}{5} - \dots$$

$$(-1)^p \frac{\cos(2p+1)\phi}{2p+1} \qquad (22)$$

Any high frequency filter causes problems if p is large. How can we talk about limit under these conditions? Even Fejer and Cesaro methods will not help. However, the solution is simple. The Gibbs phenomenon completely disappears if we take the expansion $[1 - \sin^2 \phi]^{-1/2} \cos \phi$, which gives

$$u = \cos \phi - \frac{p-1}{p+1}\left(\frac{\cos 3\phi}{3} - \frac{p-2}{p+2}\left(\frac{\cos 5\phi}{5} - \frac{p-3}{p+3}\left(\frac{\cos 7\phi}{7}\cdots \frac{1}{2p+1}\frac{\cos(2p-1)\phi}{2p-1}\cdots\right)\right)\right) \quad (23)$$

As $p \to \infty$, this expansion is the only one which converges to the expected limit at $\phi = 0$ (or $k\pi$). It is so simple that no one has perceived it!

As for the 'splines', small deviations at the given points (which could be the result of rounding) generate instabilities which are dangerously amplified, especially when we try to reduce them.

Only interpolation to digitized data, which should be used on the compact disc, is safe from these critics. All the demonstrations used in this paper are exclusively algebraic. It is reassuring to see that we can get a more satisfactory solution when we avoid complex theories by using pure algebra.

ACKNOWLEDGEMENTS

The original French manuscript was translated by Michel Menard (who used the 'Poles' during his 15 years while at Citroën), and Jack Gregory of ICAD, Cambridge, MA.

REFERENCES

1. **P. de Casteljau** *Courbes et Surfaces a Poles*, Societe Andre Citroen, 1963 (Plus several other internal notes written for the drawing school of Citroën).
2. **P. de Casteljau** *Formes a Poles*, Hermes Press, Paris, 1985.
3. **P. de Casteljau** *Shape Mathematics and CAD*, Kogan-Page, London, 1986.
4. **P. de Casteljau** *Les Quaternions*, Hermes Press, Paris, 1987.
5. **L. Ramshaw** *Blossoming: A connect-the-dots-Approach to Splines*, Digital Research Center, Palo Alto, CA, 1987.
6. **A. Kuros** *Higher Algebra*, Mir Publisher, Imported Publications, Chicago, 1989.

ABOUT THE AUTHOR

Paul de Faget de Casteljau was born in Besançon near the Swiss border on 19 November 1930. He attended catholic secondary school (St Jean de Besançon, and during the war St Caprais d'Agen) and then entered the Lycée Victor Hugo in Besançon. In 1951, after being accepted by two schools, he chose l'Ecole Normale Superieure over l'Ecole Polytechnique. In 1955, he spent his military service stationed in Algeria.

De Casteljau joined Citroën in 1958. He was involved with car body design as well as with mechanical engineering (with the camshaft of the Citroën DS21, he was able to obtain a 15% increase in power and a 10% increase in RPM). His work, although theoretical, rapidly found practical applications. Besides his work on simple and generalized polar forms, he published a study on quaternions [4] and the Euclid algorithm, which won the special prize of Concours Seymour Cray France 1987.

2. The First Years of CAD/CAM and the UNISURF CAD-System

Pierre E. Bézier

THE FIRST YEARS OF CADCAM: AN OVERVIEW

Today, CAD/CAM is part of our life, its action is fundamental in every industry: moreover, it now encompasses a large portion of our day-to-day life, from printing to art and entertainment. CAD/CAM has reached a point where the non-expert may well come to think that it is a simple technique, and he does not perceive the enormous amount of work undertaken to reach that state. Hence it seems useful to cast a glance back over the past, not for the pleasure of recalling the efforts and the successes, but to remember the reasons for having made some basic choices.

Two main facts control the development of a technique or an invention: (1) the need expressed by industry faced with difficult problems such as accuracy, efficiency, lagtime or costs; and (2) the discovery or invention of new methods which could become part of an already existing process; a good example is the laser, which was invented by scientists investigating the properties of light, and which is now used widely in industry for cutting, welding or heat treating very hard materials.

The mechanics industry always endeavoured to improve the accuracy of its products, and this may perhaps be considered the first and foremost problem. A good contact between moving parts was a compulsory condition for limiting the wear and degradation of mechanisms. Most of the surfaces require great care to be taken; planes, cylinders and cones can be defined by dimensions and limits expressed by numbers or symbols, and the measurements of such basic features became the only method for deciding whether a part was to be accepted or rejected. Regarding mechanical parts, they were mainly defined by lines and circles. However, a certain extent of their surfaces, the accuracy of which was not too critical, could only be expressed in a less precise manner, e.g. fillets, rounded corners and drafts applied to rough surfaces obtained by casting, forging or welding. Part of the definition of these surfaces was left to highly skilled operators, patternmakers, foundrymen or diesetters, and, consequently, the responsibility of defining the shape was split among a number of different people. Nevertheless, some surfaces, e.g. turbine blades, ship hulls, aircraft wings and car bodies, not representable in terms of lines and circles, required a certain degree of accuracy to be useful for practical purposes. The method used to design such surfaces was to first define cross-sections with drawings, then the templates were adjusted to the real model and interpolated by patternmakers or tool setters, and finally, parts of the model so obtained were used on copy-milling machines for manufacturing the final product. This process had two weak points: (1) the responsibility of defining a surface was divided among a number of different people – the designer traced some sections, the patternmaker interpolated them, and the final setting was left to the toolsetter . . . and none of them could be proved right or wrong in case of trouble; (2) the final 3D model could be destroyed or lost, and copies could be slightly different, hence the possible disparity among parts which had to be assembled together.

It is very likely that during World War II in a British aircraft factory, it was feared that the drawings and templates could be destroyed; hence it was compulsory to store duplicates in another location. However, instead of relying on drawings, the accuracy of which was liable to suffer from distortion, besides being far from perfect, it was preferable to store the coordinates of points located on the curves of the original drawings. These data could easily be duplicated and stored, and could not warp. Of course, this was not true CAD/CAM, but it can be considered as the first step in replacing drawings by numerical data.

In about 1955, numerically controlled (NC) machine-tools, after having remained purely experimental for some time, began to be used in industry. At that time, computers played a dual role: first, performing some kind of numerical

Fundamental Developments of Computer-Aided Geometric Modeling
ISBN 012-554765-X

calculations, such as trigonometric calculation for defining centers for drilling, reaming and tapping; and second, servo-control of the drilling or milling machine's motion. For milling, contouring was developed, where the contour curves were almost exclusively piecewise linear or circular. This came from the requirement that contouring was intended to produce mechanical parts traced with the help of a straight edge and a compass. The Automatic Programming of Tooling (APT) software was developed at the Illinois Institute of Technology Research Institute. It became rapidly evident that the definition of certain free-form curves and surfaces required a large number of circular arcs, and it would be simpler to use conoids, special ruled surfaces whose directrices are planar polynomial curves of up to twelve degrees. The description of a curve approximated by circular arcs was a time consuming procedure. To simplify computation, a parabola was used instead of a circle, since its computation was a lot faster than that of a circle. In any case, the description of space curves remained a tedious and time consuming task.

There is no doubt that the aircraft and shipbuilding industries were suffering heavily from these limitations, and they were the first to work on these problems in the late 1950s. Until that time, descriptions of curves and surfaces were Cartesian, because this fitted well with the machine tool coordinate system. However, when a curve or a surface had to be rotated in space, this entailed a large amount of computation. Therefore, it became evident that parametric definition, based on the theoretical works of Bernstein [1] and Schoenberg [2], would be more suitable. One of the first papers dealing with parametric curves and surfaces for CAD/CAM was written by James Ferguson [3], working at that time at Boeing. Ferguson's paper dealt with cubic curves and bicubic patches (see Chapter 6 in this volume). Ferguson's method was well adapted to the definition of aircraft wings and fuselages. More complicated zones such as von Karman's fillets or the cockpit were left to the shop, but this was only a small part of the entire surface.

At about the same time, a group including Ivan Sutherland were working at MIT on a project called SKETCHPAD [4]. Using a CRT and a medium-sized computer, an operator could model curves and surfaces, and store the results in a very concise manner. The naval industry also paid a great deal of attention to the development of CAD/CAM techniques as early as 1962. Some papers dealing with the mathematical lofting of ships hulls were presented at a meeting of the Society of Naval Architects and Marine Engineers. Mehlum, working at the Central Institute for Industrial Research in Oslo, developed the AUTOKON system [5], also designed specifically for the shipbuilding industry. AUTOKON not only dealt with the shape of the hulls, but with the inner frames, ribbings, staircases, scuttles, powerplants, etc., and with the behavior of a seafaring vessel in terms of stress, strain, roll period, etc.

The process of designing a car in the early 1960s consisted of the following main steps:

1. Stylists defined a general shape using small scale sketches and clay mockups.
2. Using offsets measured on the mockup, designers traced a full scale shape of the skin of the car body with the help of cross-sectional curves.
3. Plasterers built a full scale model, weighing about eight tons, starting from plywood cross-sections duplicating the curves of the drawing. This so-called 'clay model' was then lacquered, and equipped with wheels, windshield and panes, chrome trimmings, headlamps, bumpers, handles, screen wipers, etc., and it looked like a real car. The clay model was then examined by stylists and sales managers, and modified according to taste.
4. When at last the model was accepted, offsets were again measured and the final drawings were made, to full scale this time. The drawings contained not only the definition of the skin, but that of the inner panels together with the mechanical elements, e.g. frame, power plant, front and rear axles, steering mechanism, hinges, locks, rubber seals, etc. In the meantime, a model of the inside volume was built to see whether the passengers felt comfortable with elbow-room, head and leg space, etc. When the entire shape had been defined, a few prototypes were built to check stiffness, vibration and crash resistance. During this period, which could be a year or more, tooling and production specialists often suggested minor changes needed to avoid difficult and costly operations during production.

5. The drawings were finalized, and one 3D master was built as the standard for checking the press tools and stamped parts.
6. The plaster copies of the master were used for milling punches and dies on copy-machine tools. Slight changes could be inflicted either on the master or the tools at this stage at the request of the stylists or the tool specialists.

In each step of this process variations could enter the design, and the design responsibility was divided among half a dozen divisions, i.e. responsibility had practically disappeared. This whole process resulted in lengthy discussions, arguments, delays and expense; it was logical that the car industry looked for a better system.

In the early 1960s, the problem that the Ford Motor Company wanted to solve can be outlined as follows: given a clay model, form a network of curves on it defining a set of surface patches. Interpolate the network with parametric cubic patches ensuring at least slope continuity between adjacent surfaces. Two solutions were presented at about the same time. Steven A. Coons, a brilliant researcher at MIT, invented a solution known today as the Coons' patch [6] (see Chapter 8 of this volume). The solution was intended to be purely algorithmic, but it was sometimes necessary to leave some degree of freedom in order to alter the result to improve its aesthetic appearance. William J. Gordon provided a global solution [7], i.e. instead of joining patches together, he defined an interpolant as a whole, relying on the theory of Boolean sum projectors (see Chapter 7 of this volume).

The CAD Laboratory at Cambridge, UK, headed by Professor Wilkes, worked on the same subject. One of Wilkes' assistants, A. Robin Forrest, working with Steven Coons, explored parametric curves and surfaces for Computer-Aided Design [8]. The research performed by Forrest was influenced by co-workers such as Charles Lang, Malcolm Sabin and Andrew Armit (see the appropriate chapters of this volume). Richard F. Riesenfeld used B-splines for curve and surface definition after completing his PhD work at Syracuse University [9] supervised by Steven Coons (see Chapter 5 of this volume). At about the same time, FERRANTI, a Scottish company, was manufacturing electronic components defined by lines and circles. Software named Profile Data [10] was developed to control drawing and milling machines. When they were faced with the problem of free-form curves and surfaces, the tool trajectory had to be divided into a large number of small circle segments, resulting in huge storage requirements.

There is no doubt that Citroën was the first company in France that paid attention to CAD, as early as 1958. Paul de Casteljau, a highly gifted mathematician, devised a system based on the use of Bernstein polynomials [11] (see Chapter 1 of this volume). At Citroën, the accepted practice was that the drawing office first defined parts which then were transmitted to the production engineering services, who were entitled to slightly alter the shapes to make manufacturing easier. Consequently, the system devised by de Casteljau was oriented towards translating already existing shapes into patches defined in terms of numerical data. Due to Citroën's policy, the results obtained by de Casteljau were not published until 1974, and this excellent mathematician was deprived of part of the well deserved fame that his discoveries and inventions should have earned him.

In 1957, Renault investigated the possibilities of numerically controlled machine tools. They were, at that time, dedicated to drilling, reaming and tapping mechanical parts, i.e. to point-to-point operations. It was concluded that a special system would be necessary to contour-mill stamping tools, and that it should encompass the car body drawing itself. Instead of endeavouring to improve some specific steps of the process running from styling to inspection of finished parts, the basic idea was to build a system whereby definition could be transmitted from end-to-end exclusively by numbers, drawings playing only a explanatory role. The requirements of building a new system were

1. The underlying mathematics should be easily understood. Designers, stylists and technicians are mathematically not very sophisticated people, but they have an excellent knowledge of descriptive geometry, and their thinking is visual rather than abstract.
2. Space curves should be directly defined, and not only by their projections.
3. The variety of curves should be large enough to avoid definition in terms of a huge number of short segments.
4. The result should be obtained within a reasonable amount of time, i.e. a few seconds

for a curve and part of an hour for a surface, e.g. a square meter piece.

5. The styling and drawing office should be equipped not only with NC drawing boards, but with special milling machines capable of carving soft materials like styrofoam, polyurethane foam, wood, plaster, or resin at very high speed. Stylists want to investigate not only drawings before accepting a model, but they want to see and even touch a 3D model displaying color and reflection lines.

The system, initiated in 1966, was described at the Convention of the Society of Automotive Engineers, held in Detroit in 1968 [12]. It was not in full operation before the late 1970s, but it is now also applied to mechanical parts of the car (Figures 2.1–2.10 illustrate a milling machine

Figure 2.1. Prototype of a special milling machine used as a 3D tracing machine by the drawing office (1968).

Figure 2.2. Propeller (1968).

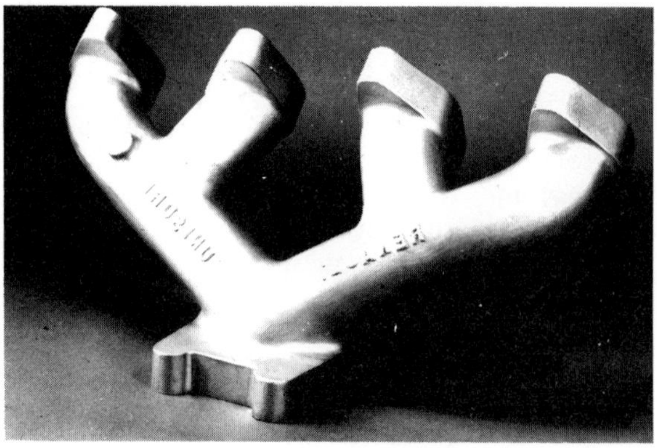

Figure 2.3. Intake manifold (1970).

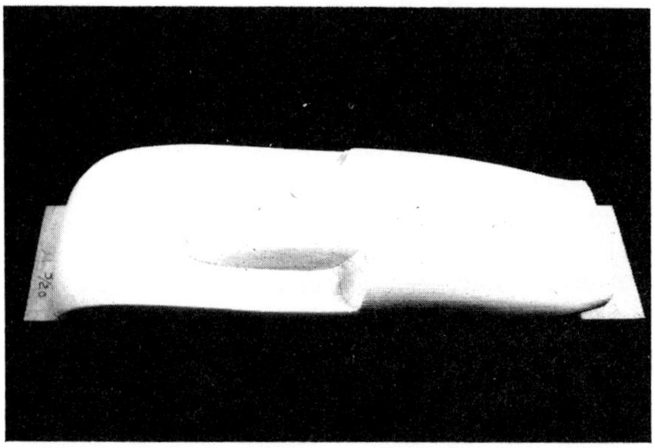

Figure 2.4. Dream car, scaled 1/8, carved in polyurethane foam (1971).

Figure 2.5. Teak wood racing boat hull (1972).

and several work-pieces designed in the early stage of UNISURF).

Professor Engeli from Switzerland developed the EUKLID system based on Bernstein polynomials, while at the University of Orsay, Paris, A. Theron and J.-M. Brun created the EUCLID system, which was capable of handling mainly analytic shapes.

All in all, CAD began to attract industry's attention in the late 1950s. Initially, a few mathematical solutions such as Cartesian polynomials and harmonic functions were investigated, but in the long-term, parametric polynomials became the foremost and quasi-only solution. CAD expansion required the development of new machines, e.g. drawing and milling machines. The most important change, however, was in the training of specialists such as designers, sylists and other technicians. Since 1975, the basic principles have not changed, but some new techniques such as non-uniform rational B-splines (NURBs) have become available in the software industry.

It is worth pointing out that the development of automatism is highly dependent on man. Those who investigated and actually created CAD had knowledge, vision, imagination, and the will to promote progress.

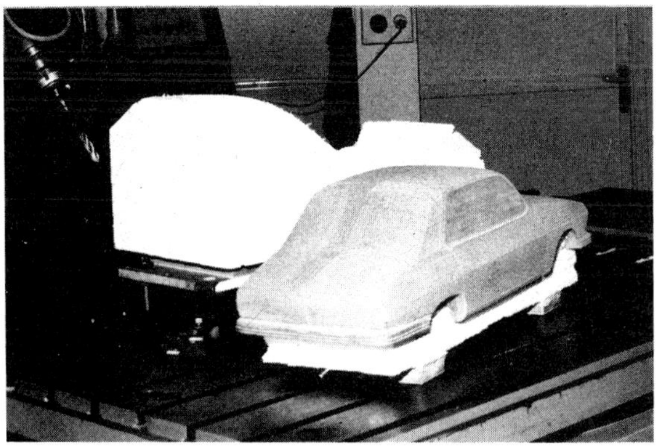

Figure 2.6. Clay model, scaled 1/8, milled on the machine shown in Figure 2.1.

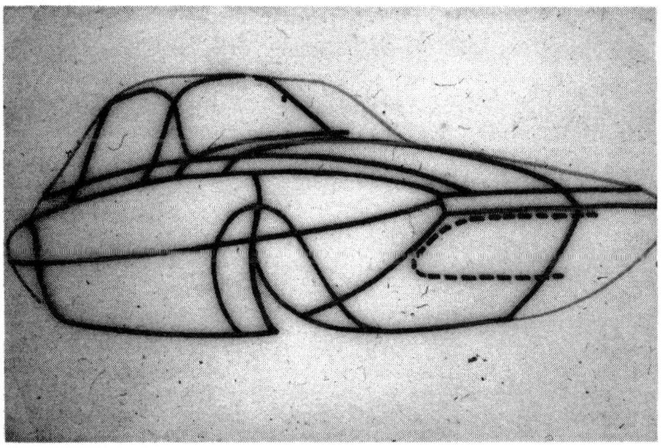

Figure 2.7. Wire frame of the FUEGO car (1973).

Figure 2.8. Rear panel of the FUEGO car. The prototype is stamped on a Kirksite puch milled on the machine shown in Figure 2.1.

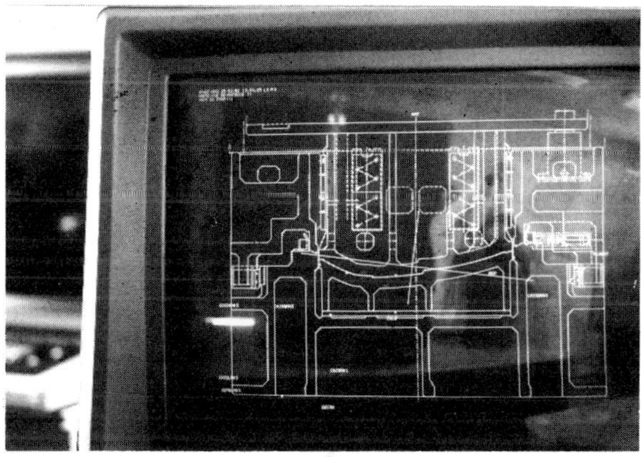

Figure 2.9. Stamping tool drawing (1974).

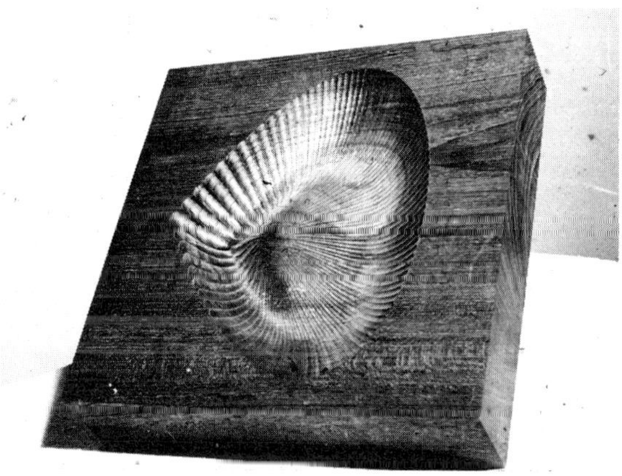

Figure 2.10. Masterpiece, one patch, carved in walnut wood.

CURVES AND SURFACES IN THE UNISURF CAD SYSTEM

Curves

Since the UNISURF system was intended for technicians, it was compulsory to found it on geometry rather than on analysis. It became obvious that the solution should be parametric rather than Cartesian. If the curve is inserted into a cube of unit length, it is easily distorted as long as the image of the cube remains a parallelepiped. To define a parametric curve, we first define some polynomial basis functions of the parameter which, by assumption, will vary between 0 and 1.

Cubic functions '*f*'

The basic curve should comply with the following requirements (see Figure 2.11):

- its start point $\mathbf{P}(0)$ should be (0, 0, 0)
- its end point $\mathbf{P}(1)$ should be (1, 1, 1)
- at $\mathbf{P}(0)$ it should be tangential to **OX**
- at $\mathbf{P}(1)$ it should be tangential to a parallel to **OZ**

To modify this curve, the designer should only define a new set of reference, say **Axyz** (see Figure 2.12) and relate it to the origin **O**. The new set can be defined by the three vectors **Ax**, **Ay** and **Az**, but it is advisable to put them end-to-end to obtain the polygon **AMNB** whose shape broadly mimics that of the curve. It should be observed that the vertices need not be coplanar, i.e. the curve is a true space curve. It is defined directly, and not only by its planar projection curves. If we denote the difference vectors **OA**, **AM**, **MN**

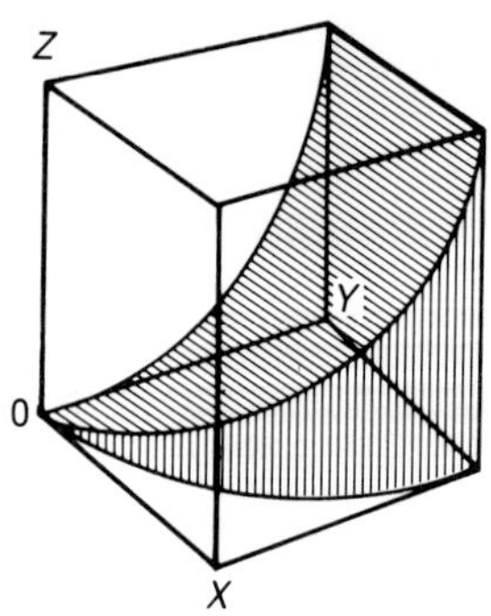

Figure 2.11. Basic curve segment.

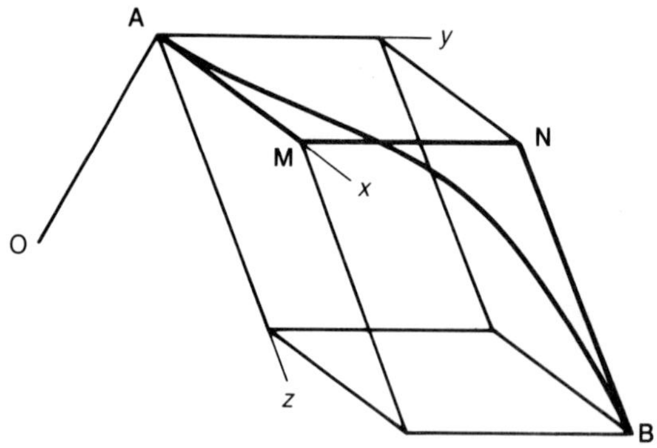

Figure 2.12. New curve defined by a new set of reference.

and **NB** by $\mathbf{a}_0$, $\mathbf{a}_1$, $\mathbf{a}_2$ and $\mathbf{a}_3$ (Figure 2.13), then a point $\mathbf{P}(u)$ on the curve can be expressed as

$$\mathbf{P}(u) = \mathbf{a}_0 f_0(u) + \mathbf{a}_1 f_1(u) + \mathbf{a}_2 f_2(u) + \mathbf{a}_3 f_3(u) \qquad (1)$$

To define the four functions 'f_i', the following requirements should be satisfied:

- **A** should coincide with $\mathbf{P}(0)$, hence

$$f_0(0)=1 \quad \text{and} \quad f_1(0)=f_2(0)=f_3(0)=0 \qquad (2)$$

- **B** should coincide with $\mathbf{P}(1)$, hence

$$f_0(1) = f_1(1) = f_2(1) = f_3(1) = 1 \qquad (3)$$

$$(\text{since } \mathbf{P}(1) = a_0 + a_1 + a_2 + a_3)$$

- at **A** the curve should be tangential to **AM**, i.e. to $\mathbf{a}_1$, hence

$$\frac{\mathrm{d}f_0}{\mathrm{d}u}(0) = \frac{\mathrm{d}f_2}{\mathrm{d}u}(0) = \frac{\mathrm{d}f_3}{\mathrm{d}u}(0) = 0 \qquad (4)$$

- at **B** the curve should be tangential to **NB**, i.e. to $\mathbf{a}_3$, hence

$$\frac{\mathrm{d}f_0}{\mathrm{d}u}(1) = \frac{\mathrm{d}f_1}{\mathrm{d}u}(1) = \frac{\mathrm{d}f_2}{\mathrm{d}u}(1) = 0 \qquad (5)$$

From (2) to (5) it follows that the functions f_0 and f_2 must satisfy four conditions, and there are only three conditions for f_1 and f_3. To define four cubic functions, two additional conditions must be added. The most convenient is to decide that the basic curve should osculate the plane **AMN** at **A** and the plane **MNB** at **B**. Hence, at $\mathbf{P}(0)$ the second derivative of f_3, and at $\mathbf{P}(1)$ the second derivative of f_1 will vanish. These conditions are summarized in Table 2.1, and the functions 'f_i' are given explicitly as

$$f_0 = 1$$
$$f_1 = u^3 - 3u^2 + 3u$$
$$f_2 = -2u^3 + 3u^2$$
$$f_3 = u^3$$

Extension to *n*-spaces

Since the vectors $\mathbf{a}_i$ need not be perpendicular to each other and of equal length, it is possible to use any number of vectors to define a curve as long as the vectors are linearly independent as long as the vectors are linearly independent in the n-space. Looking at Table 2.1, it is relatively simple to extend it, for example, to a five dimensional space (Table 2.2). For a family of $(m + 1)$ functions of degree m, the ith function takes the form

$$f_{i,m}(u) - \frac{(-u)^i}{(i-1)!} \cdot \frac{\mathrm{d}^{i-1}[(1-u)^m - 1]/u}{\mathrm{d}u^{i-1}}, \quad u \in [0, 1] \qquad (6)$$

Table 2.1.

	$u = 0$			$u = 1$		
	$f^{(0)}$	$f^{(1)}$	$f^{(2)}$	$f^{(0)}$	$f^{(1)}$	$f^{(2)}$
f_0	1	0	?	1	0	?
f_1	0	?	?	1	0	0
f_2	0	0	?	1	0	?
f_3	0	0	0	1	?	?

Bernstein polynomials

It is easier to measure the co-ordinates of the points **A**, **M**, **N** and **B** with respect to the origin than to measure the co-ordinate of the difference vectors $\mathbf{a}_0, \ldots, \mathbf{a}_3$. Denoting the position vectors of **A**, **M**, **N**, etc., by $\mathbf{s}_0$, $\mathbf{s}_1$, $\mathbf{s}_2$, etc. (see Figure 2.13), the equation of the curve becomes

$$\mathbf{P}(u) = \sum_{i=0}^{m} \mathbf{s}_i \mathbf{B}_{i,m}(u) \qquad (7)$$

where $B_{i,m}$ are the Bernstein polynomials (see Figure 2.14)

$$B_{i,m}(u) = \binom{m}{i} u^i (1-u)^{m-i} \qquad (8)$$

At first look, the change does not seem to be important, but it yields an important result: curve (7) will enjoy the variation diminishing property as the Bernstein polynomials are variation diminishing (a curve is variation diminishing if any plane intersects the curve in no more points than it intersects its control polygon). $B_{i,m}$ attains

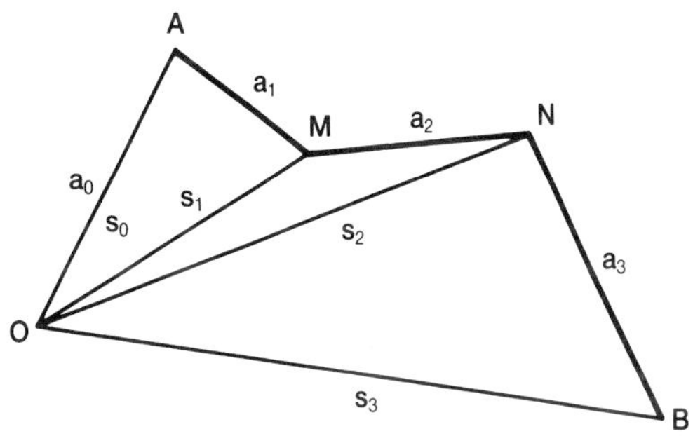

Figure 2.13. Curve defined by position vectors.

Table 2.2.

	$u = 0$					$u = 1$				
	$f^{(0)}$	$f^{(1)}$	$f^{(2)}$	$f^{(3)}$	$f^{(4)}$	$f^{(0)}$	$f^{(1)}$	$f^{(2)}$	$f^{(3)}$	$f^{(4)}$
f_0	1	0	0	?	?	1	0	0	?	?
f_1	0	?	?	?	?	1	0	0	0	0
f_2	0	0	?	?	?	1	0	0	0	?
f_3	0	0	0	?	?	1	0	0	?	?
f_4	0	0	0	0	?	1	0	?	?	?
f_5	0	0	0	0	0	1	?	?	?	?

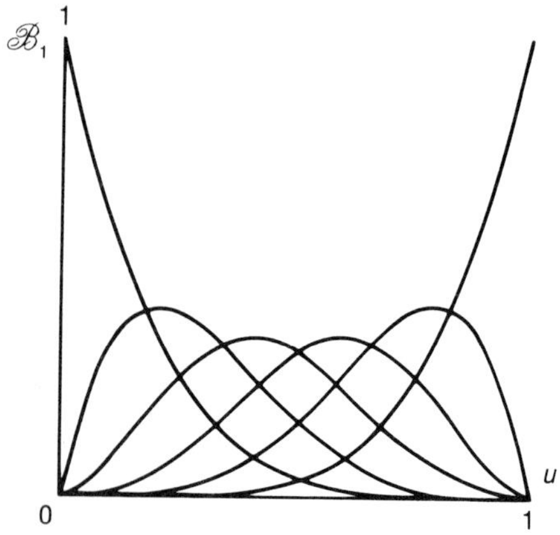

Figure 2.14. Bernstein polynomials of degree 5.

its maximum value at $u = i/m$, and this is approximately 0.3. Consequently, when an apex of the control polygon $\mathbf{s}_0, \mathbf{s}_1, \ldots$ is moved a distance δ, each point of the curve is shifted along the vector with magnitude $\delta B_{i,m}(u)$, where u is the parameter of the point. Furthermore, since the maximum of $B_{i,m}$ is about 0.3, the maximum displacement of the curve is about $\delta/3$ which happens in a region near the displaced vertex. Therefore, it is relatively easy to guess which way a polygon should be distorted to modify the curve by a certain amount.

Polynomial form

In expressions (1) or (7) we can add the vector coefficients related to the same degree of the parameter to yield

$$\mathbf{P}(u) = \sum_{i=0}^{m} \mathbf{b}_i u^i \tag{9}$$

which is the most suitable for evaluation using Horner's algorithm. There are matrices which relate the a's or s's to the b's. Since the matrix defining the Bernstein polynomials is invertible, the process works both ways.

Geometric properties

The curves previously defined enjoy several positive properties. We list below some of them which are considered important in practical implementations:

- Polygons **AMNB** and **BNMA** yield the same curve geometrically. The point $\mathbf{P}(u)$ of the first is identical to the point $\mathbf{P}(1-u)$ of the second.
- The tangents at the endpoints are collinear with $\mathbf{a}_1$ and $\mathbf{a}_m$.
- The curvatures at the endpoints depend only on the vectors $\mathbf{a}_1$ and $\mathbf{a}_2$, and $\mathbf{a}_{m-1}$ and $\mathbf{a}_m$, respectively.
- The point $\mathbf{P}(u)$ can be obtained by a simple geometric construction. Let the original polygon be **AMNRB** (Figure 2.15). The new points **A′**, **M′**, **N′** and **R′** are defined as

$$\frac{\mathbf{AA'}}{\mathbf{AM}} = \frac{\mathbf{MM'}}{\mathbf{MN}} = \frac{\mathbf{NN'}}{\mathbf{NR}} = \frac{\mathbf{RR'}}{\mathbf{RB}} = u.$$

Using the new polygon **A′M′N′R′**, points **A″**, **M″** and **N″** will be defined in the same way. Now, using **A″M″N″**, we get **A‴** and **M‴** with which P is obtained so as to satisfy $\mathbf{A'''P}/\mathbf{A'''M'''} = u$. The line **A‴M‴** is tangential to the curve at **P**, and the subpolygon **AA′A″A‴P** defines the curve between **A** and P.

Sectioning

We can express a part of a curve in the same way as the original curve, using the same expression with different control points. Let $\mathbf{P}(u_0)$ and $\mathbf{P}(u_1)$ be two points on a curve. Then the segment between them is defined as

$$\mathbf{P}'(u) = \sum_{i=0}^{m} \mathbf{b}_i[u_0 + (u_1 - u_0)u]^i \tag{10}$$

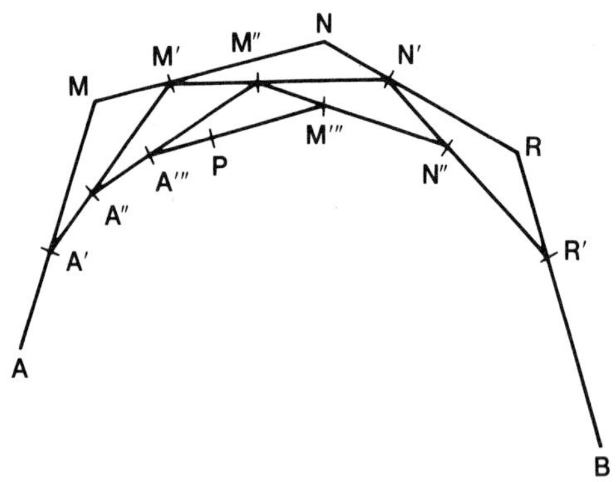

Figure 2.15. Geometric construction of a point on the curve.

The quantity between the brackets varies from u_0 to u_1 as u varies from 0 to 1, i.e. the curve (10) will sweep out the segment from $\mathbf{P}(u_0)$ to $\mathbf{P}(u_1)$. The new segment can be expressed as

$$\mathbf{P}'(u) = \sum_{i=0}^{m} \mathbf{b}'_i u^i \tag{11}$$

From the vectors $\mathbf{b}_i$ of the original curve, a matrix transformation yields the vectors $\mathbf{b}'_i$, and from the $\mathbf{b}'_i$ another matrix transformation yields the control points $\mathbf{s}'_i$.

Degree raising

It is sometimes necessary to raise the degree of the curve, either to use it for surface definition of a higher degree, or to increase the variety of shapes it can take. A degree 3 curve, for instance, can be expressed as follows (see equation (9)):

$$\mathbf{P}(u) = [1 \; u \; u^2 \; u^3] \begin{bmatrix} \mathbf{b}_0 \\ \mathbf{b}_1 \\ \mathbf{b}_2 \\ \mathbf{b}_3 \end{bmatrix}$$

Then the following expression yields the same curve

$$\mathbf{P}(u) = [1 \; u \; u^2 \; u^3 \; u^4 \; u^5] \begin{bmatrix} \mathbf{b}_0 \\ \mathbf{b}_1 \\ \mathbf{b}_2 \\ \mathbf{b}_3 \\ 0 \\ 0 \end{bmatrix}$$

From this expression, using the **b**'s matrix, we get the new control points, i.e. the s's matrix. Combining the previous two techniques, one can section a portion of a curve, raise its order, modify its shape, and finally, reinsert it into the original curve.

Hodographs

Hodographs result from the applications of nice geometric properties of the Bernstein polynomials. Let G be a curve defined by the polygon **AMNQRB** (Figure 2.16). The difference vectors

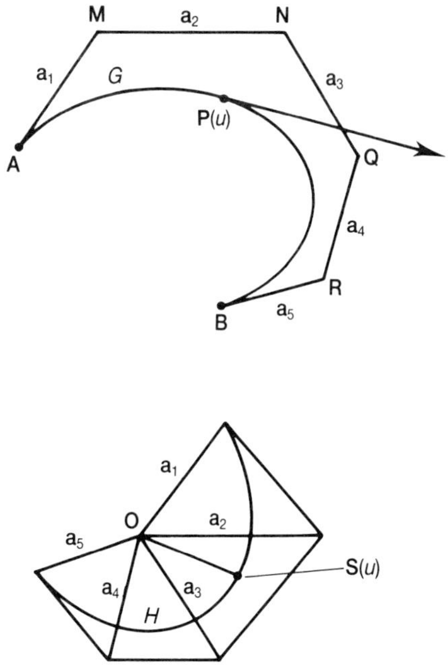

Figure 2.16. A curve and its hodograph.

$\mathbf{a}_1, \ldots, \mathbf{a}_5$, pointing from a common origin **O**, will define a curve H with control points $\mathbf{a}_1, \ldots, \mathbf{a}_5$ and whose degree is one less than that of G. It can be shown that

$$\frac{d\mathbf{P}}{du}(u) = m \cdot \mathbf{OS}\,(u)$$

where m is the degree of the original curve, i.e. derivative vectors of G are position vectors of H. Now, defining the hodograph of H, we can arrive at the second hodograph of G, and so on. From the shape of the successive hodographs, one can deduce whether the curve has an inflection point, a cusp or a flat point (a point where the curvature is zero but in the neighborhood of the point it does not change sign).

Surfaces

Principle

A surface is composed of patches connected together with slope or curvature continuity. A patch is the locus of points obtained by displacing and distorting a generatrix curve. Let the

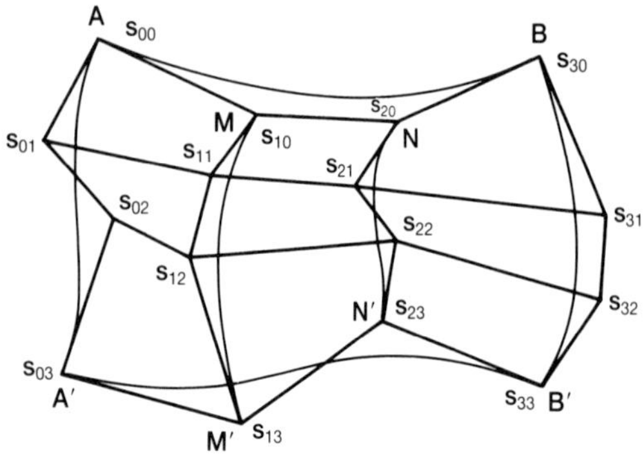

Figure 2.17. Surface defined by a control net.

boundary directrices be **AB** and **A′B′** defined by the polygons **AMNB** and **A′M′N′B′** (see Figure 2.17). We can displace **AB**, say, by displacing the corresponding control points. The trajectories of the moving **A**, **M**, **N** and **B** are defined by their polygons whose vertices can be connected to yield a net defining a patch similarly to the curve definition.

Evaluation

Let us denote the vertices of the net by $\mathbf{s}_{ij}$ (Figure 2.17). The point on the patch is given by

$$\mathbf{P}(u, v) = \sum_{i=0}^{m} \sum_{j=0}^{n} \mathbf{s}_{i,j}\mathbf{B}_{i,m}(u) B_{j,n}(v), \qquad u, v \in [0, 1] \quad (12)$$

or in power basis form

$$\mathbf{P}(u, v) = \sum_{i=0}^{m} \sum_{j=0}^{n} \mathbf{b}_{i,j}u^i v^j, \quad u, v \in [0,1] \quad (13)$$

Keeping u or v constant, we can define isoparametric lines, and it is easily seen that these lines intersect one another. Properties and techniques obtained for curves can be carried over to surfaces. For example, sectioning a patch is obtained by applying technique (10) to (13), or increasing the degree needs adding zeros to the matrix formulation of (13) and transforming back to (12). Displacing a control point a distance δ alters the shape with a distance approximately equal to $\delta/9$.

Volumes

The curve and surface definitions above can be extended to represent a volume defined by a triparametric vector-valued function

$$\mathbf{P}(u, v, w) = \sum_{i=0}^{m} \sum_{j=0}^{n} \sum_{k=0}^{p} \mathbf{b}_{i,j,k}u^i v^j w^k, \qquad u, v, w \in [0, 1] \quad (14)$$

This concept could be useful in finite elements method applied to solids. Another application is to insert an already defined object into a triorthogonal parametric space which is distorted by displacement of some of its vertices. This entails a general distorsion of the inserted object.

Blending

For scientific, technical or aesthetic reasons, continuity between adjacent curves or surfaces must be ensured. There are two main types of continuity: parametric and geometric. In some fields such as fluids dynamics, derivative (i.e. parametric) continuity is of great importance. On the other hand, in most cases geometric continuity (i.e. slope or curvature continuity) is quite sufficient. UNISURF chose to work with geometric continuity.

Continuity between curves

To ensure slope continuity between curves C and C' (see Figure 2.18), the difference vectors $\mathbf{a}_1$ and $\mathbf{a}'_1$ must be collinear. Curvature continuity is ensured if the centers of curvature of both curves at the common join coincides. This imposes conditions not only on $\mathbf{a}_1$ and $\mathbf{a}'_1$, but on $\mathbf{a}_2$ and $\mathbf{a}'_2$ (for a detailed treatment see the author's recent book [13]).

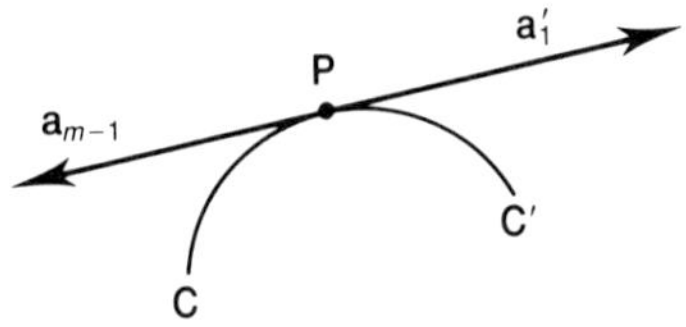

Figure 2.18. Slope continuity between curves.

Continuity between patches

UNISURF was created to solve problems related to objects having little to do with fluid dynamics, hence conditions regarding curvature continuity were not deeply investigated. The simplest solution to the slope continuity problem is similar to that of Coons. The adjacent patches must have a common boundary defined by, say, **AMNB** (see Figure 2.19). Let us denote the neighboring vertices of the respective nets by **A**′, **M**′,.. and **A**″, **M**″.... Then the condition for slope continuity can be expressed as

$$\frac{\mathbf{AA'}}{\mathbf{AA''}} = \frac{\mathbf{MM'}}{\mathbf{MM''}} = \frac{\mathbf{NN'}}{\mathbf{NN''}} = \frac{\mathbf{BB'}}{\mathbf{BB''}} = \text{const.}$$

This condition is sometimes too restrictive, especially when the joining patches are highly irregular such as those in Figure 2.20. Another less restrictive condition is obtained by requiring the partial derivative along the common boundary to be coplanar with those across this

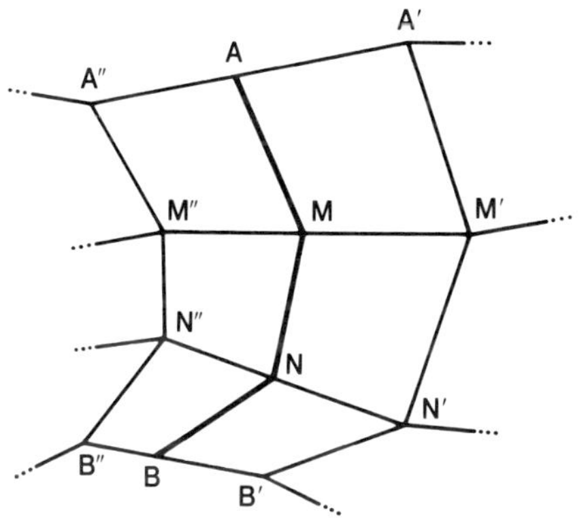

Figure 2.19. Slope continuity between patches.

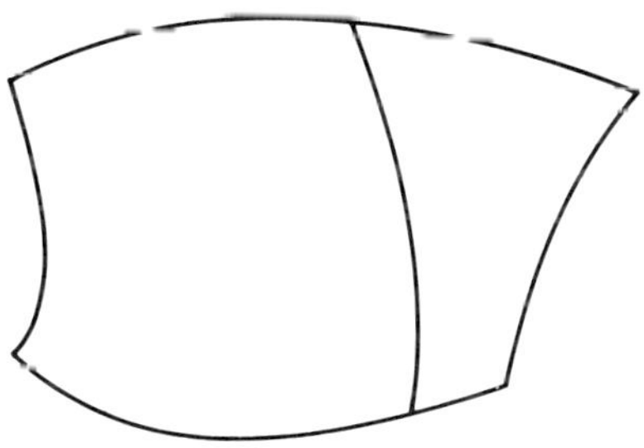

Figure 2.20. Join of irregularly shaped patches.

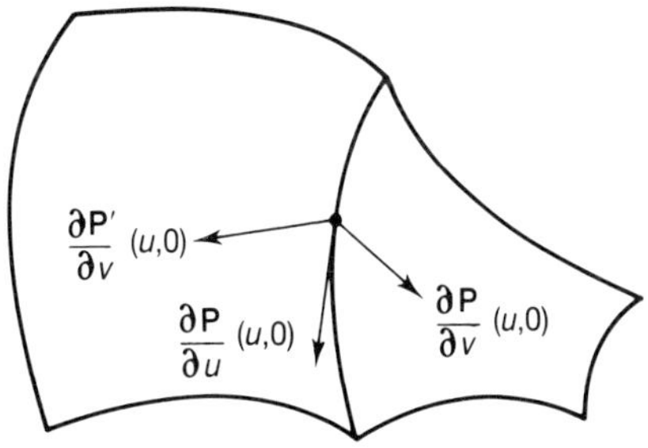

Figure 2.21. Tangent plane continuity across patches.

borderline (Figure 2.21). More precisely,

$$\frac{\partial \mathbf{P}'}{\partial v}(u, 0) = h(u)\frac{\partial \mathbf{P}}{\partial v}(u, 0) + k(u)\frac{\partial \mathbf{P}}{\partial u}(u, 0) \tag{15}$$

The functions $h(u)$ and $k(v)$ can be arbitrary. Most of the time, they are polynomials, and to ensure continuity with lateral patches, cubics are used in the form

$$h(u) = 2[h(0) - h(1)]u^3 - 3[h(0) - h(1)]u^2 + h(0)$$

where $h(0)$ and $h(1)$ are given values. A similar expression holds for $k(u)$ as well. Of course, the above process increases the degree of the patch in the u-direction by 3, and care must be taken to avoid ending up with patches with very high degree.

Transposants

In this section we give a simple definition of a curve lying on a surface. Assume that we are given a curve in the (u, v) parameter space (see Figure 2.22).

$$u = \sum_{h=0}^{p} c_h w^h \quad \text{and} \quad v = \sum_{k=0}^{p} e_k w^k$$

This curve mapped onto the patch can be expressed in the following form (see Figure 2.23):

$$\mathbf{P}(w) = \sum_{i=0}^{m}\sum_{j=0}^{n} \mathbf{b}_{i,j}\left(\sum_{h=0}^{p} c_h w^h\right)^i \left(\sum_{k=0}^{p} e_k w^k\right)^j$$

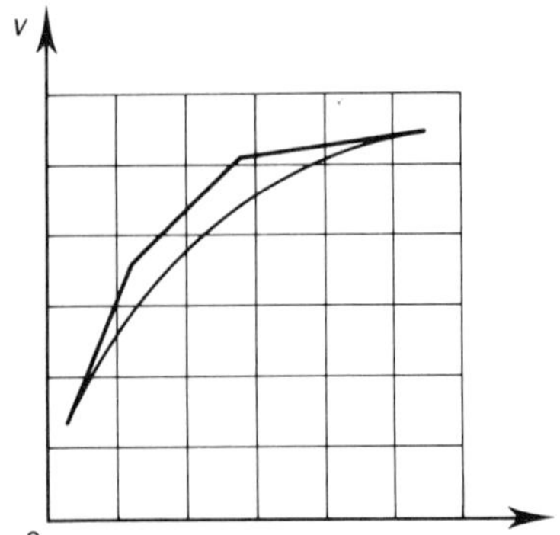

Figure 2.22. Curve defined in the (u, v) parameter space of a patch.

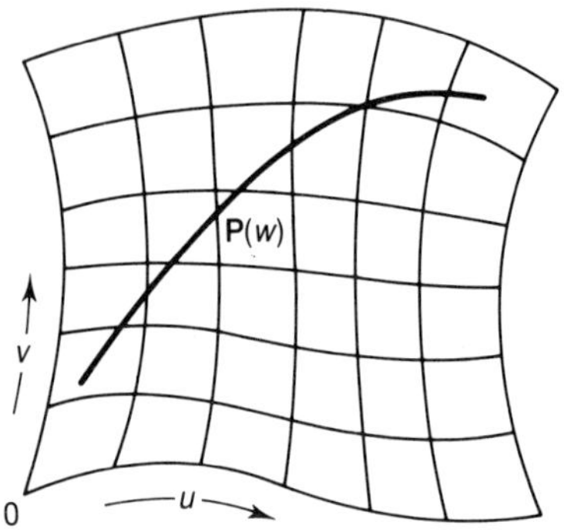

Figure 2.23. Curve on the surface obtained by mapping the parameter space curve shown in Figure 2.22.

Collecting the terms related to the same power of w yields

$$\mathbf{P}(w) = \sum_{l=0}^{(m+n)p} \mathbf{g}_l w^l \tag{16}$$

If the degree $(m + n)p$ is too high, then approximation can be used. One solution is to use the values of the function and its lower order derivatives at both ends to define a reasonable approximation. From (16) one gets

$$\mathbf{P}(0) = \mathbf{g}_0 \qquad \mathbf{P}(1) = \sum_{l=0}^{(m+n)p} \mathbf{g}_l$$

$$\frac{d\mathbf{P}}{dw}(0) = \mathbf{g}_1 \qquad \frac{d\mathbf{P}}{dw}(1) = \sum_{l=1}^{(m+n)p} l\mathbf{g}_l$$

$$\frac{d^2\mathbf{P}}{dw^2}(0) = 2\mathbf{g}_2 \qquad \frac{d^2\mathbf{P}}{dw^2}(1) = \sum_{l=2}^{(m+n)p} l(l-1)\mathbf{g}_l$$

Using these data, a degree five curve can be defined as an initial approximation. If the resulting curve differs too much from the original one, sectioning can be applied to arrive at a piecewise curve that best approximates the desired result.

Hodographs

Similar to the curves, patches can be defined representing any order of derivative with respect to one or two parameters. Hodographs can be used to detect unwanted negative-draft regions resulting from manipulation with the surface to improve manufacturing conditions.

CONCLUSIONS

The UNISURF system was developed to help car body design, stamping tool design and manufacturing. There was no need to deal with shapes defined by lines and circles – these are common among mechanical parts, but seldom used in car body design. UNISURF was designed to be used for the direct definition of shapes by stylists and designers, whilst the traditional practice was to translate an already existing hand-made full scale 3D model. It was assumed that some people would be reluctant to change a time-honored practice, hence UNISURF had to offer the capability of translating existing objects as well as creating new ones. The most important requirement was that it should be both easily understood and easily used by designers, technicians and operators having little knowledge about analysis. The second important point was that the system should be interactive, i.e. the drawing office should be equipped with a very fast milling machine capable of carving the full scale shape of a car in no more than a few days. Flexibility, interactivity and speed, together with accuracy and low costs, are the major assets of a CAD/CAM system.

REFERENCES

1 **S. N. Bernstein** Démonstration du théorème de Weierstrass fondée sur le calcul des probabilités,

Harkov Soobs. Mathem. ob-va, Vol 13 No 2, 1912, 1–2.
2 **I. J. Schoenberg** Contribution to the problem of approximation of equidistant data by analytic functions, *Q. Appl. Math.* Vol 4, 1946, 45–99.
3 **J. C. Ferguson** Multi-variable curve interpolation, *J. ACM*, Vol 11 No 2, 1964, 221–228.
4 **I. E. Sutherland** SKETCHPAD, a man-machine graphical communication system, *AFIPS Conference Proc.*, Vol 23, 1963, 329–346.
5 **E. Mehlum** *Curve and surface fitting based on variational criteria for smoothness*, Central Institute for Industrial Research, Oslo, Norway, December 1969.
6 **S. A. Coons** *Surfaces for computer-aided design of space forms*, TR MAC-TR-41, MIT, Cambridge, MA, June 1967.
7 **W. J. Gordon** Blending function methods of bivariate and multi-variate interpolation and approximation, *SIAM J. Numer. Anal.*, Vol 8 No 1, 1971, 158–177.
8 **A. R. Forrest** *Curves and surfaces for computer-aided design*, PhD Thesis, University of Cambridge, UK, 1968.
9 **R. F. Riesenfeld** *Applications of B-spline approximation to geometric problems of computer-aided design*, PhD Thesis, Syracuse University, NY, 1973.
10 **J. Ponte** Le systéme Profile Data, *Mecanique et Electricite*, August 1968.
11 **P. de Casteljau** *Formes a poles*, Note Citroën, 1959.
12 **P. E. Bézier** How Renault uses numerical control for car body design and tooling, *SAE Congress*, Paper 680010, Detroit, MI, 1968.
13 **P. E. Bézier** *The Mathematical Basis of the UNISURF CAD System*, Butterworths, London, 1986.

BIBLIOGRAPHY

Auffray and Pichouron *Définition des surfaces développables s'appuyant sur une courbe. Application a la détermination des surfaces annexes des outlis d'emboutissage*, Mémoire du C.N.A.M., 1974.

F. Bernard *Surfaces complexes lissées selon un critere d'energie minimale*, Note Avions Marcel Dassault.

P. E. Bézier Emploi de la commande numérique pour l'étude des carrosseries et des outillages, *J. SIA*, April 1968.

P. E. Bézier Définition numerique des courbes et des surfaces, *Automatisme XII*, 1967, 17–21.

P. E. Bézier *Numerical Control – Mathematics and application*, John Wiley, 1972.

P. E. Bézier *Essai de définition numérique des courbes et surfaces expérimentales*, DSc Thesis, Université de Paris, 1977.

P. E. Bézier UNISURF, from styling to tool shop, *Proc. CAPE Meeting*, North Holland, 1983.

P. E. Bézier and S. Sioussiou Semi-automatic system for defining free-form curves and surfaces, *Computer-Aided Design*, Vol 15 No 2, 1983, 65–72.

P. E. Bézier *Mathematiques et CAD – Courbes et Surfaces*, Collection Hermes, 1986.

W. Boehm Generating the Bézier points of triangular splines, *Surfaces in CAGD*, North Holland, 1983.

G. E. Farin *Bézier polynomials over triangles and the construction of C^r polynomials*, TR Brunel University, UK, 1980.

G. E. Farin Algorithms for rational Bézier curves, *Computer-Aided Design*, Vol 15 No 2, 1983, 73–77.

A. R. Forrest Interactive interpolation and approximation by Bézier polynomials, *The Computer Journal*, Vol 15 No 1, 1972, 71–78.

W. J. Gordon and R. F. Riesenfeld Bernstein–Bézier methods for the computer-aided design of free-form curves and surfaces, *J. ACM*, Vol 21 No 2, 1974, 293–310.

D. Lasser Bernstein–Bézier representation of volumes, *CAGD*, Vol 2 Nos 1–3, 1985, 145–149.

M. A. Sabin *Parametric splines in tension*, British Aircraft Corporation, July 1970.

I. J. Schoenberg Smoothing operators and their generating functions, *Bull. AMS*, 1946.

M. Véron *Contribution a l'étude des surfaces numériques UNISURF. Conditions de raccordement*, PhD Thesis, Université de Nancy, 1973.

ABOUT THE AUTHOR

Pierre E. Bézier is a retired professor of production engineering at Conservatoire National des Arts et Métiers. He attended the Ecole Nationale Supérieure des Arts et Métiers (1927–30), the Ecole Supérieure d'Electricité (1930–31), and received his DSc in mathematics from the University of Paris in 1977.

Bézier started his career at Renault as a tool setter in 1933. He became tool designer for production engineering in 1934, and director of tool design office in 1945. In 1946, as Director of Production Engineering, he was responsible for the design of the transfer lines producing most of the 4 CV mechanical parts. In 1956 he became involved in automatic storing and transportation of heavy parts, as well as loading and unloading in mass production lines. His responsibilities during 1957–60 included automatic assembly of mechanical components, and design and production of an NC drilling and milling machine, probably one of the first such machines in Europe.

Bézier started his research in CAD/CAM in 1960, when he began work on his UNISURF CAD system for free-form curves and surfaces. From 1960, his research interest focused on drawing machines, autonomous computer control, interactive design and 3D milling for manufacturing clay-models and masters. He retired from Renault in 1975, but remained active in many of his activities.

Bézier wrote four books in the field of CAD/CAM, received numerous distinctions including the Steven A. Coons award of the Association for Computing Machinery (ACM), and honorary Doctor of Engineering of the Technical University of Berlin. He is an honorary member of the American Society of Mechanical Engineers and Société Belge des Mécaniciens, ex-president of several associations including Société des Ingénieurs de l'Automobile, Ingénieurs et Scientifiques de France and Groupement pour l'Avancement de la Mécanique Industrielle, and is one of the first Advisory Editors of the *Computer-Aided Design* journal.

3 B(asic)-Spline Basics

Carl de Boor

INTRODUCTION

This chapter reviews those basic facts about (univariate) B-splines which are of interest in CAGD. The intent is to give a self-contained and complete development of the material in as simple and direct a way as possible. For this reason, the B-splines are defined via the recurrence relations, thus avoiding the discussion of divided differences which the traditional definition of a B-spline as a divided difference of a truncated power function requires. This does not force more elaborate derivations than are available to those who feel at ease with divided differences. It does force a change in the order in which facts are derived and brings more prominence to such things as Marsden's Identity or the Dual Functionals than they currently have in CAGD.

In addition, it highlights the following point: the consideration of a single B-spline is not very fruitful when proving facts about B-splines, even if these facts (such as the smoothness of a B-spline) can be stated in terms of just one B-spline. Rather, simple arguments and real understanding of B-splines are available only if one is willing to consider *all* the B-splines of a given order for a given knot sequence. Thus it focuses attention on **splines**, i.e. on the linear combination of B-splines. In this connection, it is worthwhile to stress that this essay (as does its author) maintains that the term 'B-spline' refers to a certain spline of minimal support and, contrary to usage unhappily current in CAGD, does not refer to a curve which happens to be written in terms of B-splines. It is too bad that this misuse has become current, and entirely unclear why.

The chapter deals with splines for an arbitrary knot sequence, and rarely becomes more specific. In particular, the B(ernstein–Bézier)-net for a piecewise polynomial, though a (very) special case of a representation by B-splines, gets much less attention than it deserves, given its immense usefulness in CAGD (and spline theory).

The essay deals only with spline **functions**. There is an immediate extension to spline **curves**: Allow the coefficients, be they B-spline coefficients or coefficients in some polynomial form, to be points in $\mathbb{R}^2$ or $\mathbb{R}^3$. But this misses the much richer structure for spline curves available because of the fact that even discontinuous parametrizations may describe a smooth curve.

Splines are of importance in CAGD for the same reason that they are used wherever data are to be fit or curves are to be drawn by computer: being polynomial, they can be evaluated quickly; being piecewise polynomial, they are very flexible; their representation in terms of B-splines provides geometric information and insight. See Riesenfeld's contribution in this volume for details concerning the use of splines and, especially, of their B-spline representation, in CAGD. See Cox's contribution for details concerning numerical algorithms to handle splines and their B-spline representation.

The editor of this volume has asked me to provide a careful discussion of the **place-holder** notation customary in mathematical papers on splines. This notation was invented by people who think it important to distinguish the function f from its value $f(x)$ at the point x. It is customarily used in the description of a function of *one* variable obtained from a function of *two* variables by holding one of those two variables fixed. In this essay, it appears only to describe functions obtained from others by shifting and/or scaling of the independent variable. Thus, $f(\cdot - z)$ is the function whose value at x is the number $f(x - z)$, while $g(\alpha + \beta\cdot)$ is the function whose value at t is $g(\alpha + \beta t)$, etc.

It is also worth pointing out that I have been very careful to distinguish between 'equality' and 'equality by definition'. The latter I have always indicated by using a colon on the same side of the equality sign as the term being defined. I use $D^r f$ (instead of $f^{(r)}$) to denote the rth derivative of the function f, and use Π_r to denote the collection of all polynomials of degree $\leqslant r$. The notation $\Pi_{<r}$ for the collection of all polynomials of degree $<r$ (i.e. of order r) will be particularly handy. Finally, I use a double period to indicate an interval; e.g.

Fundamental Developments of Computer-Aided Geometric Modeling
ISBN 012-554765-X

$[x\,..\,y) := \{(1-t)x + ty : 0 \leqslant t < 1\}$. This helps to distinguish the interval $(a\,..\,b)$ from the point (a, b) in the plane, or the interval $[a\,..\,b]$ from the first divided difference $[a, b]$.

B-SPLINES DEFINED

We start with a partition or **knot sequence**, i.e. a **nondecreasing** sequence $\mathbf{t} := (t_i)$. The **B-splines of order 1** for this knot sequence are the characteristic functions of this partition, i.e. the functions

$$B_{i1}(t) := X_i(t) := \begin{cases} 1, & \text{if } t_i \leqslant t < t_{i+1} \\ 0, & \text{otherwise} \end{cases} \qquad (1)$$

Note that all these functions have been chosen here to be **right-continuous.** Other choices could have been made with equal justification. The only constraint is that these B-splines should form a **partition of unity**, i.e.

$$\sum_i B_{i1}(t) = 1, \quad \text{for all } t \qquad (2)$$

In particular,

$$t_i = t_{i+1} \quad \text{implies} \quad B_{i1} = X_i = 0 \qquad (3)$$

From these first-order B-splines, we obtain **higher-order** B-splines by **recurrence**:

$$B_{ik} := \omega_{ik} B_{i,k-1} + (1 - \omega_{i+1,k}) B_{i+1,k-1} \qquad (4\text{a})$$

with

$$\omega_{ik}(t) := \begin{cases} \dfrac{t - t_i}{t_{i+k-1} - t_i}, & \text{if } t_i \neq t_{i+k-1} \\ 0, & \text{otherwise} \end{cases} \qquad (4\text{b})$$

Thus, the **second-order** B-spline is given by

$$B_{i2} = \omega_{i2} X_i + (1 - \omega_{i+1,2}) X_{i+1} \qquad (5)$$

and so consists, in general, of two nontrivial linear pieces which join continuously to form a piecewise linear function which vanishes outside the interval $[t_i\,..\,t_{i+2})$. For this reason, some call B_{i2} a **linear** B-spline. If, for example, $t_i = t_{i+1}$ (hence $X_i = 0$), but still $t_{i+1} < t_{i+2}$, then B_{i2} consists of just one nontrivial piece and fails to be continuous at the **double knot** $t_i = t_{i+1}$, as is shown in Figure 3.1.

The **third-order** B-spline is given by

$$\begin{aligned} B_{i3} &= \omega_{i3} B_{i2} + (1 - \omega_{i+1,3}) B_{i+1,2} \\ &= \omega_{i3}\omega_{i2} X_i + (\omega_{i3}(1 - \omega_{i+1,2}) \\ &\quad + (1 - \omega_{i+1,3})\omega_{i+1,2}) X_{i+1} \\ &\quad + (1 - \omega_{i+1,3})(1 - \omega_{i+2,2}) X_{i+2} \end{aligned} \qquad (6)$$

This shows that, in general, B_{i3} consists of 3 (nontrivial) quadratic pieces, and to judge from Figure 3.2, these seem to join smoothly at the knots to form a C^1 piecewise quadratic function which vanishes outside the interval $[t_i\,..\,t_{i+3})$. Coincidences among the knots $t_i, \ldots, t_{i+3}$ would change this. If, for instance, $t_i = t_{i+1} = t_{i+2}$

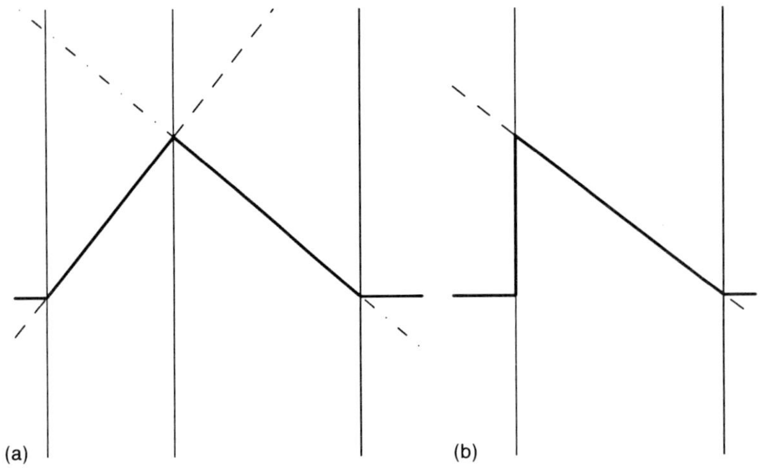

Figure 3.1. Linear B-spline with (a) simple knots, (b) a double knot.

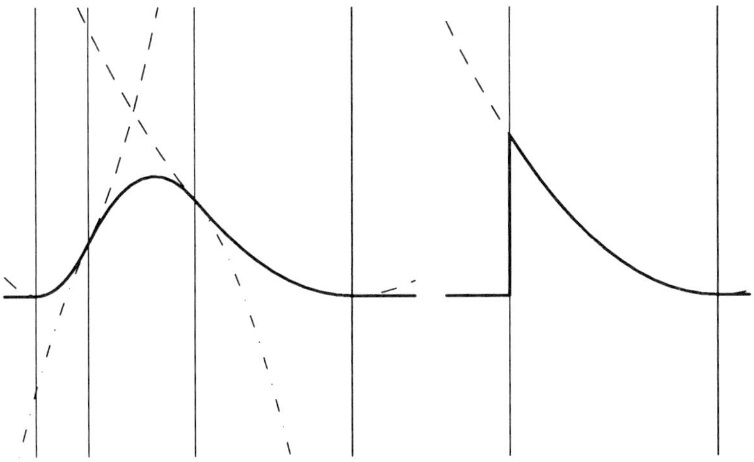

Figure 3.2. Quadratic B-spline with (a) simple knots, (b) a triple knot.

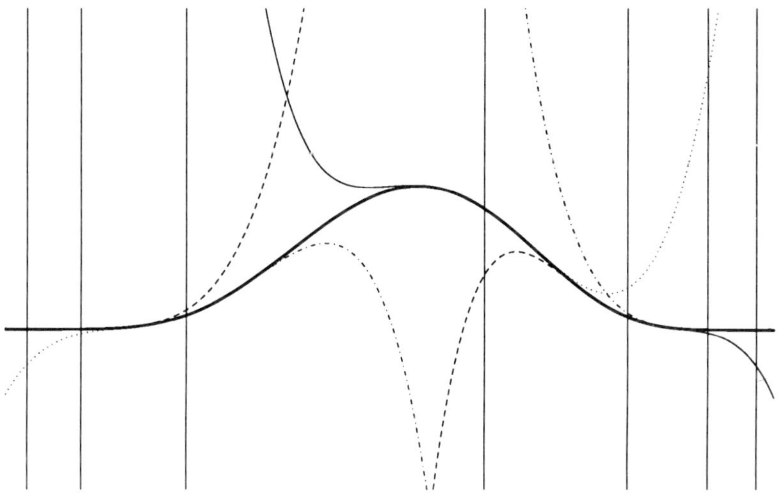

Figure 3.3. A 6th order B-spline and the six quintic polynomials whose selected pieces make up the B-spline.

(hence $X_i = X_{i+1} = 0$), then B_{i3} consists of just one nontrivial piece, fails to be even continuous at the **triple knot** t_i, but is still C^1 at the simple knot t_{i+3}, as is shown in Figure 3.2.

After $k-1$ steps of the recurrence, we obtain B_{ik} in the form

$$B_{ik} = \sum_{j=i}^{i+k-1} b_{jk} X_j \tag{7}$$

with each b_{jk} a polynomial of degree $<k$, since it is the sum of products of $k-1$ linear polynomials. Thus a B-spline of **order** k consists of polynomial pieces of **degree** $< k$. (In fact, all b_{jk} are of *exact* degree $k-1$.)

From this, we infer that B_{ik} is a piecewise polynomial of degree $< k$ which vanishes outside the interval $[t_i \,..\, t_{i+k})$ and has possible **breakpoints** $t_i, \ldots, t_{i+k}$ (cf. Figure 3.3). In particular, B_{ik} is just the zero function in case $t_i = t_{i+k}$. Also, by induction, B_{ik} is positive on the open interval $(t_i \,..\, t_{i+k})$, since both ω_{ik} and $1 - \omega_{i+1,k}$ are positive there (cf. Figure 3.4).

Further, we see that B_{ik} is completely

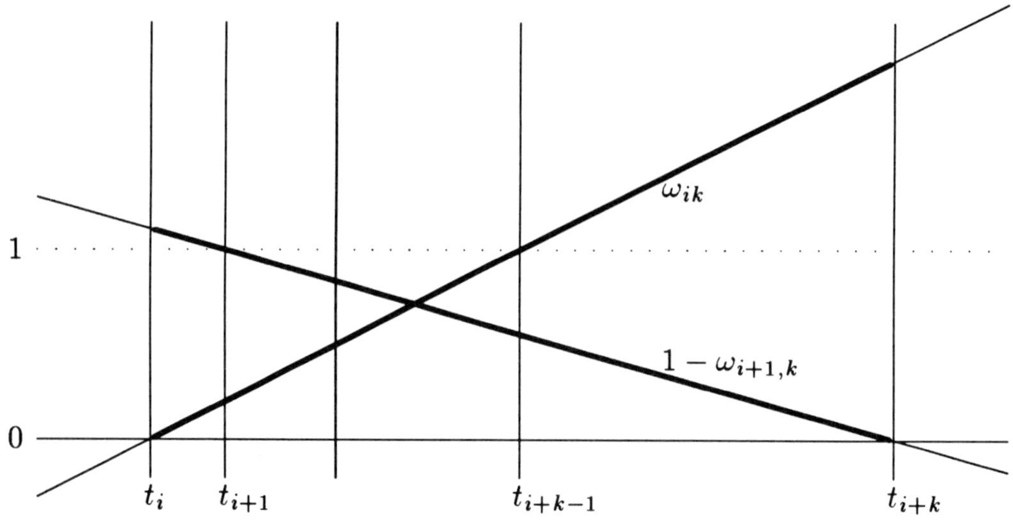

Figure 3.4. The two weight functions in (4 a) are positive on $(t_i\,..\,t_{i+k}) = \operatorname{supp} B_{ik}$.

determined by the $k+1$ knots $t_i, \ldots, t_{i+k}$. For this reason, the notation

$$B(\cdot|t_i, \ldots, t_{i+k}) := B_{ik} \tag{8}$$

is sometimes used. Other notations in use include

$$N_{ik} := B_{ik} \quad \text{and} \quad M_{ik} := (k/(t_{i+k} - t_i))B_{ik} \tag{9}$$

The many other properties of B-splines are derived most easily by considering not just one B-spline but the linear span of *all* B-splines of a given order k for a given knot sequence **t**. This brings us to splines.

SPLINES DEFINED

A **spline of order** k **with knot sequence t** is, by definition, a linear combination of the B-splines B_{ik} associated with that knot sequence. We denote by

$$S_{k,\mathbf{t}} := \left\{ \sum_i B_{ik} a_i : a_i \in \mathbb{R} \right\} \tag{10}$$

the collection of all such splines.

We have left open so far the precise nature of the knot sequence **t**, other than to specify that it be a nondecreasing real sequence. In any practical situation, **t** is necessarily a *finite* sequence. But, since on any nontrivial interval $[t_j\,..\,t_{j+1})$ at most k of the B_{ik} are nonzero, *viz.* $B_{j-k+1,k}, \ldots, B_{jk}$ (cf. Figure 3.5), it doesn't really matter whether **t** is finite, infinite, or even bi-infinite; the sum in (10) always makes pointwise sense, since on any interval $[t_j\,..\,t_{j+1})$, at most k summands are not zero.

We will pay special attention to the following two 'extreme' knot sequences, the sequence

$$\mathbb{Z} := (\ldots, -2, -1, 0, 1, 2, \ldots)$$

and the sequence

$$\mathbb{B} := (\ldots, 0, 0, 0, 1, 1, 1, \ldots)$$

A spline associated with the knot sequence $\mathbb{Z}$ is called a **cardinal** spline. This term was chosen by Schoenberg [1] because of a connection to Whittaker's Cardinal Series. This is not to be confused with its use in earlier spline literature, where it refers to a spline which vanishes at all points in a given sequence except for one at which it takes the value 1. The latter splines, though of great interest in spline interpolation, do not interest us here.

Because of the uniformity of the knot sequence $\mathbf{t} = \mathbb{Z}$, formulae involving cardinal B-splines are often much simpler than corresponding formulae for general B-splines. To begin with, *all cardinal B-splines (of a given order) are translates of one*

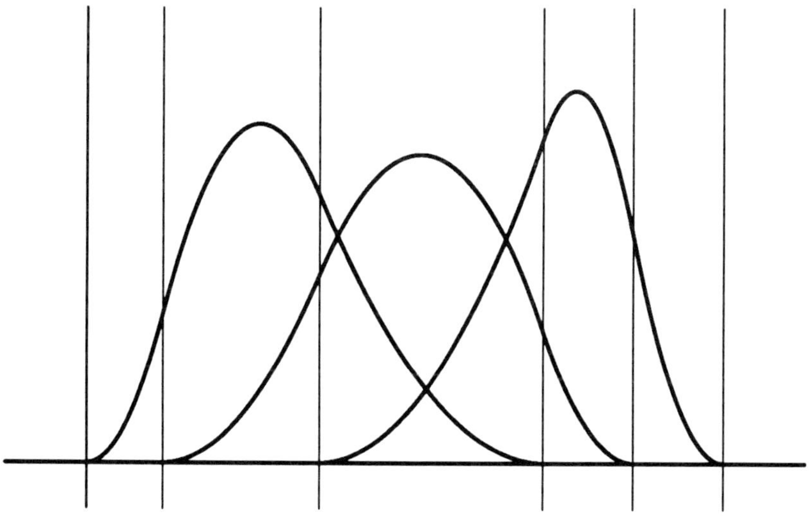

Figure 3.5. The k B-splines whose support contains $[t_j \,..\, t_{j+1})$; here $k = 3$.

another. With the natural indexing $t_i := i$, for all i, for the entries of the uniform knot sequence $\mathbf{t} = \mathbb{Z}$, we have

$$B_{ik} = N_k(\cdot - i) \tag{11}$$

with

$$N_k := B_{0k} = B(\cdot|0, \ldots, k) \tag{12}$$

The recurrence relations (4) simplify as follows:

$$(k-1)N_k(t) = tN_{k-1}(t) + (k-t)N_{k-1}(t-1) \tag{13}$$

The knot sequence $\mathbf{t} = \mathbb{B}$ contains just two points, *viz.* the points 0 and 1, but each with infinite multiplicity. The only nontrivial B-splines for this sequence are those which have both 0 and 1 as knots, i.e. those B_{ik} for which $t_i = 0$ and $t_{i+k} = 1$ (see Figure 3.6). There seems to be no natural way to index the entries in the sequence $\mathbb{B}$. Instead, it is customary to index the corresponding B-splines by the multiplicities of their two distinct knots. Precisely,

$$B_{(\mu,\nu)} := B(\cdot|\underbrace{0, \ldots, 0}_{\mu+1\,\text{times}}, \underbrace{1, \ldots, 1}_{\nu+1\,\text{times}}) \tag{14}$$

With this, the recurrence relations (4) simplify as follows:

$$B_{(\mu,\nu)}(t) = tB_{(\mu,\nu-1)}(t) + (1-t)B_{(\mu-1,\nu)}(t) \tag{15}$$

This gives the formula

$$B_{(\mu,\nu)}(t) = \binom{\mu+\nu}{\mu}(1-t)^\mu t^\nu \quad \text{for } 0 < t < 1 \tag{16}$$

for the one nontrivial polynomial piece of $B_{(\mu,\nu)}$, as one verifies by induction. The formula enables us to determine the *smoothness* of the B-splines in this simple case: Since $B_{(\mu,\nu)}$ vanishes identically outside $[0\,..\,1]$, it has exactly $\nu-1$ continuous derivatives at 0 and $\mu - 1$ continuous derivatives at 1. This amounts to ν **smoothness conditions** at 0 and μ smoothness conditions at 1. Since the order of $B_{(\mu,\nu)}$ is $\mu + \nu + 1$, this is a simple illustration of the generally valid formula

$$\#\text{smoothness conditions at knot} + \text{multiplicity of knot} = \text{order} \tag{17}$$

For fixed $\mu + \nu$, the polynomials in (16) form the so-called **Bernstein** basis (for polynomials

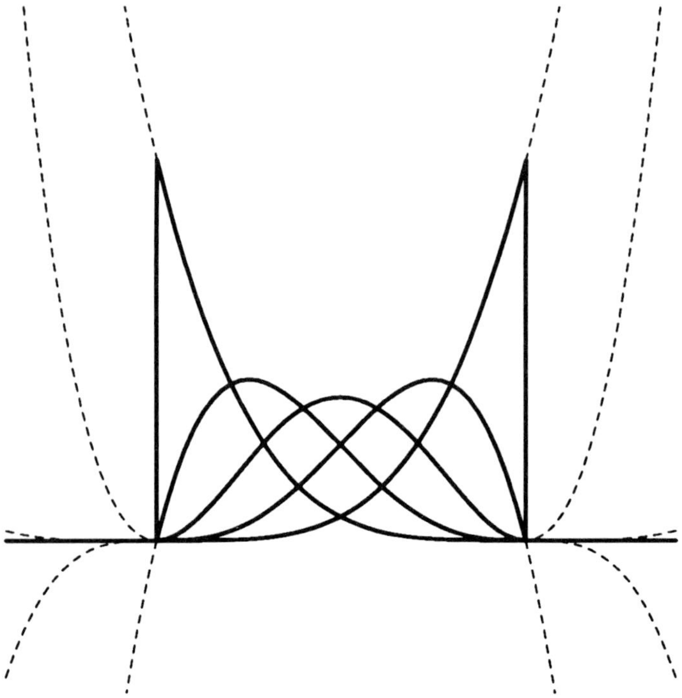

Figure 3.6. Bernstein basis of degree 4 or order 5.

of degree $\leqslant \mu + \nu$) and, correspondingly, the representation

$$p = \sum_{\mu+\nu=h} B_{(\mu,\nu)} a_{(\mu,\nu)} \tag{18}$$

is the **Bernstein** form for the polynomial $p \in \Pi_h$. In CAGD, it is more customary to refer to (18) as the **Bézier** form (for the polynomial p) or as the **Bézier polynomial**, or even the **Bernstein–Bézier** polynomial. It may be simpler to use the short **BB-form** instead.

Finally, I introduce here a **simplifying assumption.** In the next sections, I develop the basic B-spline theory by studying the spline space $S_{k,\mathbf{t}}$, i.e. the collection of all functions s of the form

$$s = \sum_i B_{ik} a_i \tag{19}$$

for a suitable coefficient vector $a = (a_i)$.

In practice, the knot sequence $\mathbf{t}$ is always finite, hence so is the sum in (19). This often requires one to pay special attention to the limits of that summation. Since I find that distracting, I will assume from now on that *the knot sequence* $\mathbf{t}$ *is bi-infinite.* This can always be achieved simply by continuing the sequence indefinitely in both directions (taking care to maintain its monotonicity) and choosing the additional B-spline coefficients to be zero.

More than that, I will assume that

$$t_{\pm\infty} := \lim_{i \to \pm\infty} t_i = \pm\infty \tag{20}$$

This assumption is convenient, since it ensures that every $\tau \in \mathbb{R}$ is in the support of some B-spline.

At times, it will be convenient to assume that

$$t_i < t_{i+k} \quad \text{for all } i \tag{21}$$

which can always be achieved by removing from $\mathbf{t}$ its ith entry as long as $t_i = t_{i+k}$. This does not change the space $S_{k,\mathbf{t}}$ since the only kth order B-splines removed thereby are zero anyway. In fact, another way to state the condition (21) is

$$B_{ik} \neq 0 \quad \text{for all } i \tag{21'}$$

THE POLYNOMIALS IN $S_{k,t}$

We show in this section that $S_{k,t}$ contains

$\Pi_{<k}$ = the collection of all polynomials of degree $<k$

and give a formula for the B-spline coefficients of $p \in \Pi_{<k}$.

We begin with **Marsden's Identity**:

Theorem 1 *For any* $\tau \in \mathbb{R}$,

$$(\cdot - \tau)^{k-1} = \sum_i B_{ik}\psi_{ik}(\tau) \tag{22a}$$

with

$$\psi_{ik}(\tau) := (t_{i+1} - \tau)\cdots(t_{i+k-1} - \tau) \tag{22b}$$

Proof We deduce from the recurrence relation (4) that for an arbitrary coefficient sequence a,

$$\sum B_{ik}a_i = \sum B_{i,k-1}((1-\omega_{ik})a_{i-1} + \omega_{ik}a_i) \tag{23}$$

On the other hand, for the special sequence

$$a_i := \psi_{ik}(\tau) := (t_{i+1} - \tau)\cdots(t_{i+k-1} - \tau)$$

(with $\tau \in \mathbb{R}$), we find for $B_{i,k-1} \neq 0$, i.e. for $t_i < t_{i+k-1}$, that

$$\begin{aligned}(1-\omega_{ik})a_{i-1} + \omega_{ik}a_i &= ((1-\omega_{ik})(t_i - \tau) \\ &\quad + \omega_{ik}\cdot(t_{i+k-1} - \tau))\psi_{i,k-1}(\tau) \\ &= (\cdot - \tau)\psi_{i,k-1}(\tau)\end{aligned} \tag{24}$$

since $(1-\omega_{ik})f(t_i) + \omega_{ik}f(t_{i+k-1})$ is the straight line which agrees with f at t_i and t_{i+k-1}, hence must equal f if, as in our case, f is linear. This shows that

$$\sum B_{ik}\psi_{ik}(\tau) = (\cdot - \tau)\sum B_{i,k-1}\psi_{i,k-1}(\tau)$$

hence, by induction, that

$$\begin{aligned}\sum B_{ik}\psi_{ik}(\tau) &= (\cdot - \tau)^{k-1}\sum B_{i1}\psi_{i1}(\tau) \\ &= (\cdot - \tau)^{k-1}\end{aligned}$$

since $\psi_{i1}(\tau) = 1$ and $\sum_i B_{i1} = 1$ (see equation 2).

Remark There may be some doubt as to why ψ_{i1} should be identically equal to 1. From the definition (22b), it would appear that ψ_{i1} is the product of *no* factors, hence, by a standard agreement concerning the empty product, equal to 1. This is the definition appropriate for use in induction arguments. Indeed, if you consider the coefficients in (23) for $k = 2$ directly, you get

$$\begin{aligned}(1-\omega_{i2})\psi_{i-1,2}(\tau) + \omega_{i2}\psi_{i2}(\tau) &= (1-\omega_{i2})\cdot(t_i - \tau) + \omega_{i2}\cdot(t_{i+1} - \tau) \\ &= (\cdot - \tau)\end{aligned}$$

which agrees with (24) for this case if we set $\psi_{i1}(\tau) = 1$.

Since τ in (22) is arbitrary, it follows that $S_{k,t}$ contains all polynomials of degree $< k$. More than that, we can even give an explicit expression for the required coefficients, as follows.

By dividing (22a) by $(k-1)!$ and then differentiating it with respect to τ, we obtain the identities

$$\frac{(\cdot - \tau)^{k-\nu}}{(k-\nu)!} = \sum_i B_{ik}\frac{(-D)^{\nu-1}\psi_{ik}(\tau)}{(k-1)!}, \qquad \nu > 0 \tag{25}$$

with Df the derivative of the function f. On using this identity in the Taylor formula

$$p = \sum_{\nu=1}^{k}\frac{(\cdot - \tau)^{k-\nu}}{(k-\nu)!}D^{k-\nu}p(\tau)$$

valid for any $p \in \Pi_{<k}$, we conclude that any such polynomial can be written in the form

$$p = \sum_i B_{ik}\lambda_{ik}p \tag{26a}$$

with λ_{ik} given by the rule

$$\lambda_{ik}f := \sum_{\nu=1}^{k}\frac{(-D)^{\nu-1}\psi_{ik}(\tau)}{(k-1)!}D^{k-\nu}f(\tau) \tag{26b}$$

Here are two special cases of particular interest. For $p = 1$, we get

$$1 = \sum_i B_{ik} \tag{27}$$

since $D^{k-1}\psi_{ik} = (-1)^{k-1}(k-1)!$, and this shows that the B_{ik} form a **partition of unity**.

Further, since $D^{k-2}\psi_{ik}$ is a linear polynomial which vanishes at

$$t_i^* := (t_{i+1} + \cdots + t_{i+k-1})/(k-1) \qquad (28)$$

we get the important identity

$$p = \sum_i B_{ik} p(t_i^*) \quad \text{for all } p \in \Pi_1 \qquad (29)$$

Remark In the *cardinal* case,

$$\psi_{ik}(\tau)/(k-1)! = \binom{i-\tau+k-1}{k-1} \qquad (22\,\mathrm{b})_{\mathbb{Z}}$$

while in the *Bernstein–Bézier* case,

$$\psi_{(\mu,\nu)}(\tau) = (-\tau)^{\mu}(1-\tau)^{\nu}$$
$$= (-1)^{\mu} B_{(\nu,\mu)} \Big/ \binom{\mu+\nu}{\mu} \qquad (22\,\mathrm{b})_{\mathbb{B}}$$

THE pp FUNCTIONS CONTAINED IN $S_{k,\mathbf{t}}$

In this section, we show that the spline space $S_{k,\mathbf{t}}$ coincides with a certain space of **pp** (:= **piecewise polynomial**) functions.

Each $s \in S_{k,\mathbf{t}}$ is pp of degree $< k$ with breakpoint sequence **t**, since each B_{ik} is pp of degree $< k$ and has breakpoints $t_i, \ldots, t_{i+k}$. In symbols,

$$S_{k,\mathbf{t}} \subseteq \Pi_{<k,\mathbf{t}} \qquad (30)$$

But $S_{k,\mathbf{t}}$ is usually a proper subspace of $\Pi_{<k,\mathbf{t}}$ since, depending on the **knot multiplicities**

$$\# t_i := \# \{t_j : t_i = t_j\} \qquad (31)$$

the splines in $S_{k,\mathbf{t}}$ are more or less smooth, while the typical element of $\Pi_{<k,\mathbf{t}}$ has a jump discontinuity at every t_i.

For the precise description, given in Theorem 2 below, of the **smoothness conditions** satisfied by the elements of $S_{k,\mathbf{t}}$, we make use of Marsden's Identity (22), since it provides us with the B-spline coefficients of various pp functions in $S_{k,\mathbf{t}}$, as follows.

Since $B_{ik}(t_j) \neq 0$ implies $\psi_{ik}(t_j) = 0$ (see Figure 3.7), the choice $\tau = t_j$ in (22) leaves only terms with support either entirely to the left or else entirely to the right of t_j (see Figure 3.8).

This implies (see also Figure 3.9) that

$$(\cdot - t_j)_+^{k-1} = \sum_{i \geq j} B_{ik}\psi_{ik}(t_j) \qquad (32)$$

with

$$\alpha_+ := \max\{\alpha, 0\} \qquad (33)$$

the positive part of the number α, i.e. $(\cdot - t_j)_+^{k-1}$ is a **truncated power** function of exact degree $k-1$. More than that, since $B_{ik}(t_j) \neq 0$ implies $D^{\nu-1}\psi_{ik}(t_j) = 0$ in case $\nu \leq \# t_j$, the same observation applied to (25) shows that

$$(\cdot - t_j)_+^{k-\nu} \in S_{k,\mathbf{t}} \quad \text{for } 1 \leq \nu \leq \# t_j \qquad (34)$$

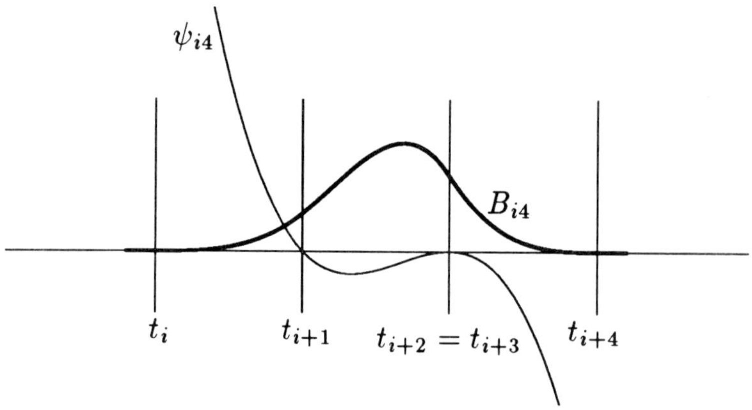

Figure 3.7. B_{i4} and ψ_{i4}; note the double knot.

Figure 3.10 illustrates further the interplay between knot multiplicity and smoothness of the resulting B-splines by showing all the relevant B-splines and truncated power functions for a certain knot sequence.

Theorem 2 *The space $S_{k,\mathbf{t}}$ coincides with the space $\tilde{S}$ of all piecewise polynomials of degree $< k$ with breakpoints t_i which are $k - 1 - \# t_i$ times continuously differentiable at t_i.*

Proof Assume without loss of generality (see the Splines Defined section on p. 30) that

$$t_i < t_{i+k} \quad \text{for all } i$$

It is sufficient to prove that for any finite interval $I := [a\,..\,b]$, the restriction $\tilde{S}_{|I}$ of the space $\tilde{S}$ to

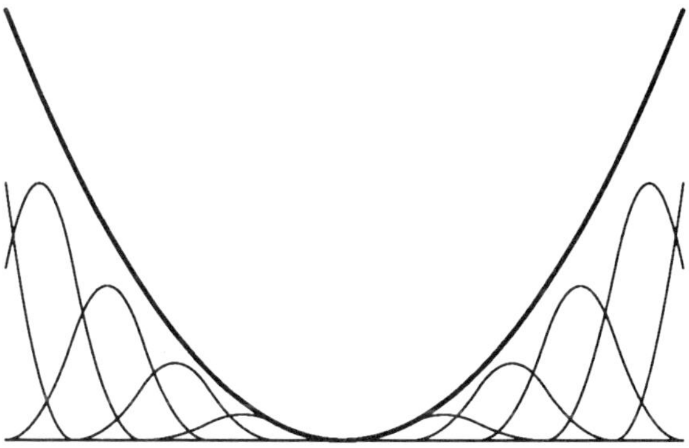

Figure 3.8. The summands $B_{i3}\psi_{i3}(t_j)$ and their sum $(\cdot - t_j)^2$.

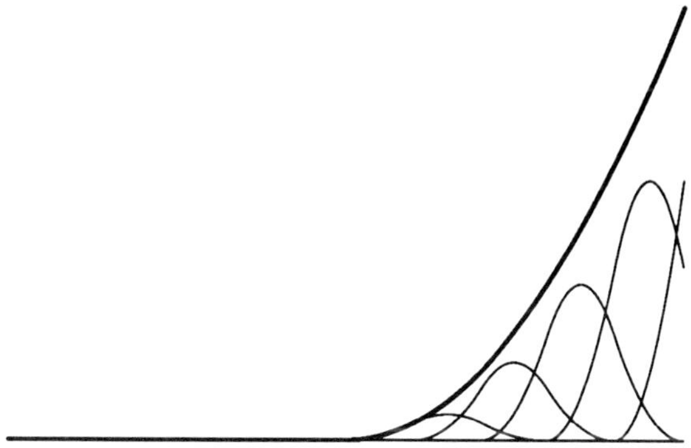

Figure 3.9. The summands $B_{i3}\psi_{i3}(t_j)$, $i \geqslant j$, and their sum $(\cdot - t_j)_+{}^2$.

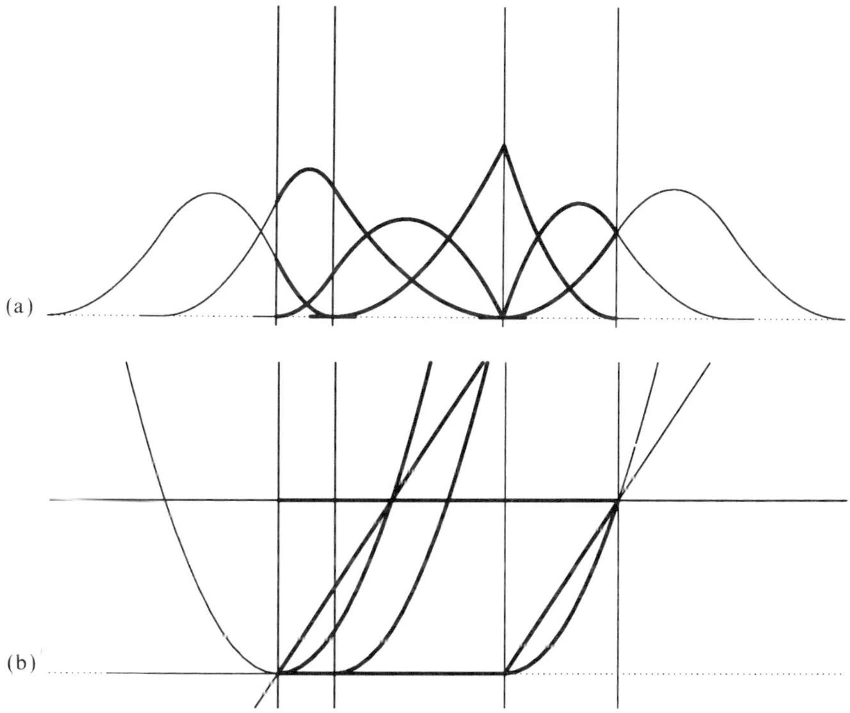

Figure 3.10. (a) The six quadratic B-splines for the case of one simple and one double interior knot; and (b) the corresponding truncated power basis.

the interval I coincides with the restriction of $S_{k,\mathbf{t}}$ to that interval. The latter space is spanned by all the B-splines having some support in I, i.e. all B_{ik} with $(t_i \,..\, t_{i+k}) \cap I \neq \emptyset$. The space $\tilde{S}_{|I}$ has a basis consisting of the functions

$$(\cdot - a)^{k-\nu}, \quad \nu = 1, \ldots, k;$$
$$(\cdot - t_i)_+^{k-\nu}, \quad \nu = 1, \ldots, \# t_i, \qquad \text{for } a < t_i < b \quad (35)$$

This follows from the observation that a pp function f with a breakpoint at t_i which is $k - 1 - \# t_i$ times continuously differentiable there can be written uniquely as

$$f = p + \sum_{\nu=1}^{\# t_i} (\cdot - t_i)_+^{k-\nu} c_\nu$$

with p a suitable polynomial of degree $< k$ and suitable coefficients c_ν. Since each of the functions in (35) lies in $S_{k,\mathbf{t}}$, by (32) and (34), we conclude that

$$\tilde{S}_{|I} \subseteq (S_{k,\mathbf{t}})_{|I} \qquad (36)$$

On the other hand, the dimension of $\tilde{S}_{|I}$, i.e. the number of functions in (35) equals the number of B-splines with some support in I (since it equals $k + \sum_{a < t_i < b} \# t_i$), hence is an upper bound on the dimension of $(S_{k,\mathbf{t}})_{|I}$. This implies that equality must hold in (36), which is what we set out to prove.

Remark The argument from linear algebra used here is the following: Suppose that we know a basis, $(f_1, f_2, \ldots, f_n)$ say, for the linear subspace F, and that we further know a sequence $(g_1, g_2, \ldots, g_m)$ whose span, G say, contains each of the f_i. Then, of course, $F \subseteq G$ and so

$$n = \dim F \leqslant \dim G \leqslant m$$

If we now know, in addition, that $n = m$, then necessarily $F = G$. Moreover, then necessarily $\dim G = m$, i.e. the sequence $(g_1, g_2, \ldots, g_m)$ must be linearly independent (since it then is *minimally spanning* for G). In our particular situation, this last observation implies that the set of B-splines having some support in I must be linearly independent over I. We pick up on this in the next section.

Remark Formula (32) provides all the information needed to deduce the divided-difference formulation for B-splines. By defining ψ_{ik}^+ to be the function which agrees with the polynomial ψ_{ik} on $[\tau_i \,..\, \infty)$ and is identically zero on $(-\infty \,..\, \tau_i)$, where τ_i is an arbitrary point in the support of B_{ik}, i.e. in $(t_i \,..\, t_{i+k})$, we are entitled to write (32) in the form

$$(t - t_j)_+^{k-1} = \sum_i B_{ik}(t)\psi_{ik}^+(t_j), \quad \text{all } t \qquad (32')$$

The function ψ_{ik}^+ agrees with ψ_{ik} at all t_j with $j > i$, and agrees with the zero function at all t_j with $j < i + k$. Since both ψ_{ik} and 0 are polynomials of degree $< k$, this implies that the kth divided difference $[t_j, \ldots, t_{j+k}]\psi_{ik}^+$ at the points $t_j, \ldots, t_{j+k}$ of ψ_{ik}^+ must be zero in case $j \neq i$. Therefore, applying this divided difference to both sides of (32′), we find that

$$[t_j, \ldots, t_{j+k}](t - \cdot)_+^{k-1} = B_{jk}(t)[t_j, \ldots, t_{j+k}]\psi_{jk}^+$$

Since ψ_{jk}^+ agrees at $t_j, \ldots, t_{j+k}$ with the kth degree polynomial $\psi_{jk} \cdot (\cdot - t_j)/(t_{j+k} - t_j)$, and this polynomial has leading coefficient $(-1)^{k-1}/(t_{j+k} - t_j)$, it follows that

$$(t_{j+k} - t_j)[t_j, \ldots, t_{j+k}](\cdot - t)_+^{k-1} = B_{jk}(t)$$

'B' STANDS FOR 'BASIC'

In this section, we discuss the **basis property** of the B-splines, as a consequence of Theorem 2 and its proof.

From the Remark following Theorem 2, we obtain the following *sharpening* of Theorem 2.

Theorem 3 *Let $I := [a \,..\, b)$ be a finite interval. Then the restrictions*

$$\{B_{ik|I} : B_{ik|I} \neq 0\} \qquad (38)$$

of those B-splines which have some support on I form a basis for the space of pp functions of degree $< k$ on I with breakpoints $\{t_i : a < t_i < b\}$ and which are $k - 1 - \# t_i$ continuously differentiable at each of their breakpoints t_i.

We conclude that the number of smoothness conditions at a knot t_i guaranteed to be satisfied by every spline in $S_{k,\mathbf{t}}$ equals $k - \# t_i$. This proves the formula

$$\begin{aligned}&\#\text{smoothness conditions at knot}\\ &\quad + \text{multiplicity of knot} \qquad = \text{order}\end{aligned} \tag{17}$$

cited earlier (in connection with the Bernstein–Bézier form).

It is worthwhile to think about this the other way around. Suppose we start off with a partition

$$a =: \xi_1 < \xi_2 < \cdots < \xi_l < \xi_{l+1} := b$$

of the interval $I := [a\,..\,b]$ and wish to consider the space

$$\Pi^{\nu}_{<k,\xi}$$

of all pp functions of degree $< k$ on I with breakpoints ξ_i which satisfy ν_i smoothness conditions at ξ_i, i.e. are $\nu_i - 1$ times continuously differentiable at ξ_i, for all i. Then a B-spline basis for this space is provided by (38), with the knot sequence $\mathbf{t}$ constructed from the breakpoint sequence ξ in the following way: To the sequence

$$(\underbrace{\xi_2, \ldots, \xi_2}_{k-\nu_2 \text{ terms}}, \underbrace{\xi_3, \ldots, \xi_3}_{k-\nu_3 \text{ terms}}, \ldots, \underbrace{\xi_l, \ldots, \xi_l}_{k-\nu_l \text{ terms}}), \tag{39}$$

adjoin at the beginning k points $\leqslant a$ and at the end k points $\geqslant b$. While the knots in (39) have to be exactly as shown to achieve the specified smoothness at the specified breakpoints, the $2k$ additional knots are quite arbitrary. They are often chosen to equal a resp. b, and this has certain advantages (among other things that of simplicity). With such a choice, it is necessary to modify the definition (1) so as to include the right endpoint, b, into the support of the rightmost nontrivial B_{i1}. In other words, if n is such that

$$t_n < t_{n+1} = b$$

then

$$B_{n1}(t) := X_n(t) := \begin{cases} 1, & \text{if } t_i \leqslant t \leqslant b \\ 0, & \text{otherwise} \end{cases} \tag{40}$$

This ensures that, in evaluating a spline or its derivatives at b, we obtain the limit from the left.

The identification of $S_{k,\mathbf{t}}$ with a certain space of pp functions allows the following conclusions of importance in calculations to be discussed later.

Corollary 1 *If $t_i < t_{i+k-1}$, then the derivative of a spline in $S_{k,\mathbf{t}}$ is a spline of degree $< k - 1$ with respect to the same knot sequence, i.e. $DS_{k,\mathbf{t}} \subseteq S_{k-1,\mathbf{t}}$.*

Proof By assumption, $\# t_i < k$, hence the pp functions in $S_{k,\mathbf{t}}$ are continuous, therefore differentiable (if we accept a possible jump at t_i in the derivative Ds of $s \in S_{k,\mathbf{t}}$ in case $\# t_i = k - 1$). Further, such a derivative Ds is pp of degree $< k - 1$ and satisfies $k - \# t_i - 1$ smoothness conditions at t_i, hence belongs to $S_{k-1,\mathbf{t}}$, by Theorem 2 or 3.

Corollary 2 *If $\hat{\mathbf{t}}$ is a refinement of the knot sequence $\mathbf{t}$, then $S_{k,\mathbf{t}} \subset S_{k,\hat{\mathbf{t}}}$.*

Proof Since $\hat{\mathbf{t}}$ is a refinement of $\mathbf{t}$, i.e. contains entries in addition to those of $\mathbf{t}$, the pp functions in $S_{k,\mathbf{t}}$ satisfy all the conditions which, by Theorem 2 or 3, characterize the pp functions in $S_{k,\hat{\mathbf{t}}}$. (But the converse does not hold, since the pp functions in $S_{k,\hat{\mathbf{t}}}$ may have more breakpoints and/or may be less smooth at some breakpoints than the pp functions in $S_{k,\mathbf{t}}$.)

These corollaries point out that it should be possible, in principle, to compute from the B-spline coefficients of a spline in $S_{k,\mathbf{t}}$ the B-spline coefficients of its derivative and its B-spline coefficients with respect to a refined knot sequence. To carry out such calculations, though, we need a means of expressing the B-spline coefficients of a spline in terms of other information, such as its values and derivatives at certain points. If the spline happens to be a polynomial, then such a formula is provided by (26). We show in the next section that the same formula works for any spline (provided we are willing to restrict the parameter τ suitably).

THE DUAL FUNCTIONALS

In this section, we prove that the formula (26) for the B-spline coefficients of a polynomial is valid for an arbitrary spline provided we restrict the parameter τ in the definition

$$\lambda_{ik}: f \mapsto \sum_{\nu=1}^{k} \frac{(-D)^{\nu-1}\psi_{ik}(\tau)}{(k-1)!} D^{k-\nu} f(\tau) \quad (26\,\text{b})$$

to the support of B_{ik}.

For this, we agree, consistent with (1), that all derivatives in (26 b) are to be taken as limits from the right in case τ coincides with a knot (except, perhaps, when τ is the right endpoint of the interval of interest, see (40)).

Theorem 4 *If τ in definition* (26 b) *of λ_{ik} is chosen in the interval $[t_i \,..\, t_{i+k})$, then*

$$\lambda_{ik}\Big(\sum_j B_{ik} a_j\Big) = a_i \quad (41)$$

Proof We prove that under the given restriction

$$\lambda_{ik} B_{jk} = \delta_{ij} := \begin{cases} 1, & \text{if } i = j \\ 0, & \text{otherwise} \end{cases} \quad (42)$$

Assume that $\tau \in [t_l \,..\, t_{l+1}) \subset [t_i \,..\, t_{i+k})$. Then (42) requires proof only for $j = l-k+1, \ldots, l$ since, for all other j, $i \neq j$ and B_{jk} vanishes identically on $[t_l \,..\, t_{l+1})$, hence also $\lambda_{ik} B_{jk} = 0$. For each of the remaining j's, let p_j be the polynomial which agrees with B_{jk} on $[t_l \,..\, t_{l+1})$ (see Figure 3.11). Then

$$\lambda_{ik} B_{jk} = \lambda_{ik} p_j$$

On the other hand,

$$p_j = \sum_{i=l-k+1}^{l} p_i \lambda_{ik} p_j \quad (43)$$

since this holds by (26 a) on $[t_l \,..\, t_{l+1})$. This forces $\lambda_{ik} p_j$, hence $\lambda_{ik} B_{jk}$, to equal δ_{ij} for $i, j = l-k+1, \ldots, l$, since, by Theorem 3 or directly from the fact that (26 a) holds for every $p \in \Pi_{<k}$, the sequence

$$p_{l-k+1}, \ldots, p_l \quad (44)$$

is linearly independent.

Remark The argument used here is that for a linearly independent sequence $(f_1, \ldots, f_n)$, the only way the equation

$$f_i = \sum_{j=1}^{n} f_j a_{ij}$$

can hold is for a_{ij} to equal 1 for $i = j$ and zero otherwise. Further, the linear independence of the sequence (44) follows from the validity of (26 a) for every $p \in \Pi_{<k}$ since that implies that the k-sequence (44) is spanning for the k-dimensional space $\Pi_{<k}$. It also follows from Theorem 3 with $I = [t_l \,..\, t_{l+1})$.

The two sequences, (B_{ik}) and (λ_{jk}), are said to be **bi-orthonormal** or **dual to each other** because they satisfy (42). For this reason, the linear functionals λ_{ik} are at times referred to as the **dual functionals** for the B-splines.

We exploit the simple formula (41) for the ith B-spline coefficient of a spline in subsequent sections, in order to derive algorithms for differentiation and knot insertion and, ultimately, to derive statements about the condition and the shape-preserving property of B-splines.

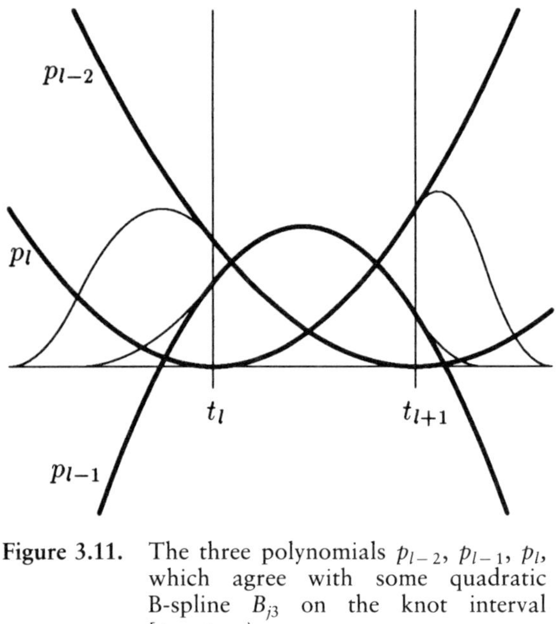

Figure 3.11. The three polynomials p_{l-2}, p_{l-1}, p_l, which agree with some quadratic B-spline B_{j3} on the knot interval $[t_l \,..\, t_{l+1})$.

CONDITION

The **condition** of a basis measures how closely relative changes in the coefficients are matched by the resulting relative changes in the element represented. The closer the match, the better

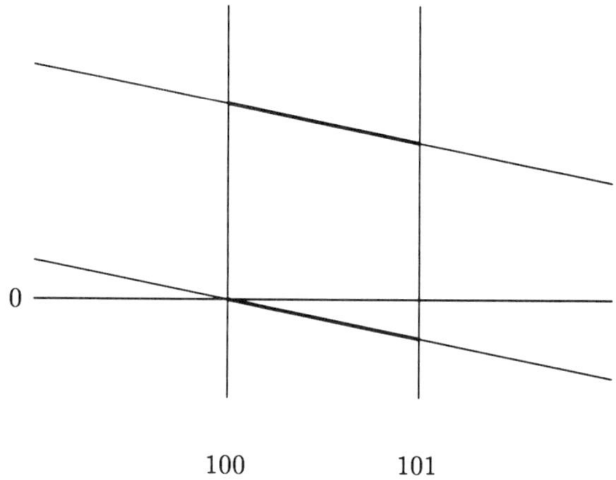

Figure 3.12. The power coefficients of the these two very different linear polynomials differ by only 0.1%.

conditioned the basis is said to be. For example, the power basis 1, t, t^2, . . . is not a good way to represent polynomials if we are interested in a positive interval $[a\,..\,b]$ with a/b close to 1. If, for example, $[a\,..\,b] = [100\,..\,101]$ then a 0.1% change in the power coefficients of the straight line $p: t \rightarrow t - 100$ can change its behaviour on $[100\,..\,101]$ by 100% (see Figure 3.12).

If we use the appropriately shifted power basis, e.g. write p in the form $p(t) = \alpha + \beta(t - 100)$, then a 0.1% change in the coefficients α, β of this form produces a 0.1% change in the polynomial on the interval $[100\,..\,101]$. The appropriately shifted power basis is often much better conditioned than the power basis. In this section, we discuss briefly the condition of the B-spline basis.

This requires us to bound the spline in terms of its B-spline coefficients and the B-spline coefficients in terms of the spline. The first turns out to be easy, while the second requires some work. Precisely, we are looking for constants $m > 0$ and M for which the inequalities

$$m \max_i |a_i| \leqslant \max_t \left|\sum_i B_{ik}(t)\,a_i\right| \leqslant M \max_i |a_i| \qquad (45)$$

hold regardless of what the coefficient vector $a = (a_i)$ might be. Since the B-splines are nonnegative and sum to 1 at any point, we have

$$\left|\sum_i B_{ik}(t)\,a_i\right| \leqslant \sum_i B_{ik}(t)\,|a_i|$$

$$\leqslant \sum_i B_{ik}(t) \max_i |a_i| = \max_i |a_i|$$

hence the second inequality always holds with $M = 1$. For the first inequality, we have to work a little harder.

Let $s := \Sigma_i B_{ik} a_i$. We know from Theorem 4 that

$$a_i = \lambda_{ik} s = \sum_{\nu=1}^{k} \frac{(-D)^{\nu-1}\psi_{ik}(\tau)}{(k-1)!} D^{k-\nu} s(\tau) \qquad (46)$$

with τ some point which we can freely choose in the interval $[t_i\,..\,t_{i+k})$. We now bound this sum in terms of $\max_t |s(t)|$.

Suppose that $\tau \in [t_l\,..\,t_{l+1}) \subset [t_i\,..\,t_{i+k})$. Then, for some const_k depending only on k, and for all $p \in \Pi_{<k}$ and all j,

$$|D^j p(\tau)| \leqslant \mathrm{const}_k (\Delta t_l)^{-j} \max_{t_l \leqslant t \leqslant t_{l+1}} |p(t)| \qquad (47)$$

The existence of such a const_k follows for the case $\Delta t_l = 1$ from the fact that $\Pi_{<k}$ is finite-dimensional, and from this it follows for arbitrary Δt_l by scaling. Since s agrees with some polynomial of degree $< k$ on $[t_l\,..\,t_{l+1})$, we conclude that

$$|D^j s(\tau)| \leqslant \mathrm{const}_k (\Delta t_l)^{-j} \max_{t_i \leqslant t \leqslant t_{i+k}} |s(t)| \qquad (48)$$

On the other hand, $\psi_{ik} = (t_{i+1} - \cdot) \ldots (t_{i+k-1} - \cdot)$ is also a polynomial of degree $< k$, and

$$\max_{t_l \leqslant t \leqslant t_{l+1}} |\psi_{ik}(t)| \leqslant \mathrm{const}'_k |\Delta t_{l^*}|^{k-1} \qquad (49)$$

for some const'_k which depends only on k and with $[t_{l^*}\,..\,t_{l^*+1})$ a largest interval of that form in $[t_i\,..\,t_{i+k})$. Therefore we choose $l = l^*$ and then obtain, from (47) with $p = \psi_{ik}$ and from (48), the bound

$$|D^{\nu-1}\psi_{ik}(\tau) D_{k-\nu} s(\tau)|$$

$$\leqslant (\mathrm{const}_k)^2\, \mathrm{const}'_k \max_{t_i \leqslant t \leqslant t_{i+k}} s|(t)|$$

Now sum these bounds over v and divide by $(k-1)!$ to obtain

$$|a_i| = |\lambda_{ik}s| \leqslant \text{const} \max_{t_i \leqslant t \leqslant t_{i+k}} |s(t)|$$

with const depending only on k.

We have proved the following:

Theorem 5 *There exists a constant D_k depending only on k so that for all knot sequences* **t** *and all $s \in S_{k,\mathbf{t}}$, and for all i,*

$$|\lambda_{ik}s| \leqslant D_k \max_{t_i \leqslant t \leqslant t_{i+k}} |s(t)| \tag{50}$$

The best value for D_k is not known exactly, but there is strong numerical evidence that $D_k \sim 2^{k-1}$. If we only consider *cardinal* splines, i.e. only uniform knot sequences, then the best value for D_k is known to be less than $(\pi/2)^k$.

Corollary *The inequalities* (45) *hold with $m = 1/D_k$ and $M = 1$.*

EVALUATION

In this section, we discuss the use of the recurrence relations (4) for the evaluation of a spline

$$s = \sum_i B_{ik}a_i \tag{51}$$

from its B-spline coefficients (a_i).

We already observed in (23) that the recurrence relations imply

$$s = \sum_i B_{ik}a_i = \sum_i B_{i,k-1}a_i^{[1]} \tag{52}$$

with

$$a_i^{[1]} := (1-\omega_{ik})a_{i-1} + \omega_{ik}a_i \tag{53}$$

Note that $a_i^{[1]}$ is not a constant, but is the straight line through the points (t_i, a_{i-1}) and (t_{i+k-1}, a_i). In particular, $a_i^{[1]}(t)$ *is a convex combination of a_{i-1} and a_i if $t_i \leqslant t \leqslant t_{i+k-1}$.*

After $k-1$-fold iterations of this procedure, we arrive at the formula

$$s = \sum_i B_{i1}a_i^{[k-1]}$$

which shows that

$$s = a_i^{[k-1]} \quad \text{on} \quad [t_i \,..\, t_{i+1})$$

Algorithm 1 *From given constant polynomials $a_i^{[0]} := a_i$, $i = j-k+1, \ldots, j$ (which determine $s := \Sigma_i B_{ik}a_i$ on $[t_j \,..\, t_{j+1})$), generate polynomials $a_i^{[r]}$, $r = 1, \ldots, k-1$, by the recurrence*

$$a_i^{[r+1]} := (1-\omega_{i,k-r})a_{i-1}^{[r]} + \omega_{i,k-r}a_i^{[r]}, \qquad j-k+r+1 < i \leqslant j \tag{54}$$

Then $s = a_j^{[k-1]}$ on $[t_j \,..\, t_{j+1})$. Moreover, for $t_j \leqslant t \leqslant t_{j+1}$, the weight $\omega_{i,k-r}(t)$ in (54) *lies between* 0 *and* 1. *Hence the computation of $s(t) = a_j^{[k-1]}(t)$ via* (54) *consists of the repeated formation of convex combinations.*

In the *cardinal* case (see the Splines Defined section esp. (11–13)), the algorithm simplifies, as follows. Now

$$s =: \sum_i N_k(\cdot - i)a_i$$

$$= \sum_i N_{k-1}(\cdot - i)a_i^{[1]}/(k-1)$$

with

$$a_i^{[1]} := (i+k-1-\cdot)a_{i-1} + (\cdot - i)a_i$$

Hence

$$s = a_j^{[k-1]}/(k-1)! \quad \text{on} \quad [j \,..\, j+1)$$

with $(54)_{\mathbb{Z}}$

$$a_i^{[r]} := (i+k-r-\cdot)a_{i-1}^{[r-1]} + (\cdot - i)a_i^{[r-1]}, \qquad j-k+r < i \leqslant j$$

In the *Bernstein–Bézier* case (see the Splines Defined section esp. (14–18)), all the nontrivial weight functions $\omega_{i,k-r}$ are the same, i.e.

$$\omega_{i,k-r}(t) = t$$

Thus, for

$$s = \sum_{\mu+\nu=h} B_{(\mu,\nu)} a_{(\mu,\nu)}$$

we get

$$s = a_{(0,0)} \quad \text{on} \quad [0..1]$$

with $\qquad (54)_{\mathbb{B}}$

$$a_{(\mu,\nu)}(t) = (1-t)a_{(\mu+1,\nu)} + ta_{(\mu,\nu+1)},$$
$$\mu+\nu = r; \quad r = h-1, \ldots, 0$$

This is **de Casteljau's Algorithm** for the evaluation of the BB-form.

DIFFERENTIATION

In this section, we derive a formula for the B-spline coefficients of the derivative of a spline in terms of the B-spline coefficients of the spline.

By Corollary 1 to Theorem 3, the derivative Ds of a spline $s \in S_{k,\mathbf{t}}$ is again a spline with the same knot sequence, but of one order lower. This means that by Theorem 4, we can compute its B-spline coefficients (a_i') by the formula

$$a_i' = \lambda_{i,k-1}(Ds)$$

provided we use $\tau \in [t_i .. t_{i+k-1})$.

To relate a' to a, we express $\lambda_{i,k-1}D$ as a linear combination of the functionals λ_{ik}, making use of the fact that λ_{ik} *depends linearly on* ψ_{ik} – recall the definition

$$\lambda_{ik}: f \mapsto \sum_{\nu=1}^{k} \frac{(-D)^{\nu-1}\psi_{ik}(\tau)}{(k-1)!} D^{k-\nu} f(\tau) \qquad (26\text{b})$$

– and that

$$(t_{i+k-1} - t_i)\psi_{i,k-1} = \psi_{ik} - \psi_{i-1,k} \qquad (55)$$

These facts imply that

$$\begin{aligned}(\lambda_{ik} - \lambda_{i-1,k})f(\tau) &= \sum_{\nu=1}^{k} \frac{(-D)^{\nu-1}(\psi_{ik} - \psi_{i-1,k})(\tau)}{(k-1)!} D^{k-\nu} f(\tau) \\ &= (t_{i+k-1} - t_i) \sum_{\nu=1}^{k-1} \frac{(-D)^{\nu-1}\psi_{i,k-1}(\tau)}{(k-1)!} \cdot D^{k-\nu} f(\tau)\end{aligned}$$

the last equality by (55), and since $D^{k-1}\psi_{i,k-1} = 0$. On the other hand, directly from the definition (26 b),

$$\begin{aligned}\lambda_{i,k-1} Df(\tau) &= \sum_{\nu=1}^{k-1} \frac{(-D)^{\nu-1}\psi_{i,k-1}(\tau)}{(k-2)!} D^{k-1+\nu} Df(\tau) \\ &= (k-1) \sum_{\nu=1}^{k-1} \frac{(-D)^{\nu-1}\psi_{i,k-1}(\tau)}{(k-1)!} \cdot D^{k-\nu} f(\tau)\end{aligned}$$

Comparison of these two displays shows that

$$\lambda_{i,k-1} D = \frac{k-1}{t_{i+k-1} - t_i} (\lambda_{ik} - \lambda_{i-1,k}) \qquad (56)$$

Assuming that $B_{i,k-1} \neq 0$, i.e. that $t_i < t_{i+k-1}$, we can choose

$$\tau \in (t_i .. t_{i+k-1}) - (t_{i-1} .. t_{i+k-1}) \cap (t_i .. t_{i+k})$$

This yields

Algorithm 2 *Compute the coefficients for* $\Sigma\, a_i' B_{i,k-1} := D\, \Sigma\, a_i B_{ik}$ *by*

$$a_i' = \frac{a_i - a_{i-1}}{(t_{i+k-1} - t_i)/(k-1)} \qquad \text{if } t_i < t_{i+k-1} \qquad (57)$$

Remark What happens when $t_i = t_{i+k-1}$? In this case, $B_{i,k-1} = 0$, hence there is no need to calculate a_i'. To be precise, in this case, the spline $s = \Sigma_i B_{ik} a_i$ may not even be continuous at t_i, therefore $(Ds)(t_i)$ makes no sense. On the other hand, the left and the right limit $(Ds)(t_i-)$ and $(Ds)(t_i+)$ always make sense, and the algorithm provides all the a_j''s needed for their calculation.

By applying the algorithm to the particular coefficient sequence $a = (\delta_{ij})$, we obtain the formula

$$\begin{aligned}DB_{ik} = {} & \frac{k-1}{t_{i+k-1} - t_i} B_{i,k-1} \\ & - \frac{k-1}{t_{i+k} - t_{i+1}} B_{i+1,k-1}\end{aligned} \qquad (58)$$

In terms of the alternative notations (9) for

B-splines, this reads

$$(DB_{ik} =)\ DN_{ik} = M_{i,k-1} - M_{i+1,k-1}$$

From this, we infer that $D(\Sigma_{i=\alpha}^{\beta} N_{ik}) = M_{\alpha,k-1} - M_{\beta+1,k-1}$. Since $\Sigma_i N_{ik} = 1$, this implies that

$$\int_{t_i}^{t_{i+k-1}} M_{i,k-1} = \int_{-\infty}^{\infty} M_{i,k-1} = 1 \qquad (59)$$

and so indicates why the particular normalization

$$M_{ik} := \frac{k}{t_{i+k} - t_i} B_{ik}$$

is of interest.

In the *cardinal* case, (57) reduces to

$$a_i' = a_i - a_{i-1} =: \nabla a_i \qquad (57)_{\mathbb{Z}}$$

and (58) reads

$$DN_k = N_{k-1} - N_{k-1}(\cdot - 1) \qquad (58)_{\mathbb{Z}}$$

On integrating this formula, we obtain

$$N_k(t) = \int_{t-1}^{t} N_{k-1}(\tau)\,d\tau \qquad (60)$$

since both sides of (60) vanish for negative t. In terms of the **convolution product**

$$(f * g)(t) := \int f(t - \tau) g(\tau)\,d\tau$$

of two functions f and g, this gives the important formula

$$N_k = N_1 * N_{k-1} \qquad (61)$$

This shows that N_k is the k-fold convolution product of N_1, i.e.

$$N_k = \underbrace{N_1 * N_1 * \ldots * N_1}_{k \text{ terms}}$$

In the *Bernstein–Bézier* case, we get

$$D \sum_{\mu+\nu=h} B_{(\mu,\nu)} a_{(\mu,\nu)} = \sum_{\mu+\nu=h-1} B_{(\mu,\nu)} a_{(\mu,\nu)} \qquad (58)_{\mathbb{B}}$$

with

$$a_{(\mu,\nu)} = (\mu + \nu + 1)(a_{(\mu+1,\nu)} - a_{(\mu,\nu+1)}) \qquad (57)_{\mathbb{B}}$$

KNOT INSERTION

In this section, we discuss the most important CAGD contribution to (univariate) spline theory, *viz.* the idea of knot insertion (also known as subdivision), particularly as introduced and practiced by Boehm; see [2] for such things as the Oslo algorithm. Since the spline order k will not change in this section, we will usually suppress it and write B_i instead of B_{ik}, ψ_i instead of ψ_{ik}, etc.

Simply put, knot insertion involves rewriting a given spline as a spline with a refined knot sequence, as can always be done by Corollary 2 of Theorem 3. Such a calculation is worthwhile, since the B-spline coefficients are nearly equal to values of the spline at known points, and this is more nearly so when the knots are closer together. Here is the precise statement.

Theorem 6 *If the spline* $s = \Sigma_i B_i a_i$ *is continuously differentiable, then*

$$|a_i - s(t_i^*)| \leqslant \text{const}\,|\mathbf{t}|^2 \sup_t |D^2 s(t)| \qquad (62)$$

with

$$t_i^* := (t_{i+1} + t_{i+2} + \cdots + t_{i+k-1})/(k-1) \qquad (28)$$

and

$$|\mathbf{t}| := \sup_i (t_{i+1} - t_i)$$

Proof Recall that

$$|a_i| = |\lambda_i s| \leqslant \text{const} \max_{t_i \leqslant t \leqslant t_{i+k}} |s(t)| \qquad (50)$$

Further, recall from (29) that

$$\lambda_i p = p(t_i^*) \quad \text{for all } p \in \Pi_1$$

Thus, choosing, in particular, $p := s(t_i^*) + (\cdot - t_i^*) Ds(t_i^*)$, i.e. the linear Taylor polynomial for s at t_i^*, we get

$$|a_i - s(t_i^*)| = |a_i - p(t_i^*)| = |\lambda_i(s - p)|$$
$$\leqslant \text{const} \max_{t_i \leqslant t \leqslant t_{i+k}} |(s-p)(t)|$$
$$\leqslant \text{const} \frac{(t_{i+k} - t_i)^2}{8} \max_{t_i \leqslant t \leqslant t_{i+k}} |D^2 s(t)|.$$

This suggests consideration of the **control polygon** (see Figure 3.13 for an example) associated with the respresentation $\Sigma_i B_i a_i$ of the spline s as an element of $S_{\mathbf{t}}$. This control polygon will be denoted by

$$C_{a,\mathbf{t}}$$

It is the broken line or piecewise linear function with vertices $P_i := (t_i^*, a_i)$. For, the theorem implies that the control polygon will be close to s if $|\mathbf{t}|$ is small. Here is the precise statement.

Corollary *Let $C_{a,\mathbf{t}}$ be the control polygon associated with the representation $\Sigma_i B_i a_i$ of the continuous spline s as an element of $S_{\mathbf{t}}$. Then*

$$\sup_t |s(t) - C_{a,\mathbf{t}}(t)| \leqslant \text{const} |\mathbf{t}|^2 \sup_t |D^2 s(t)| \tag{63}$$

Proof Let $t_i^* \leqslant t \leqslant t_{i+1}^*$ and let p be the linear polynomial which agrees with s at t_i^* and t_{i+1}^*.

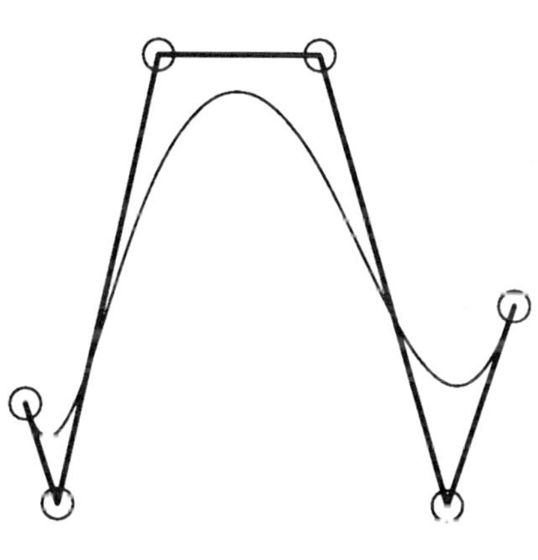

Figure 3.13. A cubic spline and its control polygon. The end knots are quadruple.

Then

$$|s(t) - p(t)|$$
$$\leqslant |t_{i+1}^* - t_i^*|^2/8 \max_{t_i^* \leqslant \tau \leqslant t_{i+1}^*} |D^2 s(\tau)|$$

while

$$|p(t) - C_{a,\mathbf{t}}(t)|$$
$$\leqslant \max\{|s(t_i^*) - a_i|, |s(t_{i+1}^*) - a_{i+1}|\}$$
$$\leqslant \text{const} |\mathbf{t}|^2 \max_\tau |D^2 s(\tau)|$$

by the theorem.

This shows that the control polygon $C_{a,\mathbf{t}}$ converges to the spline s as we refine the knot sequence $\mathbf{t}$. This is illustrated in Figure 3.14. Since the typical graphical equipment only draws broken lines anyway, this makes it attractive to construct refined control polygons for a spline.

For this, we need to know how to compute, from its B-spline coefficients a_i as an element of $S_{\mathbf{t}}$, the B-spline coefficients $\hat{a}_i$ for the spline s with respect to a refined knot sequence $\hat{\mathbf{t}}$. By Theorem 4, this is a question of comparing the corresponding $\hat{\lambda}_i$ with λ_i. Since the dual functional

$$\lambda_i : f \mapsto \sum_{\nu=1}^{k} \frac{(-D)^{\nu-1} \psi_i(\tau)}{(k-1)!} D^{k-\nu} f(\tau) \tag{26b}$$

depends linearly on ψ_i, this requires nothing more than to express

$$\hat{\psi}_i = (\hat{t}_{i+1} - \cdot) \cdots (\hat{t}_{i+k-1} - \cdot)$$

as a linear combination of the ψ_i.

This is particularly easy when $\hat{\mathbf{t}}$ is obtained from $\mathbf{t}$ by adding just one knot, say the point $\hat{t}$. Then

$$\hat{\psi}_i = \begin{cases} \psi_i, & t_{i+k-1} \leqslant \hat{t} \\ \psi_{i-1}, & \hat{t} \leqslant t_i \end{cases}$$

hence there is some actual computing necessary only for $t_i < \hat{t} < t_{i+k-1}$. For this case,

$$\alpha\psi_{i-1} + \beta\psi_i = (t_{i+1} - \cdot) \cdots (t_{i+k-2} - \cdot)$$
$$\cdot [\alpha(t_i - \cdot) + \beta(t_{i+k-1} - \cdot)]$$
$$= \hat{\psi}_i$$

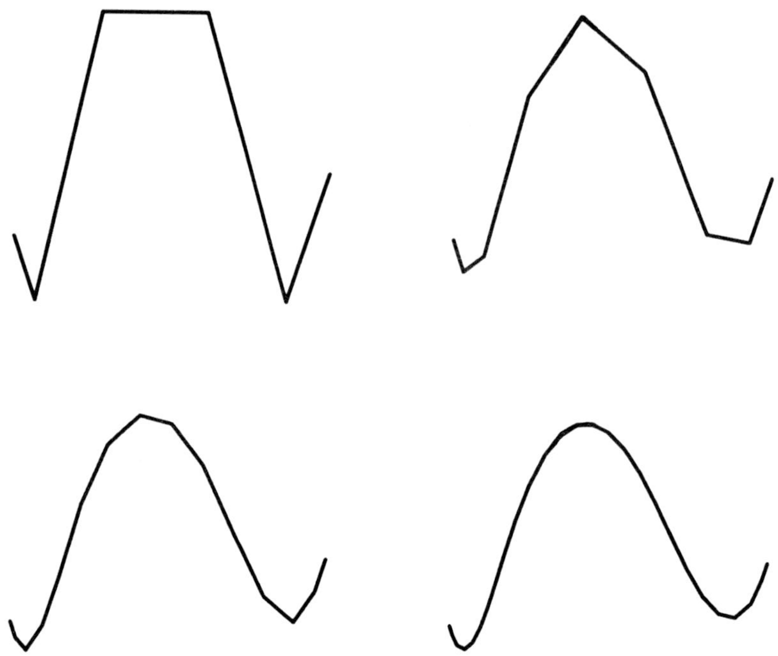

Figure 3.14. The control polygon of Figure 3.13 and three midpoint refinements.

provided $\alpha(t_i - \cdot) + \beta(t_{i+k-1} - \cdot) = (\hat{t} - \cdot)$, i.e.

$$\alpha = 1 - \omega_i(\hat{t}) \quad \text{and} \quad \beta = \omega_i(\hat{t})$$

Since $\hat{t}_i = t_i < \hat{t} < t_{i+k-1} = \hat{t}_{i+k}$, we can choose τ in the definition (26b) in the interval $(\hat{t}_i \mathinner{..} \hat{t}_{i+k}) = (t_{i-1} \mathinner{..} t_{i+k-1}) \cap (t_i \mathinner{..} t_{i+k})$. This proves

Algorithm 3 *If the knot sequence* $\hat{\mathbf{t}}$ *is obtained from the knot sequence* $\mathbf{t}$ *by addition of the point* $\hat{t}$, *then the coefficients* $\hat{a}_i$ *for the spline* s *with respect to the refined knot sequence are given by*

$$\hat{a}_i = \begin{cases} a_i & \text{if } t_{i+k-1} \leqslant \hat{t} \\ (1 - \omega_i(\hat{t}))a_{i-1} + \omega_i(\hat{t})a_i & \text{if } t_i < \hat{t} < t_{i+k-1} \\ a_{i-1} & \text{if } \hat{t} \leqslant t_i \end{cases} \tag{64}$$

Observe that $\omega_i(\hat{t}) \in [0 \mathinner{..} 1]$ when $t_i < \hat{t} < t_{i+k-1}$, and thus the coefficients $\hat{a}$ are convex combinations of the coefficients a.

This algorithm has the following very pretty graphical interpretation.

Corollary *The refined control polygon* $C_{\hat{a},\hat{\mathbf{t}}}$ *can be thought of as having been obtained by interpolation at its vertices to the original control polygon* $C_{a,\mathbf{t}}$, *i.e.*

$$C_{\hat{a},\hat{\mathbf{t}}}(\hat{t}_i^*) = C_{a,\mathbf{t}}(\hat{t}_i^*) \quad \textit{for all } i \tag{65}$$

Proof Consider the straight line $p: t \mapsto t$. It is a spline and, by (29),

$$p = \sum_i B_i t_i^*$$

i.e. (t_i^*) is its B-spline coefficient sequence with respect to the knot sequence $\mathbf{t}$. In particular, it is its own control polygon, i.e. $C_{\mathbf{t}^*,\mathbf{t}} = p$, regardless of what the knot sequence $\mathbf{t}$ might be. This implies that (64) also holds with every a replaced by t^*.

This says that the point $P_j := (\hat{t}_j^*, \hat{a}_j)$ lies on the segment $[P_{j-1} \mathinner{..} P_j]$ and cuts this segment in the ratio $(\hat{t} - t_j):(t_{j+k-1} - \hat{t})$. This is illustrated in Figure 3.15 for the control polygon of Figure 3.13.

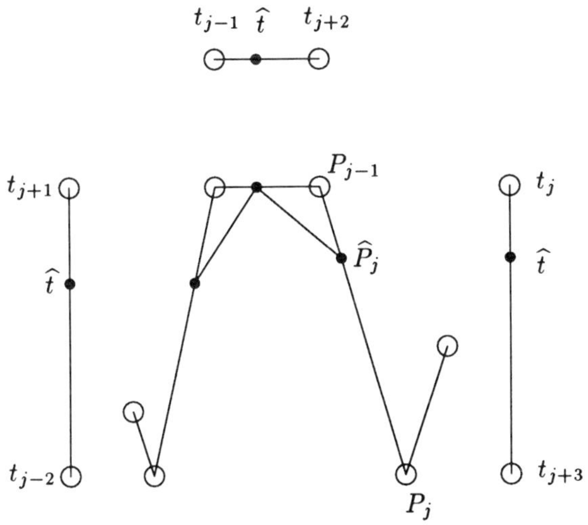

Figure 3.15. Insertion of $\hat{t} = 2$ into the knot sequence, $\mathbf{t} = (0,0,0,0,1,3,5,5,5,5)$, with $k = 4$.

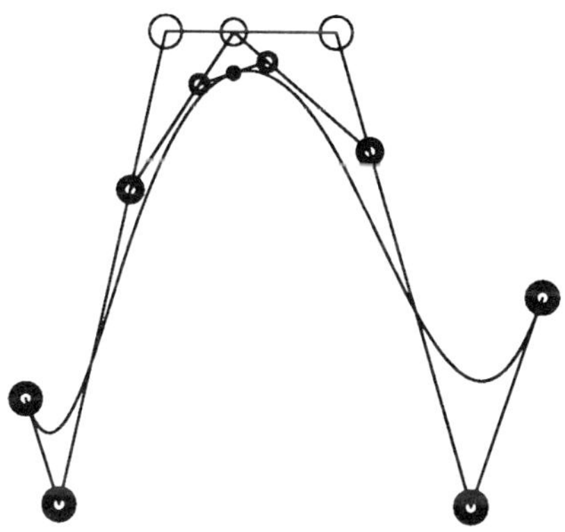

Figure 3.16. The cubic spline and its control polygon from Figure 3.13 and the sequence of control polygons generated by three-fold insertion of the same knot. (The finest control polygon differs from its predecessor only by an additional vertex point.)

If $r := \#\hat{t} \leqslant k - 1$, then, after just $(k - 1 - r)$-fold insertion of $\hat{t}$, we obtain a knot sequence $\bar{\mathbf{t}}$ in which the number $\hat{t}$ occurs exactly $k - 1$ times (see Figure 3.16 for an example). This means that there is exactly one B-spline for that knot sequence which is not zero at $\hat{t}$. Hence it must equal 1 at $\hat{t}$ and its coefficient must provide the value of s at $\hat{t}$. This makes it less surprising that the calculations in Algorithms 1 and 3 are identical.

Conversion to BB-net Let $\bar{\mathbf{t}}$ be the refined knot sequence which contains each of the knots in $\mathbf{t}$ exactly $k - 1$ times (see Figure 3.17 for an example). Then each corresponding B-spline $\bar{B}_{jk}$ is nonzero on just one knot interval, hence coincides there with a properly shifted and scaled element of the **Bernstein basis**. The k B-spline coefficients $\bar{a}_j$ associated in this way with a knot interval therefore provide the coefficients in the BB-form for the polynomial with which the spline agrees on that knot interval. The coefficient sequence $(\bar{a}_i)$, or the control polygon $C_{\bar{a},\bar{\mathbf{t}}}$, are called the **BB-net** for the given spline. It can be obtained by inserting each knot t_i of the spline $k - 1 - \#t_i$ times. The process can be speeded up slightly by inserting first every other knot, and in a second round, inserting the remaining knots. The latter insertion process is then entirely local and depends only on the ratio of the two knot intervals containing the knot being inserted.

While the formulas do simplify for the *cardinal* case, they are not of much use in that form since insertion of one knot into the sequence $\mathbf{t} = \mathbb{Z}$ would destroy the uniformity of the knot sequence. But it makes good sense to develop formulas for inserting the same number of uniformly spaced knots into every interval $[i \,.\,.\, i+1]$ since this produces again a *uniform* knot sequence. Because of its practical importance, we treat this case separately in the next section.

KNOT INSERTION FOR CARDINAL SPLINES

In this section, we consider knot refinement for cardinal splines, i.e. splines with a uniform knot

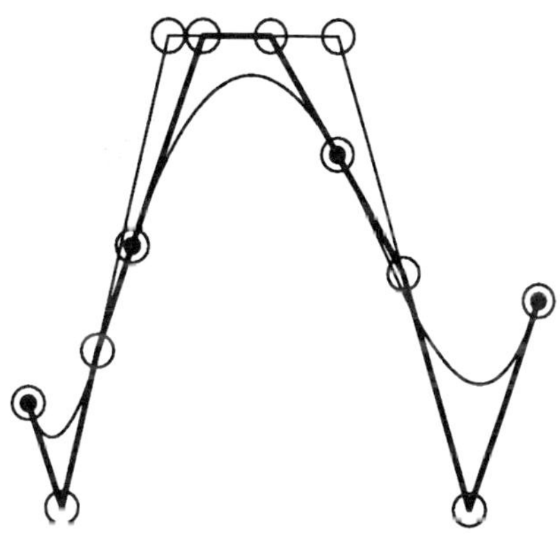

Figure 3.17. Conversion to BB-net by $(k - 2)$-fold insertion of each knot.

sequence. Here it is desirable to have the refined knot sequence again uniform. We restrict attention to the case that the given knot sequence is $\mathbf{t} = \mathbb{Z}$. That is no real restriction, since an arbitrary uniform knot sequence can always be written in the form $\alpha + \beta\mathbb{Z}$ for appropriate scalars α and β, and if s is a spline with that knot sequence, then $s(\alpha + \beta\cdot)$ is a spline with the knot sequence $\mathbb{Z}$.

If we insert $m - 1$ uniformly spaced knots into every knot interval of $\mathbb{Z}$, then the refined knot sequence is $\hat{\mathbf{t}} = \mathbb{Z}/m$. The corresponding B-splines $\hat{B}_i$ are

$$\hat{B}_i = \hat{N}_k(\cdot - i)$$

with

$$\hat{N}_k(t) := N_k(mt)$$

an appropriately scaled version of the standard cardinal B-spline N_k. This makes it trivial to determine $\hat{a}_i$ in case $k = 1$. Since

$$N_1 = \hat{N}_1 + \hat{N}_1(\cdot - 1) + \cdots + \hat{N}_1(\cdot - m + 1) \tag{66}$$

we find for this case that

$$\hat{a}_{pi+j} = a_i \quad \text{for} \quad j = 0, \ldots, m - 1$$

The formula for general order k is obtained from this with the aid of the convolution formula

$$N_k = N_1 * N_{k-1} \tag{61}$$

from page 42 as follows. We define

$$s_r := \sum_i N_r(\cdot - i)a_i =: \sum_i \hat{N}_r(\cdot - i)a_{ir}, \qquad r = 1, \ldots, k \tag{67}$$

Then $\hat{a}_i = a_{ik}$, and, from (61) and (66),

$$\begin{aligned} s_{r+1} &= N_1 * s_r \\ &= \sum_{j=1}^{m} \hat{N}_1(\cdot + 1 - j) * \sum_i \hat{N}_r(\cdot - i)a_{ir} \\ &= \sum_i \sum_{j=1}^{m} \underbrace{\hat{N}_1(\cdot + 1 - j) * \hat{N}_r(\cdot - i)}_{\hat{N}_{r+1}(\cdot - i + j - 1)/m} a_{ir} \\ &= \sum_i \sum_{j=1}^{m} (\hat{N}_{r+1}(\cdot - i)/m)\, a_{i-j+1,r} \end{aligned}$$

Here, we have used the following consequence of the convolution formula (61):

$$\begin{aligned} &\hat{N}_1(\cdot - \alpha) * \hat{N}_{r-1}(\cdot - \beta) \\ &\quad = \int N_1(m(\cdot - \tau) - \alpha)N_{r-1}(m\tau - \beta)d\tau \\ &\quad = \int N_1(m\cdot - \sigma + \beta - \alpha)N_{r-1}(\sigma)d\sigma/m \\ &\quad = \hat{N}_r(\cdot + \beta - \alpha)/m \end{aligned}$$

We conclude that

$$a_{i,r+1} := (a_{i,r} + a_{i-1,r} + \cdots + a_{i-m+1,r})/m \qquad \text{for } r > 0 \tag{68}$$

Here is the full algorithm:

Algorithm 4 *Given the B-spline coefficients* $a = (a_i)$ *of* $s \in S_{k,\mathbb{Z}}$, *its B-spline coefficients* $\hat{a} = (\hat{a}_i)$ *with respect to the refined knot sequence* $\mathbb{Z}/m$ *can be computed as follows:*

$$a_{mi+j,1} := a_i, \quad j = 0, \ldots, m - 1$$

$$a_{i,r} = \sum_{j=0}^{m-1} a_{i-j,r-1}/m, \quad r = 2, \ldots, k$$

$$\hat{a}_i := a_{ik}$$

In practice, one would use the algorithm repeatedly with $m = 2$ rather than once with a larger m. The computational cost is

$$nm(k - 1)((m - 1)A + D)$$

with n the number of coefficients to start with, and A and D the cost of one addition, respectively division. If, for example, the targeted refinement is to have $2^\mu n$ coefficients, then the cost ratio of the choice $m = 2^\mu$ *versus* the use of μ applications of the algorithm, each time with $m = 2$, is

$$\frac{2^\mu((2^\mu - 1)A + D)}{(2 + 2^2 + \cdots + 2^\mu)(A + D)} \sim 2^{\mu-1}A + D/2$$

In addition, even though the repeated application, with $m = 2$, takes roughly twice as many divisions, these are just divisions by 2.

SHAPE PRESERVATION

In this section, we use knot insertion to prove the shape preserving property of B-splines. Roughly speaking, this property says that a spline has the same shape as its control polygon. Figure 3.18 illustrates the mathematical formulation of shape preservation, i.e. the fact that any straight line crosses the spline no more often than it crosses the control polygon.

We begin with the

Convex hull property *If* $t_j \leqslant t < t_{j+1}$, *then* $s(t)$ *is a convex combination of the k B-spline coefficients* $a_{j-k+1}, \ldots, a_j$

which follows from Algorithm 1, or directly from the facts that B-splines are nonnegative and add up to 1 at every point (see (27).

For a statement of the full shape preserving property, we recall that

$$S^-(a)$$

is the standard notation for the number of (strong) sign changes in a sequence a. Thus

$$S^-(1, -1, 1, -1) = 3, \quad S^-(1, 0, 1, -1) = 1$$
$$S^-(0, 0, 0, 0) = 0$$

Theorem 7 Variation diminution $S^-(s) \leqslant S^-(a)$; *i.e. with* $x_1 < \cdots < x_r$ *arbitrary,*

$$S^-(s(x_1), \ldots, s(x_r)) \leqslant S^-(a)$$

Proof Recall from the Knot Insertion section that $s(s_1), \ldots, s(x_r)$ is a *subsequence* of the sequence $\bar{a}$ of coefficients for s with respect to the refined knot sequence $\bar{\mathbf{t}}$ which contains each x_i at least $k-1$ times. Hence it is sufficient to prove that $S^-(\bar{a}) \leqslant S^-(a)$. But this follows once we know that $S^-(\hat{a}) < S^-(a)$, with $\hat{a}$ obtained by (64), i.e. by insertion of just one knot. For this simple case, though, the conclusion is immediate if we think of the construction of $\hat{a}$ from a as occurring in two steps: In the first step, we insert $\hat{a}_i$ between a_{i-1} and a_i, and this does not increase the number of sign changes since each $\hat{a}_i$ is a *convex* combination of its neighbors

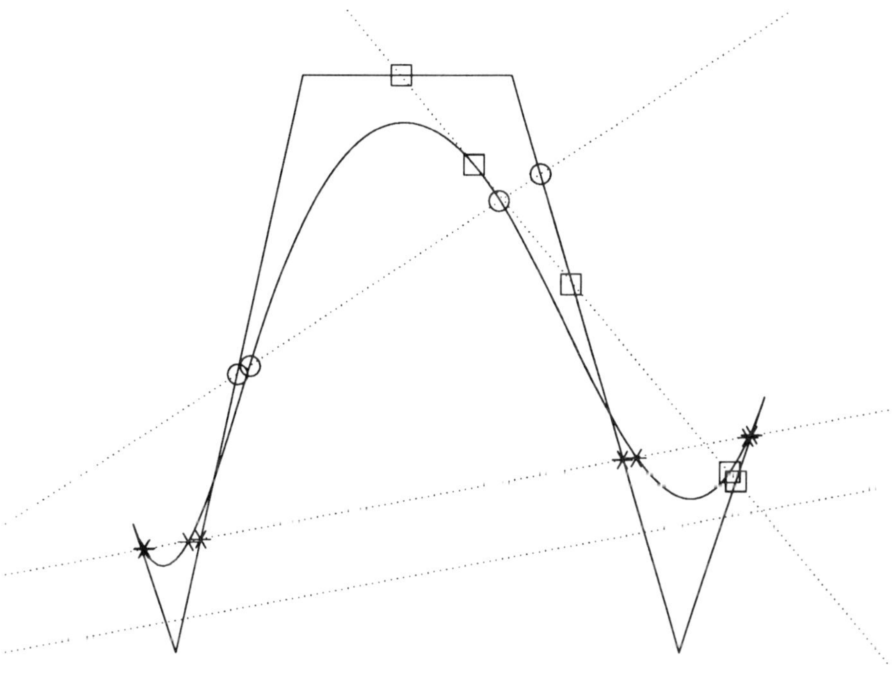

Figure 3.18. A cubic spline, its control polygon, and various straight lines intersecting them. The control polygon *exaggerates* the shape of the spline. The spline crossings are bracketed by the control polygon crossings.

a_{i-1} and a_i in that new sequence. In the second step, we pull out $\hat{a}$ as a subsequence, and this may only lower the number of sign changes.

Corollary Shape preservation *A spline crosses any straight line no more often than does its control polygon. In particular, if the control polygon is monotone (convex), then so is the spline.*

Proof Let s be the spline and p the straight line. Then $S^-(s - p)$ is the number of times the spline crosses the straight line. Since $s - p$ is a spline, this is bounded by $S^-(a - b)$, with a, b the B-spline coefficients of s resp. p with respect to **t**, and this equals the number of times the control polygon $C_{a,\mathbf{t}}$ crosses the control polygon for p. But, as we observed in the Knot Insertion section, the control polygon for the straight line p is p itself. This proves the general statement.

For the particulars, recall that a (continuous) function is monotone if and only if it crosses any constant function at most once, and that a function is convex if it crosses any straight line at most twice (dipping first below and then rising above the line in case it crosses it twice).

BACKGROUND

This chapter is a slight reworking (and correction) of the lecture notes for the first of four lectures in the course entitled 'The extension of B-spline curve algorithms to surfaces' given at *SIGGRAPH'86*. That lecture was solidly based on [3] which covers more or less the same material, in a less elaborate way and without any figures, in just seven pages.

The relevant literature on (univariate) B-splines up to about 1975 is summarized in [4], which also contains hints of the most exciting developments concerning B-splines since then: knot insertion and the multivariate B-splines. The two books on splines [5, 6] which have appeared since 1975 cover B-splines in the traditional way. As presentations of splines from the CAGD point of view, the survey article [7] and the 'Killer B's' [2, 8] are particularly recommended.

I refer you to these references and to the original papers referred to there if you are curious about just who contributed (and when and how) to the material essayed here.

ACKNOWLEDGEMENTS

The author was supported by the United States Army under Contract No. DAAL03-87-K-0030.

REFERENCES

1 **I. J. Schoenberg** Cardinal interpolation and spline functions, *J. Approximation Theory*, Vol 2, 1969, 167–206.
2 **R. H. Bartels, J. C. Beatty and B. A. Barsky** *An Introduction to Splines for Use in Computer Graphics & Geometric Modeling*, Morgan Kaufmann, 1987.
3 **C. de Boor and K. Höllig** B-splines without divided differences, in *Geometric Modeling*, G. Farin, ed., SIAM, 1987, 21–27.
4 **C. de Boor** Splines as linear combinations of B-splines, in *Approximation Theory II*, G. G. Lorentz, C. K. Chui and L. L. Schumaker, eds., 1976, 1–47.
5 **C. de Boor** *A Practical Guide to Splines*, Springer-Verlag, 1978.
6 **L. L. Schumaker** *Spline Functions*, John Wiley, 1980.
7 **W. Böhm, G. Farin and J. Kahmann** A survey of curve and surface methods in CAGD, *Computer Aided Geometric Design*, Vol 1, 1984, 1–60.
8 **R. H. Bartels, J. C. Beatty and B. A. Barsky** An Introduction to the Use of Splines in Computer Graphics, *SIGGRAPH'85*, San Francisco, 1985.

ABOUT THE AUTHOR

Carl R. de Boor is the P. L. Chebyshev Professor of Mathematics and Computer Science and the Steenbock Professor of Mathematical Sciences at the University of Wisconsin-Madison. He was Research Assistant to Garrett Birkhoff between 1959–60, an Associate Senior Research Mathematician at the General Motors Research Labs from 1960 to 1964, started his academic career at Purdue University in 1966 and has been at the University of Wisconsin since 1972. De Boor's research interests include approximation theory and numerical analysis.

Between 1956 and 1959 de Boor studied mathematics at the University of Hamburg and earned a PhD degree from the University of Michigan in 1966. He has written more than 100 technical papers, and is the author of the textbooks *Elementary Numerical Analysis* (with co-author S. E. Conte) and *A Practical Guide to Splines*.

De Boor is a member of Phi Beta Kappa, ACM's Special Interest Group on Numerical Mathematics, and the Society for Industrial and Applied Mathematics. He is an Associate Editor of *Constructive Approximation, SIAM Journal on Mathematical Analysis, Journal of Approximation Theory,* and has been on the editorial board of *Mathematics of Computation, SIAM Journal on Numerical Analysis* and is a Fellow of the American Academy of Arts and Sciences.

4 Algorithms for Spline Curves and Surfaces

Maurice G. Cox

INTRODUCTION

Many effective computer-aided design techniques concentrate on the representation and manipulation of 'objects' (simple or composite curves or surfaces in the form of mathematical functions) in terms of the vertices of a 'control' polygon or polyhedron. The designer establishes a set of vertices or 'control points' which define a representation of the object. By examining this representation using suitable software tools and analytical methods, and through the use of appropriate graphical displays, he decides the adequacy of the design. To improve the design he makes changes to individual vertices which have the effect of modifying the representation *locally*, thus giving the designer a convenient method of 'iterating' towards an acceptable form.

In an alternative approach the designer specifies a set of conditions that are to be 'satisfied' by the representation of the object. The conditions include points on the object, various derivatives at specified points, and 'shape' information such as positivity, monotonicity and convexity. A representation of the object is required that satisfies some or all of these conditions either exactly or approximately (in an appropriate sense). This paper is predominantly concerned with the latter approach, although some of the algorithms discussed relate to both approaches. An important instance of this approach arises when a designer produces a clay model of, say, a car body. An initial mathematical representation of the body shape would be given by interpolating or approximating a representative set of points on the surface of the model.

Sometimes a combination of the two approaches is appropriate. One instance arises when an object is designed using the former approach, and this object is then to be modified by augmenting it by a 'feature' defined by conditions as in the second approach. Examples in automobile engineering are a bulge in a bonnet to accommodate part of a high-powered engine, and a handle recess in a door panel.

We shall assume some familiarity with splines and their representation in terms of B-splines. See de Boor [1] for definitions and properties. We shall depart in one important respect from de Boor, *viz.* notation. The main reason is that much of the work on *algorithmic* aspects of splines employs a notation that we believe is more appropriate for software implementations. This alternative notation is adopted consistently here; it is also used in many of the references given. We indicate in Table 4.1 the main differences between the de Boor (B) and our (C) notation. Comments on the respective notations are as follows. The use of $x_{\min}$ and $x_{\max}$ rather than t_k and t_{n+1} to denote the interval endpoints is deliberately *suggestive*. It indicates explicitly the extreme values of the x-interval that are relevant to the user's problem. Despite notation B this interval is *unrelated* to the number of knots and the order of the spline. The number of knots (and the knots themselves) and the order are chosen subsequently by the user, or even determined adaptively by a knot-placement procedure. If, as in many applications, a spline interpolant or approximant is differentiated or integrated, i.e. a new spline with the same knots as before is created, but with a *different order*, notation C still holds whereas notation B has to be redefined. A similar comment applies to the case where the order is kept constant but the number of interior knots is increased, e.g. in *knot insertion* [2] and *subdivision* [3]. These considerations relate to univariate polynomial splines but extend to univariate periodic and

Table 4.1 Two notations for univariate polynomial splines

Term	de Boor (B)	This chapter (C)
Interval endpoints	t_k, t_{n+1}	$x_{\min}$, $x_{\max}$
Interior knots	$\{t_i\}_{k+1}^{n}$	$\{\lambda_j\}_1^N$
Number of interior knots	$n-k$	N
Additional knots	$\{t_i\}_1^k$, $\{t_i\}_{n+1}^{n+k}$	λ_j, $j<1$, $j>N$
Order or spline	k	n

Fundamental Developments of Computer-Aided Geometric Modeling
ISBN 012-554765-X

parametric splines and, where appropriate, to their multivariate tensor-product counterparts.

Throughout this chapter we refer to recommended software implementations of algorithms discussed, concentrating on those to which the author has contributed or of which he has intimate knowledge, *viz.* those in the NAG Library [4, 5] and DASL, the National Physical Laboratory's Data Approximation Subroutine Library [6–8]. The NAG Library provides a valuable capability for cubic and bicubic polynomial splines. DASL complements this capability by providing facilities for polynomial, periodic and parametric spline curves and polynomial and periodic spline surfaces, all of arbitrary order. The notation used in DASL documentation and in DASL routines themselves is consistent with that employed here.

Graphical illustrations are presented of results obtained using software implementations of some of the algorithms discussed. In the figures, data points are represented by small crosses or circles, and knot positions by short straight-line segments or asterisks.

REGIONS OF APPROXIMATION

In this section we describe the regions of approximation that are relevant to the various spline algorithms considered.

The line

Consider a finite interval

$$I = \{x : x_{\min} \leqslant x \leqslant x_{\max}\}$$

and its partition

$$x_{\min} < \lambda_1 \leqslant \lambda_2 \leqslant \cdots \leqslant \lambda_{N-1} \leqslant \lambda_N < x_{\max} \tag{1}$$

Subsequently, polynomial and other splines will be expressed in terms of B-splines (the chosen basis functions) defined on such a partition. The points $\lambda_1, \lambda_2, \ldots, \lambda_N$ will be termed *interior knots*. Evidently, there are N of these.[1] In order to define the full set of B-splines required, *'exterior knots'* λ_j, $j < 1$, $j > N$, satisfying

$$\begin{aligned} &\ldots \leqslant \lambda_{-1} \leqslant \lambda_0 \leqslant x_{\min}, \\ &x_{\max} \leqslant \lambda_{N+1} \leqslant \lambda_{N+2} \leqslant \ldots \end{aligned} \tag{2}$$

are introduced. The complete set of knots (interior plus exterior) is denoted by $\boldsymbol{\lambda}$. For the general knot sets considered here, and for all but periodic splines for which the following choice is not possible, there are advantages in selecting the exterior knots such that $\lambda_j = x_{\min}$, $j < 1$, and $\lambda_j = x_{\max}$, $j > N$.[2] One reason for this choice is that it is simple and independent of the other knots. Another is that it tends to optimize the numerical conditioning of spline interpolation and approximation problems (see [9] for detailed empirical evidence, and [10] for a theoretical result). A further advantage of coincident exterior knots is that their use facilitates the incorporation of derivative boundary conditions in spline approximation problems [11]. Throughout, prescribed values of $x_{\min}$, $x_{\max}$ and $\lambda_1, \lambda_2, \ldots, \lambda_N$ will be assumed to satisfy (1). λ_j is termed a *knot of multiplicity r* if $\lambda_{j-1} < \lambda_j = \lambda_{j+1} = \ldots = \lambda_{j+r-1} < \lambda_{j+r}$. If $r = 1$, λ_j is termed a *simple knot* and if $r > 1$ a *coincident or multiple knot*. Define the *subintervals*

$$I_j = x : \begin{cases} x \in [\lambda_j, \lambda_{j+1}), & j = 0, 1, \ldots, N-1 \\ x \in [\lambda_j, \lambda_{j+1}], & j = N \end{cases}$$

Note (the arbitrary decision) that, for $j < N$, interior knot λ_j is associated with subinterval I_j (rather than I_{j-1}). Also, note that in order to include the right-hand endpoint $x_{\max}$, it is considered part of I_N, which is therefore a subinterval that is closed at both ends. Finally, note that I_j is *empty* if $\lambda_j = \lambda_{j+1}$. We shall sometimes refer to a (sub)interval '*a* to *b*' rather than $[a, b]$, $[a, b)$, etc., when we do not need to be specific about whether it is closed, half-open, etc.

[1] N may be zero, i.e. there are no interior knots, with the result that a polynomial spline defined on I reduces to a single polynomial. A similar comment may be made for other splines. We shall not henceforth explicitly consider this case, but most algorithms discussed permit $N = 0$.

[2] This choice is used throughout this paper for polynomial splines. However, for splines on a uniform partition it may be desirable to choose the exterior knots at the same spacing.

For *periodic* splines, the interval of interest is bi-infinite, containing a finite interval $I = \{\theta : \theta_{\min} \leqslant \theta \leqslant \theta_{\max}\}$ which constitutes the basic period. Any value of θ is cyclically equivalent to $\theta + \ell(\theta_{\max} - \theta_{\min})$ for any integer ℓ. A partition $\theta_{\min} < \lambda_1 \leqslant \lambda_2 \ldots \leqslant \lambda_{N-1} \leqslant \lambda_N < \theta_{\max}$ provides a subdivision of the basic interval and, by periodicity, the whole of the real line. λ_j is taken as $\lambda_{j+N} - (\theta_{\max} - \theta_{\min})$ for $j < 1$ and $\lambda_{j-N} + (\theta_{\max} - \theta_{\min})$ for $j > N$. The exterior knots are defined by this extension. Note that the subinterval λ_0 to λ_1 (or λ_N to λ_{N+1}) is cyclically equivalent to the union of the subintervals λ_N to $\theta_{\max}$ and $\theta_{\min}$ to λ_1. Other properties are analogous to those for polynomial splines above.

For *parametric* splines, the interval of interest is that of the parameter, denoted here by t. The approach is similar to that for polynomial splines except that (a) the parameter t is used in place of the independent variable x, and (b) the interval of approximation is the normalized interval $[0, 1]$ in place of $[x_{\min}, x_{\max}]$.

The plane

We consider bivariate approximation over a rectangular region in the plane. In this case the above concepts and notation extend in a natural way. For polynomial splines, we consider a finite rectangular region $I \times J$, where $I = \{x : x_{\min} \leqslant x \leqslant x_{\max}\}$, $J = \{y : y_{\min} \leqslant y \leqslant y_{\max}\}$, and its partition

$$x_{\min} < \lambda_1 \leqslant \lambda_2 \leqslant \ldots \leqslant \lambda_{N_x - 1} \leqslant \lambda_{N_x} < x_{\max},$$

$$y_{\min} < \mu_1 \leqslant \mu_2 \leqslant \ldots \leqslant \mu_{N_y - 1} \leqslant \mu_{N_y} < y_{\max}$$

The points $\lambda_1, \lambda_2, \ldots, \lambda_{N_x}$ will be termed *interior x-knots* and the lines $x = \lambda_i$, $i = 1, 2, \ldots, N_x$, *interior x-knotlines*. A similar comment applies to the μ_j in the case of the y-variable. Exterior knots λ_i, $i < 1$, $i > N_x$, satisfying (2), and analogously for μ_j, are used. The complete set of y-knots will be denoted by $\boldsymbol{\mu}$.

In the periodic case the approach is as above, except that *cylindrical* coordinates are used instead of Cartesian coordinates. Thus z is used in place of x and θ in place of y. The splines considered are therefore polynomial in the first variable and periodic in the second. Extensions to *two* periodic variables can be considered analogously, using a spherical polar coordinate system, in which case the region of approximation becomes *toroidal* rather than cylindrical; we have found little practical need for such a region.

In the parametric case we consider approximations over $I \times J \equiv [0, 1] \times [0, 1]$, in variables t and u, each of which is analogous to t above. One of the variables may be periodic.

B-SPLINES

B-splines were established, for a uniform partition, by Schoenberg [12, 13] in his seminal work on spline approximation. They were not used seriously, for a general knot sequence, until a quarter of a century later.

The (normalized) B-splines

We denote by $N_{n,j}(x)$ the normalized B-spline of order n defined upon the knots $\lambda_{j-n}, \lambda_{j-n+1}, \ldots, \lambda_j$. If $\lambda_{j-n} = \lambda_{j-n+1} = \ldots = \lambda_j$, $N_{n,j}$ is taken as identically zero. Otherwise the normalized B-splines are defined through the three-term recurrence relation

$$N_{n,j}(x) = \left(\frac{x - \lambda_{j-n}}{\lambda_{j-1} - \lambda_{j-n}}\right) N_{n-1,j-1}(x) + \left(\frac{\lambda_j - x}{\lambda_j - \lambda_{j-n+1}}\right) N_{n-1,j}(x), \quad n > 1 \qquad (3)$$

with

$$N_{1,j}(x) = \begin{cases} 1, & x \in I_{j-1} \\ 0, & \text{otherwise} \end{cases} \qquad (4)$$

This recurrence relation, which was independently established simultaneously [14, 15] on both sides of the Atlantic (and also by Mansfield – see [14]), provided a breakthrough in the use of B-splines as a practical tool. Consequently, B-splines could then be evaluated by forming successively their values for orders 1, 2, ... up to the required order. The number of *flops*[3]

[3] A *flop* is a floating-point addition or subtraction, plus a floating-point multiplication or division, plus a few indexing operations.

required to form the nonzero B-splines of order n is approximately $\frac{3}{2}n^2$ [15]. Previously (see, for instance, [16,17]), B-splines were evaluated from their divided-difference definition. This approach, even for knot distributions that are not too extreme or orders that are not particularly high, can give rise to serious numerical error (see [15] for examples and an error analysis).

De Boor [14] stated that this recurrence and an associated one for evaluating a spline s from its B-spline representation appeared to be stable for orders as high as 80. Cox [15,18] gave complete floating-point error analyses which showed that the evaluation of B-splines and the B-spline representation are unconditionally stable. In particular, the values of the B-splines computed from the recurrence have *relative* errors bounded by the product of a *linear* function of n and the unit roundoff u.[4] Also the computed value of s has an *absolute* error bounded by the product of three terms: a linear function of n, the value of max $|c_j|$, and u, where the c_j denote the multipliers (coefficients) of the B-splines and the 'max' is taken over those values of j for which c_j is involved in the evaluation. In cases in which the relevant c_j all have the same numerical sign, the *relative* error is bounded by a quantity similar to that for the evaluation of the B-splines. These remarkable stability properties, virtually unsurpassed in numerical calculations, together with the compact support property, render the B-splines eminently suitable as a basis for splines. It must not be forgotten, however, that the *condition* of the B-spline basis is relevant to many calculations with splines. De Boor [19] shows that the condition of the basis grows exponentially with n (by a factor ~ 2 for each order). Despite this result, the B-splines form a well-conditioned basis for virtually all practical spline calculations.

B-spline properties

B-splines have the following valuable properties (which are not proved here – see [1] and [20] and the references therein):

Positivity:

$$N_{n,j}(x) > 0, \quad x \in (\lambda_{j-n}, \lambda_j).$$

Nonnegativity:

$$N_{n,j}(x) \geqslant 0, \quad x \in [\lambda_{j-n}, \lambda_j].$$

Compact support:

$$N_{n,j}(x) = 0, \quad x \notin [\lambda_{j-n}, \lambda_j].$$

Normalization (summation):

$$\sum_j N_{n,j}(x) \equiv 1, \quad x \in [x_{\min}, x_{\max}].$$

Normalization (integration):

$$\int_{x_{\min}}^{x_{\max}} N_{n,j}(x) = \frac{\lambda_j - \lambda_{j-n}}{n}.$$

Continuity (simple knots):

$$N_{n,j}(x) \in C^{n-2}(x_{\min}, x_{\max}).$$

The last property says that in the case of simple knots $N_{n,j}(x)$ and its derivatives of all orders up to the $(n-2)$nd are continuous everywhere within the interval of interest. The only points where discontinuities occur are the knots λ_{j-n}, $\lambda_{j-n+1}, \ldots, \lambda_j$, at each of which there is a jump in $N_{n,j}^{(n-1)}(x)$, the $(n-1)$st derivative of $N_{n,j}(x)$. $N_{n,j}^{(n-r)}(x)$ is discontinuous at a knot of multiplicity r. Any linear combination of the B-splines can only have discontinuities at the knots, at each of which it has (at least) as many continuous derivatives as the B-splines themselves. This is one of the most important advantages of using a B-spline basis.

SPLINE CURVES AND THEIR B-SPLINE REPRESENTATION

We consider four types of spline curve: polynomial, periodic, parametric, and periodic parametric. The first two are planar curves and the last two curves are in $\mathscr{R}^k$ for $k \geqslant 2$. For each we give a definition and describe its representation in terms of B-splines. This representation is used because of the many advantages it offers in terms of stability, efficiency and versatility [1,20,21].

Univariate polynomial splines

Let $s(x)$ denote a (univariate) polynomial spline curve of order n (degree $n-1$) with N interior knots $\lambda_1, \lambda_2, \ldots, \lambda_N$, which is defined over the

---

[4] u is the smallest positive machine number ε such that the computed value of $1 + \varepsilon$ exceeds unity.

interval $x_{\min} \leqslant x \leqslant x_{\max}$. If the knots are simple, s consists of a set of $N+1$ polynomial pieces of degree $n-1$ (or less), the jth of which applies to the subinterval λ_{j-1} to λ_j for $j = 1, 2, \ldots, N+1$. The jth and $(j+1)$st polynomial pieces match at $x = \lambda_j$, for $j = 1, 2, \ldots, N$, in value and all derivatives up to and including the $(n-2)$nd. If equalities are permitted among the knots, there will be correspondingly fewer polynomial pieces. Moreover, if a knot has multiplicity r, the polynomial pieces to the immediate left and right of the point match in value and all derivatives up to and including the $(n-r-1)$st.

The B-splines $N_{n,j}(x)$, $j = 1, 2, \ldots, q$, where $q = N + n$, form a basis for polynomial splines of order n (degree $n-1$) on the interval $[x_{\min}, x_{\max}]$ with interior knots $\lambda_1, \lambda_2, \ldots, \lambda_N$ [22]. Thus, if $s(x)$ is such a spline, it has the B-spline representation

$$s(x) = \sum_{j=1}^{q} c_j N_{n,j}(x), \quad x \in [x_{\min}, x_{\max}] \tag{5}$$

for certain coefficients c_j. The c_j are termed *B-spline coefficients*. From the compact support property, the value of $s(x)$ depends only on *some* of the coefficients, in fact (at most) n of them. Specifically,

$$s(x) = \sum_{j=J+1}^{J+n} c_j N_{n,j}(x), \quad x \in I_J \tag{6}$$

For specified coefficients $c_j, j = J+1, J+2, \ldots, J+n$, the computation time to evaluate s from (6) for a given value of x is about $\frac{3}{2}n^2$ flops, and therefore independent of N, the number of interior knots. However, to locate the subinterval I_J, comparisons of x must be made with a sequence of values of λ_k. The number of comparisons will in general depend on N. If, though, as is common, s is to be evaluated at a sequence of increasing or decreasing points in $[x_{\min}, x_{\max}]$, the dependence of the computation time on N is negligible. Either x lies in the same subinterval as the previous value or it lies in a neighboring subinterval. If s is to be evaluated at a 'random' point x in $[x_{\min}, x_{\max}]$, the appropriate subinterval can be located using binary search in approximately $\log_2 N$ comparisons. In practice, a combination of sequential and binary search yields an efficient process.

Figure 4.1 shows a cubic spline function with knots of various multiplicities.

Univariate periodic splines

Let $s(\theta)$ denote a (univariate) periodic spline curve of order n (degree $n-1$) with N knots $\lambda_1, \lambda_2, \ldots, \lambda_N$, which is defined over the basic interval $\theta_{\min} \leqslant \theta \leqslant \theta_{\max}$ (with the value $\theta = \theta_{\max}$ being cyclically equivalent to $\theta = \theta_{\min}$). If the knots are simple, s consists of a set of N polynomial pieces of degree $n-1$ (or less). The jth piece applies to the subinterval λ_{j-1} to λ_j, for

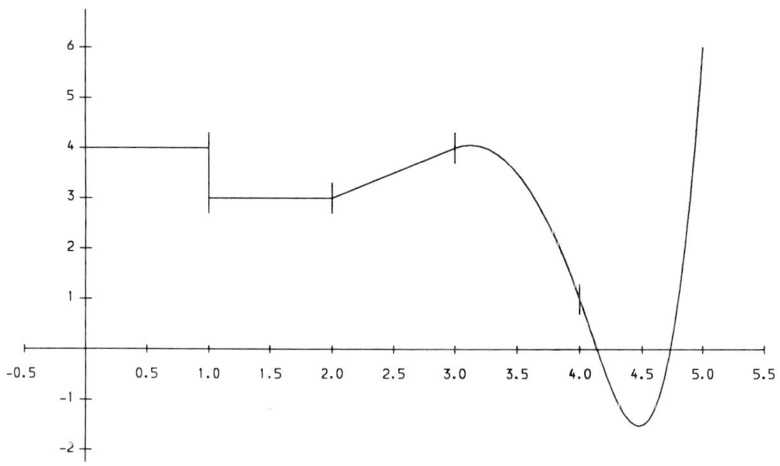

Figure 4.1. A cubic spline with knots of various multiplicities. The knots have multiplicities 4, 3, 2 and 1 at $x = 1.0, 2.0, 3.0$ and 4.0, respectively.

$j = 2, 3, \ldots, N$, and, because of periodicity, to the union of the subintervals $\theta_{\min}$ to λ_1 and λ_N to $\theta_{\max}$ (cyclically equivalent to either of the single subintervals λ_N to $\lambda_1 + (\theta_{\max} - \theta_{\min})$ or $\lambda_N - (\theta_{\max} - \theta_{\min})$ to λ_1), for $j = 1$. The jth and $(j+1)$st polynomial pieces match at $\theta = \lambda_j$, for $j = 1, 2, \ldots, N$, in value and all derivatives up to and including the $(n-2)$nd. Moreover, the value of s and *all* its derivatives at $\theta = \theta_{\max}$ agree respectively with those at $\theta = \theta_{\min}$. The statement in the preceding section relating to multiple knots applies, with an appropriate notational change, to periodic splines.

$s(\theta)$ has the B-spline representation

$$s(\theta) = \sum_{j=1}^{q} c_j N_{n,j}(\theta), \qquad \theta \in [\theta_{\min}, \theta_{\max}] \tag{7}$$

for certain coefficients c_j, where $q = N + n$ and, to ensure periodicity,

$$c_{j+N} = c_j, \quad j = 1, 2, \ldots, n \tag{8}$$

Comments similar to those for polynomial splines may be made in respect of method and efficiency of evaluation of s. For any value of θ, prior to evaluation θ is translated to the basic interval $[\theta_{\min}, \theta_{\max}]$ by argument reduction.

Parametric spline curves in $\mathscr{R}^k$

Open curves

Let $\mathbf{s}(t) = (s_1(t), s_2(t), \ldots, s_k(t))$ denote a parametric spline curve of order n (degree $n-1$) in k dimensions with N knots $\lambda_1, \lambda_2, \ldots, \lambda_N$, which is defined over the interval $0 \leqslant t \leqslant 1$. For convenience the knots will here be assumed to be simple, and therefore $s_h(t)$, for $h = 1, 2, \ldots, k$, consists of a set of $N+1$ polynomial pieces of degree $n-1$ (or less) in the variable t. The jth piece applies to the subinterval λ_{j-1} to λ_j, for $j = 1, 2, \ldots, N+1$. (λ_j is taken as zero for $j < 1$ and unity for $j > N$.) The jth and $(j+1)$st polynomial pieces match at $t = \lambda_j$, for $j = 1, 2, \ldots, N$, in value and all derivatives up to and including the $(n-2)$nd. Normally, the parametric spline itself has the same properties as its component parts.

The parametric spline has the B-spline

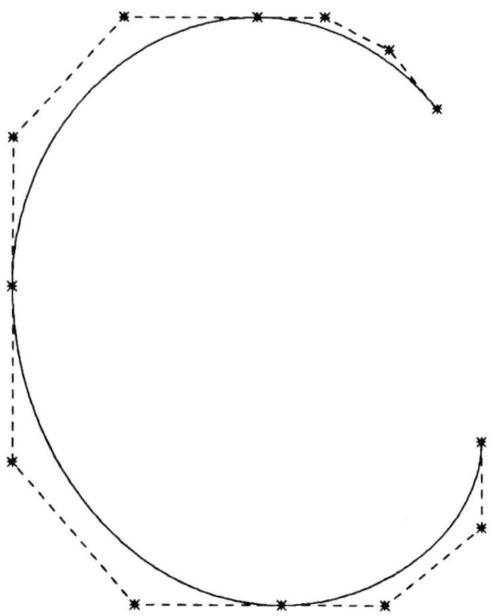

Figure 4.2. A parametric cubic spline function s in $k = 2$ dimensions with three double knots [23]. The B-spline coefficients are depicted as points (the *control points* in the CAD literature) in the space of the (two) variables of s. The points are connected by straight lines to form the so-called *control polygon*.

representation

$$s_h(t) = \sum_{j=1}^{q} c_{j,h} N_{n,j}(t), \quad t \in [0, 1], \qquad h = 1, 2, \ldots, k \tag{9}$$

for certain coefficients $c_{j,h}$, where $q = N + n$. Comments similar to those for polynomial splines may be made in respect of method and efficiency of evaluation of one component of the spline. However, once the n B-splines that are nonzero in a subinterval I_j, for a given $t \in I_j$, have been determined, evaluation of $\mathbf{s}(t)$ requires only nk flops.

Figure 4.2 shows a parametric cubic spline function in $k = 2$ dimensions.

Closed curves

Let $\mathbf{s}(t) = (s_1(t), s_2(t), \ldots, s_k(t))$ denote a parametric periodic spline curve of order n (degree $n-1$) in k dimensions with N knots $\lambda_1, \lambda_2, \ldots, \lambda_N$, which is defined over the interval

$0 \leqslant t \leqslant 1$ (with the value $t = 1$ being cyclically equivalent to $t = 0$). Again, the knots will be assumed simple and therefore $s_h(t)$, for $h = 1, 2, \ldots, k$, consists of a set of N polynomial pieces of degree $n - 1$ (or less) in the variable t. The jth piece applies to the subinterval λ_{j-1} to λ_j, *for* $j = 2, 3, \ldots, N$, and, because of periodicity, to the union of the subintervals 0 to λ_1 and λ_N to 1 (cyclically equivalent to either of the single subintervals λ_N to $\lambda_1 + 1$ or $\lambda_N - 1$ to λ_1), for $j = 1$. (λ_j is taken as $\lambda_{j+N} - 1$ for $j < 1$ and $\lambda_{j-N} + 1$ for $j > N$.) The above continuity properties for parametric splines apply. The parametric periodic spline has the B-spline representation (9) for certain coefficients $c_{j,h}$, where $q = N + n$ and, to ensure periodicity,

$$c_{j+N,h} = c_{j,h}, \quad j = 1, 2, \ldots, n, \quad h = 1, 2, \ldots, k$$

Comments similar to those for periodic splines may be made in respect of method, and to those for nonperiodic parametric splines in respect of efficiency of evaluation of the spline.

SPLINE SURFACES AND THEIR B-SPLINE REPRESENTATION

We define three classes of spline surface, polynomial, periodic and parametric, and give their representations in terms of B-splines. The extension to periodic parametric, as in the univariate case, is straightforward and for brevity not made here. The treatment is of *bivariate* tensor-product splines; adaptation to the multivariate case is straightforward but notationally more cumbersome. Tensor-product splines are a natural generalization of univariate splines in that for a fixed value of y the bivariate spline $s(x, y)$ is simply a univariate spline in x, and conversely a univariate spline in y for a fixed value of x.

Bivariate polynomial splines

Let $s(x, y)$ denote a (bivariate) polynomial spline surface of order n_x (degree $n_x - 1$) in x and order n_y (degree $n_y - 1$) in y with N_x knots $\lambda_1, \lambda_2, \ldots, \lambda_{N_x}$ in x, and N_y knots $\mu_1, \mu_2, \ldots, \mu_{N_y}$ in y, which is defined over the rectangular region $x_{\min} \leqslant x \leqslant x_{\max}$, $y_{\min} \leqslant y \leqslant y_{\max}$. If all knots are simple, s consists of a set of $(N_x + 1)(N_y + 1)$ polynomial 'patches' of degree $n_x - 1$ (or less) in x and degree $n_y - 1$ (or less) in y. The patch numbered (i, j) applies to the region bounded by the rectangle λ_{i-1} to λ_i in x and μ_{j-1} to μ_j in y (λ_i is taken as $x_{\min}$ for $i < 1$ and $x_{\max}$ for $i > N_x$, and μ_j as $y_{\min}$ for $j < 1$ and $y_{\max}$ for $j > N_y$). Patch (i, j) matches its neighboring patches on its bounding x-knotlines in value and all derivatives with respect to x up to and including the $(n_x - 2)$nd, and on its bounding y-knotlines in value and all derivatives with respect to y up to and including the $(n_y - 2)$nd. If equalities are permitted among the knots, there will be correspondingly fewer polynomial patches. Moreover, if a knot in x is repeated exactly r times, the polynomial patches neighboring this value of x will match them in value, and in all derivatives with respect to x up to and including the $(n_x - r - 1)$st. A similar comment applies to a y-knotline.

s has the B-spline representation

$$s(x, y) = \sum_{i=1}^{q_x} \sum_{j=1}^{q_y} c_{i,j} N_{n_x,i}(\boldsymbol{\lambda}; x) N_{n_y,j}(\boldsymbol{\mu}; y) \tag{10}$$

where the $N_{n_x,i}(\boldsymbol{\lambda}; x)$, $i = 1, 2, \ldots, q_x := N_x + n_x$, denote the B-splines of order n_x in x with the above x-knotlines, and the $N_{n_y,j}(\boldsymbol{\mu}; y)$, $j = 1, 2, \ldots, q_y := N_y + n_y$, the B-splines of order n_y in y with the above y-knotlines. We represent the $q_x \times q_y$ matrix of B-spline coefficients $\{c_{i,j}\}_{i=1;j=1}^{q_x;\ q_y}$ by C, and let q denote the total number of coefficients $q_x q_y$. We see that s is represented as a linear combination of all products of the B-splines of order n_x in x with x-knotlines $\boldsymbol{\lambda}$ and the B-splines of order n_y in y with y-knotlines $\boldsymbol{\mu}$. An individual basis function of s is typified by $N_{n_x,i}(\boldsymbol{\lambda}; x)N_{n_y,j}(\boldsymbol{\mu}; y)$. The properties and advantages of the univariate B-spline representation extend in a natural way to the bivariate form.

Bivariate periodic splines

Let $s(z, \theta)$ denote a (bivariate) periodic spline surface of order n_z (degree $n_z - 1$) in z and order n_θ (degree $n_\theta - 1$) in θ, that is periodic in θ, with N_z knots $\lambda_1, \lambda_2, \ldots, \lambda_{N_z}$ in z, and N_θ knots μ_1,

$\mu_2, \ldots, \mu_{N_\theta}$ in θ, which is defined over the basic rectangle $z_{\min} \leqslant z \leqslant z_{\max}$, $\theta_{\min} \leqslant \theta \leqslant \theta_{\max}$ (with the value $\theta = \theta_{\max}$ being cyclically equivalent to $\theta = \theta_{\min}$). If all knots are simple, s consists of a set of $(N_z + 1)N_\theta$ polynomial 'patches' of degree $n_z - 1$ (or less) in z and degree $n_\theta - 1$ (or less) in θ. Similar comments to those in the previous section can be made about these patches, bearing in mind that one variable is periodic, and the effect of multiple knots.

Bivariate parametric splines

Let $(s_x(t, u), s_y(t, u), s_z(t, u))$ denote a parametric spline surface of order n_t in t and order n_u in u, with N_t knots $\lambda_1, \lambda_2, \ldots, \lambda_{N_t}$ in the parameter t, and N_u knots $\mu_1, \mu_2, \ldots, \mu_{N_u}$ in the parameter u, which is defined over the rectangular region $0 \leqslant t \leqslant 1$, $0 \leqslant u \leqslant 1$. Simple knots will be assumed, in which case $s_\alpha(t, u)$, where α denotes x, y or z, consists of a set of $(N_t + 1)(N_u + 1)$ polynomial patches of degree $n_t - 1$ (or less) in t and degree $n_u - 1$ (or less) in u. The parametric spline surface has the B-spline representation

$$s_\alpha(t, u) = \sum_{i=1}^{q_t} \sum_{j=1}^{q_u} c_{\alpha_{i,j}} N_{n_t,i}(\boldsymbol{\lambda}; t) N_{n_u,j}(\boldsymbol{\mu}; u)$$

$$\alpha \equiv x, y, z \qquad (11)$$

for certain coefficients $c_{\alpha_{i,j}}$, $\alpha \equiv x, y, z$, where the $N_{n_t,i}(\boldsymbol{\lambda}; t)$, $i = 1, 2, \ldots, q_t := N_t + n_t$, denote the B-splines of order n_t in t with the above t-knots, and the $N_{n_u,j}(\boldsymbol{\mu}; u)$, $j = 1, 2, \ldots, q_u := N_u + n_u$, the B-splines of order n_u in u with the above u-knots.

SPLINE CURVE INTERPOLANTS

We outline in this section algorithms for interpolation by univariate polynomial, periodic and parametric splines. The algorithms apply to general distributions of data points. The methods considered do not require derivatives to be specified at the interval endpoints.

Univariate polynomial splines

The univariate polynomial spline interpolation problem[5] may be stated as follows. Given m data points (x_i, f_i), $i = 1, 2, \ldots, m$, with $x_1 < x_2 < \ldots < x_m$, determine a spline function $s(x)$ satisfying the conditions

$$s(x_i) = f_i, \qquad i = 1, 2, \ldots, m \qquad (12)$$

There are three choices to make: the order n of the spline, the number N of knots, and the knot set $\boldsymbol{\lambda}$. The choice $n = 4$ (*cubic* splines) is suitable for many purposes; this order is sufficiently high to give reasonable continuity (in value, first derivative and second derivative, in the case of simple knots) and to give s sufficient flexibility to adapt to a variety of underlying functions, but low enough not to introduce undesirable oscillations. In certain situations a different order is more appropriate. The value of N is given immediately, since for general data the number of parameters q $(= N + n)$ of s must match the number of points m; hence $N = m - n$.

Schoenberg and Whitney [24] provide necessary and sufficient conditions for existence and uniqueness of s: a unique data point whose abscissa lies strictly within the support of the B-spline can be associated with each B-spline in the basis. These conditions can rapidly be tested within a spline interpolation algorithm before attempting to form and solve the defining equations (16) below. They can be stated as

$$x_i < \lambda_i < x_{i+n}, \quad i = 1, 2, \ldots, m - n \qquad (13)$$

Despite this *mathematical* result it is desirable that these conditions are in a sense *well-satisfied*, else s may unnecessarily have unreasonable behavior. In this respect, the choice

$$\lambda_j = \begin{cases} x_{j+r}, & \text{if } n = 2r \quad \text{(i.e. } n \text{ even)} \\ (x_{j+r-1} + x_{j+r})/2, & \text{if } n = 2r - 1 \quad \text{(i.e. } n \text{ odd)} \end{cases} \qquad (14)$$

for $j = 1, 2, \ldots, N$, is recommended. In the important case of cubic splines this choice means that knots are positioned at all data points apart

[5] Schumaker [16] made the comment that this particular interpolation had been all but neglected. Previously, seemingly all the effort in designing spline interpolation algorithms was for splines of even order with a knot placed at each x_i and $n - 2$ specified endpoint derivative values.

from the first two and the last two. Thus, in this case, the cubic arcs of which s is composed are such that the first arc spans the first three data points, the last arc spans the last three data points, and each of the intermediate arcs spans an interval between neighboring data points. Since it appears to be satisfactory for general use, this choice is adopted in DASL [7, 8]. There are special circumstances in which other choices are appropriate. For instance, suppose it is required to mimic a gradient discontinuity in the underlying function at some point. In the case $n = 4$, for example, the λ_j may then be defined as above except that λ_{k-1} and λ_{k+1}, for some particular choice of k $(1 < k < N)$, are then replaced by the value of λ_k. The Schoenberg–Whitney (S–W) conditions (13) are still satisfied, and the effect is to introduce a triple knot at $x = \lambda_k$, thus providing a discontinuity in s' at this point.

The use of the B-spline representation (5) of s together with (12) yields the defining equations

$$\sum_{j=1}^{q} c_j N_{n,j}(x_i) = f_i, \quad i = 1, 2, \ldots, m \tag{15}$$

for the B-spline coefficients c_j, $j = 1, 2, \ldots, q$ $(=m)$, or, in matrix form,

$$Ac = f \tag{16}$$

where A is the $m \times m$ matrix with $a_{i,j} = N_{n,j}(x_i)$, $f = (f_1, f_2, \ldots, f_m)^T$ and $c = (c_1, c_2, \ldots, c_m)^T$. The matrix[6] A is *banded*, of bandwidth n, each row containing at most n nonzero entries (occurring in contiguous positions) as a result of the compact support of the B-splines.

A is totally positive, i.e. all its minors are nonnegative [26]: a linear system with such a coefficient matrix can be solved safely using Gaussian elimination without pivoting [27]. An algorithm for solving this system of equations, which exploits the structure of A, is outlined by Cox [20]. Because pivoting is unnecessary, the elimination can be carried out *in situ*, and thus no additional array space to accommodate fill-in is required. In fact the matrix A need not be formed *in toto* and then factorized; it is computationally convenient to form and process sequentially successive rows of (A, f) using elementary transformations [28], stabilization being unnecessary because of the above property. This sequential method of solution is compatible with that considered subsequently for spline *approximation*, for which the savings in computer memory can be considerable. The storage requirements for the implementation of this approach in DASL is essentially that for the upper triangular system to which (16) is reduced. The matrix in this system has bandwidth n and thus approximately mn storage locations are required. The solution of the equations takes $O(mn^2)$ flops. Since the B-spline values required in forming the equations can also be determined in $O(mn^2)$ flops, the total time to construct s is $O(mn^2)$. In particular, for a *fixed* order of spline (e.g. order 4, cubic), the time is proportional to the number of data points. A backward error analysis [67] of the complete problem, i.e. the formation and solution of the defining equations, shows that the computed solution represents an exact interpolant for a neighboring point set. The bound derived for the departure of this set from the given point set is independent of the condition of A.

Univariate periodic splines

The approach to periodic spline interpolation differs from that of the previous section in the following way: m distinct data points denoted by (θ_i, f_i), $i = 1, 2, \ldots, m$, are given, and now it is required to solve equations (15), with the $N_{n,j}$ defined on a periodically extended knot set, and with θ_i in place of x_i, subject to the conditions (8). These conditions can be used to eliminate n of the coefficients, thus reducing the problem to the solution of m equations in N unknowns. As a result we must specify m interior knots. If $m \leqslant 2n$ the coefficient matrix A is generally full; otherwise A is bordered[7] with bandwidth n and borderwidth $n - 1$. There is no straightforward counterpart of the S–W conditions (13) for periodic spline interpolation.

[6] As a consequence of the ordering of the x_i the matrix is also in standard form [25], i.e. the column position of the first nonzero entry in each row is a nondecreasing function of row number.

[7] A bordered (or augmented band) matrix of bandwidth p and borderwidth p' is identical to a band matrix of bandwidth p except that the rightmost p' column contain nonzero elements.

However, Schumaker [29, pp 297–308] provides conditions that apply to cases where m is odd. The choice

$$\lambda_j = \begin{cases} \theta_j, & \text{if } n \text{ is even} \\ (\theta_j + \theta_{j+1})/2 & \text{if } n \text{ is odd} \end{cases}$$

where $\theta_{m+1} = \theta_1 + (\theta_{max} - \theta_{min})$, which is adopted in DASL, is satisfactory for many purposes.[8] For this choice, A is invertible and hence a periodic spline interpolant exists and is unique. The matrix A is not in general totally positive so, to ensure stability, Gaussian elimination with pivoting, stabilized elementary transformations or orthogonal transformations should be employed. A strategy employing orthogonal transformations in the form of Givens plane rotations [25] is used in the DASL implementation of a periodic spline interpolation algorithm. A method which employs full matrix or structured matrix techniques is used according to whether $m <$ or $\geqslant 2n$. Again, the number of flops is proportional to mn^2.

Parametric spline curves in $\mathscr{R}^k$

The approach of the previous two sections can, together with a choice of parametrization, straightforwardly be applied to parametric spline curves in $\mathscr{R}^k$. We distinguish between open and closed curves, these being the parametric counterparts of polynomial and periodic splines.

For open curves, the parametric spline interpolation problem can be stated as follows. Given m *ordered* points $\mathbf{x}_i \equiv (x_{i,1}, x_{i,2}, \ldots, x_{i,k}) \in \mathscr{R}^k$, $i = 1, 2, \ldots, m$, with $\mathbf{x}_{i-1} \neq \mathbf{x}_i$, $i = 2, 3, \ldots, m$, determine a vector-valued spline function $\mathbf{s}(t) \in \mathscr{R}^k$ satisfying the conditions

$$\mathbf{s}(t_i) = \mathbf{x}_i, \quad i = 1, 2, \ldots, m \tag{17}$$

In (17) the t_i represent a suitable set of parameter values associated with the $\mathbf{x}_i$. A convenient parametrization, used in DASL, is *normalized cumulative Euclidean distance* given by $t_i = d_i/d_m$, $i = 1, 2, \ldots, m$, where

$$d_1 = 0; \quad d_i = d_{i-1} + \|\mathbf{x}_i - \mathbf{x}_{i-1}\|_2, \quad i = 2, 3, \ldots, m$$

Knots may sensibly be chosen in t in the same way as they are in x for polynomial splines. The form (9) is used for each component of $\mathbf{s}$, the coefficients in which can be determined by the univariate polynomial spline interpolation method outlined earlier. The coefficient matrix A is common to all k calculations. Thus the problem can be reduced to a system of linear algebraic equations with k right-hand sides, with obvious gains in efficiency. The above choice of parameterization is usually effective. It is influenced, however, by the scales of the axes in the k-dimensional axis system. The choice is such, though, that spurious behavior (e.g. cusps or loops) is unlikely. The use of certain other parameterizations, including uniformly-spaced values of t_i, may be unsatisfactory in this respect [1, pp 316–318].[9] If the use of cumulative Euclidean distance is considered inappropriate, cumulative arc length may be used instead. Of course, arc length is unknown *a priori*, but it may be calculated (by numerical quadrature) following computation of an interpolant based on chord-length parametrization. If necessary, the process can be repeated; for practical purposes it normally stabilizes in a small number of iterations.

The above approach is modified slightly if a closed-curve interpolant is sought. It is then required that $\mathbf{s}(t_i) = \mathbf{x}_i$, $i = 1, 2, \ldots, m+1$, where $\mathbf{x}_{m+1} = \mathbf{x}_1$. The most significant changes to the above approach are that knots may be selected in t as they are in θ for periodic splines, and that each component of $\mathbf{s}$ is now a periodic spline, to which the method of the previous subsection is applied.

Figure 4.3 shows a periodic parametric spline interpolant to data representing a closed curve.

SPLINE CURVE APPROXIMANTS

In this section we consider methods for computing least-squares (ℓ_2-norm) approximants to data by polynomial, periodic and parametric splines. Approximation in other norms is possible but computationally much less efficient for large problems because matrix structure

[8] We choose the basic period $[\theta_{min}, \theta_{max}]$ such that $\theta_{min} < \theta_1$ and $\theta_m < \theta_{max}$.

[9] Epstein [30] has proved that parametric cubic splines with simple knots have a continuously varying tangent line if the parametrization is based on a norm (which includes the above chord-length parametrization).

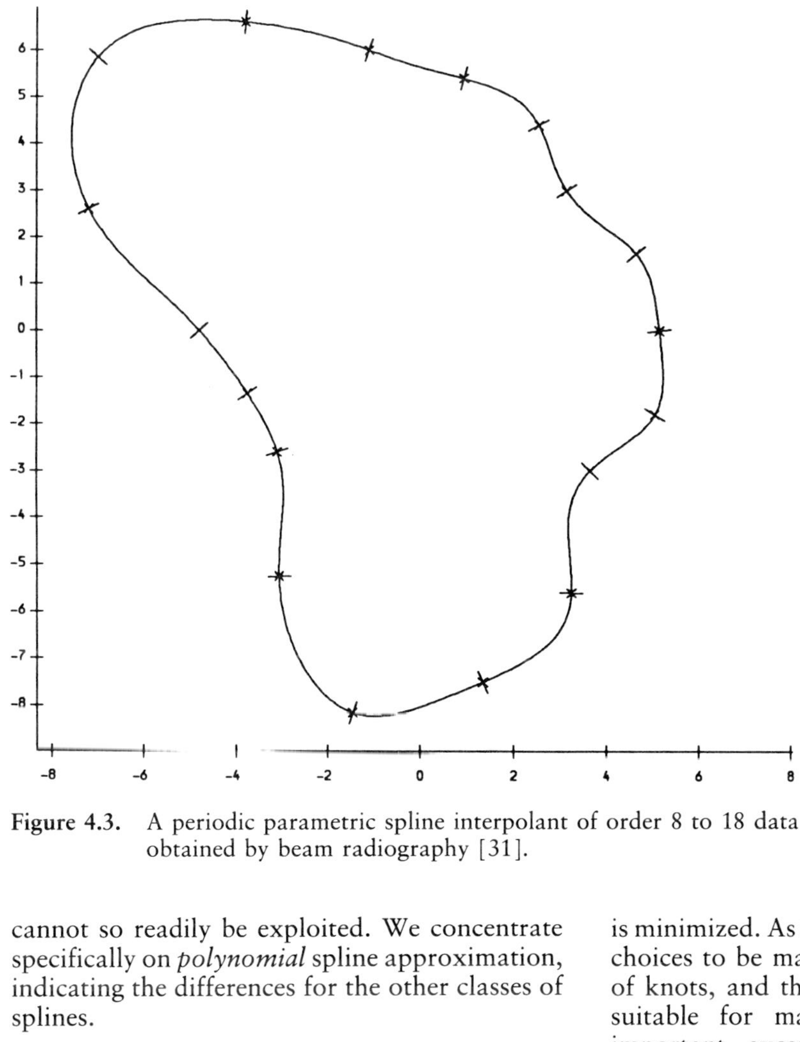

Figure 4.3. A periodic parametric spline interpolant of order 8 to 18 data points representing a lung section obtained by beam radiography [31].

cannot so readily be exploited. We concentrate specifically on *polynomial* spline approximation, indicating the differences for the other classes of splines.

Univariate polynomial splines

The univariate least-squares polynomial spline approximation problem may be stated as follows (note the similarities with interpolation). Given m (not necessarily distinct) data points (x_i, f_i), $i = 1, 2, \ldots, m$, and corresponding weights[10] w_i, $i = 1, 2, \ldots, m$, determine a spline function $s(x)$ such that the weighted residual sum of squares $\sum_{i=1}^{m} e_i^2$, where

$$e_i = w_i\{f_i - s(x_i)\}, \quad i = 1, 2, \ldots, m$$

is minimized. As for interpolation, there are three choices to be made: the order n, the number N of knots, and the knot set[11] $\boldsymbol{\lambda}$. Again, $n = 4$ is suitable for many problems, but there are important exceptions. In contrast to interpolation, in which N is decided by the quantity of data and the order of the spline, i.e. $N = m - n$, for approximation we prefer, if possible, to use a relatively small number of knots and so, normally, $N < m - n$ and, ideally, $N \ll m - n$. Generally, knots should be more greatly concentrated in regions where the function underlying the data is more 'severe' [32].

[10] Each w_i is permitted to be any real number. The solution is unchanged if a weight is negated, and a zero weight means that the corresponding data point has no effect.

[11] In this section we shall assume that the knots are to be selected in advance so that the residual sum of squares is to be minimized with respect to the *linear* parameters (here the B-spline coefficients) of s. Later we consider the knots themselves as additional minimization variables.

The problem is now to solve the above minimization problem in which, using the B-spline representation of s

$$e_i = w_i\left\{f_i - \sum_{j=1}^{q} c_j N_{n,j}(x_i)\right\}, \quad i = 1, 2, \ldots, m$$

In matrix terms, the problem becomes

$$\min_c e^{\mathrm{T}}e, \quad e = W(f - Ac)$$

where $e = (e_1, e_2, \ldots, e_m)^{\mathrm{T}}$, $W = \mathrm{diag}\{w_1, w_2, \ldots, w_m\}$, $f = (f_1, f_2, \ldots, f_m)^{\mathrm{T}}$, $c = (c_1, c_2, \ldots, c_q)^{\mathrm{T}}$, and A is the $m \times q$ matrix with $a_{i,j} = N_{n,j}(x_i)$. A is banded, of bandwidth n, as for interpolation. If the matrix WA has full rank, the solution to this problem is defined by the normal equations $Hc = g$, where $H = A^{\mathrm{T}}W^2A$, $g = A^{\mathrm{T}}W^2f$, and may be computed as follows. First, form H (a matrix of bandwidth $2n - 1$) and g. Second, determine the Cholesky decomposition

$$H = R^{\mathrm{T}}R \tag{18}$$

where R, the Cholesky factor, is a band upper triangular matrix of order q and bandwidth n. Finally, solve the resulting lower and upper triangular systems $R^{\mathrm{T}}\theta = g$ and $Rc = \theta$. An algorithm based on this approach is given by de Boor [1]. Alternatively, and more reliably but computationally less efficiently, solution can be effected by orthogonal triangularization. With this approach orthogonal transformations are applied to WA to reduce it to triangular form R (which is *mathematically* equal to the Cholesky factor provided in the normal-equations approach). The same transformations are applied to Wf to produce a vector θ (again mathematically the same vector as before), and the upper triangular system $Rc = \theta$ is then solved. Again, full account can be taken of the structure of A. Details are given by Cox [25]. The method described there is based on the use of Givens rotations to effect the orthogonal transformations. This approach permits successive rows of A and f to be formed and processed, and requires minimal memory. Since their introduction [33, 34] for general matrices and their specialization to band [25, 35] and bordered band systems [25], Givens-based methods seem to be becoming 'standard' for a wide class of linear least-squares problems in general and least-squares spline approximation problems in particular. A backward error analysis [9] shows that the computed vector c corresponds to a spline approximant that is exact for a neighboring point set. In this case the bound (inevitably) depends on the condition of A. Total computation time is again essentially $O(mn^2)$. Note that this time is independent of the number of knots. Thus, for given data a spline of a fixed order with many knots can be determined as efficiently as one of the same order with just a few or even no knots.

Figure 4.4 portrays a polynomial spline approximant to photon count data.

For certain positions of the knots (when the number of knots in certain regions is 'large' for the number of distinct data points there), the spline will not necessarily be unique. It is possible to test for this eventuality before attempting to determine s. The spline is unique if and only if there exists a subset of q distinct data points (with nonzero weights) for which a spline *interpolant* with the specified knots can be computed. Thus the test is based on determining whether there exists such a subset satisfying the S–W conditions (cf. (13)).[12] If these conditions are not satisfied an *infinity* of least-squares solutions exists. In DASL a particular solution is computed by incorporating appropriate *resolving constraints* [34]. The constraints are selected such that the solution has desirable geometric properties [20]. Note that the minimum norm solution (in our case that which minimizes $c^{\mathrm{T}}c$), often used for rank-deficient least-squares problems, does not possess this property (for a simple example, see Cox [36]). *Numerical* rank deficiency can arise, i.e. cases in which A is close to a rank deficient matrix but for which the corresponding S–W conditions are (not necessarily just) satisfied. In such an instance, the strategy used in DASL is to detect such cases, using a suitable numerical tolerance, and then to proceed as if A *were* rank deficient.

Univariate periodic splines

The approach of the previous section extends straightforwardly to univariate periodic spline approximation. Data provided are a set of points

[12] Cox and Hayes [35] provide an algorithm for determining whether such a subset exists.

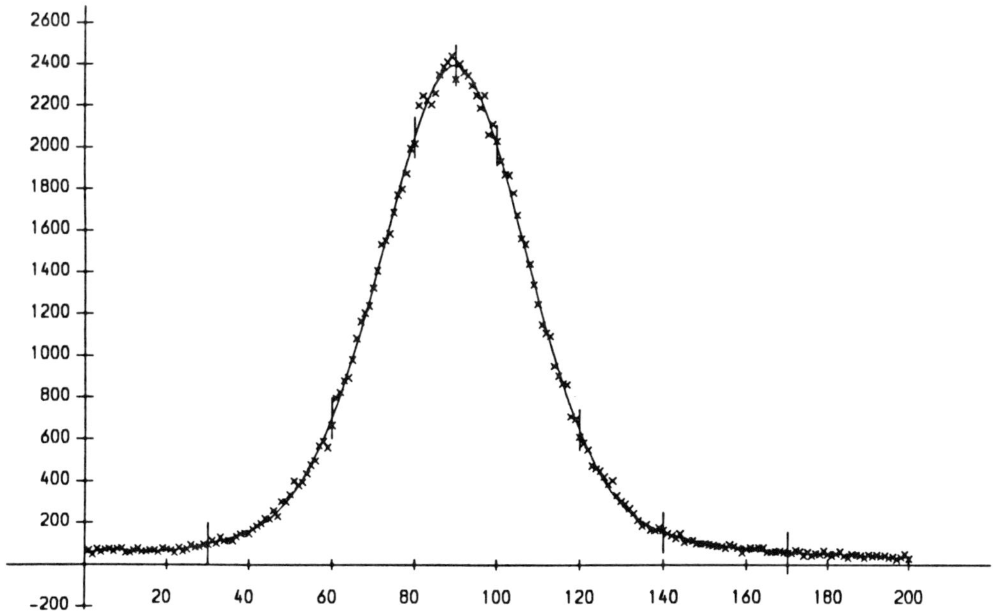

Figure 4.4. A least-squares polynomial spline approximant of order 6 with 8 non-uniform knots to 200 uniform data points obtained from a photon counting experiment [8].

(θ_i, f_i) and weights w_i, $i = 1, 2, \ldots, m$. Knots are selected as before, the B-spline representation (7) of a periodic spline employed, and the problem formulated in matrix terms. The matrix A is now bordered, with bandwidth n and border width $n - 1$. The least-squares solution is computed taking advantage of this structure [25]. No counterpart of the S–W conditions exists, and so the algorithm must identify cases of rank deficiency or near-rank deficiency; the method used in DASL is similar to that for polynomial splines (see previous section).

Parametric spline curves in $\mathscr{R}^k$

The approach to least-squares parametric spline approximation follows that of the earlier section on parametric spline curves. Here, however, equations (17) are overdetermined and are solved by least squares. A variant of the method of the previous two subsections is used according to whether the underlying curve is open or closed. In either case the least-squares system to be solved has k righthand sides. The DASL implementations provide knots that are a generalization of those recommended for interpolation. These often prove to be satisfactory, but the choice may be overridden.

Figure 4.5 shows a periodic parametric spline approximant to data representing a closed curve.

SPLINE SURFACE INTERPOLANTS

We outline algorithms for bivariate spline interpolation to data specified on a rectangular mesh. The mesh lies in the space of the independent variables for polynomial and periodic spline interpolation and in parameter space for parametric spline interpolation. Only *tensor-product* spline interpolation is considered here. This form of interpolation is generally inappropriate for scattered data. There are many other approaches (not using B-splines) for such data (see Franke [37]).

Bivariate polynomial spline

Given data values $F = \{f_{h,k}\}_{h=1;k=1}^{m_x;\ m_y}$, specified at the vertices of an $m_x \times m_y$ rectangular mesh $\{x, y : x = x_h, y = y_k\}_{h=1;k=1}^{m_x;\ m_y}$, the problem is to construct a bivariate polynomial spline interpolant $s(x, y)$ such that

$$s(x_h, y_k) = f_{h,k}, \quad h = 1, 2, \ldots, m_x,$$
$$k = 1, 2, \ldots, m_y$$

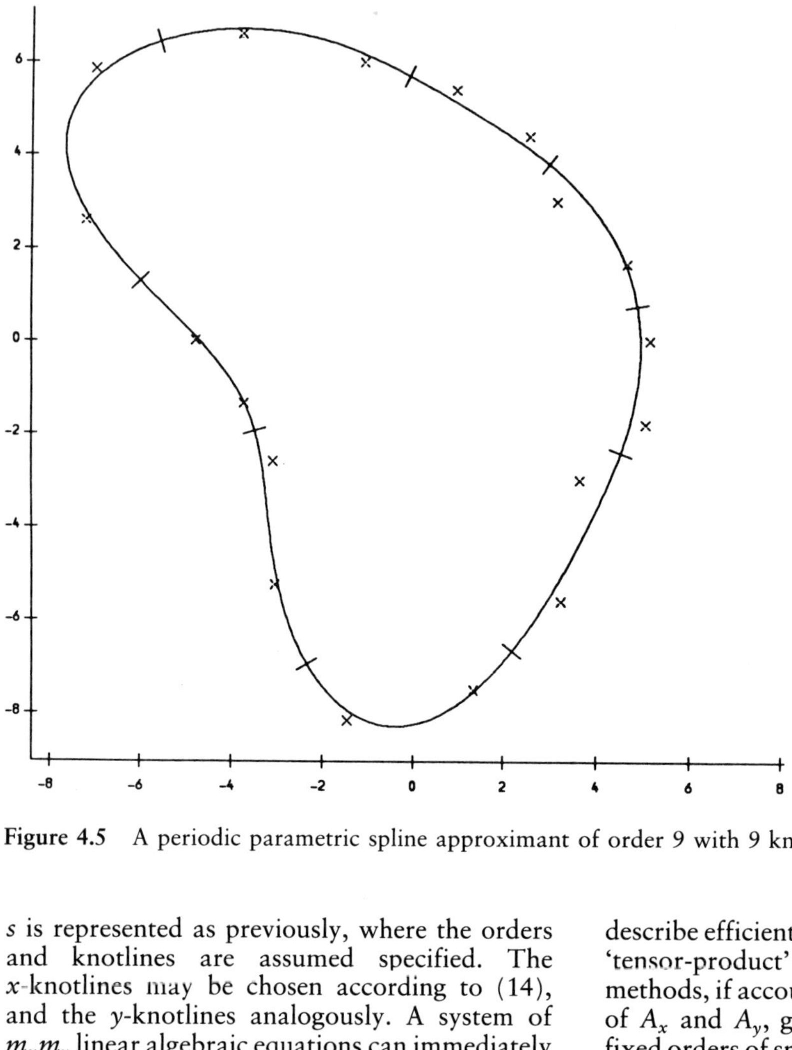

Figure 4.5 A periodic parametric spline approximant of order 9 with 9 knots to the data of Figure 4.3.

s is represented as previously, where the orders and knotlines are assumed specified. The x-knotlines may be chosen according to (14), and the y-knotlines analogously. A system of $m_x m_y$ linear algebraic equations can immediately be constructed which define the coefficients C. However, the work involved in solving this system is enormously reduced if advantage is taken of separability. The defining equations can be written as $A_x C A_y^T = F$ and therefore solved in two stages, through an intermediate $m_x \times m_y$ matrix B:

$$A_x B = F, \quad A_y C^T = B^T \tag{19}$$

The matrix A_x is of order m_x, of bandwidth n_x, and contains the B-spline basis functions in x evaluated at the x-meshlines, and A_y similarly for y. Thus each of the above two stages corresponds to the solution of a univariate polynomial spline interpolation problem, the only exception being that there are multiple 'right-hand sides'. De Boor [38] and Cox [39] describe efficient methods for implementing such 'tensor-product' schemes. The use of these methods, if account is taken of the band structure of A_x and A_y, gives an execution time that for fixed orders of spline is $O(m)$, where $m = m_x m_y$.

Bivariate periodic splines

The formulation for bivariate periodic spline interpolation follows closely that above for the polynomial case. The major difference is that one of the two systems in (19) is *bordered*. Again, for fixed orders of spline, $O(m)$ flops are required.

Bivariate parametric splines

We do not concentrate specifically on bivariate parametric spline surface *interpolation* in this paper. Some of the statements in a following

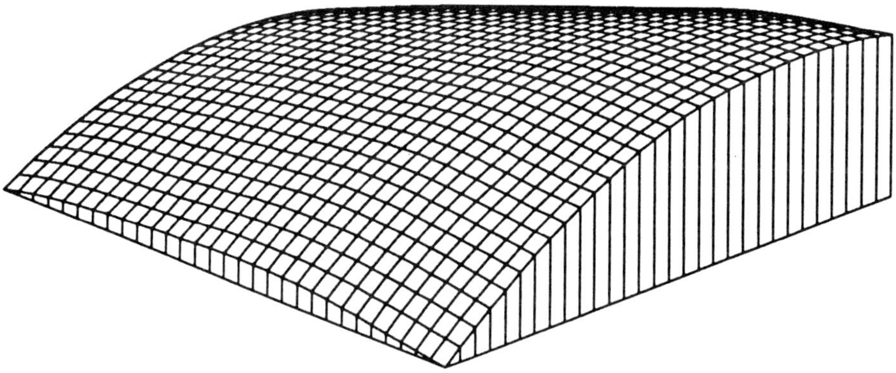

Figure 4.6. An isometric projection of a bicubic polynomial spline interpolant to flatness calibration data specified at the vertices of a 7 × 7 rectangular mesh [40].

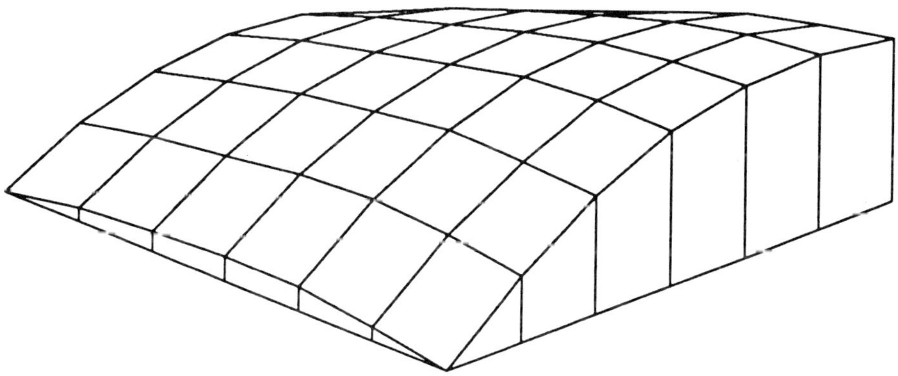

Figure 4.7. An isometric projection of the data from which Figure 4.6 was computed. The data is connected by straight lines for purposes of visualization. The figure was formed by determining a bilinear spline interpolant.

section on the corresponding approximation problem can be specialized to interpolation.

Figure 4.6 shows a bicubic polynomial spline interpolant to flatness calibration data. Figure 4.7 shows by means of its bilinear spline interpolant the data from which Figure 4.6 was computed.

SPLINE SURFACE APPROXIMANTS

We outline algorithms for least-squares bivariate spline approximation by polynomial and parametric splines. We do not explicitly consider periodic splines since their treatment follows straightforwardly from the polynomial spline case together with the univariate periodic spline concepts.

Bivariate polynomial splines

We distinguish between two cases: data on a mesh and scattered data.

Data on mesh

In the case of data on a mesh the problem is as posed previously, except that instead of an interpolant a bivariate polynomial spline approximant $s(x, y)$ is sought such that the residual sum of squares[13] $\sum_{h=1}^{m_x} \sum_{k=1}^{m_y} e_{h,k}^2$ is

[13] Note that a weight matrix cannot be incorporated because, unless it has appropriate structure, separability is no longer possible. If general weights are required the method of the following section can be applied.

minimized, where

$$e_{h,k} = f_{h,k} - s(x_h, y_k), \quad h = 1, 2, \ldots, m_x, \quad k = 1, 2, \ldots, m_y$$

The problem separates, as for interpolation, and is a special case of least-squares approximation by a tensor product of two sets of general basis functions, as analyzed by Greville [41]. It reduces to the least-squares solutions of overdetermined systems of the form (19), where now A_x is $m_x \times q_x$, of bandwith n_x, and A_y is $m_y \times q_y$, of bandwidth n_y. As for interpolation, for fixed orders of spline these systems can be solved in $O(m)$ time. If no account if taken of separability, $O(qm)$ time, where $q = q_x q_y$, is required. For a problem of quite modest size the corresponding computation times could be 10 seconds and 1 hour. Separability also implies that the S–W conditions may be tested independently for each variable, and that cases of rank deficiency may be treated as for univariate polynomial splines.

Figure 4.8 displays a bicubic polynomial spline approximant to photodiode data.

Scattered data

There appears to be greater concentration in the literature on the development of algorithms for scattered data *interpolation* than for approximation. The reason may be due either to the fact that it is easier to characterize the solution to an interpolation problem (for example, no decisions have to be made in respect of choice of approximation norm or value of a smoothing parameter), or to a belief that interpolation problems occur more frequently in practice. The former reason is indeed valid but the second, we maintain, is untenable. In our opinion, most sets of scattered data are the result of measurement and therefore subject to error; hence approximation methods should often be more appropriate. In any particular circumstance, unless there are overriding reasons to the contrary, an approximant should be sought that reproduces the data to the accuracy warranted by the data. (Similar comments can to a great extent be made in respect of univariate interpolation.)

The problem does not separate as in the case of data on a mesh; the resulting formulation is a relatively large matrix problem. However, the matrix A is still banded, with bandwidth $n_y q_x - q_x + n_x$ (cf. Cox [42]), but there are only $n_x n_y$ nonzero elements in each row. It is not straightforward to take significant account of this fine structure. In forming the Cholesky factor R of A, most of the structure within the band is lost. See Cox [42] for examples. The situation is markedly different from the

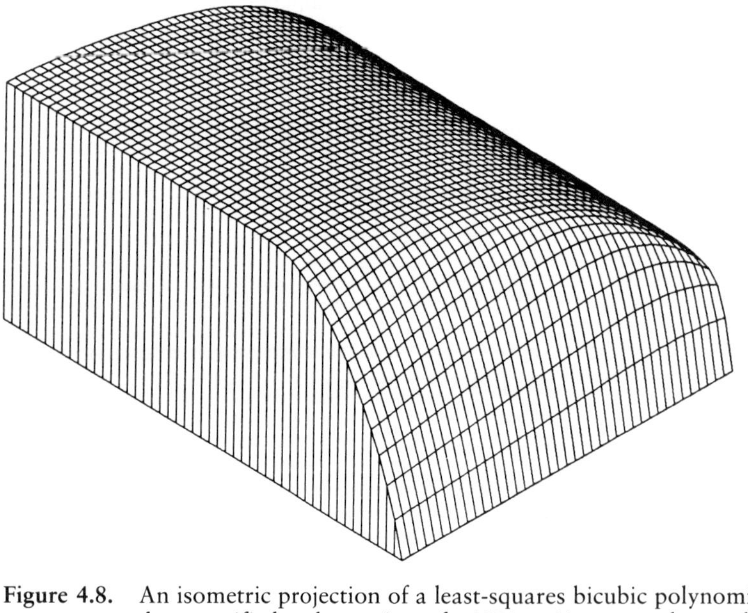

Figure 4.8. An isometric projection of a least-squares bicubic polynomial spline approximant s to photodiode data specified at the vertices of a 1024×41 rectangular mesh. s has 50 x-knotlines and 2 y-knotlines [8].

univariate case in that the computation time now depends on the number of knots. For some of these problems iterative methods, such as the Lanczos-based algorithm *LSQR* of Paige and Saunders [43], may be competitive. For a class of model problems, Cox [42] shows that iterative methods have smaller memory requirements for bivariate problems, and are more efficient in both space and time for higher-dimensional problems. An important consideration, however, in many fields of application is that iterative methods are unable readily to provide the statistical information associated with the solution that is embodied in R (although *LSQR* delivers crude estimates of the variances of the coefficients).

We consider the least-squares approximation of 'scattered' data (x_i, y_i, f_i), $i = 1, 2, \ldots, m$, by a spline s in the form (10). It is necessary to select the x- and y-orders, n_x and n_y, respectively, of s, and the numbers, N_x and N_y, of the x- and y-knotlines, and their locations $\lambda_1, \lambda_2, \ldots, \lambda_{N_x}$ (in x) and $\mu_1, \mu_2, \ldots, \mu_{N_y}$ (in y). Clearly, this is a much harder task than in the case of univariate spline fitting. However, in respect of orders, the choice $n_x = n_y = 4$, giving a bicubic spline, is frequently satisfactory. In practice, if there is sufficient data, it is often sensible to choose initially a small number of uniformly-spaced knots in each of the x- and y-variables. A bivariate spline is then computed for this choice and, by examining the residuals and suitable graphical displays of the spline surface, knots may be added, deleted or adjusted until an acceptable approximation is achieved. Automatic methods, with varying degrees of success, depending on the data, can be constructed, as discussed later.

There is no counterpart of the S–W conditions in the case of scattered data. Thus the algorithm used to determine s must be sufficiently versatile to cope reliably with cases of nonuniqueness, which is very common in such problems. It can arise when there are panels within the knot mesh which contain no or very few points. In cases of nonuniqueness it is necessary to select a particular solution from the space of possible solutions, as in the univariate case. The NAG Library routine, for bicubic splines, selects the minimum norm solution, and the DASL routine, for general orders, determines a geometrically invariant solution. The NAG routine uses a method due to Hayes and Halliday [44], modified by Anthony and Cox to take account of structure in A, and the DASL routine uses the method described by Cox [21].

Figure 4.9 shows a bicubic polymomial spline approximant to values from an exponential function. The values were obtained by random sampling over a domain consisting of a square with a hole.

Bivariate parametric splines

The bivariate parametric spline approximation problem can be posed as follows. Given m points (x_i, y_i, z_i) and associated parameter values t_i, u_i, $i = 1, 2, \ldots, m$, determine polynomial spline functions $s_x(t, u)$, $s_y(t, u)$ and $s_z(t, u)$ such that the residual sum of squares is minimized for each variable x, y and z. In the case of the x-variable, it is required to minimize $\Sigma_{i=1}^{m} e_{x_i}^2$, where $e_{x_i} = x_i - s_x(t_i, u_i)$, and similarly for the y- and z-variables. Major difficulties are the choice of parameter values t_i and u_i and of t- and u-knotlines. If suitable choices can be made, with the same knotlines for all three variables, the method of the previous subsection, with three right-hand sides, can be applied. In the remainder of this section we outline methods proposed to tackle the first difficulty. Many methods are tailored for particular applications rather than purporting to be general approaches.

Parametric spline surface approximation is generally a much harder problem than the corresponding curve case. Normally, the data to be presented by a parametric spline curve is *ordered* in a natural way, and thus a sensible parametrization such as that mentioned earlier can readily be used. Even if the data is not provided in order, it is not usually difficult to arrange it sensibly, although there may be problems if the underlying curve intersects itself (or comes close to doing so), or if the noise in the data is comparable to or greater in magnitude than the (ordered) interpoint spacing. In the case of data representing a surface, there is no immediate concept of ordering. It is therefore necessary to (attempt to) impose some structure on the data. Structure may already exist of course: previous work with similar data sets may suggest an assignment of t, u parameter values to the data points; the data points may have been gathered in x, y, z space in some 'orderly' way (e.g. by measuring systematically points at the vertices of a curvilinear mesh laid down on the surface of a clay model). In all cases it will be

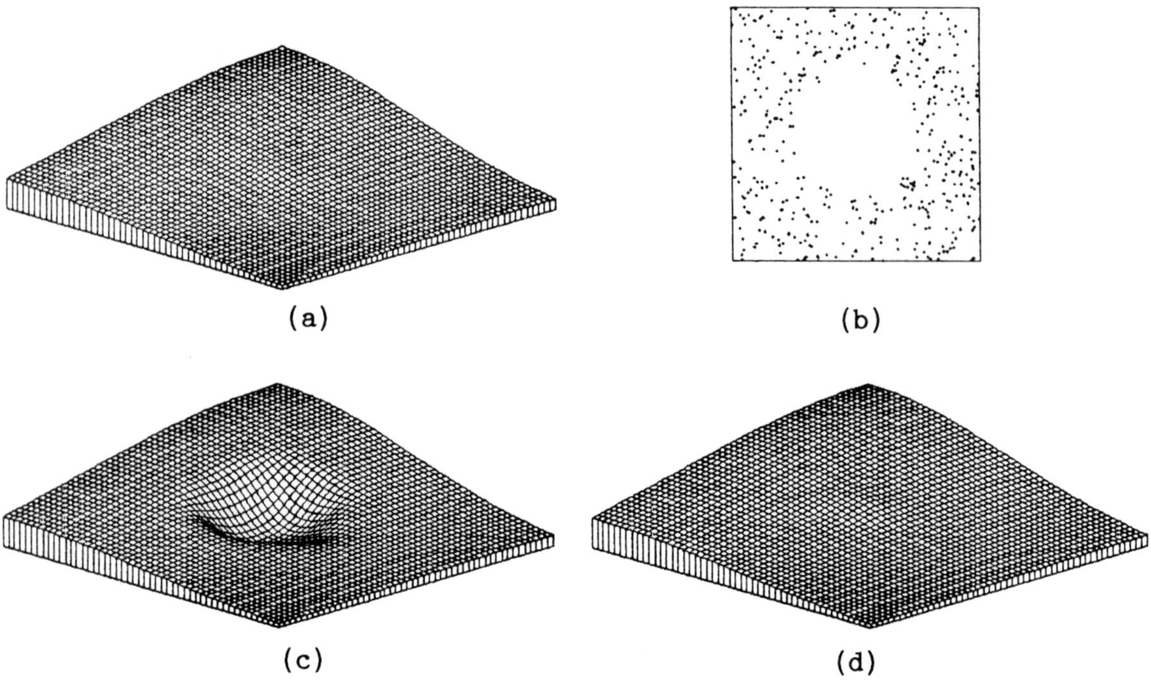

Figure 4.9. (a) $f(x, y) = \exp(-x^2 - 2y^2)$, a 'test' function defined over the square $S = [0, 1] \times [0, 1]$, (b) the locations of 500 data points obtained by sampling f uniformly over $S - O$, where O is the circle with radius $\frac{1}{4}$ and center $(\frac{1}{2}, \frac{1}{2})$, (c) a least-squares bicubic spline approximant to f based on nine uniformly spaced x-knotlines and nine uniformly spaced y-knotlines using the NAG bicubic spline fitting routine (which employs the minimum norm criterion), (d) as (c) but using the DASL bivariate spline fitting routine (which employs a geometrically invariant criterion) [8].

necessary to provide *an* initial parametrization. A surface can then be fitted based on this parametrization and then, as in the curve case, refinement to the parametrization made and a new approximation obtained as necessary.

Available algorithms for bivariate parametric spline approximation are few. The SSR1 system developed by Computer Aided Manufacturing International, Inc [45] provides a basic facility, but 'smoothing' takes place in one parametric direction only. A method due to Kjellander [46] is designed to provide a smooth approximation to a regular network of curves, but the user is required to interact with the software to make various point adjustments. It is possible that Kjellander's method will not give good results unless the parametric mesh lines are reasonably well aligned with the lines of principal curvature of the surface. Earnshaw and Yuille [47] use least-squares parametric spline surface approximation for representing ship hull forms. Data are provided on a series of parallel plane cross-sectional curves determined from data points by least squares, with parameter values assigned using approximate arclengths. A parametric net is then constructed by deriving curves orthogonal to the first set which interpolate them at uniformly spaced parameter values. Spline surface fitting is then carried out using this parametric net.

Recent work by Pratt [48] goes some way towards a full surface approximation system. It assumes that the data points are arranged along 'tracks' on some measured surface. Chord-length parametrization is used along each track, and linear interpolation between these curves is used to assign (t, u) parameter values to all data points. An approximating bicubic spline surface is obtained by applying the NAG bicubic spline routine mentioned in the previous section. The overall method is iterated in respect of the parametrization to improve the quality of the least-squares fit. Pratt remarks, however, that "despite the apparent success of the method described, this work has prompted more questions than it has answered". Clearly, much remains to be done to develop a universal algorithm for parametric surface approximation. For further details of the above methods see Pratt [48].

CURVILINEAR KNOT MESHES

Tensor product splines cannot cope effectively with data sampled from certain types of function, such as a function that contains a narrow curved ridge but elsewhere is relatively flat. The difficulty arises for the following reason. A plane section of the function that is parallel to the x-axis contains a sharp narrow peak where it intersects the ridge, but is fairly smooth elsewhere. To obtain a bivariate spline that satisfactorily approximates this section a greater density of knots will be required in the neighborhood of the peak. However, this argument can also be applied to other plane sections. Consequently, knots will be required at a fine spacing throughout the region. Although the method described in the section on scattered data would hopefully deliver an approximant s that was free from the spurious extrema that might arise if the basic or minimum norm solutions were determined, s would contain very many coefficients, perhaps even more than the number of data points m.

Hayes [49] introduced *curved knotlines* to handle such problems efficiently. His approach can also treat a class of problems in which the *order* of a discontinuity in s varies across the surface. Normally, a tensor product spline is defined (as in previous sections) in terms of a rectangular mesh of knotlines, one set parallel to the y-axis and the other to the x-axis. The B-spline basis in x at any point (x_0, y_0) depends only on the value of x_0 and the x-knotlines, and that in y on y_0 and the y-knotlines. In the case of curved knotlines the dependence is stronger. Consider a curvilinear mesh of knotlines. Denote the ith x-knotline by $x = \lambda_i(y)$ and the jth y-knotline by $y = \mu_j(x)$. The mesh is required to be topologically equivalent to a rectangular mesh with the exception that neighboring lines are permitted to merge or coalesce. Consider the point (x_0, y_0) in terms of this mesh. The B-splines in x are computed in terms of knots given by the values $\lambda_i(y_0)$ and those in y in terms of the values $\mu_j(x_0)$. With this exception the computation proceeds as for a rectangular mesh of knotlines. A choice of curved knotlines which follow the ridge (i.e. are relatively close in the region of the ridge, but more widely spread elsewhere) enables the curved ridge problem to be handled satisfactorily.

No algorithm for bivariate spline approximation with curved knotlines appears to be generally available. Results obtained with pilot software developed at NPL clearly indicate the advantages of such an algorithm. Some of these results are given in Cox [21]. A major reason for the absence of a general algorithm is the difficulty of designing a suitable interface with the user. In addition to his providing the data points the user also has to provide $N_x + N_y$ functional forms describing the knotlines. Additionally, the boundaries of the region could also be curved, i.e. $y_{\min}$ and $y_{\max}$ functions of x, and $x_{\min}$ and $x_{\max}$ functions of y.[14] Design of a suitable interface should prove of significant value in CAD and other areas.

Figure 4.10 depicts bicubic polynomial spline approximants with and without curved knotlines to data representing a narrow curved ridge.

MANUAL AND AUTOMATIC KNOT SELECTION

Manual knot placement

Although the choice of a suitable number of uniformly spaced knots can be effective for a range of univariate and bivariate spline approximation problems, there are many circumstances where a nonuniform knot distribution pays handsome dividends. An important instance arises when approximating data describing a function consisting of a sharp peak and smooth tails. The use of a few knots concentrated near the maximum, with the remaining knots more widely separated, typically yields a superior approximation. An often-quoted example is the titanium heat data considered by de Boor [1]. Generally, sensible choices for the number and positions of the interior knots of s may often be estimated in any particular instance by examining the shape of the underlying curve as indicated by the data. In general, more knots will

[14] Alternatively, the knotlines and boundaries could be specified by sets of points which are then interpolated or approximated by splines, using one of the methods described in this paper.

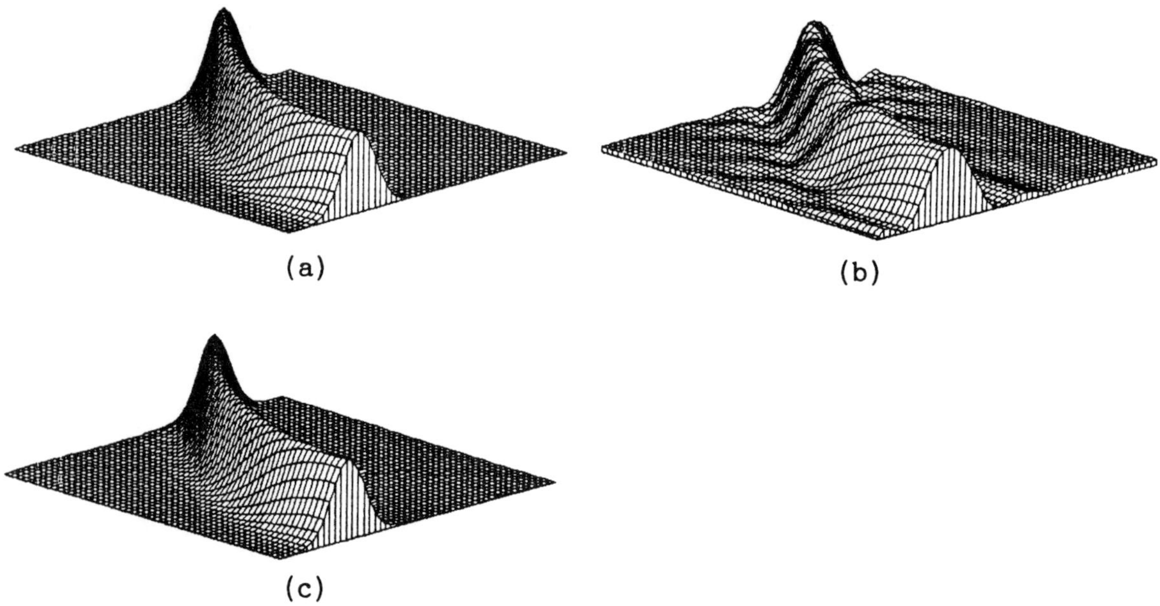

Figure 4.10. (a) A function f containing a narrow curved ridge that cannot efficiently be approximated by a conventional bivariate spline, (b) a least-squares bicubic spline approximant to f based on nine uniformly spaced x-knotlines and nine uniformly spaced y-knotlines, (c) as (b) but obtained with a pilot version of a routine permitting curved knotlines. The approximant in (c) is based on five curved knotlines a constant x-distance apart [8].

be needed in regions where the behavior is changing rapidly and fewer where it is relatively smooth.

A 'manual' strategy for obtaining a least-squares univariate spline approximant is as follows [32]:

1. Position an initial set of knots in accordance with the above 'criterion'.
2. Obtain an approximating spline s based on the 'current' knots using the method of the section on univariate polynomial splines.
3. Examine key parameters and other features of the approximant, such as the root mean square residual and the individual values of the residuals e_i, and the behavior of s in regions where there are few data points or in the neighborhood of special features such as inflexion points and extrema.[15]
4. In regions where the approximant is inadequate, introduce additional knots, perhaps after adjusting the positions of existing ones. In regions where the approximant is 'too good', i.e. where it follows the data values so closely that the spline has oscillations with amplitudes of the order of the noise level of the data, remove a number of knots, again adjusting the positions of existing ones if necessary.
5. Repeat as necessary from step 2.

With a little experience, many data sets can be satisfactorily approximated after having made say three or four passes through the process. This approach was used effectively as part of an interactive cubic spline fitting system developed some 20 years ago at NPL. The system included a large (4ft × 3ft) laser-driven screen which enabled six sizeable pictures to be displayed simultaneously. A typical mode of operation was as follows. Display the given data as the first picture. Estimate an initial number of knots and their locations, entering their values at the keyboard. Determine an approximation with these knots and display it together with the data

[15] For the degree of precision normally associated with empirical data a graphical form of output, in which the data points, the approximating spline and the knot positions are displayed, is particularly useful at this stage.

as the second picture on the screen. On inspection of the results, adjust the knots as in the above strategy and repeat as necessary. The system offered straightforward facilities for adding and deleting knots, for requesting spline computations and for displaying the data and approximants.

Automatic knot placement

Rather than the 'user' having to select a suitable set of knots, it would clearly be desirable to have available an automatic method which in some sense chose optimal or at least 'good' knot positions in any particular instance. A number of workers have examined this important problem.

Most of the work on knot placement addresses the univariate case but there are some algorithms for bivariate splines. One of the earliest algorithms is due to de Boor and Rice [50]. It computes successively least-squares cubic splines for $N = 1, 2, \ldots$ interior knots. For each choice of N the construction of the spline involves the solution of a nonlinear least squares problem with N knots as variables. Initial estimates of the knots at each stage are taken as the knots provided by the previous stage, together with an additional knot inserted in the region where the contribution to the residual sum of squares is greatest.

Another type of algorithm, which is based on a theoretical expression for the distance of a spline from the 'underlying' function, is due to de Boor [51]. For a function $f \in C^{(n)}[x_{\min}, x_{\max}]$, a bound for the distance $|f(x) - s(x)|$ can be expressed in terms of $f^{(n)}(\xi_x)$, where ξ_x is 'local to' x, and the local knot spacing. Thus, there is a choice of knots for which the extreme departures of s from f are equalized. For any given choice of N the algorithm constructs approximately such a set of knots. It is based on the iterative estimation of $f^{(n)}(x)$.

A third type of algorithm successively inserts knots in regions where the contribution to the residual sum of squares is greatest. One such algorithm, which is available in the NAG Library, is due to Dierckx [52]. In his algorithm, the residual sum of squares is augmented by a smoothing term (cf. Reinsch [53]).

Recent work involves fitting polymomials to (overlapping) subsets of the data to obtain information about the local behavior of the underlying function. On the basis of this information knots are located at hopefully appropriate positions. A spline based on these knots is computed and examined for acceptability. Knots may be added, deleted or re-distributed using a suitable strategy to improve the approximation.[16] Methods of this type are considered by Cox, Harris and Jones [54, 55]. The approach seems promising but further testing and enhancements are necessary.

Two algorithms for the bivariate case, for meshed and scattered data, are also due to Dierckx [56, 57], and available in the NAG Library. They are similar in concept to the univariate algorithm.

Automatic knot placement is one of the most challenging problems in practical data approximation by splines; the development of generally effective algorithms would prove to be of very significant practical value. A major difficulty is that the residual sum of squares as a function of the knots may possess local minima [58], some of which may be unsatisfactory. In principle, though, minimizing the residual sum of squares, if a global solution could be guaranteed, would provide the optimal solution. It does not follow, however, that this solution is necessarily satisfactory to the 'user'.

Figure 4.11 shows a cubic spline approximant, with knots computed automatically, to data descriptive of the response of an oscilloscope.

SHAPE CONTROL USING CONSTRAINED SPLINES

It is frequently required that an approximant inherits characteristics of the function underlying the data so that it is physically feasible. Approximation of the data alone will not guarantee this. In this section we discuss algorithms for computing spline interpolants and approximants to specified data that also satisfy prescribed shape conditions. Most approaches for determining shape-preserving or constrained splines can be classified as local methods for interpolation and global methods for approximation.

[16] Alternatively, as in the approach of de Boor and Rice [50], the knots so obtained may be optimized numerically.

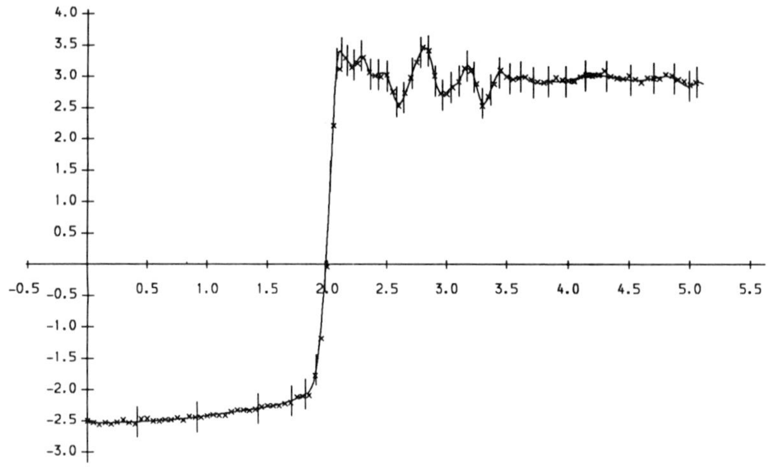

Figure 4.11. A cubic spline approximant to 512 data points descriptive of the response of a fast-rise time oscilloscope to a step input function. The knots were computed automatically [54, 55]. For clarity only every fifth data point is portrayed.

In the case of univariate interpolation, given data points (x_i, f_i), $i = 1, 2, \ldots, m$, with $x_1 < x_2 < \ldots < x_m$, local methods typically construct an arc over each subinterval between consecutive data points using the specified function values at those points and adjacent points. For instance, if $2 < i < m$,[17] a cubic arc $g_i(x)$, say, can be determined over $[x_{i-1}, x_i]$ such that $g_i(x_j) = f_j$, $j = i - 1, i$, and $g_i'(x_j) = d_j$, $j = i - 1, i$, where d_j is an estimate of f_j', the first derivative of the underlying function at x_j, obtained from the points (x_k, f_k), $k = j - 1, j, j + 1$. For any choice of the d_j, the resulting sequence of arcs forms a C^1 spline s over $[x_1, x_m]$.[18] The freedom of choice in the d_j permits values to be determined such that s has the required property,[19] e.g. positivity, monotonicity or convexity, or possibly some or all of these. Routines for monotonic piecewise cubic interpolation, using a method due to Fritsch and Butland [59], are available in the NAG Library.

We now consider methods for constrained *approximation* by a spline in its B-spline form (5). The approach is that of the first paragraph of the section on univariate least-squares spline approximation. Now, however, we augment the problem statement by certain shape-preserving conditions. Some such conditions can be implemented as a *finite* set of constraints. For example, if s is to be convex, we require $s''(x) \geqslant 0$ for $x \in [x_{\min}, x_{\max}]$. In the case $n = 4$ (cubic splines), $s''(x)$ is piecewise linear and the condition therefore holds if and only if $s''(\lambda_j) \geqslant 0$, $j = 0, 1, \ldots, N$. In terms of the B-spline representation (5) of s, we can, using the compact support property of the B-splines, express these constraints as

$$\sum_{k=j+1}^{j+3} c_k N''_{4,k}(\lambda_j) \geqslant 0, \quad j = 0, 1, \ldots, N \tag{20}$$

where the derivative values in (20) are calculated by recurrence [14]. If a least-squares solution is required, in this case and in general it is necessary to minimize a sum of squares of linear functions subject to linear inequalities: this is (a special case of) a *quadratic program* [60]. A generalization is a formulation in terms of the p-norm, in which case it is required to minimize $\|e\|_p$ for a suitable value of p (usually 1, 2 or ∞), where $e = (e_1, e_2, \ldots, e_m)^T$, $e_i = w_i(f_i - s(x_i))$, subject to linear constraints on s to impose the required shape. A practical advantage of the choice $p = 1$ (suitable if the

[17] The outlined approach is modified slightly if $i = 2$ or $i = m$.

[18] If required it can be converted to B-spline form (5) with a *double* knot at each interior x_i.

[19] A solution will not exist if the data is inconsistent with the requirement. Moreover, if a solution does exist, the choice of d_i is rarely unique: different choices may make discernable changes to the appearance of the interpolant.

data are likely to contain 'wild' points) or $p = \infty$ (appropriate if the data errors are rectangularly distributed, as in a conventional mathematical table) is that the problem can be formulated as a *linear* program.[20] Linear programming algorithms specially tailored for the solution of 1- and ∞-norm approximation problems are available in the NAG Library based on the work of Barrodale and Roberts [61] and Barrodale and Phillips [62].

In the above case, convexity, the constraints (20) are both necessary and sufficient [63]. In general, it is possible to bound any derivative of s in terms of a finite set of linear inequalities in c. The resulting constraints are normally merely sufficient, but the approach can nevertheless be used to obtain an approximant that is not in general optimal but often 'good'. Hanson [64] describes an iterative scheme which heuristically but effectively treats constraints that cannot *a priori* be replaced by a finite set of inequality conditions.

There are many types of constrained spline approximation problems, including point constraints, line constraints, boundary constraints, and local and global constraints. Generally, the B-spline representation is a very suitable tool for expressing such constraints in a form that permits useful algorithms to be constructed. A method with a considerable degree of generality for solving bivariate spline approximation problems where functional forms for s at the boundary and derivatives normal to the boundary of a rectangular region are specified is given by Anthony and Cox [65]. This method has similarities to the blending-function methods introduced by Coons [66] and since used widely by the CAD fraternity.

CLOSING REMARKS

This chapter has described some of the algorithmic aspects of spline interpolation and approximation in one and two independent variables. A number of excellent algorithms have been developed in this area, with a considerable proportion readily available through recognized software libraries. Most of the algorithms are well supported by sound mathematical theory, but further theoretical advances will be necessary to give scientists and design engineers a more complete set of software tools for working with splines.

[20] Linear programming algorithms are simpler and more compact than quadratic programming algorithms. In contrast, it is more straightforward to solve *unconstrained* linear least-squares problems than unrestricted 1- and ∞-norm problems.

ACKNOWLEDGEMENTS

G. T. Anthony and P. M. Harris made valuable comments on a draft of this chapter. I am also grateful to P. M. Harris for assisting with the figures.

REFERENCES

1 **C. de Boor** *A Practical Guide to Splines,* Springer-Verlag, New York, 1978.
2 **W. Boehm** Inserting new knots into B-spline curves, *Comput.-Aided Design,* Vol 12, 1980, 199–201.
3 **E. Cohen, T. Lyche and R. F. Riesenfeld** Discrete B-splines and subdivision techniques in computer aided geometric design and computer graphics, *Comput. Graphics Image Process.,* Vol 14, 1980, 87–111.
4 **B. Ford, J. Bentley, J. J. du Croz and S. J. Hague** The NAG Library 'machine', *Software – Practice and Experience,* Vol 9, 1979, 56–72.
5 **J. G. Hayes** NAG algorithms for the approximation of functions and data, In *Algorithms for Approximation,* J. C. Mason and M. G. Cox, eds., Clarendon Press, Oxford, 1987, 653–668.
6 **M. G. Cox** Topic libraries for mathematical computation, *Software – Practice and Experience,* Vol 15, 1985, 397–411.
7 **G. T. Anthony and M. G. Cox** The National Physical Laboratory's Data Approximation Subroutine Library, in *Algorithms for Approximation,* J. C. Mason and M. G. Cox, eds, Clarendon Press, Oxford, 1987, 669–687.
8 **M. G. Cox** The NPL Data Approximation Subroutine Library: current and planned facilities, *NAG Newsletter,* Vol 2, 1987, 3–16.
9 **M. G. Cox** *Numerical methods for the interpolation and approximation of data by spline functions,* PhD thesis, City University, London, 1975.
10 **J. Kozak** *On the choice of the exterior knots in the B-spline basis for a spline space,* Technical Report 2148, University of Wisconsin, 1980.
11 **M. G. Cox** The incorporation of boundary conditions in spline approximation problems, in *Seventh Manitoba Conference on Numerical*

Mathematics and Computing, D. McCarthy and H. C. Williams, eds, University of Manitoba, Winnipeg, 1978, 73–105.

12 **I. J. Schoenberg** Contributions to the problem of approximation of equidistant data by analytic functions. Part A – On the problem of smoothing or graduation. A first class of analytic approximation formulae, *Qu. Appl. Math.,* Vol 4, 1946, 45–99.

13 **I. J. Schoenberg** Contributions to the problem of approximation of equidistant data by analytic functions. Part B – On the problem of osculatory interpolation. A second class of analytic approximation formulae, *Qu. Appl. Math.,* Vol 4, 1946, 112–141.

14 **C. de Boor** On calculating with B-splines, *J. Approx. Theory,* Vol 6, 1972, 50–62.

15 **M. G. Cox** The numerical evaluation of B-splines, *J. Inst. Math. Appl.,* Vol 10, 1972, 134–149.

16 **L. L. Schumaker** Some algorithms for the computation of interpolating and approximating spline functions, in *Theory and Applications of Spline Functions,* T. N. E. Greville, ed., Academic Press, New York, 1969, 87–102.

17 **J. G. Herriott and C. H. Reinsch** *Algol 60 procedures for the calculation of interpolating natural spline functions,* Technical Report STAN-CS-71-200, Computer Science Department, Stanford University, 1971.

18 **M. G. Cox** The numerical evaluation of a spline from its B-spline representation, *J. Inst. Math. Appl.,* Vol 21, 1978, 135–143.

19 **C. de Boor** On local linear functionals which vanish at all B-splines but one, in *Theory of Approximation with Applications,* A. G. Law and B. N. Sahney, eds., Academic Press, New York, 1976, 120–145.

20 **M. G. Cox** Practical spline approximation, in *Notes in Mathematics 965: Topics in Numerical Analysis* P. R. Turner, ed., Springer-Verlag, Berlin, 1982, 79–112.

21 **M. G. Cox** Data approximation by splines in one and two independent variables. in *The State of the Art in Numerical Analysis,* A. Iserles and M. J. D. Powell, eds, Oxford University Press, Oxford, 1987, 111–138.

22 **H. B. Curry and I. J. Schoenberg** On Pólya frequency functions. IV: The fundamental spline functions and their limits, *J. d'Analyse Math.,* Vol 17, 1966, 71–107.

23 **M. G. Cox and P. M. Harris** *The approximation of a composite cubic Bézier curve by a composite quadratic Bézier curve,* Technical Report DITC 147/89, National Physical Laboratory, Teddington, UK, 1989.

24 **I. J. Schoenberg and A. Whitney** On Pólya frequency functions III, *Trans. Am. Math.,* Vol 74, 1953, 246–259.

25 **M. G. Cox** The least squares solution of overdetermined linear equations having band or augmented band structure, *IMA J. Numer. Anal.,* Vol 1, 1981, 3–22.

26 **C. de Boor and R. DeVore** A geometric proof of total positivity for spline interpolation, *Math. Comput.,* Vol 45, 1985, 497–504.

27 **C. de Boor and A. Pinkus** Backward error analysis for totally positive linear systems, *Numer. Math.,* Vol 27, 1977, 485–490.

28 **J. H. Wilkinson** *The Algebraic Eigenvalue Problem,* Clarendon Press, Oxford, 1965.

29 **L. L. Schumaker** *Spline Functions: Basic Theory,* John Wiley, New York, 1981.

30 **M. P. Epstein** On the influence of parametrization in parametric spline interpolation, *SIAM J. Numer. Anal.,* Vol 13, 1976, 261–268.

31 **J. Van de Geijn** Computational methods in beam therapy planning, in *Computer Programs in Biomedicine,* 1972, 153–168.

32 **M. G. Cox** A survey of numerical methods for data and function approximation, in *The State of the Art in Numerical Analysis,* D. A. H. Jacobs, ed., Academic Press, London, 1977, 627–668.

33 **W. M. Gentleman** Least squares computations by Givens transformations without square roots, *J. Inst. Math. Appl.* Vol 12, 1973, 329–336.

34 **W. M. Gentleman** Basic procedures for large, sparse or weighted linear least squares problems, *Appl. Statist.,* Vol 23, 1974, 448–454.

35 **M. G. Cox and J. G. Hayes** *Curve fitting: a guide and suite of algorithms for the non-specialist user,* Technical Report NAC 23, National Physical Laboratory, Teddington, UK, 1973.

36 **M. G. Cox** Linear algebra support modules for approximation and other software, in *Scientific Software Systems,* J. C. Mason and M. G. Cox, eds, Chapman and Hall, London, 1990, 21–29.

37 **R. Franke** *A critical comparison of some methods for interpolation of scattered data,* Technical Report TR-NPS-53-79-003, Naval Postgraduate School, Monterey, California, 1979.

38 **C. de Boor** Efficient computer manipulation of tensor products, *ACM Trans. Math. Software,* Vol 5, 1979, 173–182.

39 **M. G. Cox** Versatile parameter lists for scientific library routines, in *Production and Assessment of Numerical Software,* M. A. Hennell and L. M. Delves, eds., Academic Press, London, 1980, 249–262.

40 **K. G. Birch and M. G. Cox** *Calculation of the flatness of surfaces; a least-squares approach,* Technical Report MOM 5, National Physical Laboratory, Teddington, UK, 1973.

41 **T. N. E. Greville** Note on fitting of functions of several independent variables, *J. Soc. Ind. Appl. Math.,* Vol 9, 1961, 109–115.

42 **M. G. Cox** Direct versus iterative methods of

solution for multivariate spline-fitting problems, *IMA J. Numer. Anal.*, Vol 2, 1982, 73–81.
43 **C. C. Paige and M. A. Saunders** LSQR: an algorithm for sparse linear equations and sparse least squares, *ACM Trans. Math. Software,* Vol 8, 1982, 43–71.
44 **J. G. Hayes and J. Halliday** The least-squares fitting of cubic spline surfaces to general data sets, *J. Inst. Math. Appl.*, Vol 14, 1974, 89–103.
45 **CAM-I, Inc.** *User documentation for the CAM-I sculptured surfaces system,* Technical Report PS-77-SS-06, CAM-I, Inc., Arlington, TX, 1977.
46 **J. A. P. Kjellander** Smoothing of bicubic parametric spline surfaces, *Comput.-Aided Design,* Vol 15, 1983, 288–293.
47 **J. L. Earnshaw and I. M. Yuille** A method of fitting parametric equations for curves and surfaces to sets of points defining them approximately, *Comput.-Aided Design,* Vol 3, 1971, 19–22.
48 **M. J. Pratt** Smooth parametric surface approximations to discrete data, *Computer-Aided Geometric Design,* Vol 2, 1985, 165–171.
49 **J. G. Hayes** New shapes from bi-cubic splines, in *Proc. CAD 74,* IPC Business Press, Guildford, 1974.
50 **C. de Boor and J. R. Rice** *Least squares cubic spline approximation II – variable knots,* Technical Report CSD TR 21, Purdue University, 1968.
51 **C. de Boor** Good approximation by splines with variable knots II, in *Numerical Solution of Differential Equations, Lecture Notes in Mathematics No. 363,* G. A. Watson, ed., Springer-Verlag, Berlin, 1974, 12–20.
52 **P. Dierckx** *An improved algorithm for curve fitting with spline functions,* Technical Report TW54, Katholieke Universiteit Leuven, 1981.
53 **C. H. Reinsch** Smoothing by spline functions, *Numer. Math.*, Vol 10, 1967, 177–183.
54 **M. G. Cox, P. M. Harris and H. M. Jones** *Strategies for knot placement in least-squares data fitting by splines,* Technical Report DITC 101/87, National Physical Laboratory, Teddington, UK, 1987.
55 **M. G. Cox, P. M. Harris and H. M. Jones** A knot placement strategy for least squares spline fitting based on the use of local polynomial approximations, in *Algorithms for Approximation II,* J. C. Mason and M. G. Cox, eds., Chapman and Hall, London, 1990, 37–45.
56 **P. Dierckx** An algorithm for surface fitting with spline functions. *IMA J. Numer. Anal.*, Vol 1, 1981, 267–283.
57 **P. Dierckx** A fast algorithm for smoothing data on a rectangular grid while using spline functions, *SIAM J. Numer. Anal.*, Vol 19, 1982, 1286–1304.
58 **M. G. Cox** Curve fitting with piecewise polynomials, *J. Inst. Math. Appl.*, Vol 8, 1971, 36–52.
59 **F. N. Fritsch and J. Butland** A method for constructing local monotone piecewise cubic interpolants, *SIAM J. Statist. Sci. Comput.*, Vol 5, 1984, 300–304.
60 **P. E. Gill, W. Murray and M. H. Wright** *Practical Optimization,* Academic Press, London, 1981.
61 **I. Barrodale and F. D. K. Roberts** Solution of an overdetermined system of linear equations in the ℓ_1-norm (Algorithm 478), *Commun. ACM,* Vol 17, 1974, 319–320.
62 **I. Barrodale and C. Phillips** Solution of an overdetermined system of linear equations in the Chebyshev norm (Algorithm 495), *ACM Trans. Math. Software,* Vol 1, 1975, 264–270.
63 **M. G. Cox and H. M. Jones** Shape-preserving spline approximation in the ℓ_1 norm, in *Algorithms for Approximation,* J. C. Mason and M. G. Cox, eds., Clarendon Press, Oxford, 1987, 115–129.
64 **R. J. Hanson** *Constrained least squares curve fitting to discrete data using B-splines – a users guide,* Technical Report SAND78-1291, Sandia Laboratories, 1979.
65 **G. T. Anthony and M. G. Cox** The fitting of extremely large data sets by bivariate splines, in *Algorithms for Approximation,* Clarendon Press, Oxford, 1987, 5–20.
66 **S. A. Coons** *Surfaces for computer aided design,* Technical Report, Mech. Eng. Dept., MIT Design Division, Boston, MA, 1964.
67 **M. G. Cox** An algorithm for spline interpolation, *J. Inst. Math. Appl.* Vol 15, 1975, 95–108.

ABOUT THE AUTHOR

Maurice G. Cox obtained a BSc in Applied Mathematics at the City University, London, in 1963. Following graduation he worked for Associated Electrical Industries (now General Electric Company), Rugby. He was a consultant mathematician there until 1968 when he joined the National Physical Laboratory (NPL), Teddington, as a Senior Scientific Officer. He became Principal Scientific Officer in 1971, obtained an external PhD on algorithms for spline functions at City University in 1975, and was further promoted to Senior Principal Scientific Officer in 1984. He was appointed Visiting Professor in Computational Mathematics at Kingston University in 1992.

His major areas of research have been spline approximation and overdetermined linear algebraic systems. Professor Cox has concentrated on the design and development of efficient algorithms in these areas; software implementations of his algorithms form the core of the interpolation and curve and surface fitting chapters of the NAG Library. He played the major role in designing and promoting the NPL Data Approximation Subroutine Library, which is now available from NAG Ltd. For virtually all his algorithms Professor Cox has provided a complete floating-point error analysis to demonstrate their stability properties.

Professor Cox led NPL's Mathematical Software Group from 1984 to 1991 and is now responsible for longer-term mathematical research and application at NPL. Moreover, in recent years he has taken a strong interest in Algorithms for Metrology, a high-profile NPL project for which he is responsible. This project benefits from his extensive experience in the theory and practice of approximation.

He has applied his research to many scientific and engineering disciplines. He led a recently-completed major multinational project for the European Commission concerned with the preparation of reference algorithms for the approximation of fundamental geometrical forms in dimensional metrology. He is also very active in related standardization activities, both at national and international level.

Professor Cox has over 80 publications on approximation, linear algebra, metrology and related topics. He was a founder member of the Institute of Mathematics and its Applications (IMA) and is a Fellow of the Institute.

5 Modeling with NURBS Curves and Surfaces

Richard F. Riesenfeld

INTRODUCTION

There are two fundamental issues surrounding the choice of a representation scheme which are of primary concern. One is primarily a mathematical issue, that is, as a theoretical matter, what is the general class of objects which can be represented exactly by a particular mathematical scheme. One can refer to this property as the power of the representational scheme. Nonuniform rational B-splines have become rather widespread and popular in recent years, partly because they provide a flexible and general representation for a large class of interesting objects. More recently, there has been a greater use of rational splines because they provide the capability to represent all conic shapes exactly. Quadratic shapes are common and important in engineering and manufacturing, so that ability to represent them exactly is both attractive and important.

The second fundamental aspect of choosing a mathematical representation has to do with the ability to devise an effective and intuitive user interface for exploiting the mathematical potential which the representation theoretically possesses. This is quite another question from the first question discussed above. Here we are concerned with the user's ability to actually specify a shape even though we are assured that appropriate coefficients for such a shape do, in fact, theoretically exist. In the spirit we can distinguish the *B-spline Basis* from the *B-spline Method* for designing curves and surfaces. The former is simply the mathematical entity, the specific bases which entail some favorable computational characteristics. The latter refers to the Bézier-style method of specifying shape by manipulating a control polygon in order to use an intuitive shape-based interface to specify coefficients of a B-spline basis.

Important work has transpired in recent years to make B-splines more tractable as a mathematical representation because we have developed more tools for the user to invoke in specifying shape at a much higher level than the control polygon level. These methods involve tools sometimes called shape operators, where one specifies something much more natural and physically intuitive, like a Warp or a Bend, to imply the coefficient changes that are appropriate to a whole collection of vertices. In fact, in some systems such shape operators have become the preferred method of specifying B-spline shape coefficients. Other tools for providing wholesale simultaneous specification for a collection of B-spline coefficients is to use parametrically defined primitives, like rounded boxes, cylinders, and the like. In this way we are using names of known things to give specification to the associated B-spline representation. Such methods have greatly contributed to the ease and popularity of B-spline representations.

Also, CAD/CAM software developers and users have become much more sophisticated in the last two decades. As people involved with CAD/CAM have become more knowledgeable on spline theory and computation, they have been more willing to incorporate such technology in their efforts. While B-spline methods can be significantly more complex than Bézier methods, they provide greater power and flexibility. As the understanding of B-splines has become more broadly based, their adoption has become more widespread.

And finally, one must recognize the tremendous change that the computing environments have undergone in the last couple of decades. We are working in environments that provide from more sophisticated software languages and environments which significantly facilitate implementing a B-spline based approach to shape representation, and we enjoy vastly improved computational resources which make practical schemes which were only theoretically attractive a decade ago. Further, there is even direct hardware support for evaluating B-splines available on the newest engineering workstations, which further facilities their use.

Displaying and visualizing B-splines have become a routine matter at this point. This was still a major area of research a decade ago. Many

The colour plates for this chapter appear between pp. 210 and 211.

Fundamental Developments of Computer-Aided Geometric Modeling
ISBN 012-554765-X

applications areas avoided using splines because there was not a good manner available to coupling spline technology with high quality raster graphics. The important progress that has occurred in this direction has been key for promoting the wider application use of B-spline methods.

All of these factors have contributed to the current state of wide acceptance and use of B-splines for making computer models of geometric shapes. This paper goes into further detail on some important issues involved in modeling with B-splines. It is hoped that the sections that follow give insight and understanding in explaining why B-splines have grown to become the most standard way of modeling today instead of an esoteric and highly specialized method of several years ago.

DEFINITIONS AND PROPERTIES OF B-SPLINE CURVES AND SURFACES

Splines became computationally tractable when de Boor [1] and Cox [2] introduced stable and efficient algorithms for computing B-splines. The author first proposed the notion of using this computational approach as a means of extending Bézier's Method for designing with polynomial curves and surfaces to B-spline curves and surfaces in his thesis [3]. From the spline basics presented in other parts of this book [1,2], we can define a straightforward generalization to curves and surfaces as follows:

Definition Let $t = \{t_i\}_{i=0}^{N}$ be a sequence of nondecreasing real numbers. To define a piecewise polynomial curve of degree $k - 1$, called $\gamma(t)$, with domain $[t_{k-1}, t_{N-k})$,

1. For a given $t \geqslant t_{k-1}$, find J such that $t \in [t_J, t_{J+1})$.
2. Define $P_i^{[0]} = P_i$.
3. Let

$$P_i^{[p]} = \frac{t - t_i}{t_{i+k-p} - t_i} P_i^{[p-1]} + \frac{t_{i+k-p} - t}{t_{i+k-p} - t_i} P_{i-1}^{[p-1]},$$

$$i = J, J-1, \ldots, J-k+1+p.$$

4. Then, $\gamma(t) = P_J^{[k-1]}$.

Looking at the generating algorithm, we see that the curve will have the form $\gamma(t) = \Sigma P_i B_{i,k}(t)$, where each $B_{i,k}(t)$ is a piecewise polynomial, and exactly k of them are nonzero on each open interval between knots. The B-spline functions described in de Boor's paper [1] have been shown to be correct functions. Their individual properties have already been shown.

Definition The *tensor product B-spline surface* is defined as the piecewise polynomial surface

$$S(u, v) = \sum_{i=0}^{m} \sum_{j=0}^{n} P_{i,j} B_{j,k_v}(v) N_{i,k_u}(u)$$

where B_{j,k_v} are the B-splines of order k_v on the knot vector $\boldsymbol{v} = \{v_0, \ldots, v_{n+k_v}\}$ in the v direction, and N_{i,k_u} are the B-splines of order k_u on the knot vector $\boldsymbol{u} = \{u_0, \ldots, u_{m+k_u}\}$ in the u direction.

Note that the order of the tensor product used may be different in the u direction than in the v direction, as well as having distinct knot vectors. For example, one could have open quadratic in u and floating cubic in v. Further, for any fixed value of u, say $u = r$, $S(r, v)$ is a B-spline curve of order k_v with knot vector $\boldsymbol{v}$. Similarly, for a fixed value of v, say $v = r$, $S(u, r)$ is a B-spline curve of order k_u with knot vector $\boldsymbol{u}$. Remember that if $\boldsymbol{u}$ and $\boldsymbol{v}$ are two knot vectors, of length $m + k_u + 1$ and $n + k_v + 1$, respectively, then the domain over which the surface is defined is $[u_{k_u-1}, u_{m+1}] \times [v_{k_v-1}, v_{n+1}]$.

Some very strong geometric properties can be associated with the above definition. It is important to observe that a B-spline curve or surface essentially is given as a convex combination of its control points. Although the weighting coefficients of the control points are functions of some parametric domain, either an interval or a rectangle, they are always nonnegative and sum identically to 1. So for any fixed parameter values, the coefficients specify a point on a curve or surface as a convex combination of the control points. From this we get the equivalent statement that a B-spline curve or surface must always lie within the convex hull of its defining polygon. This has become known as the Convex Hull Property for B-spline curve and surface, an unusual approximation property enjoyed by both Bézier–Bernstein and B-spline methods.

Other extraordinary geometric properties are available from similar simple observations. Since B-splines form a local basis, with only a small number of basis functions being nonzero at any point in the associated knot vector over which they are defined, we can recognize an ever stronger Local (or Strong) Convex Hull Property. Since most weighting coefficients, or blending functions, are zero for any particular parametric value on a curve or surface, it must be true that the curve or surface is confined to an even smaller convex hull than the entire set of control points. Only those control points associated with nonzero coefficient functions can contribute to the definition of the curves or surface. Therefore we have a local convex hull consisting of only neighboring control points. For a cubic curve, for example, only four successive control points are involved in the definition of a particular polynomial span of a B-spline. Alternatively, we can say that each polynomial span of a B-spline is contained within the convex hull of four consecutive control points. Analogously, a grid of 16 local control points completely defines each bicubic patch of a cubic B-spline surface. This is due to the fact that only the basis functions in a small neighborhood of any point on a B-spline curve or surface are nonzero.

Geometrically, this means that moving a control point on a B-spline curve or surface can cause only a local perturbation beyond which the remainder of the curve or surface remains identically unchanged. This is a property that makes B-spline curves and surfaces particularly attractive for computer aided geometric design, and probably accounts for much of their widespread use today. One often wants to make a local change without concern that it can propagate throughout a design in a global manner.

Affine invariance comes as a straightforward corollary. Once again we notice that a B-spline curve or surface can be viewed as a geometric entity defined as a convex combination relative to some control points. Now suppose that some fixed affine transformation is applied to each of the control points which preserves the shape of the control polygon. It easily follows that the B-spline curve or surface definition is similar relative to the transformed points, i.e. the shape of the parametric curve or surface has been affinely transformed, but not fundamentally affected, if the shape of the polygon has been preserved. This property is a direct consequence of the compound nature of parametric B-spline curves and surfaces. The curve or surface is defined locally relative to the control points, and the control points are then positionally defined and oriented in the embedding space. If the control points are removed in a shape preserving manner, then so is the curve or surface.

One more unusual shape related property makes Bézier–Bernstein and B-spline methods rather attractive for geometric modeling, and that is the variation diminishing property. As an approximation scheme, Bernstein or B-spline approximation produces an approximant which crosses the origin no more often than the primitive function it is approximating. This can be extended to show that it is true for any straight line as well, so the approximant intersects any straight line no more often than the primitive function itself. In the context of a Bézier or B-spline method, this has the consequence that the curve defined by a control polygon intersects any straight line no more often than the polygon itself. This property is sometimes summarized by stating that the Bézier or B-spline spline curve exhibits no more 'wiggles' than the control polygon itself. All of these characterizations are related to the variation diminishing property of B-spline approximation, which derives from the more theoretical property of 'total positivity', a subject well beyond the scope of this paper.

There is considerable freedom in defining a knot vector that gives rise to a set of basis functions, even after the degree and number of basis functions for the curve (or surface) are established. In fact, so long as the values of the knots in the ordered collection are nondecreasing and the number of knots with the same real value does not exceed one more than the degree of the piecewise polynomials, the collection forms a legal knot vector which, in principle, would suffice for generating a curve. In general, the notion of trying fully to exploit this freedom is considered too complex for a designer, but there are some fundamental choices that have become somewhat popular and standard. A uniform B-spline with knots at the integers, or otherwise evenly spaced, is the very simplest way to define a B-spline curve or surface. It also gives rise to basis functions that are translates of each other so that it is not necessary to recompute them. In fact, they can be stored in a table in advance and not recomputed with each invocation of an evaluation routine.

With the increasing power of dedicated engineering workstations, it is increasingly popular to allow for the more general case involving nonuniform knot placement. In this case, a knot vector must only satisfy the restrictions on multiplicity and monotonicity. A major impetus for using nonuniform knot vectors, called nonuniform B-splines for short, is that they arise naturally as a result of spline manipulations like trying to compute the intersection of two splines or tightly approximating a spline by a piecewise polynomial. Such applications often involve knot insertions, which give rise to nonuniform splines even if one begins the process with uniform splines. Since applications are increasingly supportive of such rendering and modeling operations, we are seeing a significant gain in the use of nonuniform splines.

As the accompanying papers in this volume by de Boor [1] and Cox [2] indicate, there are a variety of treatments for the end conditions necessary for terminating a knot vector in order to define properly a B-spline defined over a finite domain. This choice is somewhat independent of whether the internal knot spacing is constant or variable. In earlier uses it was quite common to use a tabled set of basis functions, and then simply increase the multiplicity of the control point, instead of the knot, to achieve an interpolating effect relative to the endpoints of the curve. While this does force interpolation with curve tangency, a property to which users have become accustomed from use of Bézier methods, it does have certain ensuing inelegant geometric properties. The curve has flattened ends characterized by embedded line segments for the curve case or embedded bilinears for the surface case in that region. Use of multiple knots at the endpoints of the knot vector instead of multiple control points gives a much more desirable geometry result, albeit at somewhat greater computational expense. (But in today's engineering economics superior geometric qualities are usually worth some additional computation.) To achieve the proper definition of the B-spline basis for a given degree N, we introduce $N + 1$ multiple knots at either end of the knot vector. Internal knots are then added to provide for sufficiently many basis functions to satisfy the requirement that each control point be blended by an individual basis function. The internal knots can be evenly spaced, or spaced in some proportion to the length of the sides of the defining control polygon. As long as a scheme is consistent, it is not strongly critical for shape description how the choices are, in fact, made.

There is a final issue of defining periodic *versus* nonperiodic curves. Actually periodic curves have a conceptual, if not computational, convenience that one does not have to be concerned with end conditions. The curve is defined in a periodic manner, so the issue of defining end conditions never arises.

Definition A *periodic B-spline curve* is a curve defined with parametric domain the unit circle. The mapping from the real line is made by the function circle: $\boldsymbol{R} \to \{\text{unit circle}\}$ such that

$$\text{circle}(t) = \exp\left(2\pi i \frac{t - t_0}{t_{n+1} - t_0}\right)$$

There are $n + 1$ distinct basis functions over this domain. We say that $u = v \bmod a$ if $u = v + ma$ where m is integer valued. This mapping to the circle is equivalent to setting $a = t_{n+1} - t_0$ and applying the modular function.

The curve can be thought of as being defined (parametrically) on a unit circle, with the knots being points on the unit circle. Thus, using appropriate modulo arithmetic, the matter of dealing with end conditions is avoided. This approach is highly suitable for defining closed curves that are actually intended to be viewed as periodic, like the cross-section of a cam shaft, for example. There is no naturally distinguished initial point or end point to such a curve, and therefore a CAD scheme should not force one on the user. If a nonperiodic curve, one with naturally occurring endpoints, is desired, then one of the earlier alternates is the usual method of definition. Either multiple coefficients (i.e. control points) or multiple knots are introduced at the endpoints. Thus we have discussed the major choices associated with defining a specific B-spline curve or surface.

Refinement

Refinement, the process of expressing a particular B-spline in terms of a related spline that has additional knots (i.e. a refined knot vector) has evolved into a powerful technique. It comes up in several applications, as well as in theoretical contexts. This process is also referred to as *knot insertion,* since new knots are inserted into the original knot vector to produce a proper refinement. It is sufficiently important that we provide a formal definition before discussing it further.

Consider two knot vectors $\boldsymbol{\tau}$ and $\boldsymbol{t}$ where $\boldsymbol{\tau}$ is a subset of $\boldsymbol{t}$, that is $\boldsymbol{t} = \{t_i\} = \boldsymbol{\tau} \cup \boldsymbol{\eta}$ where $\{\eta_i\}_{i=1}^{l}$ for some values η_i. If $B_{i,k}(t) = B_{i,k,\tau}(t)$ are the B-splines of order k over the knot vector $\boldsymbol{\tau}$, and $N_{j,k}(t) = N_{j,k,t}(t)$ are the B-splines of order k over the knot vector $\boldsymbol{t}$, then it is clear that $span\{B_{i,k}\} \subset span\{N_{j,k}\}$. $\boldsymbol{t}$ is called a *refinement* of the knot vector $\boldsymbol{\tau}$. We further require that the multiplicity of any knot in both $\boldsymbol{\tau}$ and $\boldsymbol{t}$ be no greater than k. Thus, if $\gamma(t) = \Sigma_i P_i B_{i,k}(t)$, then there exist coefficients D_j such that $\gamma(t) = \Sigma_j D_j N_{j,k}(t)$. The problem is to compute the D_j's. Note that the curve γ is unchanged. Hence even though the new refined basis functions $\{N_{j,k}\}$ have higher order discontinuities at the values of $\boldsymbol{\eta}$ than $\{B_{i,k}\}$, the *curve* does not. If $\eta_j \notin \boldsymbol{\tau}$, then γ is the same polynomial segment on both sides of $\gamma(\eta_j)$, and $\gamma(\eta_j)$ is called a *pseudoknot*.

Since all splines that can be written as a linear combination of the $B_{i,k}$'s can also be written as a linear combination of the $N_{i,k}$'s, certainly the almost trivial case can be done. That is, for each i and each k, there exist scalar coefficients to $N_{j,k}$ which depend on $\boldsymbol{\tau}$ and $\boldsymbol{t}$, call them $\alpha_{i,k}(j) = \alpha_{i,k,\tau,t}(j)$, such that $B_{i,k}(t) = \Sigma_{j=0}^{m} \alpha_{i,k}(j) N_{j,k}(t)$. Substituting this into the original formula for γ yields

$$\gamma(t) = \sum_{i=0}^{n} P_i B_{i,k}(t)$$

$$= \sum_{i=0}^{n} P_i \sum_{j=0}^{m} \alpha_{i,k}(j) N_{j,k}(t)$$

Rearranging the summations yields,

$$= \sum_{j=0}^{m} \left[\sum_{i=0}^{n} P_i \alpha_{i,k}(j) \right] N_{j,k}(t)$$

Hence, $D_j = \Sigma_{i=0}^{n} P_i \alpha_{i,k}(j)$.

The $\alpha_{i,k}(j) = \alpha_{i,k,\tau,t}(j)$ are called *discrete B-splines of order k on* $\boldsymbol{t}$ *with knots* $\boldsymbol{\tau}$. We can show that

$$\alpha_{i,k,\tau,t}(j) = \begin{cases} w_{i,k}(t_{j+k-1})\alpha_{i,k-1})(j) & \\ \quad + (1 - w_{i+1,k}(t_{j+k-1}))\alpha_{i+1,k-1}(j) & k > 1 \\ 1 & \text{if } \tau_i \leqslant t_j < \tau_{i+1} \text{ and } k = 1 \\ 0 & \text{otherwise} \end{cases}$$

where

$$w_{i,k,\tau}(x) = \begin{cases} \dfrac{x - \tau_i}{\tau_{i+k-1} - \tau_i}, & \text{if } \tau_{i+k-1} > \tau_i \\ 0 & \text{otherwise} \end{cases}$$

It immediately follows that

1. $\tau_J \leqslant t_j < \tau_{J+1}$, then $\alpha_{i,k}(j) = 0$ for $i \leqslant J - k$ and $i > J$.
2. For $0 \leqslant j \leqslant m$, let J be such that $\tau_J \leqslant t_j < \tau_{J+1}$. Then $\alpha_{i,k}(j) = 0$ for $i \notin \{J - k + 1, \ldots, J\}$. That is, for any j there are at most k discrete B-splines which can be nonzero.
3. $\alpha_{i,k}(j) > 0$.
4. $\sum_{i=0}^{n} \alpha_{i,k}(j) = 1, \quad \tau_{k-1} \leqslant t_j < \tau_{n+1}$

The ability to add flex to the representation through refinement means that by giving up some higher order continuity, the designer gets unlimited flexibility for use in hierarchical design. This ability to add flex and use it in hierarchical design is an important requirement for the shape operators to act properly.

Further, if we consider a sequence of refinements of a knot vector, so that in each successive one the maximal distance between knots is smaller and that maximal value converges to zero, it can be shown [4] that the piecewise linear functions formed by connecting the successive control points converge to the curve. One method of using this result is to implement subdivision of NURBS for rendering.

There are efficient algorithms for implementing both the general case [5] and the special case for single knot insertion [6]. The various efficiencies are analyzed in [7].

Subdivision of B-spline curves/surfaces

The algorithm for subdivision can be an application of the rule for insertion of a single multiple knot (which effects splitting the curve into two spline subcurves) or it can occur with multiple instances of multiple knots (splitting the curve into several spline subcurves). Unlike the Bézier case, the curves are defined only over the domain specified by their new knot vectors.

Suppose an open B-spline curve, γ of order k is defined by knot vector $\boldsymbol{\tau}$ and control polygon $\boldsymbol{P} = \{P_0, \ldots, P_n\}$. Further suppose that one wants to subdivide the curve into two open B-spline curves, γ_Q and γ_R of order k at the middle of its parametric range. Since the domain of γ is from $\tau_0 = \tau_{k-1}$ to $\tau_{n+1} = \tau_{n+k}$, the subdividing point will be $t^* = (\tau_0 + \tau_{n+k})/2$. The subdivision process must first determine the refined knot vector, then refine the curve using the Oslo Algorithm, and then split the curve. Note that γ_Q and γ_R are well defined open B-spline curves of order k.

One of the refinement algorithms mentioned above would be appropriate for implementing subdivision. Whether the general method or the special case (single knot inserted) depends on whether the designer needs general refinement to support design operators and on whether the subdivision is used on surfaces or just curves [7].

DATA FITTING TECHNIQUES

There are several approaches to data fitting with splines, some based on interpolation and some based on least squares techniques. Within the interpolation group are those methods which interpolate discrete data (points, derivatives, etc.), and those which interpolate curve forms. Adaptive least squares techniques are the more traditional methods which start with a simple form and add knots based on some error criteria and those techniques which start with a complex spline representation and try to *remove* knots to get a simpler representation based on error and ranking criteria. We shall briefly discuss the generic categories.

Interpolation to a grid of data points

This is the most familiar problem seen by researchers. Suppose there is an $M \times N$ grid of data points $\{R_{i,j}\} = \{(x_{i,j}, y_{i,j}, z_{i,j})\}$. The problem is to find a (unique) B-spline surface $\sigma(u, v)$ so that $\sigma(u_m, v_n) = R_{m,n}$. This statement of the problem has implicitly imposed the constraint that the data is parametrized so that each row (column) has one of the parameter values held constant. In general this is written,

$$R_{m,n} = \sigma(u_m, v_n) = \sum_i \sum_j P_{i,j} B_{j,k_v,\boldsymbol{\nu}}(v_n) N_{i,k_u,\boldsymbol{\mu}}(u_m)$$

The orders in each direction and the knot vectors to be used must be specified in advance by the user or through some estimation routines. Notice we can write the inner sum as

$$Q_i^n = \sum_j P_{i,j} B_{j,k_v,\boldsymbol{\nu}}(v_n) \tag{1}$$

for each i and n, then the interpolation can be written, for fixed n,

$$R_{m,n} = \sum_i Q_i^n N_{i,k_u,\boldsymbol{\mu}}(u_m) \quad m = 0, \ldots, M-1 \tag{2}$$

But this has just reduced the surface interpolation problem to an example of curve interpolation. As long as $u_i \in \{t : N_{i,k_u,\boldsymbol{\mu}}(t) > 0\}$, $i = 0, \ldots, M-1$ [8], then the system in (2) can be uniquely solved. This system can be applied for each $n = 0, \ldots, N-1$ until all the coefficients Q_i^n are obtained. Then, as long as $v_j \in \{t : N_{i,k_v,\boldsymbol{\nu}}(t) > 0\}$, the system in (1) can be solved.

This result is most frequently used in a slightly modified form to obtain the *complete spline interpolant*. The complete spline interpolant uses as knot vectors

$$\mu_i = \begin{cases} u_0 & \text{for } 0 \leqslant i < k_u \\ u_{i+1-k_u} & \text{for } k_u \leqslant i < M + k_u - 2 \\ u_{M-1} & \text{for } M + k_u - 2 \leqslant i < M + 2k_u - 2 \end{cases} \tag{3}$$

$$\nu_j = \begin{cases} v_0 & \text{for } 0 \leqslant i < k_v \\ v_{i+1-k_v} & \text{for } k_v \leqslant i < N + k_v - 2 \\ v_{N-1} & \text{for } N + k_v - 2 \leqslant i < N + 2k_v - 2 \end{cases} \tag{4}$$

There are two extra B-spline functions which result from this assignment, as well as a reordering of the association between basis function and data point. In curve interpolation, the end tangent values are assigned to be the extra data. In surface interpolation, this is not so easily performed. While equations (2) can be solved using the data tangents, there are no such values to insert into equation (1). Estimation routines are most often used.

Lofting and skinning

One of the oldest such methods which has existed since the early days of CAGD is the 'lofting' operator, a function which derives its name from the large architectural spaces (lofts) where ship hull designers traditionally were located. Lofting deals with fitting a smooth patch through two parametric boundary curves that are to become embedded on two parametrically opposite sides of a patch or surface. The lofted surface $L(s, t)$ has the form

$$L(s, t) = b_1(s)\gamma_1(t) + b_2(s)\gamma_2(t)$$

where the boundary curves are γ_1 and γ_2, and the blend curves are b_1 and b_2, where $b_1, b_2 \geqslant 0$ and $b_1 + b_2 = 1$. It is an example of equation (1), and cuts out the first part of interpolation, assuming the curves which are found with equation (2) are already known.

A ruled surface (i.e. when $b_1 = (1 - s)$ and $b_2 = s$), is perhaps the simplest lofted surface. Although it is not motivated by geometrical considerations, a topological homotopy function from elementary topology is exactly the same as a lofting function. Recall that a homotopy is simply a parametrized surface that transitions smoothly between two space curves, precisely the role of a lofted surface.

A simple generalization of this is an operation that generates a surface that interpolates a family of parametrically parallel curves. This is exactly the operation that is done by traditional lofters when they seek a surface that passes smoothly through the cross-sectional curves of a ship's hull. A corresponding problem exists in defining the fuselage of an airplane from its cross-sectional curves. This can be thought of as a higher level design operation, especially if it is used to determine a tensor product spline surface. It is an efficient method of specifying several coefficients simultaneously through a single, coherent and geometrically meaningful and intuitive operation. Conceptually, this operation is rather straightforward, but its application is rather more complicated. It is assumed that each of the curves is a parametric B-spline curve of the same order over the same knot vector. However, interpolation with piecewise polynomials can result in shapes which have undesirable characteristics away from the constrained points or curves. In those cases, intuition and the interpolation operator do not give the same surface. It is sometimes possible to improve results by reparametrizing or adding knots and constraint conditions. However, dealing with nonintuitive interpolating surfaces continues to be an area of research. Most recently, attempts have been made to use piecewise cubic Bézier curves constrained to be G^1 where they meet, as the blending functions, instead of B-splines. This requires estimating the respective tangent directions and relative magnitudes, which is a whole different area of research.

Coons-type operators

The traditional Coons' Patch, or boolean sum surface, harks back to the earliest days of CAD itself. It was motivated by the requirement of interpolating a smooth surface through four connected boundary curves. Considerable freedom is allowed in the interior of such a patch after it meets the constraint of interpolating the boundary curves, as well as some optional normal derivative information. If we let $h_{i,j}(u)$ be the cubic polynomial such that $h_{i,j}(k) = \delta_{k,i}\delta_{0,j}$, $h'_{i,j}(k) = \delta_{k,i}\delta_{1,j}$, and let $\sigma(s, t)$ represent the function whose boundary curves and cross boundary derivatives will be used as the boundary information to form C. The cubically blended Coons' patch is defined as

$$\begin{aligned} C(s, t) = {} & \sum_{i=0}^{1} h_{i,0}(t)\sigma(s, i) + \sum_{i=0}^{1} h_{i,1}(t)\frac{\partial\sigma(s, i)}{\partial t} \\ & + \sum_{i=0}^{1} h_{i,0}(s)\sigma(i, t) \\ & + \sum_{i=0}^{1} h_{i,1}(s)\frac{\partial\sigma(i, t)}{\partial s} - T(s, t) \end{aligned} \tag{5}$$

where

$$
\begin{aligned}
T(s,t) = {} & \sum_{i=0}^{1}\sum_{p=0}^{1} h_{i,0}(s)h_{p,0}(t)\sigma(i,p) \\
& + \sum_{i=0}^{1}\sum_{p=0}^{1} h_{i,0}(s)h_{p,1}(t)\frac{\partial\sigma(i,p)}{\partial t} \\
& + \sum_{i=0}^{1}\sum_{p=0}^{1} h_{i,1}(s)h_{p,0}(t)\frac{\partial\sigma(i,p)}{\partial s} \\
& + \sum_{i=0}^{1}\sum_{p=0}^{1} h_{i,1}(s)h_{p,1}(t)\frac{\partial^2\sigma(i,p)}{\partial s\partial t}
\end{aligned}
\tag{7}
$$

The *bilinearly blended* Coons' Patch uses the simple linear blending functions, and does not require normal derivative information. This general form allows σ to be any parametric bivariate function which has well defined cross boundary derivatives. Inspired by the widespread utility and applicability of the Coons' Patch even in a highly restricted form, a 'boolean sum' operator can be based on the idea of generally filling in a spline surface in a region bounded by four curves [9], where the blending functions are written as spline blending functions.

Using the homotopy analogy, a boolean sum surface is a double homotopy. As we remarked above, a loft is a homotopy between two curves. A boolean sum surface satisfies the homotopy requirements in both directions, so it is a special kind of function simultaneously homotopic in both variable directions. Building such functions constructively was studied by Coons and Gordon in the early years of CAGD.

In many instances during a design process, the critical features of a region are the boundary curves, with the interior being rather flexible in its specification. The boolean sum operator is a useful tool in a modeling world where one is constantly called upon to impose a system of rectangular, tensor product patches on objects not naturally modeled in this 'topology'.

When the boundary curves and cross boundary derivatives (as well as corner twists) are constrained to lie on a bicubic polynomial surface, $C(s,t)$ and $T(s,t)$ reduce to the same surface, the *original* surface, and the form is much simpler. However, that is not true in the case that the boundary curves and the cross boundary derivatives lie on a tensor product spline surface which is at least bicubic. In that case, however, the boolean sum, $C(s,t)$ is a tensor product spline surface which *does not* reproduce the original surface, although it does reproduce the boundaries and the cross boundary derivatives. It can be represented as such by simply refining the h functions appropriately.

Adaptive least squares

Methods for fitting curves to data points have traditionally attempted to reduce the error in one of the L^p norms, using more and more polynomial segments until a fit results that does not exceed some given error bound [10–12]. The number and location of the knots is determined adaptively from the data, and some methods determine new parametrizations [12, 13] based on the data. These methods extend directly to grids of data by treating one parameter at a time (as in the case of interpolation), and requiring that the error be below the tolerance on every curve. That results in a collection of curves, each with the same knot vector, ready to be interpolated, or to have the same least squares methods applied to them to obtain a surface.

Data reduction

In an alternative approach Lyche and Mørken [14–16] developed an algorithm that begins with a piecewise linear B-spline with open end conditions passing through the data points to be approximated. That curve is re-represented as a curve piecewise degree $k-1$, through degree raising, with knots of multiplicity $k-1$ at each data point. The degree-raised curve is then used as the initial curve and knots are removed, creating a true degree $k-1$ spline, according to some global requirements until the maximum error tolerance is met. The final spline representation is the 'data reduced' spline. During knot removal, some or all of the multiple knots are removed.

The basic algorithm has three distinct sections. The first deals with determining the importance of each knot in the curve based on the approximate distance between the actual curve and a curve defined on the knot sequence without the knot in question. The second part of the algorithm is concerned with removing the ranked knots in the order that will perturb the curve the least amount. The final part calculates an approximation to the original curve over the new knot vector. These parts are termed *rank, remove* and *approximate,* respectively. Interesting results have been obtained so far.

The algorithm requires access to the complete data set before beginning its ranking effort. A modification of the basic approach has been proposed by Banks and Cohen [17] which can be applied interactively to data that are being fed to the algorithm incrementally, and arrive at an approximation in realtime.

AB INITIO DESIGN

B-splines provide a geometrically intuitive basis, one in which the coefficients (i.e. control points), have a predictable and understandable relationship to the shape of the curve or surface. This is an unusual property in approximation schemes, one which makes B-splines extremely attractive for geometric design. Further, it is an approximation scheme in which an individual control point has only a local effect, and does not perturb a design beyond the localized region of influence. This is a direct consequence of the local approximation of B-splines mentioned in the previous section.

From a strictly theoretical point of view, approximation schemes, along with their associated design schemes, can be classified as either local or global. A scheme which does not have its effect restricted to a local region, is a global scheme. From a practical standpoint, however, we must be concerned with the extent of influence that a local scheme might have. For example, moving a control point of a cubic B-spline curve only affects four local spans of the B-spline, with the more remote spans being less strongly affected. This is the minimum effect that a changed control point can have. But there may be reasonable design scenarios where we desire a large region of effect, but we do not want a global change. For example, suppose that two neighboring control points are modified on a cubic B-spline curve. Then the modified region on the curve will consist of a local region comprised of five contiguous spans. The three central spans will be influenced by both neighboring control points, and each will individually affect a separate span on either side of the three spans with overlapping influence.

The result of an operation applied to a spline surface is thus defined as the spline surface defined by the transformed control points together with the original knot vectors and surface orders. More formally, the original surface is of the form

$$\sigma(u, v) = \sum_{i=0}^{m} \sum_{j=0}^{n} P_{i,j} N_{i,k}(u) N_{j,l}(v) \tag{8}$$

and

$$\sigma_T(u, v) = \sum_{i=0}^{m} \sum_{j=0}^{n} T[P_{i,j}] N_{i,k}(u) N_{j,l}(v) \tag{9}$$

is the result of applying the transformation T to the spline surface S.

In the case where T is an *affine* transformation, this strategy produces exactly the same surface which would have been produced if T had been applied to each point on the surface. Equation (9) can be rewritten for linear T as

$$\sigma_T(u, v) = T\left[\sum_{i=0}^{m} \sum_{j=0}^{n} P_{i,j} N_{i,k}(u) N_{j,l}(v)\right] \tag{10}$$

Most of the choices for T that will be discussed as interesting possibilities for surface modification operators are nonlinear. For some transformations, most notably those inspired by characteristics of the spline representation, the notion of an 'exact' application of the transformation suggested by equation (10) may not be meaningful. In the warp operation, the smoothing properties of the splines are intrinsically involved in the definition of the transformation. In these cases, refinement is in a sense a parameter of the transformation that affects the result. It is the particular choice for the representation of the spline (the amount of refinement) that determines whether the blending from the part of the surface affected by the transformation to the unaffected part will be quite smooth or rather sharp. The refinement in some cases may largely determine the region over which the transformation affects the surface.

For other transformations, the 'exact' solution suggested by equation (10) does have an intuitive meaning, but cannot in general be represented as a spline surface. An important example is the offset operator. It is not possible in general to represent the offset surface of a spline surface as a spline surface. To maintain the closure property, yet allow important operations like offset, it is necessary to produce an approximation to the true surface in equation (10).

With the previous discussion, we have tried to motivate and explain a mechanism for how one might use B-spline control points to modify regions of a curve larger than the minimum that is affected by a single control point. One can simply move a neighboring collection of control points in order to operate on a larger, but still local, region. Now one is challenged to do this in an orderly and intuitive way. One natural approach is to try to mimic the behavior of common phenomena in our life, actions about which we already have considerable intuition.

This setting has given rise to a collection of 'shape operators'. They perform the analog of an intuitive physical action with which we are all familiar. Instead of a bump, or warp, one could analogously define a flatten operator. With a flatten operator one could align a neighboring collection of control points to be collinear according to the positioning of a flattening line or surface. This would act much like a board being pressed against clay to produce a flattened region in a defined local domain.

A bend operator can be similarly defined. One could provide a center for the bend, a local region to which the bend should be applied, and a couple of lines indicating the final positions after a bend is effected. In other words, one must specify the extent, or full excursion, for a bend operation. With this information, it is straightforward to move control points in a manner that simulates the physical effects of a bend. The radius of the bend and how fast the effects of the bend fall off are parameters for experimentation.

One can carry the idea of 'Shape operators' to considerable generality. Nearly any physical analog can give rise to a related spline shape operator, like twist, or thicken, fillet, or the like.

Surfaces from curves

Sweeps

A sweep surface, or 'sweep', can be defined, in very general terms, as the generation of a curve, surface or volume by moving a geometric object (a point, curve, surface or volume) in 3-space. The union of all points in 3-space that are occupied by the object as it is moved, becomes the resulting sweep entity. This idea can be further generalized by allowing the object that is being swept to deform as it is moved. A familiar example using this kind of sweep construction is a torus, which is defined as a profile circle swept through a trajectory circle. This is the ordinary cross product definition of a torus from elementary topology. The 'space curve sweep' is an abstraction of a snug washer that can slide along a flexible rod, in such a manner that it always remains normal to the rod. At every point, the washer always remains in the moving plane normal to the tangent vector of the curve defined by the rod.

We consider the generation of surfaces by sweeping one 3-space curve (called the **cross-section curve**) along another 3-space curve (called the **axis curve**). This sweep paradigm is attractive in that it reduces the problem of surface design to the simpler problem of curve design. A wide variety of shapes can be described intuitively in terms of a (possibly deformable) cross-section moving along a central axis. There are certainly many examples in practice for which such a generalization is already quite useful, and provides a fairly powerful design paradigm. Suppose one wants to define a shape like a drawer handle, which is topologically similar to a half torus. This kind of sweep operation is very useful for defining shapes that are essentially forms that can be thought of as bent pipes, or bent extrusions with more complicated profile sections. There are manifold occurrences of these shapes in the everyday world of mechanical design.

Bloomenthal [18] represents this form as

$$\boldsymbol{\sigma}(u, v) = \gamma(u) + [\boldsymbol{X}(u) \quad \boldsymbol{Y}(u) \quad \boldsymbol{Z}(u)]\boldsymbol{p}(v) \tag{11}$$

where $p(v) = (x(v), y(v), z(v))$ designates the *profile* or *cross-section* curve, $\gamma(u)$ is a parametrically defined general 3-space curve called the **axis curve** along which the cross-section curve will be swept, and $\boldsymbol{X}(u)$, $\boldsymbol{Y}(u)$, $\boldsymbol{Z}(u)$ are three vector valued functions that together specify three ortho-normal vectors at each value of the parameter u in the domain of $\gamma(u)$. (In equation (11) each of these functions is taken as a column vector.) These functions provide orientation information along the axis curve. $[\boldsymbol{X}\ \boldsymbol{Y}\ \boldsymbol{Z}] = [\boldsymbol{X}(u)\ \boldsymbol{Y}(u)\ \boldsymbol{Z}(u)]$ will be referred to as a *local coordinate frame* or *local orientation frame*.

Equation (11) specifies rigid body motion of the cross-section curve along the axis curve. At a given value of u, equation (11) specifies an affine transformation of the cross-section curve, with translation given by $\gamma(u)$ and rotation given by $[\boldsymbol{X}(u)\ \boldsymbol{Y}(u)\ \boldsymbol{Z}(u)]$.

A further useful generalization with apparent utility is one in which the cross-section is allowed to vary as a function of the distance along the extruded direction. In this manner one can define a duct or channel that smoothly blends from a square hole to a round hole, for example. Note that such shapes cannot be described as surfaces of revolution. This 'blended sweep' can be combined with the space sweep to produce a highly general compound sweep. It can be represented as [18]

$$\sigma(u, v) = \gamma(u) + [\boldsymbol{X}(u)\ \boldsymbol{Y}(u)\ \boldsymbol{Z}(u)]\boldsymbol{p}(uv)$$

Other useful generalizations of sweeps exist. It is not necessary conceptually to make the profile curve planar. Suppose one wants to define a duct that connects two nonplanar openings. This situation makes physical sense, and one ought to be able to define a mathematical operation that captures this notion. The above equation need only modify the 0 to be a $z(u, v)$ function.

For example, the Frenet frame is frequently thought of as mathematically the correct coordinate system to use for defining sweep operations. However, the normal flips sides when a curve has an inflection point. Similar difficulties can arise for space curves with high torsion. The normal can rotate rapidly causing the Frenet frame to gyrate wildly. The Frenet frame is not defined for a linear segment of a curve, either, although the geometrical concept of a sweep still seems to make intuitive sense there. The proper CAGD generalization is still a matter of research, but one can certainly develop heuristic techniques to deal with the difficulties mentioned [18, 19].

Extrusions

The 'extrusion operation' is a simple sweep where γ, the axis curve, is a straight line and the coordinate frame is a fixed vector, with its $\boldsymbol{Z}$ element in the direction of the axis curve. An extrusion can be represented as a simple lofting between $\boldsymbol{p}(v)$ and $\boldsymbol{d} + \boldsymbol{p}(v)$, where $\boldsymbol{d}$ is a vector in the direction of the axis line with magnitude the length of the extrusion. Then the extruded surface is defined as the lofted surface, usually linearly ruled, between the two instances of the end profile curves. The extruded B-spline surface is defined as only the surface lofted between the profile curves. It is always ruled in the direction of the extrusion. In the world of freeform surfaces, only the surface is defined. Note that an open curve can be extruded by this definition. The result is simply an open, ruled surface defined by the profile curve, not the boundary of a solid object.

Another useful generalization is to allow the profile curve to be in a plane that is not orthogonal to the trajectory curve. A simple practical case of this occurs if one wants to model a piece of French bread that has come from cutting a loaf with a series of parallel cuts that are on a bias, that is, not orthogonal to the longitudinal axis of the loaf. This is the usual way of cutting French bread. Each piece is a simple extrusion of a profile curve that lies in parallel planes not orthogonal to the major axis of the loaf. Such shapes frequently occur in mechanical design as well.

Extruded lettering of various kinds is tediously pervasive in computer generated television logos and many advertisements. This has become one of the most common examples of the use of extrusion to generate shapes.

Surface of revolution

The widespread use of surfaces of revolution in practical situations has made it one of the most basic and natural design tools. Mathematically, the user specifies a profile curve and an axis of revolution, and then asks for a surface of revolution. Since the circle can be represented as a rational spline curve, then the surface of revolution can be represented as a spline if the profile curve is also a spline curve. This simple operation is highly useful, for many objects have radial symmetry, in the first approximation.

Surfaces from surfaces or deformations

To meet these various design requirements, the approach for adapting operators to the B-spline representation and for developing new operators for B-splines is to define the operations on the *control points* of the spline surface. Typically, each existing control point is transformed by the operation to a new control point. The actual transformation may also involve extra derived surface information and other parameters of the operator, but it will depend also on the geometric position of the original control point. This approach will ensure that many of the design considerations above are met automatically. Closure (and to some extent the unification goal) is obviously insured, since the result is a control mesh for a spline surface which uses the same parametric information as the original. Similarly, since the parametric information does not change and the control mesh has the same dimensions, data complexity is maintained at the same level as the original surface. Although interaction speed is dependent on the actual operation, it is clear that it is closely related to the number of parameters (control points) in the original surface.

A major motivation for such operators arises from frustration with the tedium of the interactive control point movement paradigm for objects with large numbers of defining points, whether surface control meshes, polyhedral data, or sampled point data. An operator that simultaneously affects many points has the possibility of providing a useful higher level operation. One reason for wishing to affect many points at once is efficiency and ease for the designer. Another motivation is the difficulty in figuring out how much, and in what direction, to perturb individual points to achieve a regional effect.

Physical analogs

There is a collection of useful shape operators for splines which were developed in [9] including 'taper', 'bend', 'flatten' and 'warp', each of whose names associates them with an intuitive action. The taper operator applies a scaling function to a surface in one direction. 'Bend' is the analog of the physical operation. One simply has to specify the final bend angle and how sharp (how local) the bend should be. 'Flatten' has the effect of applying a flat plane to a clay model, for example. Flatten has the effect of depressing a surface in a specified direction by a variable, unimodal, well function. The physical analog similarly motivates 'warp', which has an opposite effect of flatten in the sense that flatten can be used to undo a warp. Warp has the effect of puffing up a region by pushing with a (virtual) thumb. The 'bulge' operation of Shirman and Sequin [20] is also in this group of highly intuitive shape operators, and has similar motivation.

All of these operators can be implemented in a direct and straightforward manner by simply applying the corresponding actions to the control points of the B-splines themselves. All of these operators exploit the local support property of B-splines. Deciding on which control points these operators should be applied makes the specification difficult. The relationship between visual effect and parameter values, or even geometry, is poorly understood. The operators only become truly tractable in an interactive environment where one can apply the operations for one set of parameters, see the effect, and then reapply them again for slightly different values until the desired results are achieved. The ability of the user to apply them interactively until the right shape is defined is very important. Much of the scientific contribution in developing them is actually in identifying them and their importance rather than in their particular implementation.

Distortion operators

Many aesthetically appealing shapes are defined by a variety of transformations that we can collectively call 'distortion' operations [21, 22]. These are operations that can mildly deform a model, or can be applied in ways that create Salvador Dali-type images. They can take a well constructed model of a glass bottle, say, and make it look as though it has been exposed to intense heat. Other operations apply curvilinear transformations to make objects take on a serpentine form or shape. So far these deformation operations have tended to be more applicable to the artist domain, partly because such tools are somewhat hard to control precisely for many mechanical applications. Perhaps with greater experience, these operators will find more widespread use in mechanical design.

Offset

The offset operator is important for the process of defining a spline model of a mechanical part, and also for defining the surfaces needed for defining a process plan for manufacturing as well. An offset surface, in the simplest form, is simply a translate of a surface by a fixed amount in a fixed given, typically normal, direction. If the unit normal to a surface σ is defined as

$$n = \left(\frac{\partial \sigma}{\partial u} \times \frac{\partial \sigma}{\partial v}\right) \Big/ \left\| \frac{\partial \sigma}{\partial u} \times \frac{\partial \sigma}{\partial v} \right\|$$

then the offset surface of distance d from σ is defined as $\sigma(u, v) + dn(u, v)$. However, the offset surface has self-intersections and loops if the radius of curvature is less than the offset distance. If a surface neighbors itself (nonlocally) by less than twice the offset distance, its offset surface will clearly intersect itself.

An 'offset surface' is a concept that comes up naturally in developing a process plan for many manufacturing processes. When one is concerned with milling a surface with a ball end cutter, one usually has to specify the machining operation in terms of a surface through which the tool center passes. When using either a flatend or ball mill, the 'tool path surface' is one defined as a surface that is offset by a fixed amount at each point of the surface in the direction normal to the original surface. This form of the 'normal offset operator' produces a translate only when the original surface is planar. Normal offset surfaces also arise in manufacturing processes like plastic injection molding for which shrinkage is a factor. The mold has to be larger than the designed artifact to account for final shrinkage in the cooling stage of the process.

No real process can tolerate the mathematical offset surface with its loops and self-intersections, so research continues into modified offset surfaces which satisfy real process needs. If loops are generated in the offset toolpath space for a milling machine, they should be eliminated to save valuable machine time. Detecting such loops, however, is not simple. Additionally, the offset of a B-spline surface is not another tensor product B-spline surface. In a system that supports only tensor product B-spline representations, it then becomes necessary to define an approximate normal offset surface that comes close enough to the correct offset in order to serve the intended purposes. Spline approximations can be derived from the original surface with sufficient accuracy to meet the needs of the processes.

REFINEMENT IN SUPPORT OF MODELING TECHNIQUES

It has often been reiterated that the local properties of B-splines make them very useful for representing geometry. Their ability to support general, stable, rapid refinement algorithms help make them supportive of hierarchical design.

The classic example is when the gross shape of an object has been defined, but the detail is still missing. Then the surface is refined in that region to allow finer design, while still retaining the continuity of parametric derivatives of appropriate orders.

Since many design operators are applied to the *control points* of the B-spline, the number and density of control points is one of the controlling factors in the final effect of the design operators. For example, suppose a cylinder is defined as a ruled surface (represented as a single cubic in the ruled direction) connecting the two end circles. Trying to bend such a representation is an exercise in frustration, since there are only two intermediate points which can be adjusted. However, if the surface is refined in that direction and then bent, the effects of the bend can be very finely tuned. Warp operations have the similar need, to be able to refine the surface 'enough' to attain the desired result.

Refinement is also used to support the Coons' boolean sum operator. Frequently, the curves obtained from designing the boundaries do not have compatible knot vectors, and so refinement can be applied so that the knot vectors match. The cubic blending functions are also refined to the appropriate knot vectors, and then the whole surface is represented in its tensor product B-spline form. Finally refinement, and its special case of subdivision, are used throughout the adaptive rendering process, as will be shown below.

Booleans and trimmed surfaces

From the nonspline culture of constructive solid geometry (CSG) models, which uses algebraic assembly of simple shapes as its main modeling operator, we inherit the notion of performing boolean operations with geometric entities. In the simple, primitive world, a pipe can be defined as the difference of two concentric cylinders, one with slightly smaller radius. To implement volumetric algebra in the primitive world one can store a parsed, unevaluated expression tree, called a CSG tree, and then compute analytic intersections and perform the evaluations only as required. While this task is not simple, it is still much easier than trying to work with objects whose boundaries are defined by sculptured spline surfaces. Regardless, the idea of union, intersection, and difference is attractive and powerful, and can be extended to objects made up of spline surfaces. Moreover, using Boolean operations to specify shape is a paradigm that users have accepted in the world of simple, primitive building blocks, and also want to have available in a spline-based system. In the more complex environment of spline objects the intuitive meaning of Boolean operations is relatively clear. Sometimes, however, the designer wants to perform Booleans on objects that are only partially specified. Then ambiguities can arise when surfaces are not parts of closed objects. The mathematical and computational aspects become more problematic; and strategies must be developed to overcome them [23].

First we should restate the context of the problem. It is generally implicitly understood that when Boolean operations are applied to objects defined by splines, then we are working with a boundary representation of the objects. A full implementation of a Boolean operation is technically complex. It often makes perfect sense to *trim* one surface by another surface even if no boolean operation is indicated. At this point, it should be noted that a *trim* operation has taken us outside the world of tensor product B-spline surfaces defined over rectangular regions. If we wish to continue to perform modeling operations with the results of a cut operation, and hence for any boolean operation, as they can all be implemented using combinations of cuts, we must allow for the 'trimmed surface' data type. More formally, a trimmed surface is an ordinary tensor product (B-spline) surface that has a restricted parameter domain.

The cut or *trimming* operation, which requires finding surface-surface intersections is too formidable to implement in an analytic manner, so it is usually approached by invoking one or more numerical methods. Two fundamental approaches based on subdivision or incremental techniques are typically used in practice. Subdivision methods tend to be more stable and are nearly certain not to miss any parts of the solution curve, but are also relatively slow. Incremental methods involving standard Newton iteration provide a faster, but less certain alternative. Hybrid schemes are obviously possible, as well. In general this is a difficult computational problem. There are fundamental issues to discover concerning basic representations for geometric entities, algorithms that operate on such entities, and methods for making geometric decisions with greater assurance of correctness in light of known numerical inaccuracies [24].

A B-spline curve intersection algorithm

An important problem is to find the intersections of two curves. Even for explicit polynomial curves, this cannot be done analytically for curves of degree higher than 4, i.e. for arbitrary polynomials of degree five and higher, no general formula can be given expressing the roots in terms of combinations of radicals of rational functions of the coefficients. Particular functions may be amenable to such a decomposition. For parametrically defined curves, one wants to find the actual coordinates of the intersection and also find the distinct parametric values associated with that point on each curve.

Newton–Raphson style numerical algorithms have certain appeal, but difficulties in finding starting points, insuring convergence, and being sure that all roots have been found have all argued against algorithms based solely on numerical methods. Algorithms based solely on subdivision methods can guarantee convergence, up to floating point accuracy, but can be quite slow. So, sometimes, the two methods are combined, i.e. subdivision is used until good starting points can be found for the numerical methods, and until the numerical methods can be guaranteed to converge.

Refining and subdividing surfaces

Let $\sigma(u, v) = \Sigma\Sigma P_{i,j} B_{i,k_u,\tau_u}(u) B_{j,k_v,\tau_v}(v)$ with knot vectors τ_u and τ_v, respectively. Performing refinement at several values of u can be accomplished by merely forming the correct $\mathbf{t}_u$ knot vector. For example, to subdivide simultaneously at r different values, the appropriate $\mathbf{t}$ knot vector would be $\{\tau\}_u + \{c_{u_1,j}, \ldots, c_{u_r,j}\}_{j=0}^{n_u}$. The one would consider the 'curves' $\Sigma P_{i,j} B_{i,k_u}$ and subdivide by refining into

$\Sigma_m D_{m,j} B_{m,k_u}$, where B are the B-splines over the refined knot vector, yielding

$$\sigma(u, v) = \sum \sum D_{m,j} B_{m,k_u}(u) N_{j,k_v}(v)$$

As has been shown [7], a modified matrix approach is more efficient for computation. We denote by

$$\boldsymbol{B}_{k_r,\tau_r}(r) = [B_{0,k_r,\tau_r}(r)\, B_{1,k_r,\tau_r}(r) \ldots B_{m_r,k_r,\tau_r}(r)]$$

$$\boldsymbol{N}_{k_r,t_r}(r) = [N_{0,k_r,t_r}(r)\, N_{1,k_r,t_r}(r) \ldots N_{m'_r,k_r,t_r}(r)]$$

where $r = u$ or $r = v$ as appropriate.

We have shown

$$\sigma(u, v) = \sum_{i=0}^{m} \sum_{j=0}^{n} P_{i,j} B_{j,k_v,\tau_v} v B_{i,k_u,\tau_u}(u)$$

which in turn can be written

$$\sigma(u, v) = \boldsymbol{B}_{k_u,\tau_u}(u) \boldsymbol{P} \boldsymbol{B}^t_{k_v,\tau_v}(v)$$

But,

$$\boldsymbol{B}_{k_u,\tau_u}(u) = \boldsymbol{N}_{k_u,t_u}(u) A^t_{\tau_u,t_u}$$

where

$$A^t_{\tau_u,t_u} = \begin{bmatrix} \alpha_{0,k_v,\tau_v,t_v}(0) & \alpha_{0,k_v,\tau_v,t_v}(1) & \ldots & \alpha_{0,k_v,\tau_v,t_v}(n') \\ \alpha_{1,k_v,\tau_v,t_v}(0) & \alpha_{1,k_v,\tau_v,t_v}(1) & \ldots & \alpha_{1,k_v,\tau_v,t_v}(n') \\ & & \vdots & \\ \alpha_{n',k_v,\tau_v,t_v}(0) & \alpha_{n',k_v,\tau_v,t_v}(1) & \ldots & \alpha_{n',k_v,\tau_v,t_v}(n') \end{bmatrix}$$

Also,

$$\boldsymbol{B}^t_{k_v,\tau_v} = A_{\tau_v,t_v} \boldsymbol{N}^t_{k_v,t_v}$$

where

$$A_{\tau_v,t_v} = \begin{bmatrix} \alpha_{0,k_v,\tau_v,t_v}(0) & \alpha_{0,k_v,\tau_v,t_v}(1) & \ldots & \alpha_{0,h_v,\tau_v,t_v}(n') \\ \alpha_{1,k_v,\tau_v,t_v}(0) & \alpha_{1,k_v,\tau_v,t_v}(1) & \ldots & \alpha_{1,k_v,\tau_v,t_v}(n') \\ & & \vdots & \\ \alpha_{n',k_v,\tau_v,t_v}(0) & \alpha_{n',k_v,\tau_v,t_v}(1) & \ldots & \alpha_{n',k_v,\tau_v,t_v}(n') \end{bmatrix}$$

If

$$\boldsymbol{P} = \begin{bmatrix} P_{\mu-k_u+1,\nu-k_v+1} & P_{\mu-k_u+1,\nu-k_v+2} & \ldots & P_{\mu-k_u+1,\nu} \\ P_{\mu-k_u+2,\nu-k_v+1} & P_{\mu-k_u+2,\nu-k_v+2} & \ldots & P_{\mu-k_u+2,\nu} \\ & & \vdots & \\ P_{\mu,\nu-k_v+1} & P_{\mu,\nu-k_v+2} & \ldots & P_{\mu,\nu} \end{bmatrix}$$

then we can write

$$\sigma(u, v) = \boldsymbol{N}^t_{k_u,t_u} A^t_{\tau_u,t_u} \boldsymbol{P} A_{\tau_v,t_v} \boldsymbol{N}_{k_v,t_v}$$

We can see from this form that the new mesh, the $\boldsymbol{D}$ matrix is written $\boldsymbol{D} = A^t_{\tau_u,t_u} \boldsymbol{P} A_{\tau_v,t_v}$.

The more general capability would be to introduce more flexibility into the surface representation by refining the knot vectors in certain areas. The procedure outlined above for calculating the alpha matrices and then performing matrix multiplication to evaluate the new mesh points is equally applicable for straightforward refinement.

If, instead, it is necessary to subdivide a surface into four subsurfaces, the strategy developed below incorporates the matrix approach.

The method of calculation then for rendering, refinement in support of hierarchical design, and in support of intersection operators is to calculate the discrete splines for each of the tensor product directions, the refinement vectors in each direction being decided by the type of problem at hand. Then perform the matrix multiplication indicated above to get the $\boldsymbol{D}$ matrix, the mesh coefficients in the new tensor product basis. Note that if no refinement is wanted in a particular direction, then it is unnecessary to apply the operator, and only one matrix multiplication is performed.

Rendering open surfaces

The rendering of B-spline surfaces, as well as curves, is important. The algorithm below outlines such a procedure. Again, the surface σ carries all the information of mesh, orders, and knot vectors:

```
Procedure SplineSurfgen (σ)
    Are the points of the mesh of σ
    within tolerance?
    if yes,
        Draw_mesh (σ)
```

```
        else begin
            SplineSurfsplit
            (σ, σ1, σ2, σ3, σ4)
            SplineSurfgen(σ1)
            SplineSurfgen(σ2)
            SplineSurfgen(σ3)
            SplineSurfgen(σ4)
        end {else}
    end
```

Since subdivided curves meet at their endpoints, it is easy to find piecewise linear approximations to each 'flat' subdivided curve, which when unioned form a piecewise linear curve. In fact the curve ψ^k is such a curve for the Bézier case, and the control polygon D is such a curve for the general B-spline curve. The situation is more complicated for surfaces.

The decision of whether the mesh for a subdivided surface is 'flat enough' is clearly more complicated than the decision for a curve, since each quadrilateral region of the mesh, $P_{i,j}$, $P_{i+1,j}$, $P_{i,j+1}$, and $P_{i+1,j+1}$, which corresponds to the linear segments of the curve, is *not* planar (linear)! One strategy is to use the convex hull test, i.e. the test requires that the convex hull must be 'flat enough' by some measuring. Finding the convex hull for a parametric surface is a far more complicated and time consuming task than for curves. An alternative is to find a bounding box for the convex hull, which, of course, also includes the surface, and test the bounding box. Since bounding boxes are quite easy to find this test is not onerous. Unfortunately, the bounding box can be substantially larger than the convex hull, and so the algorithm might subdivide the surface much more than is really necessary. However, suppose it has been determined that the subdivided piece of surface is 'flat enough'. Then one must still determine further attributes to make it feasible to render this surface on a raster display. In particular, most graphical rendering algorithms [25] require that the database be polygons. While one might want to imitate the curve case and use the B-spline mesh, that is impossible since while the whole mesh may be 'flat enough' to approximate the surface, each quadrilateral is not planar. The person wanting to render sculptured surfaces must determine planar pieces. Even if only the four corner points of the submesh are used, it is still impossible to fit just one plane through them. Also, the user must somehow determine a strategy which will be consistent from one subdivided piece of its neighbor, so a least square fit is not suitable. Sometimes people just treat the quadrilaterals as if they were planar polygons. Sometimes this strategy is successful; other times the method fails. Certainly, this is not a general strategy. When two sides along a subboundary are subdivided to a different level, the black hole problem can result. Along one side of the boundary, a single line segment represents the edge, but along the other side, two or more line segments might represent the edge. This is called the 'black hole problem', and is shown in Figure 5.1.

To determine a consistent planar fitting capability based on just three of the points at a time always gives planar fits based on the data. There are many which work very well.

ALPHA_1, A NURBS BASED MODELING SYSTEM

The Alpha_1 project started in 1980. An initial major goal of the Alpha_1 project was to produce a spline based, boundary representation solid modeler that could be used to model the large class of sculptured mechanical parts which were not well represented as Computational Solid Geometry (CSG) models. Objects which fall into this category include turbine blades, airplanes, helicopters, automobiles, sewing machine parts, shoes, automobile engine blocks, silverware, and so on. Even parts that have, at first glance, a simple geometry sometimes cannot be accurately modeled by a CSG approach.

Alpha_1 is also intended to serve as an experimental testbed on which new modeling ideas and techniques can be easily mounted and evaluated. For example, it should be easy to insert new surface representations into the system as they are developed. The initial surface representation upon which the Alpha_1 system was built, and the surface type supported by the entire system, is the trimmed parametric tensor product nonuniform rational B-spline (NURB) surface.

The Alpha_1 system provides a number of capabilities. A short listing of the features of Alpha_1 includes

- an interactive, extensible geometric editor
- drafting style geometric construction operators

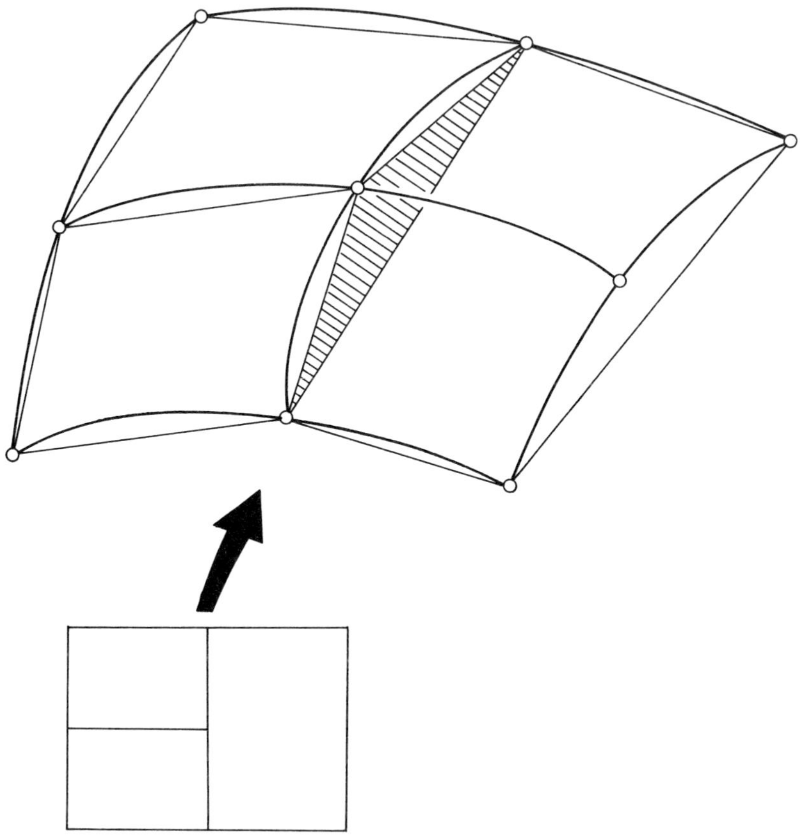

Figure 5.1. Adjacent subdivided subsurfaces with their corner points.

- high-level shape modification and design operators
- parametric design
- automatic creation of primitive objects
- the ability to perform set operations to combine objects into more complex assemblies
- high quality graphical renderings of object models
- computation of simple mass properties of modeled objects
- automatically generated N/C cutter paths.

Designing in Alpha_1

One of the goals of the Alpha_1 geometric modeling system has been to move the *design* of geometric objects to a higher level, i.e. many designers are not familiar with the mathematics of the representations used in the modeling systems they use, and have little intuition as to what blending function coefficients will lead to the shape they have in mind, so requiring them to directly specify blending function coefficients is less than ideal for them. The goal of higher level operators is to take information meaningful to the designer and return the mathematical representation, using the blending functions that the designer has chosen.

The goal of most geometric design is a realizable object, and in the style of design described in this paper, that means designing objects bounded by NURBS surfaces. Hence, curves are created as a means to attain boundary descriptions.

Compass and ruler construction style operators have been traditional draftsman's tools, and were the first type of operations in computer aided drafting systems. These operators are *curve* creation and movement operators dealing with points, lines and arcs, generally in a plane. Other styles of curve creation use the *interpolation* or *data fitting* and the *approximation* paradigms. In some sense, the NURBS curves approximate the piecewise linear curve formed by connecting the coefficients sequentially. So, any sequence of specified points can be used either as data for interpolation or as the coefficients for the B-splines. Curves can be used to construct surfaces in many ways, including all of the *curve-to-surface* (i.e. lofting, skinning, Coons type operators, extrusions, generalized sweeps) methods discussed above.

To simplify the task of working with NURB surfaces, high level shape modification and design operators have been created in Alpha_1. Initial surfaces can be created through the methods discussed above, through interpolation of grids of data, or through using the routines for automatically creating the basic primitives of CSG provided in Alpha_1. Since these basic primitives, such as boxes, spheres, ellipsoids, cylinders, cones and tori can be represented as objects bounded by NURBS surfaces, once created, the shape modification operators can also be applied to them. As we have seen, the goal of these operators is to move and specify groups of coefficients in predictable and acceptable ways.

We discuss the design development of several models in what follows and how the high level operators supported their creation.

Plate 1 shows some concentric surfaces of revolution, which are then associated with each other in a *group* operation. The flat radial structures with large holes through them are *instances* of an identical structure which are again associated with a *group* operation. In Plate 2 we can see a basic form of extrusion, a ring defined with a surface of revolution, and some boolean operations providing the detailed definition. The bottle in Plate 3 is formed by taking a standard bottle, a surface of revolution, and applying a *flatten* operation three times to the sides. Then the flattened sides are dented inwardly with a *warp*.

The composite image in Plate 4 shows a virtual

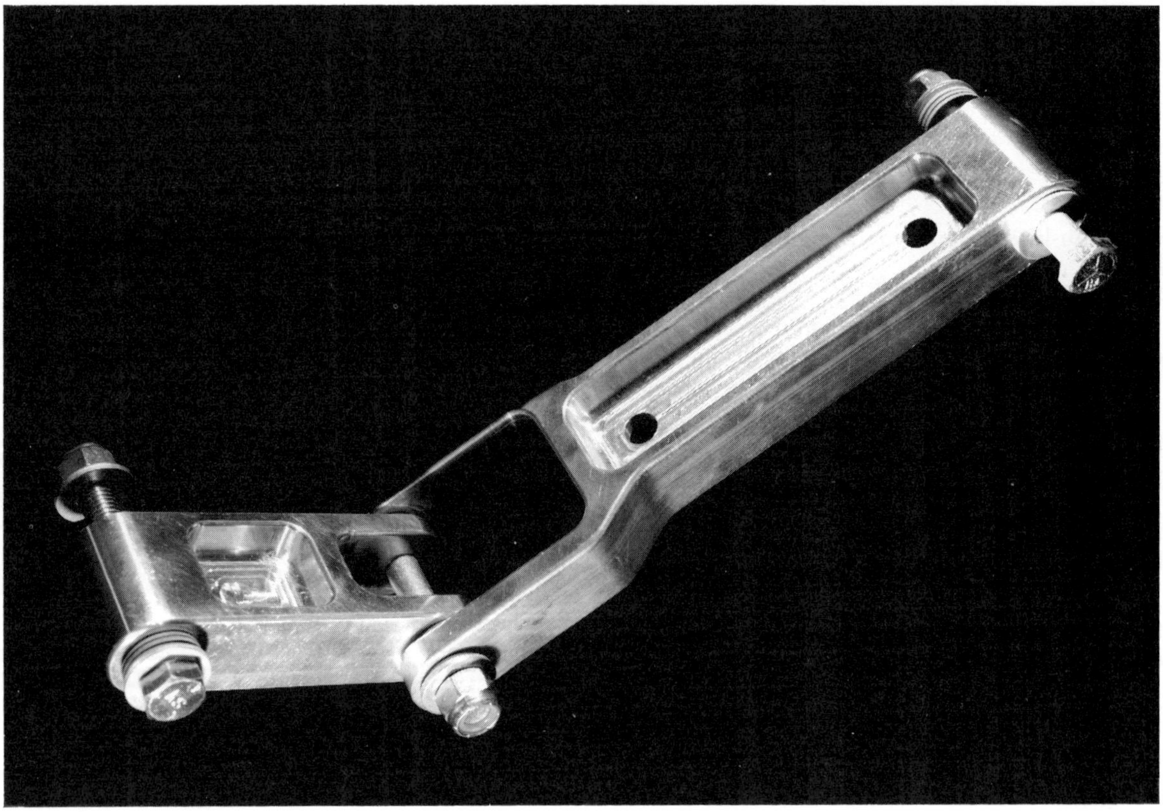

Figure 5.2 Complex mechanical assembly.

model of a spoon in the upper right, and then the actual metal mold that was used to make injected plastic models. The spoon is a rich example of shape operators. It was formed with warps, an extrude for the handle, followed by a taper and a bend applied to the handle, and finally a *thicken* used to make a surface into a closed object.

The gears shown in Plate 5 show a parametric modeling process that is suggestive of manufacturing. The gear blank is a surface of revolution, from which the excess material was removed as it is moved through space. A gear is a parametric model type that can be respecified in different variations simply by providing the four characteristic numbers that define a typical gear of this kind.

The turbine blade root (Plate 6) and the composite blade structure (Plate 7) illustrate rather complicated geometry. The root is *trimmed* with carefully specified trimming surfaces. The basic root geometry is given as the intersection of two not quite orthogonal extruded forms. The upper portion of the blade, the aerodynamic surface, is defined as a *loft* through a large number of cross-sections. Finally, the upper part and the lower part are unioned together to form the final part in Plate 7.

The hydrodynamic surface in Plate 8 is blended to the hub in a complex blending operation. While many blends are quite simple in application, others can be rather complex. This part indicates a more complex one between two sculptured surfaces.

Figures 5.2 and 5.3 show increasingly complex mechanical assemblies comprised of multiple parts. This is typical for many applications in mechanical design. Simply being able to specify a single part is not powerful enough for many design domain. Instead, we require the capability of specifying large numbers of related parts, and constraining their function and behavior relative

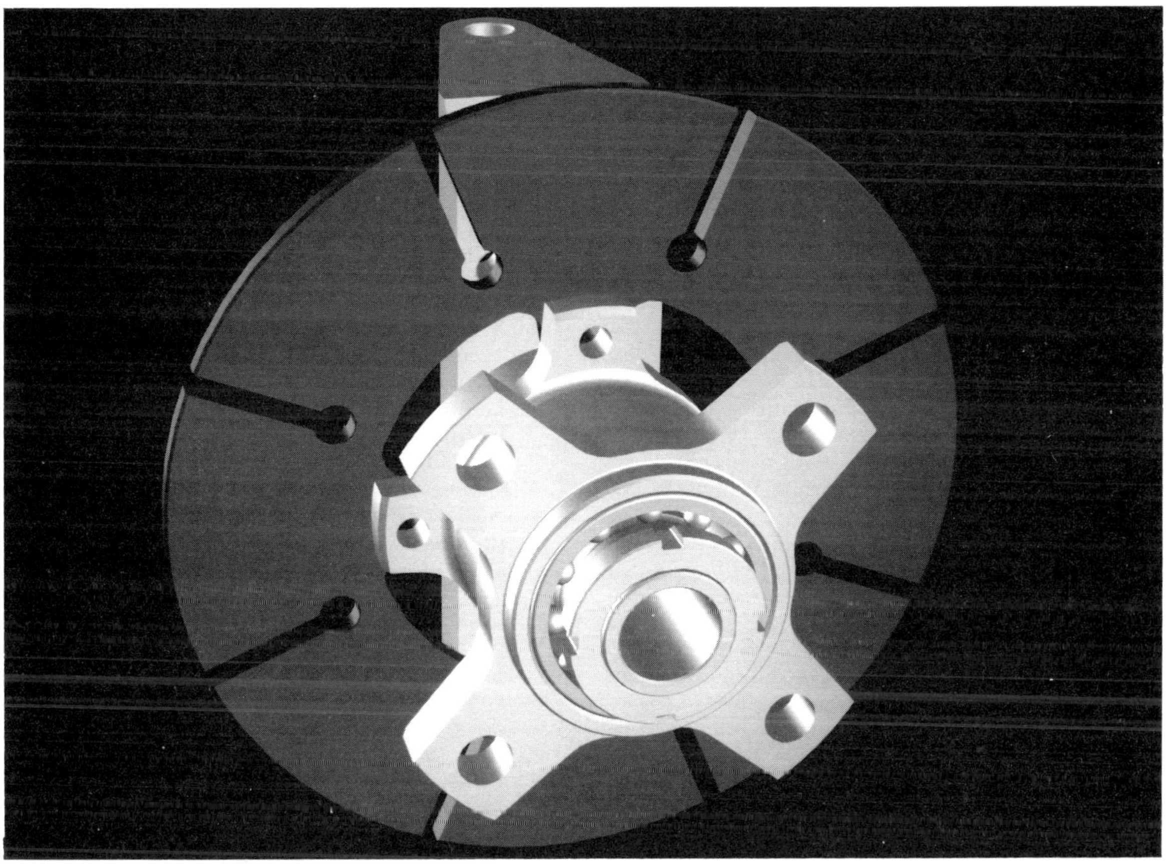

Figure 5.3. Brake disc assembly.

to each other. Feature design is extensively used in these last two examples. Pockets are defined as features and manufactured as such. Similarly for the hub and brake disc assembly in Figure 5.3, feature hole patterns, slots, and the like, are specified to make a repetitive design more straightforward to complete. The manufacturing is then optimized with respect to the features and their constituent subprocesses like spot drilling, for example.

ACKNOWLEDGEMENTS

I am grateful for the encouragement, patience, and editorial improvements provided by Les Piegl. Appreciation is due to Elaine Cohen and the Alpha_1 Group for their help and illustrative material. This work was supported in part by DARPA (N00014-88-K-0689). All opinions, findings, conclusions or recommendations expresed in this document are those of the author and do not necessarily reflect the views of the sponsoring agency.

REFERENCES

1 **C. de Boor** B(asic)-spline basics, This volume, pp 27–49.

2 **M. G. Cox** Algorithms for spline curves and surfaces. This volume, pp 51–76.

3 **R. F. Riesenfeld** *Applications of B-spline Approximation to Geometric Problems of Computer-Aided Design,* PhD thesis, Syracuse University, May 1973. (Available as TR No UTEC-CSc-73-126, UUCS.)

4 **E. Cohen and L. L. Schumaker** Rates of convergence of control polygons, *Comput.-Aided Geometric Design,* Vol 2, September 1985, 229–235.

5 **T. Lyche and K. Morken** Making the Oslo algorithm more efficient, *SIAM J. Numer. Anal.,* Vol 23, June 1986, 663–675.

6 **W. Boehm** Inserting new knots into B-spline curves, *Comput.-Aided Design,* Vol 12 No 4, July 1980, 199–201.

7 **E. Cohen, T. Lyche and K. Morken** Knot line refinement algorithms for tensor product B-spline surfaces, *Comput.-Aided Geometric Design,* Vol 2, April 1985, 133–139.

8 **L. L. Schumaker** *Spline Functions: Basic Theory,* John Wiley, Chichester, 1981.

9 **E. S. Cobb** *Design of Sculptured Surfaces Using the B-spline Representation,* PhD thesis, University of Utah, June 1984.

10 **P. Dierckx** Algorithms for smoothing data with periodic and parametric splines, *Comput. Graphics and Image Processing,* Vol 20 No 2, October 1982, 171–184.

11 **M. Plass and M. Stone** Curve-fitting with piecewise parametric cubics, *Computer Graphics,* Vol 17, July 1983, 229–239.

12 **C. L. F. O'Dell** *Approximating Data with Parametric B-splines,* Master's thesis, University of Utah, December 1985.

13 **E. Cohen and C. L. O'Dell** *Mathematical Methods in Computer Aided Geometric Design,* Academic Press, London, 1989, 155–166.

14 **T. Lyche and K. Morken** *A data reduction strategy for splines,* Research Report 107, University of Oslo, February 1987.

15 **T. Lyche and K. Morken** *A Discrete Approach to Knot Removal and Degree Reduction Algorithms for Splines,* Clarendon Press, Oxford, 1987.

16 **T. Lyche and K. Morken** Knot removal for parametric B-spline curves and surfaces, *Comput.-Aided Geometric Design,* Vol 4, November 1987, 217–230.

17 **M. J. Banks and E. Cohen** Realtime B-spline curves from interactively sketched data, *Computer Graphics,* Vol 24 No 2, March 1990, 99–107.

18 **M. Bloomenthal** *Approximation of sweep surfaces by tensor product B-splines,* Technical Report, Computer Science, University of Utah, 1988.

19 **B. D. Donahue** *Modeling Complex Objects with Generalized Sweeps,* Master's thesis, University of Utah, 1985.

20 **L. A. Shirman and C. H. Sequin** Local surface interpolation with shape parameters between adjoining Gregory patches, *Comput.-Aided Geometric Design,* Vol 7 No 5, 1990, 375–388.

21 **A. Barr** Global and local deformations of solid primitives, *Computer Graphics,* Vol 18, July 1984, 21–30.

22 **T. W. Sederberg and S. R. Parry** Free-form deformations of solid geometric models, *Computer Graphics,* Vol 20, August 1986, 151–160.

23 **S. W. Thomas** *Modeling Volumes Bounded by B-spline Surfaces,* PhD thesis, University of Utah, June 1984.

24 **C. M. Hoffman** *Geometric and Solid Modeling,* Morgan Kaufman, New York, 1988.

25 **J. D. Foley and A. Van Dam** *Fundamentals of Interactive Computer Graphics,* Addison-Wesley, New York, 1982.

ABOUT THE AUTHOR

Richard F. Riesenfeld majored in mathematics, and received an AB degree from Princeton University in 1966, where he emphasized the areas of probability, statistics and computing. This led to graduate work in mathematics at Syracuse University, and he concentrated on pure mathematics for three years leading to an MA degree in 1969.

While at Syracuse, he first became attracted to the area of computer aided geometric design and computer graphics when taking a course from Steve Coons. Subsequent study with Robin Forrest and Charles Lang during a visit with the CAD Group at Cambridge University strengthened the interest in CAD. When Bill Gordon joined the Syracuse University faculty, Riesenfeld began to study spline applications in geometric modeling with him. This alignment of interests resulted in some joint research activity with Gordon, under whom he did his doctoral thesis research introducing the B-spline method for computer-aided design. This thesis has become a standard reference in the field, as B-splines have become more ubiquitous. In 1973 he received a PhD from Syracuse University in computer science.

Since 1972, Riesenfeld has been a member of the Computer Science faculty at the University of Utah. He is co-head of the computer-aided design and computer graphics group. In 1980, together with Elaine Cohen, he initiated the Alpha_1 Project. The Alpha_1 System is a trimmed NURBS-based geometric modeling system used for research and application. As a unified approach, the system uses rational B-splines exclusively to represent all objects in the system. Alpha_1 represented one of the first attempts at unifying the CSG modeling style with the complex surface style of modeling. The system has provided a useful testbed for much experiment work in this area aimed at advancing the field of spline modeling. The system is now beginning to gain wider use beyond Utah.

In 1980 Riesenfeld worked with Elaine Cohen and Tom Lyche to produce the Oslo Algorithm, which offers a theoretical and computational basis from which to effect refinement and subdivision of B-spline curves. The power of this approach has become widely recognized with subdivision and refinement becoming a fundamental instrument in spline modeling. More recently, his interest has carried him closer to the integration of the design and manufacturing processes.

Having remained on the Utah faculty since 1972, Riesenfeld has been active in the field as a researcher and consultant. He has published many articles on spline related topics in CAD/CAM, and has served on several editorial boards. During his tenure at Utah he has served as a previous chair of the Department of Computer Science for six years.

6 F-Methods for Free-Form Curve and Hypersurface Definition

James C. Ferguson

INTRODUCTION

The F-techniques were developed at Boeing during the early 1960s for general engineering curve and surface definition of aerovehicular shapes.

The primary intention of this chapter is to present (1) a review of F-methods as they were essentially originally written, (2) some shape classification techniques, and (3) outlines of interesting applications and research directions.

Background

The history of F-curve, surface, and hypersurface techniques is rooted in the Master Dimensions Group of Boeing Company (Seattle, early 1960s). Bill Hoar led that group, along with Mark Danielsen. Both were extraordinarily gifted in both engineering science and management, an enviable combination of talent that easily drew the best from each member of that team. When hired (April 1960) I was 'applications green', straight out of college, and grounded in pure mathematics. These qualities ordinarily were not sought by real world engineering projects. However, Bill and Mark were launching a novel and very geometric approach to automatic aircraft surface lofting and analysis, and my background in differential geometry and vector analysis seemed to fit well. As will be mentioned, it was this very education that inspired development of the F-techniques.

F-methods (this terminology was coined by coworkers) arose, along with a number of other important geometric methods, as a result of a challenge posed by Bill to solve the general 'mesh of points' problem: determine a smooth surface, friendly for use in a design/manufacturing/computing environment, that interpolates a given matrix of points in space. Within this lattice, certain subsets formed nodes on control curves, themselves forming a bidirectional network of space curves. Thus certain topological relations were implied. Here, the control curves were assumed neither planar nor single-valued off any coordinate plane. At that time, solutions to this problem were being eagerly and independently pursued by several organizations around the country. The progression toward the parametric polynomial interpolation approach at Boeing included both a combination of personal insight and experience gained by daily interaction with all members of the team. In this sense, the F-methods must rest as a group accomplishment, and the true pioneers must include all of the early MDG members as well as several persons in the Numerical Control Group.

A model, known as the 'flying carpet' (see Figure 6.1) was circulating nationally as an open challenge to interpolate and mill. Several of the control curves were not planar, and the composite set was not regularly oriented with respect to any coordinate plane. We were first to discover and announce a solution [1–5] – it is interesting to note Gellert's original use of currently common terminology. An account of

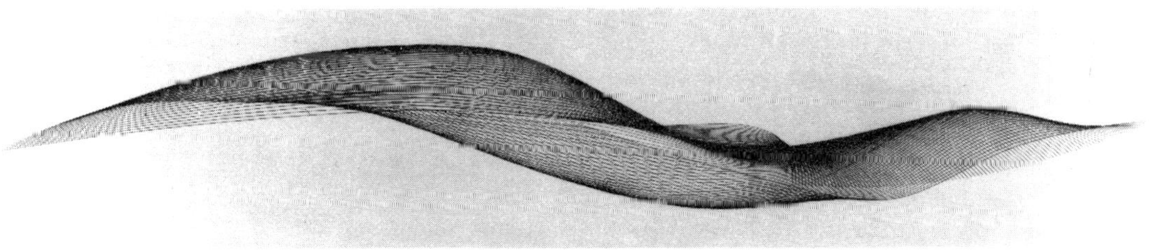

Figure 6.1. The 'flying carpet' model [1].

Fundamental Developments of Computer-Aided Geometric Modeling
ISBN 012-554765-X

how dramatic this unveiling was is given in Carlberg and Childs [2] (Ed Carlberg directed the NCG at Boeing). Boeing made history by uncovering the 'flying carpet' surface, milled in a heavy block of aluminum, at the end of a national conference during which the impossibility of this achievement had been predominantly vocalized. Apparently, a state of shock arose when our model was revealed. Thus began a rapid and persistent spawning of ingenious generalizations (e.g. Coons, Bézier, Forrest, Gordon, Riesenfeld) and applications that have driven CAD/CAM mathematics and algorithms for years. Typical applications for parametric F-methods quickly came from the major auto, aerospace, and shipbuilding industries. These covered diverse areas of engineering including aerodynamics analysis, structural design, stress analysis, computer-aided wind tunnel analysis, turbine blade design, auto body design, wave impact analysis, hypersonic flow, etc.

A progress report, Hoar (internal Boeing communication, 1962), was prepared to summarize our group activity at that time. One paragraph of that note read,

> "The publication of D2-22026, 'Multi-Variable Curve Fitting', dated August, 1962 culminates several months of effort to establish a satisfactory surface fit through a given array of points in space. This surface has a seemingly endless variety of applications to Aero-Space problems. It is believed that this surface fitting technique is not duplicated elsewhere in the industry."

The mathematics of differential geometry underwrote the parametric technique and carried our successes, mainly since (1) analytic surface points, normals and, curvatures could be quickly obtained to any degree of accuracy without numerical iteration, (2) surfaces could easily be delineated or bounded, (3) reorientation (under Euclidean motion) of the basic geometric data produced congruent interpolation surfaces, (4) surface patches could be pieced together to form a smooth conglomerate, (5) for normal engineering applications, surface shape could be predicted and controlled, and (6) various surface degeneracies could be handled and integrated within a network of F-patches. F-volumes (3D hypersurfaces) were ahead of their time, and were not exploited as much as the F-surfaces. This was mainly due to relatively low computer power in those days. F-volumes unfortunately thus fell into obscurity until the 1970s, when solid modeling became popular.

F-curves

The search for a general surface interpolation scheme initially began as a search for a general space curve interpolation method. After all, the control points in the mesh were arranged with the idea of developing a surface that interpolated an intersecting network of curves. Lofting methods, used for interpolating piecewise conic planar section curves in one direction, had been one of the common techniques for surface development, and thus the lofting concept also influenced our direction.

The basic problem for space curves was to find a curve segment that interpolated two given points while remaining tangent to two given lines at these points. This setup attended to three things:

1. It allowed a convex segment.
2. It allowed a singly inflected segment.
3. It allowed development of a chain of segments smoothly interpolating a set of distinct ordered points.

The logical choice for the simplest mathematical representation for such an interpolating curve was a parametric cubic since it involved four vector-valued coefficients sufficient to balance the given Hermite conditions.

Thus (see Figure 6.2a), let $\mathbf{T}_0$ and $\mathbf{T}_1$ denote direction vectors at two given points $\mathbf{P}_0$, $\mathbf{P}_1$, respectively. Then the curve segment can be written as $\mathbf{P}(u) = (x(u), y(u), z(u))$, where,

$$\mathbf{P}(u) = \sum_{0}^{3} \mathbf{A}_k u^k, \quad 0 \leqslant u \leqslant 1$$

The conditions

$$\mathbf{P}(0) = \mathbf{P}_0, \mathbf{P}(1) = \mathbf{P}_1, \mathbf{P}'(0) = \mathbf{T}_0,$$

and

$$\mathbf{P}'(1) = \mathbf{T}_1$$

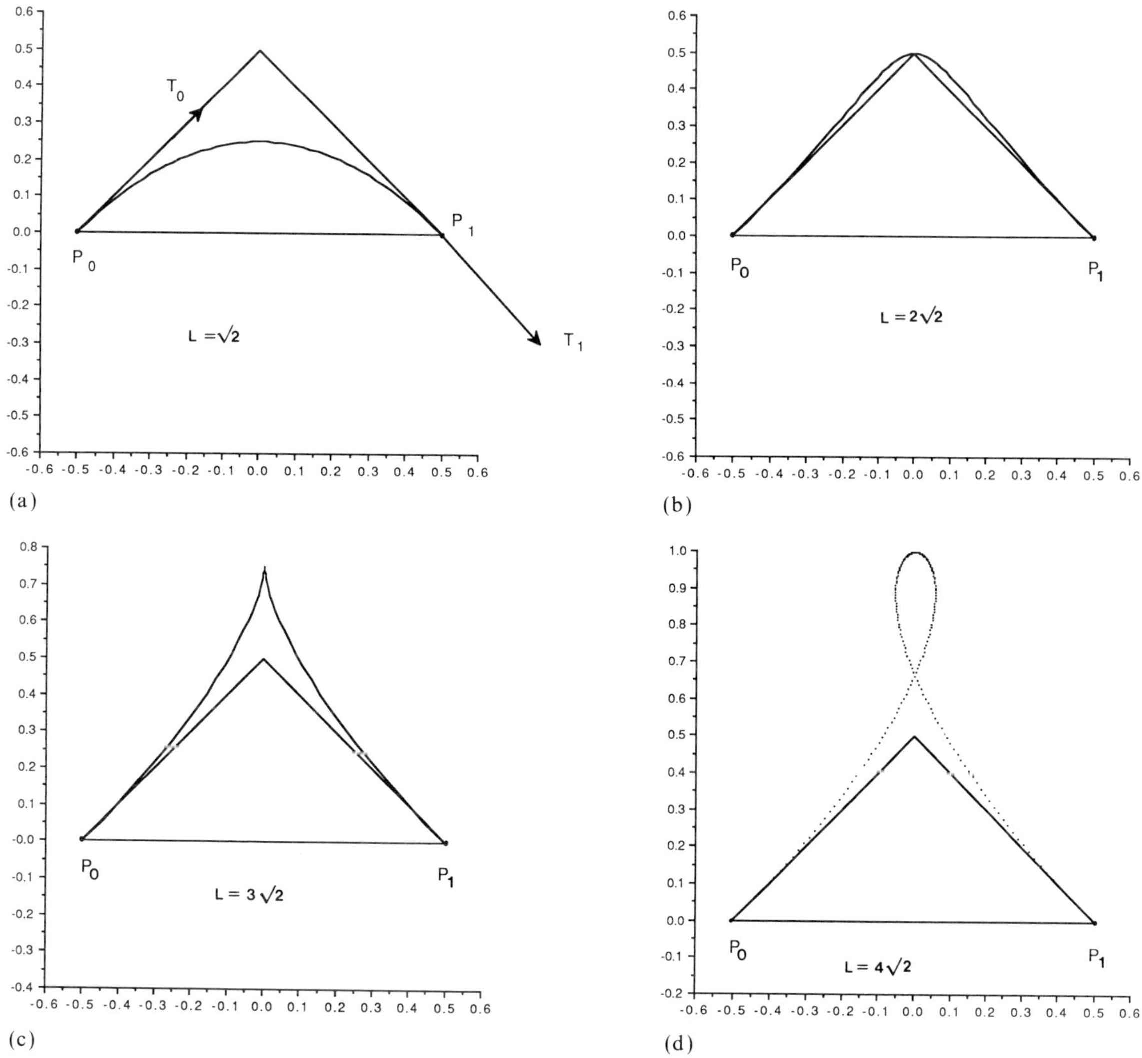

Figure 6.2. F-curve segment with tangent lengths (a) $\sqrt{2}$, (b) $2\sqrt{2}$, (c) $3\sqrt{2}$, (d) $4\sqrt{2}$.

give the following values for the $\mathbf{A}_k$,

$$\mathbf{A}_0 = \mathbf{P}_0$$
$$\mathbf{A}_1 = \mathbf{T}_0$$
$$\mathbf{A}_2 = 3(\mathbf{P}_1 - \mathbf{P}_0) - 2\mathbf{T}_0 - \mathbf{T}_1$$
$$\mathbf{A}_3 = 2(\mathbf{P}_0 - \mathbf{P}_1) + \mathbf{T}_0 + \mathbf{T}_1$$

Let, l_0, l_1 be the respective lengths of $\mathbf{T}_0$, $\mathbf{T}_1$. It was very apparent that these parameters could be 'tuned' in a variety of ways depending on the application. For example, one might want to find the values of l_0, l_1 that give rise to a segment having the most desirable flow characteristics in a fluid physics problem. Or, interest might be in finding values for l_0, l_1 that optimize a profile within an optical lens, i.e. we essentially look for external solutions, within the class of parametric cubics, to problems involving optimization of curve functionals.

In fact, one of the shapes a single segment can assume looks like an airfoil (Figure 6.3), and it is interesting to pose the question what are the aerodynamic properties of such curves as functions of l_0, l_1? Perhaps this might be too simplistic. If a specific chord value is given then

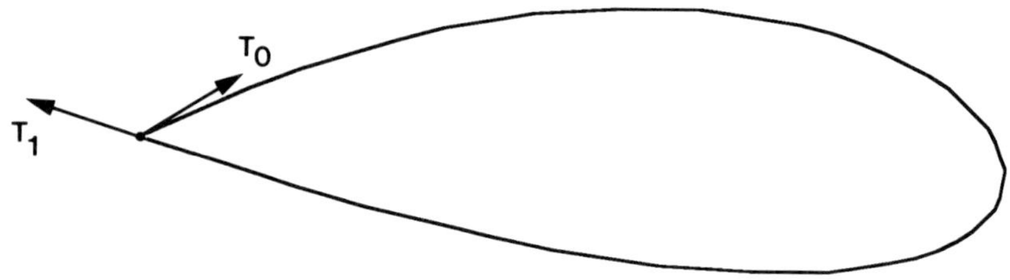

Figure 6.3. A tear drop shaped F-curve segment.

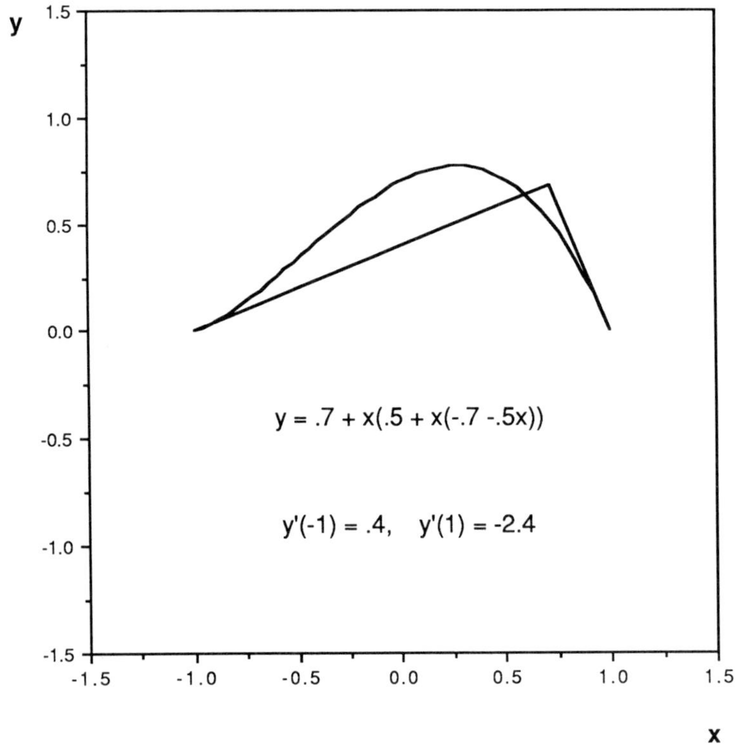

Figure 6.4. Scalar cubic with forced inflection.

we need two segments, at least. Then the above question has more meaning, since we can play with four parameters instead of two, assuming fixed tangent directions at the hard points. Although posed in the early 1960s, study of this problem is still wanting. And, in general, using these parametric segments, with their rich variety of shapes and with their simple format, as function classes in a wide variety of approximation problems is fertile territory for research. An example is given subsequently for script character generation.

Figure 6.2a shows a convex segment through two points separated by 1 unit. The values of l_0, l_1 equal $\sqrt{2}$ for this frame. It is a shape one might expect for this arrangement of points and tangent lines. Parenthetically, if one uses a scalar-valued cubic, i.e. part of a hypothetical cubic spline, the resulting shape can be inflected (Figure 6.4), contrary to what one normally expects. In fact, this example clearly exposes the occasional bad behavior of scalar cubic splines. Therefore, we can immediately sense some of the power of the vector-valued cubic: it has richer

Figure 6.5. A single inflected F-curve segment.

Table 6.1. Shape/tangent length relationship for a special case F-curve (see Figures 6.2a–d)

Tangent length	Shape
$0 < l_0 \leqslant 3\sqrt{2}/2$	convex
$3\sqrt{2}/2 < l_0 < 3\sqrt{2}$	doubly inflected
$l_0 = 3\sqrt{2}$	cusp
$l_0 > 3\sqrt{2}$	loop

shape variety while needing only the arithmetic involved with cubic polynomials.

It is useful to see further examples of shapes for this curve segment. Consider the sequence of planar segments in Figure 6.2a–d. Increasing the values of l_0, l_1 pulls the curve closer to the respective tangent lines. This action also simultaneously enlarges the segment. Notice the transition of shape as it moves within and through the 'characteristic triangle' by increasing values of l_0, l_1:

convex → doubly inflected → cusped → looped
(6.2a) (6.2b) (6.2c) (6.2d)

Normally, one is only interested in the family of either (1) convex shapes when the point/tangent arrangement is as suggested in Figure 6.2a, or (2) singly inflected shapes, as indicated in Figure 6.5. Thus, a question immediately arises. What are the values of l_0, l_1 that separate the shapes we have seen? This question has been addressed by several workers [6–14], depending on the format used to express the curve (Hermite, Bézier, or B spline, polynomial or rational-based). It is instructive to give a partial answer for the simple case when the data are as depicted as in Figure 6.2a, where the tangents are of equal length and both make 45° angles with a unit length chord. Subsequently we will summarize a complete analysis for planar parametric cubic curves. The dependence of shape on tangent length for this special instance is given in Table 6.1.

F-curve shape analysis

This section classifies planar F-curves and is based on work by Ferguson and Miller [7, 9, 15].

Let each plane curve be determined by two distinct points, **Q** and **R**, and two tangent vectors, **T** and **S**, applied at **Q**, **R**, respectively. Also, assume these data are in 'characteristic' position (Figure 6.6), where

$$\mathbf{Q} = (-c/2, 0), \quad \mathbf{R} = (c/2, 0),$$
$$\mathbf{T} = (t_1, t_2) \quad \text{and } \mathbf{S} = (s_1, s_2)$$

and with $c > 0$. Then an equation for the curve is $\mathbf{P}(u) = (x(u), y(u))$, where

$$x(u) = (-2c + s_1 + t_1)u^3 + (3c - 2t_1 - s_1)u^2 + t_1 u - c/2$$

$$y(u) = (t_2 + s_2)u^3 - (2t_2 + s_2)u^2 + t_2 u,$$
$$0 \leqslant u \leqslant 1$$

and where $\mathbf{P}(u)$ is the position vector to a variable point. Such a curve is an algebraic plane curve of at most the third degree. Our main interest is in the portion of the curve between **Q** and **R** when $0 \leqslant u \leqslant 1$.

In the following we adopt the terminology that a curve $\mathbf{P}(u)$ has a *loop* if it possesses a self-intersection point that is associated with two distinct values of u. It has a *cusp* if it contains a point at which $x'(u)$ and $y'(u)$ vanish together. Also, a *flat point* is one at which the curvature vanishes. We proceed to give two central

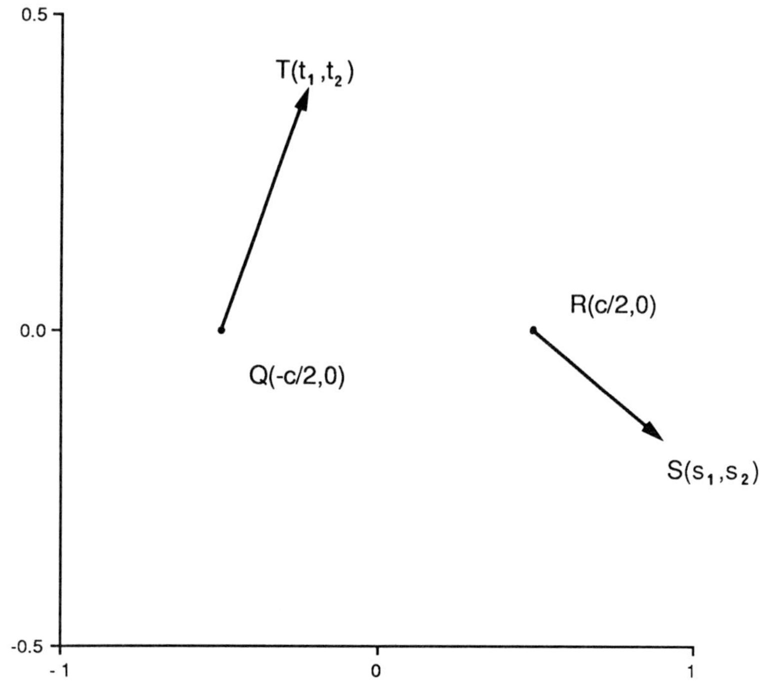

Figure 6.6. Data defining a canonical F-curve segment.

theorems. First assume t_2 is nonzero and define,

$$\sigma := s_2/t_2$$
$$\alpha := (s_1t_2 - s_2t_1)/ct_2$$
$$A := (1 - \sigma - \alpha)$$
$$B := (\alpha - 2)$$
$$C := (3 - \alpha)$$
$$G(\alpha, \sigma) := \alpha(\alpha - 4) + 4\sigma(\alpha - 3) = 4AC - 3B^2$$

Theorem 1 *If A is nonzero, then the curve* $\mathbf{P}(u)$ *has a loop if and only if* $G(\alpha, \sigma) > 0$. *The self-intersection point is* $\mathbf{P}(u_1) = \mathbf{P}(u_2)$ *where* $u_1, u_2 = (-B \pm \sqrt{G})/2A$. *Also, the curve* $\mathbf{P}(u)$ *has a cusp if and only if* $G(\alpha, \sigma) = 0$ *(the cusp point is at* $u = -2C/3B$*). If* $A = 0$, *then the curve* $\mathbf{P}(u)$ *has no loop or cusp points.*

Proof The arguments (see [9]) use methods from elimination theory and are not repeated here.

Theorem 2 *If A is nonzero, then the curve* $\mathbf{P}(u)$ *has exactly two flat points if and only if* $G(\alpha, \sigma) < 0$. *These points are given by*

$$u_1, u_2 = (-3B \pm \sqrt{-3G})/6A$$

If $A = 0$, *then the curve* $\mathbf{P}(u)$ *has exactly one flat point at* $u = -C/3B$, *provided B is nonzero.*

Proof The arguments (see [9]) involve examination of the expression for curvature.

Both of the above two theorems apply to $\mathbf{P}(u)$ with no restrictions on the variable u. In each theorem the special case $\alpha = 2$, $\sigma = -1$ occurs when $A = 0 = B$. The curve for this case is at most a parametric quadratic, and thus has no self-intersection, cusp or flat (unless a line) points.

These results can be used to provide a classification of parametric cubic curves. Figure 6.7a depicts global results, i.e. for the unbounded u-interval, where the boundary ($G = 0$) between looped and inflected curves identifies cusped curves. In the practical case of characterizing shape over the interval $0 < u < 1$, we must explore the conditions under which self-intersection and flat points are interior to this interval. This can be done in a direct manner by

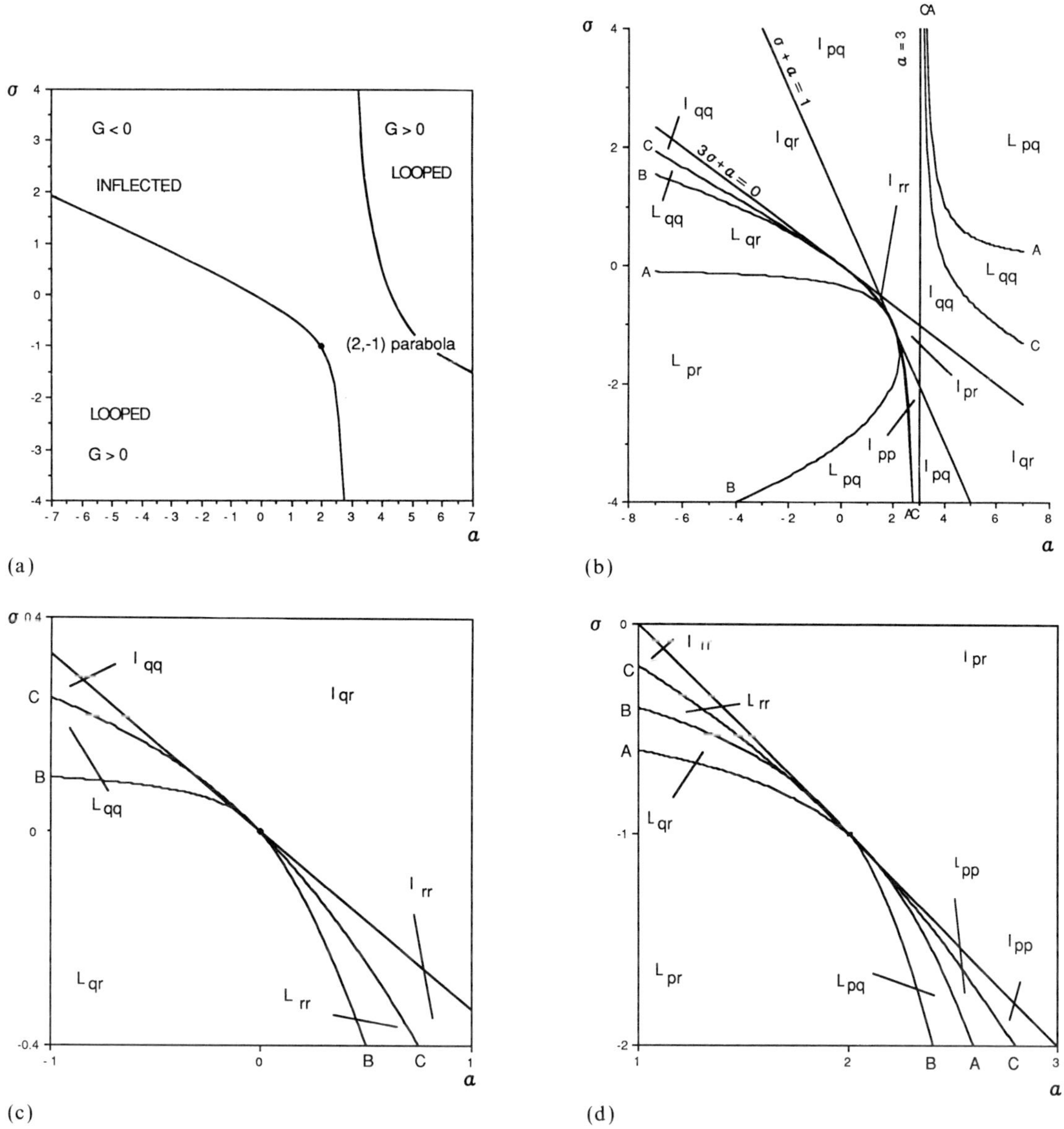

Figure 6.7. (a) Shape parameter regions for unconstrained u-value, (b) full display of shape parameter regions, (c) enlargement of regions around (0, 0), (d) enlargement of regions around (2, −1).

restricting the appropriate values of u_1, u_2 to the open interval (0, 1). The results are summarized in Figures 6.7b–d for t_2 nonzero, wherein one finds the α, σ plane divided into regions that are labelled either with a doubly subscripted I or L, indicating inflections or loops, respectively. (In Figure 6.7: curve A is $(3-\alpha)\sigma + 1 = 0$; curve B is $\sigma(3+\sigma) + \alpha = 0$; curve C is $G = 0$.) Any α, σ values falling in one of these areas means the curves $\mathbf{P}(u)$ having these values are looped (L) or inflected (I). If $\mathbf{P}(u)$ has a loop then the subscripts indicate where the u-values of the self-intersection point are located with respect to the interval (0, 1), where p means $u < 0$, q means $0 < u < 1$, and r means $u > 1$. Similarly, if $\mathbf{P}(u)$ is inflected, the subscripts of I indicate where the

u-values of the flat points are located with respect to the interval $(0, 1)$. Special consideration occurs along the line $\alpha + \sigma = 1$, where a single inflection occurs in $(0, 1)$ for $\alpha > 3$ or for $\alpha < 1.5$, and no inflection occurs in $(0, 1)$ for $1.5 \leqslant \alpha \leqslant 3$. When $t_2 = 0$ and s_2 nonzero, one gets the corresponding behavior of the curves $\mathbf{P}(u)$ by examining the $\sigma = 0$ axis, replacing α by t_1/c, u by $(1 - u)$, and interchanging subscripts p, r. For $s_2 = 0 = t_2$, we obtain a straight line joining points $\mathbf{Q}$ and $\mathbf{R}$. Further comments on this are given below.

Several remarks on Figures 6.7b–d are in order. First, Figures 6.7c and d consist of two graphs depicting, qualitatively, blowups of neighborhoods of two special points in Figure 6.7b. This gives greater clarity to the various curve crossings and tangencies involved. Curves $\mathbf{P}(u)$ with a cusp have α, σ values on $G(\alpha, \sigma) = 0$. This line forms the boundary between looped and inflected curves $\mathbf{P}(u)$. If we restrict our attention to the curve segment $\mathbf{P}(u)$, $0 < u < 1$, then we notice that it has

(i) a loop if $G > 0$ and either $\sigma > 0$, $\alpha + \sigma(3 + \sigma) \geqslant 0$ or $\alpha > 3$, $\sigma(3 - \alpha) + 1 \geqslant 0$;

(ii) exactly one flat point if either $3\sigma + \alpha > 0$, $\alpha \leqslant 3$ or $\alpha > 3$, $3\sigma + \alpha \leqslant 0$ or $3\sigma + \alpha = 0$, $\alpha < 0$;

(iii) neither inflections, loops, nor cusps if $3\sigma + \alpha \leqslant 0$, $0 < \alpha \leqslant 3$ or $\alpha + \sigma(3 + \sigma) < 0$, $\alpha \leqslant 0$.

Also, on the boundary lines between regions we can infer loop or flat conditions at the $u = 0, 1$ end points, which is an interesting check on subscript continuity when passing from one type of curve to another.

A further illustration of the use of Figure 6.7 comes from early applications of parametric cubics in industry. For some time a 'rule of thumb', concerning the choice of the end point tangent components, has been the requirements

$$s_1^2 + s_2^2 \leqslant c^2 \quad \text{and} \quad t_1^2 + t_2^2 \leqslant c^2$$

By observation and experience, it was found that the fulfillment of these conditions gives 'nice' curve segments $\mathbf{P}(u)$, $0 \leqslant u \leqslant 1$. If we consider s_1, $t_1 > 0$, then the geometries involved are like those of Figures 6.2a and 6.5, and (as we will see later) the above 'rule of thumb' guarantees that these segments are not looped or cusped. This alone is a strong reason for calling such segments 'nice'; however, more is really desired. For instance, one usually wants, for Figure 6.2a, a convex curve segment. Examination of Figure 6.7 shows that this occurs when $0 < \alpha \leqslant 3$, $3\sigma + \alpha \leqslant 0$, since in this case we are considering s_1, t_1, $t_2 > 0$ and $s_2 < 0$ (i.e. $\sigma < 0$, $\alpha > 0$). On the other hand, it is clear that the 'rule of thumb' might indeed not always give a convex segment. The techniques of this example can easily be applied to other basic point/tangent arrangements.

A short remark about when $x(u)$ is monotonic (originally analyzed by Rowin [6]) is appropriate to amplify the comment above concerning the 'rule of thumb'. By examining $x'(u) = 0$, we construct Figure 6.8 as we did for Figure 6.7. The symbolism R_{jk} is used where, as before, $j, k = p, q$ or r and relates the roots of $x'(u) = 0$ to the open interval $(0, 1)$. Within the ellipse $x'(u)$ is never zero, since the discriminant of $x'(u)$ is negative. If we recall the 'rule of thumb' conditions, and if again we regard s_1, $t_1 > 0$, then we have s_1/c, $t_1/c < 1$ and hence s_1, t_1 fall in the region R_{pr} meaning there are no roots of $x'(u) = 0$ for $0 < u < 1$. Hence, the curve segment $\mathbf{P}(u)$, $0 < u < 1$, cannot be looped or hooked. Also, if $s_2 = 0 = t_2$, then the resulting straight line has no kinks (i.e. it does not double back on itself) in the interval $(0, 1)$ if $x(u)$ is monotonic here. Such conditions can occur interior to the chord or exterior to it, where in the latter case the segment graph actually covers a length greater than the chord, implying $|x(u)| > c/2$ for some $0 < u < 1$. In Figure 6.8 the special case $t_1 = c = s_1$ is such that $x'(u)$ is nonzero, although at other points of the ellipse $x'(u)$ can be zero.

Other examples on the use of Figure 6.7 and the above results follow. Assume s_1, t_1, $t_2 > 0$ and $s_2 < 0$:

(a) Figure 6.2a: $t_1 = t_2 = s_1 = 1$, $s_2 = -1$. Then, $\sigma = -1$, $\alpha = 2$, $A = G = 0$. This is the special case mentioned above (parametric quadratic) and has constant sign curvature.

(b) Figure 6.2b: $t_1 = t_2 = s_1 = 2$, $s_2 = -2$. Then, $\sigma = -1$, $\alpha = 4$, $A = -2$, $G = -4$. A is nonzero and $G < 0$, again implying there are two flat points. From Figure 6.7,

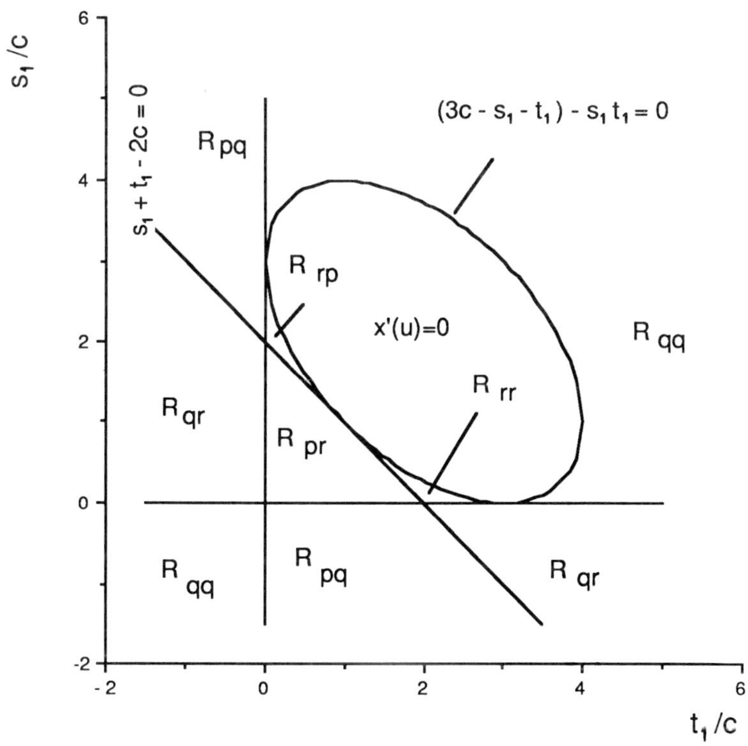

Figure 6.8. Monotonicity regions for $x(u)$.

$\alpha = 4$, $\sigma = -1$ is in an I_{qq} region. Thus the flat points are at values of u in the interval $0 < u < 1$. In particular, from Theorem 2

$$u_1 = (-6 + 2\sqrt{3})/(-12)$$
$$= 1/2 - \sqrt{3}/6$$

and

$$u_2 = 1/2 + \sqrt{3}/6$$

(c) Figure 6.2d: $t_1 = t_2 = s_1 = 4$, $s_2 = -4$. Then, $\sigma = -1$, $\alpha = 8$, $A = -6$, $G = 12$. A is nonzero and $G > 0$ imply that the curve has a loop. From Figure 6.7, $\alpha = 8$, $\sigma = 1$ is in an L_{qq} region, meaning $\mathbf{P}(u)$, $0 < u < 1$, has a loop. The self-intersection point is given by either $u_1, u_2 = 1/2 \pm \sqrt{3}/6$.

(d) Figure 6.5: $t_1 = t_2 = s_2 = 1$, $s_1 = 2$. Then, $\sigma = 1$, $\alpha = 1$, $A = -1$, $G = -11$. Theorem 2 implies the curve is singly inflected on (0, 1) at $u =$ slightly less than 1/2. Also, examination of Figure 6.7 corroborates these statements by placing α, σ in an I_{pq} region.

These local results can be used for automatically producing shape preserving composite F-curves, where such properties as monotonicity, convexity, minimal oscillation, positivity and functional minimization are considered. For example, see Ferguson [16] and Ferguson and Pruess [17].

Parallel analyses can be made using the Bézier form that involves a Bernstein representation. It should be noted that visual inspection of the associated control polygons does not always give precise information as to the shape characteristics of the curve. For example, a looped polygon can be associated with a doubly inflected curve (Figure 6.2b has an associated looped Bézier control polygon). This remark also applies to B-spline curves (Wang [12] treats the general shape questions for this representation).

Approximation using F-curve composites

It is desired to approximate a given ordered sequence of N distinct noncolinear points $\{\mathbf{P}_k\}_N$, $N > 2$, by a cubic parametrized curve (CPC) having initial point $\mathbf{P}_1$ and final point $\mathbf{P}_N$ (Figure 6.9). Several such sequences are independently approximated as subsequences of another and joined to form CPC-chains, maintaining either curve continuity or tangency. Efficient organization of the data into CPC units requires an awareness of the several shape varieties these curves can assume. An iterative least squares technique is given for the necessary determination of the parameters involved, and an interesting application of these equations is given for script character generation. The material is based on Ferguson [15]. A related reference is Plass and Stone [18].

The general CPC segment is defined as an F-curve with the representation

$$\mathbf{P}(u) = (2u+1)(u-1)^2\,\mathbf{P}_1 + (3-2u)u^2\mathbf{P}_N + l_1 u(u-1)^2\mathbf{T}_1 + l_N u^2(u-1)\mathbf{T}_N$$

where $0 \leqslant u \leqslant 1$. Assume further that the unit vectors $\mathbf{T}_1$, $\mathbf{T}_N$ are given. With this information the quantities l_1 and l_N are determined such that the resulting CPC best fits the points in the following sense. Let

$$u_k := \sum_{2}^{k} \frac{|\mathbf{P}_j - \mathbf{P}_{j-1}|}{\sum_{2}^{N} |\mathbf{P}_j - \mathbf{P}_{j-1}|}, \qquad k = 2, \ldots, N-1$$

and define

$$F(l_1, l_N) := \sum_{2}^{N-1} |\mathbf{P}(u_k) - \mathbf{P}_k|^2$$

Then compute those values of l_1, l_N that satisfy

$$\frac{\partial F}{\partial l_1} = 0 = \frac{\partial F}{\partial l_N}$$

Noting the identity

$$|\mathbf{P}(u_k) - \mathbf{P}_k|^2 = (\mathbf{P}(u_k) - \mathbf{P}_k)\cdot(\mathbf{P}(u_k) - \mathbf{P}_k)$$

writing this as a function that is quadratic in l_1, l_N, and using it in the above partial derivative equations, gives the following two equations in

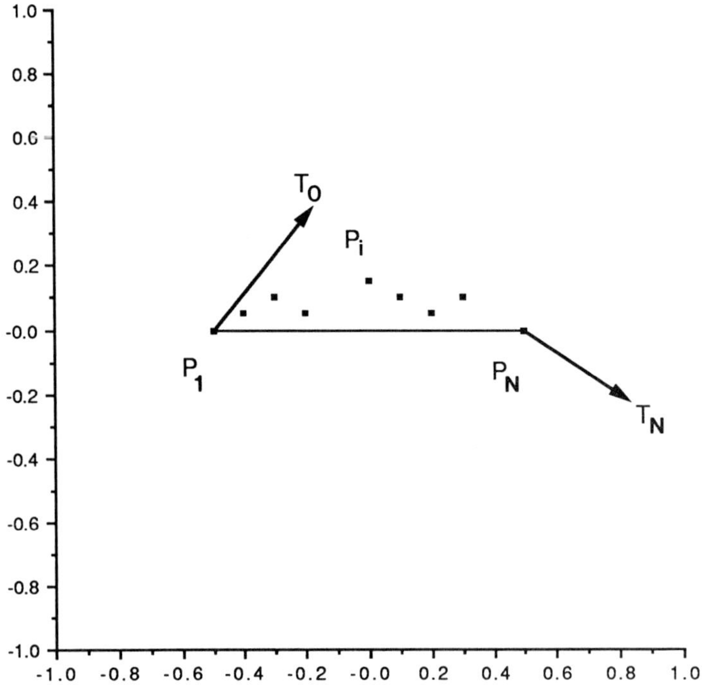

Figure 6.9. Approximation to data by an F-curve segment.

the unknowns l_1, l_N (all summations are from $k = 2, N - 1$):

$$l_1[\Sigma u_k^2(u_k - 1)^4] + l_N(\mathbf{T}_1 \cdot \mathbf{T}_N)[\Sigma u_k^3(u_k - 1)^3] = a$$

$$l_1(\mathbf{T}_1 \cdot \mathbf{T}_N)[\Sigma u_k^3(u_k - 1)^3] + l_N[\Sigma u_k^4(u_k - 1)^2] = b$$

where

$$a = \mathbf{T}_1 \cdot [\Sigma u_k(u_k - 1)^2 \mathbf{P}_k] - (\mathbf{T}_1 \cdot \mathbf{P}_1)\Sigma u_k(2u_k + 1)(u_k - 1)^4 - (\mathbf{T}_1 \cdot \mathbf{P}_N)\Sigma u_k^3(3 - 2u_k)(u_k - 1)^2$$

and

$$b = \mathbf{T}_N \cdot [\Sigma u_k^2(u_k - 1)\mathbf{P}_k] - (\mathbf{T}_N \cdot \mathbf{P}_1)\Sigma u_k^2(2u_k + 1)(u_k - 1)^3 - (\mathbf{T}_N \cdot \mathbf{P}_N)\Sigma u_k^4(u_k - 1)(3 - 2u_k)$$

The values for l_1, l_N obtained by solving these two linear equations (assume exist and unique) are used to obtain another set of numbers u_k^*, $k = 2, \ldots, N - 1$, better than the above u_k in the sense that they are computed to satisfy

$$f_k(u) := (\mathbf{P}(u) - \mathbf{P}_k) \cdot \mathbf{P}'(u) = 0$$

This is accomplished by iterating

$$u_k^{(n+1)} = u_k^{(n)} - f_k(u_k^{(n)})/f_k'(u_k^{(n)}),$$

where $u_k^{(1)} = u_k$

Now, define u_k^* as the value to which this sequence converges (assume u_k is a sufficiently close estimate of u_k^* to ensure convergence). Then redefine $u_k = u_k^*$, $k = 2, \ldots, N - 1$, and again compute the corresponding l_1, l_N. Repeat the above iteration until some measure of error in the tangent lengths is small enough.

Next, an application of this procedure to script character generation (and in general, to font definition) is given.

Each character is geometrically analyzed for an efficient way of representing it by a chain of CPCs. Suppose a decision as to the number of these necessary has been made, the basis of which is intimately related to the character topology as well as to the variety of CPCs available. For instance, the letter 'a' (see Figure 6.10) can be segmented into two CPCs, one of

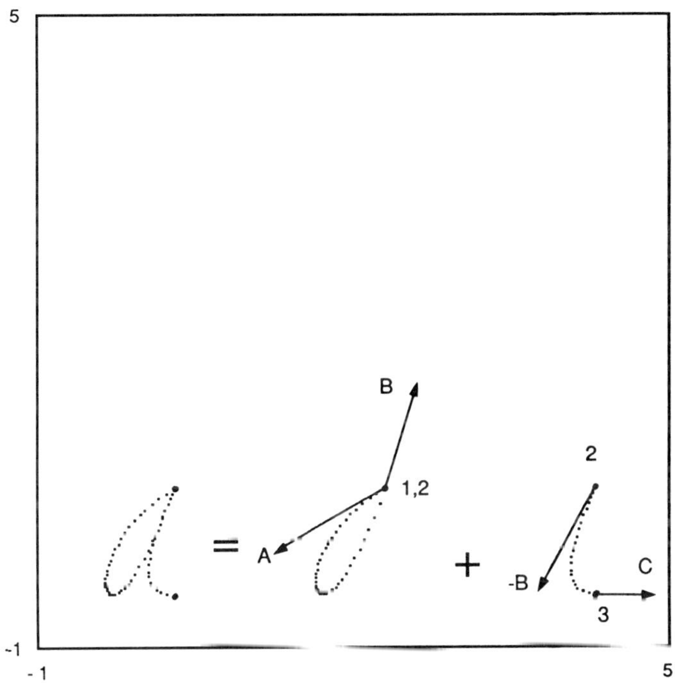

Figure 6.10. Splitting characters into F-curve segments.

a teardrop shape, the other with distinct end points. Then each CPC must be defined, up to determining appropriate tangent lengths, by two end points and the associated unit tangents. Choice of these data depends on whether one desires to join a CPC continuously or tangentially to another. The first CPC (see Figure 6.10) is defined by two coincident points, 1 and 2, and two unit tangents **A** and **B**. This part of the character is traced from point 1 in the direction **A** toward point 2 as u increases from 0 to 1. The second CPC (see Figure 6.10) is defined by points 2, 3 and tangents $-$**B**, **C**. Notice that, purposely, the two CPCs are tangent at point 2, even though the adjacent lengths might differ as a result of using separate minimizations. A number of control points are then selected for each CPC, and are used to adjust the relative magnitudes in the desired composite shape. The CPC-chain is determined once the four tangent lengths have been computed, and thus completes the definition of the letter 'a'.

Figure 6.11. The first four characters described by composite CPCs.

Table 6.2. Parameters for the script character 'a' (see Figure 6.10)

No. CPCs	2	
Link	1	2
$\mathbf{P}_1$	(0.3, 0.5)	(0.3, 0.5)
$\mathbf{P}_N$	(0.3, 0.5)	(0.3, -0.5)
$\mathbf{T}_1$	(-0.87416, -0.48564)	(-0.33872, -0.94089)
$\mathbf{T}_N$	(0.33872, 0.94089)	(1, 0)
l_1	3.53	0.76
l_N	6.15	1.46

Table 6.3. Parameters for the script character 'b' (see Figure 6.11)

No. CPCs	3		
Link	1	2	3
$\mathbf{P}_1$	(−0.5, −0.5)	(−0.1, −0.4)	(0.4, −0.1)
$\mathbf{P}_N$	(−0.1, −0.4)	(0.4, −0.1)	(0.5, 0.3)
$\mathbf{T}_1$	(0.6, 0.8)	(0.74651, −0.66537)	(0.06238, 0.99805)
$\mathbf{T}_N$	(0.74651, −0.66537)	(0.06238, 0.99805)	(0.75926, 0.65079)
l_1	11.85	0.78	2.72
l_N	5.07	0.50	2.17

Table 6.4. Parameters for the script characters 'c' and 'd' (see Figure 6.11)

	c	d	
No. CPCs	1	2	
Link	1	1	2
$\mathbf{P}_1$	(0.3, 0.2)	(0, 0.35)	(−0.1, −0.4)
$\mathbf{P}_N$	(0.2, −0.3)	(−0.1, −0.4)	(0.2, −0.5)
$\mathbf{T}_1$	(0, 1)	(−0.93888, −0.34426)	(0.74154, 0.67091)
$\mathbf{T}_N$	(0.75926, 0.65079)	(0.74154, 0.67091)	(0.25163, −0.96782)
l_1	3.85	2.19	6.09
l_N	4.60	1.14	9.02

Results of following this technique for the first four written English letters are given in Figure 6.11, and in Tables 6.2–6.4.

F-hyperpatch

This section discusses N-dimensional F-Surfaces, and follows the material given in Ferguson [1]. (Also see Coons [19] and Forrest [20] for alternative approaches.)

An N-dimensional hyperpatch resides in $N+1$ space, and is defined on an N-dimensional unit cube $[0, 1]^N$ (Cartesian product of N unit intervals) in the parameter space. The parametric function defined on this cube is recursively constructed from $(N+1)$-dimensional position and tangent data at the corners using lower dimensional constructs. Let $\mathbf{P}_{k1\ldots kN}$ denote the corner points, and let $\mathbf{T}^j_{k1\ldots kN}$ denote the respective tangent vectors at these points, where j identifies the independent variable u_j of the curve to which the tangent vector applies. Here, $km = 0$ or 1 and $j = 1, \ldots, N$. For example, $N = 1$ treats two points $\mathbf{P}_0$, $\mathbf{P}_1$ and two tangent vectors $\mathbf{T}^1_0$ and $\mathbf{T}^1_1$ applied at these points, respectively. For another example, let $N = 2$. Then the corner points are $\mathbf{P}_{00}$, $\mathbf{P}_{10}$, $\mathbf{P}_{01}$, $\mathbf{P}_{11}$ and the tangent vectors are $\mathbf{T}^1_{00}$ and $\mathbf{T}^2_{00}$, $\mathbf{T}^1_{10}$ and $\mathbf{T}^2_{10}$, $\mathbf{T}^1_{01}$ and $\mathbf{T}^2_{01}$, $\mathbf{T}^1_{11}$ and $\mathbf{T}^2_{11}$ applied, respectively at like indexed points (Figure 6.12). The specific nature of the parametric function that interpolates these data is defined next.

General equation

First, let d_k denote a partial differentiation interpolation operator with respect to u_k defined subsequently. Then the F-hyperpatch is given by

$$\mathbf{P}(u_1, \ldots, u_N) = [\mathbf{P}(u_1, \ldots, u_{N-1}, 0), \mathbf{P}(u_1, \ldots, u_{N-1}, 1), d_N\mathbf{P}(u_1, \ldots, u_{N-1}, 0), d_N\mathbf{P}(u_1, \ldots, u_{N-1}, 1)] \cdot \mathbf{F}(u_N) \quad \text{for } N > 1$$

where $0 \leqslant u_k \leqslant 1$ and it is understood that

$$\mathbf{P}(u_1) = [\mathbf{P}_0, \mathbf{P}_1, \mathbf{T}^1_0, \mathbf{T}^1_1] \cdot \mathbf{F}(u_1)$$

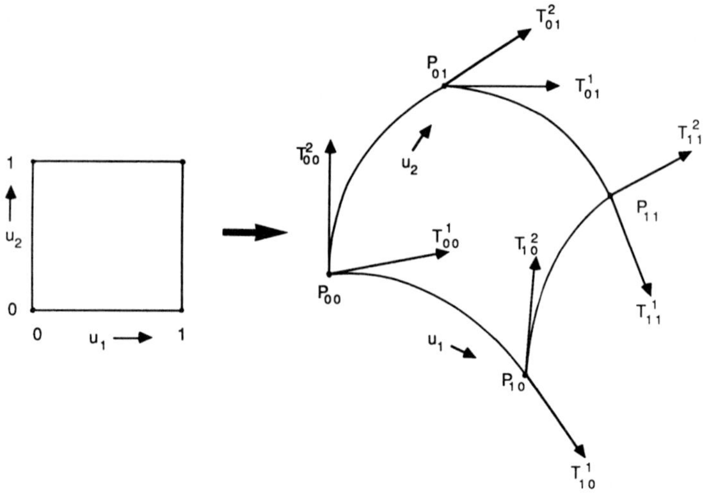

Figure 6.12. Data defining a two dimensional F-hyperpatch.

Here, $\mathbf{F}(u_k) = [(2u_k + 1)(u_k - 1)^2, u_k^2(3 - 2u_k), u_k(u_k - 1)^2, u_k^2(u_k - 1)]$ and **P**, **T** vectors have $N + 1$ coordinates.

The embedded hyperpatch and tangent interpolation operators for general indexes are needed to expand the above definition. These relations are, respectively,

$$\begin{aligned}\mathbf{P}(u_1, \ldots, u_k, i_{k+1}, \ldots, i_N) \\ = [&\mathbf{P}(u_1, \ldots, u_{k-1}, 0, i_{k+1}, \ldots, i_N), \\ &\mathbf{P}(u_1, \ldots, u_{k-1}, 1, i_{k+1}, \ldots, i_N), \\ &d_k\mathbf{P}(u_1, \ldots, u_{k-1}, 0, i_{k+1}, \ldots, i_N), \\ &d_k\mathbf{P}(u_1, \ldots, u_{k-1}, 1, i_{k+1}, \ldots, i_N)] \cdot \mathbf{F}(u_k), \\ &k > 1\end{aligned}$$

and

$$\begin{aligned}d_m\mathbf{P}(u_1, \ldots, u_n, j_{n+1}, \ldots, j_N) \\ = [&d_m\mathbf{P}(u_1, \ldots, u_{n-1}, 0, j_{n+1}, \ldots, j_N), \\ &d_m\mathbf{P}(u_1, \ldots, u_{n-1}, 1, j_{n+1}, \ldots, j_N)] \\ &\cdot G(u_m), \quad n > 1\end{aligned}$$

where

$$G(u_m) = [(2u_m + 1)(u_m - 1)^2, u_m^2(3 - 2u_m)]$$

Again, it is understood that

$$\begin{aligned}\mathbf{P}(u_1, k2, \ldots, kN) \\ = [&\mathbf{P}_{0k2\ldots kN}, \mathbf{P}_{1k2\ldots kN}, \mathbf{T}^1_{0k2\ldots kN}, \\ &\mathbf{T}^1_{1k2\ldots kN}] \cdot F(u_1)\end{aligned}$$

and

$$\begin{aligned}d_m\mathbf{P}(u_1, j2, \ldots, jN) \\ = [\mathbf{T}^m_{0\,j2,\ldots,jN}, \mathbf{T}^m_{1\,j2,\ldots,jN}] \cdot G(u_1)\end{aligned}$$

One-dimensional case: Curve

The F-curve expression for $N = 1$ is

$$\begin{aligned}\mathbf{P}(u_1) &= [\mathbf{P}(0), \mathbf{P}(1), d_1\mathbf{P}(0), d_1\mathbf{P}(1)] \cdot F(u_1) \\ &= [\mathbf{P}_0, \mathbf{P}_1, \mathbf{T}^1_0, \mathbf{T}^1_1] \cdot F(u_1)\end{aligned}$$

Two-dimensional case: Surface

The F-surface expression for $N = 2$ is

$$\begin{aligned}\mathbf{P}(u_1, u_2) \\ = [&\mathbf{P}(u_1, 0), \mathbf{P}(u_1, 1), d_2\mathbf{P}(u_1, 0), \\ &d_2\mathbf{P}(u_1, 1)] \cdot F(u_2)\end{aligned}$$

where

$$\mathbf{P}(u_1, 0) = [\mathbf{P}_{00}, \mathbf{P}_{10}, \mathbf{T}^1_{00}, \mathbf{T}^1_{10}] \cdot \mathbf{F}(u_1)$$

$$\mathbf{P}(u_1, 1) = [\mathbf{P}_{01}, \mathbf{P}_{11}, \mathbf{T}^1_{01}, \mathbf{T}^1_{11}] \cdot \mathbf{F}(u_1)$$

$$d_2\mathbf{P}(u_1, 0) = [\mathbf{T}^2_{00}, \mathbf{T}^2_{10}] \cdot \mathbf{G}(u_2)$$

$$d_2\mathbf{P}(u_1, 1) = [\mathbf{T}^2_{01}, \mathbf{T}^2_{11}] \cdot \mathbf{G}(u_2)$$

(see Figures 6.12, 6.13 and 6.14).

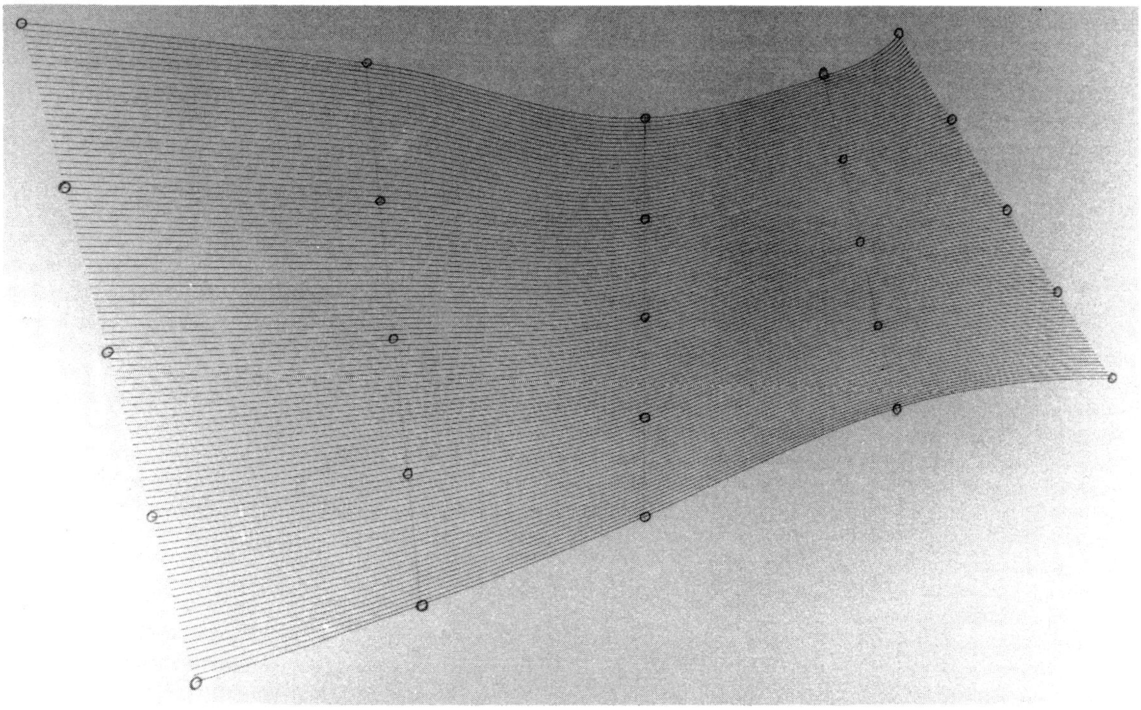

Figure 6.13. Composite F-surface obtained by interpolating through 25 data points [1].

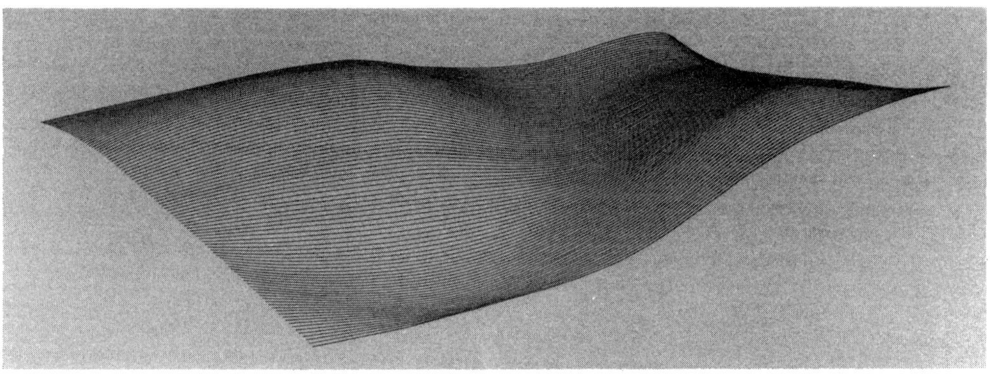

Figure 6.14. A different view of the F-surface in Figure 6.13 [1].

Three-dimensional case: Volume

A final example is the F-Volume where $N = 3$ (Figure 6.15: for convenience, vector names on the parametric cube are the same as those in 4-space),

$$\mathbf{P}(u_1, u_2, u_3) = [\mathbf{P}(u_1, u_2, 0), \mathbf{P}(u_1, u_2, 1), d_3\mathbf{P}(u_1, u_2, 0), d_3\mathbf{P}(u_1, u_2, 1)] \cdot \mathbf{F}(u_3)$$

where

$$\mathbf{P}(u_1, u_2, 0) = [\mathbf{P}(u_1, 0, 0), \mathbf{P}(u_1, 1, 0), d_2\mathbf{P}(u_1, 0, 0), d_2\mathbf{P}(u_1, 1, 0)] \cdot \mathbf{F}(u_2)$$

$$\mathbf{P}(u_1, u_2, 1) = [\mathbf{P}(u_1, 0, 1), \mathbf{P}(u_1, 1, 1), d_2\mathbf{P}(u_1, 0, 1), d_2\mathbf{P}(u_1, 1, 1)] \cdot \mathbf{F}(u_2)$$

$$\mathbf{P}(u_1, 0, 0) = [\mathbf{P}_{000}, \mathbf{P}_{100}, \mathbf{T}^1_{000}, \mathbf{T}^1_{100}] \cdot \mathbf{F}(u_1)$$

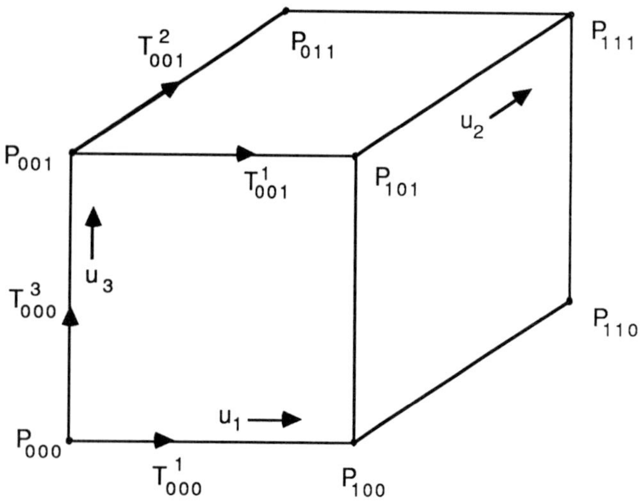

Figure 6.15. Three dimensional parameter domain for 3D F-hyperpatch.

$$\mathbf{P}(u_1, 1, 0) = [\mathbf{P}_{010}, \mathbf{P}_{110}, \mathbf{T}^1_{010}, \mathbf{T}^1_{110}]\cdot\mathbf{F}(u_1)$$

$$\mathbf{P}(u_1, 0, 1) = [\mathbf{P}_{001}, \mathbf{P}_{101}, \mathbf{T}^1_{001}, \mathbf{T}^1_{101}]\cdot\mathbf{F}(u_1)$$

$$\mathbf{P}(u_1, 1, 1) = [\mathbf{P}_{011}, \mathbf{P}_{111}, \mathbf{T}^1_{011}, \mathbf{T}^1_{111}]\cdot\mathbf{F}(u_1)$$

$$d_3\mathbf{P}(u_1, u_2, 0) = [d_3\mathbf{P}(u_1, 0, 0), d_3\mathbf{P}(u_1, 1, 0)]\cdot\mathbf{G}(u_1)$$

$$d_3\mathbf{P}(u_1, u_2, 1) = [d_3\mathbf{P}(u_1, 0, 1), d_3\mathbf{P}(u_1, 1, 1)]\cdot\mathbf{G}(u_1)$$

$$d_3\mathbf{P}(u_1, 0, 0) = [d_3\mathbf{P}(0, 0, 0), d_3\mathbf{P}(1, 0, 0)]\cdot\mathbf{G}(u_1) = [\mathbf{T}^3_{000}, \mathbf{T}^3_{100}]\cdot\mathbf{G}(u_1)$$

$$d_3\mathbf{P}(u_1, 1, 0) = [\mathbf{T}^3_{010}, \mathbf{T}^3_{110}]\cdot\mathbf{G}(u_1)$$

$$d_3\mathbf{P}(u_1, 0, 1) = [\mathbf{T}^3_{001}, \mathbf{T}^3_{101}]\cdot\mathbf{G}(u_1)$$

$$d_3\mathbf{P}(u_1, 1, 1) = [\mathbf{T}^3_{011}, \mathbf{T}^3_{111}]\cdot\mathbf{G}(u_1)$$

$$d_2\mathbf{P}(u_1, 0, 0) = [\mathbf{T}^2_{000}, \mathbf{T}^2_{100}]\cdot\mathbf{G}(u_1)$$

$$d_2\mathbf{P}(u_1, 1, 0) = [\mathbf{T}^2_{010}, \mathbf{T}^2_{110}]\cdot\mathbf{G}(u_1)$$

$$d_2\mathbf{P}(u_1, 0, 1) = [\mathbf{T}^2_{001}, \mathbf{T}^2_{101}]\cdot\mathbf{G}(u_1)$$

$$d_2\mathbf{P}(u_1, 1, 1) = [\mathbf{T}^2_{011}, \mathbf{T}^2_{111}]\cdot\mathbf{G}(u_1)$$

Note: The position **P** and tangent **T** geometry has four coordinates.

Connectivity

Adjacent hyperpatches that share position and tangent data can form a network that is geometrically smooth. Such assemblies should be defined so that adjacent chord vectors along control curves are of relatively uniform length, otherwise distorted interpolation surfaces might arise (i.e. relative point spacing is a consideration). This circumstance usually is not a distraction for most engineering applications. One common and practical method is to assign the minimum (of two) chord lengths to the tangent at a node along a control curve. Alternatives to equally lengthed entrant and exit tangent assignments across nodes exist, but are not dealt with here. These comments apply generally to B-spline and Bézier forms, although the situation might be more subtle, since the simplicity in the input data act as a camouflage, especially when these forms are used for interpolation during interactive design applications. Further information can be found in Lichten [21] or Faux and Pratt [22].

REFERENCES

1 **J. Ferguson** Multivariable curve interpolation, *J. ACM*, Vol 11, 1964, 221–228.

2 **E. F. Carlberg and J. J. Childs** Boeing expands numerical control contouring capability, *Machine and Tool Blue Book*, Vol 58, 1963, 1.

3 **J. Ferguson** *Multivariable curve fitting*, Boeing Co., Seattle, #D222026, 1962. (Also, in expanded form: D2-22504, April 1963.)

4 **J. Ferguson** *Multivariable curve interpolation,* MS thesis, University of Washington, 1963.
5 **G. O. Gellert** Geometric computing, electronic geometry for semi-automated design, *Machine Design,* parts 1/2, March/April 1965, 152–159/94–100.
6 **M. Rowin** *Reversals of F-curves,* Master Dimensions Group, Boeing Co., Seattle, WA, 1964.
7 **J. Ferguson** *Form, characterized in a special class of parametrized curves,* Applied Math. Group, TRW, Inc., CA, Report #3122-31, 1966.
8 **J. Ferguson** *Form, characterized in a special class of parametrized curves – II,* TRW Inc., Report #3122-3-237, 1967.
9 **J. Ferguson and K. Miller** *Characterization of shape in a class of third degree algebraic curves,* TRW Inc., Report #5322-3-5, 1969.
10 **A. R. Forrest** *The twisted cubic curve,* CAD Group Doc. No. 50, University of Cambridge, UK, November 1970.
11 **A. R. Forrest** *Shape classification of the non-rational twisted cubic curve in terms of Bézier polygons,* CAD Group Doc. No. 52, University of Cambridge, UK, 1970.
12 **C. Y. Wang** Shape classification of the parametric cubic curve and parametric B-spline cubic curve. *Comput.-Aided Design,* Vol 13, 1981, 199–206.
13 **J. Jortner** *Curves and Surfaces for Computer Graphics,* PhD thesis, Louisiana State University, 1986.
14 **J. A. Roulier** Bézier curves of positive curvature, *Comput.-Aided Geometric Design,* Vol 5, 1988, 59–70.
15 **J. Ferguson** *Approximation of discretely defined space curves by cubic parametrized segments,* TRW Inc., Report #3122-79, 1966.
16 **J. Ferguson** *Shape preserving parametric cubic curve interpolation,* PhD thesis, University of New Mexico, Albuquerque, 1984.
17 **J. C. Ferguson and S. Pruess** Shape-preserving interpolation by parametric piecewise cubic polynomials, *Comput.-Aided Design,* Vol 23 No 7, 1991, 98–505.
18 **M. Plass and M. Stone** Curve fitting with piecewise parametric cubics, *Computer Graphics,* Vol 1 No 3, July 1983, 229–239.
19 **S. Coons** *Surfaces for computer-aided design of space forms,* Design Division, Mechanical Engineering Department, MIT, January 1964.
20 **A. R. Forrest** *Coons surfaces and multivariable functional interpolation,* CAD Group, University of Cambridge, UK, July 1970.
21 **L. Lichten** *Studies in the Computer-Aided Design of Connected Surfaces,* UCLA PhD Dissertation, 1976.
22 **I. D. Faux and M. J. Pratt** *Computational Geometry for Design and Manufacture,* Ellis Horwood, Chichester, 1979.

ABOUT THE AUTHOR

James C. Ferguson is currently working as a mathematician and numerical geometry consultant: Suite 2582, 1420 N.W. Gilman, Issaquah, WA. His responsibilities include developing algorithms for geographical information systems, 3D computer graphics and model development. He spent the late 1970s and early 1980s at Los Alamos National Lab in New Mexico as a staff scientist working on graphics, modeling and numerical algorithm development for support of nuclear reactor safety and particle transport applications. Ferguson received an Advanced Study Fellowship from Los Alamos to pursue his PhD studies, and finished the degree in 1984. His thesis covered shape-preserving curve interpolation written under Steve Pruess, with guidance from Alex Stone and Richard Allen (all at the University of New Mexico at that time). Significant support was also given by distinguished scientists from Los Alamos, including G. Milton Wing (general analysis), Mike Steuerwalt (numerical analysis), and James Hyman (shape preserving interpolation). Previous years were spent in a variety of research, consulting and industrial settings. Significant among these was TRW Systems in Redondo Beach, CA during the last half of the 1960s. Here he worked for the well-known Applied Mathematics Group directed by Oliver Smith. Much of his work there related to numerical techniques for interpolation, approximation, optimization and contouring. In particular, he produced a complete shape analysis for planar parametric cubics and published these results as internal reports. Ferguson's industrial career began at Boeing Company in Seattle, WA in 1960 after attending Reed College and the University of Washington. In the Master Dimensions Group at Boeing, he developed new curve and surface definition techniques that significantly influenced the mathematics used in the CAD/CAM industry. These results were developed during 1961–62, and were published as internal Boeing documents, as an MS thesis, and as a formal journal paper.

Ferguson's technical interests range from cosmology, to strange attractors, through nanotechnology and molecular biology. He is a member of the American Mathematical Society, the Society of Industrial and Applied Mathematics, the Institute of Electrical and Electronics Engineers, and the Association for Computing Machinery.

7 Sculptured Surface Definition via Blending-Function Methods

William J. Gordon

Preface

I was pleased to be asked by Les Piegl to contribute a chapter to this volume. Although Dr Piegl suggested that I provide a historical overview of the development of surfacing techniques, after much thought I have elected to dwell on my own contributions. Full books could be written on the contributions of my friends and colleagues P. Bézier, C. de Boor, G. Birkhoff and S. A. Coons, with long chapters devoted to the contributions of many others who I will not attempt to enumerate.

This paper is a retrospective personal exposition inasmuch as it represents a summary of the results of my own research in the years 1967–73. It concentrates on CAD/CAM applications, and falls short of a full discussion of these same techniques to finite element and finite difference applications. However, these applications are rather straightforward and can be inferred from references [1] and [2].

General Motors implemented the so-called 'Gordon Surface' in the late 1960s, and it quickly became the Corporate CAD/CAM standard. This paper is my first openly published account of the mathematics which underlie the 'Gordon Surface' technique. The key to this is the proper parametrization of constituent network curves and the use of cubic splines as blending functions.

The body of the paper represents a concrete application of a general theory of multivariate interpolation and approximation. For the details of this, see elsewhere [3].

INTRODUCTION

This paper is concerned with the problem of mathematically interpolating a smooth, free-form ('sculptured') surface through a topologically rectangular network of curves in 3-space. In practice, the curves of the skeletal network are the result of a laborious design process, and the designer simply seeks to obtain a smooth surface which inherits all of the essential characteristics of this curve collection. In other words, the shape of the surface is thought of as being more or less uniquely determined by the given curves; alterations in the surface shape are accomplished by changing the shapes of the input curves.

We wish to emphasize that the curves of the given network may be of any functional form whatsoever, including parametric polynomials, splines, rational functions, or simply 'point-defined' representations obtained, for example, by direct digitization of curves from engineering drawings. Indeed, the key to the practical implemtation of the surfacing techniques to be discussed is that the associated formulas contain arbitrary curve forms which are unspecified until the surface equation is to be evaluated. At that time, the computer program must identify and be able to evaluate the functional form of each curve. In other words, a computer program for the implementation of blended bivariate interpolation should be written so that the univariate functions needed to define the curves of the network are treated as dummy argument functions which are called at the time of program execution.

PARAMETRIC SURFACES AND BLENDED INTERPOLATION

In problems of geometric design, one is generally concerned with parametrically defined surfaces of the general form

$$U(s,t) = \begin{pmatrix} X(s,t) \\ Y(s,t) \\ Z(s,t) \end{pmatrix}$$

in which the three components $X(s,t)$, $Y(s,t)$ and $Z(s,t)$ of the vector function $U(s,t)$ are

The colour plate for this chapter appears between pp. 210 and 211.

Fundamental Developments of Computer-Aided Geometric Modeling
ISBN 012-554765-X

single-valued functions of the two independent variables (parameters) s and t. For each point (s, t) in the parameter domain, there corresponds a unique point on the parametric surface whose Cartesian coordinates are $x = X(s, t)$, $y = Y(s, t)$ and $z = Z(s, t)$. As the parameters s and t vary over their allowable domain, the vector-valued function $\mathbf{U}(s, t)$ generates a surface in Euclidean 3-space. For our purposes, it will suffice to assume that the domain of the variables s and t is the unit square $S = [0, 1] \times [0, 1]$.

In this section we are concerned with the problem of interpolating a free-form surface $\mathbf{U}(s, t)$ through a given skeletal network of curves in space.

A general twisted curve in Euclidean 3-space can be written parametrically as a vector $(x(s), y(s), z(s))$ where s is any parameter which is *monotone* as a function of the arc length along the curve. Suppose we are given two families of intersecting space curves $\{\mathbf{G}_i(t)\}_{i=0}^{M}$ and $\{\mathbf{F}_j(s)\}_{j=0}^{N}$ defined respectively over the constant parameter lines $0 = s_0 < s_1 < \ldots < s_M = 1$ and $0 = t_0 < t_1 < \ldots < t_N = 1$ in the s, t domain $[0, 1] \times [0, 1]$

$$\mathbf{G}_i(t) = \begin{pmatrix} g_i^1(t) \\ g_i^2(t) \\ g_i^3(t) \end{pmatrix}, \quad \mathbf{F}_j(s) = \begin{pmatrix} f_j^1(s) \\ f_j^2(s) \\ f_j^3(s) \end{pmatrix} \qquad (2)$$

where the superscripts 1, 2 and 3 on the functions $g_i(t)$ and $f_j(s)$ represent the x, y and z components, respectively. Since the curves are assumed to intersect, we have the *compatibility conditions*

$$\mathbf{U}_{ij} = \mathbf{G}_i(t_j) = \mathbf{F}_j(s_i) \qquad (3)$$

or, more explicitly,

$$\mathbf{U}_{ij} = \begin{pmatrix} g_i^1(t_j) \\ g_i^2(t_j) \\ g_i^3(t_j) \end{pmatrix} = \begin{pmatrix} f_j^1(s_i) \\ f_j^2(s_i) \\ f_j^3(s_i) \end{pmatrix} = \begin{pmatrix} u_{ij}^1 \\ u_{ij}^2 \\ u_{ij}^3 \end{pmatrix} \qquad (4)$$

Our objective is to construct a surface $\mathbf{U}(s, t)$ which interpolates all $M + N + 2$ of the given space curves, i.e.

$$\mathbf{U}(s_i, t) = \mathbf{G}_i(t) \quad (i = 0, 1, \ldots, M) \qquad (5\text{a})$$

$$\mathbf{U}(s, t_j) = \mathbf{F}_j(s) \quad (j = 0, 1, \ldots, N) \qquad (5\text{b})$$

Clearly, equations (5) will be satisfied if we can construct scalar-valued bivariate functions $U^1(s, t)$, $U^2(s, t)$ and $U^3(s, t)$ which are, respectively, the x, y and z components of the vector-valued function $\mathbf{U}(s, t)$, such that these scalar functions interpolate the respective scalar-valued univariate components of the given space curves $\{\mathbf{G}_i\}$ and $\{\mathbf{F}_j\}$.

Thus, we see that in suitable parametric coordinates s and t, the problem of interpolating a free-form surface through an intersecting skeletal network of space curves can be reduced to the following *scalar-valued* bivariate interpolation problem, in which we drop the superscripts denoting the three coordinate components: *Construct a bivariate function $U(s, t)$ which interpolates the $M + N + 2$ univariate functions $\{g_i\}_{i=0}^{M}$ and $\{f_j\}_{j=0}^{N}$:*

$$U(s_i, t) = g_i(t) \quad (i = 0, 1, \ldots, M) \qquad (6\text{a})$$

$$U(s, t_j) = f_j(s) \quad (j = 0, 1, \ldots, N) \qquad (6\text{b})$$

A general solution to this problem is provided by the following theorem. Much of the remainder of this paper is concerned with interpretations, variations and applications of this result. For a considerably more formal treatment, see [3–5].

Theorem 1 *If $\{\phi_i(s)\}_{i=0}^{M}$ and $\{\psi_j(t)\}_{j=0}^{N}$ are any two sets of functions which satisfy the* **cardinality** *properties*

$$\phi_i(s_k) = \begin{cases} 0 & \text{if } k \neq i \\ 1 & \text{if } k = i \end{cases} \qquad (7\text{a})$$

$$\psi_j(t_\ell) = \begin{cases} 0 & \text{if } \ell \neq j \\ 1 & \text{if } \ell = j \end{cases} \qquad (7\text{b})$$

and if $\{g_i(t)\}_{i=0}^{M}$ and $\{f_j(s)\}_{j=0}^{N}$ are any two collections of functions such that the following **compatibility** *relations are satisfied*

$$g_i(t_j) = f_j(s_i) \quad \text{for all } i \text{ and } j$$

then the bivariate function $U(s, t)$ given by

$$U(s, t) = \sum_{i=0}^{M} g_i(t)\phi_i(s) + \sum_{j=0}^{N} f_j(s)\psi_j(t) - \sum_{i=0}^{M} \sum_{j=0}^{N} u_{ij}\phi_i(s)\psi_j(t) \qquad (8)$$

solves the network interpolation problem, i.e.

$$U(s_k, t) = g_k(t) \quad (k = 0, 1, \ldots, M) \qquad (9\text{a})$$

$$U(s, t_\ell) = f_\ell(s) \quad (\ell = 0, 1, \ldots, N) \qquad (9\text{b})$$

Proof To prove the theorem, it will suffice to verify the conclusions (9a) and (9b). To this end, we evaluate the function $U(s, t)$ along the line $s = s_k$

$$U(s_k, t) = \sum_{i=0}^{M} g_i(t)\phi_i(s_k) + \sum_{j=0}^{N} f_j(s_k)\psi_j(t) - \sum_{i=0}^{M}\sum_{j=0}^{N} u_{ij}\phi_i(s_k)\psi_j(t)$$

By using the relations (7a), this reduces to

$$U(s_k, t) = g_k(t) + \sum_{j=0}^{N} f_j(s_k)\psi_j(t) - \sum_{j=0}^{N} u_{kj}\psi_j(t)$$

and, noting from (3) that $u_{kj} = f_j(s_k)$, we obtain

$$U(s_k, t) = g_k(t) + \sum_{j=0}^{N} u_{kj}\psi_j(t) - \sum_{j=0}^{N} u_{kj}\psi_j(t) = g_k(t)$$

which establishes (9a). By evaluating $U(s, t)$ along the line $t = t_\ell$, we verify (9b) similarly.

It is important to recognize that Theorem 1 is true for *any* compatible families of functions $\{g_i\}$ and $\{f_j\}$, and for *any* sets of functions $\{\phi_i(s)\}_{i=0}^{M}$ and $\{\psi_j(t)\}_{j=0}^{N}$ provided only that they satisfy (7a) and (7b), respectively. Hence, the theorem is a purely algebraic result which makes no assumptions regarding the specific functional forms or analytic properties of the functions appearing in expression (8).

The functions $\phi_i(s)$ and $\psi_j(t)$ in (8) are called *blending-functions* since they play the role of blending together the various curves comprising the network. Although the blending-functions are quite arbitrary, the kinds of functions which immediately come to mind as being practical candidates are:

1. Polynomials (Figure 7.1a)

$$\phi_i(s) = \prod_{k \neq i} \left(\frac{s - s_k}{s_i - s_k} \right) \qquad (10)$$

2. Step functions (Figure 7.1b)

$$\phi_i(s) = \begin{cases} 1 & \text{for } s \in ((s_{i-1} + s_i)/2, (s_i + s_{i+1})/2) \\ 0 & \text{otherwise} \end{cases} \qquad (11)$$

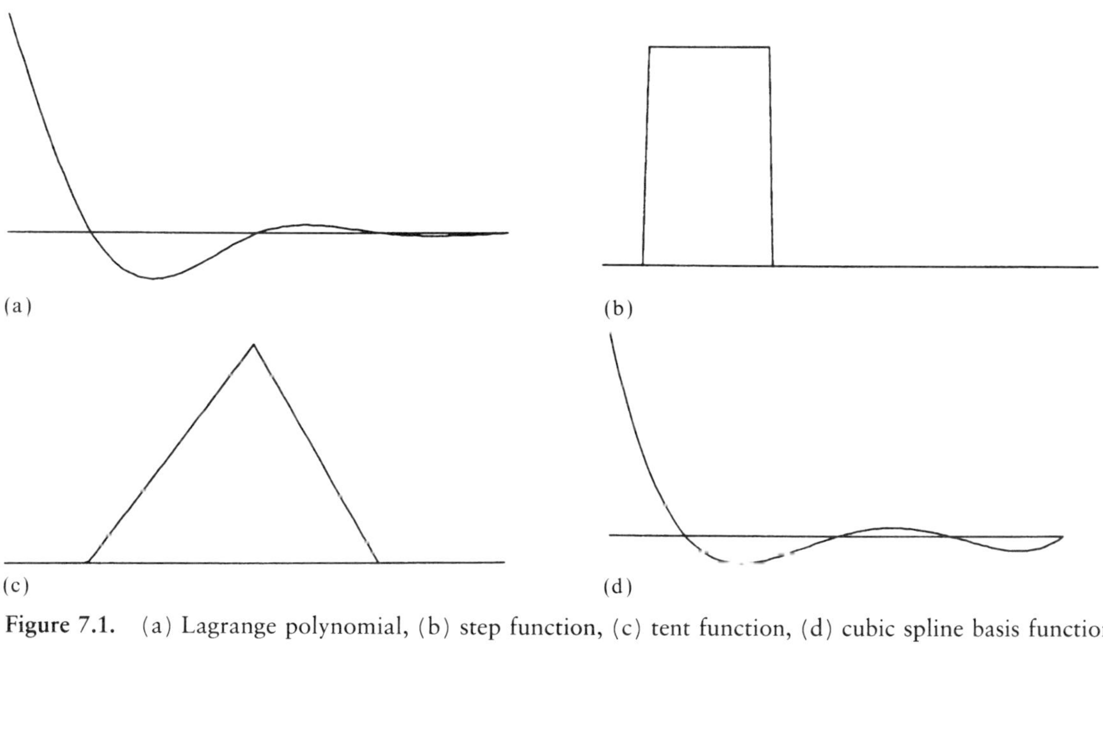

Figure 7.1. (a) Lagrange polynomial, (b) step function, (c) tent function, (d) cubic spline basis function.

3. 'Tent' functions (i.e. piecewise linear) (Figure 7.1c)

$$\phi_i(s) = \begin{cases} (s-s_{i-1})/(s_i-s_{i-1}) & \text{for } s \in (s_{i-1}, s_i) \\ (s_{i+1}-s)/(s_{i+1}-s_i) & \text{for } s \in (s_i, s_{i+1}) \\ 0 & \text{otherwise} \end{cases} \quad (12)$$

4. Piecewise polynomials, e.g. cubic splines (Figure 7.1d) or Hermite-type polynomials (see below).

We note that the surface equation (8) can be interpreted as the sum of three simple surfaces, i.e.

$$U(s, t) = V_1(s, t) + V_2(s, t) - T(s, t) \quad (13)$$

where

$$V_1(s, t) = \sum_{i=0}^{M} g_i(t)\phi_i(s) \quad (14\,a)$$

$$V_2(s, t) = \sum_{j=0}^{N} f_j(s)\psi_j(t) \quad (14\,b)$$

$$T(s, t) = \sum_{i=0}^{M} \sum_{j=0}^{N} u_{ij}\phi_i(s)\psi_j(t) \quad (14\,c)$$

Let us examine each of these three surfaces separately.

Lemma 1 *Let* $\{g_i(t)\}_{i=0}^{M}$ *be any collection of functions defined over the lines* $s = s_i$ $(i = 0, 1, \ldots, M)$, *and let the set of blending-functions* $\{\phi_i(s)\}_{i=0}^{M}$ *satisfy the conditions* (7 a) *of Theorem 1. Then, the surface* $V_1(s, t)$ *given by* (14 a) *interpolates the family of curves* $g_i(t)$ $(i = 0, 1, \ldots, M)$

$$V_1(s_k, t) = g_k(t) \quad (k = 0, 1, \ldots, M) \quad (15)$$

The lemma is easily verified by simply evaluating the function $V_1(s, t)$ along the line $s = s_k$ and using properties (7 a). Of course, an analogous statement applies to the second surface $V_2(s, t)$, i.e. $V_2(s, t)$ interpolates the family of curves $\{f_j(s)\}_{j=0}^{N}$ provided that the blending-functions $\{\psi_j(t)\}_{j=0}^{N}$ have the properties (7 b)

$$V_2(s, t_\ell) = f_\ell(s) \quad (\ell = 0, 1, \ldots, N) \quad (16)$$

Formulas (14 a) and (14 b) are the mathematical equivalents of the drafting technique known as 'surface lofting' which is widely used in the aircraft and shipbuilding industries.

Lemma 2 *If* $\{\phi_i(s)\}_{i=0}^{M}$ *and* $\{\psi_j(t)\}_{j=0}^{N}$ *satisfy* (7 a) *and* (7 b), *then the ('tensor product') function* $T(s, t)$ *interpolates the* $(M+1)\cdot(N+1)$ *scalar values* u_{ij} *given at the mesh-points* (s_i, t_j) $(i = 0, 1, \ldots, M; j = 0, 1, \ldots, N)$

$$T(s_i, t_j) = u_{ij} \quad (i = 0, 1, \ldots, M; j = 0, 1, \ldots, N) \quad (17)$$

Figure 7.2 shows a compatible network of space curves. In Figures 7.3a–d, step functions (and in Figures 7.4a–d tent functions) were used to obtain lofted, tensor product and Gordon surfaces. Notice that the tent-blending substantially improves the tensor product as well as the Gordon surface interpolants. Figures 7.5a–f show cubic spline-blended lofted, tensor product and Gordon surfaces. In Figures 7.5d and 7.5f, the interpolants are compared to the original network of curves (note the difference between the tensor product and the Gordon surface).

Most of the existing literature on bivariate interpolation is concerned with tensor product (alias cross product, Cartesian product, Kronecker product) techniques. We will examine the relation of these product schemes to more general blending-function techniques in the following sections.

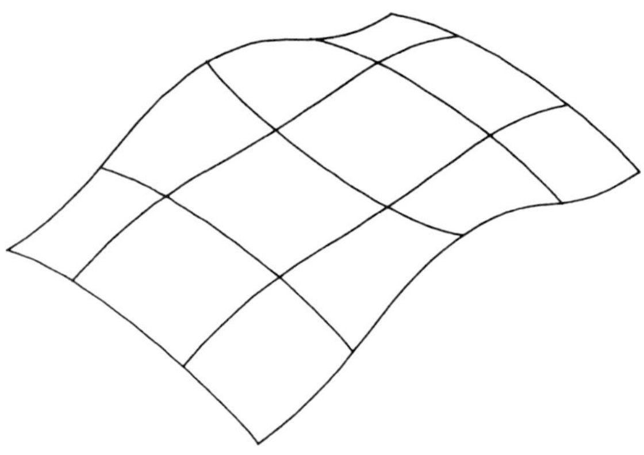

Figure 7.2. Compatible network of curves.

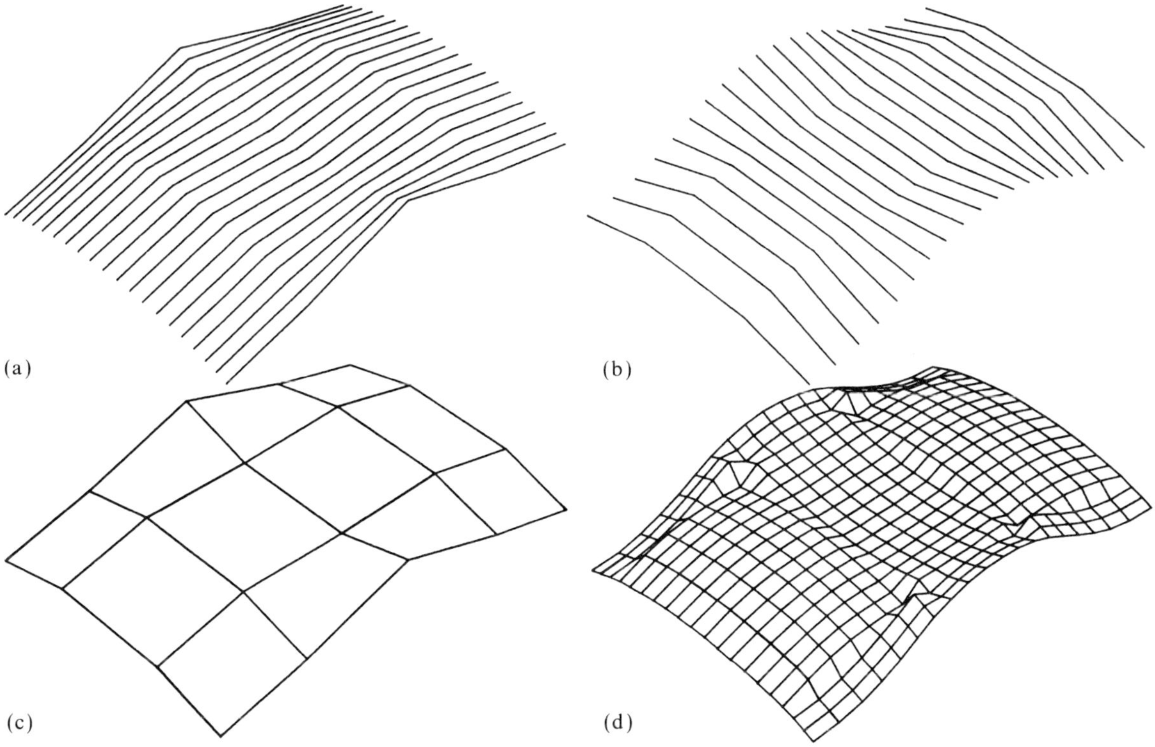

Figure 7.3. Step function blended (a) s-directional lofted surface interpolant, (b) t-directional lofted surface interpolant, (c) tensor product surface interpolant, (d) Gordon surface interplant.

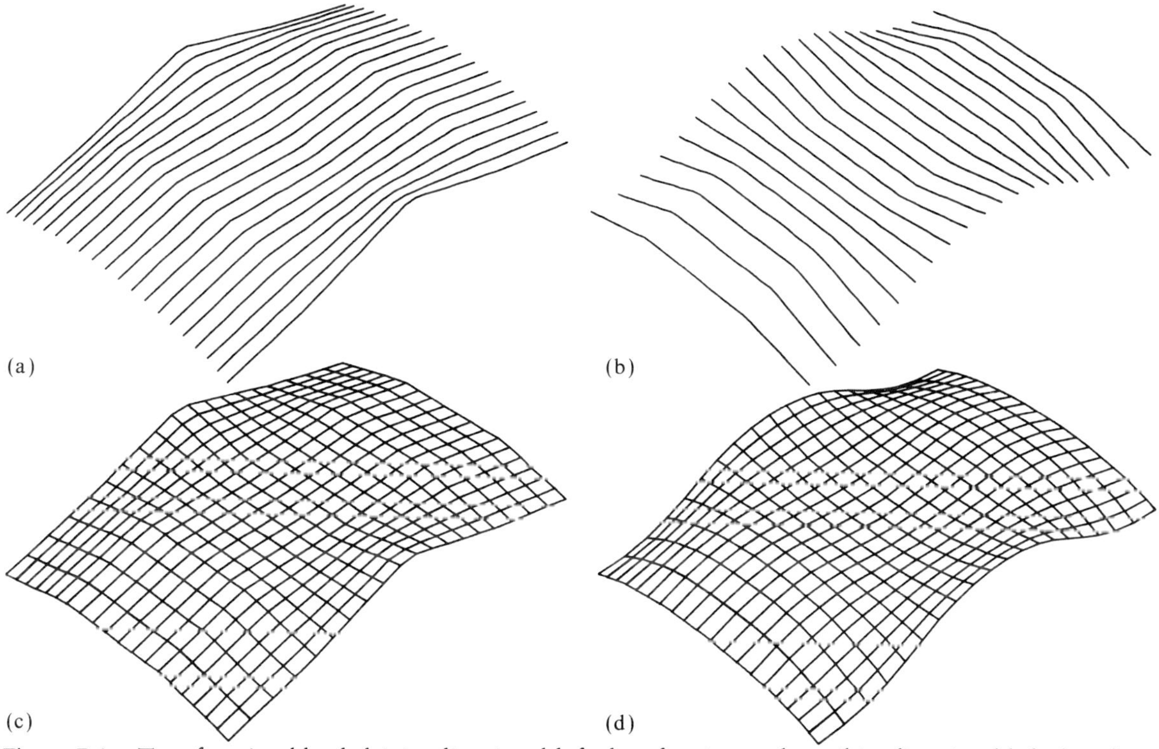

Figure 7.4. Tent function blended (a) s-directional lofted surface interpolant, (b) t-directional lofted surface interpolant, (c) tensor product surface interpolant, (d) Gordon surface interpolant.

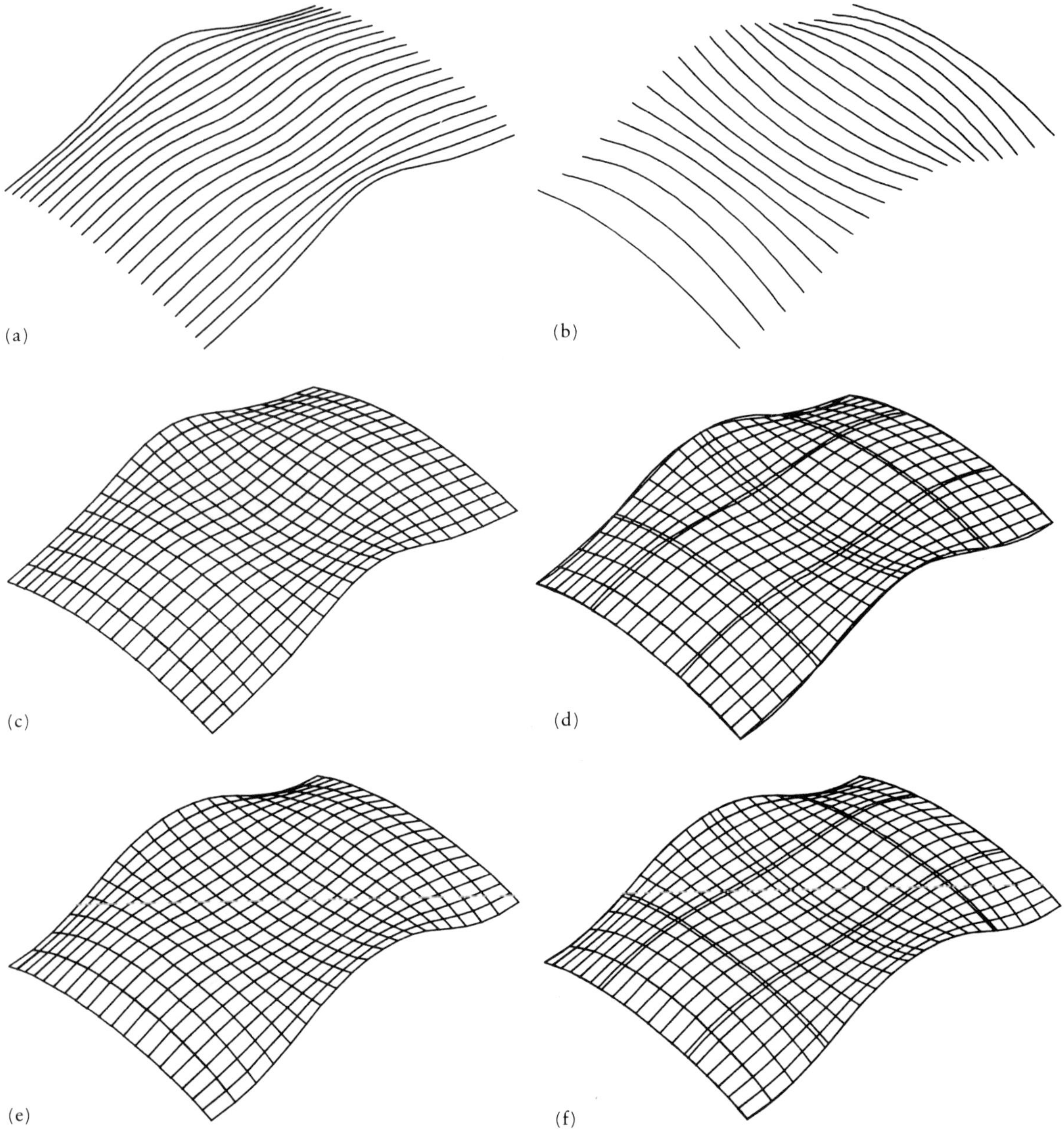

Figure 7.5. Cubic spline-blended (a) s-directional lofted surface interpolant, (b) t-directional lofted surface interpolant, (c) tensor product surface interpolant; (d) tensor product surface compared to the original network of curves; (e) cubic spline-blended Gordon surface interpolant, (f) Gordon surface compared to the original network of curves.

GLOBAL METHODS OF SURFACE BLENDING

In this and the following sections we shall adopt a somewhat different point of view regarding problems of surface interpolation. Namely, we shall assume that the univariate functions $g_i(t)$ and $f_j(s)$ to which we interpolate with the bivariate function $U(s, t)$ have actually been extracted from some (usually hypothetical) function $F(s, t)$. This is the point of view of mathematical *analysis*, where one is primarily concerned with questions involving the error made by approximating the *primitive function* $F(s, t)$ by the *synthesized function* $U(s, t)$. Even though the notion of an underlying primitive function may seem totally irrelevant in problems of free-form sculptured surface *design*, it nevertheless provides a mathematically and notationally more appealing and succinct way of describing techniques of bivariate interpolation and approximation.

To begin, let us paraphrase Theorem 1 in terms of the hypothetical primitive function $F(s, t)$:

Theorem 2 *Let $F(s, t)$ be an arbitrary*[1] *bivariate function and let $\{\phi_i(s)\}_{i=0}^{M}$ and $\{\psi_j(t)\}_{j=0}^{N}$ be two families of blending-functions satisfying* (7a) *and* (7b). *Then the bivariate function $U(s, t)$ given by*

$$U(s,t) = \sum_{i=0}^{M} F(s_i, t)\phi_i(s) + \sum_{j=0}^{N} F(s, t_j)\psi_j(t) - \sum_{i=0}^{M}\sum_{j=0}^{N} F(s_i, t_j)\phi_i(s)\psi_j(t) \quad (18)$$

interpolates $F(s, t)$ along the lines $s = s_i$ $(i = 0, 1, \ldots, M)$ and $t = t_j$ $(j = 0, 1, \ldots, N)$

$$U(s_i, t) = F(s_i, t) \quad (i = 0, 1, \ldots, M) \quad (19\text{a})$$

$$U(s, t_j) = F(s, t_j) \quad (j = 0, 1, \ldots, N) \quad (19\text{b})$$

In the previous section we have already remarked on the enormous freedom one has in choice of the blending-functions. Now, we shall consider two *global* schemes (in contrast to the *local* 'patch surface' methods of the next section) for solving the network interpolation problem. The first of these global methods uses polynomials of the form (10) as blending-functions and the second uses cubic splines. The polynomial scheme has the advantage that the blending-functions are known explicitly, but the disadvantage that the resulting surface may be undulating – a reflection of the well-known characteristic of univariate interpolation by polynomials of fairly high degree. On the other hand, even though the blending-functions of cubic spline type must be computed (i.e. they have no explicit formula), this modest amount of computation is usually rewarded by the fact that the spline-blended surface is, in a sense to be explained below, the 'smoothest' surface obtainable – again a reflection of the well-known smoothness properties of univariate cubic splines.

Consider first the case of a polynomially blended surface. The following result is a special case of a theorem established in [5].

Corollary 1 *If the blending-functions of Theorem 2 are taken to be the polynomials for Lagrange interpolation at the sets of points $\{s_i\}_{i=0}^{M}$ and $\{t_j\}_{j=0}^{N}$, i.e.*

$$\phi_i(s) = \prod_{k \neq i} \left(\frac{s - s_k}{s_i - s_k}\right) \quad (i, k = 0, 1, \ldots, M) \quad (20\text{a})$$

$$\psi_j(t) = \prod_{\ell \neq j} \left(\frac{t - t_\ell}{t_j - t_\ell}\right) \quad (j, \ell = 0, 1, \ldots, N) \quad (20\text{b})$$

then the function $U(s, t)$ given by (18) *is the* **unique** *function which interpolates $F(s, t)$ along the lines $s = s_i$ $(i = 0, 1, \ldots, M)$ and $t = t_j$ $(j = 0, 1, \ldots, N)$ and satisfies the partial differential equation*

$$\frac{\partial^{M+N+2}}{\partial s^{M+1}\partial t^{N+1}} U(s, t) = 0 \quad (21)$$

To verify (21), one uses the implied hypotheses that $\phi_i^{(M+1)}(s) = 0$ and $\psi_j^{(N+1)}(t) = 0$ (since they are polynomials of degree M and N, respectively). More generally (cf. [5]), if the $\{\phi_i(s)\}_{i=0}^{M}$ and $\{\psi_j(t)\}_{j=0}^{N}$ in Theorem 2 are bases for the null spaces of arbitrary ordinary differential operators D_s and D_t, respectively,

[1] At the expense of mathematical rigor, we have throughout this paper omitted precise statements of differentiability hypotheses, since the requisite assumptions are, in all cases, obvious from the subsequent formulae. Hence, by 'arbitrary' we mean sufficiently differentiable for the purpose at hand, but of otherwise unspecified functional form. In Theorem 2, for example, all that is required of $F(s, t)$ is that it be a continuous function.

then the interpolating function $U(s, t)$ given by (18) is in the infinite dimensional null space of the partial differential operator D_sD_t, i.e. $D_sD_t[U] = 0$. In Corollary 1, $D_s = d^{M+1}/ds^{M+1}$ and $D_t = d^{N+1}/dt^{N+1}$.

Corollary 2 *Let $F(s, t)$ be an arbitrary bivariate function, and let $\phi_i(s)$ and $\psi_j(t)$ be as in Corollary 1. Then the tensor product function $T(s, t)$ given by*

$$T(s, t) = \sum_{i=0}^{M} \sum_{j=0}^{N} F(s_i, t_j)\phi_i(s)\psi_j(t) \tag{22}$$

is the unique **bipolynomial** *of degree M in s and N in t which interpolates $F(s, t)$ at the $(M + 1)\cdot(N + 1)$ mesh-points (s_i, t_j)*

$$T(s_i, t_j) = F(s_i, t_j) \quad \begin{pmatrix} i = 0, 1, \ldots, M \\ j = 0, 1, \ldots, N \end{pmatrix} \tag{23}$$

In general, a *tensor product interpolation scheme of the form* (23) can be devised using any two sets of blending-functions $\{\phi_i(s)\}$ and $\{\psi_j(t)\}$ which are bases for finite dimensional linear spaces Φ_M and Ψ_N. For it is easy to show that the product of two such bases (i.e. the set of $(M + 1)\cdot(N + 1)$ functions $\{\phi_i(s)\psi_j(t)\}_{i=0,j=0}^{M,\ N}$) is a basis for the $(M + 1)\cdot(N + 1)$ dimensional tensor product space $\Phi_M \otimes \Psi_N$. The notion of bivariate tensor product interpolation is classic (e.g. Taylor's formula for functions of two variables is of tensor product form, cf. [3]) and serves as the basis for many of the conventional interpolation schemes, such as those discussed in [6–8].

The important distinction between blending-function methods and tensor product interpolation methods cannot be overemphasized: It is that the former interpolates to *arbitrary curves* given along two families of lines $s = s_i$ $(i = 0, 1, \ldots, M)$ and $t = t_j$ $(j = 0, 1, \ldots, N)$, and obviously, therefore, interpolates at the $(M + 1)\cdot(N + 1)$ mesh-points (s_i, t_j). The latter, on the other hand, interpolates *only* at the $(M + 1)\cdot(N + 1)$ mesh-points (s_i, t_j).

As an aside, we note that an interesting interpretation of tensor product formulas such as the bipolynomial (22) is that it is a polynomial blend of polynomial interpolants. To construct the function $T(s, t)$ of (22), we may first construct the $M + 1$ univariate polynomials $p_i(t)$ of degree N which (for fixed $i = 0, 1, \ldots, M$) interpolate to the data points $\{t_j, F(s_i, t_j)\}_{j=0}^{N}$ along the lines $= s_i$. These polynomials are given in terms of blending-functions $\{\psi_j(t)\}$ as

$$p_i(t) = \sum_{j=0}^{N} \alpha_{ij}\psi_j(t) \tag{24}$$

in which the coefficients α_{ij} are precisely the values $F(s_i, t_j)$

$$\alpha_{ij} = F(s_i, t_j) \tag{25}$$

Then, for each fixed value of $j = 0, 1, \ldots, N$, we form the polynomial which *interpolates the $M + 1$ coefficients* α_{ij} $(i = 0, 1, \ldots, M)$ of the polynomials $p_i(t)$. The net result of these constructions is the bipolynomial $T(s, t)$ given by

$$\begin{aligned} T(s, t) &= \sum_{i=0}^{M} \phi_i(s)\left(\sum_{j=0}^{N} \alpha_{ij}\psi_j(t)\right) \\ &= \sum_{i=0}^{M} \sum_{j=0}^{N} \alpha_{ij}\phi_i(s)\psi_j(t) \qquad (26) \\ &= \sum_{i=0}^{M} \sum_{j=0}^{N} F(s_i, t_j)\phi_i(s)\psi_j(t) \end{aligned}$$

which, of course, is the same as the function $T(s, t)$ of (22). This construction is intimately related to the procedure of surface lofting alluded to earlier.

We shall next consider the special case of Theorem 2 in which we take the blending-functions $\{\phi_i(s)\}$ and $\{\psi_j(t)\}$ to be cubic splines. Our purpose here is to summarize the principal results concerning spline-blended surfaces. A complete and formal mathematical development of these notions can be found in [4].

A *cubic spline function* with knots at the internal mesh-points $s_1, s_2, \ldots, s_{M-1}$ of the interval $[s_0, s_M]$ is a cubic polynomial function in each subinterval $[s_i, s_{i+1}]$, and which is continuous in value, slope and curvature across the knots $s = s_i$ $(i = 1, 2, \ldots, M - 1)$. As is well-known, univariate interpolation with cubic splines is possible under a variety of specified conditions (cf. [6]). Of particular importance in applications is 'free-endpoint' interpolation with cubic splines, as described in the following lemma:

Lemma 3 *Let $f(s)$ be an arbitrary function of s defined for $s \in [s_0, s_M]$. Then, there exists a* *unique cubic spline $\phi(s)$ with joints at $s_1, s_2, \ldots, s_{M-1}$ which interpolates $f(s)$ at the $M+1$ points $\{s_i\}_{i=0}^{M}$*

$$\phi(s_i) = f(s_i) \quad (i = 0, 1, \ldots, M) \tag{27}$$

and whose second derivative $\phi''(s)$ vanishes at the endpoints $s = s_0$ and $s = s_M$

$$\phi''(s_0) = \phi''(s_M) = 0 \tag{28}$$

Moreover, $\phi(s)$ is given explicitly by

$$\phi(s) = \sum_{i=0}^{M} f(s_i)\phi_i(s) \tag{29}$$

where the set $\{\phi_i(s)\}_{i=0}^{M}$ is the unique basis of cubic splines with the properties that

$$\phi_i(s_k) = \begin{cases} 0 & \text{if } i \neq k \\ 1 & \text{if } i = k \end{cases}$$
$$\phi_i''(s_0) = \phi_i''(s_M) = 0 \tag{30}$$

for $i = 0, 1, \ldots, M$.

A very important characteristic of cubic splines is their so-called 'minimum norm' property [6], which states that among all twice continuously differentiable functions which interpolate to $f(s)$ at the points $\{s_i\}_{i=0}^{M}$, the cubic spline (29) is the unique function which minimizes the functional $J(u)$ given by

$$J(u) = \int_{s_0}^{s_M} (u''(s))^2 \, ds \tag{31}$$

This functional measures the linearized *strain energy* of the function $u(s)$ considered as a thin flexible beam subjected to mechanical constraints equivalent to those of (27) and (28). If the total linearized strain energy $J(u)$ is used as a measure of the overall (i.e. global) smoothness of interpolation, then the cubic spline must be said to be the 'smoothest' function which interpolates $f(s)$ at the points $s = s_i$. Because of their *global* smoothness, and because cubic splines are everywhere continuous in value, slope and curvature, they are an excellent class of functions for practical applications – especially when as in many design problems, smoothness of the curves and surfaces is an important consideration.

To apply the notion of spline-blending to the network interpolation problem of the second section, we also need to consider a second collection of cubic splines $\{\psi_j(t)\}_{j=0}^{N}$ with the analogous properties that

$$\psi_j(t_\ell) = \begin{cases} 0 & \text{if } j \neq \ell \\ 1 & \text{if } j = \ell \end{cases} \tag{32}$$

$$\psi_j''(t_0) = \psi_j''(t_N) = 0 \tag{33}$$

By taking the blending-functions in Theorem 2 to be the cubic splines $\{\phi_i(s)\}$ and $\{\psi_j(t)\}$ just described, we obtain the following result, the details of the proof of which may be found in [4].

Corollary 3 *If the blending functions $\{\phi_i(s)\}_{i=0}^{M}$ and $\{\psi_j(t)\}_{j=0}^{N}$ of Theorem 2 are the 'free-endpoint' cubic splines defined above, then the spline-blended function $U(s,t)$ given by* (18) *is the unique function which satisfies* (19a) *and* (19b) **and** *minimizes the functional*

$$\mathscr{J}(U) = \int_{s_0}^{s_M} \int_{t_0}^{t_N} (U^{(2,2)}(s,t))^2 \, ds\, dt \tag{34}$$

where $U^{(2,2)} = \partial^4 U/\partial s^2 \partial t^2$. In this sense, the spline-blended surface is the 'smoothest' surface which interpolates the curve network.

Without going into the details of its construction, we want also to consider a related problem of surface interpolation which is of substantial practical interest (e.g. in smoothly concatenating a surface with an already existing surface) and which is easily solved by spline-blending. This is the problem of interpolating a surface through a network of curves *and* satisfying specified boundary conditions on the normal derivatives ('surface tangents') of the surface along the exterior perimeter of the rectangular domain $\mathscr{R} = [s_0, s_M] \times [t_0, t_N]$. (Note that we are not given normal derivatives across internal mesh-lines, but only on the perimeter of $\mathscr{R}$.) We can state this result somewhat heuristically as follows. For a more precise statement and formulas, see section 2 of [4].

Corollary 4 *Let $F(s,t)$ be an arbitrary function defined for $(s,t) \in \mathscr{R} = [s_0, s_M] \times [t_0, t_N]$. Then, with suitable cubic spline blending-functions, one can construct a spline-blended surface*

$U(s, t)$ which interpolates to the $M + N + 2$ plane sections of $F(s, t)$

$$U(s_i, t) = F(s_i, t) \quad (i = 0, 1, \ldots, M) \tag{35 a}$$

$$U(s, t_j) = F(s, t_j) \quad (j = 0, 1, \ldots, N) \tag{35 b}$$

and *interpolates to the first normal derivative $\partial F/\partial n$ on the perimeter of $\mathcal{R}$, i.e.*

$$\left.\begin{aligned} U_s(s_0, t) &= F_s(s_0, t) \\ U_s(s_M, t) &= F_s(s_M, t) \end{aligned}\right\} \quad \text{for all } t \in \{t_0, t_N\}$$
$$\tag{36}$$
$$\left.\begin{aligned} U_t(s, t_0) &= F_t(s, t_0) \\ U_t(s, t_N) &= F_t(s, t_N) \end{aligned}\right\} \quad \text{for all } s \in \{s_0, s_M\}$$

Moreover, this surface is everywhere continuous in value, slope and curvature in both the s and t directions,[2] and is the smoothest surface which satisfies (35 a), (35 b) and (36) in the sense that it minimizes the functional $\mathcal{J}(U)$ of (34).

In the special case where all of the curves of the given network are themselves univariate cubic splines with joints at $\{s_i\}_{i=1}^{M-1}$ and $\{t_j\}_{j=1}^{N-1}$, the spline-blended surface $U(s, t)$ reduces to a *bicubic spline* function of the type first studied by de Boor [8]. Such bicubic splines are tensor product functions and can be expressed in the form of formula (22).

In the case where only four boundary curves are given (i.e. no curves are specified over lines internal to the square S but just on the four edges of S) a spline-blended surface reduces to a *cubically-blended Coons' patch* to be studied in the next section (formula (45)). In that case, the cubic splines apropos to Corollary 4 are just ordinary cubic polynomials, *viz.* the polynomials of (46) below.

Figure 7.6 shows a parametric surface to be interpolated using the methods discussed above. Figure 7.7 shows just the boundaries of this surface to be used to obtain lofted, tensor product and Gordon surface interpolants applying Lagrange polynomials as blending-functions

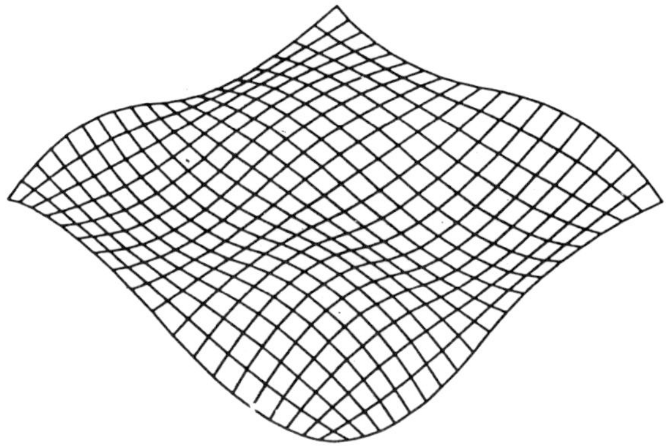

Figure 7.6. Parametric surface (primitive function) to be interpolated.

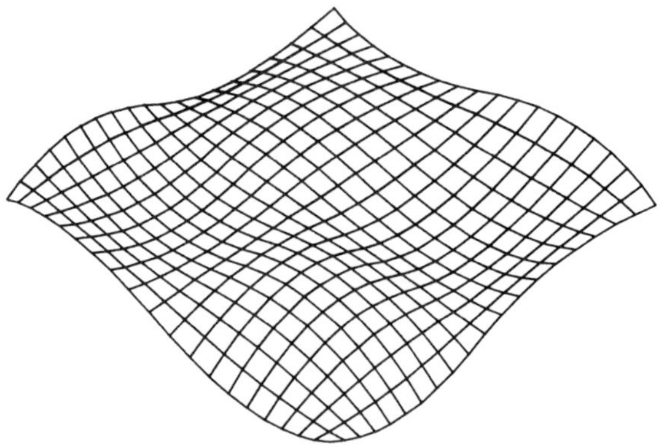

Figure 7.7. Special network of curves (the boundaries) sampled from the surface in Figure 7.6.

(Figures 7.8a–d). In Figure 7.9 we sampled more curves to obtain a better approximation of the original surface. Figures 7.10a, 7.11a, 7.12a and 7.13a show cubic spline-blended lofted, tensor product and Gordon surface interpolants, whereas in Figures 7.10b, 7.11b, 7.12b and 7.13b, these interpolants are compared to the original surface to check the value of the interpolation. Note that the tensor product surface is the worst, whereas the Gordon surface is the best interpolant.

The method of cubic spline surface blending just outlined was developed by the author in the late 1960s, and has for more than two decades been the standard technique used at General Motors for the mathematical representation of exterior automobile panels. As indicated in the first section, there are certain preprocessing and

[2] Assuming, of course, that the constituent univariate functions $F(s_i, t)$, $F(s, t_j)$, $F_s(s_0, t)$, etc., in equations (35) and (36) are continuous in value, slope and curvature.

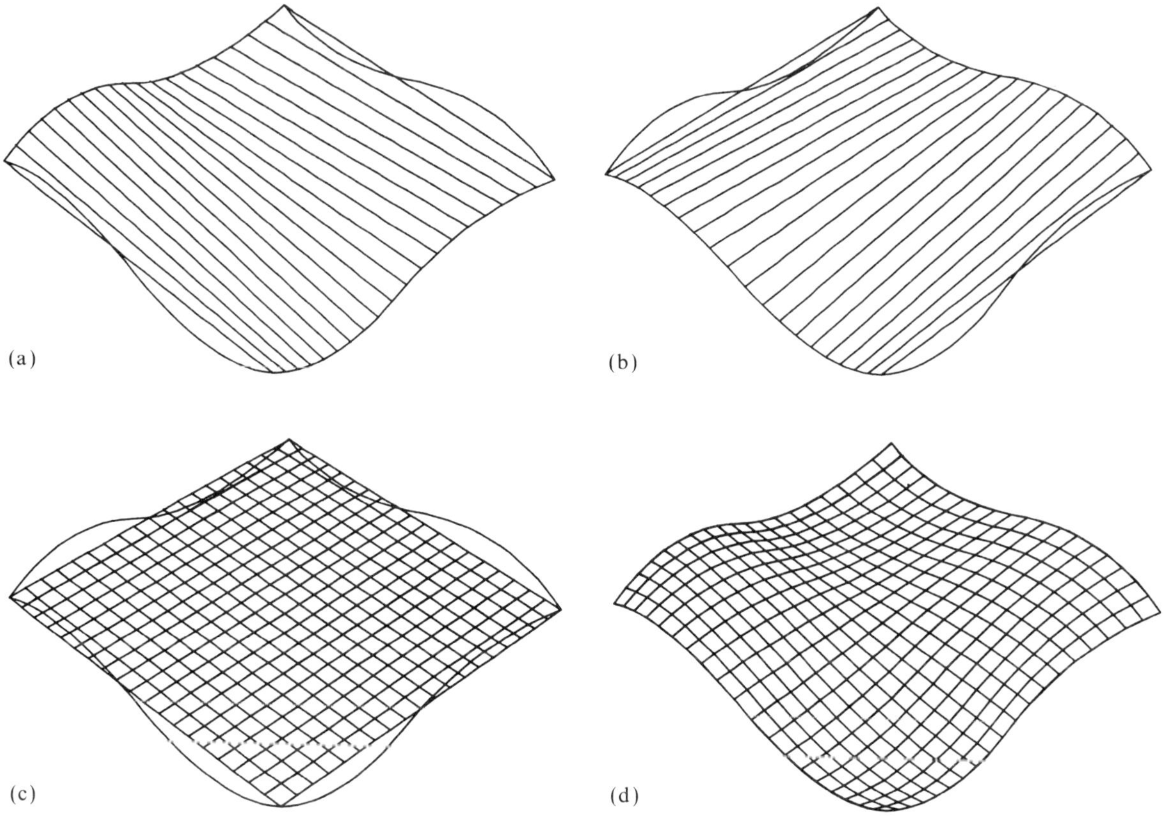

Figure 7.8. Lagrange polynomial blended (a) s-directional, (b) t-directional lofted surface interpolants, (c) tensor product, (d) Gordon surface interpolants.

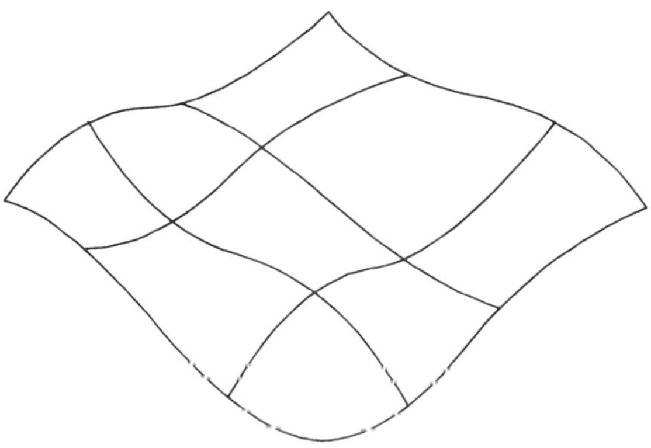

Figure 7.9. Curve network sampled from the surface in Figure 7.6.

data management problems which must be handled by auxiliary subroutines in any such practical implementation. For instance, in practice one is seldom given a network of curves which is *a priori* properly parametrized. Thus, a preprocessor is required which outputs two families of curves $\{G_i(s)\}$ and $\{F_j(t)\}$ which are properly parametrized by the normalized variables s and t. Another difficulty which often arises is that *a priori* the curves of the skeletal network may not be compatible, as assumed by equation (3) and the subsequent interpolation formulas. Hence, one requires a preprocessing operator which 'meshes' a skeletal network in which $G_i(t_j) \neq F_j(s_i)$ by moving each curve a minimal amount to force compatibility. Unfortunately, space does not allow a full discussion of these or other practical details in the present paper.

It should also be remarked that the very same methods described above for the mathematical definition of 3D *shell structures* are also applicable to *planar mapping problems* in which one is interested in constructing univalent (one-to-one) maps of the unit square onto an arbitrary (bounded and simply connected) region $\mathscr{R}$ in the x,y-plane. Such mapping

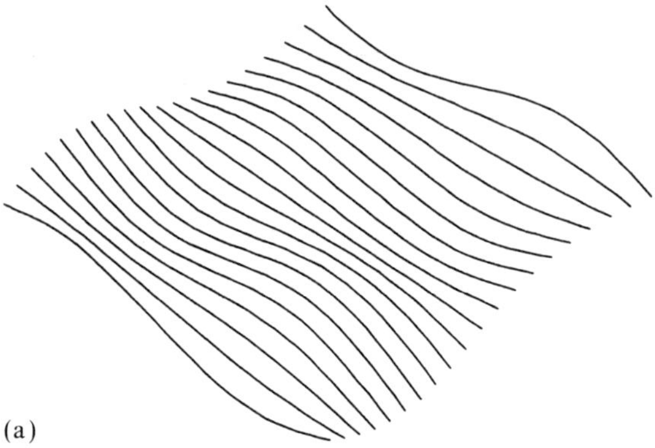

(a)

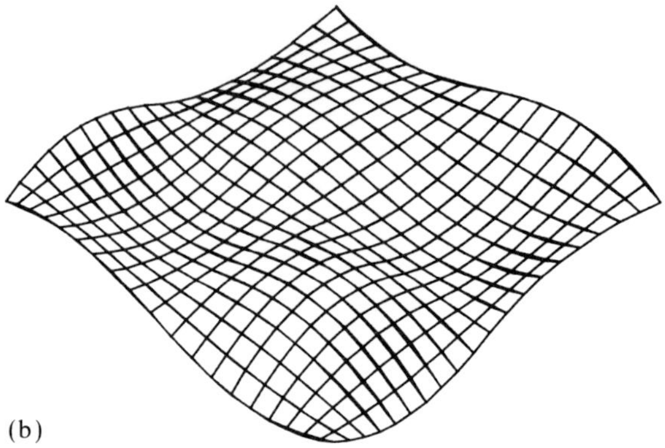

(b)

Figure 7.10. (a) Cubic spline blended s-directional lofted surface interpolant, (b) lofted surface in (a) compared to the original surface in Figure 7.6.

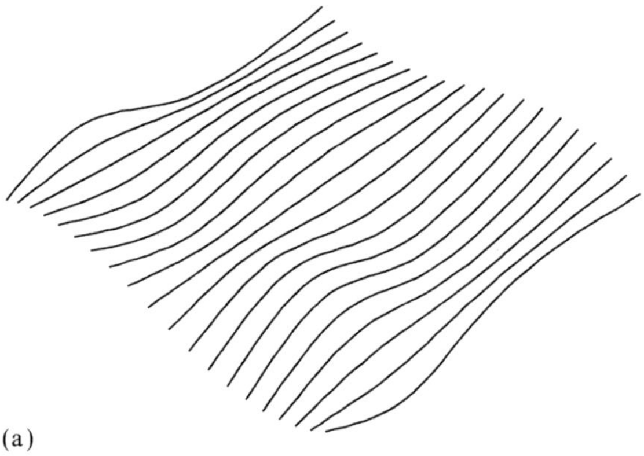

(a)

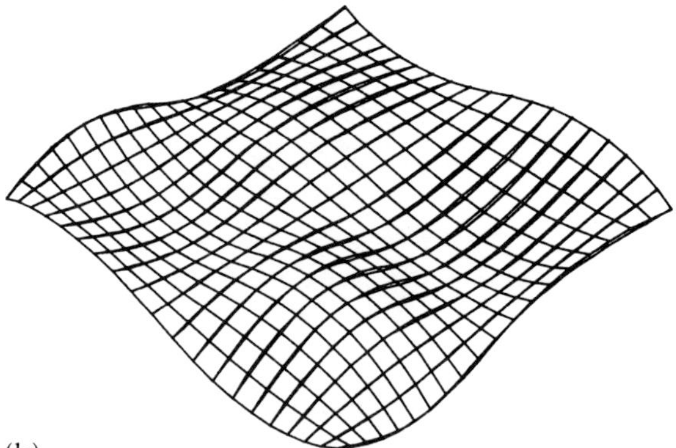

(b)

Figure 7.11. (a) Cubic spline blended t-directional surface interpolant, (b) lofted surface in (a) compared to the original surface in Figure 7.6.

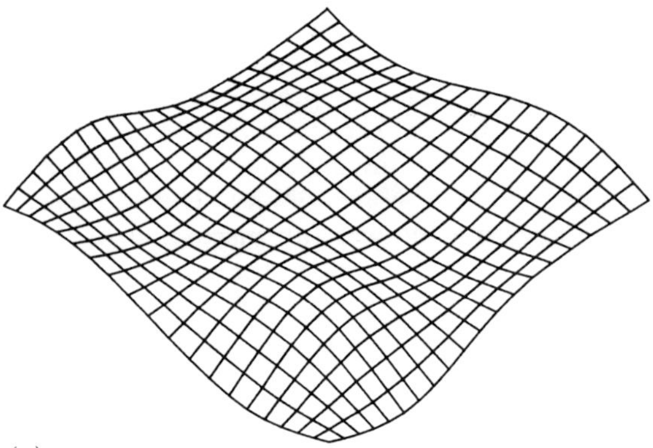

(a)

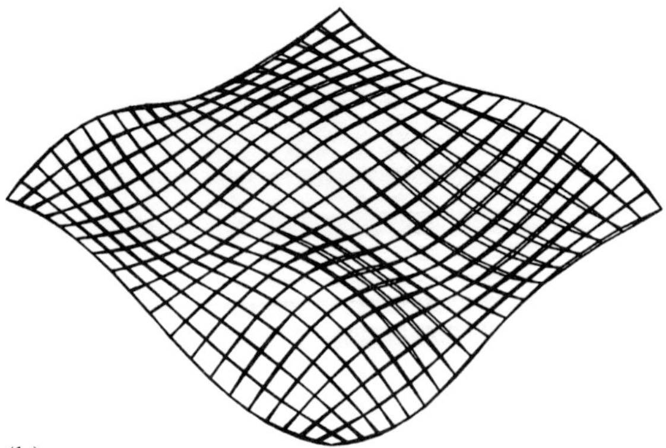

(b)

Figure 7.12. (a) Cubic spline blended tensor product surface interpolant, (b) tensor product surface in (a) compared to the original surface in Figure 7.6.

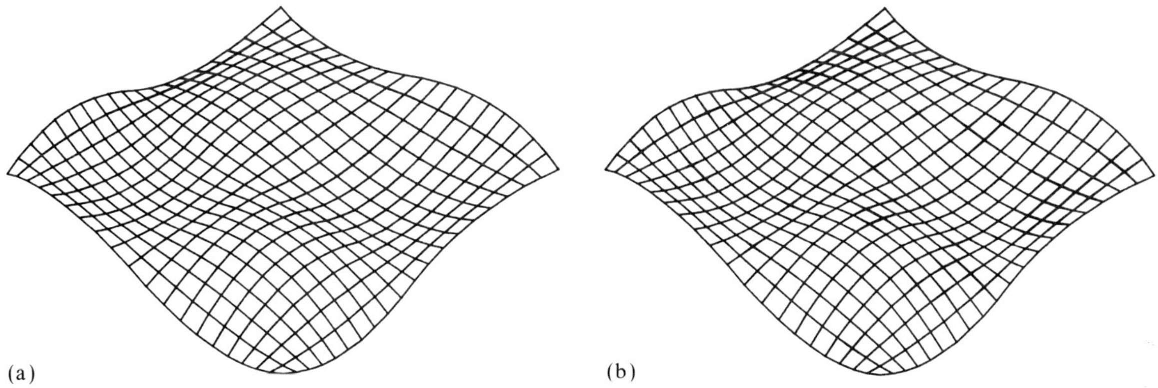

(a) (b)

Figure 7.13. (a) Cubic spline blended Gordon surface interpolant, (b) Gordon surface in (a) compared to the original surface in Figure 7.6.

problems and their relation to the geometric aspects of the finite element method were explored by Gordon and Hall [1].

LOCAL METHODS OF SURFACE INTERPOLATION – THE COONS PATCHES

Both of the 'surfacing techniques' (i.e. bivariate interpolation schemes) considered in the previous section – Lagrange-type polynomial blending and spline blending – are global methods in the sense that a perturbation of any curve in the network will, at least to a small degree, affect the entire surface. Thus, such surfacing methods are sometimes referred to as 'elastic', a term first suggested by Coons. This is because the polynomials and the splines used there have as their 'support' the entire interval $[s_0, s_M]$ (or $[t_0, t_N]$ as the case may be). In other words, the functions $\phi_i(s)$ are non-zero in every subinterval $[s_k, s_{k+1}]$ of the parameter range $[s_0, s_M]$. Since they automatically provide smooth transitions across mesh-lines and use only information which is normally available to the designer, such global methods are usually better suited to problems of computer-aided design of *smooth* surfaces than are the local patch methods discussed in this section. In contrast, the local methods described below require *ad hoc* techniques for estimating surface tangents across patches or cross derivatives (e.g. $U^{(1,1)} \equiv U_{st}$) at interior corners – concepts with which the traditionally trained designer is unfamiliar.

However, whether one uses local or global methods of surface development or surface description is so dependent upon the specific application and the criteria of smoothness and differentiability which must be met (not to mention the considerations of computer cost) that we hesitate to be overly definitive in our recommendations.

By a 'patch method' or local method of surface interpolation we mean a method for which the blending-function $\phi_i(s)$ vanishes identically outside some neighborhood of the point $s = s_i$ $(i = 0, 1, \ldots, M)$, and similarly for the blending-functions $\{\psi_j(t)\}$. For example, the 'tent functions' of formula (12) are of this type, since $\phi_i(s) \equiv 0$ for all s outside the interval $[s_{i-1}, s_{i+1}]$. So also are the step functions of (11). Patch type surfacing methods are sometimes referred to as 'plastic', since local perturbations in the curve network do not propagate throughout the entire surface.

The local schemes discussed here (i.e. the Coons schemes) are most conveniently studied by using the unit square $S = [0, 1] \times [0, 1]$ as the normalized domain of definition. This is easy to arrange, since any rectangular parameter domain $R_{ij} = [s_i, s_{i+1}] \times [t_j, t_{j+1}]$ can be affinely mapped onto the unit square by the transformations $s \mapsto \alpha_1 s + \alpha_2$ and $t \mapsto \beta_1 t + \beta_2$ where $\alpha_1 = 1/(s_{i+1} - s_i)$, $\alpha_2 = s_i/(s_i - s_{i+1})$, and $\beta_1 = 1/(t_{j+1} - t_j)$, $\beta_2 = t_j/(t_j - t_{j+1})$.

The methods developed by the late Steven Coons for surface interpolation were first described by him in [9], and later by others [3–5, 10, 11]. They solve the following class of problems: Along the four edges of the unit square one is given the values and the first $p-1$ normal derivatives of a bivariate function $F(s,t)$, and the problem is to construct a surface $U(s,t)$ which interpolates these conditions. In [9], Coons showed how to *recursively* construct the solutions to this class of problems for arbitary $p = 1, 2, \ldots$, using polynomials of degree $2p-1$ as blending-functions. Later, Birkhoff and Gordon [11] provided a *characterization* of these surfaces by proving the following result:

Theorem 3 *The Coons surface $U(s,t)$ which interpolates to $F(s,t)$ and the first $p-1$ normal derivatives $\partial^{p-1}F/\partial n^{p-1}$ on the boundary of $[0,1] \times [0,1]$ is the* **unique** *solution to the generalized 'draftsman's equation'*

$$\frac{\partial^{4p}}{\partial s^{2p}\partial t^{2p}} U(s,t) = 0 \tag{37}$$

The error $F - U$ in approximating F by its interpolant U is given by

$$F(s,t) - U(s,t) = \int_0^1 \int_0^1 G_{2p}(s;\xi)G_{2p}(t;\eta)F^{(2p,2p)}(\xi,\eta)\,\mathrm{d}\xi\,\mathrm{d}\eta \tag{38}$$

where $G_{2p}(s;\xi)$ and $G_{2p}(t;\eta)$ are Green's functions for the ordinary differential operators $\mathrm{d}^{2p}/\mathrm{d}s^{2p}$ and $\mathrm{d}^{2p}/\mathrm{d}t^{2p}$. This error is bounded by (*cf.* [11])

$$|F(s,t) - U(s,t)| \leqslant \frac{1}{(2p!)^2} \max_{(s,t)\in S} |F^{(2p,2p)}(s,t)| \tag{39}$$

The simplest Coons surface is the *linearly blended* function of the following corollary. Formula (41) is easily derived from the general network solution given by (18) by taking $M = N = 1$, $s_0 = t_0 = 0$ and $s_1 = t_1 = 1$ and using the *linear blending-functions*

$$\begin{aligned} \phi_0(s) &= 1 - s & \psi_0(t) &= 1 - t \\ \phi_1(s) &= s & \psi_1(t) &= t \end{aligned} \tag{40}$$

Corollary 5 *Let $F(s,t)$ be an arbitrary bivariate function defined for $(s,t) \in [0,1] \times [0,1]$. The function*

$$\begin{aligned} U(s,t) = {} & (1-s)F(0,t) + sF(1,t) \\ & + (1-t)F(s,0) + tF(s,1) \\ & - (1-s)(1-t)F(0,0) \\ & - (1-s)tF(0,1) \\ & - s(1-t)F(1,0) - stF(1,1) \end{aligned} \tag{41}$$

is the unique function which satisfies the boundary conditions

$$\begin{aligned} U(0,t) &= F(0,t) \\ U(1,t) &= F(1,t) \\ U(s,0) &= F(s,0) \\ U(s,1) &= F(s,1) \end{aligned} \tag{42}$$

and satisfies the partial differential equation

$$\frac{\partial^4}{\partial s^2 \partial t^2} U(s,t) = 0 \tag{43}$$

Note that the last group of four terms in (41) is the four-parameter *tensor product bilinear* function

$$\begin{aligned} T(s,t) = {} & (1-s)(1-t)F(0,0) \\ & + (1-s)tF(0,1) \\ & + s(1-t)F(1,0) + stF(1,1) \end{aligned} \tag{44}$$

which interpolates to $F(s,t)$ at the *four corners* of the unit square.

The original constructions given by Coons [9] use the notion of *correction surfaces* to satisfy normal derivative interpolation conditions. The procedure is to first construct the blinearly blended surface, (41) above, which interpolates to the values $F(0,t)$, $F(1,t)$, $F(s,0)$ and $F(s,1)$ on the perimeter of the square, and to subtract the normal derivatives of this first surface from the given normal derivative conditions. Then, one uses cubic blending functions to interpolate a surface which vanishes on the boundary and whose normal derivatives 'correct' for the error in the derivatives of the first surface. The process continues in this way until after p stages one has constructed the functions $U_0(s,t)$, $U_1(s,t)$, $\ldots$, $U_{p-1}(s,t)$, the sum of which provides the complete solution to the original problem. This procedure is precisely the bivariate analog to performing $2p-1$ degree univariate Hermite (two endpoint) interpolation to conditions $f(0)$, $f'(0)$, ..., $f^{(p-1)}(0)$ and $f(1)$, $f'(1)$, ..., $f^{(p-1)}(1)$ by first constructing a straight line through $f(0)$ and $f(1)$, then constructing a cubic correction for first derivatives, quintic correction for second, etc. Whereas this is a valuable and quite instructive point of view, it is needlessly complicated for practical applications, since one can straightaway construct the complete solution, as we indicate below.

For example, the next corollary gives the explicit solution to the problem of interpolating to F and its normal derivatives using cubic polynomial blending-functions. Explicit formulas for related schemes of higher order can be found in [5].

Corollary 6 *Let $F(s, t)$ be arbitrary, and define $U(s, t)$ by*

$$\begin{aligned} U(s,t) = {} & \phi_0 F(0,t) + \phi_1 F(1,t) \\ & + \phi_2 F_s(0,t) + \phi_3 F_s(1,t) \\ & + \psi_0 F(s,0) + \psi_1 F(s,1) \\ & + \psi_2 F_t(s,0) + \psi_3 F_t(s,1) \\ & - \phi_0\psi_0 F(0,0) - \phi_0\psi_1 F(0,1) \\ & - \phi_0\psi_2 F_t(0,0) - \phi_0\psi_3 F_t(0,1) \\ & - \phi_1\psi_0 F(1,0) - \phi_1\psi_1 F(1,1) \\ & - \phi_1\psi_2 F_t(1,0) - \phi_1\psi_3 F_t(1,1) \\ & - \phi_2\psi_0 F_s(0,0) - \phi_2\psi_1 F_s(0,1) \\ & - \phi_2\psi_2 F_{st}(0,0) - \phi_2\psi_3 F_{st}(0,1) \\ & - \phi_3\psi_0 F_s(1,0) - \phi_3\psi_1 F_s(1,1) \\ & - \phi_3\psi_2 F_{st}(1,0) - \phi_3\psi_3 F_{st}(1,1) \end{aligned} \tag{45}$$

in which the blending functions are the cardinal functions for univariate cubic Hermite interpolation

$$\begin{aligned} \phi_0(s) &= s^2(2s-3)+1 \\ \phi_1(s) &= -s^2(2s-3) \\ \phi_2(s) &= s(s-1)^2 \\ \phi_3(s) &= s^2(s-1) \\ \psi_0(t) &= t^2(2t-3)+1 \\ \psi_1(t) &= -t^2(2t-3) \\ \psi_2(t) &= t(t-1)^2 \\ \psi_3(t) &= t^2(t-1) \end{aligned} \tag{46}$$

Then, the function $U(s, t)$ interpolates to $F(s, t)$ and the first normal derivative of $F(s, t)$ on the boundary of the unit square, i.e. $U(s, t)$, satisfies conditions (42) **and** *satisfies*

$$\begin{aligned} U_s(0,t) &= F_s(0,t) \\ U_s(1,t) &= F_s(1,t) \\ U_t(s,0) &= F_t(s,0) \\ U_t(s,1) &= F_t(s,1) \end{aligned} \tag{47}$$

Moreover, U is the unique function which interpolates F and $\partial F/\partial n$ and satisfies the eighth order hyperbolic partial differential equation

$$\frac{\partial^8}{\partial s^4\,\partial t^4} U(s,t) = 0 \tag{48}$$

The last group of 16 terms in (45), that is, the function $T(s, t)$

$$\begin{aligned} T(s,t) = {} & \phi_0\psi_0 F(0,0) + \phi_0\psi_1 F(0,1) \\ & + \phi_0\psi_2 F_t(0,0) + \phi_0\psi_3 F_t(0,1) \\ & + \phi_1\psi_0 F(1,0) + \phi_1\psi_1 F(1,1) \\ & + \phi_1\psi_2 F_t(1,0) + \phi_1\psi_3 F_t(1,1) \\ & + \phi_2\psi_0 F_s(0,0) + \phi_2\psi_1 F_s(0,1) \\ & + \phi_2\psi_2 F_{st}(0,0) + \phi_2\psi_3 F_{st}(0,1) \\ & + \phi_3\psi_0 F_s(1,0) + \phi_3\psi_1 F_s(1,1) \\ & + \phi_3\psi_2 F_{st}(1,0) + \phi_3\psi_3 F_{st}(1,1) \end{aligned} \tag{49}$$

is the unique 16-parameter *bicubic Hermite* interpolant to values of F, F_s, F_t and F_{st} at the four corners of the unit square. As with most *tensor product* type methods of interpolation, the original source of this scheme is shrouded in the folklore of approximation theory. During the past two decades, however, it has received a great deal of attention because of its simplicity and its usefulness in CAD/CAM applications. In the same sense that the local Coons' patches are a special case of the global spline-blended surfaces described in the previous section, so too the local finite dimensional scheme of bicubic Hermite interpolation is a special instance (i.e. no knots) of de Boor's global scheme of bicubic spline interpolation [8].

It is important that one fully appreciate the basic distinction between the cubically blended Coons' patch (45) *and the bicubic polynomial of* (49). The former interpolates to eight *arbitrary univariate functions* (i.e. $F(0, t)$, $F(1, t)$, etc., in formula (45)), whereas the latter interpolates only to 16 *scalar values* given at the corners of the square. The Coons' surface reduces to the bicubic Hermite function in the very special case where all eight of the functions $F(0, t)$, . . ., $F_t(s, 1)$ in (45) are cubic polynomials. This is precisely the special case which is implemented in many CAD/CAM systems and mistakenly referred to as a Coons' patch.

In applying the Coons' techniques to the problem of surface interpolation to a curve network, one encounters a number of practical difficulties associated with the joining of adjacent patches. By using $2p - 1$ degree polynomials for blending-functions over each subrectangle $R_{ij} = [s_i, s_{i+1}] \times [t_j, t_{j+1}]$, adjacent patches turn out to be joined together with continuity of derivatives up to and including order $p - 1$. For example, by using the transformed version of formula (41) for interpolating to the four curves over the boundary of the subrectangle R_{ij}, one obtains a composite surface which is continuous only in value in going from one patch to another. With cubic blending (i.e. formula (45) transformed to R_{ij}) the patches are connected with continuity of value and slope, but not curvature. However, implementation of formula (45), or a higher order Coons' surface, requires some knowledge of the behavior of the normal derivative (i.e. 'surface tangent') along the perimeter of *each* subrectangle $R_{ij} = [s_i, s_{i+1}] \times [t_j, t_{j+1}]$ and, in practice, *this information is seldom available to a designer.* This difficulty has been treated by a number of special *ad hoc* techniques which have been described to the author in private communications; however, none of these seems very satisfactory.

Moreover, continuity properties alone cannot insure satisfactory 'smoothness' of surface interpolation. As we implied in the previous section, the intuitive notion of smoothness is a combination of local properties (e.g. the degree of differentiability) and global properties, such as gross undulations of the surface. Hence, even if one can devise mathematically acceptable methods for estimating normal derivatives, the resulting surface may still have such aesthetically objectionable features as 'flat spots', 'waves', etc.

Spline-blending, on the other hand, provides a very 'smooth surface', but circumvents all of the derivative estimation problems associated with the local patch methods. For example, by using cubic splines as blending-functions we obtain a solution to the network interpolation problem which is *continuous in value, slope and curvature* across the boundaries of the subrectangles R_{ij}, and which is also *globally smooth* in the sense of minimizing the functional (34). Equally important, we need not estimate the normal derivatives along the edges of interior subrectangles R_{ij}; and, we can either use or not use normal derivative information on the boundary of $\mathscr{R} = [s_0, s_M] \times [t_0, t_N]$ – depending upon whether or not it is an explicit part of the problem specification (see [4]).

The above-mentioned advantages of (cubic) spline-blended surfaces over Coons' patches are especially important when, as in automobile exterior design, surface smoothness and tight design tolerances are factors of overriding importance. Nevertheless, there are many applications in which this is *not* the case and for which the previously mentioned *ad hoc* methods for obtaining the requisite derivative information along patch boundaries and at corners do produce acceptably smooth and accurate surface representations. In such cases, the local patch methods are to be recommended inasmuch as they are computationally simpler and, hence, surface evaluations are more efficient.

One should note, however, that each patch of a cubic spline-blended surface (i.e. each surface segment determined by the values of s and t in a subrectangle $R_{ij} = [s_i, s_{i+1}] \times [t_j, t_{j+1}]$) is (*a posteriori*) of the *same functional form* as a cubically blended Coons' patch. This observation follows from the uniqueness result of equations (37) and (48) and the fact that cubic splines are merely piecewise cubic polynomials concatenated with C^2 continuity from segment to segment. Thus, for surfaces consisting of a great many patches which will be evaluated, manipulated, sliced into sections, used to prepare NC cutting tapes, etc., we have found that it is often worth the computational expense of *a posteriori converting* from the global database representation associated with cubic spline blending to the equivalent local, patch-by-patch representation using the ordinary cubic Hermite cardinal functions of (46). This conversion from global to equivalent local representation of spline-blended surfaces is totally analogous to the well-known and computationally efficient procedure of *a posteriori* representing a univariate cubic spline with knots at $s_1, s_2, \ldots, s_{M-1}$ as a series of M cubic polynomial segments with spans $[s_i, s_{i+1}]$.

In this chapter I have tried to bring out the distinctions and the relationships between local and global blending-function and tensor product

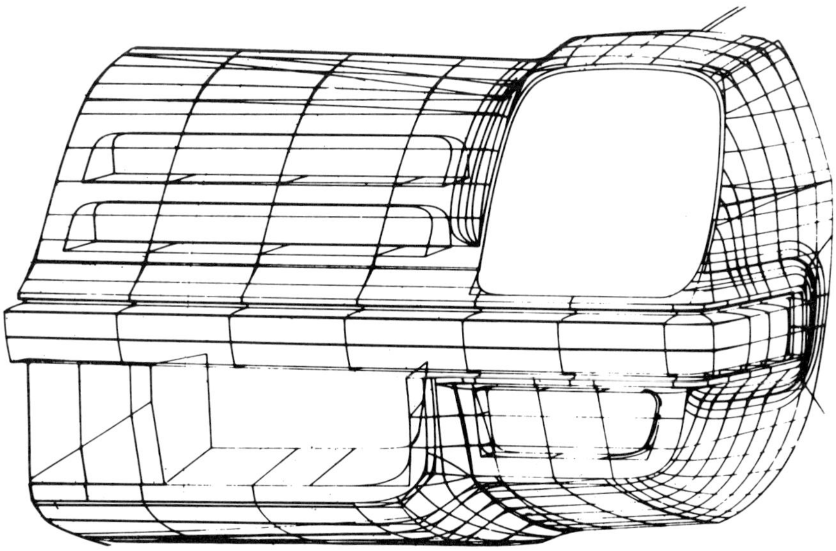

Figure 7.14. Design of the fascia of a bumper using Gordon surfaces (Courtesy of General Motors Design Staff).

surface interpolation methods; and, guided by my own experience with CAD/CAM systems techniques in use at General Motors and other industrial organizations, to provide guidelines in determining which of the various methods are best suited to one's particular needs. As final examples Figure 7.14 shows the design of a 'fascia' (the bumper area of a car) where Gordon surfaces suited the best to satisfy various requirements, whereas in Plate 9 the latest GM design car called 'Buick Wild Cat' can be seen designed by uniform B-splines.

REFERENCES

1 **W. J. Gordon and C. A. Hall** Construction of Curvilinear Coordinate Systems and Applications to Mesh Generation, *International J. Numer. Methods on Engineering*, Vol 7, 1973, 461–477.

3 **W. J. Gordon** Distributive Lattices and the Approximation of Multivariate Functions, *Proceedings of the Symposium on Approximation with Special Emphasis on Spline Functions*, University of Wisconsin, May 5–7 1969, 223–277.

4 **W. J. Gordon** Spline-Blended Surface Interpolation Through Curve Networks, *J. Math. Mech.* Vol 18 No 10, 1969, 931–952.

5 **W. J. Gordon** Blending-Function Methods of Bivariate and Multivariate Interpolation and Approximation, *SIAM J. Numerical Analysis*, Vol 8, 1971, 158–177.

6 **J. H. Ahlberg, E. N. Nilson and J. L. Walsh** *The Theory of Splines and Their Applications*, Academic Press, New York, 1967.

7 **A. C. Ahlin** A Bivariate Generalization of Hermite's Interpolation Formula, *Math. Comp.*, Vol 18, 1964, 264–273.

8 **C. de Boor** Bicubic Spline Interpolation, *J. Math. Phys.*, Vol 41, 1962, 212–218.

9 **S. A. Coons** *Surfaces for Computer-Aided Design of Space Forms*, Project MAC, Design Div., Dept. of Mech. Engineering, MIT, 1964. (Revised to MAC-TR-41, June, 1967. Available through CFSTI, Sills Building, 5285 Port Royal Road, Springfield, VA 22151.)

10 **D. V. Ahuja and S. A. Coons** Geometry for Construction and Display, *IBM Systems Journal*, Vol 7 No 3–4, 1968, 188–205.

11 **G. Birkhoff and W. J. Gordon** The Draftsman's and Related Equations, *J. of Approximation Theory*, Vol 1, 1968, 199–208.

ABOUT THE AUTHOR

William J. Gordon received his Doctorate of Science in Applied Mathematics from Brown University in 1965. His undergraduate work in mathematics and physics was done at the University of Pittsburgh. Following Brown, he joined the research staff of General Motors.

P. J. Davis, G. Birkhoff, Michael Stirling and H. L. Garabedian were very influential in his post-doctoral development and early research.

By 1967, H. Burchard at GM had satisfactorily solved the problem of smooth interpolation to curve data extracted from a 3D clay model of a full-size automobile. The logical next step was to develop a fool-proof method of passing a smooth surface through a network of curves in Euclidean 3-space. Based upon the work of C. de Boor and S. A. Coons, in 1967 Gordon successfully developed and implemented into GM's CAD/CAM system the interpolatory technique known as 'spline-blending'. The practical, parametric implementation of this technique – which is the subject of the present paper – is termed the 'Gordon Surface'.

Dr Gordon has been Professor of Mathematics and Computer Science at Drexel University in Philadelphia for the past ten years. In addition to General Motors, he worked for Boeing and Westinghouse Corporations, and served on the faculties of the Universities of Detroit, Syracuse and Utah. From 1976–78 he served as Liaison Scientist for the U.S. Office of Naval Research in London. He is currently in semi-retirement from Drexel.

8 Coons' Patches and Convex Combinations

Robert E. Barnhill

INTRODUCTION AND EARLY HISTORY

Computer Aided Geometric Design (CAGD) is the representation and approximation of curves, surfaces, and volumes in an interactive computer graphics environment. CAGD may be considered to have been launched by two pioneers: S. A. Coons and P. Bézier. Coons' surfaces [1] and Bézier's surfaces [2, 3] each consists of a network of 'patches' which have a rectilinear topology. Coons' patches match exactly certain information, whereas Bézier's patches have the different flavor that some data are matched exactly and the rest are approximated. Thus Coons' patches are a form of interpolation, and Bézier's patches are a form of approximation, which corresponds to a high level of Davis' dichotomy of interpolation and approximation [4]. This paper will concentrate on Coons' patches. Bézier patches are described by their creator in Chapter 2 of this volume.

Both Coons and Bézier were working in engineering environments when they discovered their surface models. In order for mathematicians to analyze their methods, the underlying mathematical structures had to be recognized. Gordon [5] discovered that Coons' patches have the powerful underlying algebraic structure of forming distributive lattices. Gordon described Coons' patches as Boolean sums of lower-dimensional 'projectors' which were themselves interpolants to lower-dimensional information. He also created 'Gordon surfaces', which he describes in Chapter 7 of this volume.

A generalization of Coons' original patches was necessary for those situations in which four-sided topology is not applicable. Barnhill *et al.* [6] initiated triangular Coons' patches for the case of arbitrarily located information. These triangular patches have a more complicated data structure through which they solve the more complex problem of interpolation to more general data. The generalization to general triangles was accomplished (but not published) by Little [7].

Some personal history: I was introduced to Coons' patches by Bill Gordon during a summer at the General Motors Research Labs in 1969. I introduced these surface methods into my graduate classes at the University of Utah thereafter. In 1971/72 I met John Gregory during a sabbatical year at Brunel University. In 1974 John Gregory and Lois Mansfield both visited Utah for several months, and 'Gregory's Square' (see below) emerged. Also in 1974, Rich Riesenfeld and I created the term 'Computer Aided Geometric Design' and hosted a corresponding conference at Utah [8]. CAGD has boomed ever since.

We now turn to Coons' patches *per se*.

RECTANGULAR COONS' PATCHES

Bilinearly blended Coons' patches

Construction

Coons' patches were created to solve the problem of interpolation to position and tangent information in networks of curves with a rectilinear structure [1], as in Figure 8.1. Gordon [5] recognized the Boolean structure of Coons' patches. We can use Gordon's result to quickly write the bilinearly blended Coons' patch as follows: The most used interpolant in science is univariate linear interpolation. If we denote the linear interpolation operator by P_1, then linear interpolation to a 'primitive' function $f(t)$ on $[0, 1]$ is given by

$$(P_1f)(t) = (1 - t)f(0) + tf(1)$$

See Figure 8.2 for the corresponding geometry. For free form curves in parametric form, we think of the function f as the vector function consisting of the coordinate functions.

Historically, bivariate interpolants have been built up from univariate interpolants, frequently by means of tensor products. For example, with

The colour plates for this chapter appear between pp. 210 and 211.

Fundamental Developments of Computer-Aided Geometric Modeling
ISBN 012-554765-X

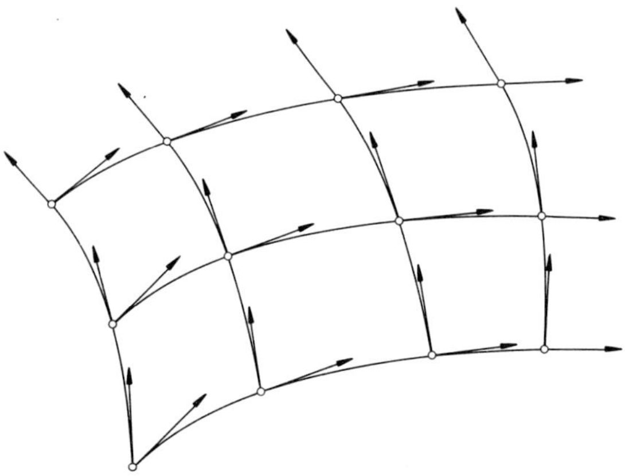

Figure 8.1. Position and tangent data in networks of curves with a rectilinear structure.

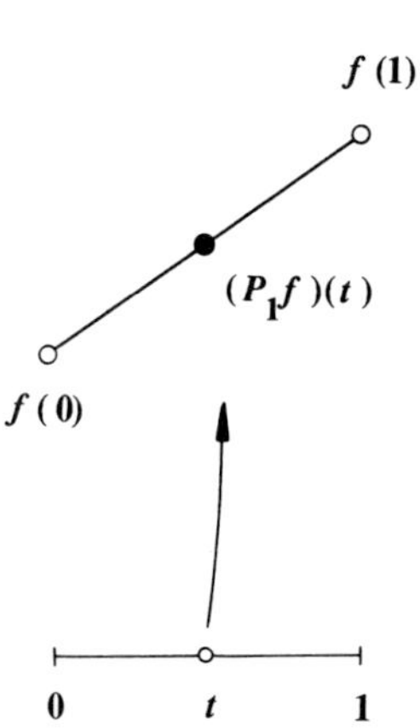

Figure 8.2 Parametric univariate linear interpolation.

$F = F(u, v)$, given the data $F(0, 0)$, $F(0, 1)$, $F(1, 0)$ and $F(1, 1)$, the following bilinear function interpolates to these data:

$$BF(u, v) = (1-u)(1-v)F(0,0) + (1-u)vF(0,1) + u(1-v)F(1,0) + uvF(1,1)$$

$BF(u, v)$ is linearly ruled along parallels to the uv-axes. $BF(u, v)$ is the tensor product of univariate linear interpolation in the variables u and v (see Figure 8.3):

$$P_1P_2F = BF$$

Coons' generalization of interpolation to four points is to interpolate to four curves of data. (Gordon evocatively named this 'transfinite interpolation'.

Now, S. A. Coons [1] developed an interpolating surface as follows: Suppose we are given two curves in space, each of which corresponds to a fixed value of one of the parameters, say v. We form the *lofted* surface (or *lofting interpolant*) by using the appropriate linear operator as before, e.g.

$$P_2F = (P_2F)(u, v) \qquad (1)$$

is a linear interpolant in v (notice that the subscript indicates the interpolating variable).

$$P_2F = (1-v)F(u, 0) + vF(u, 1) \qquad (2)$$

Also, notice that $F(u, 0)$ and $F(u, 1)$ are curves in the two planes $v = 0$ and $v = 1$. This is also called a *linearly ruled surface* (see Figure 8.4).

Next, suppose that instead of two boundary curves, we have four boundary curves, arranged so that the endpoints match to form a 'rectangular' surface. Now, examine the difference of surface F and the lofting interpolant P_2F. (Here, we assume that $F = F(u, v)$.) It is apparent that the lofting interpolant is exact along $v = 0$ and $v = 1$. Along $u = 0$ and $u = 1$,

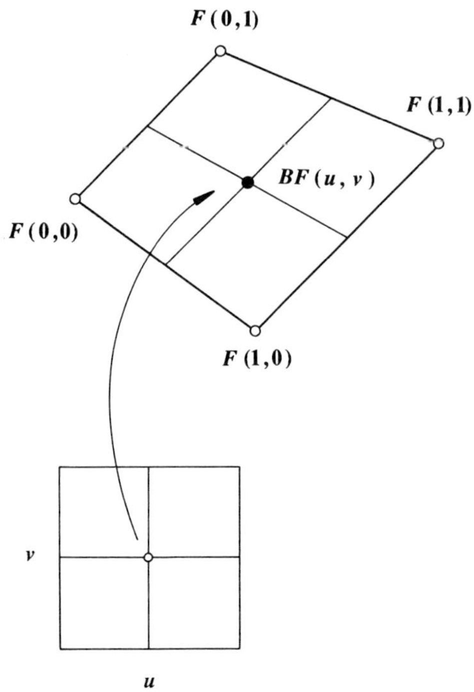

Figure 8.3. Bilinear interpolation to positional data at four corners.

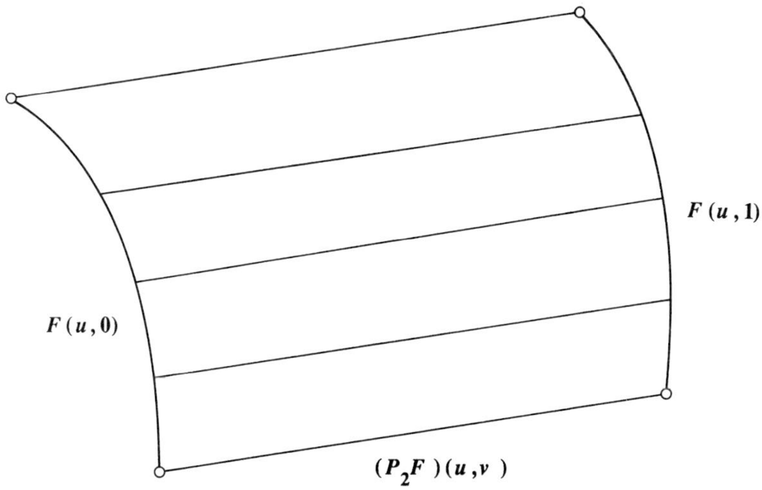

Figure 8.4. The lofting interpolant P_2F defined in equation (2).

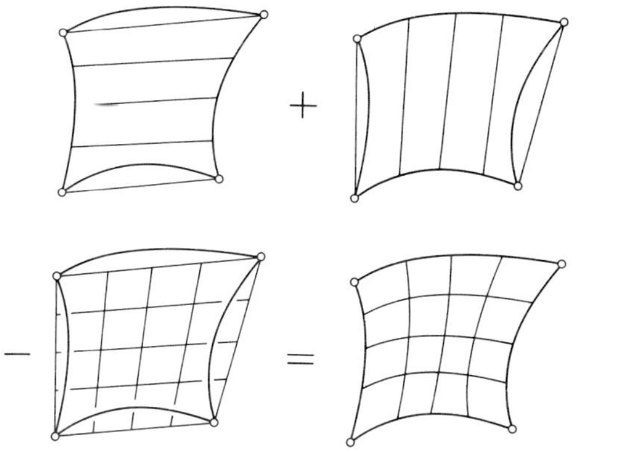

Figure 8.5. Geometry corresponding to the algebra of the Boolean sum defined in equation (3).

we have the curves minus the linear interpolant. We can then apply the same procedure to these remainder curves by forming $P_1(F - P_2F)$. We find that the desired surface is

$$\begin{aligned} P_2F + P_1(F - P_2F) &= P_2F + P_1F - P_1P_2F \\ &= P_1F + P_2F - P_1P_2F \\ &= (P_1 + P_2 - P_1P_2)F \\ &= (P_1 \oplus P_2)F \qquad (3) \end{aligned}$$

$$P_1 \oplus P_2 = P_1 + P_2 - P_1P_2$$

is called a *Boolean sum*. This type of interpolant is known as a *Coons' patch* (see Figure 8.5).

Boolean sum properties can be related via a lattice structure. For example, if we draw the lattice structure of the relation $\subseteq$, we have

$$\begin{array}{ccc} & A \cap B & \\ \swarrow & & \searrow \\ A & & B \\ \searrow & & \swarrow \\ & A \cup B & \end{array}$$

This type of analysis was used by Gordon [5] to explain certain properties of Coons' patches.

Notice that with a Coons' patch we are interpolating whole curves of information. Interpolants of this type are called *transfinite*.

Discretization of boundary curves

Suppose that instead of using the actual boundary curves (or because they are not actually available), we use a discretization as follows. Let

$$\tilde{F}(u, 0) \equiv (1 - u)F(0, 0) + uF(1, 0) \qquad (4)$$

with similar formulae for the other boundary curves. We leave it as an exercise to show that $(P_1 \oplus P_2)\tilde{F} = P_1P_2F$, i.e. we recapture the bilinear interpolant.

Other discretizations are possible. If, for example, we use a quadratic interpolant to the boundary curve for $u = 0, \frac{1}{2}, 1$, and similarly for the other curves, we would get the *quadratic discretization of the bilinearly blended Coons' patch*.

The points on the domain which determine the discretization used form what is called a *stencil*. This arises in finite element methods as well. Originally, these were found to have higher accuracy than expected, and so came to be called 'serendipity elements'.

The question of smoothness conditions can also be considered, but for now we simply state that across patch boundaries, the surface is in the class C^0 if the boundary curves match, but it is not in the class C^1. We will discuss this more later.

Consider the remainder function

$$\begin{aligned} RF &= F - (P_1 \oplus P_2)F \\ &= (I - (P_1 \oplus P_2))F \end{aligned} \tag{5}$$

where I = the identity operator. Now, let

$$\mathscr{R}_i \equiv I - P_i, \quad i = 1, 2$$

Then

$$\begin{aligned} I - (P_1 \oplus P_2) &= I - (P_1 + P_2 - P_1P_2) \\ &= (I - P_1) - P_2(I - P_1) \\ &= (I - P_1)(I - P_2) \\ &= \mathscr{R}_1\mathscr{R}_2 \end{aligned} \tag{6}$$

Thus, the error of the Boolean sum is the product of the errors of the two lofting interpolants, e.g. if the P_i's are linear interpolants, then $\mathscr{R}_i = O(h^2)$ and $\mathscr{R}_1\mathscr{R}_2 = O(h^4)$ (h is a 'step length'). This explains the serendipitous nature of the Boolean sum.

Polynomial precision sets

Consider the precision set of P_1. When is $P_1(u^iv^j) = u^iv^j$? Since P_1 is linear in u, we have $P_1(u^iv^j) = u^iv^j$ for $i = 0, 1$. What if $F(0, v) = f(v)$? We get $P_1F = f(v)$, clearly. In fact, the precision set is $f_1(v) + f_2(v)u$, and for polynomials it is $a + bu$. If we graph this in ij-space, we get nodes at $i = 0, 1, j = 0, 1, 2, \ldots$. Then, for P_2, we get nodes at $j = 0, 1, i = 0, 1, 2, \ldots$. Finally, for $P_1 \oplus P_2$, we get the union of the above two sets. Notice that the 'first' point not in the set is $(2, 2)$, which corresponds to u^2v^2. This implies $O(h^4)$ accuracy, as determined before.

The generalization to higher order is obvious.

Compatibility conditions

Suppose that the four corner values $F(0, 0)$, $F(0, 1)$, $F(1, 0)$ and $F(1, 1)$ are not uniquely defined. This frequently arises when the data come from experimental measurements which are independent. The Boolean sum is the following:

$$\begin{aligned} (P_1 \oplus P_2)F = {} & (1 - u)F(0, v) + uF(1, v) \\ & + (1 - v)F(u, 0) + vF(u, 1) \\ & - [(1 - u)P_2F(0, v) + uP_2F(1, v)] \end{aligned}$$

What does $P_2F(0, v)$ mean?

$$\begin{aligned} P_2F(0, v) &\equiv \lim_{u \to 0} P_2F(u, v) \\ &= \lim_{u \to 0} [(1 - v)F(u, 0) + vF(u, 1)] \\ &= (1 - v) \lim_{u \to 0} F(u, 0) \\ &\quad + v \lim_{u \to 0} F(u, 1) \end{aligned}$$

Therefore,

$$\begin{aligned} (P_1 \oplus P_2)F = {} & (1 - u)F(0, v) + uF(1, v) \\ & + (1 - v)F(u, 0) + vF(u, 1) \\ & - \Big[(1 - u)(1 - v) \lim_{u \to 0} F(u, 0) \\ & + (1 - u)v \lim_{u \to 0} F(u, 1) \\ & + u(1 - v) \lim_{u \to 1} F(u, 0) \\ & + uv \lim_{u \to 1} F(u, 1) \Big] \end{aligned}$$

There are asymmetries in the tensor product term, so $(P_1 \oplus P_2)F$ may not interpolate F on the boundary of S. Let's check the interpolation. Let $u = 0$:

$$\begin{aligned} & (P_1 \oplus P_2)F(0, v) \\ & \quad = F(0, v) + (1 - v) \lim_{u \to 0} F(u, 0) \end{aligned}$$

$$+ v \lim_{u \to 0} F(u, 1)$$

$$- \left[(1 - v) \lim_{u \to 0} F(u, 0) + v \lim_{u \to 0} F(u, 1) \right]$$

$$= F(0, v)$$

The case of $u = 1$ is analogous:

Let $v = 0$:

$$(P_1 \oplus P_2) F(u, 0)$$

$$= (1 - u) \lim_{v \to 0} F(0, v) + u \lim_{v \to 0} F(1, v) + F(u, 0)$$

$$- \left[(1 - u) \lim_{u \to 0} F(u, 0) + u \lim_{u \to 1} F(u, 0) \right]$$

$$= F(u, 0) \quad \text{iff} \lim_{v \to 0} F(0, v) = \lim_{u \to 0} F(u, 0)$$

$$\text{and} \quad \lim_{v \to 0} F(1, v) = \lim_{u \to 1} F(u, 0)$$

The case of $v = 1$ is analogous.

Let us analyze this discrepancy further:

$$\lim_{v \to 0} F(0, v) = \lim_{v \to 0} P_1 F(0, v) \equiv \alpha$$

$$\lim_{u \to 0} F(u, 0) = \lim_{u \to 0} P_2 F(u, 0) \equiv \beta$$

Therefore:

$$(P_1 \oplus P_2) F$$

$$= P_1 F + P_2 F - [1 - u \quad u] \times \begin{bmatrix} \lim_{u \to 0} F(u, 0) & \lim_{u \to 0} F(u, 1) \\ \lim_{u \to 1} F(u, 0) & \lim_{u \to 1} F(u, 1) \end{bmatrix} \times \begin{bmatrix} 1 - v \\ v \end{bmatrix}$$

Isolate the problem:

Assume that the only discontinuity is at $(0, 0)$ and try to fix it. If this can be done, then the other three corners will be analogous.

The creative step: Assume that the 2×2 matrix is redefined in the form

$$[1 - u \quad u] \begin{bmatrix} T_{00}(u, v) & T_{01}(u, v) \\ T_{10}(u, v) & T_{11}(u, v) \end{bmatrix} \begin{bmatrix} 1 - v \\ v \end{bmatrix}$$

where the function $T_{00}(u, v)$ has the properties necessary that the resulting interpolant – call it PF – interpolates. Now

$$PF = P_1 F + P_2 F - [1 - u \quad u] \times \begin{bmatrix} T_{00}(u, v) & \cdot \\ \cdot & \cdot \end{bmatrix} \begin{bmatrix} 1 - v \\ v \end{bmatrix}$$

so that

$$(PF)(0, v) = (P_1 F)(0, v) + (P_2 F)(0, v) - [T_{00}(0, v) \quad T_{01}(0, v)] \times \begin{bmatrix} 1 - v \\ v \end{bmatrix}$$

$$= F(0, v) + (1 - v) \lim_{u \to 0} F(u, 0) + v \lim_{u \to 0} F(u, 1) - (1 - v) T_{00}(0, v) - v T_{01}(0, v)$$

Hence, we require:

$$\lim_{u \to 0} F(u, 0) = T_{00}(0, v)$$

Next,

$$(PF)(u, 0) = (P_1 F)(u, 0) + (P_2 F)(u, 0) - [1 - u \quad u] \begin{bmatrix} T_{00}(u, 0) \\ T_{10}(u, 0) \end{bmatrix}$$

$$= (1 - u) \lim_{v \to 0} F(0, v) + u \lim_{v \to 0} F(1, v) + F(u, 0) - (1 - u) T_{00}(u, 0) - u T_{10}(u, 0)$$

Hence, we require:

$$\lim_{v \to 0} F(0, v) = T_{00}(u, 0)$$

Let $T_{00}(u, v) = (u\alpha + v\beta)/(u + v)$. The compatibly corrected interpolant is

$$B\,F(u, v) = [1-u \quad u]\begin{bmatrix} F(0, v) \\ F(1, v) \end{bmatrix} + [F(u, 0) F(u, 1)]\begin{bmatrix} 1-v \\ v \end{bmatrix}$$
$$- [1-u \quad u]\begin{bmatrix} \dfrac{uP_1F(0,0) + vP_2F(0,0)}{u+v} & \dfrac{uP_1F(0,1) + (1-v)P_2F(0,1)}{u+1-v} \\ \dfrac{(1-u)P_1F(1,0) + vP_2F(1,0)}{1-u+v} & \dfrac{(1-u)P_1F(1,1) + (1-v)P_2F(1,1)}{1-u+1-v} \end{bmatrix}\begin{bmatrix} 1-v \\ v \end{bmatrix}$$

where

$$P_1F(0,0) \equiv \lim_{v \to 0} P_1F(0, v) = \lim_{v \to 0} F(0, v)$$

$$P_2F(0,0) \equiv \lim_{u \to 0} P_2F(u, 0) = \lim_{u \to 0} F(u, 0)$$

This is an interpolant that interpolates to four completely unrestricted boundary curves.

Continuity of Coons' patches

Before discussing continuity of Coons' patches, we note that P_1 and P_2 are *projectors*. A projector is a linear idempotent operator, i.e. $P^2 = P$.

Now, we remarked that in a piecewise sense, bilinearly blended Coons' patches are only of the class C^0. We would prefer to have C^1 continuity (at least). To do this, we need to use cubic Hermite lofting interpolants in the formula $P_1 \oplus P_2$. For the bivariate case, these are of the form

$$P_1F = h_0(u)F(0, v) + h_1(u)F(1, v) + \bar{h}_0(u)F_u(0, v) + \bar{h}_1(u)F_u(1, v) \tag{7}$$

where $F = F(u, v)$ and

$$h_0(t) = 1 - 3t^2 + 2t^3$$
$$h_1(t) = 3t^2 - 2t^3$$
$$\bar{h}_0(t) = t - 2t^2 + t^3$$
$$\bar{h}_1(t) = -t^2 + t^3$$

(Note that we sometimes use numerical subscripts to indicate the derivatives, such as $F_{1,0}$ in place of F_u.) Note that P_1 is an operator, and not a functional, as the variable u appears on the right-hand side. Now,

$$P_2F = h_0(v)F(u, 0) + h_1(v)F(u, 1) + \bar{h}_0(v)F_v(u, 0) + \bar{h}_1(v)F_v(u, 1) \tag{8}$$

See Figure 8.6 for a graphical illustration. Forming the tensor product term we get,

$$\begin{aligned} P_1P_2F &= P_1(h_0(v)F(u, 0) + h_1(v)F(u, 1) \\ &\quad + \bar{h}_0(v)F_v(u, 0) + \bar{h}_1(v)F_v(u, 1)) \\ &= h_0(u)[h_0(v)F(0, 0) + h_1(v)F(0, 1) \\ &\quad + \bar{h}_0(v)F_v(0, 0) + \bar{h}_1(v)F_v(0, 1)] \\ &\quad + h_1(u)[h_0(v)F(1, 0) \\ &\quad + h_1(v)F(1, 1) + \bar{h}_0(v)F_v(1, 0) \\ &\quad + \bar{h}_1(v)F_v(1, 1)] \\ &\quad + \bar{h}_0(u)[h_0(v)F_u(0, 0) \\ &\quad + h_1(v)F_u(0, 1) \\ &\quad + \bar{h}_0(v)\frac{\partial}{\partial u}F_v(u, 0)|_{u=0} \\ &\quad + \bar{h}_1(v)\frac{\partial}{\partial u}F_v(u, 0)|_{u=0}\Big] \\ &\quad + \bar{h}_1(u)\Big[h_0(v)F_u(1,0) + h_1(v)F_u(1,1) \\ &\quad + \bar{h}_0(v)\frac{\partial}{\partial u}F_v(u, 0)|_{u=1} \\ &\quad + \bar{h}_1(v)\frac{\partial}{\partial u}F_v(u, 0)|_{u=1}\Big] \end{aligned} \tag{9}$$

It is especially important that we have terms involving $\partial/(\partial u)F_v$ and $\partial/(\partial v)F_u$ instead of F_{uv} and F_{vu}, because we are unsure of the equality of mixed second partials at this point.

If we let $\boldsymbol{h}(u) = [h_0(u)h_1(u)\bar{h}_0(u)\bar{h}_1(u)]$, and similarly for $\boldsymbol{h}(v)$, and recognize the inner

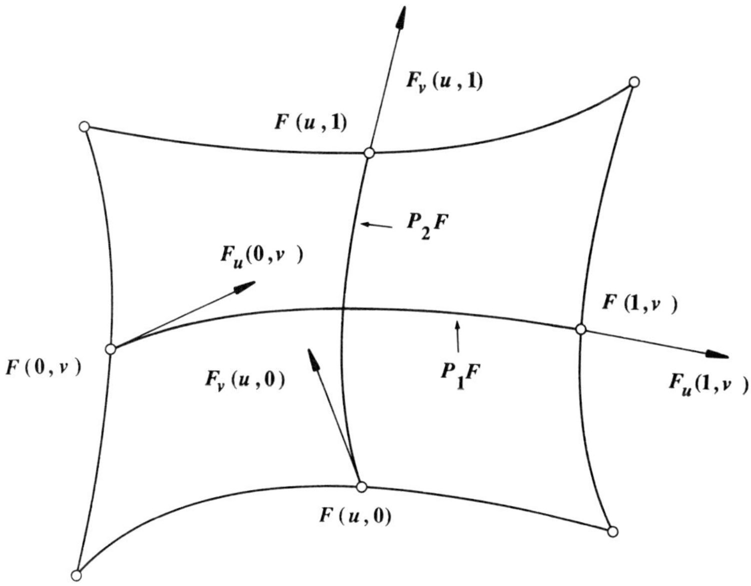

Figure 8.6. Positional and tangent ribbon data for the lofting interpolants defined in equations (7) and (8).

products, then we have

$$P_1P_2F = \boldsymbol{h}(u)M\boldsymbol{h}^T(v) \tag{10}$$

where

$$M = \begin{bmatrix} F(0,0) & F(0,1) & F_v(0,0) & F_v(0,1) \\ F(1,0) & F(1,1) & F_v(1,0) & F_v(1,1) \\ F_u(0,0) & F_u(0,1) & \frac{\partial}{\partial u}F_v(u,0)|_{u=0} & \frac{\partial}{\partial u}F_v(u,1)|_{u=0} \\ F_u(1,0) & F_u(1,1) & \frac{\partial}{\partial u}F_v(u,0)|_{u=1} & \frac{\partial}{\partial u}F_v(u,1)|_{u=1} \end{bmatrix} \tag{11}$$

The tensor product part is clearly bicubic. We have 16 pieces of data (or *degrees of freedom*). We also have the transfinite information.

A graphical interpretation of the data actually interpolated is given in Figure 8.7.

We are said to have a *ribbon* of information, called the *cross-boundary tangents*. We interpolate to the boundary curves and then the *tangential* derivatives (derivatives along the curve in the direction of the curve) are interpolated automatically.

Now, we can clearly see that if two patches share cross-boundary information (ribbons), we have C^1 continuity across patches.

The $F_{1,1} = F_{uv}$ derivatives are called *twists*. Are these unique? For example, consider the curves that define the boundaries, and rather than transfinite derivative information (ribbons), we are given derivatives in both u and v directions at the corners. We no longer have F_u, F_v at arbitrary points, but only at the corners. Can we approximate these? Let

$$\tilde{F}_u(0,v) \equiv (1-v)F_u(0,0) + vF_u(0,1) \tag{12}$$

This is *semi-discretization*, that is, discretizing derivatives but not positions. We can similarly discretize the other derivatives and develop

$$(P_1 \oplus P_2)\tilde{F} \tag{13}$$

This interpolant is useful in networks of patches with only positional data, because derivative data at corners, as in equation (12), are implied by positional data from orthogonal sides of the domain (see Figure 8.8).

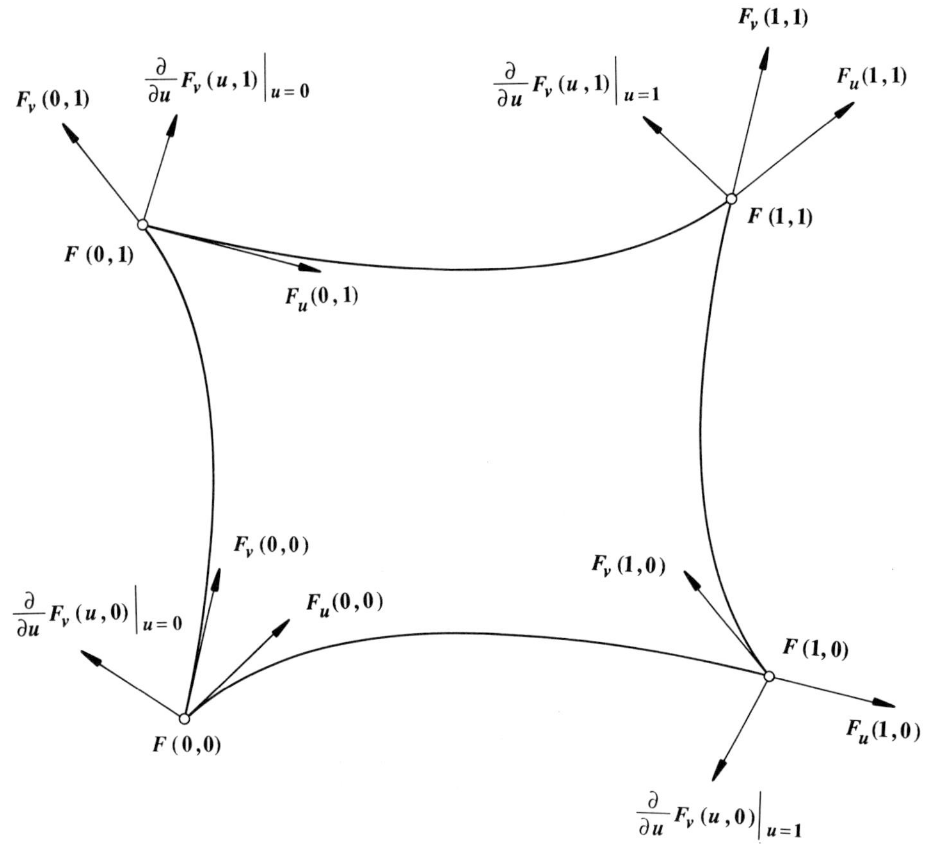

Figure 8.7. Data interpolated by the tensor product defined in equation (9).

Bicubically blended Coons' patches

Construction

We now consider the *bicubically blended Coons' patch*. This is defined as

$$(P_1 \oplus P_2)F = P_1F + P_2F - P_1P_2F = \boldsymbol{h}(u)\begin{bmatrix} F(0,v) \\ F(1,v) \\ F_u(0,v) \\ F_u(1,v) \end{bmatrix}$$

$$+ [F(u,0)F(u,1)F_v(u,0)F_v(u,1)]\boldsymbol{h}^{\mathrm{T}}(v) - \boldsymbol{h}(u)M\boldsymbol{h}^{\mathrm{T}}(v) \tag{14}$$

where

$$M = \begin{bmatrix} F(0,0) & F(0,1) & F_v(0,0) & F_v(0,1) \\ F(1,0) & F(1,1) & F_v(1,0) & F_v(1,1) \\ F_u(0,0) & F_u(0,1) & \frac{\partial}{\partial u}F_v(0,0) & \frac{\partial}{\partial u}F_v(0,1) \\ F_u(1,0) & F_u(1,1) & \frac{\partial}{\partial u}F_v(1,0) & \frac{\partial}{\partial u}F_v(1,1) \end{bmatrix} \tag{15}$$

or

$$M = \begin{bmatrix} F & F_v \\ F_u & F_{uv} \end{bmatrix}$$

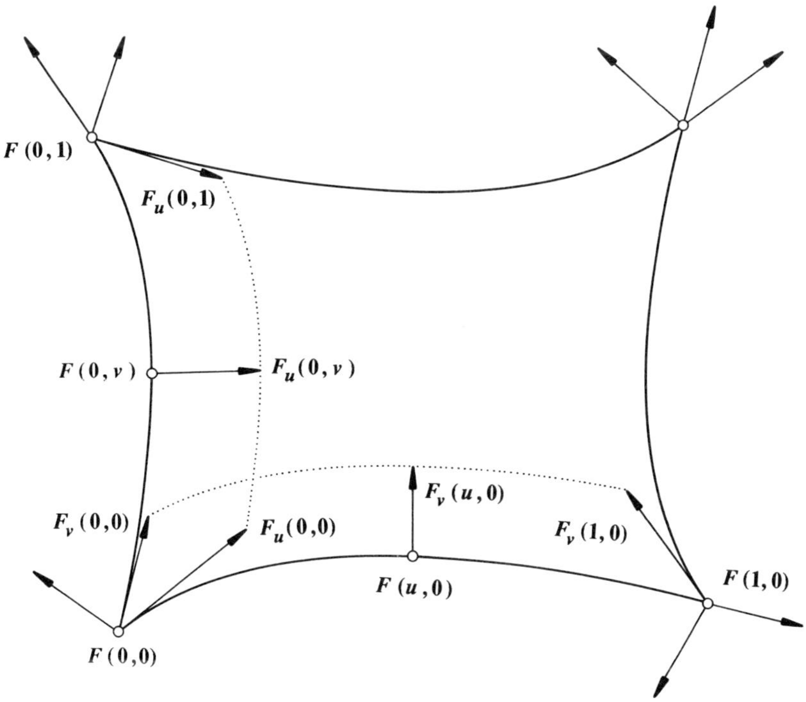

Figure 8.8. A geometric interpretation of the linear approximation of cross boundary derivatives.

where each submatrix is evaluated at

$$\begin{matrix} (0,0) & (0,1) \\ (1,0) & (1,1) \end{matrix}$$

Notice that the pattern of subscripts matches the pattern of evaluation, so we can use this as a mnemonic device.

Twist uniqueness in bicubically blended Coons' patches

Barnhill, Gregory and Mansfield investigated the question of twist uniqueness for these types of patches. We follow their development. First, consider discretization as an example and examine the twists at $(u, v) = (0, 0)$.

$$\begin{aligned} F_u(0, v) &\approx \tilde{F}_u(0, v) \\ &\equiv (1 - v)F_u(0, 0) + vF_u(0, 1) \\ F_v(u, 0) &\approx \tilde{F}_v(u, 0) \\ &\equiv (1 - u)F_v(0, 0) + uF_v(1, 0) \end{aligned}$$

See Figure 8.8 for a geometric interpretation of these approximations. Now, from examining the derivatives of these approximations, we see that

$$\begin{aligned} \frac{\partial}{\partial v} \tilde{F}_u(0, v)|_{v=0} &= F_u(0, 1) - F_u(0, 0) \\ \frac{\partial}{\partial u} \tilde{F}_v(u, 0)|_{u=0} &= F_v(1, 0) - F_v(0, 0) \end{aligned} \tag{16}$$

which are clearly not equal (see Figure 8.9). Why is this the case? In most situations, all we have are the boundary curves (not bivariate functions) and the cross-boundary derivatives. We then interpolate in some manner, as above, and have the above type of inequality.

Since, in general, we do not have twist information, we must approximate the twists. Several methods are in common use, and we will consider some of them shortly. However, note first that Forrest has found that the simplistic solution of setting all twists equal to zero causes *pseudoflats* in the surface. These are regions where it appears as if a thumb had been pushed into the surface, flattening it out. We will consider better methods below.

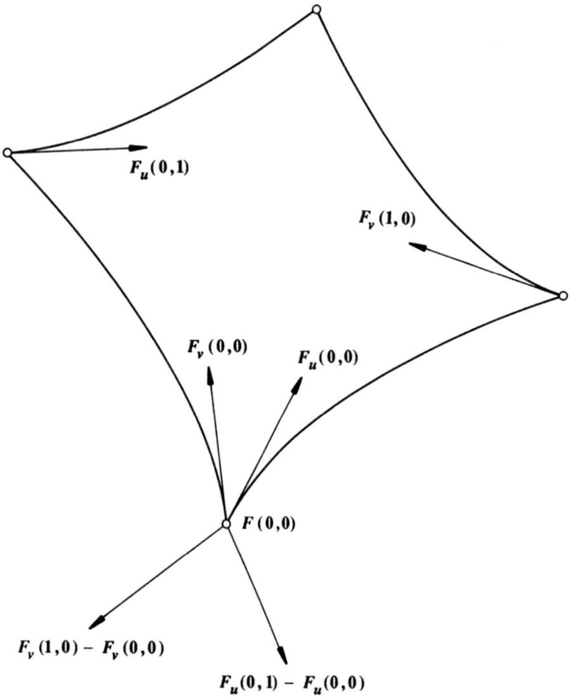

Figure 8.9. Unequal twists, from a semi-discretization, corresponding to equation (16).

This 'flattening' can be seen as follows: let us approximate the surface near (0, 0). For small Δt we have

$$h_0(\Delta t) = 1 - 3\Delta t^2 + 2\Delta t^3 \approx 1$$

$$h_1(\Delta t) = 3\Delta t^2 - 2\Delta t^3 \approx 0$$

$$\bar{h}_0(\Delta t) = \Delta t - 2\Delta t^2 + \Delta t^3 \approx \Delta t$$

$$\bar{h}_1(\Delta t) = -\Delta t^2 + \Delta t^3 \approx 0$$

Substitution into the surface form yields

$$\begin{aligned}(P_1 \oplus P_2)F(\Delta u, \Delta v) = F(0,0) &+ F_u(0,0)\Delta u \\ &+ F_v(0,0)\Delta v \\ &+ F_{uv}(0,0)\Delta u \Delta v\end{aligned}$$

Now if $F_{uv} = 0$, then the surface point lies in the plane spanned by F_u and F_v. (We are indebted to the Editor for this paragraph.) See Figure 8.10 for a geometric interpretation.

What functions have zero twists? Gerald Farin [10] points out that those surfaces with identically zero twists are the translational surfaces. Normalized to (0, 0), such a surface is

$$\mathbf{t}(u, v) = \mathbf{c}_1(u) + \mathbf{c}_2(v) - \mathbf{a}$$

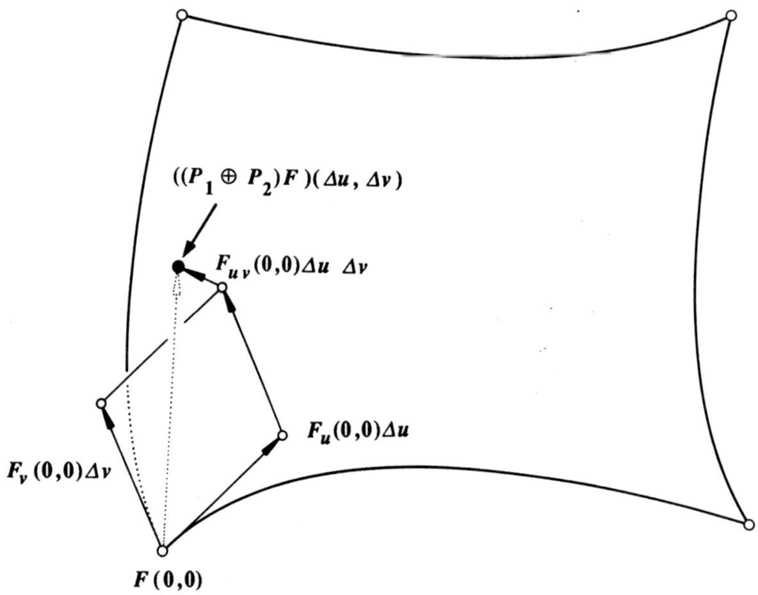

Figure 8.10. A geometric interpretation of pseudo-flats.

where $\mathbf{a} = \mathbf{c}_1(0) = \mathbf{c}_2(0)$. An additional interesting property of translational surfaces is that the bilinearly blended Coons' patch can be written as the bilinear blend of translational surfaces for the four corners of the unit square [10].

Interpolation properties of $(P_1 \oplus P_2)F$

We must, of course, verify that our proposed interpolant actually does interpolate the values needed. To this end, it is quite easy to see that, for example, $(P_1 \oplus P_2)F(0, v) = F(0, v)$

$$
\begin{aligned}
(P_1 \oplus P_2)F|_{u=0} &= [1, 0, 0, 0]\begin{bmatrix} F(0, v) \\ F(1, v) \\ F_u(0, v) \\ F_u(1, v) \end{bmatrix} \\
&\quad + [F(0, 0)F(0, 1)F_v(0, 0)F_v(0, 1)]\boldsymbol{h}^{\mathrm{T}}(v) \\
&\quad - [1, 0, 0, 0]M\boldsymbol{h}^{\mathrm{T}}(v) \\
&= F(0, v) \\
&\quad + [F(0, 0)F(0, 1)F_v(0, 0)F_v(0, 1)]\boldsymbol{h}^{\mathrm{T}}(v) \\
&\quad - [F(0, 0)F(0, 1)F_v(0, 0)F_v(0, 1)]\boldsymbol{h}^{\mathrm{T}}(v) \\
&= F(0, v)
\end{aligned}
$$

from the δ_{ij} properties of the basis polynomials,

$$
h_i(j) = \bar{h}_i'(j) = \delta_{ij} = \begin{cases} 1 & \text{if } i = j \\ 0 & \text{otherwise} \end{cases}
$$

and $\bar{h}_i(j) = 0$ for $i, j = 0, 1$. Next, we look at the cross-boundary derivative interpolation properties

$$
\begin{aligned}
\frac{\partial}{\partial u}(P_1 \oplus P_2)F|_{u=0} &= \boldsymbol{h}'(u)\begin{bmatrix} F(0, v) \\ F(1, v) \\ F_u(0, v) \\ F_u(1, v) \end{bmatrix} \\
&\quad + \left[F_u(0, 0)F_u(0, 1)\frac{\partial}{\partial u}F_v(u, 0)|_{u=0} \frac{\partial}{\partial u}F_v(u, 1)|_{u=0}\right]\boldsymbol{h}^{\mathrm{T}}(v) \\
&\quad - \boldsymbol{h}'(u)M\boldsymbol{h}^{\mathrm{T}}(v) \\
&= [0, 0, 1, 0]\begin{bmatrix} F(0, v) \\ F(1, v) \\ F_u(0, v) \\ F_u(1, v) \end{bmatrix} \\
&\quad + \left[F_u(0, 0)F_u(0, 1)\frac{\partial}{\partial u}F_v(u, 0)|_{u=0} \frac{\partial}{\partial u}F_v(u, 1)|_{u=0}\right]\boldsymbol{h}^{\mathrm{T}}(v) \\
&\quad - [0, 0, 1, 0]M\boldsymbol{h}^{\mathrm{T}}(v) \\
&= F_u(0, v) \\
&\quad + \left[F_u(0, 0)F_u(0, 1)\frac{\partial}{\partial u}F_v(u, 0)|_{u=0} \frac{\partial}{\partial u}F_v(u, 1)|_{u=0}\right]\boldsymbol{h}^{\mathrm{T}}(v) \\
&\quad - \left[F_u(0, 0)F_u(0, 1)\frac{\partial}{\partial u}F_v(0, 0) \frac{\partial}{\partial u}F_v(0, 1)\right]\boldsymbol{h}^{\mathrm{T}}(v) \\
&= F_u(0, v)
\end{aligned}
$$

Analogously,

$$
\frac{\partial}{\partial v}(P_1 \oplus P_2)F|_{v=0} = F_v(u, 0) + \boldsymbol{h}(u) \times \begin{bmatrix} 0 \\ 0 \\ \frac{\partial}{\partial v}F_u(0, v)|_{v=0} - \frac{\partial}{\partial u}F_v(u, 0)|_{u=0} \\ \frac{\partial}{\partial v}F_u(1, v)_{v=0} - \frac{\partial}{\partial u}F_v(u, 0)|_{u=1} \end{bmatrix}
$$

Continuity class

Consider two adjacent bicubically blended Coons' patches as pictured, with $F = F^i$, $i = 1, 2$ on regions 1 and 2 parametrized respectively by $0 \leqslant u \leqslant 1, 0 \leqslant v \leqslant 1$ and $0 \leqslant u \leqslant 1, 0 \leqslant v^* \leqslant 1$ with common boundary $v = 0$ and $v^* = 1$.

Although we know that

$$\left.\frac{\partial(P_1 \oplus P_2)F^1}{\partial v}\right|_{v=0} = F^1_{0,1}(u,0) + [\bar{h}_0(u)\bar{h}_1(u)]$$

$$\times \begin{bmatrix} \frac{\partial}{\partial v}F^1_{1,0}(0,v)|_{v=0} - \frac{\partial}{\partial u}F^1_{0,1}(u,0)|_{u=0} \\ \frac{\partial}{\partial v}F^1_{1,0}(1,v)|_{v=0} - \frac{\partial}{\partial u}F^1_{0,1}(u,0)|_{u=1} \end{bmatrix}$$

and

$$\left.\frac{\partial(P_1 \oplus P_2)F^2}{\partial v^*}\right|_{v^*=1} = F^2_{0,1}(u,0) + [\bar{h}_0(u)\bar{h}_1(u)]$$

$$\times \begin{bmatrix} \frac{\partial}{\partial v^*}F^2_{1,0}(0,v^*)|_{v^*=1} - \frac{\partial}{\partial u}F^2_{0,1}(u,1)|_{u=0} \\ \frac{\partial}{\partial v^*}F^2_{1,0}(1,v^*)|_{v^*=1} - \frac{\partial}{\partial u}F^2_{0,1}(u,1)|_{u=1} \end{bmatrix}$$

could the interpolant be C^1? That is, could

$$\left.\frac{\partial(P_1 \oplus P_2)F^1}{\partial v}\right|_{v=0} = \left.\frac{\partial(P_1 \oplus P_2)F^2}{\partial v^*}\right|_{v^*=1} ?$$

Now $F^1_{0,1}(u,0) = F^2_{0,1}(u,0)$ by the definition

$$F_{0,1} = \begin{bmatrix} F^1_{0,1} \; on\, 1 \\ F^2_{0,1} \; on\, 2 \end{bmatrix}$$

In addition, this definition also implies that

$$\frac{\partial}{\partial u}F^1_{0,1}(u,0)|_{u=0} = \frac{\partial}{\partial u}F^2_{0,1}(u,1)|_{u=0}$$

Similarly, at $u = 1$

$$\left.\frac{\partial(P_1 \oplus P_2)F}{\partial v^*}\right|_{v^*=1} = F_{0,1}(u,1) + [\bar{h}_0(u)\bar{h}_1(u)]$$

$$\times \begin{bmatrix} \frac{\partial}{\partial v^*}F_{1,0}(0,v^*)|_{v^*=1} & - \frac{\partial}{\partial u}F_{0,1}(u,1)|_{u=0} \\ \frac{\partial}{\partial v^*}F_{1,0}(1,v^*)|_{v^*=1} & - \frac{\partial}{\partial u}F_{0,1}(u,1)|_{u=1} \end{bmatrix}$$

What about

$$\frac{\partial}{\partial v}F^1_{1,0}(0,v)|_{v=0} = \frac{\partial}{\partial v}F^2_{1,0}(0,v^*)|_{v^*=0}?$$

If the function

$$F_{1,0}(0,v) \equiv \begin{bmatrix} F^1_{1,0}(0,v), 0 \leqslant v \leqslant 1 \\ F^2_{1,0}(0,v^*), 0 \leqslant v^* \leqslant 1 \end{bmatrix}$$

is C^1, then equality holds.

Therefore we can say that the overall interpolant is C^1 iff $F_{1,0}(0,v)$ and $F_{1,0}(1,v)$ are C^1.

This is a surprising result: We do not have interpolation to the C^1 data, but we do have a C^1 interpolant. This result has sometimes obscured the lack of interpolation to the C^1 data.

Historical notes: A major consideration in piecing together surface patches is the smoothness between adjacent patches. Motivated by Nielson's [11] nu-splines, I invented the term 'visual continuity' to refer to smoothness between curve segments or surface patches. For curves, visual continuity means, for example, tangent vector continuity, curvature vector continuity, in general, smoothness with respect to the rth derivative with respect to arc length. For surfaces, visual continuity means tangent plane continuity, etc. I taught this in the Utah CAGD class for years. (A standard homework question was: What is visual continuity? For example, B. Barsky took the class and subsequently developed his beta-splines.) Visual continuity is now called 'geometric continuity' to indicate its intrinsic geometric nature. Unfortunately, the newer term loses the computer graphics connotation of the original.

After the fact, I re-read Coons' [1] 'little red book' and found that he discussed 'tangent plane continuity' for his bicubically blended patches. Since these patches are parametrically smooth, they are tangent plane continuous. However, a point is that Coons recognized that visual continuity, rather than parametric continuity, is actually the important concept for smoothness between surface patches.

Gregory's square

As noted above,

$$\frac{\partial}{\partial v}(P_1 \oplus P_2)F|_{v=0} = F_v(u,0) + \boldsymbol{h}(u)$$

$$\times \begin{bmatrix} 0 \\ 0 \\ \dfrac{\partial}{\partial v} F_u(0, v)|_{v=0} - \dfrac{\partial}{\partial u} F_v(u, 0)|_{u=0} \\ \dfrac{\partial}{\partial v} F_u(1, v)|_{v=0} - \dfrac{\partial}{\partial u} F_v(u, 0)|_{u=1} \end{bmatrix}$$

We must introduce some sort of correction, known as a *compatibility correction,* to our scheme. One possibility would be

$$\frac{(P_1 \oplus P_2) + (P_2 \oplus P_1)}{2}$$

but this is quickly determined not to work. We next consider the possibility of using a 'convex combination' of the two lofting interpolants, of the form

$$\alpha(P_1 \oplus P_2) + \beta(P_2 \oplus P_1), \quad \alpha + \beta \equiv 1 \qquad (17)$$

We need only examine the twist partition of the matrix M. By symmetry arguments, we need only examine one of the terms. The problem can now be stated as

Problem *Let the twist partition of the Coons' patch scheme be denoted by*

$$\begin{bmatrix} \dfrac{\partial}{\partial u} F_v(0, 0) & \dfrac{\partial}{\partial u} F_v(0, 1) \\ \dfrac{\partial}{\partial u} F_v(1, 0) & \dfrac{\partial}{\partial v} F_v(1, 1) \end{bmatrix} = \begin{bmatrix} T_1 & T_2 \\ T_3 & T_4 \end{bmatrix} \qquad (18)$$

and similarly define the twist partition for the Coons' patch with the differentiation done in the opposite order as

$$\begin{bmatrix} \dfrac{\partial}{\partial v} F_u(0, 0) & \dfrac{\partial}{\partial v} F_u(0, 1) \\ \dfrac{\partial}{\partial v} F_u(1, 0) & \dfrac{\partial}{\partial u} F_u(1, 1) \end{bmatrix} = \begin{bmatrix} T'_1 & T'_2 \\ T'_3 & T'_4 \end{bmatrix} \qquad (19)$$

Then form the terms

$$G_1 = \bar{h}_0(u)\bar{h}_0(v)[\alpha(u, v)T_1 + \beta(u, v)T'_1] \qquad (20)$$

and similarly for G_2, G_3 and G_4 so that the scheme so created yields the correct values for the mixed second partials regardless of the order of differentiation.

As one solution to just one of the four elements, we use

$$G_1 = \bar{h}_0(u)\bar{h}_0(v)\left[\frac{u}{u+v} T'_1 + \frac{v}{u+v} T_1\right] \qquad (21)$$

This is one element of what is called *Gregory's square*, a solution to the problem of devising an interpolant that interpolates to the two twists at the corners. Note the non-polynomial nature of the correction term.

If we set $u = 0$, we have $u/(u + v) = 0\ \forall v \neq 0$, and if we set $v = 0$, we have $u/(u + v) = 1$ $\forall u \neq 0$. But, what must we do to make the terms correct if $u = v = 0$? What happens at other values of (u, v)?

Before discussing Gregory's square in more detail, we introduce the following notation: Let

$$t^1_{00} \equiv \frac{\partial}{\partial v} F_{1,0}(0, v)|_{v=0}$$

$$= \frac{\partial}{\partial v}\frac{\partial}{\partial u} P_1 F(u, v)|_{u=0}|_{v=0}$$

and similarly for all permutations of the superscripts and subscripts, where the subscripts indicate the geometric location and the superscripts indicate the projector used. For example,

$$t^2_{00} \equiv \frac{\partial}{\partial u} F_{0,1}(u, 0)|_{u=0}$$

$$= \frac{\partial}{\partial u}\frac{\partial}{\partial v} P_2 F(u, v)|_{v=0}|_{u=0}$$

Now, the compatibly corrected Coons' patch is of the form

$$GF = P_1F + P_2F - TF \qquad (22)$$

where

$$TF = \boldsymbol{h}(u)M\boldsymbol{h}(v)^T \qquad (23)$$

with

$$M = \begin{bmatrix} F & F_{0,1} \\ F_{1,0} & \begin{pmatrix} T_1(u,v) & T_2(u,v) \\ T_3(u,v) & T_4(u,v) \end{pmatrix} \end{bmatrix} \tag{24}$$

Now, let

$$\mathscr{T}_1(u,v) = h_0(u)\mathrm{h}_0(v)\,T_1(u,v) \tag{25}$$

$\mathscr{T}_1$ must have the following properties:

1. $\mathscr{T}_1 \equiv 0$ for $u = 0, 1$ and $v = 0, 1$.
2. $\dfrac{\partial \mathscr{T}_1}{\partial u}|_{u=1} = \dfrac{\partial \mathscr{T}_1}{\partial v}|_{v=1} = 0$, assuming $\dfrac{\partial \mathscr{T}_1}{\partial u}|_{u=1}$ is bounded (similarly for the other partial).

Now, what exactly does the partial $\dfrac{\partial \mathscr{T}_1}{\partial u}|_{u=0}$ represent? If we differentiate equation (22) with respect to u, and evaluate it at $u = 0$, we find (after simplification)

$$\frac{\partial G\,F}{\partial u}|_{u=0} = \frac{\partial F}{\partial u}|_{u=0} + \frac{\partial P_2 F}{\partial u}|_{u=0} - [F_{1,0}(0,0)\,F_{1,0}(0,1)\,T_1(0,v)\,T_2(0,v)]\,\boldsymbol{h}(v)^T \tag{26}$$

Now, the $\dfrac{\partial P_2 F}{\partial u}|_{u=0}$ term is, as previously discussed,

$$\frac{\partial P_2 F}{\partial u}|_{u=0} = \boldsymbol{h}(v)\begin{bmatrix} F_{1,0}(0,0) \\ F_{1,0}(0,1) \\ \dfrac{\partial}{\partial u}F_{0,1}(u,0)|_{u=0} \\ \dfrac{\partial}{\partial u}F_{0,1}(u,1)|_{u=0} \end{bmatrix} \tag{27}$$

Inserting (24) into (23), we have

$$\frac{\partial G\,F}{\partial u}|_{u=0} = \frac{\partial F}{\partial u}|_{u=0} + \boldsymbol{h}(v)\begin{bmatrix} 0 \\ 0 \\ \dfrac{\partial}{\partial u}F_{0,1}(u,0)|_{u=0} - T_1(0,v) \\ \dfrac{\partial}{\partial u}F_{0,1}(u,1)|_{u=0} - T_2(0,v) \end{bmatrix} \tag{28}$$

which implies

$$\begin{aligned} &\frac{\partial}{\partial u}F_{0,1}(u,0)|_{u=0} - T_1(0,v) \\ &\quad = t_{00}^2 - T_1(0,v) = 0 \\ &\frac{\partial}{\partial u}F_{0,1}(u,1)|_{u=0} - T_2(0,v) \\ &\quad = t_{01}^2 - T_2(0,v) = 0 \end{aligned} \tag{29}$$

Examining the first of these two equations shows that $T_1(0,v) = t_{00}^2$, and by symmetry, $T_1(u,0) = t_{00}^1$. It is now possible to see how to create $T_1(u,v)$ so that these conditions hold. For example, notice that by using combinations of $u/(u+v)$ and $v/(u+v)$, we can create a T_1 which has these properties:

$$T_1(u,v) = \frac{u}{u+v}t_{00}^1 + \frac{v}{u+v}t_{00}^2 \tag{30}$$

Notice that this is a convex combination of the two functions of u and v. The other $T_i(u,v)$'s are created similarly:

$$\begin{aligned} T_2(u,v) &= \frac{ut_{01}^1 + (1-v)t_{01}^2}{u + (1-v)} \\ T_3(u,v) &= \frac{(1-u)t_{10}^1 + vt_{10}^2}{(1-u)+v} \\ T_4(u,v) &= \frac{(1-u)t_{11}^1 + (1-v)t_{11}^2}{(1-u)+(1-v)} \end{aligned} \tag{31}$$

The twist partition of M is sometimes referred to as a *variable twist partition.*

As an additional property, the boundedness of T_1 ensures that $T\,F(0,0) = \boldsymbol{h}(0)M\boldsymbol{h}(0)^T = F(0,0)$.

Historical notes: The derivation of Gregory's square given above differs from Gregory's original development, in which a correction term for the bicubically blended Coons' patch was found. The above derivation assumes a certain structure, namely, that a solution to the incompatibility problem can be found by replacing the twist partition in the bicubically blended Coons' patch. The proof that the originally derived compatibility correction is of this form was pointed out to me by Frank Little [7].

Gregory's square can be viewed as the bicubically blended Coons' patch with a different twist partition. Mike Dincau at Lockheed therefore called it a 'variable twist partition', connoting that the Gregory Square twists are functions rather than numbers. (I like this language.)

Discretizations

The transfinite data in $G\,F$ are the positions and the normal derivatives around the boundary of the square S. These may come from transfinite information or may come from discretizations.

Example 1 Semi-discretization
The positions are transfinite and the normal derivatives are finite dimensional. This example could arise from positions that are (univariate) splines, but where little information about the normal derivatives is known.

Subexample 1 Cubic Hermite Normal Derivatives
Let

$$F_{1,0}(0,v) \simeq \tilde{F}_{1,0}(0,v) \equiv h_0(v)F_{1,0}(0,0) + h_1(v)F_{1,0}(0,1) + \bar{h}_0(v)t_{00}^1 + \bar{h}_1(v)t_{01}^1$$

and similarly for $F_{1,0}(1,v)$, $F_{0,1}(u,0)$, $F_{0,1}(u,1)$:

$$\tilde{F}_{1,0}(1,v) = h_0(v)F_{1,0}(1,0) + h_1(v)F_{1,0}(1,1) + \bar{h}_0(v)t_{10}^1 + \bar{h}_1(v)t_{11}^1$$

$$\tilde{F}_{0,1}(u,0) \equiv h_0(u)F_{0,1}(0,0) + h_1(u)F_{0,1}(1,0) + \bar{h}_0(u)t_{00}^2 + \bar{h}_1(u)t_{10}^2$$

$$\tilde{F}_{0,1}(u,1) \equiv h_0(u)F_{0,1}(0,1) + h_1(u)F_{0,1}(1,1) + \bar{h}_0(u)t_{01}^2 + \bar{h}_1(u)t_{11}^2$$

and let F and $\tilde{F}$ have the same positional values. Then

$$G\tilde{F} = [h_0(u)h_1(u)]\begin{bmatrix}F(0,v)\\F(1,v)\end{bmatrix} + [F(u,0)F(u,1)]\begin{bmatrix}h_0(v)\\h_1(v)\end{bmatrix} - [h_0(u)h_1(u)]\begin{bmatrix}F(0,0)F(0,1)\\F(1,0)F(1,1)\end{bmatrix} + [\bar{h}_0(u)\bar{h}_1(u)]$$
$$\times\begin{bmatrix}\dfrac{vt_{00}^1+ut_{00}^2}{u+v} & \dfrac{(1-v)t_{01}^1+ut_{01}^2}{u+1-v}\\ \dfrac{vt_{11}^1+(1-u)t_{11}^2}{1-u+v} & \dfrac{(1-v)t_{11}^1+(1-u)t_{11}^2}{2-u-v}\end{bmatrix} \times\begin{bmatrix}\bar{h}_0(v)\\\bar{h}_1(v)\end{bmatrix}$$

Notice the reversal of weighting functions in the variable twist partition. The precision of this interpolant is bicubic.

Subexample 2 Zero Twists
If all the twists are zero, then $P\,\tilde{F}$ depends only on the positional data and on the positional blending functions h_0 and h_1. The zero twists drastically affect the precision, which is

$$\{\textit{linear polynomials}\} \cup \{u^i, 0<i\} \cup \{v^j, 0<j\}$$

This precision is determined by considering those monomials which have zero twists.

Subexample 3 Linear Normal Derivatives
Since twists are usually not known, and setting them equal to zero produces pseudo-flats, an alternative is to specify that the normal derivatives be linear. That is, let

$$F_{1,0}(0,v) \simeq \tilde{F}_{1,0}(0,v) \equiv (1-v)F_{1,0}(0,0) + vF_{1,0}(0,1)$$

$$F_{1,0}(1,v) \simeq \tilde{F}_{1,0}(1,v) \equiv (1-v)F_{1,0}(1,0) + vF_{1,0}(1,1)$$

$$F_{0,1}(u,0) \simeq \tilde{F}_{0,1}(u,0) \equiv (1-u)F_{0,1}(0,0) + uF_{0,1}(1,0)$$

$$F_{0,1}(u,1) \simeq \tilde{F}_{0,1}(u,1) \equiv (1-u)F_{0,1}(0,1) + uF_{0,1}(1,1)$$

The eight twists are calculated directly from these discretizations, e.g.

$$t_{00}^1 \equiv \frac{\partial \tilde{F}_{1,0}(0, v)}{\partial v}\Big|_{v=0}$$

$$= -F_{1,0}(0, 0) + F_{1,0}(0, 1)$$

$$t_{00}^2 \equiv \frac{\partial \tilde{F}_{0,1}(u, 0)}{\partial u}\Big|_{u=0}$$

$$= -F_{0,1}(0, 0) + F_{0,1}(1, 0)$$

The precision of this discretization is that subset of the precision set of the undiscretized interpolant whose members have linear cross boundary derivatives.

The transfinite interpolant had precision $S = \{u^i v^j : 0 \leqslant i \leqslant 3 \text{ or } 0 \leqslant j \leqslant 3\}$. A function in S such that, e.g. $F_{1,0}(0, v)$ and $F_{1,0}(1, v)$ are linear in v, is in $S_D^1 \equiv \{u^i v^j : 0 \leqslant j \leqslant 1\} \cup \{v^2\}$. The discretizations of $F_{0,1}(u, 0)$ and $F_{0,1}(u, 1)$ have the precision set $S_D^2 \equiv \{u^i v^j : 0 \leqslant i \leqslant 1\} \cup \{u^2\}$. The overall precision set of the discretizations is

$$S_D^1 \cap S_D^2 = \{\textit{quadratic polynomials}\}$$

The precision set of the overall interpolant is the intersection of the precision of the transfinite interpolant and the discretization precision, which equals $\{\textit{quadratic polynomials}\}$.

Example 2 Discretization of Positional Data
Let the positional data be discretized as a cubic Hermite. Then (at least) three possibilities are immediately conceivable: (1) Let the normal derivatives be cubic Hermite, which leads to a 20 degree of freedom interpolant; (2) let the normals be cubic Hermite but assume that the twists are uniquely defined, which leads to a 16 degree of freedom interpolant; (3) let the normals be linear, which leads to a 12 degree of freedom interpolant. We derive the formula for (1) only. The others are analogous. Cubic Hermite positions means that

$$\begin{aligned} F(0, v) &\simeq \tilde{F}(0, v) \\ &\equiv h_0(v) F(0, 0) + h_1(v) F(0, 1) \\ &\quad + \bar{h}_0(v) F_{0,1}(0, 0) \\ &\quad + \bar{h}_1(v) F_{0,1}(0, 1) \end{aligned}$$

etc.

$$G\tilde{F} = [h_0(u) h_1(u) \bar{h}_0(u) \bar{h}_1(u)] \times \begin{bmatrix} F(0,0) & F(0,1) & F_{0,1}(0,0) & F_{0,1}(0,1) \\ F(1,0) & F(1,1) & F_{0,1}(1,0) & F_{0,1}(1,1) \\ F_{1,0}(0,0) & F_{1,0}(0,1) & \dfrac{v t_{00}^1 + u t_{00}^2}{u+v} & \dfrac{(1-v) t_{01}^1 + u t_{01}^2}{u+1-v} \\ F_{1,0}(1,0) & F_{1,0}(1,1) & \dfrac{v t_{10}^1 + (1-u) t_{10}^2}{1-u+v} & \dfrac{(1-v) t_{11}^1 + (1-u) t_{11}^2}{2-u-v} \end{bmatrix} \times \begin{bmatrix} h_0(v) \\ h_1(v) \\ \bar{h}_0(v) \\ \bar{h}_1(v) \end{bmatrix}$$

The 20 degrees of freedom are the four positions, eight slopes, and eight twists.

Brown's Square

The twist incompatibilities in Coons' patches arise from the tensor product $P_1 P_2 F$. That is, incompatibility problems arise from taking the composition of the lower dimensional projectors. Thus we could consider constructing interpolants by taking linear combinations of the projectors. This topic is taken up later in the section on convex combinations. Here we remark that Brown's Square was a first example of this technique.

Brown's Square: C^1 case
We seek to avoid the question of incompatibilities that arise in tensor product terms of the form $P_1 P_2$. Let P_1 and P_2 be the univariate cubic Hermite operators, as before. Then consider

$$BF \equiv \alpha(u, v) P_1 F(u, v) + \beta(u, v) P_2 F(u, v)$$

where

$$\alpha(u, v) \equiv \frac{\dfrac{1}{u^2(1-u)^2}}{\dfrac{1}{u^2(1-u)^2} + \dfrac{1}{v^2(1-v)^2}}$$

and $\beta(u, v) \equiv \alpha(v, u)$. A useful form of Brown's C^1 Square is

$$\alpha(u, v) = \frac{v^2(1-v)^2}{u^2(1-u)^2 + v^2(1-v)^2}$$

$B\,F$ interpolates to F and to $(\partial F)/(\partial n)$ all around the boundary of S.

Check:

$$(B\,F)(0, v) = P_1F(0, v) = F(0, v)$$
$$(B\,F)(u, 0) = P_2F(u, 0) = F(u, 0)$$

$$\frac{\partial B\,F}{\partial u}\Big|_{u=0} = \alpha(0, v)\frac{\partial P_1F}{\partial u}\Big|_{u=0} + \frac{\partial \alpha}{\partial u}\Big|_{u=0} P_1F(0, v)$$
$$+\beta(0, v)\frac{\partial P_2F}{\partial u}\Big|_{u=0} + \frac{\partial \beta}{\partial u}\Big|_{u=0} P_2F(0, v)$$
$$= \frac{\partial P_1F}{\partial u}\Big|_{u=0} = F_{1,0}(0, v)$$

$\partial B\,F/\partial v|_{v=0}$ is similarly calculated.

Thus the defining properties of α and β are that

$$\alpha\Big|_{\substack{u=0\\u=1}} \equiv 1, \quad \alpha\Big|_{\substack{v=0\\v=1}} \equiv 0, \quad \frac{\partial \alpha}{\partial n}\Big|_{\partial S} \equiv 0$$

where $\partial/\partial n$ is the cross boundary derivative.

The zero properties

$$\alpha\Big|_{\substack{v=0\\v=1}} \equiv 0 \equiv \frac{\partial \alpha}{\partial v}\Big|_{\substack{v=0\\v=1}}$$

imply that $\alpha(u, v)$ is of the form

$$\alpha(u, v) = v^2(1-v)^2H(u, v)$$

for same fraction H. The requirement that

$$\alpha(u, v) + \beta(u, v) \equiv 1$$

and the fact that β's properties are dual to α's, i.e.

$$\beta\Big|_{\substack{u=0\\u=1}} \equiv 0 \equiv \frac{\partial \beta}{\partial u}\Big|_{\substack{u=0\\u=1}}$$

imply that the definition

$$\alpha(u, v) \equiv \frac{v^2(1-v)^2}{u^2(1-u)^2 + v^2(1-v)^2}$$

and $\beta(u, v) = \alpha(v, u)$ is a solution.

Twist approximations

Bicubic patches and bicubically blended Coons' patches both contain twists, the (1, 1) derivatives. Twists *per se* are rarely supplied by a user, although they may be implied by given transfinite cross-boundary derivatives. In the latter case, two distinct twists are implied and Gregory's Square is applicable. Historically, users have avoided twists by setting them equal to zero. (This is sometimes hidden in the formulas.) In the mid-1970s I studied finite element methods, particularly during years at Brunel University and the University of Dundee. In finite element analysis, there are 'serendipity elements', which are methods with unexpectedly high accuracy. The explanation for the higher accuracy is the higher polynomial precision. Most finite elements can be used as interpolants. 'Adini's Rectangle' is a serendipitous 12 degree of freedom interpolant that is C^0 over a network of rectangular patches. The 12 degrees of freedom are the positions and tangents at the four corners of a rectangular element. Jim Brown and I had the idea of calculating the twist of Adini's Rectangle for use as a twist approximation. (The first possibility is to use the twist of a bilinear interpolant to positional data; Adini's Rectangle is an upscale form.) Of course, a vertex common to four adjoining rectangles would thus have four twists, so these are suitably averaged. The details appear in [12]. These are now called 'Adini's twists' in the CAGD literature. We gave the formulas for uniform domains: Boehm *et al.* [13] give the formula for possibly scaled rectangular domains.

Figures 8.11–8.20 illustrate a sequence of Coons' patch possibilities. A Coons' patch with zero twists is shown in Figure 8.11 and the Coons' patch with exact twists is shown in Figure 8.12. Gregory's Square and Brown's Square to the same data are shown in Figures 8.13 and 8.14, respectively. Boolean sums tend to give 'fuller' surfaces than convex combinations. This tendency is somewhat discernible from Figures 8.13 and 8.14. Figures 8.15–8.20 represent the differences between various patches, scaled to emphasize the differences.

Curvature analysis

Twists are (1, 1) derivatives, i.e. second derivatives. (Twists are not geometric invariants

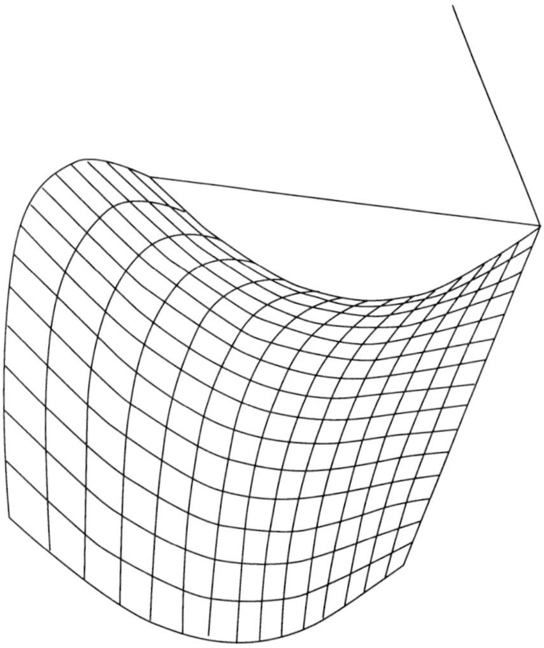

Figure 8.11. A Coons' Patch with zero twists.

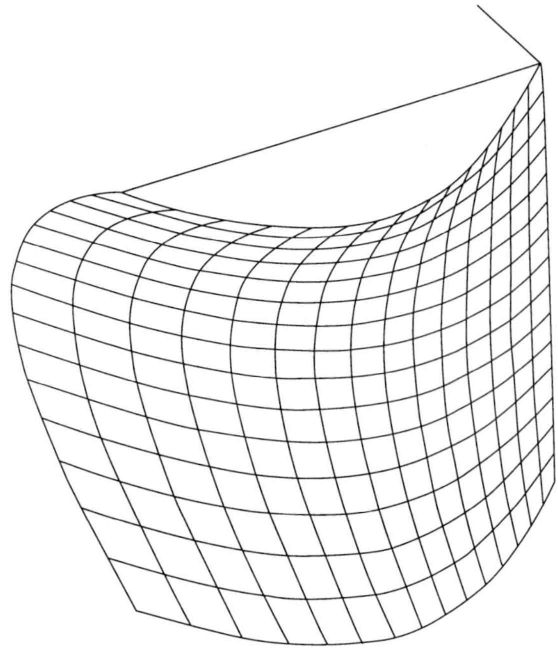

Figure 8.13. Gregory's Square to the same data.

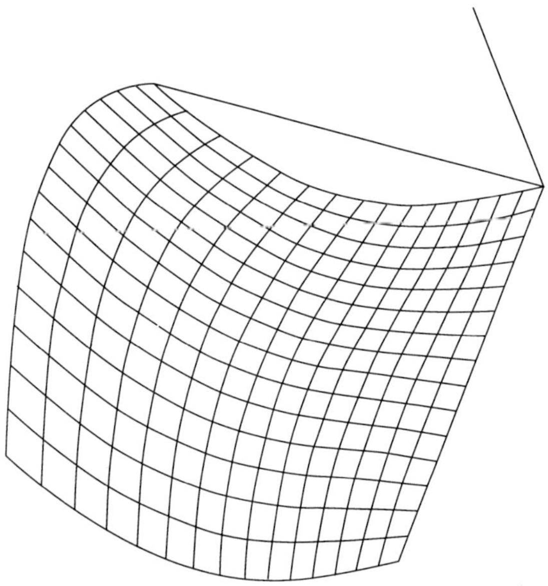

Figure 8.12. The Coons' Patch in Figure 8.11 with exact twists.

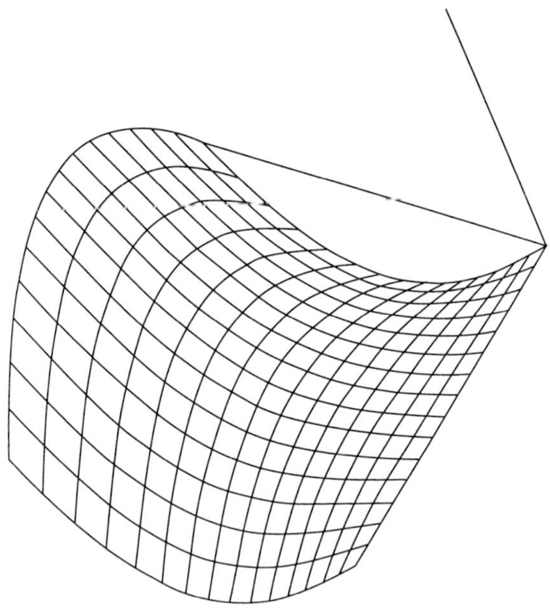

Figure 8.14. Brown's Square to the same data.

in that a reparametrization changes the twists [14].) Detection of different twists by viewing surfaces *per se* is difficult. Thus Barnhill and Farin had the idea of using surface curvatures, which involve second derivatives, to study twist approximations. This study included considering various curvatures for surface interrogations. The three curvature measures of mean curvature, Gaussian curvature, and the average of the absolute curvatures when used together provide

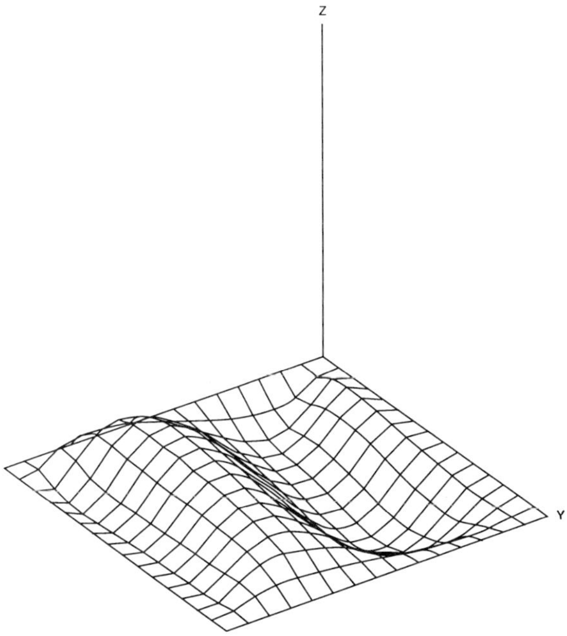

Figure 8.15. Gregory's Square minus Brown's Square, multiplied by 20.

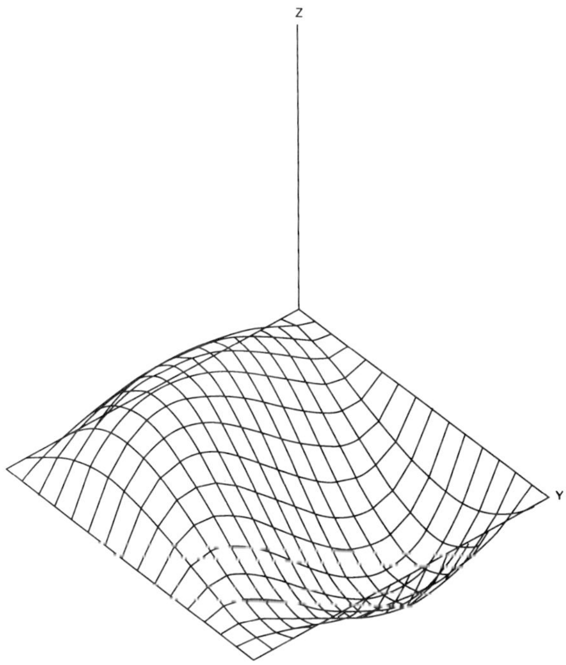

Figure 8.16. Gregory's Square minus the exact primitive function, multiplied by 5.

a good curvature analysis for surface interrogation. (See [15] and the references therein.) Curvature analysis is a modern tool that is just coming into real practical utility.

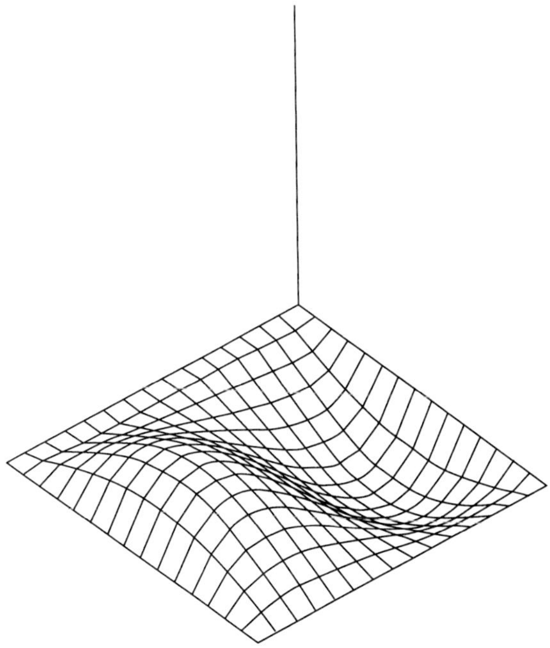

Figure 8.17. Gregory's Square minus the zero twist patch, multiplied by 5.

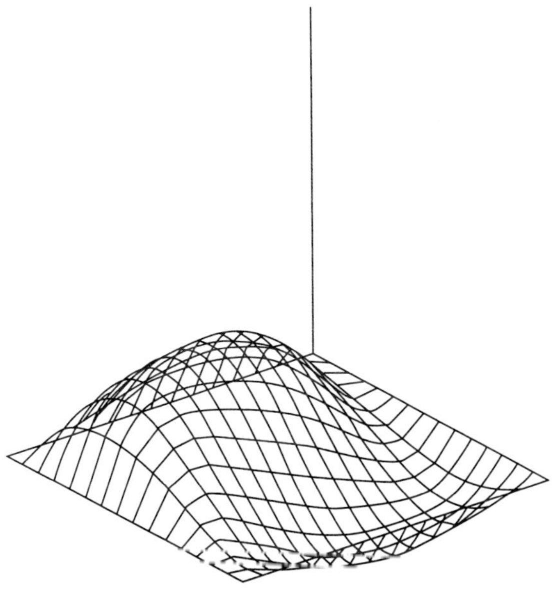

Figure 8.18. Brown's Square minus the exact primitive function, multiplied by 5.

The following pictures from Fayard [16] show some of the possibilities for surface interrogation using curvature analysis. A 'trunk lid' and its three curvatures, Gaussian, mean and average of the absolute curvatures, is displayed in Figure 8.21. Next the trunk lid (of length 760

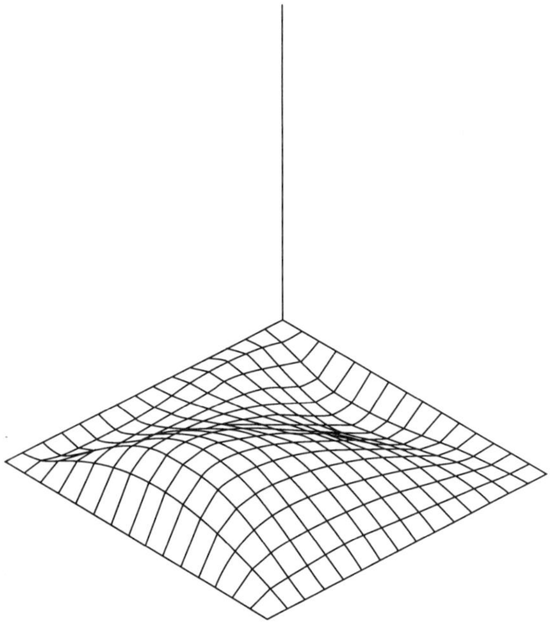

Figure 8.19. Brown's Square minus the zero twist patch, multiplied by 5.

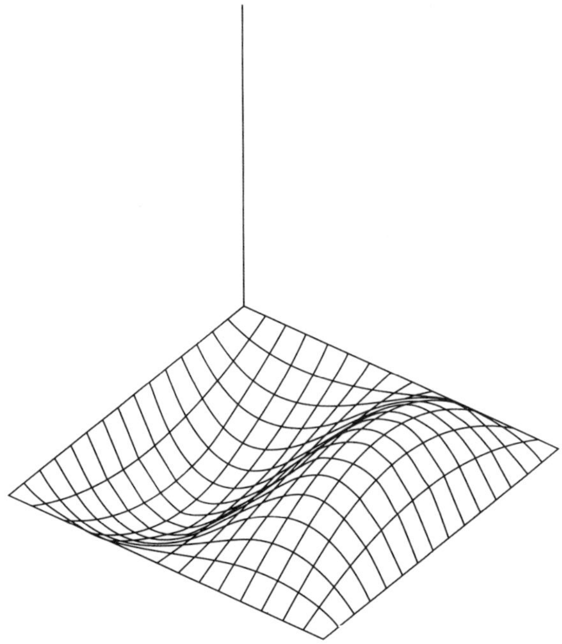

Figure 8.20. The exact primitive function minus the zero twist patch, multiplied by 5.

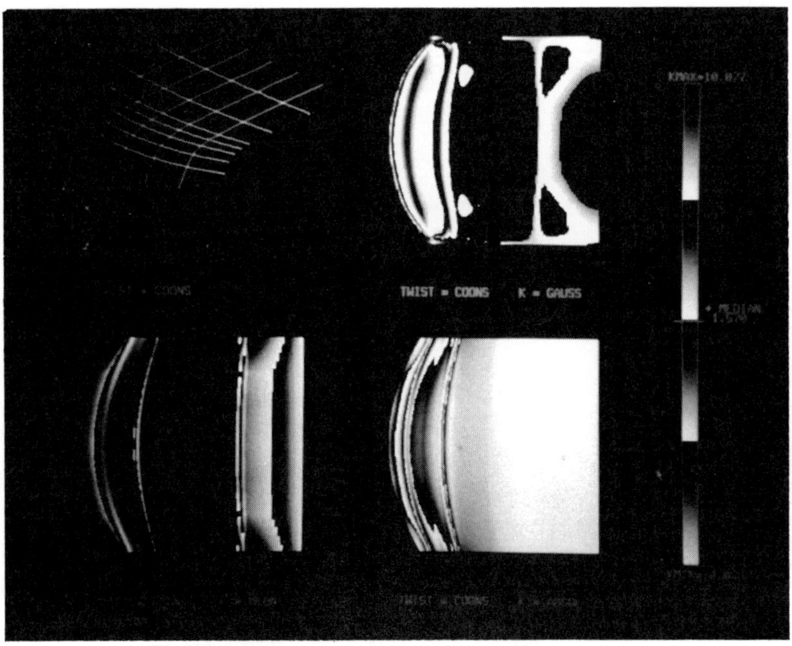

Figure 8.21. A 'trunk lid' and its three curvatures: Gaussian, mean and average of the absolute curvatures.

centimeters) is perturbed by 0.1 millimeter. It looks the same as the unperturbed trunk lid. But the three curvatures are considerably different, as shown in Figure 8.22.

The use of curvature analysis to determine the quality of twist approximations is the next topic to be illustrated. The Gaussian curvature of the trunk lid for four twist approximations is

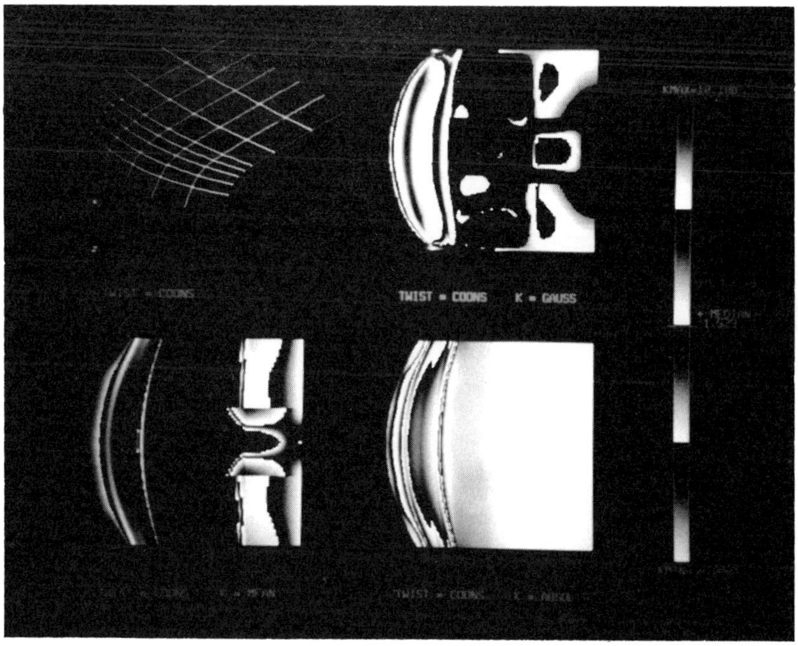

Figure 8.22. A 'perturbed' trunk lid and its three curvatures.

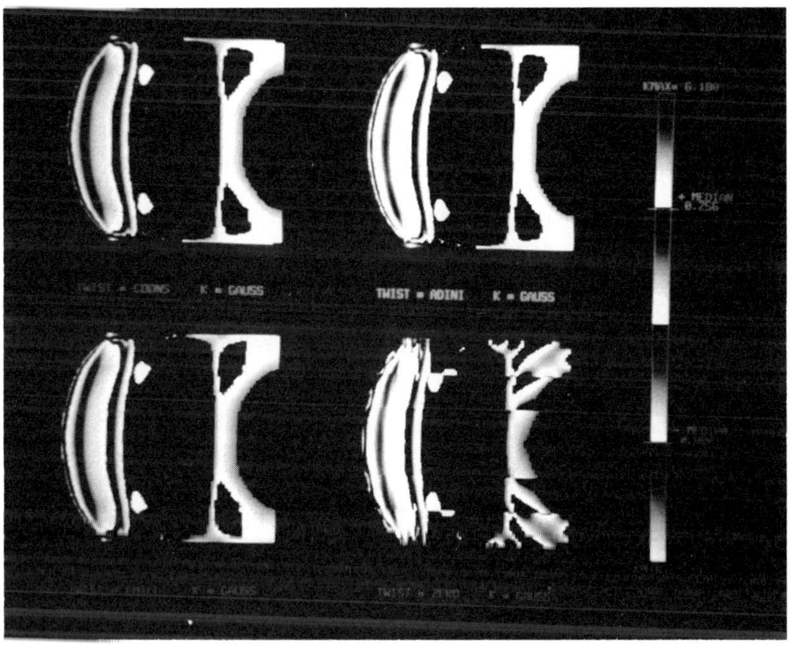

Figure 8.23. Gaussian curvature of the trunk lid for four twist approximations.

illustrated in Figure 8.23. In particular, we see that the curvature analysis indicates that using zero twists leads to poor results. In this and the following pictures, the 'Coons' curvature is the curvature of a bilinear patch (see [12]), 'FMILL' is discussed in Boehm *et al.* [13] and 'Adini' curvature comes from the Adini's Rectangle discussed above [12].

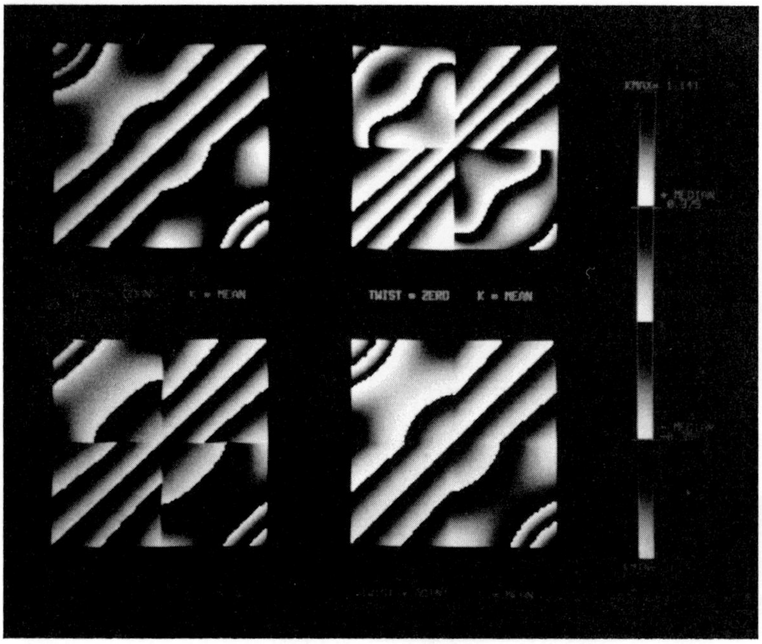

Figure 8.24. An alternative rendering, 'gray scale bands', for the example in Plate 11.

Reflection lines provide a different surface interrogation tool. For the surface shown in Plate 10 the 'horizontal' and 'diagonal' reflection lines are depicted on the bottom row's left and right sides, respectively. The mean curvature of this surface is used to study the quality of the four twist approximations in Plate 11. Only the Adini's twists yield good results for this surface. If 'gray scale bands' are used for the example in Plate 11, the result is shown in Figure 8.24. Color pictures allow a continuous change of the variables whereas gray scales allow sharp demarcations but do not specify the actual values from the 'color bar' shown on the right side of each picture. The Gaussian curvature for the same data and approximations is shown in Figure 8.25. Considerable insight into a surface can be obtained from studying such pictures.

Generalizations

The compatibly corrected bicubically blended Coons' patch can be generalized in several ways. We shall discuss two: higher degree bivariate interpolants and multidimensional Coons' patches.

Higher order incompatibilities

We gave the compatibly corrected bilinearly blended and bicubically blended Coons' patches above. Frank Little generalized Gregory's Square to take care of incompatibilities in position and tangent data [17]. A next case is biquintically blended Coons' patches. My 1981/82 CAGD class (their names are listed in the reference) and I developed a C^2 Gregory's Square and C^2 Brown's Square [18]. The composition of univariate quintic Hermite interpolants leads to incompatibilities involving derivatives of order two, three and four. (Lower order incompatibilities may also be considered, as above.)

Compatibly corrected interpolants are proper generalizations: if there are no incompatibilities, then the interpolants reduce to the usual 'uncorrected' schemes. Thus, from a scientific viewpoint, the compatibly corrected interpolants should always be used. However, many users are reluctant to program the more complicated compatibly corrected interpolants. This reluctance is not well founded because, for example, in the bicubic case, the 'variable twist partition' can be dealt with independently of the rest of the scheme. Since twists usually must be

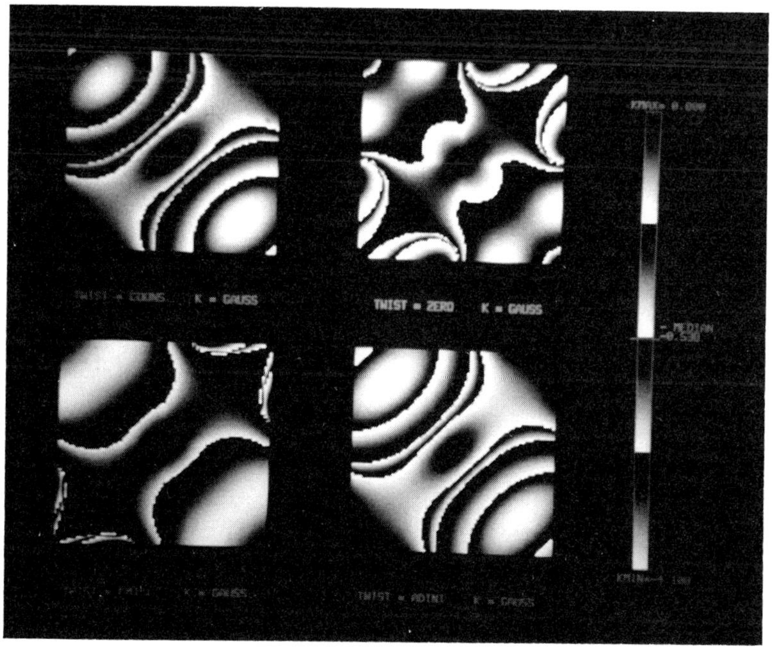

Figure 8.25. Gaussian curvature used for the example in Plate 11.

generated anyway, the change in data structure is not significant.

Higher dimensional patches

Bivariate Coons' patches are useful for three-dimensional surface design. Higher dimensional Coons' patches can also be useful. For example, trivariate Coons' patches can be used to map the unit cube into other solids. (This 'domain deformation' had been used by Bruce Brown with Coons' Patches and Tom Sederberg with Bézier Patches.) The utility of such mappings is that data defined on a somewhat arbitrary domain can sometimes be mapped back to (or at least associated with) data on a standard domain such as a cube, for which interpolants are well known. I became interested in the trivariate Coons' patch in connection with 'surfaces defined on surfaces' [19], a major research area *per se*. I initiated the trivariate generalization, and then collaborated with Andrew Worsey [20], who also continued this generalization [21].

TRIANGULAR COONS' PATCHES: THE BBG INTERPOLANT

In the summer of 1969, Garrett Birkhoff, Bill Gordon and I generalized Coons' patches to a triangular domain. This research eventually appeared in 1973 [6]. Our idea was to use lofting interpolants to two sets of boundary data, followed by an appropriate Boolean sum of the lofting interpolants. Later, Frank Little generalized the domain from a 'standard' triangle to an arbitrary triangle via barycentric coordinates and appropriate differentiation rules as follows (see [18, 22]):

If V is an arbitrary point in triangle $V_1V_2V_3$, then it can be represented as $V = b_1V_1 + b_2V_2 + b_3V_3$ where $b_1 + b_2 + b_3 = 1$ and the b_i are non-negative. The b_i are called the barycentric coordinates of V with respect to triangle $V_1V_2V_3$ and are the relative areas of the subtriangles illustrated in Figure 8.26. V may also be considered as a point (x, y). Then a function $F - F(x, y)$ can be differentiated as follows:

$$F(x, y) = F(b_1V_1 + b_2V_2 + b_3V_3) = \hat{F}(b_1, b_2, b_3) \tag{32}$$

A directional derivative is given by

$$\frac{\partial F}{\partial e} = \sum_{i=1}^{3} \frac{\partial \hat{F}}{\partial b_i}\frac{\partial b_i}{\partial e} \tag{33}$$

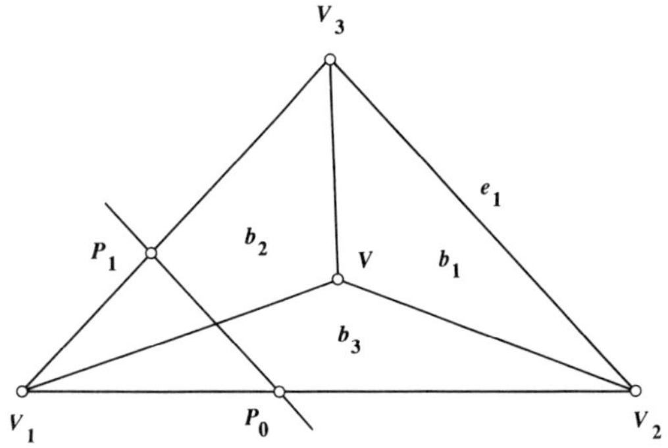

Figure 8.26. The barycentric coordinates of the point V and also the points P_0 and P_1 used in the definition of the BBG lofting interpolant and defined in equation (31).

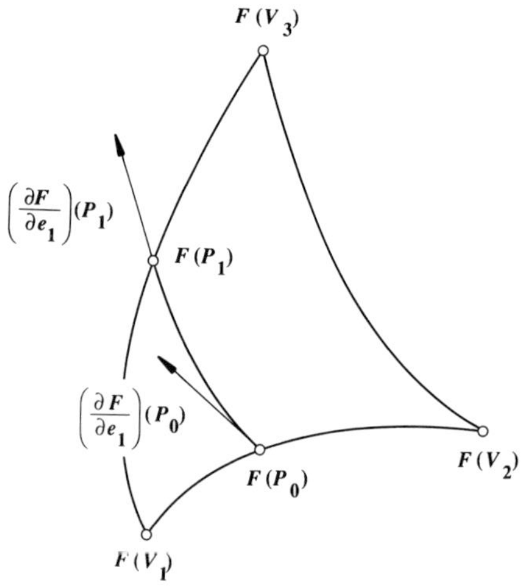

Figure 8.27. The data used in the definition of the lofting interpolant P_1F in the Barnhill, Birkhoff and Gordon triangular Coons' Patch.

where e is an arbitrary direction. Equations (32) and (33) enable us to derive triangular Coons' patches of several types, such as Barnhill, Birkhoff, Gordon triangles, Nielson's radial interpolants, and others, all of which can be discretized to finite dimensional and other kinds of interpolants.

Next we define the BBG interpolant for an arbitrary triangle.

P_iF is defined as the cubic Hermite interpolant to C^1 data parallel to edge e_i, $i = 1, 2, 3$. With the notation

$$P_0 = b_1V_1 + (1 - b_1)V_2$$

and (34)

$$P_1 = b_1V_1 + (1 - b_1)V_3$$

we can write P_1F as

$$P_1F = h_0\left(\frac{b_3}{1-b_1}\right)F(P_0) + h_1\left(\frac{b_3}{1-b_1}\right)F(P_1) + \bar{h}_0\left(\frac{b_3}{1-b_1}\right)(1-b_1)\left(\frac{\partial F}{\partial e_1}\right)(P_0) + \bar{h}_1\left(\frac{b_3}{1-b_1}\right)(1-b_1)\left(\frac{\partial F}{\partial e_1}\right)(P_1) \tag{35}$$

The data used in this lofting interpolant are shown in Figure 8.27.

A compatibility corrected BBG scheme is the following [18]:

$$\begin{aligned}
[P_3 \oplus (P_1 \oplus P_2)]F
&= h_0\left(\frac{b_1}{1-b_2}\right)F(b_2V_2 + (1-b_2)V_3) \\
&+ h_1\left(\frac{b_1}{1-b_2}\right)F((1-b_2)V_1 + b_2V_2) \\
&+ \bar{h}_0\left(\frac{b_1}{1-b_2}\right)(1-b_2) \\
&\times \frac{\partial F}{\partial e_2}(b_2V_2 + (1-b_2)V_3) \\
&+ \bar{h}_1\left(\frac{b_1}{1-b_2}\right)(1-b_2) \\
&\times \frac{\partial F}{\partial e_2}((1-b_2)V_1 + b_2V_2) \\
&+ h_1\left(\frac{b_3}{1-b_1}\right)F(b_1V_1 + (1-b_1)V_3) \\
&+ \bar{h}_1\left(\frac{b_3}{1-b_1}\right)(1-b_1) \\
&\times \frac{\partial F}{\partial e_1}(b_1V_1 + (1-b_1)V_3) \\
&- h_1\left(\frac{b_3}{1-b_1}\right)\Big[h_0(b_1)F(V_3)
\end{aligned}$$

$$+ h_1(b_1)F(V_1) + \bar{h}_0(b_1) \times \frac{\partial F}{\partial e_2}(V_3) + \bar{h}_1(b_1)\frac{\partial F}{\partial e_2}(V_1)\Big]$$

$$- \bar{h}_1\left(\frac{b_3}{1-b_1}\right)(1-b_1) \times \Big\{(-b_1)\Big[h_0'(b_1)F(V_3) + h_1'(b_1)F(V_1) + \bar{h}_0'(b_1)\frac{\partial F}{\partial e_2}(V_3) + \bar{h}_1'(b_1)\frac{\partial F}{\partial e_2}(V_1)\Big]$$

$$+ \bar{h}_0(b_1)\frac{\partial F}{\partial e_2}(V_3) + \bar{h}_1(b_1)\frac{\partial F}{\partial e_2}(V_1) + h_0(b_1)\frac{\partial F}{\partial e_1}(V_3) - h_1(b_1)\frac{\partial F}{\partial e_3}(V_1) + \bar{h}_0(b_1) \times \frac{\partial^2 F}{\partial e_1\,\partial e_2}(V_3) - \bar{h}_1(b_1)\frac{\partial^2 F}{\partial e_3\,\partial e_2}(V_1)\Big\}$$

$$+ \bar{h}_1\left(\frac{b_2}{1-b_3}\right)(1-b_3)\bar{h}_1(b_3) \times \left[\frac{\partial^2 F}{\partial e_2\,\partial e_1}(V_3) - \frac{\partial^2 F}{\partial e_1\,\partial e_2}(V_3)\right]$$

Suppose that a user has arbitrarily spaced data of the form $\{x_i, y_i, F(x_i, y_i)\}_{i=1}^{n}$ and that he wants a C^1 surface interpolating to these data. Such a C^1 surface can be constructed as follows:

1. Triangulate the $\{(x_i, y_i)\}_{i=1}^{n}$.
   (a) Obtain a quick first triangulation.
   (b) Optimize this triangulation.
2. Estimate the gradient $\nabla F(x_i, y_i)$ at the data points.
3. Use a C^1 scheme which requires C^1 data.

Step 3 can be achieved by discretizing the transfinite BBG scheme. We accomplish this discretization by assuming that F is cubic along each edge of the triangle, and its normal derivative is linear there. Little has worked out the following discretization of the transfinite BBG scheme:

$$[P_3 \oplus P_2 \oplus P_1 \oplus C]\tilde{F} = CF + 4b_1b_2b_3\sum_{i=1}^{3} w_i N_i$$

where

$$w_i = b_j b_k\left(\frac{1}{1-b_j} + \frac{1}{1-b_k}\right)$$

and

$$N_i = \frac{\partial(F - CF)}{\partial n_i}\bigg|_{m_i}$$

i, j, k all different, n_i being the direction normal to edge e_i and m_i the midpoint of e_i. Here

$$CF = \sum_{i=1}^{3} b_i^2\left[(3-2b_i)F(V_i) + b_j\frac{\partial F(V)}{\partial(V_j - V_i)} + b_k\frac{\partial F(V_i)}{\partial(V_k - V_i)}\right]$$

$$\frac{\partial F}{\partial n_i}\Big|_{m_i} = \frac{1}{2}[\nabla F(V_j) + \nabla F(V_k)]\cdot \boldsymbol{n}_i$$

and

$$\frac{\partial CF}{\partial n_i}\Big|_{m_i} = \left\{e_i\cdot e_j\left[\frac{3}{2}F(V_j) + \frac{1}{4}\frac{\partial F}{\partial e_i}(V_j)\right] + e_i\cdot e_k\left[\frac{3}{2}F(V_k) - \frac{1}{4}\frac{\partial F}{\partial e_i}(V_k)\right] + e_i\cdot e_i\left[-\frac{\partial F}{\partial e_j}(V_j) - \frac{\partial F}{\partial e_k}(V_k)\right]\right\}\Big/ ((\bar{e}_i\cdot e_k)\|e_i\|)$$

where $i = 1, 2, 3$, $j = i \bmod 3 + 1$ and $k = j \bmod 3 + 1$, and if $e = (a, b)$, then $\bar{e} = (-b, a)$.

Triangular interpolants have been generalized to tetrahedral for trivariate data. Barnhill and Little [23] developed the trivariate BBG and Nielson's radial interpolant. The development uses the permanence principle for Boolean sums, incidentally.

ADDITIONAL BOOLEAN SUM TOPICS

Barnhill and Gregory [9] investigated the interpolant and precision properties of the Boolean sum $P \oplus Q$ in terms of its constituents P and Q. This search led to the Barnhill/Gregory Theorem:

Duality theorem for Boolean sums *The Boolean sum $P \oplus Q$ has (at least) the interpolant properties of P and (at least) the function precision of Q.*

(They were constructing complicated triangular interpolants which led to the creation of additional useful theorems about Boolean sums. We are only using their simplest and most basic result here.)

This theorem can be used in (at least) two ways.

1. Take the Boolean sum of an interpolation scheme P which does interpolate but has poor polynomial precision with a scheme Q which has the desired polynomial precision but need not interpolate.
2. Build up a scheme from simpler schemes analogously to the Newton form of the (univariate) interpolating polynomial. That is, P_n is the interpolating polynomial operator to $n+1$ data, then

$$\begin{aligned} P_n \oplus P_{n-1} &= P_n + P_{n-1} - P_n P_{n-1} \\ &= P_n + P_{n-1} - P_{n-1} \\ &= P_n \end{aligned}$$

because $P_{n-1}F$ is in the precision set of P_n. We may view

$$P_n = P_n \oplus P_{n-1} = P_{n-1} + (I - P_n)P_{n-1}$$

as P_{n-1} plus a correction term. Using Davis' [4] language, this is a permanence principle.

Barnhill and Farin [24] used this idea to build up an explicit representation of a (bivariate) C^1 polynomial defined over a triangle. More precisely, the C^1 quintic Q_{21} was built up using Coons' and Bézier's representations, respectively.

I had Poeppelmeier [25] build up interpolants to arbitrary bivariate data by taking the Boolean sum $S \oplus B$ where S is Shepard's Formula [26] and B is the piecewise defined Barnhill-Gregory nine-degrees-of-freedom C^1 triangular interpolant. Poeppelmeier's scheme is an instance of both (1) and (2) above.

We state the general result as follows. Let $SF = \sum_{i\in\alpha} u_i L_i F$ and $BF = \sum_{j\in\beta} v_j M_j F$, where SF and BF are both in cardinal form, i.e. S interpolates to the linear functionals L_i with basis functions u_i, i in an index set α, and M_j, v_j and β_j are analogous. Then the Boolean sum $(S \oplus B)F$ has the following properties:

1. $S \oplus B$ has at least the interpolation properties of S.
2. $S \oplus B$ has at least the function precision of B.
3. $S \oplus B$ is of the (simpler) form

$$(S \oplus B)F = BF + \sum_{i\in\alpha-(\alpha\cap\beta)} u_i[L_iF - L_iBF]$$

A simpler example is S equal to Shepard's Formula S_0 for four points and L equal to linear interpolation for the first three of the four points. Then $(S_0 \oplus \mathrm{L})F = LF + w_4[F_4 - LF_4]$ where w_4 is the fourth Shepard weight function and G_4 means evaluation of the function G at the fourth point. We observe that the Boolean sum $S_0 \oplus L$ interpolates at all four points, has linear polynomial precision, and its computation involves the addition of only one term to the (simpler) function LF, which is a type of permanence principle.

A second example in Poeppelmeier [25] uses S_0 and the Barnhill/Gregory scheme B defined over a subset of the points that define S_0. The Boolean sum $(S_0 \oplus B)F$ is C^1, in addition to having properties (1)–(3) above. This smoothness is an instance of the following general result:

Smoothness of Boolean sums
The smoothness of a general Boolean sum $P \oplus Q$ is as follows. Let $PF \in C^p$ and $QF \in C^q$ for all relevant F. Assume that PQF (or at least $P(I-Q)F$) can be formed. Then

$$(P \oplus Q)F \in C^{\min\{p,q\}}$$

(For additional information, see Barnhill and Stead [27].)

CONVEX COMBINATIONS

Convex combinations as an alternative to Boolean sums

We mentioned above that forming the composition of univariate cubic Hermite interpolants introduces (1, 1) derivatives, called 'twists'. It is obvious that the composition of two interpolants containing derivative functionals always involves derivatives of higher order than the projectors themselves. We pondered this fact – and what to do about it – at the University of Utah in the mid-1970s. We assembled ingredients as follows: In Philip Davis' book, we read of convex hulls. (See, in particular, his example of finding the convex hull of planar points with nails and string.)

Definitions The convex hull (CH) of a given set S is the intersection of all convex sets containing S [4]. A set is convex if it contains all of its line segments, that is, a and b in S implies that $(1-t)a+tb$ is in S for $0 \leqslant t \leqslant 1$. Then I made the definition: A convex combination (CC) of $\{a_i\}_{i=1}^n$ is a linear combination of the form $\Sigma_{i=1}^n \alpha_i a_i$ where all the $\alpha_i \geqslant 0$ and $\Sigma_{i=1}^n \alpha_i \equiv 1$ [17].

For 15 years or so, students in my CAGD classes have proved that CH = CC. (This problem is usually given in the Bézier approximation part of the course.) Some approximations are convex combinations, but others are not. Examples of these possibilities are B-splines and Lagrange interpolation, respectively. We learned later that convex combinations are also called 'partitions of unity' in other literature.

Convex combination theorems

Several interpolants, such as Brown's Square and Gregory's simplex interpolants were developed in the 1970s. In 1978 Frank Little and I were working on triangular interpolants such as the BBG interpolants and Nielson's radial interpolant. I developed the following (heretofore unpublished) theorem in an effort to systematize this research (as always, it appeared in the CAGD class):

Convex combination theorem (for C^1 triangular interpolants) *Let F_1, F_2 and F_3 be defined on triangle T and $F_1 = F_2 = F_3 = F$ on ∂T (the boundary of T). Let α, β and γ be functions defined on T and have the following properties:*

$$\alpha \equiv 1 \quad on\ e_1$$
$$\beta \equiv 1 \quad on\ e_2$$
$$\gamma \equiv 1 \quad on\ e_3$$

and $\alpha + \beta + \gamma - 1$, where the e_i are T's edges. Form

$$P \equiv \alpha F_1 + \beta F_2 + \gamma F_3$$

Then $P = F$ on ∂T and moreover, $DP = DF_i$ on e_i, $i = 1, 2, 3$ respectively.

General case of convex combination theorem (for triangular interpolants) Given F_i defined on e_i for $i = 1, 2, 3$ and α, β, γ such that

$$\alpha \equiv 1 \quad on\ e_1$$
$$\beta \equiv 1 \quad on\ e_2$$
$$\gamma \equiv 1 \quad on\ e_3$$

and $\alpha + \beta + \gamma \equiv 1$ on ∂T. Let

$$PF = \alpha F_1 + \beta F_2 + \gamma F_3$$

Then

$$D^n PF|_{e_i} = D^n F_i|_{e_i}$$

if:

1. $D^i F_1 = D^i F_2 = D^i F_3 \equiv D^i F \quad \text{on } \partial T$

   $i = 0, 1, \ldots, m, \quad m < n$

2. $D^j \alpha = D^j \beta = D^j \gamma \equiv 0 \quad \text{on } \partial T$

   $j = 1, \ldots, n-m-1, \quad -1 \leqslant m$

In fact, $D^K PF|_{e_i} = D^K F_i|_{e_i}\ \forall k \leqslant n$.

These theorems enabled one to combine various interpolants to produce interpolants with appropriate smoothness but without forming Boolean sums. (Also see [28].) Generalizations to interpolants over domains other than triangles are straightforward.

For additional information about convex combinations see [29]. The very important triangular Shepard's method, due to Little, is described in his article.

CONCLUSION

The original Coons' patches were developed as solutions to concrete problems in engineering design. But a rich and remarkable plethora of ideas have come from this beginning, including: (1) Boolean sums, both for description of old methods and for construction of new methods; (2) transfinite triangular interpolants (and their connection to finite element analysis); (3)

multidimensional interpolants; (4) convex combinations as alternatives to Boolean sums. The totality of these ideas provides powerful methods for solving many problems in science and engineering.

ACKNOWLEDGEMENTS

I would like to thank the Editor, Les Piegl, for inspiring me to write this article. I would like to thank the Editor for Figures 8.1–8.10, 8.26 and 8.27, Steve Watson for Figures 8.11–8.20 and Larry Fayard for Plates 10 and 11 as well as Figures 8.21–8.29. This research was supported by DoE grant DE-FG-02-87ER25041 and by NSF grant DMC-8807747 to Arizona State University.

REFERENCES

1 **S. A. Coons** *Surfaces for computer aided design,* Mechanical Engineering Department, MIT, 1964 (revised 1967). (Available as AD 663 504 from the National Technical Information Service, Springfield, VA 22161.)

2 **P. Bézier** Définition numérique des courbes et surfaces, *Automatisme,* Vol 11, 1966, 625–632.

3 **P. Bézier** Définition numérique des courbes et surfaces (II), *Automatisme,* Vol 12, 1967, 17–21.

4 **P. J. Davis** *Interpolation and Approximation,* Dover, New York, 1975.

5 **W. J. Gordon** Distributive lattices and the approximation of multivariate functions, in *Approximations with Special Emphasis on Splines,* I. J. Schoenberg, ed., University of Wisconsin Press, Madison, 1969.

6 **R. E. Barnhill, G. Birkhoff and W. J. Gordon** Smooth interpolation in triangles, *J. Approx. Theory,* Vol 8 No 2, 1973, 114–128.

7 **F. F. Little** Private communication, 1978.

8 **R. E. Barnhill and R. F. Riesenfeld eds.** *Computer Aided Geometric Design,* Academic Press, New York, 1974.

9 **R. E. Barnhill and J. A. Gregory** Polynomial interpolation to boundary data on triangles, *Mathematics of Computation,* Vol 29, 1975, 726–735.

10 **G. Farin** *Curves and Surfaces for Computer Aided Geometric Design* (second edition), Academic Press, San Diego, 1990.

11 **G. M. Nielson** Some piecewise polynomial alternatives to splines under tension, in *Computer Aided Geometric Design,* R. E. Barnhill and R. F. Riesenfeld, eds. Academic Press, New York, 1974.

12 **R. E. Barnhill, J. H. Brown and I. M. Klucewicz** A new twist for computer aided geometric design, *Computer Graphics and Image Processing,* Vol 8, 1978, 78–91.

13 **W. Boehm, G. Farin and J. Kahmann** A survey of curve and surface methods in CAGD, *Comput. Aided Geometric Design,* Vol 1, 1984, 1–60.

14 **I. D. Faux and M. J. Pratt** *Computational Geometry for Design and Manufacture,* Halsted Press, New York, 1979.

15 **R. E. Barnhill, G. Farin, L. Fayard and H. Hagen** Twists, curvatures and surface interrogation, *Comput.-Aided Design,* 1987, 341–346.

16 **L. J. Fayard** *Curvature for surface interrogation.* Masters thesis, Computer Science Department, Arizona State University, 1988.

17 **R. E. Barnhill** Representation and approximation of surfaces, in *Mathematical Software III,* J. R. Rice, ed., Academic Press, New York, 1977.

18 **R. E. Barnhill** Computer aided surface representation and design, in *Surfaces in Computer Aided Geometric Design,* R. E. Barnhill and W. Boehm, eds., North-Holland, Amsterdam, 1983.

19 **R. E. Barnhill and T. A. Foley** Methods for constructing surfaces on surfaces, in *Geometric Modeling,* H. Hagen and D. Roller, eds. Springer, New York, 1991.

20 **R. E. Barnhill and A. J. Worsey** Smooth interpolation over hypercubes, *Comput.-Aided Geometric Design,* Vol 1, 1984, 101–113.

21 **A. J. Worsey** C^2 interpolation over hypercubes, in *Surfaces in CAGD '84,* R. E. Barnhill and W. Boehm, North-Holland, Amsterdam, 1985.

22 **R. E. Barnhill** Coons' patches, *Computers in Industry,* Vol 3 No 2, 1982, 37–43.

23 **R. E. Barnhill and F. F. Little** Three- and Four-dimensional surfaces, *Rocky Mountain J. Math.* Vol 14, 1984, 77–102.

24 **R. E. Barnhill and G. Farin** C^1 quintic interpolation over triangles: two explicit representations, *International J. Numer. Methods in Engineering,* Vol 17, 1981, 1763–1778.

25 **C. C. Poeppelmeier** *A Boolean sum interpolation scheme to random data for computer aided geometric design,* Masters thesis, Computer Science Department, University of Utah, 1975.

26 **D. Shepard** A two-dimensional interpolation function for irregularly-spaced data, *Proceedings Association for Computing Machinery National Conference,* 1968, 517–524.

27 **R. E. Barnhill and S. E. Stead** Multistage trivariate surfaces, *Rocky Mountain J. Math.,* Vol 14, 1984, 103–118.

28 **G. J. Herron** *Triangular and Multisided Patch Schemes,* PhD thesis, Department of Mathematics, University of Utah, 1979.

29 **F. F. Little** Convex combination surfaces, in *Surfaces in Computer Aided Geometric Design,* R. E. Barnhill and W. Boehm, eds., North-Holland, Amsterdam, 1983.

ABOUT THE AUTHOR

Robert E. Barnhill received his doctorate in mathematics, with a numerical analysis specialization, at the University of Wisconsin in 1964. His undergraduate work in mathematics (minor: sociology) was at the University of Kansas. He was always interested in mathematics and became interested in computers working for Paul Gilles and the physical chemists at the University of Kansas. He joined the mathematics faculty at the University of Utah in 1964, where he remained until 1986. During this time he spent a year at Brown University, three years in the United Kingdom (two years at Brunel University and one year at the University of Dundee), and shorter periods at the General Motors Research Labs and at the Office of Naval Research. In 1986 he accepted a new kind of challenge by becomining the Chair of the Computer Science Department at the Arizona State University. He has served as Interim Vice-President for Research at ASU from 1991–1993.

In the 1970s, Dr Barnhill merged his interest in surfaces and computer graphics by naming and helping to establish *Computer Aided Geometric Design,* commonly referred to as CAGD, as a scientific discipline. CAGD is now known worldwide. To provide a high quality publication outlet, he helped form the research journal *Computer Aided Geometric Design.* Both at Utah and at Arizona State, he formed CAGD Groups, emphasizing a team approach to research and development.

Dr Barnhill has brought many young researchers into the field, e.g. while at Utah he supervised 19 PhD dissertations, 16 Master theses, and was a consultant on another 10 dissertations and theses at other universities. Many of his students are themselves well known. Dr Barnhill has had federal research funding throughout his scientific career. He has emphasized the applications of research, which has led to many industrial, government and university collaborations.

Preston Hammer, his advisor at Wisconsin, and Philip Davis at Brown had particular impact on his way looking at the world of research. However, many others have had a significant influence.

9 Boundary Modeling

Ian C. Braid

The last 20 years have seen engineering drawings giving way gradually to computer models of one kind and another that serve as descriptions of mechanical components and assemblies. One approach has been to record the bounding surfaces of solids in a piecewise manner. This technique, known as 'boundary modeling', underlies many of today's commercial systems for design of solid objects.

Sometimes boundary models have evolved from wireframe and surface models used in drafting systems; in other instances they have come about as a result of research into the general problem of representing and manipulating solids in a computer.

This chapter considers the evolution of boundary modeling, particularly the progress of research and its subsequent commercial exploitation by groups at Cambridge in England.

EARLY DAYS

In 1965, Professor M. V. Wilkes, Director of the University Mathematical Laboratory, invited Charles Lang, an engineering graduate of the University then working with Doug Ross at MIT, to return to Cambridge to set up a research team, the CAD Group, to study computer graphics and computer-aided design [1]. The Laboratory had recently acquired a PDP7 computer and DEC 340 display. One of the first tasks for the new group was to make a connection between the PDP7 and the Titan mainframe computer. Other early projects included ASP [2], a data structure package that found application in computer-aided design systems in Germany and Japan, a systems-programming language SAL [3], and a device-independent graphics package GINO [4] later developed and sold commercially by the Cambridge CAD Centre (an institute established by the British government in 1968 to promote the use of computer-aided design by industry). The CAD Group went on to study parametric curves and surfaces – the mathematics underlying them [5], their interactive design [6], and the production of physical models of surfaces cut in soft materials using a lightweight three-axis milling machine [7] constructed in the Laboratory. It also collaborated with Professor O. R. Frisch of the University Physics Department in the development of a very large screen laser display [8], which although it did not fulfill our hopes of a drawing-board sized screen for computer-aided design, found application in cartography.

In keeping with its practical orientation, the group had many contacts in industry, and was host at various times to perhaps a dozen industrial researchers who came for a few terms to pursue problems they had brought with them, or to assist in the current activities of the Group. One influential contact was that with Dr D. T. N. Williamson, research director to Molins Ltd., manufacturers of cigarette-making machinery (and developers of System-24, a forerunner of modern automated machining systems). Williamson pointed out that the ordinary bread-and-butter components of mechanical engineering were well worth studying, and helped us to see that these humble piece parts had a complexity, albeit different from that of doubly-curved surfaces, that might profit from the application of computer techniques.

In October 1969, I joined the group after undergraduate studies in mechanical engineering at Melbourne University, and four years of work in industry when my interests and job had shifted steadily towards computing. A *Scientific American* article by Ivan Sutherland [9] had revealed the possibilities of computer-aided design. Having first cleared an obligatory hurdle (to pass the one-year diploma course in computing with distinction), I arrived in the CAD Group hoping to work on computer-aided mechanism design. Charles Lang explained that it would be helpful if first we learnt how to model the components of a mechanism, and so my research topic was set.

From the start, the aim was to build up volumes by combining other volumes in a manner comprehensible to practising designers.

The colour plate for this chapter appears between pp. 210 and 211.

Fundamental Developments of Computer-Aided Geometric Modeling
ISBN 012-554765-X

Various schemes were tried. The first relied on juxtaposing many identical small volumes to make larger shapes. Representing each unit volume by setting a bit would allow general volumes to be combined using Boolean operations on bits. However, this scheme demanded too much storage space for adequate resolution (as engineers expect a range of 1 in 10^5, it seemed that 10^{15} bits might be required). There was also the expense of re-creating models after rotation, and the difficulty of deriving line drawings from such a model.

A second scheme envisaged stacks of blocks of variable size, again juxtaposed, each primitive being represented once and then being instanced repeatedly in a directed graph of transformations (Figure 9.1). The idea of using linear transformations and homogeneous coordinates came from Roberts [10]. However, there were at least two problems: the user might have difficulty creating, understanding and editing the directed graphs of transformations, and drawing bodies without showing hidden lines and 'cracks' between volumes would be expensive and could jeopardize interactive performance.

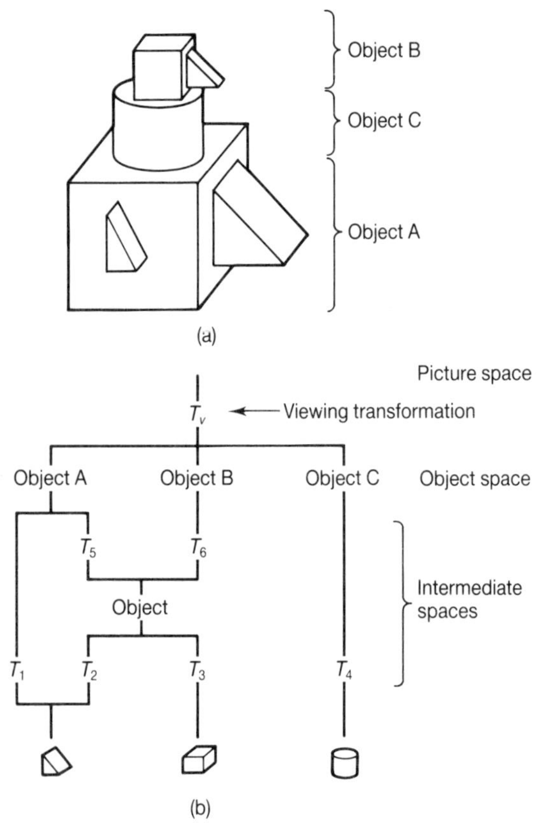

Figure 9.1. (a) A shape made by juxtaposing volumes of various sizes, (b) corresponding directed graph of transformations.

One possible solution to the latter problem envisaged an intermediate form consisting of just the outer surfaces of the juxtaposed bodies. The intermediate form would be stored as a collection of faces, edges and vertices, with external 'cracks' and all interior surfaces, edges and vertices, omitted. It would need to change when the directed graph was altered, but at least would speed up creation of a new picture following a change of view.

The next step was to find a structure to represent the primitive volumes at the leaf nodes of the directed graph. The same structure – of faces, edges and vertices – turned out to be suitable for representing the primitives. At this point it became apparent that the troublesome directed graph could be discarded, as primitives could be combined straight away into the intermediate form as soon as they had been positioned next to each other.

This became the basis of the final scheme which, by a roundabout route, had come to be what would now be termed a boundary modeler, with the 'intermediate form' or boundary model as its data. A vestige of the directed graph remained: face equations, i.e. surfaces, were held as references to surfaces of instanced and transformed primitives. It was hoped to preserve the advantage of the directed graph in grouping primitives together, by employing a suitable input language perhaps augmented by a system of constraints.

In the data structure (Figure 9.2), planes were held in implicit form ($f(x, y, z) = 0$), and cylindrical surfaces and straight and conic curves in rational parametric form. General curves of intersection between cylinders were not handled at all. Data structure entities were grouped into simple compact lists. As only 12 k (or exceptionally 16 k) of 48-bit words was available for program and data, pointers were represented as indices into the compact lists, and were stored as small integers packed into 9 bits of 12-bit quarter words. Thus there could be at most 512 of each entity (faces, edges and so on) in a body. The program itself was written in SAL (see above), plus a small amount of Fortran for routines making heavy use of floating-point arithmetic, and incorporated a simple overlay scheme.

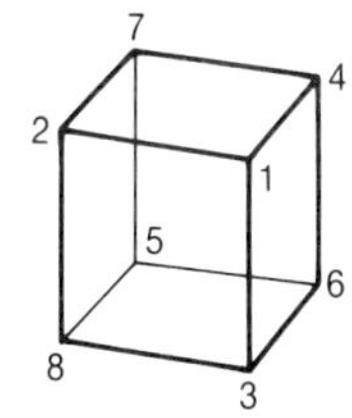

T_1

1·	0·	0·
0·	1·	0·
0·	0·	1·
0·	0·	0·

Unmodified face equation (stored once) $Ax+By+Cz+D=0$

	A	B	C	D
1	1·	0·	0·	−1·
2	0·	1·	0·	−1·
3	0·	0·	1·	−1·
4	−1·	0·	0·	−1·
5	0·	−1·	0·	−1·
6	0·	0·	−1·	−1·

Vertices

1	1·	1·	1·
2	−1·	1·	1·
3	1·	−1·	1·
4	1·	1·	−1·
5	−1·	−1·	−1·
6	1·	−1·	−1·
7	−1·	1·	−1·
8	−1·	−1·	1·

Faces → Edge lists

	Faces		Edge lists			
1	1	1	1	2	3	5
2	2	1	8′	6′	5′	4′
3	3	1	1′	6	7′	12
4	4	1	7	8	9	11
5	5	1	2′	12′	11′	10′
6	6	1	10	9′	4	3′

Reference to T_1

Reference to equation of *unmodified* face (face type)

Straight edges

	1st vertex	2nd vertex
1	1	3
2	3	6
3	6	4
4	7	4
5	4	1
6	1	2
7	8	2
8	2	7
9	7	5
10	6	5
11	5	8
12	8	3

Figure 9.2. The Build 1 data structure for a cube.

With a data structure established, the next task was to develop a method for gluing objects together at juxtaposed flat faces. Given two faces, with surfaces coplanar and of opposite orientation, the gluing routine compared all edges in one face with all edges of the other. Edges were split as soon as intersections were found, and hence subsequent comparisons had to be done with the pieces of the original edges resulting from the splitting. A second phase reconnected the edge pieces into new faces. Determination of multiply-connected faces required a point-in-face test to arrange hole and peripheral loops correctly. The fact that the area of a hole loop had to be less than the area of its enclosing peripheral loop, less the area of hole loops already assigned, reduced the number of calls to the test. A final pass over the data removed deleted entities and compressed the compact lists; it also removed 'smooth' edges where faces met without change to slope (Figure 9.3).

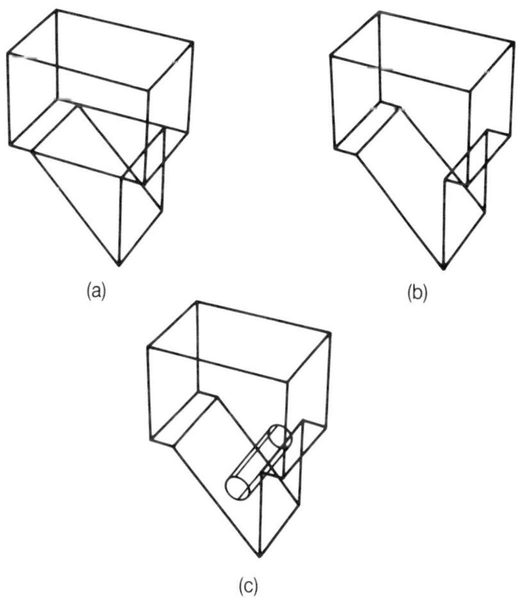

Figure 9.3. Gluing objects together at touching flat faces, (a) before, (b) after, (c) a negative cylinder 'glued' to a positive body.

With gluing working, it was then realized that if one object were to be negated, it would be possible to use gluing to make holes in bodies – at least in certain cases such as a hole entering and leaving a body through parallel flat faces where the ends of a negative cylinder could be glued to the parallel faces.

By this stage, one could construct quite a range of bodies, but the limitations of gluing were irksome. Building the commonly-required prismatic shapes by gluing blocks, sectors and blends together was laborious where it was not impossible. So a method of defining and sweeping profiles was added to construct such shapes directly and quickly.

Even then there was still no way of making a hole that entered and left a body through non-parallel faces, or was only partly contained by the body. A third operation was devised to allow the subtraction of a (transformed) cylinder or block from a general body. The complexities already met when gluing bodies at touching faces recommended starting with an operation simpler than general subtractions of arbitrary bodies.

A form of boxing speeded up the process. Both bodies were transformed to bring the cylinder (or cube) to the origin, to align it with the axes, and to make it of unit size. Faces of the arbitrary body found not to cut a sphere enclosing the unit cylinder or cube were marked, as were their edges (Figure 9.4). Detailed comparisons were then made between the unmarked edges of each body and the unmarked faces of the other, and intersections were recorded in a separate temporary list as well as by modifying the data structure of the bodies. A further stage re-assembled the pieces to form the resulting shape, using complicated algorithms that attempted to deal with any combination of intersecting or coinciding faces. As with gluing, it was found, unsurprisingly, that by negating argument bodies or result bodies or both, the subtraction procedure could be made to perform unions and intersections as well.

Looking back at this early enterprise [11] one can discern shortcomings not always evident then. In the Boolean algorithms, a great many particular cases were identified and dealt with, but there was no certainty that all had been enumerated, let alone solved (this was noted at the time). The problems of working with finite precision, and the need for care in numerical techniques, were hardly addressed at all. There was a dim appreciation that tangent intersections caused difficulties, but the problem was not tackled. No thought was given to the possibility of non-manifold shapes arising, or to the closure of the set of solids that could be built using the implemented operations (which clearly were very far from complete). A desire to protect the engineering user from the abstractions of Boolean algebra led to a deliberate underplaying of the notion that shapes could be expressed as Boolean combinations of point sets in three-dimensional space. In the implementation, the data structure gave rapid access from faces to edges and on to vertices, but access from vertices back to edges and thence to faces required searching.

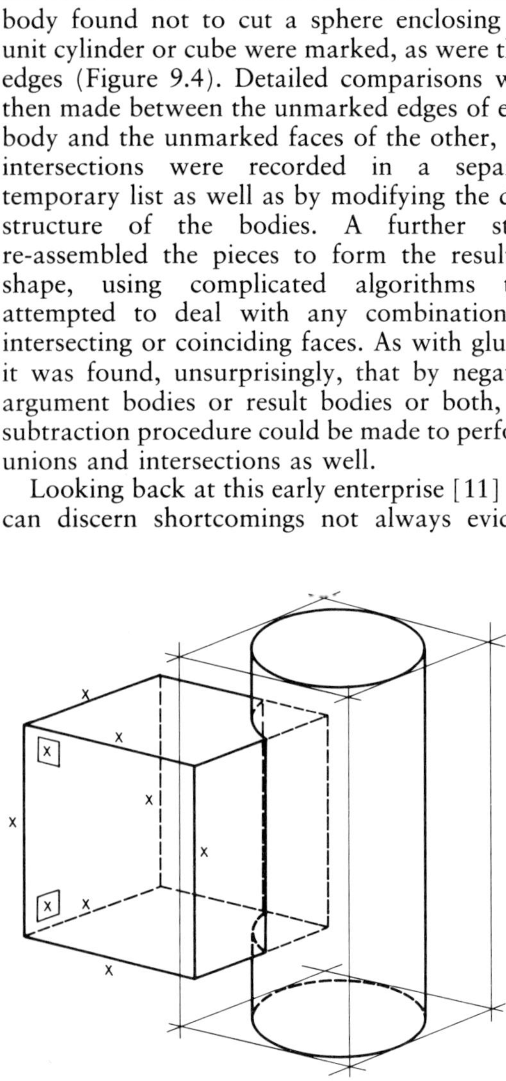

Figure 9.4. Boxing and marking of faces and edges. (Faces and edges found, from boxes around cylinder and block, to be non-intersecting are marked 'x'.)

Figure 9.5 shows a component modeled with the prototype system (termed 'Build', later 'Build 1'). Watkins' algorithm produced the picture as scan lines which were then drawn on the 340 display. Hidden-line pictures could also be made using a program developed by Alan Grayer, a research student and member of the CAD Group. Build also made accurate volume calculations using line integrals evaluated along the edges of each face. Permanent storage of models relied on writing the explicit face-edge-vertex structure to disc, or alternatively, on saving an implicit description in the form of the input commands used to create the part.

The first published account of this investigation was given at the 1973 Prolamat Conference held in Budapest [12]. That exciting conference brought together for the first time many researchers with similar interests in 3D modeling.

One speaker was Professor N. Okino of the University of Hokkaido. Since the mid-1960s he had been developing a modeler TIPS. It was what would now be called a CSG (constructive solid geometry) system in which shapes were recorded implicitly as Boolean combinations of primitive volumes [13]. The primitives were composed of intersections of half-spaces and could be unbounded (as were Build's negated bodies). Primitives could even be given as Fortran functions which took a point (x, y, z) as input and returned an approximate signed distance

Figure 9.5. A shape modeled with Build 1.

indicating the displacement of the point from the boundary, and whether it was within, outside, or on the boundary of the primitive. General shapes were recorded as positive or negative instances of primitives, which were combined by taking in effect the union of the positive primitives less the union of the negative primitives, as the resulting shape. Thus some care was needed to ensure that the expected result materialized. Whenever an output such as a picture, a mass property or machining data, was wanted, it was computed direct from the stored implicit description. An auxiliary model employing a regular array of cells in 3D together with a steepest-descent method using the distance function mentioned above, speeded up the process of finding points on the boundary of the resulting shape. Although the system was rather slow, it had few limitations and covered a far wider range of surface shapes than did other modelers at that time.

Others whom we met for the first time in Budapest included Professor M. Hosaka of Tokyo University, Professor G. Spur and Dr F.-L. Krause of the Technical University of West Berlin, and Professor H. B. Voelcker of Rochester University. All these institutions were developing research groups that would make significant contributions to the field of geometric modeling (see Chapter 11 of this book for an account of the PADL Group at Rochester).

There were others active in the field but not present at the conference. One was Dr M. Engeli of Fides Rechenzentrum, Zurich who had developed an Algol-like language in which geometric constructs and hence shapes could be expressed. Originally his system, Euklid [14] modeled only polyhedral shapes, and was notable for offering a rich set of sweeping operations, as well as for general Boolean operations in polyhedra. It has since been steadily improved, and now stands out for its ability to model and machine difficult doubly-curved shapes such as those found in forged and die-cast components.

Not to be confused with Engeli's system is the French modeler Euclid developed at this time by Dr J.-M. Brun at the University of Grenoble [15]. It too employed a specialized language, but based on Fortran augmented for geometric modeling. Its data structure, being intended only for polyhedra, was very concise and the modeler was fast. Euclid has become a leading European commercial modeler.

Turning from Europe to the United States, we find two more early modelers, neither intended for mechanical engineering. In one case, Professor C. M. Eastman at Carnegie Mellon University [16] created a modeler for architecture. His team paid more attention to topological complexity and to long-term storage of models than was common at the time. They envisaged models consisting of large numbers of separate components – even tens or hundreds of thousands as would be required to represent complete buildings, and so were led to consider the interplay of modelers and databases. They were also one of the first to make a clear distinction between the geometric and connective (also termed 'topological') data in a boundary model. By delaying the binding of geometric data to the topological data, they obtained models with fixed topology and variable geometry.

The second developer was B. G. Baumgart at Stanford University, who in a modeler constructed in support of vision research, made some of the most influential advances in boundary modeling [17]. One of these was the data structure which he termed a 'winged-edge' structure (Figure 9.6). Consisting of fixed-length records containing data and pointers, it gave rapid access between faces, edges and vertices. Most of the pointers were carried by edges. An edge referred not only to the vertices at its ends, and to the faces to either side, but also contained pointers to the neighboring edges met going clockwise or

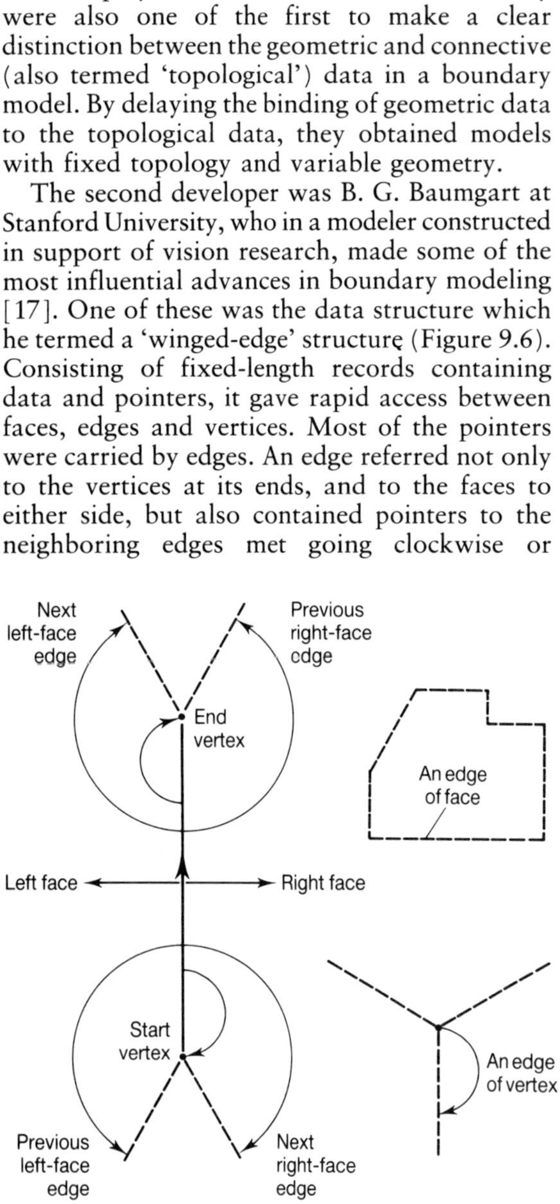

Figure 9.6. Baumgart's winged-edge data structure.

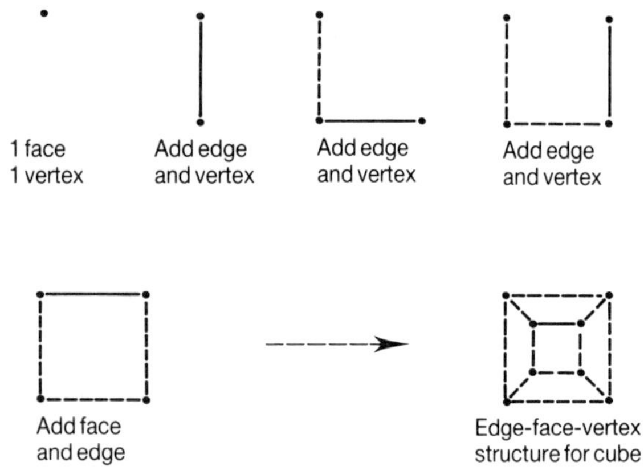

Figure 9.7. Using Baumgart's Euler operators to make the topology of a cube.

anti-clockwise round either face of the edge. The pointers to the four neighboring edges were the 'winged-edge' pointers. If a face carried a pointer to one of its edges, and likewise a vertex, the winged-edge pointers gave rapid access to the other edges of the face or vertex.

Baumgart's second innovation concerned the arrangements for creating and changing the data structure. He developed a small number of low-level routines for building up or dismantling the embedded graph formed by the edges of the solid. Having observed that the numbers of faces, edges and vertices of a body obeyed Euler's rule (faces + vertices = edges + 2), he provided a primitive consisting of a face and vertex only, together with two operations termed Euler operations: to add an edge and vertex, and to add an edge and face. This was a considerable advance on Build 1 where the six primitives had been constructed by hand – a laborious and error-prone business (Figure 9.7).

CONSOLIDATION AND APPLICATIONS

In 1973, Cambridge University replaced its Titan computer with an IBM machine. Before this happened, Alan Grayer [18] had created the only application of Build 1 in which prismatic shapes designed with Build, were cut on the model-making machine in an automatic process. His algorithms for generating cutter paths from boundary models ran effectively in a time proportional to model complexity. The face-adjacency information in the boundary model facilitated checks required in the machining process, for example, to discover quickly the faces bordering a face being machined and whether they did so at convex or concave edges (whence appropriate machining instructions could be generated).

The demise of Titan removed from the CAD Group any temptation to continue work on Build 1, and so design of son-of-Build began. This enjoyable period spanned much of 1974 and the early part of 1975 when I could devote all my time to planning and prototyping the new modeler (I succeeded Charles Lang as leader of the Group in May 1975).

It has been said [19] that where the first version of a computer program suffers the omissions of its creators such as cases not dealt with or data structures of insufficient richness, the second version is likely to err the other way with too much generality leading to too large a system, with too much complexity and too little efficiency. Build 2 was no exception.

Several aims guided the second round of modeling. We wished to handle the majority of engineering components by providing a much greater range of curves and surfaces, including doubly-curved surfaces such as spheres and toruses, and also free-form surfaces represented in parametric form. The modeler should be capable of representing and manipulating parts of greater topological complexity than hitherto. We hoped to arrive at more convenient and rapid means of building and changing models. Another aim was to develop applications based on the modeler, and to construct the modeler so that other programs could easily interface to it. A version of Build 1 organized as a subroutine library callable by Fortran routines had shown in a limited way what might be done.

At the same time, an attempt was made to understand and define mathematically the Boolean operations on solids used in Build 1 [20]. It was evident that the negate operation did not quite correspond to taking the complement of a (closed) point set, since the boundary remained. So rather than using open and closed point sets, we adopted a scheme based on three-valued logic. Although it found limited application in Build 2 which employed three-valued logicals and operators, it did not lead to a comprehensive theory on which to base robust and comprehensive Boolean algorithms.

It was decided to use a graph of nodes and pointers as the basis for the topological data structure so as to avoid the awkward compact lists of Build 1, and to give a flexible basis for experiment. News of Baumgart's work came soon after this decision, and his scheme of a winged-edge data structure was speedily adopted, and generalized to encompass objects containing multiply-connected faces.

The terminology of face-edge-vertex and surface-curve-point for topological and corresponding geometric entities was settled at this time, and came to be generally adopted.

On the geometric side, the choice was made for quadric surfaces, encouraged by Levin's [21] report of techniques for classifying quadrics and their curves of intersection. Toruses and parametric surfaces were put off until later, although it was hoped to organize the modeler in such a way that further surface types could be added without too much difficulty.

Another question to be settled was the language of implementation. The CAD Group had always counted itself fortunate to live in a computing milieu. This helped especially in the 'experimental' side of our work – in this case, in the construction of large, interactive, graphical programs which were then by no means as commonplace or as easily built as today. We were conscious of the need for good computing tools, and were fortunate that some such as the Titan multiple-access operating system already existed, and were able to build others (such as ASP, SAL and GINO, already referred to above).

The experience of creating Build 1 did not encourage further programming in assembler or Fortran. Luckily, others in the Laboratory were building a compiler for Algol-68 [22], so we decided to try that language for the new modeler. Although in the early days of the compiler, debugging was not without its perils (we sometimes had to resort to core dumps), the language turned out to be a huge asset. Wholesale changes to data structures, unthinkable for Build 1, could be undertaken knowing that the compiler would detect most errors. High-level data constructs coupled with programmer-defined operators invaluable for vector and matrix arithmetic and for geometry, made for much easier development and improvement of algorithms. A further advantage was that the arrangements for separate compilation together with the readability of the code meant that a handful of programmers could work together on a substantial program, aided by our simple source management system and little documentation beyond the source itself.

The first months were spent understanding what could be done with the language and in extending and implementing Baumgart's data structure and Euler operators. These were based on a more general form of Euler's equation

$$\text{vertices} + 2 \times \text{faces} - \text{edges} - \text{loops} = 2 \times (\text{bodies} - \text{genus})$$

where bodies stood for the total number of shells (connected sets of faces) of all bodies and genus for the sum of the genus over all bodies [23]. At this stage, the program was neither graphical nor interactive. Short sequences of compiled commands written in Algol-68 sufficed to control execution.

The Euler operators were immediately put to work in creating primitive wires and sheets, and in sweeping wires into sheets and sheets into solids. A 'wire' was the name given to an object composed of one or more edges embedded in a single face. The edges might be connected into a tree structure; more often they formed a simple open chain. As soon as an open chain became closed by the addition of a further edge, a second face came into being and each edge then belonged to two faces back to back, the whole forming a degenerate solid termed a lamina. A face of a lamina could be selected and swept along a vector; further calls to Euler operators built lateral faces corresponding to the surfaces swept out by each edge of the swept face. Wires and laminae could also be swept around an axis to create solids of revolution. Besides wires and laminae, both recorded as degenerate solids, Build 2 permitted free-standing geometry which was used to specify points and curves when making profiles with the Euler operators that added an edge and vertex or edge and face (Figure 9.8).

Sweeping was termed a *compound Euler operation*. Another such operation chamfered or rounded an edge or vertex, replacing it with a planar, cylindrical or toroidal face (Figure 9.9). Although these operations had limitations, they nonetheless provided a quick means of making certain changes to bodies, changes that were tedious with Boolean operations where the modifying primitives had first to be carefully shaped and positioned. It also foreshadowed the development of more general blending operations [24].

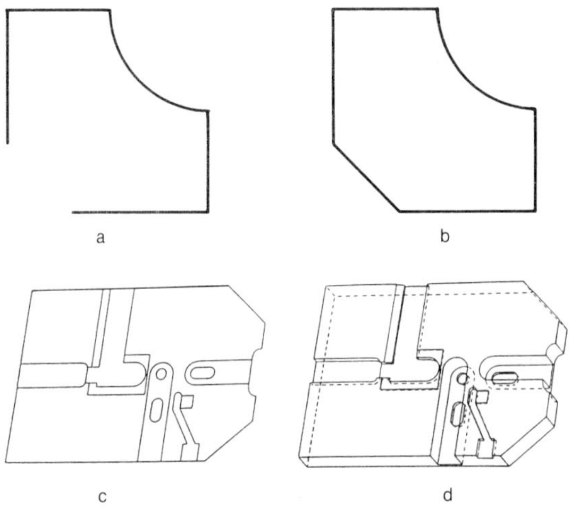

Figure 9.8. Making and sweeping profiles. (a) A wire body, (b) a sheet body, (c) a more complex sheet body, (d) after sweeping by various amounts, to make a solid.

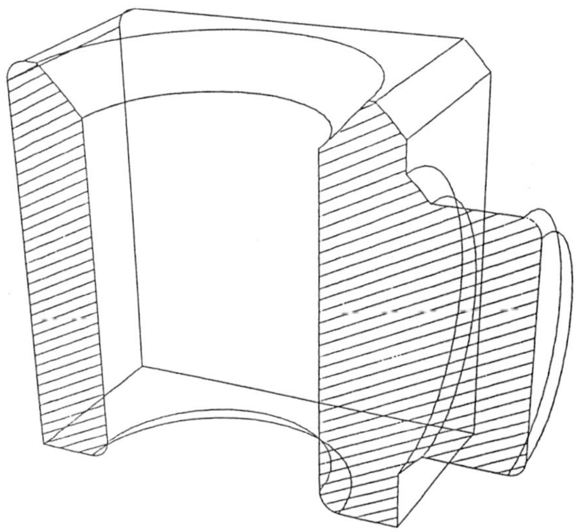
Figure 9.9. Chamfering and rounding edges.

Build 2 demonstrated other operations which had a local effect on the boundary of a body. One was termed *tweaking* in which the geometry of a few faces of the boundary was altered, for example, to give a draft to a part that was to be withdrawn from a mold. The basic operation allowed the surface of one or more faces to be altered, either by linear transformations or by changing the type of face, for example, from planar to cylindrical, and its bounding edges and vertices recomputed. Although tweaking had originally been tried as a more rapid alternative to Boolean operations, it was soon apparent that its real advantage lay in giving the user means to make certain kinds of model changes that were very difficult to carry out with Boolean operations. Of course, without further checks, tweaking and chamfering could give rise to self-intersecting boundaries that could cause subsequent Boolean operations to fail. In practice, users seemed happy to put up with a little danger to enjoy the convenience of this and other local operations. Those who were familiar with surface design systems had already experienced similar freedom in modifying bounding surfaces of bodies. When similar operations arrived in Romulus (see below), they came with the option of automatically detecting self-intersecting boundaries (Figure 9.10).

Assemblies were modeled by generalizing bodies to consist of either a collection of faces or a collection of instances of other bodies. By labeling assemblies as conjunctive or disjunctive, and letting instances optionally negate the bodies referred to, the same structure was made to record general Boolean expressions in implicit form [25]; indeed, it could represent a model at any stage of evaluation from fully implicit to fully explicit. In practice, users preferred an implicit record in the form of textual command sequences which could be easily read, understood and edited.

The Boolean operations of Build 2 occurred in two phases: first the boundaries of the two bodies being united, intersected or subtracted were compared, and their intersections noted; and second, the boundaries were divided at their intersections, portions of the boundaries being discarded and the remainder re-connected to form the boundaries of the resulting bodies.

Most of the work occurs in the first phase where in principle every face of one body must be compared with every face of the other. Although cases can be constructed in which every such comparison does give rise to an intersection, usually rather few comparisons result in actual intersections. Thus it is important to design a Boolean algorithm to avoid unproductive face-to-face comparison as far as possible. A second question for the algorithm designer is how to record the face-to-face intersections that do occur, and how to make that information available to the relinking phase to follow.

Build 2 based the first phase on face-to-face

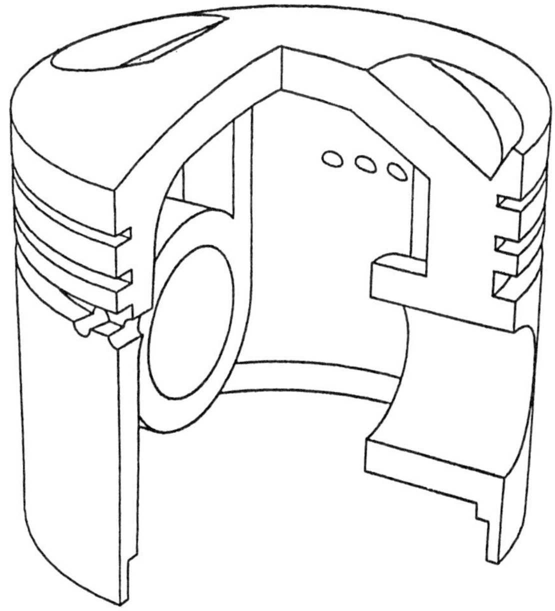

(a)

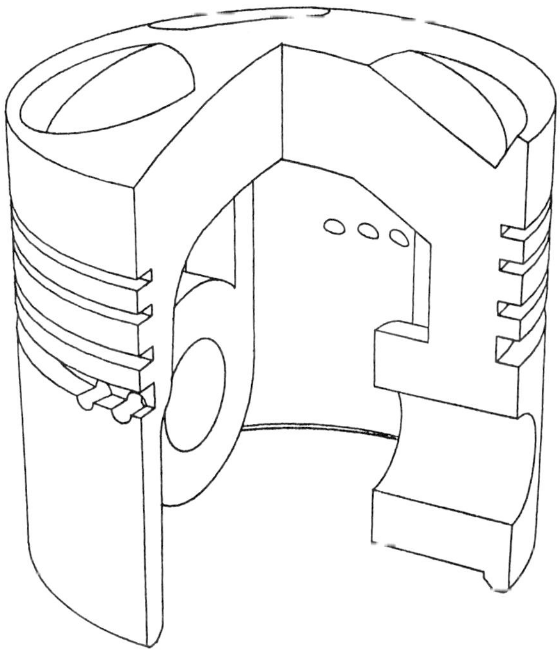

(b)

Figure 9.10. Tweaking faces and copying features to change a body (using Romulus). (a) Before, (b) after.

comparisons rather than on face-to-edge comparisons as in Build 1. Faces were not split during the face-to-face comparison phase and so retained their identity throughout the first phase. By contrast, Romulus used Euler operators to imprint new edges and hence to split faces immediately, and then made subsequent comparisons using the descendant faces. Build 2 did divide the edges of each body where they were found to pass through the boundary of the other body, at the same time marking the edge pieces as 'to be kept' or 'to be deleted', determined by noting whether each edge piece was inside or outside the other body, and by taking account of the kind of Boolean operation.

At each face-to-face intersection, the curve of intersection of the surfaces of the two faces were found, then the portions of that curve lying within each face were determined by intersecting the curve with the edges of the face, and lastly, the regions of overlap of the curve portions in one face with those in the other face were discovered. For each region of overlap, a new edge was made and partially linked into the model. The ends of the edge were at vertices new or old, the loop pointers referred to the loops of the original faces, and the winged-edge pointers were left unset. In this way the original faces kept their identity while they were being compared.

The second phase completed the linking of new edges into the structure, setting the winged-edge pointers at their ends, using the loop pointers of edges, new or old, and the 'to be kept' 'to be deleted' marks to determine how the pointers should be set. It then grouped edges new or old into new loops by starting at a new edge and following the winged-edge pointers while at the same time setting the loop pointers of the edges to refer to new loops. The new loops were associated with the original faces of which they formed a part. The next step formed loops new or old into new faces, using a point-in-face test to determine the relative enclosures of loops. Lastly, faces were collected into shells by traversing their loops and edges to find all connected faces. Braid [26] gives a fuller description of the Build 2 Booleans at that date.

Further changes occurred later. In particular, the algorithms were substantially modified so that faces need not be plane-projectible, and thus, for example a cylinder no longer had to include one or more edges (props) joining its top and bottom circular edges. This change followed a recognition that artifacts such as props were undesirable since they did not correspond to any physical property of the shape being modeled. Such a change required a generalization of Euler's rule to allow closed edges with no vertex and closed surfaces with no vertex.

The goal of making the Boolean operations general was reached. A single algorithm catered for unions, intersections and subtractions of pairs of bodies whose faces could intersect in curves or could coincide in common surfaces. Eventually it became fairly reliable, thanks especially to Alan Smith and Bernard Solomon, and detected but did not handle cases leading to non-manifold bodies.

A system of bounding spheres placed around faces or groups of faces, shells or bodies, speeded up Booleans. For example, when seeking the intersection of two faces, one from each body, the spheres enclosing each face were compared to see whether they overlapped or not (if the distance between their centers exceeds the sum of their radii, the faces must be disjoint – a fact that can be discovered without calculating square roots if the squares of the two quantities are used). If this test did not show the faces to be disjoint, the curves of intersection of their surfaces were sought. Since general curves of intersection of quadric surfaces were approximated by fitted Bézier curves (see below), it was important to limit the curves created to the region of interest, which in this case could be taken as the region of overlap of the spheres enclosing the two faces. The next step looked for intersections between the edges of a face and the curves of intersection, and again, an edge could be tested against the sphere surrounding the other face and only if it was not disjoint, need be made the more expensive test of an edge against an intersection curve (or equivalently, against the surface of the other face).

Face groups were intended to act both as a means of representing features of a boundary as specified by the user, and as a way of denoting sub-regions of a body boundary, which could then be enclosed by spheres and so provide a hierarchy of spheres with the result, it was hoped, of further speeding up face-to-face comparisons. In fact, by trying to do two tasks at once, they failed to do either well. There was the further difficulty that no good way was found of automatically grouping faces into hierarchies of propinquitous faces. The asymmetry of comparing each face of one body with all faces of the other seemed unsatisfactory, at least from an aesthetic point of view. Recently, Guibas and Seidel [27] have described an algorithm that treats the faces of each body in an even-handed way.

So far as geometry was concerned, the choice of general quadrics for Build 2 turned out to be unwise. Although mathematical discriminants exist that will determine whether a quadric surface is a plane, sphere, cylinder, hyperbolic paraboloid, and so on, and similarly classify curves of intersection of quadrics, the discriminants do not relate to physical distance measures. In practice, one wishes to regard a spherical surface as planar if *over a face* say, it departs from planarity by no more than a standard length discriminant. There was also the difficulty that storing a quadric by its coefficients could lead to very small or very large numbers in the case of the coefficients of the second-degree terms ([28] discusses these points in detail). The Romulus approach to storing surfaces by their 'obvious' properties such as a center line and radius for a cylinder, seems to work much better.

A small number of operations performed geometric processing throughout the modeler. A single dyadic operator took two geometric entities (point, curve or surface) and returned their geometric intersection. Other operators gave the distance of a point from a surface or derivatives at a point on a curve or surface. Building on the work of Levin [21], Solomon developed methods for intersecting quadric surfaces and approximating their quartic curves of intersection by Bézier curves. These methods relied on the fact that any curve of intersection of two quadric surfaces is also the curve of intersection of a ruled quadric surface with another quadric (not ruled in general). Thus points along the curve of intersection could be found by repeatedly intersecting generators of the ruled surface, i.e. straight lines, with the general quadric. Although this step is straightforward and requires merely one square root to be found, the overall problem of producing the generators so as to find points on the curve only within a predetermined region of space, is more complex. There is also the requirement to generate the points sensibly spaced along the curve so that the curve interpolated through them, in this case a B-spline, approximates the true curve of intersection within a given tolerance.

Solomon also went some way towards adding B-spline parametric surfaces, and finding their intersections with quadric surfaces and with one another. Later, Várady used the same framework to investigate the use of double quadratic surfaces [29] in a modeler (he had only to supply the small number of standard intersection and interrogation routines). It was fortunate that the Boolean algorithms did not make heavy use of a distance-of-point-from-surface test, and hence the cost in speed of incorporating parametric surfaces was not as high as it might have been.

Besides being a basis for experiments in boundary modeling, Build 2 was the basis for applications. Hillyard [30] used the modeler for an investigation into dimensions and tolerances. He regarded a dimension as a constraint that fixed the relative position of two or more topological elements: for example, two edges might be constrained to be parallel and a certain distance apart, or all the vertices of a face might be required to lie in a plane. His method formulated each dimensional constraint as a relationship between the coordinates of the vertices of the topological elements concerned. A fully-dimensioned body gave rise to a system of non-linear simultaneous equations that related values of dimensions to vertex coordinates. Using the current vertex coordinates as a set of starting values, an iterative procedure solved for new coordinate positions when dimensions were changed. The method discovered objects under- or over-dimensioned, and from the differential forms of the equations deduced a sensitivity analysis for tolerances by showing how small changes in dimensions contributed to displacements at vertices (Figure 9.11). To implement it with Build 2, Hillyard defined a new data structure type for a constraint (holding *inter alia* pointers to the topological elements constrained, and the type, value, tolerance and measurement direction of the dimension). He also added other routines that built and solved the set of equations, calling on lower-level Build routines to read or change the model.

At MIT, Gossard and his students have taken this approach much further. For example, they have developed methods of practical efficiency and scope by utilizing better numerical techniques and by extending the class of objects to include those bounded by curved surfaces [31].

The second application of Build 2 was to parts classification and coding [32]. Once again, both model and modeler were extended by the addition of further data types and procedures,

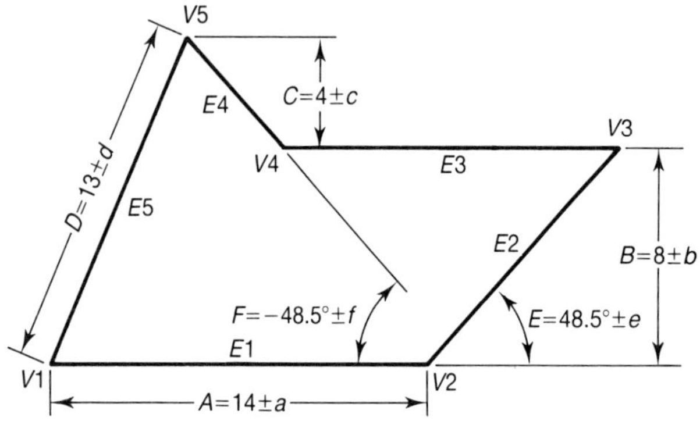

a

$$\begin{bmatrix} i & 0 & 0 & 0 & 0 \\ j & 0 & 0 & 0 & 0 \\ 0 & j & 0 & 0 & 0 \\ i & -i & 0 & 0 & 0 \\ \frac{1}{2}j & \frac{1}{2}j & \frac{1}{2}j & \frac{1}{2}j & 0 \\ p_1 & -p_1 & p_2 & -p_2 & 0 \\ 0 & 0 & \frac{1}{2}j & \frac{1}{2}j & -j \\ p_3 & 0 & 0 & 0 & -p_3 \\ p_1 & p_1-p_4 & p & p_4 & 0 \\ p_1 & -p_1 & 0 & p_5 & -p_5 \end{bmatrix} \cdot \begin{bmatrix} d_1 \\ d_2 \\ d_3 \\ d_4 \\ d_5 \end{bmatrix} = \begin{bmatrix} 0 \\ 0 \\ 0 \\ u_a \\ u_b \\ \theta \\ u_c \\ u_d \\ u_e \\ u_f \end{bmatrix}$$

$p_1 = -i/A,\quad p_2 = i/(x_3 - x_4),\quad p_3 = (x_5 - x_1, y_5 - y_1)/D$
$p_4 = w/w^2$ where $w = (y_3 - y_2, x_2 - x_3)$,
$p_5 = w/w^2$ where $w = (y_5 - y_4, x_4 - x_5)$.

b

Figure 9.11. A system of constraints and their equations. (a) 2D object with constraints (*A* to *F*) and point variables (*V*1 to *V*5), (b) constraint equations relating (small) changes in dimensional constraints (u_a to u_f) to displacements (*d*1 to *d*5) at the points.

in this case to recognize and represent part features, and to generate classification codes from the recognized features (the source of the part code generator was itself created by a program which took as input a description of feature combinations and their corresponding part codes). The class of parts was limited to rotational and nearly-rotational parts. Although an explicit definition of what should constitute a feature was desirable, it was decided for this first attempt to let the feature be defined implicitly by the recognizing procedure. Recognition utilized model properties such as whether the loops of multiply-connected (plane-projectible) faces were 'inner' or 'outer', and whether edges were convex or concave. The result was a feature-recognizer and part code generator that was very fast (one or two seconds CPU on an IBM370). Of course, when the aim is merely to break down a population of parts into groups and sub-groups of manageable size, the choice of features is not as critical as when features must relate to function or manufacture.

In a third application of Build 2, Wördenweber [33] investigated the generation of finite-element (FE) meshes in two and three dimensions, from Build models. In this case, it was necessary in effect to create a new modeler for FE meshes. The Build model first had its curved edges approximated by polygons, then each face was faceted with triangles, and lastly an assembly of juxtaposed tetrahedra was created from the faceted solid using a small but powerful set of operators (Figure 9.12). The same operators, and their inverse forms, were available for adaptive mesh refinement (or coarsening) in response to error data which would have to be output by the solution process. Once again, the application created data structures of its own, and these

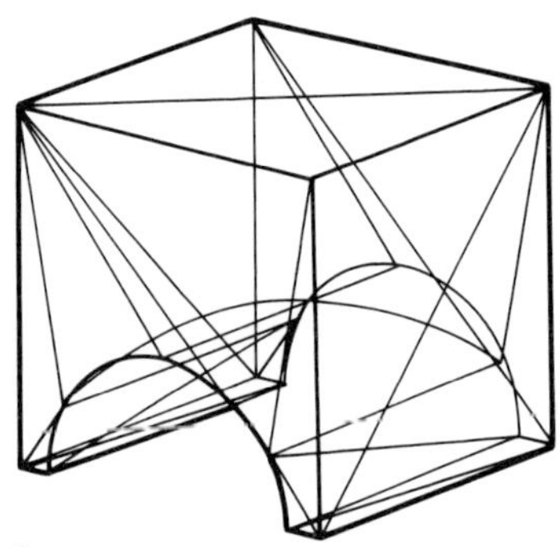

a

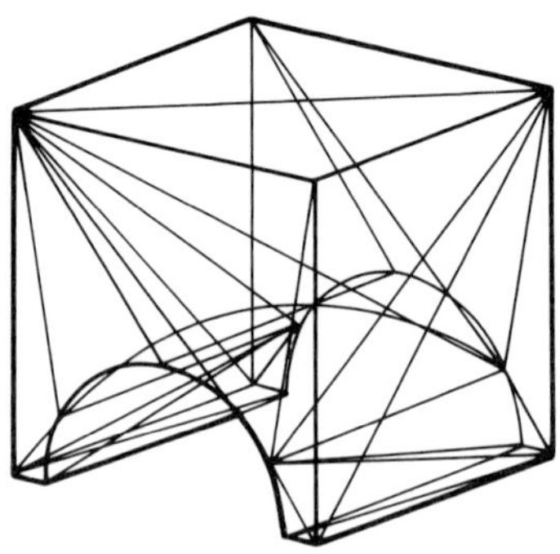

b

Figure 9.12. A finite-element mesh created from a Build 2 object. (a) After surface triangulation, (b) after volume triangulation.

contained both their own data and references to the Build data.

This section will conclude by mentioning some exact boundary modelers developed by other research groups over the period when Build 2 came into existence. In West Germany, a group at the Ruhr-Universität Bochum made the PROREN2 system, while at the Technical University of Berlin another group constructed the COMPAC system. PROREN2 [34] used a compact hierarchical data structure where bodies referred to faces, faces to edges, and edges to vertices (but not *vice versa*). Profiles were not made directly as in Build 1 and 2, but were made by using Boolean operations on primitives in two dimensions. The 3D Boolean operations required the user to distinguish two cases: two bodies could be glued at touching faces, or they could be combined by Boolean operations provided no pair of faces coincided, i.e. could be glued. The surface types were planes, cylinders, cones and spheres. PROREN2 had a well-deserved reputation for speed, and was certainly faster than Build 2. Later the group at Bochum added toroidal surfaces and experimented with free-form surfaces [35]. PROREN2 has become available as a commercial product.

COMPAC [36] was also written in Fortran, and had similar restrictions on its Boolean operations. In its early days, COMPAC was run offline from a file of commands that was edited and rerun each time. Although this might be thought to have detracted from its use as an interactive design tool, it is fair to say that for designing parts of any complexity, modelers have more often than not been run in this way. The COMPAC data structure did not have a separate data type for loops of edges; instead, it allowed positive and negative faces so that a face with a hole in it would be represented by a positive face plus a negative face (it might be better to describe this structure as allowing loops but not faces). The data structure was implemented using the ring-data structure package ASP developed, but little used, by the CAD Group at Cambridge. In its original class of surfaces and their subsequent generalization, COMPAC was similar to PROREN2.

In this brief summary of boundary modeling research, the work of one more research group must not be overlooked. This is the team at Tokyo University started by Professor Hosaka and continued by Professor Kimura. They have constructed several systems, and have followed a course somewhat similar to Build 2 in that they have used Euler operations and have also investigated local operations. Their boundary modeler was one of the first to include free-form surfaces [37].

REALIZATION IN COMMERCIAL PRODUCTS

One of the pleasures of working in an applied subject like geometric modeling comes from seeing research ideas carried through from prototypes to practically useful products. Allied to this is the stimulus of the demands that soon flow from real users of any modeler. Many researchers are content to let others make and market products based on their results, while they turn to new pastures. But there is the option, especially if research funding and university prospects are uncertain, for the researchers to follow the development through, and themselves to undertake the commercial development of the research ideas. In any case, four members of the CAD Group, Charles Lang, Peter Veenman, Alan Grayer and myself, founded Shape Data Ltd. in 1974. The company had a so-called 'soft' start [38], and used consultancy and bespoke software development rather than outside finance to grow the company and fund the making of a product called Romulus.

Though it followed many of the methods tried out in Build 1 and 2, Romulus, designed in the first instance largely by Alan Grayer and Martyn Horner, differed significantly. From the start Romulus had to be a product rather than a piece of research apparatus. Even if at first its operations were limited in scope, they had to be reliable, predictable and stable. Its language was Fortran as it was to be portable and integrated into other software also likely to be in Fortran. The data structure (Figure 9.13) was a variant of Baumgart's winged-edge model in which edges referred to their adjacent faces via co-edges (some termed 'edge uses' [39]). This simplified and speeded up the common case of accessing all edges of a face, and avoided the marking of wire edges (i.e. edges having the same face either side) that had been necessary in Build when traversing the edges of a loop. Romulus was stricter than Build in that it used Euler operators to make almost all model changes, including those needed in Boolean operations. More important was Alan Grayer's insistence from the start on careful numerical coding of the algorithms which led to Romulus becoming eventually an unusually reliable boundary modeler.

The earliest versions of Romulus created models directly by means of 'edge-based' construction. A user began by making an 'acorn', i.e. a vertex embedded in an infinite surface. From it, he could scribe edges to make a finite face from which the surface geometry of the face was found. He could scribe further edges across the face provided they lay on the surface of the face; alternatively, he could suspend the face geometry by setting the face to be 'rubber' and could then create edges freely in three dimensions (they remained topologically within the face). This process was likened to putting up the metal frame inside a tent; as the edges in the data structure were always embedded, a notional surface existed even if its actual shape was not yet defined.

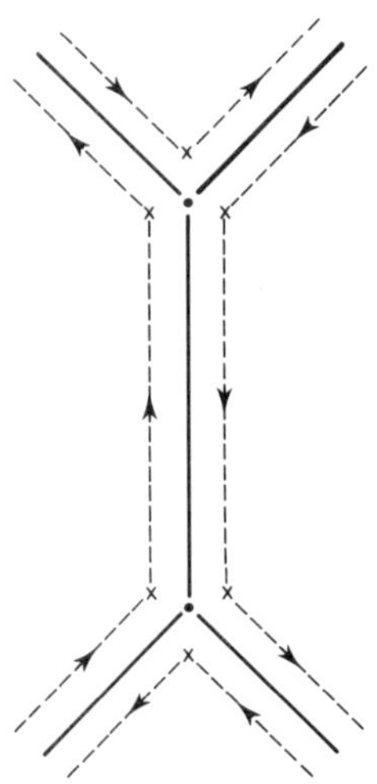

Figure 9.13. Edges and co-edges in Romulus. ●: vertex; ×: co-vertex; ——: edge; – – –: co-edge.

There was also a sweep operation that could be applied to any face of a body, with the option of a sweep-and-pierce by means of which a closed profile could be scribed within one face of a cube, say, and be swept perpendicularly through the cube until it met and pierced the opposite face. Early Romulus (c. 1979) already had an operation to section a body and then cross-hatch section faces (see Figure 9.14). Its surface types were planes and circular cylinders; curves could be straight lines, circles or ellipses.

Romulus gradually gained in power as cones, spheres and toruses joined its list of surface types, and general curves of intersection were included in its range of curves. Judy Broadway added Boolean operations in their initial form. Bernard Solomon and David Jackson, amongst others, contributed to their extension and improvement so that the handling of tangencies became more robust and other awkward cases such as subtracting a body from itself were encompassed. The solution adopted for non-manifold bodies was to get the modeler to recognize such cases and to refuse to make them.

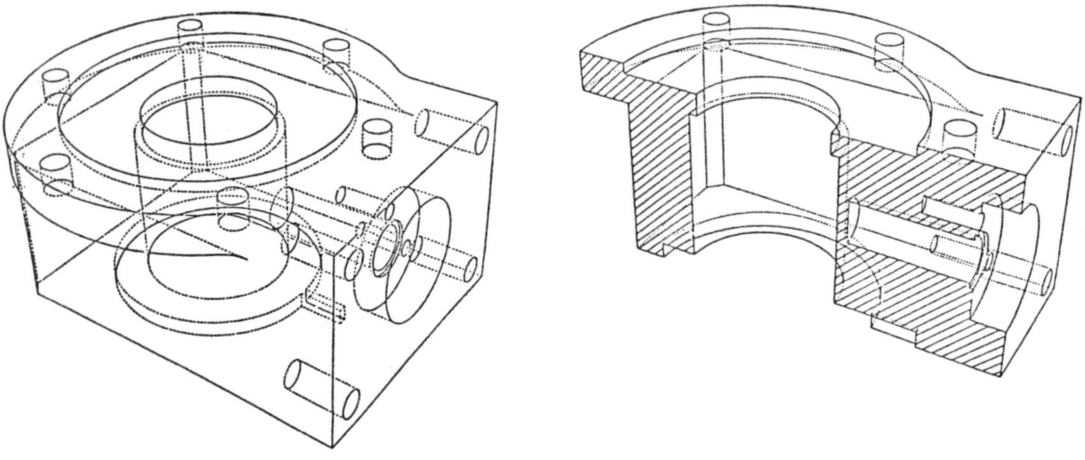

Figure 9.14. An early Romulus object.

The modeler acquired further local operations including topology-changing operations on features (recorded as sets of faces). A feature such as a hole or slot could be pulled off a body, the remaining faces of which would then 'grow', if possible, to cover over the missing portion of the bounding surface of the body. A similar operation permitted a feature to be copied so as to form a body composed of the faces in the feature plus new faces corresponding to the pieces of surface that would have been grown had the feature been pulled off. Users liked these operations but some found the difficulty of knowing when they would succeed or not, trying. Also there seemed to be no good way to characterize the operations theoretically. One suggestion was to regard feature-deletion, for example, as the removal of one or more half-spaces from a CSG tree, but then how has a unique CSG tree to be derived from the boundary model?

Models need attributes if they are to represent more than just shape information. Romulus allowed for attributes to be attached to model entities, and moreover, updated attributes as the models changed. A few simple rules governed the updating process that controlled what was to happen to an attribute when its owning entity altered (e.g. was transformed), divided into pieces or merged with another entity.

The original aim had been to offer Romulus as a systems component rather than as a stand-alone program. In practice, its test harness became part of the product, and communicating programs controlled it by sending commands in the same way that a user would. Eventually a satisfactory system design was arrived at where the modeler presented an interface in the form of a collection of subroutines. Its design followed the Applications Interface proposed in 1980 by Shape Data in a contract for the international research association CAM-I [40]. Any errors in the modeler were caught within the kernel, and the model was returned to its state at entry to the kernel routine in which the error occurred. The interface also noted in a journal values of arguments given to it and returned by it, so that a modeler malfunction could be reported to the originators and reproduced by them. As a further service, after each modeling operation, Romulus left a record of all model changes so that the calling system could inspect it and update other data such as a graphic data structure, needing to be kept in step with the model. The modeler backed up the model before making changes to it so that the user could quickly undo and re-do one or more modeling operations. It could also generate high-quality, structured graphic output for inclusion in design documents such as drawings.

Romulus has formed the nucleus of a number of complete CAD/CAM systems, and several thousand seats for systems incorporating it have been sold. In common with other solid modeling systems, it has been most successful at user sites where the shapes or product being designed happen to match the capabilities of the modeler, where there is a clear pay-off such as the avoidance of clashing components or reduction in material quantities for a component, and most important, where there is at least one able enthusiast for solids technology [41]. Plate 12 shows an example of a topologically complex assembly modeled with Romulus. From a commercial point of view, its greatest drawback has been the lack of free-form surfaces. Shape Data's new modeler, Parasolid, written in pre-processed C to a specification similar to Romulus, makes good that deficiency and also offers much improved performance.

Yet another descendant of Build and Romulus is ACIS, a kernel modeler developed by Alan Grayer, Charles Lang and the author at their company Three-Space Ltd. in Cambridge. The ACIS model is rich enough to encompass wireframe, surface, solid and cellular models, as well as non-manifold solids and incomplete models. Its geometry covers common implicit surfaces and also non-uniform rational B-spline curves and surfaces managed by a bought-in geometry package. Use of an efficient implementation language (C++) and careful algorithm design have given worthwhile gains in performance, especially in the maintenance of speed as model complexity rises. The object-oriented nature of C++ allows the interface between ACIS and the calling system to be wide and flexible. This is particularly helpful for application-defined attributes, where the application can decide what data and pointers attributes should carry, and what action should occur when model entities that own attributes change (i.e. are created, merged, split or deleted). By this means it is hoped that ACIS will be able to model not just shapes, but products [42].

Besides Build and its descendants, Cambridge has been the home of several other modeling systems that should be mentioned, even though they do not fall into the category of exact boundary modelers. One of the earliest was DUCT, a system originated in the University Engineering Department [43] and now a commercial product. Like Engeli's EUKLID system, DUCT is particularly suited to design and machining of the complex surfaces that occur in parts made with dies and molds. The same university department also saw the development of methods of curve and surface design based on differential geometry [44]. Two commercial products centered on polyhedral modelers should not be overlooked. Medusa, devised by Newell, Sancha and others at Cambridge Interactive Systems, has been used widely in Europe, while Microsolid, developed at Perspective Design by Horner and Broadway, formerly of Shape Data, has been influential in making solid modeling available on personal computers.

Similar examples of the spread of boundary modeler technology have occurred in many other countries. There is room here to do no more than mention some cases. The Geomap modeler at Tokyo University [37] has spawned commercial modelers made and used in Japanese corporations. In Germany, two university modelers, COMPAC and PROREN2, have themselves appeared as products.

Another approach, more popular in the US, has been to take the surfaced wireframe models of existing drafting systems as the basis for solid models, and to provide Boolean operations on them. This gives users a smoother transition to solid modeling and also offers an alternative when the modeler cannot produce the desired model: the user can create it directly by building and surfacing a wireframe model. Such modelers can also represent incomplete models, that is, models whose boundary is not fully specified. Though this goes against the completeness sometimes claimed as an advantage of volume modeling, it fits well with practical design where parts are frequently left in a semi-specified state during the design process. Care has to be taken to ensure that the exacting requirements of model consistency necessary for Boolean operations to work well, are met by wireframe and surfacing routines originally written in more lenient circumstances.

A variant of this route, sometimes followed when adding modeling capability to surface packages, is to facet surfaces before performing Boolean operations on them. Clearly, the logic of the Boolean operations is simpler if they are confined to polyhedra, and one avoids complex geometric calculations, but the resulting models are more suitable for making pictures than for other purposes such as machining that demand an accurate model.

The popularity of parametric, more especially rational parametric, curves and surfaces, in systems for designing free-form curves and surfaces, has prompted the development of modelers that use a uniform basis of parametric curves and surfaces for all their geometry. Riesenfeld's Alpha_1 is an early example of the genre, and has been followed by others such as those from corporations such as ATP, PRIME and Integraph. While geometric uniformity is attractive esthetically and may promise compact and tidy software, the need for speed and reliability, especially in the presence of tangent intersections, leads ineluctably to the adoption of more specific forms for common surfaces such as planes and cylinders.

Two more kinds of boundary modeler deserve mention before concluding this section. Polyhedral boundary modelers such as Euclid in France and the US GEOMOD and ARIES systems, have played an important role in making solid modeling widely available. They have often found a place in conceptual design where packing problems occur, and where early visualization of preliminary designs is wanted. The models produced can be used when preparing data for finite element analysis, for example, but for detail design, it is usual to resort to a drafting system. Polyhedral models have seldom been satisfactory as a basis for generating cutter paths for numerically-controlled machining of components.

All the boundary modelers discussed so far have found application in designing assemblies of moderate complexity, i.e. having not more than a few tens of components. For really large assemblies such as chemical plant and power stations, a different kind of modeler has arisen. These attend to the problems of storing permanently models composed of thousands of components. They may provide for storage of models at varying levels of detail, and for access based on spatial or topological criteria. Models must be organized so that many designers can work on them at once, and specialized routines may be needed so that graphic output can be produced in a reasonable time. Example of these systems include the Cambridge CAD Centre's PDMS and the Medusa assembly modeler.

REFLECTIONS AND CONCLUSIONS

Boundary modeling has advanced by a variety of routes until now there are modelers capable of supporting real design. Sometimes, perhaps more often than one imagines, solid models have given significant help to the design process, and in the transfer of data from design to analysis and manufacture. The processes of engineering design are not so very different from one company to another, so it seems reasonable to suppose that one day the technique will be widely used, and will come to be seen as being just as essential to mechanical design as computer methods are now in the field of electronics.

Yet one cannot but admit to some dissatisfaction from the research point of view. Boundary models are redundant, and so when held in computers of finite precision are inevitably self-inconsistent to a small extent (the model tolerance). Today's modelers use double precision so that model inconsistencies are small compared to the accuracy expected of the model by the user. Their Boolean algorithms give priority to some parts of models over others (e.g. to surfaces as compared to curves) and iterate when need be, to improve the accuracy of data taken from the model. Nonetheless, it is not difficult to make boundary modelers fail when model details approach the model tolerance in size. One has only to try subtracting a copy of a body from itself, having first rotated it by a small angle. For an angle so small that the model and its copy differ only by amounts substantially less than the model tolerance, all is well. Likewise, the Boolean operation will work if the two bodies differ by amounts materially greater than the model tolerance. But for some angles in between, the Boolean will fail. Perhaps this unsatisfactory state of affairs explains why we still await a thorough account of Boolean operations on boundary models.

These problems are beginning to receive attention. One recent experiment [45] illustrates a method for finding Voronoi diagrams, that is proof against ill-conditioned input data. The diagrams produced may be locally in error, but the algorithm is able to operate successfully even on diagrams in which such errors occur. Moreover, as the model precision is increased, the model comes steadily closer to representing the true result. In effect, the class of entities modeled has been enlarged so that it encompasses any result produced by the algorithms employed. A similar method will deal with non-manifold shapes: rather than letting the modeler fail or having it refuse to continue, the model is enriched so that it can represent non-manifold bodies, and the algorithms are generalized to work with such bodies.

A more elegant solution is to avoid creating redundant data at all. The method of constructive geometry (CSG) does this by taking as its primary model of a shape, a (regularized) Boolean combination of point sets in three dimensions (see Chapter 11 by Voelcker and Requicha). Problems of small discriminants are still there, but arise in the interpretation of the primary model rather than in its creation, and so have more limited consequences. On the other hand, the user is constrained to make models that are solid at every stage in their design, and has a smaller repertoire of model-changing operations than are usual with boundary modelers. The extension of CSG methods to shapes bounded by free-form parametric surfaces has not been easy, since such surfaces by themselves do not in general bound point sets in three dimensions (although by addition of further surfaces they can be made to do so).

The desire for models that can represent incompletely specified designs and construction data such as center lines and surfaces, and the response from modeler-suppliers, is evidence of a general need for computer assistance not merely in representing the result of the design process, but for modeling the data and relationships germane to the process itself. The recent appearance of systems that can model constraints or record the logic of design in various ways, testifies to the strength of this need. Now modeler developers have the task of representing and manipulating these abstract relationships and of making them evident and comprehensible to the user – more difficult than in the case of models restricted to wires, surfaces and solids which all have visual equivalents in the physical world.

In conclusion, one can say that although boundary modeling has advanced in 20 years to the point of real if not yet widespread utility, the story is certainly not over.

ACKNOWLEDGEMENTS

Thanks are due to Hewlett-Packard Ltd. and Woods Ltd. for the picture of an assembly modeled using Romulus within the HP ME-30 system.

It will be evident even from this short and partial account that the development of this branch of geometric modeling has been the work of many people. Besides those mentioned in the text, I would like to pay tribute to all who have worked unsung within commercial enterprises on geometric modelers. The steady stream of large and small improvements they have made has done much to make modelers useful, reliable and fast.

REFERENCES

1 **M. V. Wilkes** *Memoirs of a computer pioneer,* MIT Press, 1985.

2 **C. A. Lang and J. C. Gray** ASP – a ring-implemented associative structure package, *Commun. ACM,* Vol 11 No 8, 1968, 550–555.

3 **C. A. Lang** SAL – systems assembly language, *Proceedings Spring Joint Computer Conference,* 1969, 543–555.

4 **P. A. Woodsford** The design and implementation of the GINO 3D graphics software package, *Software – Practice and Experience,* Vol 1 No 4, 1971, 335–365.

5 **A. R. Forrest** *Curves and surfaces for computer-aided design,* PhD thesis, University of Cambridge, 1968.

6 **A. P. Armit** *Computer systems for interactive design of three-dimensional shapes,* PhD thesis, University of Cambridge, 1970.

7 **C. A. Lang** A three-dimensional model making machine, in J. Hatvany, ed., *Computer languages for numerical control,* North-Holland, 1973.

8 **P. A. Woodsford** The HRD-1 laser display system, *Computer Graphics,* Vol 10 No 2, 1976, 68–73.

9 **I. E. Sutherland** Computer inputs and outputs, *Scientific American,* Vol 215 No 3, 1966, 86–96.

10 **L. G. Roberts** *Machine perception of 3D solids,* Technical Report No 315, Massachusetts Institute of Technology, 1965.

11 **I. C. Braid** *Designing with volumes,* PhD thesis, University of Cambridge, 1973.

12 **I. C. Braid and C. A. Lang** Computer-aided design of mechanical components with volume building bricks, in J. Hatvany, ed., *Computer languages for numerical control,* North-Holland, 1973.

13 **N. Okino, Y. Kakazu and H. Kubo** TIPS-1: technical information processing system for computer-aided design, in Hatvany, J., ed., *Computer languages for numerical control,* North-Holland, 1973

14 **M. Engeli and V. Hrdliczka** *EUKLID – eine Einführung,* Fides Rechenzentrum, Zurich, 1974.

15 **J.-M. Brun** *EUCLID: A manual,* Équipe Graphique du Laboratoire d'Information pour la Mechanique et les Sciences de l'Ingénieur, LIMSA, Orsay, France, 1976.

16 **C. M. Eastman** General purpose building description systems, *Comput.-Aided Design,* Vol 8 No 1, 1979, 17–26.

17 **B. G. Baumgart** *Geometric modelling for computer vision,* PhD thesis, Computer Science Department, Stanford University report STAN-CS-320, 1974.

18 **A. R. Grayer** *A computer link between design and manufacture,* PhD thesis, University of Cambridge, 1978.

19 **F. P. Brookes** *The mythical man-month,* Addison-Wesley, New York, 1975.

20 **I. C. Braid** The synthesis of solids bounded by many faces, *Commun. ACM,* Vol 18 No 4, 1975, 209–216.

21 **J. Z. Levin** A parametric algorithm for drawing pictures of solid objects composed of quadric surfaces, *Commun. ACM,* Vol 9 No 10, 1976, 555–563.

22 **S. R. Bourne, A. D. Birrell and I. Walker** *ALGOL68C reference manual,* Computer Laboratory, University of Cambridge, 1975.

23 **I. C. Braid, R. C. Hillyard and I. A. Stroud** Stepwise construction of polyhedra in geometric modelling, in K. W. Brodie, ed., *Mathematical methods in computer graphics and design,* Academic Press, 1980.

24 **A. P. Rockwood and J. C. Owen** Blending surfaces in solid modeling, in G. E. Farin, ed., *Geometric modeling: algorithms and new trends,* Society for Industrial and Applied Mathematics, Philadelphia, 1987, 367–383.

25 **I. C. Braid** On storing and changing shape information, *Computer Graphics,* Vol 12 No 3, 1978.

26 **I. C. Braid** *Notes on a geometric modeller,* CAD Group document 101, Computer Laboratory, University of Cambridge, 1979.

27 **L. J. Guibas and R. Seidel** Computing convolutions by reciprocal search, *Proceedings 2nd Annual ACM Symposium on Computational Geometry,* 1986, 90–99.

28 **B. J. Solomon** *Surface intersections for solid modelling,* PhD thesis, University of Cambridge, 1985.

29 **G. E. M. Jared and T. Várady** Synthesis of volume modelling and sculptured surfaces in BUILD, *Proceedings CAD '84,* IPC Science and Technology Press, 1984.

30 **R. C. Hillyard** *Dimensions and tolerances in shape design,* PhD thesis, University of Cambridge, 1978.

31 **R. A. Light and D. C. Gossard** Modification of

geometric models through variational geometry, *Comput.-Aided Design,* Vol 14 No 4, 1982, 209–214.
32 **L. K. Kyprianou** *Shape classification in computer-aided design,* PhD thesis, University of Cambridge, 1980.
33 **B. Wördenweber** *Automatic mesh generation of 2 and 3 dimension curvilinear manifolds,* PhD thesis, University of Cambridge, 1981.
34 **N. Bargelé** *PROREN2 – Einen Graphik Software zur dreidimensionalen rechnerinternen Darstellung von Bauteilgeometrien,* PhD thesis, Ruhr-Universität Bochum, 1978.
35 **J.-F. Grätz** *Modellalgorithmen zur dreidimensionalen Geometriefestlegung Komplexer Bauteile mit beliebiger Flächenbegrenzung in der rechnerunterstützten Konstruktion,* PhD thesis, Ruhr-Universität, Bochum, 1983.
36 **G. Spur, F.-L. Krause and J. J. Harder** The Compac solid modeller, *Computers in Mechanical Engineering,* Vol 1 No 2, 1982, 44–53.
37 **F. Kimura** Geomap-III: designing solids with free-form surfaces, *IEEE Computer Graphics and its Application,* Vol 4 No 6, 1984, 58–72.
38 **Segal, Quince and Partners** *The Cambridge Phenomenon,* Brand Bros. & Co., London, 1985.
39 **K. Weiler** Edge-based data structures for solid modeling in curved-surface environments, *IEEE Computer Graphics and its Applications,* Vol 5 No 1, 1985, 21–40.
40 **CAM-I Inc.** *The CAM-I Applications Interface Specification* (volumes I and II), CAM-I Inc., Arlington, TX 76011, USA, 1986.
41 **R. A. O'Neill** CAMX at Otter Controls, *Comput.-Aided Design,* Vol 17 No 3, 1985, 149–152.
42 **I. C. Braid** Improving product models and kernel modellers, in T. Sata, ed., *Organization for Engineering Knowledge for Product Modeling in Computer Integrated Manufacturing,* Elsevier, 1989, 275–302.
43 **A. R. Johnson and D. P. Sturge** The DUCT system of design and manufacture for patterns, moulds and dies, *Proceedings 20th International Machine Tool Design and Research Conference,* 1979, 17–24.
44 **A. W. Nutbourne and R. R. Martin** *Differential geometry applied to the design of curves and surfaces,* Ellis Horwood, Chichester, 1988.
45 **K. Sugihara and M. Ira** *Geometric algorithms in finite precision arithmetic,* Research Memorandum RM 88-10, Department of Mathematical Engineering & Information Physics, Faculty of Engineering, University of Tokyo, 1988.
46 **B.-J. Gausemeier** *Eine Methode zur rechnerorientierten Darstellung technischer Objekte im Maschinenbau,* PhD thesis, TU Berlin, 1977.

ABOUT THE AUTHOR

Ian C. Braid studied mechanical engineering at Melbourne University and spent several years in industry before moving to Cambridge University. After gaining a diploma in computer science, he joined the Computer-Aided Design Group in the Computer Laboratory, and made one of the first investigations into solid modeling, which resulted in 1973 in his PhD thesis *Designing with volumes.* From 1975 to 1980 the CAD Group under his leadership continued to work on solid modeling and its applications. Having been a founder director of Shape Data Limited in 1974, he joined the company full time in 1980, and became responsible for the continuing technical development of the Romulus kernel solid modeler. He held this position until 1985, when he left to help set up Three-Space Limited, where he has carried out consulting assignments and has contributed to the development of a new modeler, ACIS. He is a chartered engineer and member of the Institution of Mechanical Engineers, London.

10 Conceptual Modeling in Design

Charles M. Eastman

INITIAL CONCEPTION OF COMPUTER-AIDED ARCHITECTURAL DESIGN

Man has used drawings for thousands of years to characterize the spatial configuration of a design. It is not surprising, then, that when the initial melding of computer graphics with dedicated computing was achieved to produce computer-aided design (CAD), the first generation of products were oriented toward drawings. While the object being represented was limited, the goals were ambitious: to communicate design ideas through drawings to the computer, which would analyze and check them and fill in details. The goal was an 'electronic design assistant' [1,2].

In practice, communicating a drawing to a computer was (and is) arduous and time consuming. The computer's initial ability to respond to a design was limited. The area where the computer was able to contribute was in the production drawing phase of design practice. Operations that allowed manipulation of multiple lines at a time provided some productivity benefits. Data sharing through the use of layers, the addition of intelligence to entities allowing parts accounting, design rule checking and automation of application interfaces, provided a range of advantages over manual drafting. The electronic environment of CADD also allowed easier project management and control. Such were the primary benefits of CAD, as initially developed.

Early drafting systems were expensive and used by only a few US architectural firms. Early results were mixed, with a few installations driving their purchasers into bankruptcy.

Even before CAD, a number of analysis programs were developed that allowed quick and accurate evaluation of a building design for structural performance, of a site plan for cut-and-fill, and calculation of energy loads. Three-dimensional modeling and perspective generation was demonstrated on the very first interactive CAD system [3]; the computer could also be used for rendering.

Thus, the late 1960s was a world populated with a growing number of applications potentially useful in architectural design. Each of the programs, however, required separate data preparation. The cost of data entry was so great that programs were only used in critical areas, as hand calculations had been used earlier. CAD was just another application supporting drafting. The larger potential benefits of CAD, espoused by Coons and J. C. R. Licklider, were not realized.

EVOLUTIONARY GOAL OF BUILDING MODELING

In the early 1970s, I too embarked on the development of various application level design tools [4]. After doing such work, it seemed to me that developing further computer-based tools would have little impact on the practice of architecture. The problem was that the benefit of using any one program was lost because of its data input requirements. At the same time, much of the information needed for drafting and for energy, structural, lighting and other analyses were the same. If a building description could be defined once and organized so that the data could be automatically extracted to send to a variety of programs, then the benefits of the separate applications would be much better realized. This was the concept of *building modeling*.

As the idea of building modeling evolved, it seemed to me and my students that design information ideally would be organized cohesively by components, so that geometry, material, and other properties of components, could be entered only once and be used multiple times.

A property used in many analyses and the primary focus of drawings was *geometry*. Needed was a canonical representation of geometry that could be used to both guarantee its integrity and present different views for use in drawings and analyses. I developed an initial direction in 1972, based on horizontal lofting [5], and presented an outline of the approach at

Fundamental Developments of Computer-Aided Geometric Modeling
ISBN 012-554765-X

a symposium at Cambridge University. While on that trip, I met Ian Braid and saw a demonstration of Build, the first boundary representation solid modeler [6]. Upon return I also learned of Baumgart's work [7], and it was obvious that these were better approaches than the one I had initiated.

BDS (Building Description System)

Our initial modeling strategy was that design information should be organized cohesively by components, so that geometry, material and other properties of components could be entered at one time and be logically related. Geometry was the most dominant property in building design. Modeling it, so that it could be shared in creating multiple drawings and in sending data to analyses, was an obvious critical task.

Modeling of closed 3D shapes was required, so that the consistency of different geometric views could be verified.

Based on the work of Braid and Baumgart, my group, consisting of Joe Lividini, Doug Stoker and myself, set out to develop an initial building modeler at Carnegie-Mellon University, based on boundary representation solids modeling. We called it Building Description System (BDS) [8–10]. A shape was defined as a topology, an expression set for assigning coordinates to vertices, and a set of parameters. Shape topologies were defined using a subset of the Euler operators, as defined in Baumgart's thesis [7]. Our implementation was described in Stoker [11]. Expression Sets assigned coordinates to vertices, allowing some to be parametrized. A Form was the name given to a shape template, consisting of a Topology, Expression Set, and a set of parameters. The vertices were used to compute the surfaces of the shape, when they were needed for computations.

Topologies, Expression Sets, and Shapes were all separate entities and identified by name. Thus a topology could be shared across multiple shapes, which saved storage [10]. If a new topology was created in editing operations, such as by the spatial set operators, it would have the name of the shape. Instances were identified by subscripts and were defined by assigning a transform to a FORM. A building was a composition of several thousand such shape Instances [12].

BDS incorporated a limited implementation of the spatial (regularized) set operators that ignored coincident faces. It also incorporated a display operation that cut sections through an assembly of shapes and, if desired, showed the new surfaces defined by the cut. Sections could be applied to perspectives as well as orthographic views, as shown in Figure 10.1. Graphic display was limited to line drawings.

Forms and their instances were stored together in a record format on disc as they were created; thousands could be easily defined. They were read, one form at a time, and all the selected instances displayed, resulting in several object instances being displayed per second. All interaction was by keyboard or from command files.

BDS was written in BLISS on a DEC-10 [13], driving a Three Rivers display [14]. A version was also implemented for use with Tektronix displays. Work was begun on a PDP-11 version, but it was never completed; 32K memory was too tight for the task. Extension of BDS to include material properties was superseded by the desire to resolve more fundamental issues.

In the papers on BDS, I postulated the benefits of a building model [8]:

- design refinements would be facilitated, because a change could be made once to impacted components, with all information regarding those components updated automatically;
- designers could work directly in three-dimensions, gaining a fuller insight into the spatial and visual aspects of their work;
- use of analyses and other types of applications would be facilitated because the data they require could be extracted and prepared automatically;
- drawing production could be automated, allowing any type of view to be created on demand, reducing time and costs;
  - — integrity of information would be enhanced:
  - — certain kinds of checking, such as spatial conflict testing, could be automated;
  - — by eliminating the redundancy of information across multiple representations, it could be guaranteed that all design information was consistent.

Most of the objectives are valid today and still largely unrealized.

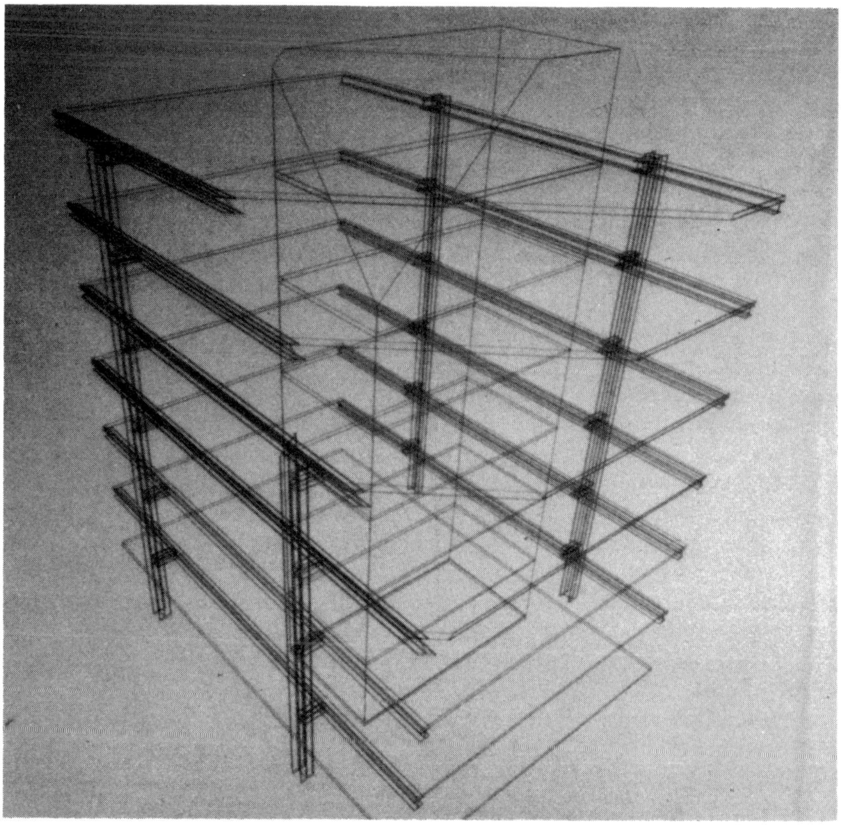

Figure 10.1. An early perspective image from BDS, showing the section cut capability incorporated in the view operation.

During this time, we learned of similar approaches used with industrialized building systems, especially the British hospital systems. For a pre-fabricated building system, the components could be defined and their properties of interest, their geometry and graphic representation in different types of drawings could all be pre-entered into an electronic building parts library [15, 16]. CAD provided the ability to arrange the components and extract the appropriate information for presentation in different drawings or as input to analyses. Several of these systems included conceptual design tools for laying out the organizational units of a hospital on the appropriate grid and for comparing the design to the program [17].

In the US, there was little interest in building systems. The challenge was to allow users to define custom components as well as their arrangement. A paper I wrote in 1976 titled 'General Purpose Building Description Systems' delineated these distinctions [9].

BDS generated significant excitement among those learning of its capabilities. That a computer 'database' could replace drawings was considered architectural blasphemy. BDS was described to architects in an early paper in the American Institute of Architects Journal [18].

GLIDE (Graphical Language for Interactive Design)

Though BDS's focus was solids modeling, these capabilities were limited. The basic data structure for geometry defined a face as a single loop. Faces with holes in them required an 'umbilical cord' connecting the hole to the outside loop. This condition led to many special cases in the Boolean and other editing operations.

It was also apparent that more powerful and general computing language capabilities were needed to define complex designs, such as buildings. We had in mind automatic detailing routines, such as framing, and mechanical system layout. The spatial set operators suggested that shapes could be treated as data types, just like numbers, and that in a language for design, it should be possible to compute shapes in the same manner as it is to compute numbers. Also, all instances of a shape in BDS had to be of the same geometric size. This made the definition of certain elements, such as the variable length studs on a gable end wall of a building, extremely tedious to define.

In 1976/77, my group at Carnegie-Mellon both grew and lost members. Kevin Weiler took on a staff position and Max Henrion, Gilles Lafue, Robert Thorton and Mark Birnbaum were students. This group undertook the development of such a design language, which later was named GLIDE [19]. The goals of GLIDE were to provide an interpretive programming language whose global data were permanent. Geometry operations were built into the language. Vertices, edges, faces, loops, topologies and shape data types were language primitives. The Euler operators, spatial set operators and transforms were provided for manipulating shapes. Planer and cylindrical surface types were supported. Internally, GLIDE used a winged-edge data structure [20]. Non-spatial attributes could be associated with a shape.

In GLIDE, shape classes were user defined and called *Forms*. Data objects created from Forms were called *Instances*. A Form consisted of a user defined topology, created using a full implementation of the Euler operators [21], and a geometry consisting of vertex assignments. A Form also incorporated a set of attributes. All vertex coordinates were accessed indirectly through pointers. Some vertex coordinates were set up to point to stored literals, others could be set up to point to attribute values. In effect, the shapes of Instances could be parametrized, allowing each to vary in size. These parametrized shape classes could be defined by users. Other attributes could be used to specify material properties. Default values were defined for all attributes. Location was also an attribute. Thus some attributes of a Form were system defined, others were user defined.

An Instance was defined as a Form, a transform and a sequence of name value pairs of new attribute values. Instances were identified by index and users could specify the starting index, allowing parts to be defined by floor number or other controlled scheme. Checks were made that different sequences of indices for a form did not overlap. Attributes could be added to a form after it was created. Instances could have their attribute values changed, including shape attributes and location.

The major step achieved in GLIDE was that these capabilities were embedded in an interpretive programming language. GLIDE had a Pascal-like syntax and would execute any program whose outer block was closed. All procedures were permanently stored to disk and could be called directly or nested in other procedures.

Users could write programs that computed rather complex assemblies of shapes or procedures that detailed an existing shape. Figure 10.2a shows a procedure calling the Euler operators to define an extrusion topology. Figure 10.2b then uses the topologies to define Forms, consisting of a parametrized shape and material attributes. These are then embedded in another procedure that defines a composition in the form of a structural grid, based on dimensions and sizes provided by the user in Figure 10.2c. Figure 10.2d shows the execution dialogue of that program and Figure 10.2e shows the display of the result.

The shapes of Forms were defined in their own coordinate system, selected to facilitate their parametrization. Their Instances could be located relative to the project or to any other Instance. Thus GLIDE supported multiple coordinate systems. This capability turned out to be very important for automatic detailing. The detailing program only had to know about the Form type it was designed to operate on and the Form's coordinate system and could generate its layout ignoring how the Instances it worked on were placed in the world. This is how framing details as shown in Figure 10.3 were developed. Detailing that had to refer to multiple shapes grouped those shapes as children of the same parent shape. All detailing was done in the parent's coordinate system. Some of these details were presented in Eastman [22].

The programmability of GLIDE provided a powerful design environment. Within it, designing consisted of programming the composition of a layout. In classes, students wrote programs for wood and steel framing, piping design, structural steel detailing, and so forth, without prior experience in programming. A nice idea proposed by Robert Thornton and implemented was that a camera was a special object whose shape is a view frustum. It held parameters

```
.RUN GLIDE

Welcome to GLIDE:    Version 1.4 November 77
Type project name, without extension: T5
Continuing with existing project.

! SHAPES.GLI:  Several useful shape procedures;

! To create a topology with n quadrilateral sides and;
! 2 n-sided ends;

TOPO PROCEDURE EXTRUDE(REAL N)=BTOPO
! BTOPO is initialized with a body made up of a single vertex;
! and face;
     CMV (N-1),(N+1);    ! creates a series of edge-vertex pairs;
                        ! chained end-to-end;
     CFE 1,N,2;         ! creates an edge-face pair by connecting;
                            ! the two given vertices;
     CMV N,1;                ! creates the top edge ring;
     CFE (2*N),(N+1),(N+2);  ! creates the top face;
     ! create side faces;
     FOR I FROM 2 TO N DO BEGIN
          CFE I,(I+N),(I-1) END;      CLOSE;
     ETOPO;

TOPO TOP3=EXTRUDE 3;
TOPO TOP4=EXTRUDE 4
TOPO TOP6=EXTRUDE 6;
TOPO TOP10=EXTRUDE 10;
TOPO TOP12=EXTRUDE 12;
```

(a)

```
>?>
> GET F2.GLI
>!   routine to create a structural frame;
>!   first declare those attribs and other things needed;
> REAL ATTRIB STIFFNESS;
> FORM PROCEDURE STRUCTEL(REAL X,Y,Z)=BFORM
    > EXTERNAL TOPO TOP4;
    > EXTERNAL REAL ATTRIB LEN,STIFFNESS;
    > LEN_Z;
    > STIFFNESS_1;
    > SHAPE_BPOLY TOP4;
         > VX[2 TO 3,6,7]_X;
         > VZ[3,4,7,8]_LEN;
         > VY[5 TO 8]_Y;
         > EPOLY;
    > EFORM;
>
```

(b)

```
>! now for the main program;
> PROCEDURE STRUCTFRAME = BEGIN
    > GLOBAL REAL NAISLES,NBAYS,NFLRS,WAISLES,WBAYS,HTFLRS;
    > GLOBAL REAL COLSIZ;
    > TEXT MES;
    > EXTERNAL PROCEDURE SFRAME1;
    > EXTERNAL PROCEDURE SFRAME2;
    > MES_'THIS ROUTINE BUILDS A STRUCTURAL GRID';
    > WRITE MES;
    > MES_'THERE ARE N+1 ROWS OF COLS. FOR N BAYS.';
    > WRITE MES;
    > MES_'SAME FOR AISLES.';
    > WRITE MES;
    > MES_'ENTER NO. OF BAYS. THESE GO DOWN Z-AXIS';
    > WRITE MES;
    > NBAYS_READREAL;
    > MES_'ENTER NO. OF AISLES. THESE GO DOWN X-AXIS';
    > WRITE MES;
    > NAISLES_READREAL;
    > MES_'ENTER NO. OF FLOORS. THESE GO UP Y-AXIS';
    > WRITE MES;
    > NFLRS_READREAL;
    > MES_'NOW DIMENSIONS ARE NEEDED';
    > WRITE MES;
    > MES_'ENTER CTR. TO CTR. WIDTH OF BAYS';
    > WRITE MES;
    > WBAYS_READREAL;
    > MES_'ENTER CTR. TO CTR. WIDTH OF AISLES;
    > WRITE MES;
    > WAISLES_READREAL;
    > SFRAME1();
    > SFRAME2();
    > END;


> PROCEDURE SFRAME1 = BEGIN
    > EXTERNAL FORM PROCEDURE STRUCTEL(REAL X,Y,Z);
    > EXTERNAL REAL NFLRS,HTFLRS,NAISLES,WAISLES,WBAYS,NBAYS;
    > EXTERNAL REAL COLSIZ;
    > TEXT MES;
    > MES_'ENTER FLR. TO FLR. HEIGHT';
    > WRITE MES;
    > HTFLRS_READREAL;
    > MES_'ENTER UNIT DIMENSION OF STRUCTURAL ELEMENT';
    > WRITE MES;
    > COLSIZ_READREAL;
    > MES_'COLUMNS ARE IN FORM COL';
    > WRITE MES;
    > GLOBAL FORM COL = STRUCTEL(COLSIZ,COLSIZ,HTFLRS);
    > FOR I FROM 0 TO (NFLRS-1) DO
      > FOR J FROM 0 TO NBAYS DO
        > FOR K FROM 0 TO NAISLES DO
          > COPY COL=<WAISLES*K,(HTFLRS*I),(WBAYS*J+COLSIZ)\90>;
    > END;
>
> ! procedure to put in beams;
> PROCEDURE SFRAME2 = BEGIN
 >EXTERNAL REAL NFLRS,HTFLRS,NAISLES,WAISLES,WBAYS,NBAYS
 >EXTERNAL REAL COLSIZ;
 >EXTERNAL FORM COL;
 >EXTERNAL REAL ATTRIB LEN;
 >TEXT MES;
 >MES_'NOW BEAMS ARE LAID OUT IN BOTH DIRECTIONS';
 >WRITE MES;
 >MES_'BEAMS ARE CREATED IN FORM COL TOO';
 >WRITE MES;
 >FOR I TO NFLRS DO
  >FOR J FROM 0 TO (NBAYS-1) DO
   >FOR K FROM 0 TO (NAISLES-1) DO
    >BEGIN
     >COPY COL=<WAISLES*K,(HTFLRS*I-COLSIZ),(WBAYS*J+COLSIZ);
          >       LEN_WBAYS-COLSIZ>;
     >COPY COL=<WAISLES*K+COLSIZ,(HTFLRS*I-COLSIZ),
          >       (WBAYS*J+COLSIZ)\0,-90;LEN_WAISLES-COLSIZ>;
    >END;
 >FOR P TO NFLRS DO
  >BEGIN
   >FOR L FROM 0 TO (NBAYS-1) DO
    >COPY COL=<WAISLES*NAISLES,(HTFLRS*P-COLSIZ);
         >       (WBAYS*L+COLSIZ),LEN_WBAYS-COLSIZ>;
   >FOR M FROM 0 TO (NAISLES-1) DO
    >COPY COL=<WAISLES*M+COLSIZ,(HTFLRS*P-COLSIZ),
         >       (WBAYS*NBAYS+COLSIZ)\0,-90;LEN_WAISLES-COLSIZ>;
  >END;
 >END
>
```

(c)

Figure 10.2. (a) An extrude operation of GLIDE for making topologies, (b) GLIDE procedure creating rectangular structural elements. The user procedures that compose structural elements into a structural grid, (d) user dialog with the resulting programs, (e) image of the resulting composition shown in Figures 10.2a–d.

```
.RU GLIDE

Welcome to GLIDE:  Version 1.4 November 77
Type project name, without extension: T5
Continuing with existing project.

>STRUCTFRAME;
THIS ROUTINE BUILDS A STRUCTURAL GRID
THERE ARE N+1 ROWS OF COLS. FOR N BAYS.
SAME FOR AISLES.
ENTER NO. OF BAYS. THESE GO DOWN Z-AXIS
< 4
ENTER NO. OF AISLES. THESE GO DOWN X-AXIS
< 2
ENTER NO. OF FLOORS. THESE GO UP Y-AXIS
< 2
NOW DIMENSIONS ARE NEEDED
ENTER CTR. TO CTR. WIDTH OF BAYS
< 145
ENTER CTR. TO CTR. WIDTH OF AISLES
< 136.5
ENTER FLR. TO FLR. HEIGHT
< 140
ENTER UNIT DIMENSION OF STRUCTURAL ELEMENT
< 11
COLUMNS ARE IN FORM COL
global var. COL already in symbol table*user execution warning 222
```

```
new copy =  1
new copy =  2
new copy =  3
new copy =  4
new copy =  5
new copy =  6
new copy =  7
new copy =  8
new copy =  9
new copy = 10
new copy = 11
new copy = 12
new copy = 13
new copy = 14
new copy = 15
new copy = 16
new copy = 17
new copy = 18
new copy = 19
new copy = 20
new copy = 21
new copy = 22
new copy = 23
new copy = 24
new copy = 25
new copy = 26
new copy = 27
```

(d)

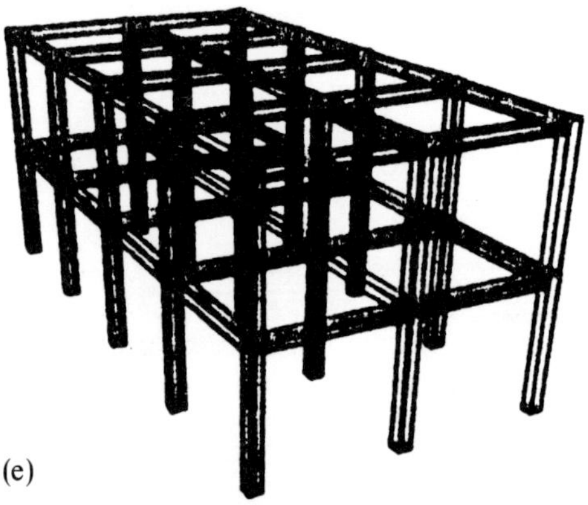

(e)

Figure 10.2 *continued*

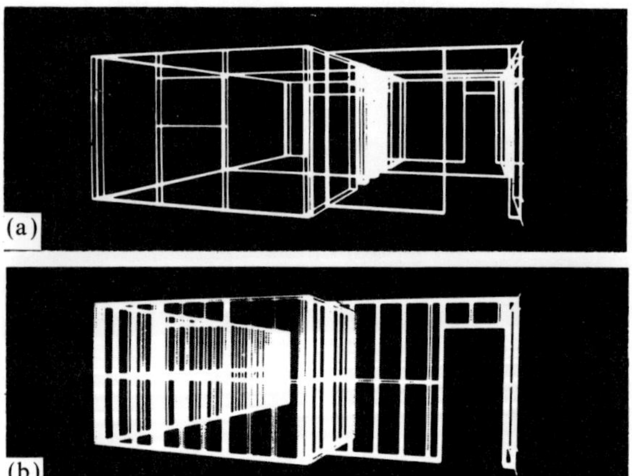

Figure 10.3. (a) A set of rectangular wall segments created in GLIDE that are then framed using a simple automatic framing procedure (b).

regarding clipping and other view parameters. Cameras were object types and used as arguments for display operations. A display was created by turning on a camera. No part of GLIDE involved direct graphical manipulation.

GLIDE supported the definition of thousands of shapes, all as parameterized instances of user-defined Forms. New Forms and programs could be added at any time, providing schema extensibility. It supported automatic detailing and even allowed the embedding of simple evaluation functions into the procedures that created Instances.

Another fact about GLIDE was that all names were global. It became apparent from using GLIDE that a multi-user system required a segmented name space. Like BDS, GLIDE incorporated a buffer space where objects from disk were placed to be worked on. Changes were immediately written back to disk. Thus any error in a design program could corrupt the design file by creating only some of the Forms and Instances desired. From these kinds of errors, we learned about the issue of database transaction managemement.

I considered GLIDE to be a database management system, being a language whose global data was permanent. It did not support many of the standard database support functions, such as concurrency, recovery or queries, but we learned quickly about these issues and existing techniques to deal with them. My understanding of some of these issues are addressed in [23].

After several semesters of GLIDE use, additional criteria were defined that would further the capabilities of a design database. We found that objects must be organized in some form of structure allowing differentiated access, not in GLIDE's flat structure. There was also a need to create and store user views of data in other than graphical formats, for analysis interfaces. Better facilities were needed to query and access data. We also concluded that the shape modeling scheme should be extendable; sophisticated users should be able to define new surface types or shape editing operations.

Moreover, in GLIDE a shape was defined at the time an object (Instance) was created. The shape location was the object's location. It could be modified, but each object had only one shape value. A shape could not be deleted without deleting the object. We found that during design development there is a need to define objects and possibly locate them, even if they do not have a specific shape. It should also be possible to both delete a shape from an object and assign another one and allow an object to have multiple shapes, such as one of the rough-cut form and another that is fully detailed. That is, an object is distinct from its geometry, or as they say today, the issue is *object identity*.

We also learned that the variables corresponding to the relevant physical properties of objects could not be fixed at the time the objects were initially declared or instanced. There was the need to add properties incrementally, as the specific functions of the objects and the types of evaluations needed for them were identified. These attributes also were grouped by the functions for which they allowed evaluation. There was a need to group them to support addition and local addressing.

As we developed programs that computed Instances, it was also clear that there was a need to maintain the relations that were used to compute the instances originally. If one of the parameters of the design procedure changed, it usually became necessary to re-compute the Instances. In a CAD database, this re-computation should be automatic. Hewitt's PLANNER program at MIT showed the way to using constraints to maintain relations between objects [24]. In GLIDE we had implemented constraints embedded in operators. We found that understanding the behavior of autonomous constraints in complex designs was extremely difficult; constraint management tools were also needed.

GLIDE was an excellent tool for teaching computational aspects of design. It supported exploration of design development and detailing and raised deep questions about the structure of design information. GLIDE also was implemented in BLISS on a DEC-10. The graphics on the Three Rivers display was limited to line drawings, again with sectioning and back plane removal. It was distributed to about a dozen organizations.

GLIDE II

These insights (and funding by the US Army Corps of Engineers to build a portable design database system) led to the definition of GLIDE II and its implementation [25]. As students graduated, the GLIDE II team changed to include Kevin Weiler and Mark Birnbaum, John Callen, Greg Glass, Yehuda Kalay, Ali Kutay, Robert Thornton and Yacob Yasky. GLIDE II was defined by developing both the database language and a schema supporting building design in parallel. The database language provided the structure for defining the schema. New needed schema structures could be added after the database was loaded with values.

GLIDE II was also an interpretive programming language with permanence. Syntactically, it was a superset of Pascal and built by extending a P-code interpreter.

Shape modeling capabilities were built on top of GLIDE II, so they could be extended or modified. Thus the Euler operators, shape definition and transform operators were all written in GLIDE II (the spatial set operators were never implemented). This potentially allowed new surfaces to be added as abstract data types; they consisted of a data structure and a set of operations for manipulating the new surface type.

Separate name spaces were added, called *Frames*. Frames were similar to directories and organized hierarchically. They held both objects and operators and allowed structuring of abstract data types into a hierarchical database schema. Accessing procedures or variables in other frames required their path name as a prefix.

Some example code listings from GLIDE II are shown in Figure 10.4. Figure 10.4a is the listing for an extrusion topology. Figure 10.4b shows the coordinate assignment procedure and then its use in defining a parameterized rectangular shape.

```
FUNCTION extrusion(n:INTEGER):BPTR;
{ function which returns an n-sided extruded body }
VAR  bd : bptr;  { body pointer }
     fd : FPTR;  { face pointer }
     rd : RPTR;  { ring pointer }
     ebeg1,ebeg2,elast,newe,ehold1,ehold2 : EPTR; { edge pointers }
     vbeg1,vbeg2,vlast,vhold1,vhold2 : VPTR; { vertex pointers }
     i : INTEGER;
BEGIN
     IF n > 2 THEN
       BEGIN
         cme(n-1,NIL,NIL,CW,vbeg1,ebeg1,vlast,elast);
         { create a chain of vertex edge pairs to form the top }
         { ring of vertices }
         me(vbeg1,ebeg1,CCW,vlast,elast,CW,elast,rd,fd);
         { create a new edge to close the top }
         vhold1 := vlast; { store last vertex }
         ehold1 := elast; { store last edge }
         cme(n,vbeg1,ebeg1,CCW,vbeg2,ebeg2,vlast,elast);
         { create bottom ring of vertices }
         me(vbeg2,ebeg2,CW,vlast,elast,CCW,newe,rd,fd);
         { close bottom face }
         vhold2 := vlast;  { store last vertex }
         ehold2 := newe;   { store last edge }
         FOR i := 1 TO (n-1) DO
           BEGIN
             me(vhold2,ehold2,CW,vhold1,ehold1,CCW,newe,rd,fd);
             { connect top and bottom }
             ehold1 := ccwe(ehold1,vhold1,NIL);
             { return the next counterclockwise edge }
             ehold2 := cwe(ehold2,NIL,vhold2);
             { return the next clockwise edge }
             vhold1 := overt(ehold1,vhold1);
             { return the opposite vertex of vhold1 }
             { lying on ehold1 }
             vhold2 := overt(ehold2,vhold2);
             { return the opposite vertex of vhold2 }
             { lying on ehold2 }
           END;
         bd := ebptr(ebeg2);
         closetope(bd);
         extrusion := bd
       END ELSE
         BEGIN
           WRITELN(tty,'ERROR in extrusion: edge num. para. wrong');
           extrusion := NIL
         END
END;  { extrusion }
```

(a)

```
PROCEDURE assignvert(shape:gptr; vid1,vid2,vid3,vid4:REAL;
                     dim:xyz; varid:CHAR; ptr:^REAL: lit:REAL);
{ assign 4 vertices a value, either a literal or pointer }
BEGIN
          setvert(shape,vid1,dim);  setvert(shape,vid2,dim);
          setvert(shape,vid3,dim);  setvert(shape,vid4,dim);
          { access vertices and dimension on shape }
          assigncord(shape,varid,ptr,lit);
          { assign all previously accessed vertex coordinates }
          { in shape the value defined by ptr or lit }
END;   { assignvert }


FUNCTION rectangle(xmin,xmax,ymin,ymax,zmin,zmax:REAL):gptr;
{ creates a rectangular solid relative to some predefined origin.
  checks that the parameters are valid, eg not inverted.
  returns a shape record.  creates the appropriate topo, if it
  does not exist. }
VAR  rectemp : gptr;
BEGIN
     IF topo4 = NIL then topo4 := extrusion(4);
     { creates topo4 if not existing }
     IF  (xmin<xmax)and(ymin<ymax)and(zmin<zmax) THEN
         BEGIN
         rectemp := inishape(topo4);
         assignvert(rectemp,2,3,6,7,x,'a',NIL,xmax);
         assignvert(rectemp,1,4,5,8,x,'b',NIL,xmin);
         assignvert(rectemp,5,6,7,8,y,'c',NIL,ymax);
         assignvert(rectemp,1,2,3,4,y,'d',NIL,ymin);
         assignvert(rectemp,3,4,7,8,z,'e',NIL,zmax);
         assignvert(rectemp,1,2,5,6,z,'f',NIL,zmin);
         closeshape(rectemp);
         rect := rectemp;
         END
     ELSE BEGIN
           writeln('illegal shape params. in rectangle');
           rect := NIL
         END
END;  { function RECT }
```

(b)

Figure 10.4. (a) Programming style of GLIDE II: extrusion procedure, making a topology, (b) assignment of vertices and the creation of a rectangular solid.

Such programs were thought not to be the products of end use designers, but rather systems people supporting design. The use of a common environment, however, supported the smooth transition between what is commonly viewed as system development and design. This smooth transition, where designers start with high level tools, but then have the opportunity to move into the incremental definition of new tools, was an important concept behind the development of GLIDE and GLIDE II.

GLIDE II supported design alternatives. At any point in time, a set of design operations could be re-directed to a new file. Work then would proceed, using the data in that file as primary. A user could return to the original file and create other alternatives, creating multiple branches in a parent-child relation. Secondary alternatives could be created from the first set, resulting in a tree of alternative files. Name searches within the GLIDE II runtime environment were from the current leaf node up the alternative tree. Once a base design or alternative had child alternatives, it could not be written to. Any alternative could be merged into its parent, which destroyed its siblings [23].

GLIDE II's Frame structure facilitated serious work in schema organization. In construction, a single fixed schema is not practical, as many different technologies can be used to solve the same problem (steel, concrete or wood structures). Thus a semi-hierarchical structure in which detail type definitions and operations are loaded (from stored libraries) as one designs seemed to be the desired schema definition strategy. Each higher level Frame need not know from what or how the performance data was computed. This structure was documented in Eastman [26].

This form of extensibility was outlined in Eastman [23] along with its limitations. Frames offered no clear way to add new types of performance after the schema was initialized. Other language extensions were needed to resolve this limitation.

Frames did not support inheritance or instantiation. They were more like modules in Modula. Instances of form types were created within them. A form of polymorphism called type unions was developed so that existing types could have new semantics added to them. Thus a building or other high level data object could have new variables appended to receive additional values. A set of appended variables could correspond to a class of performance. Similar appendages could be associated with other data objects up or down the tree, which would allow new detail properties to be assigned, then aggregated up to the top level.

Our motivation was to allow new performance properties to be added to a form after it was created. This type of extensibility involved three levels of complexity: unions of types, unions of newly created instances (without values) and unions that extended an existing instance already loaded with values. We only got around to the first two levels of implementation, though the third level is what is commonly called for in design. Type unions were our own pre-cursor to inheritance capabilities as implemented in object-oriented programming. The need in design databases to inherit new data fields after an object is instanced is still a capability missing from available systems.

GLIDE II allowed very flexible development of design models, organized in complex schemas. But this flexibility came at significant cost. While GLIDE could create objects at the speed of several a second, GLIDE II took serveral seconds to create each object. Some layout procedure would create hundreds of objects at a time, resulting in performance that really was not interactive, but batch operation oriented. Like GLIDE, GLIDE II wrote directly to disk, without transaction management.

GLIDE II was used as a testbed for development of serious design models, both for research and for the Army Corps of Engineers. A significant building model was implemented [27]. It consisted of a hierarchical Frame structure that decomposed a building into systems and components [26]. Systems and objects were defined at multiple levels of abstraction. Performance properties were aggregated upward from parts to assemblies. The system architecture diagram for a Corps of Engineers program is shown in Figure 10.5. Some of the research was documented in Yasky [28] and his thesis [29]. Graphical interaction with GLIDE II was attempted using an Evans and Sutherland Picture System. While models of geometry could be developed on that system or the VAX used for GLIDE II, proper communication between them was never achieved. The Army Corps of Engineers were never able to see images generated by the models their funding paid for. GLIDE II was made available in the public domain and was picked up by several commercial firms. It was developed into at least one commercial product.

Solids modeling was a very important topic at SIGGRAPH and among industrial groups. Industrial consortia, such as Computer Aided Manufacturing International (CAM-I), held workshops on solids modeling. The various researchers got to know each other well, and various rivalries emerged. This was especially between the boundary representation modelers and those doing constructive solids geometry. Only much later were the relative contributions and complementarity of a parametric representation and an evaluated surface model fully appreciated.

FORM:ULAE™ [1]

In 1983, I took a leave of absence from Carnegie-Mellon University to attempt production development of an engineering design environment. I co-founded Formative Technologies, Inc.

[1] FORM:ULAE™ is a registered trademark of FORMTEK. The information describing FORM:ULAE is used with the permission of FORMTEK.

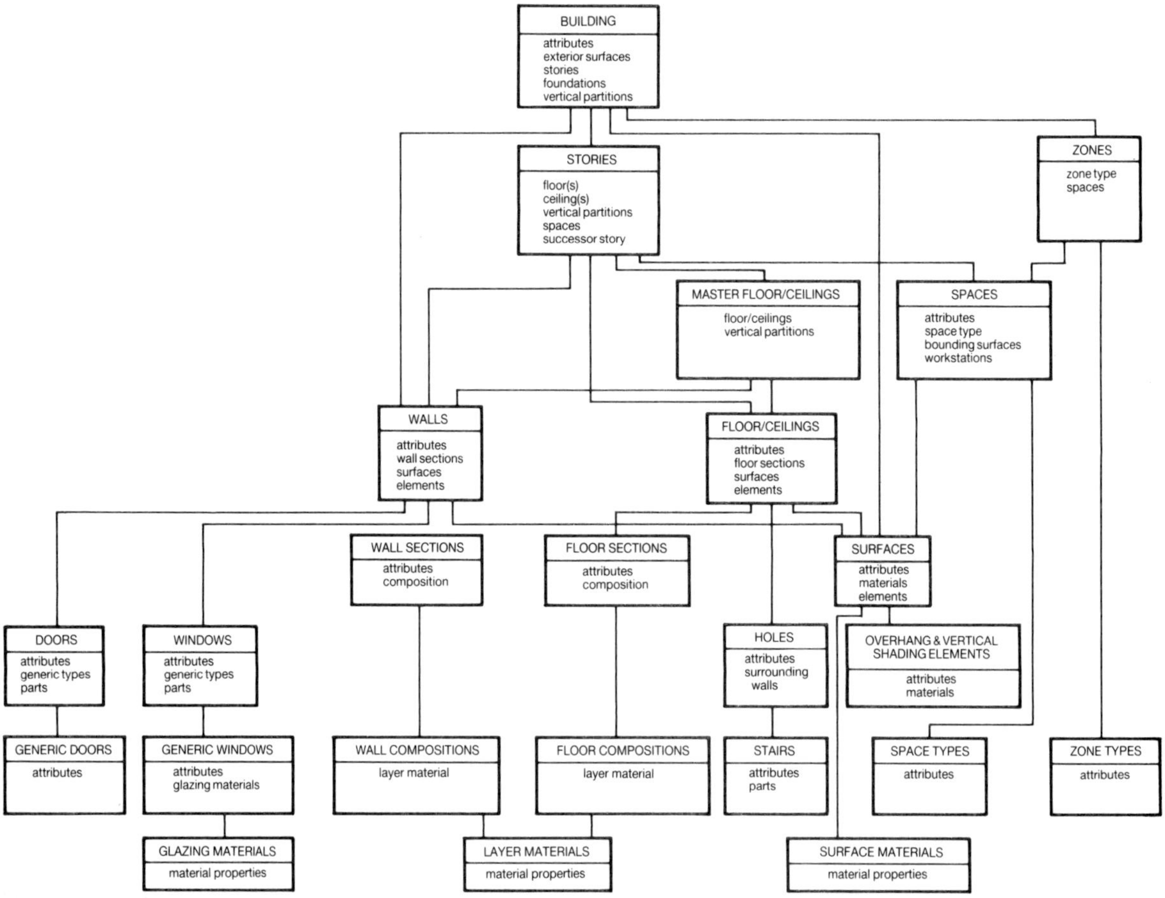

Figure 10.5. System architecture defined for building design evaluation for the US Army Corp of Engineers. Only parts were implemented.

(FORMTEK) with the plan to implement a production engineering database system and a suite of application software on top of it. The initial application was a drafting product, developed before the database to generate incremental income.

The FORMTEK engineering database followed the line of earlier efforts: it was a language with permanence and was called FORM:ULAE™. Syntactically, it was a superset of Modula-II, with extensions supporting object-orientedness, multiple inheritance and constraint management [30, 31]. Both objects and modules were supported, as alternative means to embed code from other sources. Constraints were simply pointers to procedures, which were written in a form that allowed them, when invoked, to access other variables identified by name and position. It was expected that most would be used to compute some functions of data within their own data object and pass the result to a parent data object (or set a flag in the parent).

The planned structure for developing models within FORM:ULAE™ relied on the semi-hierarchical structure developed in GLIDE II. Three structures of data objects were designed that corresponded to the aggregation needed of particular performance properties. These hierarchies were defined for each property, not by physical object. Each data object had internal constraints and could pass the result of internal calculations or testing to its parent, allowing integrity maintenance up any tree. Downward propagations were design actions and would be implemented by the general procedures (methods) used for designing. A physical object node was expected to incorporate property objects from multiple hierarchies; changes to such a node, possibly determined by the performance requirements in one hierarchy, affects multiple hierarchies. This was an important effect to capture.

The FORM:ULAE^{TM} environment was planned as a semi-distributed environment, with a check-out, check-in file server [30]. While the distributed versions would operate like a language with permanence, the file server would include a full transaction manager that would never over-write itself. Thus each update of the library would be an application level transaction which could be undone later, if necessary.

Alternatives were to be supported as a variation on this strategy. Roll-backs to earlier design states were also supported by this scheme.

In late 1985, FORM:ULAE^{TM} was running in prototype mode. However, its development was put on hold because FORMTEK determined that its introduction at that time was not commercially attractive. It was never re-started. From the perspective of the current time, shortcomings of its design are apparent. However, its features are similar to those of the current generation of object-oriented database systems. A few problems were never resolved: the ability to do constraint management on a distributed database is still a significant research issue.

The research experiences gained from BDS, GLIDE, GLIDE II and FORM:ULAE^{TM} not only provided a wealth of insight into the proper direction of design databases, it educated a generation of future researchers who would later take positions further developing the technology. There were only a few similar efforts in engineering databases [32, 33]. The architectural application area also had a few serious efforts, particularly Borkin's group at Michigan [34] and Bijl's at Edinburgh [35], as well as the industrialized building system oriented CAD packages in the UK.

STRUCTURE OF BUILDING MODELS

The challenge that has faced architectural CAD research for the last ten years has been how to develop the technology that supports building modeling for custom, one-off designs that incorporate a mixture of off-the-shelf and custom components and a project unique mix of use criteria. My group's work at Carnegie-Mellon was directed toward answering this question. While many major issues and details have been worked out, there still remains much to do, as of today, early 1993. Only a few prototype systems exist to allow research exploration of these issues. A few systems are commercially available that provide support for building modeling but they have not been particularly successful in the marketplace. Part of the unmet challenge involves re-defining design practices so as to take advantage of the computer as a new design medium and information environment.

Part of the challenges are peculiar to architectural and building practices: the small size of firms; the heavy reliance on the cultural practices of the local construction industry; and the sharing of design responsibilities between designers, consultants and contractors. These issues will require special adaptation of information handling methods to these organizational structures and practices.

Below, I attempt to summarize the lessons learned regarding how a building model should be organized.

Scope of building modeling

The information needed to design, construct, operate and maintain a building over its lifetime is enormous. If we consider design from the architect's standpoint, part of the necessary design information for a building are the context, expected use, social or organizational processes, and other inputs to the design process to which the building is a response. For construction, construction practices and processes, both explicit and implicit, are required. For operations and maintenance, we need usage and maintenance schedules.

Traditionally, the explicit scope of design information supplied to the contractor has been *a state description of the building to be provided at the completion of construction*. While this is only a small part of the potential scope of a building model, it is an ambitious enough first step for us to address.

Construction is a craft; people make buildings. Drawings are the most effective means known to convey layout and composition information between people. Thus the main product of architectural CAD for some time will continue to be drawings. A major requirement of architectural CAD is the ability to produce drawings incorporating existing formats and conventions.

Organization of information within a building model

Designers rely on multiple representations that each partially describe the objects making up the design. Layout drawings of plan, section and elevation, area calculations, written specifications, are all partial descriptions of the evolving design. In general, a representation allows abstracting from the total design information that which allows evaluation of particular properties; its value is in allowing direct expression of these properties and derivation of certain relations between collections of properties. In the database sense, a representation is a view.

New objects must be easily created within any view, with the necessary linkage structure to relate the initial view with others added later for other purposes. Physical objects must be defined separately from their shape, so that an abstract depiction of an object can have different shapes and other properties assigned to it.

New property types need to be added, as additional performances become relevant during the design process. It would be impractical to carry all possible properties of objects just for the possibility of their use. Properties consist of a set of variables describing a class of behavior and these are associated to the operations that can assign values, modify them and check their correctness. The incremental addition of properties was the reason we provided union types in GLIDE II and inheritance in FORM:ULAE™.

During design, objects are decomposed into their parts. This is to specify how the higher level object is to be composed. Each decomposed object is a small design problem. Objects are also aggregated to allow verification of the functioning of the higher level object.

The description of the design at various levels of aggregation is required to several reasons. At a high level of aggregation, global variables are assigned values using estimation (and experience) for planning purposes. Also, different applications require information about the design at different levels of aggregation. For example, bills of material need to know the material quantities in a wall, but an energy analysis needs only the aggregate thermal performance of the wall. Thus within any technology, there is the need to aggregate information from potentially diverse objects below it in the hierarchy.

Aggregations take place in different hierarchies. For example, a DC transformer may be part of the electrical and the control systems and contribute to performance evaluation in both hierarchies. At a higher level of aggregation, a wall may be part of both the space enclosure and the structural hierarchies. We came to believe that all individual physical properties could be managed within a strict hierarchical structure. Each property type is aggregated hierarchically, but the multiple properties of a single object are aggregated in multiple property hierarchies. Physical objects consist of a collection of physical properties and means to maintain the integrity between them. Physical objects are embedded in a possibly complex directed, acyclic graph. The directionality of the graph comes from the direction of aggregation of various properties. We came to call this structure an *abstraction hierarchy* [26]. Figure 10.6 is an example of an early schema developed to demonstrate these ideas.

At the top of an abstraction hierarchy is the global description of the object being defined. We considered a building to be made up of a predefined set of systems, each responsible for one or more specific properties. For us, a typical building consisted of a site, structure, mechanical equipment, space enclosure, external shell and spaces. These systems each have their own performance requirements, some handled by consulting engineers. Each system can be resolved by one or a mixture of technologies. Sometimes, one technology resolves multiple systems. For example, alternative technologies for a structural system might be steel frame, bearing wall, poured-in-place concrete, etc. Each technology defines abstractly some components and their necessary properties. Implementation-wise, we take a technology to correspond to a sub-schema, composed as an abstract data type [36].

These functional properties will eventually map onto unique physical objects or alternatively, several functional properties may map onto a single physical object. This last case shows how a single object may have multiple functions. Thus a stair bannister may also be a heating radiator or a balcony railing may also be a rain gutter. Such multi-functionality is a crucial capability

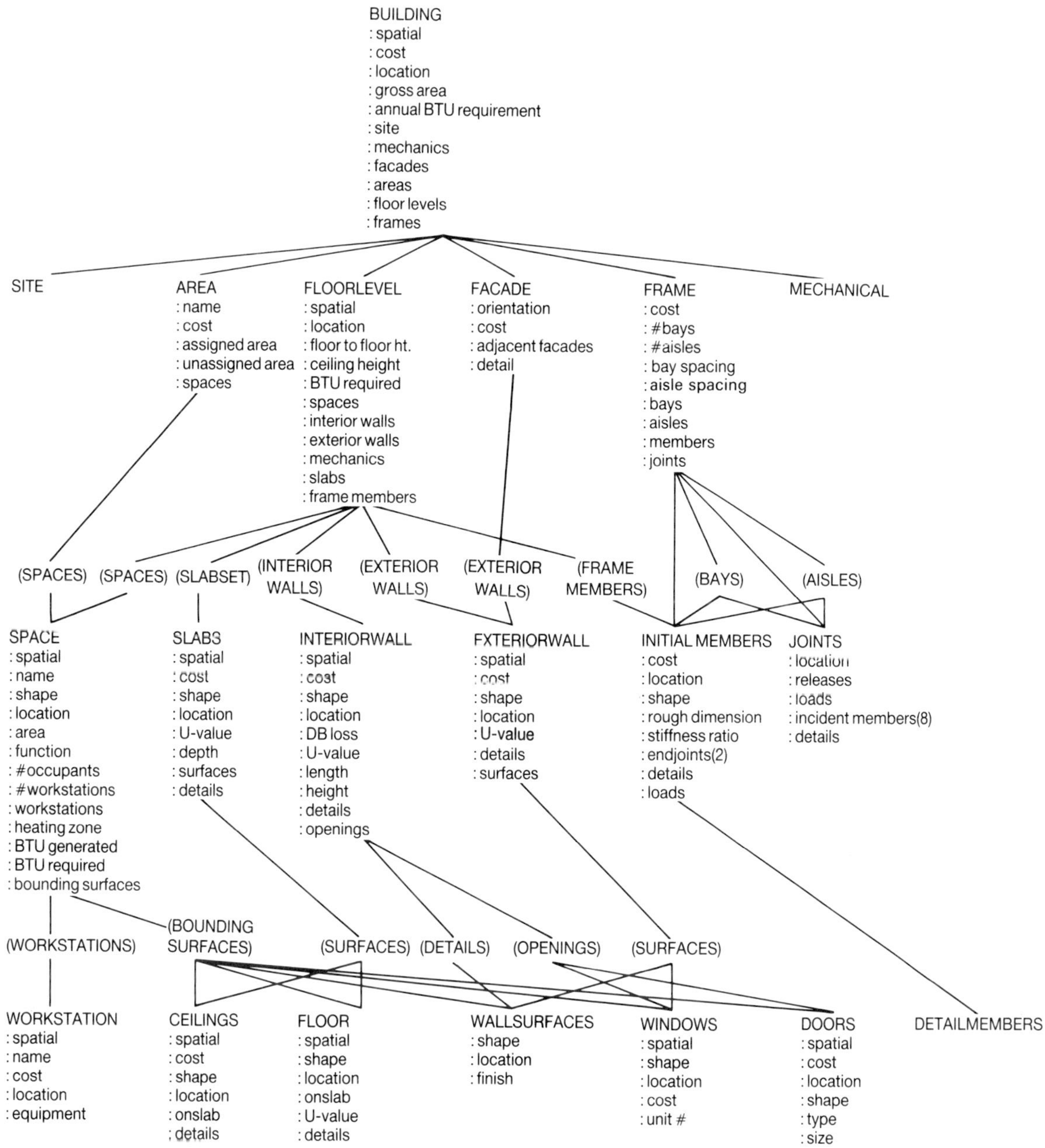

Figure 10.6. Diagram of a building model schema developed in GLIDE II, incorporating different levels of abstraction. Different technologies could be plugged into the high level functional requirements.

for creative design. Objects should be accessible by their different names as well as their different functions and properties.

A detailed data structure developed about this time for a wall system is shown in Figure 10.7. It treats a wall as a core made up of multiple layers and a surface finish. Various entities may be part of the wall, such as doors and windows. Other parts pass through a wall, such as pipes and conduit. While the parts move with the wall, additional work is associated with entities passing through the wall.

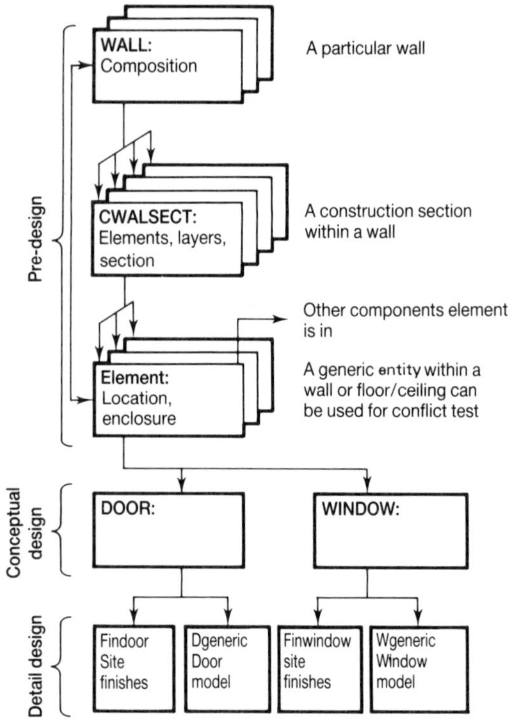

Figure 10.7. The GLIDE II database-schema for a wall.

Constraints

As design actions are made, they respond to various relations. Occasionally these relations are defined explicitly, as in certain spacing rules. More often, however, they are implicit and embedded in either complex code or the head of the designer. An example here would be the alternative fire exit requirements for a multi-storey building. Later, as decisions are made in the layout of the building, occupancy changes may lead to corridor or door size changes, because of the fire codes. The effect is that one change results in other changes, possibly propagating in a chain effect. Like others, I call these relations *constraints*. Constraints operate in multiple directions, depending on what is being modified and what checked. The appropriate representation of constraints is an open issue.

Some constraints are based on the logical correctness of the data, e.g. its integrity. An example is that when a wall is moved, then the volumes of spaces on both sides of the wall also change. If these changes were not made, the data would be logically inconsistent. Some changes are the result of design intentions. For example, pipes may be routed through the moved wall. When the wall moves, the pipes must be re-routed. We believed strongly that a major goal of a building model should be to incorporate such constraints and to manage them and if possible to update them automatically. If automatic updating is not possible, the relations should be flagged for a human to deal with. Thus management of design integrity is a fundamental issue in design research.

I have come to believe that design deals with three types of constraint structure. One structure is defined by the traditional representations used for composing or analyzing objects so as to *satisfy various functional performances*. This constraint structure defines the relation between the different properties of various objects and various kinds of functional performance. The second structure *covers the different properties of a single or hierarchically organized set of objects*. It requires that they all must be consistent. The third structure is the one imposed by the designer, in terms of shared definition of materials, spatial organization, details or other *imposed criteria*. See Figure 10.8.

Until recently, most research efforts have been directed to the first structure: the analysis of a composition's behavior and developing means to explicate the structure imposed by various functions. A building database explicitly adds the second structure for managing the multiple descriptions or objects. The third structure, consisting of the constraints imposed by the designer for coherence, ease of manufacture of other self-imposed objectives, is new to the formal definition of design models.

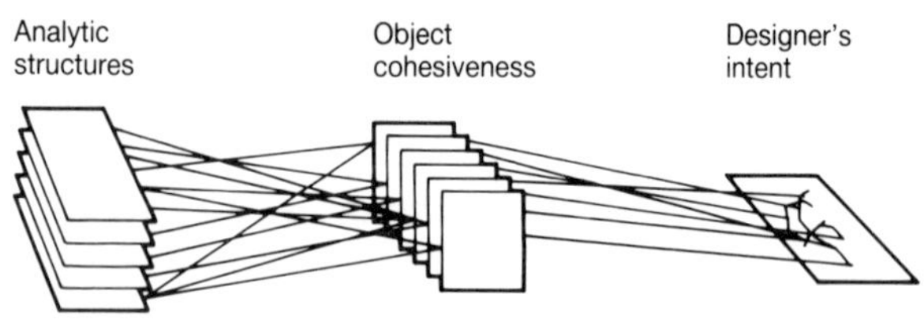

Figure 10.8. The three logical structures embedded in building design models: functional performances, qualitative performances and the integrity of a single object.

Special concerns regarding the geometry of buildings

A unique quality of architecture is that the spaces created by solid objects are as important as the solids and carry a large number of functional properties. A good AEC shape model must allow explicit representation of spaces as well as solids. We are interested in manipulating spatial properties as well as the properties of solids. There are obviously a large number of integrity conditions across the properties of the spaces and their bounding solids, not only regarding shape properties, but also lighting, acoustics and energy properties. This special property of architecture has not been adequately responded to in commercial CAD systems and only a few research efforts have addressed these issues [37,38].

Our efforts at building modeling assumed that all graphic information could be extracted as sections or projections of a 3D model of the building [9]. Several commercial systems came to market embedding the same assumption. These treatments of geometry failed as production solutions because it was inherently too restrictive. Components are often represented in symbolically, in not-to-scale drawings, center-lines and other forms of representation, as well as sections of a 3D model and attributes [39]. A solid model as the basis for generating all geometry turned out to be too limited; needed was the ability to carry multiple shapes and presentation data for an object.

Drawings and analysis input are defined as queries that select objects with particular properties and/or that are located within a specified region. This information is then formatted into a particular report. Drawings are graphic reports. Annotations, dimensions and other presentation information can be stored with the objects and keyed to a particular type of drawing. In some cases, the geometric information is computed from projections of 3D shapes.

This outline of building modeling describes those features identified during and after the development of GLIDE and GLIDE II and FORM:ULAE^{TM}. Many of these database organization issues apply to other types of design models, beyond buildings. Much further development and testing is required before production building models can become practical.

SOFTWARE TOOLS FOR BUILDING MODELING

In parallel with these considerations of the structure of a building model, we also investigated development of the necessary system tools for defining databases of complex physical systems. BDS, GLIDE and GLIDE II were all prototype implementations of such systems.

In design it is necessary for multiple applications to read and modify design data. It is very risky that all such modification operations be implemented multiple times in each of the applications. It is much preferred for the operators to be implemented once and provided centrally to all applications. Centralizing the actions both reduces the total amount of code and allows control of the semantics of the model. Abstract data types have thus been central to our thinking [36,40].

Because design decisions are selected among different technologies, it was clear that a single schema for storing design data was impractical and that dynamic schema definition was required [23].

There is an intrinsic partial precedence ordering to design data that imposes a partial ordering on design actions. This ordering imposes a dependency chain when iterations of actions are made [41,42]. We outlined a means to globally structure constraints within a design as semantic transactions [43]. Semantic transactions impose a rigorous structure for the management of constraints.

The development of property hierarchies, including methods to create, modify, and evaluate them, involved capturing aspects of the behavior of the physical world. These property hierarchies are the foundations on which different design models are built. They are combined with physical object descriptions into an AND- and OR-graph structure described in Eastman [44]. Physical objects are the union of a set of properties. Not adequately worked out are methods for maintaining the integrity of these different property descriptions that describe a single object.

There is much current interest in the development of object-oriented databases [33, 45]. Object-orientedness, to my understanding, incorporates two types of structures: abstract data types and method and data inheritance. Abstract data types are a basic means to provide data independence and to incorporate procedural semantics into databases. Our experience with schema definition suggests that inheritance plays a similar important role in the refinement of design information.

In the not too distant future, design will become a mixture of machine operations and human controlled ones. The human controlled operation sequences that are frequently repeated will become automated. Practical accomplishment of such automation requires the development of a design and computational framework allowing the information relevant to a decision to be accessed and complex assessments to be made. Languages for encoding such information are required. I am encouraged by certain developments in shape grammars that provide these capabilities and package them in a form useable by architects and other types of designers [46].

CONCLUSION

The work reported here built upon current computer language and semantic concepts avidly discussed in classes and seminars at Carnegie-Mellon University. We applied the newest concepts to the research we were undertaking, sometimes extending the concepts, occasionally through misinterpretation. It was a heady time.

I strongly believe that computer tools will be highly integrated with theories of design, not in any restricted sense, but in the sense that only by gaining a deep understanding of design will we ever develop appropriate design tools. I believe that this two-way dialogue, between the abstract information concepts of computer science and the challenges of creative design, is enriching on both sides.

I remember ten years ago presenting the prospect of modeling as a means of design, rather than multi-overlay drafting. The idea then was revolutionary, completely outside the culture of architectural practice. Now the possibility is becoming widely known and initial systems aspiring to these capabilities, if not realizing them, are on the market. In the next few years, I expect more suitable system configurations to emerge on which to do design modeling.

The means of architectural development now used, based on drawings, has existed for hundreds of years. The computer is slowly allowing development of the concepts and tools needed to create a new means of architectural development, based on integrated models. Many more issues must be resolved before such building models can have a real impact on the practice of architecture and building.

The work reviewed here emerged from a highly collaborative environment. Many people deserve credit for these results. While I have attempted to identify as many of them as I can recall, I probably have forgotten others. My apologies.

REFERENCES

1 **J. C. R. Licklider** Man-Computer symbiosis, *IRE Trans. on Human Factors,* Vol 1 No 1, 1960, 4–11.

2 **S. A. Coons** An outline for the requirements for a computer-aided design system, *Proceedings Spring Joint Comput. Conf.,* Spartan Books, Washington, DC, 1963.

3 **T. I. Johnson** SKETCHPAD-III: A Computer Program for Drawing in Three-Dimensions, *Proceedings Spring Joint Comput. Conf.,* Spartan Books, Washington, DC, 1963.

4 **C. M. Eastman** Preliminary report on a language for general space planning, *Commun. ACM,* Vol 15 No 2, 1972, 76–87.

5 **C. M. Eastman** An interrogation language for building descriptions, in *Models and Systems in Architecture and Planning,* D. Hawkes, ed., Construction Press, London, 1972.

6 **I. C. Braid** *Designing with Volumes,* Cantab Press, Cambridge University, 1973.

7 **B. G. Baumgart** *Geometric Representation for Computer Vision,* Stanford University Computer Science Dept., AIM-249, STAN-CS-74-463, October 1974.

8 **C. M. Eastman, J. Lividini and D. Stoker** A database for designing large physical systems, *AFIPS Press, NCC,* 1975, 603–611.

9 **C. M. Eastman** General purpose building description systems, *Comput.-Aided Design,* Vol 8 No 1, 1976, 17–26.

10 **C. M. Eastman** The concise structuring of geometric data for computer aided design, in *Data Structures for Pattern Recognition and Computer Graphics,* A. Klinger, K. S. Fu and T. Kunii, eds., Academic Press, New York, 1976.

11 **D. Stoker** *CRIPL-edge data structure,* unpublished report, Institute of Physical Planning, Carnegie-Mellon University, 1974.

12 **BDS-10 User's Manual** Institute of Physical Planning, Carnegie-Mellon University, Pittsburgh, PA, 1976.

13 **W. A. Wulf, D. Russell and N. Haberman** BLISS: a language for systems programming, *Commun. ACM*, Vol 14 No 2, December 1971, 780–790.

14 **B. Rosen** The architecture of a high-performance graphics display terminal, *Soc. for Info. Displays, Symposium*, New York, May 1973.

15 **E. M. Hoskins** Computer aids in system building, in *Computer-Aided Design*, J. Vliestra and R. Wielinga, eds., North-Holland, Amsterdam, 1973.

16 **J. W. Paterson** An integrated CAD system for an architects' department, *Comput.-Aided Design*, Vol 6 No 1, 1974, 25–31.

17 **M. A. Meager** The application of computer aids to hospital building, in *Computer-Aided Design*, J. Vliestra and R. Wielinga, eds., North-Holland, Amsterdam, 1973.

18 **C. M. Eastman** The use of computers instead of drawings in building design, *J. Amer. Inst. of Architects*, March 1975, 46–50.

19 **C. M. Eastman and M. Henrion** GLIDE: a language for design information systems, *SIGGRAPH'77 Proceedings Computer Graphics*, Vol 11 No 2, 1977, 24–33.

20 **A. Baer, C. Eastman and M. Henrion** Geometrical modeling: A survey, *Comput.-Aided Design*, Vol 11 No 5, 1979, 253–272.

21 **C. M. Eastman and K. Weiler** Geometric modeling using the Euler operators, *Proceedings NSF Conf. on CAD/CAM*, D. Gossard, ed., MIT Press, 1979.

22 **C. M. Eastman** The design of assemblies, *SAE International Congress and Expo*. Tech. Paper No. 810197, 1981.

23 **C. M. Eastman,** Systems facilities for CAD databases, *Proceedings 17th Design Auto. Conf.*, 1980, 50–56.

24 **C. Hewitt** *Description and theoretical analysis (using schemata) of PLANNER: a language for proving theorems and manipulating models of a robot*, AI Memo No. 251, Project MAXC, MIT Cambridge, MA, April 1972.

25 **C. M. Eastman and R. Thornton** *A report on the GLIDE II language definition*, Institute of Physical Planning, Carnegie-Mellon University, March 1979.

26 **C. M. Eastman** Representation of design problems and maintenance of their structure, in *Artificial Intelligence and Pattern Recognition in Computer Aided Design*, J. C. Latombe, ed., North-Holland, Amsterdam, 1978.

27 **CAD-Graphics Lab** *The integrated building model and database schema for the Second Phase of Integrated CAEADS*, Cad-Graphics Laboratory, Institute of Building Sciences, Carnegie Mellon University, 1981.

28 **Y. Yasky** Transforming a set of building drawings into a consistent database, *Proceedings CAD80 Conf.*, IPC Press, London, 1980, 101–108.

29 **Y. Yasky** *A consistent database for an integrated design CAAD system: fundamentals for an automated design assistant*, unpublished PhD thesis, Dept. of Architecture, Carnegie-Mellon University, 1981.

30 **C. M. Eastman and A. Kutay** Functional specification of a distributed engineering data management system, *Proceedings ASCE Conf.*, San Francisco, CA, 1984.

31 **C. M. Eastman and A. Kutay** Specification of FORM:ULAE: a distributed engineering data management system, *Proceedings ASCE Conf.*, San Francisco, CA, 1985.

32 **K. R. Dittrich, W. Gotthard and P. Lockeman** Complex entities for engineering applications, *Proceedings 5th Entity-Telationship Conf.*, North-Holland, Amsterdam, 1986.

33 **D. S. Batory and W. Kim** Modeling concepts for VLSI CAD Objects, *ACM Trans. Database Syst.*, Vol 10 No 3, 1985, 322–346.

34 **H. J. Borkin, J. McIntosh, P. McIntosh and J. Turner** *ARCH:MODEL: Geometric Modeling Relational Database System*, Architectural Research Laboratory, University of Michigan, 1982.

35 **A. Bijl and G. Shawcross** Housing site layout system, *Comput.-Aided Design*, Vol 7 No 1 1975, 2–10.

36 **J. V. Guttag** Abstract data types and the development of data structures, *Commun. ACM*, Vol 20 No 6, June 1977, 396–404.

37 **K. Aggar** Data structures for function and geometrical description of buildings, *Proceedings CAD'76*, IPC Press, UK, 1976, 87–95.

38 **C. Yessios** The computability of void architectural modeling, in *Computability of Design*, Y. Kalay, ed., Wiley Interscience, New York, 1987, 141–172.

39 **C. M. Eastman** Fundamental problems in the development of computer-based architectural design models, in *Computability in Design*, Y. Kalay, ed., John Wiley, 1987.

40 **D. Parnas** On the criteria to be used in decomposing systems into modules, *Commun. ACM*, Vol 15 No 12, 1973, 1053–1058.

41 **C. M. Eastman** Database facilities for engineering design, *Proceedings IEEE*, Vol 69 No 10, 1981, 1249–1263.

42 **G. Lafue and C. Eastman** Semantic integrity transactions in design databases, in *File Structure and Databases for Design*, J. Encarnacao and F. L. Krause, eds., North-Holland, Amsterdam, 1982.

43 **A. Kutay and C. M. Eastman** Transaction management in engineering databases, *SIGMOD Conf. on Engineering DB*, 1983.

44 **C. M. Eastman** Automatic composition in design, in *Design Theory and Methodology*, S. Newsome, W. Spillers and S. Finger, eds., Springer-Verlag, New York, 1989.

45 **T. M. Atwood** An object-oriented DBMS for design support applications, *Proceedings IEEE COMPINT*, New York, 1985, 299–307.
46 **G. Stiny** *Pictorial and Formal Aspects of Shape and Shape Grammars*, Birkhauser, Basel, Switzerland, 1975.

ABOUT THE AUTHOR

Charles M. (Chuck) Eastman became intrigued with the possibilities of computers helping to manage the complexity of information facing architects while he was a graduate student at the University of California, Berkeley. While there he put himself through school designing houses in the Bay area.

He taught at the University of Wisconsin for one year, then was offered a joint appointment in Computer Science and Architecture at Carnegie-Mellon University, Pittsburgh. He spent the next 20 years in Pittsburgh, as an Assistant, Associate and full Professor in the two departments. While there, he founded and directed the Institute of Physical Planning jointly with the School of Urban and Public Affairs. This was succeeded by founding and directing the Institute for Building Sciences and the CAD Graphics Laboratory. He also initiated the PhD program in Architecture. During this time, he undertook research on cognitive processing of design, automated design, solids modeling and design databases. He published over 60 papers and edited two books. Chuck has collaborated in research with many types of designers, including civil, chemical and mechanical engineers and product designers. He has consulted for companies making buildings, cars, airplanes and VLSI chips.

In 1983, he founded Formative Technologies Inc., which has become a producer of optical scanning, editing, archiving and distributing systems for engineering data.

In 1987, he moved to Los Angeles and became Professor at the Graduate School of Architecture and Urban Planning, University of California. His current interests include solids modeling, design theory and the conceptual modeling of buildings. He has been on the Editorial board of *Computer-Aided Design* journal since 1973.

11 Research in Solid Modeling at the University of Rochester: 1972–87

Herbert B. Voelcker and Aristides A. G. Requicha

PROLOGUE

The Production Automation Project (PAP) was founded at the University of Rochester in 1972 and disbanded in 1987. During its 15-year life, the PAP contributed to the foundations of solid modeling, developed core technology for CSG modeling systems and applications of solid modeling, and provided codifications of the field that are reflected today in the research literature, textbooks, and university courses. This chapter summarizes the evolution and technical character of those contributions.

The PAP was led by one or the other of us through its life, and was a satisfying group to lead. It had *élan* and a distinctive working style, due in no small measure to our belief that we were doing important work and were, in the 1970s, in continuing contention for leadership with similarly strong, spirited groups at Cambridge, Hokkaido, Berlin, and a few other places. We have chosen to organize this chapter on partially historical lines to convey some of the conceptual dynamism of those years, and to provide an archival record for the PAP.

The Context section below summarizes the evolution of solid modeling, the American R&D scene, and the chronology of PAP projects. The succeeding section, which is the body of the chapter, summarizes technical themes and important concepts, algorithms, and systems as they emerged in the PAP. Throughout, history 'stops' c. 1986/87, and issues that have been dispatched in subsequent years are left open or dangling.

CONTEXT

The emergence of solid modeling[1]

The inventions of the 1950s – NC machine tools, early industrial robots, primitive graphic displays, and of course viable digital computers – spawned in the 1960s and 1970s four streams of development aimed at representing 3D objects: wireframes, polygonal schemes, sculptured surfaces, and solid modeling. Development within each stream proceeded largely independently for almost two decades, but in the 1980s the streams began to merge conceptually and technologically.

The birth of solid modeling may be traced to three projects in the early 1960s: Roberts at MIT, Gutterman at Sandia, and Luh and Krolak at IBM.[2] These projects were remarkably prescient, but they were a decade too early and largely failed. In the late 1960s and early 1970s a first generation of experimentally viable systems appeared with a much more international flavor: in England, Braid's Build-1 system; on the continent, Compac by Spur *et al.* in Berlin, Brun's Euclid in France, and Engeli's Euklid in Switzerland; in Japan, Okino's TIPS-1 at Hokkaido and Hosaka's Geomap at Tokyo; in the USA, Synthavision at MAGI and (under a different name) at the Army Ballistic Research Lab, Shapes at the MIT Draper Lab, Eastman's Glide-1 at Carnegie Mellon, and PADL-1 at Rochester. These systems were also prescient, and recognized the importance of Boolean operations, but most were quirky and executed Booleans properly only under restrictive conditions with *ad hoc* algorithms.

In the 1970s, theoretical foundations and systematic algorithms for solid modeling began to appear, and the first generation of experimental systems was superseded by a second generation of much better experimental systems – Build-2, later versions of Euclid, and PADL-2, among others; some of these systems furnished the techniques, and in some cases core code, for a first generation of commercial and industrial

The colour plates for this chapter appear between pp. 210 and 211.

[1] See Requicha [1,2] for an extended discussion of the material in this section, and for references.

[2] The earliest conceptual study with solid foundations is Price's [3] interesting proposal for a Design Machine.

Fundamental Developments of Computer-Aided Geometric Modeling
ISBN 012-554765-X

systems that appeared in the early 1980s (Romulus, GMSolid, Unisolids, Bravo and others). These early industrial systems were slow and limited in functionality, but they provided *new* functionality (automatic mass properties and shading, primarily) and were received initially with enthusiasm, because much better systems were expected in the later 1980s. It is fair to say that by the later 1980s the early enthusiasm had waned because much better systems were not appearing, and did not seem imminent. The problems we thought solvable in the early 1980s turned out to be harder than expected.

The American R&D environment

The launching of the Soviet satellite Sputnik-1 in 1957 triggered in America a wave of enthusiasm for science and engineering that peaked in the later 1960s, just prior to the Apollo moon landing, and then faded amidst the counter-culture fads of the 1960s and 1970s. In the 1970s, American science and engineering were in retreat, and there was growing pressure to do applied research relevant to national needs – initially the environment and alternative energy sources, with the latter triggered by two oil crises. Industrial productivity and international competitiveness were first tagged as lower priority needs, but their importance grew gradually in the 1970s and then swiftly in the 1980s, as America's trade balance deteriorated.

The National Science Foundation established the RANN (Research Applied to National Needs) Program in the early 1970s to address these various needs. RANN supported a short-lived consortium called the Automation Research Council, and – pivotally – a program launched in 1972 by Bernard Chern that sowed the seeds for the ambitious NSF programs in design and manufacturing that appeared a decade later. It is fair to say that Chern's early grantees – Barash, Binford, Mergler, Nevins, Paul, Rosen, Voelcker, Wang, Whitney – set the character of much of American automation and modeling research for almost a decade. RANN, however, was a controversial program because it promoted applied science in an agency with pure science traditions, and because it tried to use aerospace-industry management techniques. RANN lasted for about five years, but the beachhead it built in the NSF for problem-oriented R&D survived, and influenced strongly the reorganization of the NSF Engineering Directorate in 1984. Chern founded a new NSF Division for Design and Manufacturing (and, briefly, Computer Engineering) as part of that reorganization.

On the industrial side, the CAM-I consortium was incorporated in 1972 to replace and broaden the old APT Long Range Program, and CAM-I's Geometric Modeling Project stimulated important discussions and experimentation in solid modeling in the 1970s. CAM-I never attracted enough subscription income to support serious R&D, and thus became mainly a forum, but it was a useful forum. ARPA (now DARPA – the Defense Advanced Research Project Agency) was an active promoter through the 1960s and 1970s of research in computer graphics, robotics, networking, AI, and VLSI, but it did not promote work in manufacturing and related modeling technologies until the mid- and later 1980s.

Two of the most serious problems we and other university-based research groups faced in the 1970s were the lack of archival (or indeed any) journals in which to publish research on modeling and automation, and the lack of recognized peer groups whose blessings were needed to legitimize such research in university tenure cases. We spent many hours doing battle with editors and reviewers, and some talented junior people were warned-off areas that today are 'hot'. But time cures many things, and by the later 1980s these problems were much less onerous.

Historical summary of the PAP

The Production Automation Project (PAP) was established in 1972 by Herb Voelcker, then a professor of electrical engineering at Rochester, to develop automatic programming systems for numerically controlled (NC) machine tools. Ari Requicha joined the project in October of 1973, but influenced it earlier. (*Why* these two communication engineers decided to enter the world of industrial automation is explained elsewhere [4]). Figure 11.1 provides a graphical summary of the PAP's history.

While the PAP's initial goal was automatic NC programming, the staff were neophytes in NC and CAD, and so focussed initially on surveying relevant technologies. Surveys conducted in 1972–74 [5–7] showed that major advances in NC programming and mechanical CAD were

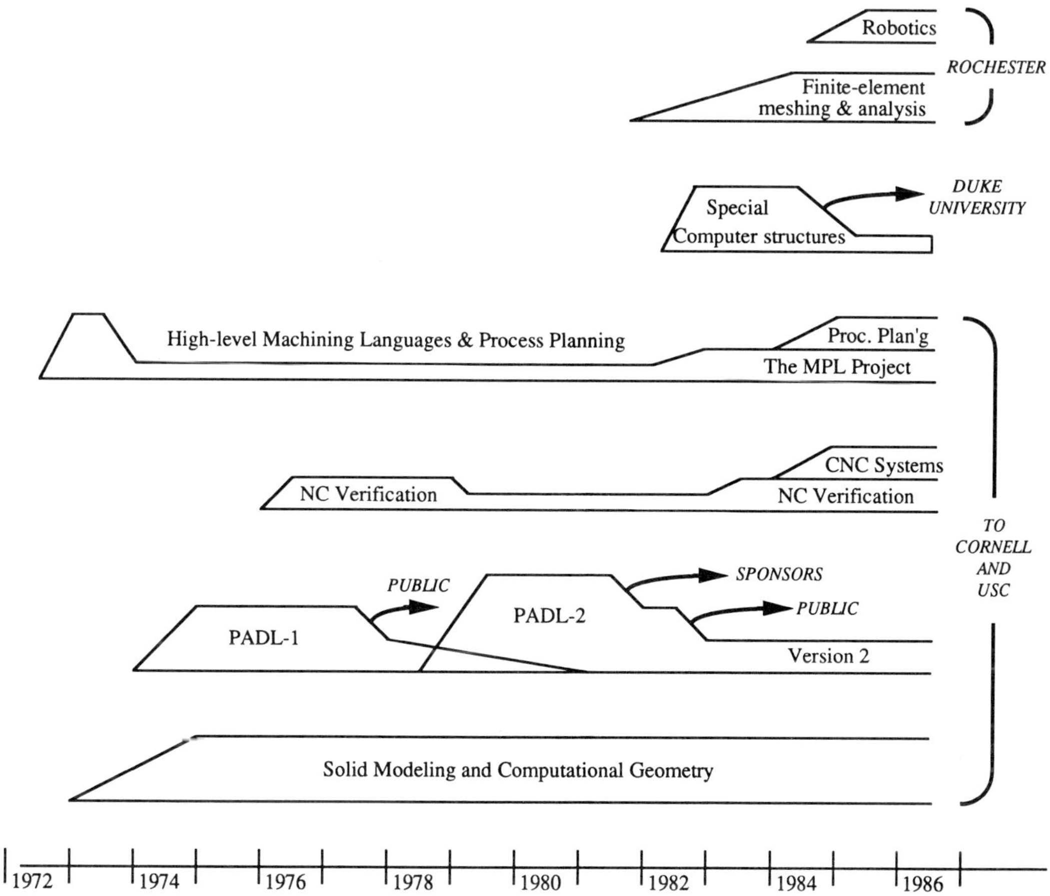

Figure 11.1 PAP history.

impossible at that time, because there were no mathematical and computational means for describing mechanical parts unambiguously. The PAP therefore redirected most of its effort to what is now called *solid modeling*. Important theoretical results were in hand by 1975, and a prototype modeling system, PADL-1[3], with a limited set of very powerful capabilities, was demonstrated at a CAM-I meeting in 1976, and made publicly available 15 months later. PADL-1 proved popular in both industrial and educational circles, and we expected industry to take over the development of solid modeling. This did not happen, for reasons summarized in Voelcker [8], and thus we were led into a second round of modeling-system work. This yielded a new wave of theoretical findings and, in 1981, a first PADL-2 system [9]. A public version of PADL-2 was ready by mid-1982 [10], and the system then acquired a life of its own, which we summarize briefly in a later section.

As PADL-1 approached completion in 1976, we tried to return to machining research through a study of modeler-based NC verification, but were drawn back into modeling research by the PADL-2 project. It was only in 1982 – ten years after the PAP's founding – that we were able to restart machining research with adequate modeler support. In that year we launched four projects: two in machining, one in finite elements (automatic mesh generation), and one in special computer structures to improve the performance of solid modeling systems. The first three prospered, and were followed a few years later by a small robotics initiative. The computer-structures project yielded in late 1982 an initial design for an ingenious ray-casting computer,

[3] PADL is an acronym for Part and Assembly Description Language.

but the team leader – Gershon Kedem – left Rochester in 1984 to seek better microelectronic facilities at Duke University and environs.

The PAP 'trifurcated' in 1986, and was formally disbanded in 1987. Herb Voelcker and two senior staff members – Gene Hartquist and Rich Marisa – moved to mechanical engineering at Cornell. Ari Requicha moved to computer science at the University of Southern California, and two assistant professors remained at Rochester to continue finite-element and robotics research. The trifurcation was orderly and amicable, with equipment and residual funds being split equitably, and Cornell assuming responsibility for disseminating the PADL-2 system and PAP documents. The PAP did not die: it was 'killed' for a number of reasons, with the primary being 'It was time'. The PAP had done good work but was going stale, and new venues, new faces, and new research alignments were in order.

TECHNICAL THEMES, CONCEPTS, ALGORITHMS, AND SYSTEMS

The PAP's work can be described in terms of eight themes:

- Models and representations for solids
- Tolerances
- Algorithms
- Applications
- Systems
- Codification
- Technology transfer
- Facilities and organization.

Each is addressed below in a subsection. Interplay between the various theme activities was a critical catalyst, in that progress or problems in one area stimulated work in others. Figure 11.2 shows the main interactions over four of the themes in the early years, when

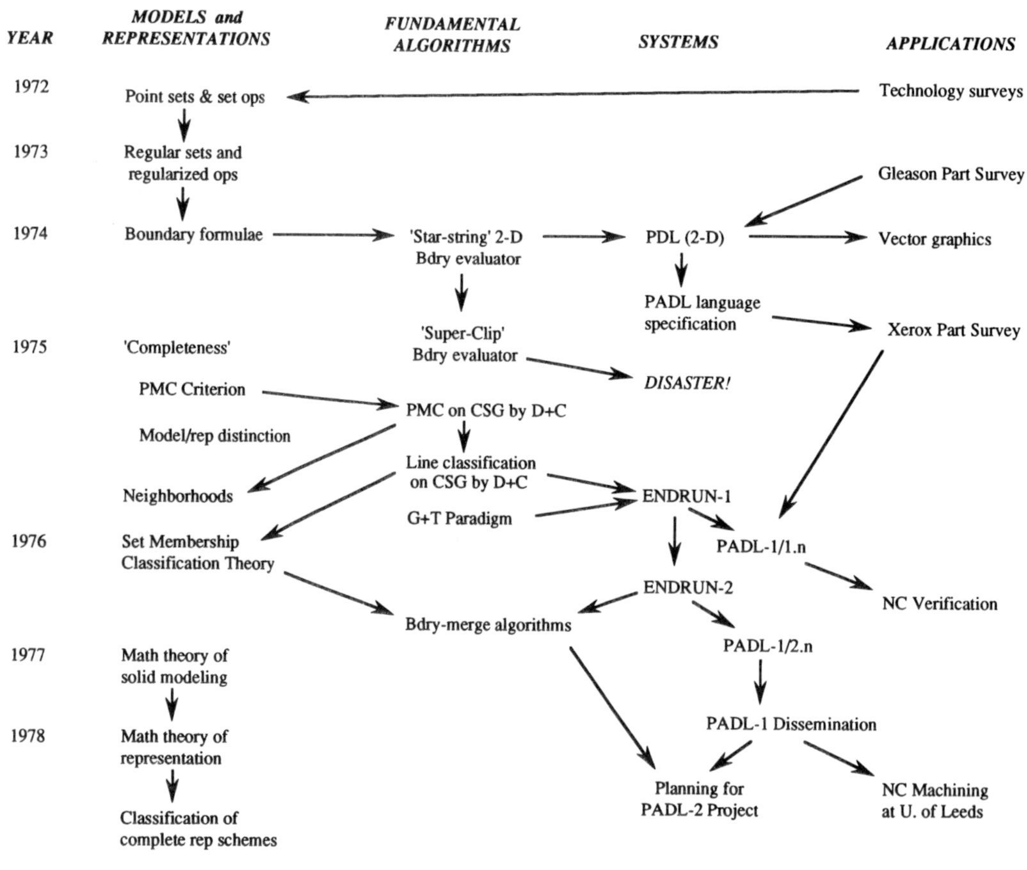

Figure 11.2. Research interactions in the early years.

foundations were being laid and codification, tolerancing, and technology transfer were not yet major concerns.

Models and representations for solids[4]

Getting started

It was evident from the outset that mechanical part modeling would be a critical issue, and we launched low-key research while doing the technology surveys noted earlier. The obvious candidates for models of mechanical parts were sets of points in E^3, and our initial focus on machining led us to view Boolean operations – union, intersection, and especially *difference*, as a mechanism for modeling machining – as essential. That much was easy; the first hard question was 'What and how do we compute?'

We seized intuitively on boundaries of sets because, in 2D examples, these drove the vector displays of that era. (The distinctions between modeling and graphics were fuzzy, and only gradually did we learn to separate the two cleanly.) By early 1973 we were experimenting on paper with such conjectures as

$$\partial(A \cap B) ?=? (\partial A \cap B) \cup (\partial B \cap A) \qquad (1)$$

where A and B are closed sets of points, ∂ denotes boundary, and $\cup$ and $\cap$ are conventional set operators. Figure 11.3a shows that (1) seems to be true, whereas Figure 11.3b implies that it may

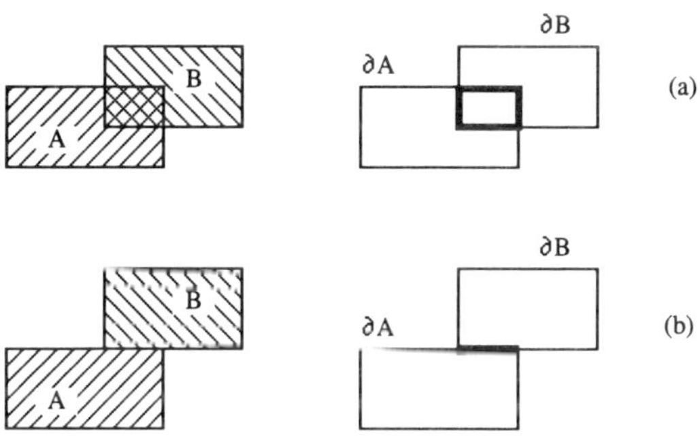

Figure 11.3. Ostensible boundary 'pathologies' for the solid $A \cap B$.

[4] 'Solid' generally refers to an ideal-form rigid solid. There are subtleties here that we shall not address, e.g. are rigid solids deformable?

be false . . . because sets that 'just touch' should have no 'volume' and hence no boundary. We added terms to (1) to handle 'boundary pathologies' as we discovered them, and essentially derived versions of the correct boundary equations 'experimentally', but we did so with the uneasy feeling that we were 'hacking' at issues that deserved formal analysis. ('Boundary pathologies' clearly motivated the restrictions in Ian Braid's Build system, which had just been described at the second Prolamat conference [11] but we could find no corresponding restrictions in the (opaque) description of the TIPS-1 system described at the same conference [12].)

Regular sets, regularized operators, and boundary formulae

By good fortune, our sole graduate student at the time, Jon Shopiro, was a mathematician by undergraduate training, and he led us to mathematical literature on *regular sets*.[5] The essential notions we developed in 1973/74 were these.

1. Solids may be modeled with closed regular sets. Henceforth all sets will be assumed closed unless stated otherwise. Regular (closed) sets satisfy

$$X = k\,i\,X \qquad (2a)$$
$$= r_n X \qquad (2b)$$

where X is regular in E^n, i denotes interior, and k is a topological closure 'operator' which, in essence, adds limit points to iX; r_n is a regularization operator denoting the ordered application of i then k in a topological space of dimension n. Intuitively, a regular set has a homogeneous 'solid' interior covered with a tight bounding 'skin'. r_n regularizes a non regular set by trimming subsets that have no interiors in the usual topology of E^n ('dangling edges' and the like), and then supplying a boundary for the open set that remains.

[5] Shopiro also constructed the popular example – see Figure 2 in Voelcker [13] – showing that the edges of a polyhedron generally do not provide an unambiguous representation of the polyhedron.

2. Regular sets are not algebraically closed under conventional set operations. For example, in Figure 11.3b, $X = A \cap B$ is a line; because a line has no interior in the usual topology of E^2, (2) fails. Algebraic closure struck us as an essential characteristic of a reliable modeling system, and so we defined *regularized* operators $\cap_n$, $\cup_n$ and $-_n$, with

$$A \langle op \rangle_n B \equiv r_n(A \langle op \rangle B) \quad (3)$$

to force closure.[6] We convinced ourselves that regular sets and regularized operators have the same algebraic properties as ordinary sets and set operators, but we did not establish this fact formally until 1977, in the first major wave of codification [14].

3. Boundary formulae for regularized compositions of regular sets followed relatively easily once the basic definitions were in hand. By mid-1974 we had, for regular sets A and B,

$$\partial(A \cap_n B) = (\partial A \cap iB) \cup (\partial B \cap iA) \cup (\partial A \cap \partial B \cap k(iA \cap iB)) \quad (4)$$

plus similar formulae for $\cup_n$ and $-_n$ [14]. The third term in (4), and its analogs in the other boundary formulae, provided mechanisms for dealing with 'boundary pathologies'. How to use them computationally was less obvious at the time.

In mid-1974 we decided to try to build a system, and this led to work on algorithms, languages, and other matters summarized later (see Figure 11.2). Basic work on models and representations did not resume until the spring of 1975.

Basic issues in modeling and representation

In early 1975 we began to attend meetings of the newly formed CAM-I Geometric Modeling Project (GMP), and also taught an experimental course in geometric modeling and graphics. The GMP meetings became exercises in frustration, in part because there was no agreed terminology in which to discuss technical issues. (Many members were unaware of how little was known about geometric and solid modeling; they were wont to construct 'wish lists' of capabilities without appreciating the technical implications of their wishes.) Our contributions were (1) the notion of 'geometric completeness', and (2) some terminology – 'boundary representations' (B-reps) for the vertex/edge/face structures, which were then called 'evaluated' or 'explicit' models of solids, and 'constructive solid geometry' (CSG) for the Boolean-composition-of-primitives systems which were then called 'unevaluated' or 'procedural' models. There was abroad a general view, perhaps stemming from the pervasiveness of drafting, that B-reps were the only 'real' models, and that others, e.g. CSG, could not support useful computation until they were 'evaluated' (converted to B-reps).

'Completeness' was put forward in part to defuse this issue, and to reduce the confusion between representational *efficiency* and *sufficiency*. The 1975 definition, recast slightly in modern terminology:

> A representation of a solid is *geometrically complete* if it contains enough information to enable any well defined[7] geometric property to be computed automatically (at least in principle).

This notion found moderate support in the GMP, and led naturally to the question: how does one establish or test for completeness? We suggested the *point membership classification* (PMC) criterion[8] shown in Figure 11.4, and shortly thereafter assigned the writing of a PMC algorithm as a homework problem in a course we were teaching. The students failed to solve the problem, but we managed to. From the ashes of that problem grew a powerful family of divide-and-conquer classification algorithms (see Figure 11.2) that is summarized later. The development of those algorithms in 1975–77 posed some new modeling and representational issues involving neighborhoods and regularization

---

[6] PAP documents used an asterisk superscript, e.g. $\cap^*$, to denote regularization, with the topology inferred from the context. The *n*-subscript notation used here to denote regularization helps to clarify the description of algorithms.

[7] 'Well defined' means defined by a computable function.

[8] A simpler version made only IN|OUT distinctions. The rationale was based on using point classification to construct the indicator (function) for the defined set.

in relative topologies. For brevity, we shall not describe the issues here because they are mentioned later and are covered in the literature [14–17].

The PMC criterion for completeness drew even warmer support in CAM-I circles than the completeness notion itself (perhaps because it was more tangible), and the criterion persisted for about two years – much to our embarrassment, because we knew within a few weeks of proposing the criterion that it was bogus. The fatal flaw: how does one decide whether the system's answers are *correct*? Once the flaw was understood, it was a relatively short route to the meta-mathematical scheme – see Figure 11.5 – widely accepted today.

The key distinction in Figure 11.5 is between *models* (of physical entities) and *representations* (of models). Models in **M** are representation-free (existential) elements whose mathematical properties are chosen to capture important physical properties. Representations in **R** or **R′** are symbol structures – boundary or CSG representations, for example – that designate or 'name' elements in **M**. The associations between **M**-elements and **R**-elements are determined by a *representation scheme s*, which is a relation (a set of ordered pairs) $s:\mathbf{M} \rightarrow \mathbf{R}$. Important properties flow from this conception, as we shall explain below, with completeness of a representation $r \in \mathbf{R}$ being definable as the mapping of r through s^{-1} to only *one* $m \in \mathbf{M}$.

Once this view is taken, the key issues devolve to defining **M**, and then formally characterizing known representations and representation schemes (and perhaps inventing new ones).

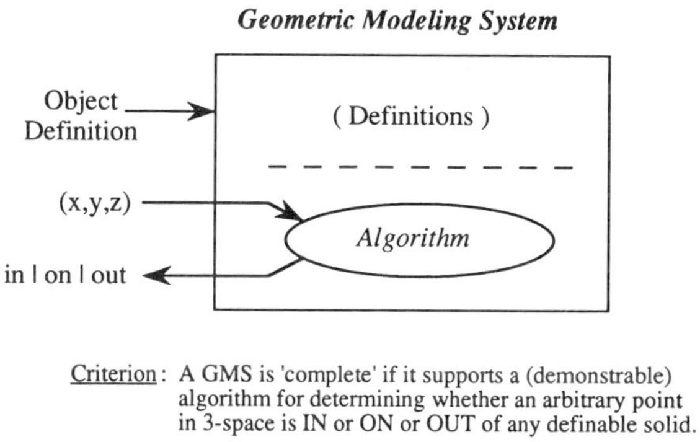

Figure 11.4. A putative but inadequate completeness criterion (PMC, 1975).

Mathematical models for rigid solids

The elements of **M** clearly are sets of points, but to model rigid solids one must restrict the admissible sets by mathematical criteria that capture the *physical* essence of 'rigid solid'. By late 1975 we had settled on the following list of requirements:

entity:	set of points in E^n
rigidity:	shape invariance (fixed distances and signed angles between all pairs and triples of points)
solidity:	homogeneous interior, in the sense of (2)
boundedness:	finite size
closure under Boolean operations:	regularized operations, in the sense of (3)

with the last being a matter of expediency. (We wanted a theory of solid modeling broad enough to model processes with set operations.) With these constraints, **M** becomes a space of compact

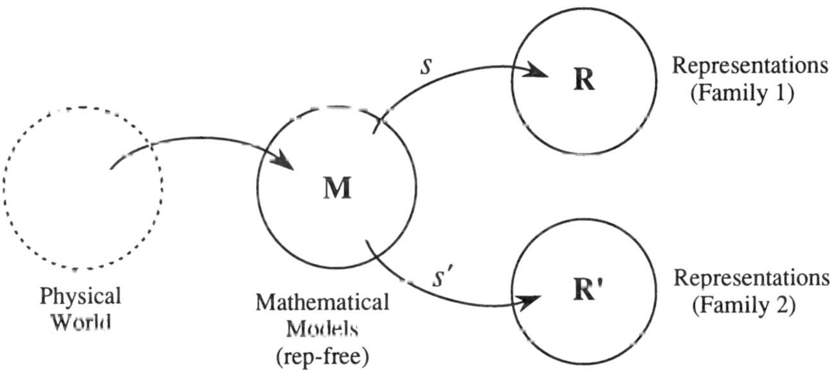

Figure 11.5. Models, representations, and representation schemes.

regular sets, with associated operations consisting of rigid motions (translation, rotation) and regularized set union, intersection, and difference. Our working tools to establish this were taken mainly from general (point-set) topology. We could prove that regularized CSG representations on bounded primitives were complete and not unique, but we could not say much about boundary representations . . . and there matters rested for a year or more.

By mid-1976 the PADL-1 system (described below) was providing reassuring proof that our approach to modeling through regularized CSG was mathematically consistent and computationally viable, and so basic work on modeling and representation was restarted to complete and document the underlying theory summarized in Figure 11.5. Requicha drew results from combinatorial topology to add two important properties to the foregoing list of properties to be captured mathematically in **M**:

finite describability:	finite triangulability, in the sense of simplicial complexes
boundary determinism:	a set X in E^n is determined uniquely by ∂X, its boundary.

The first was motivated in part by computability concerns, and the second by the common-sense requirement that a 'real' solid *should* be determined by its boundary.

Compact regular sets do not meet these new criteria, and so a smaller class – *semi-analytic* compact regular sets – was defined and dubbed *r-sets*. Semi-analytic (or semi-algebraic) sets can be thought of as Boolean compositions of halfspaces whose boundaries are defined by analytic or algebraic functions, and thus do not vary 'wildly'. Intuitively, *r*-sets are solid polyhedra with curved faces [18]. Boundary determinism had been proved, insofar as we knew, only for sets of dimension $\leqslant 3$.

The selection of *r*-sets as the elements of **M** was the final stage in our effort to model rigid solids; we directed our efforts thereafter to other matters – representations, algorithms, systems, and so forth. R-sets have come to be quite widely accepted as thus-far adequate models, with only one variant proposed in subsequent years.[9]

A competing theory of modeling emerged informally in about 1980 from the B-rep school of system builders, in which solids are required to be bordered 3-manifolds with unbordered 2-manifold boundaries [19–21]. These 'manifold models' were motivated by a popular family of data structures and structure-building algorithms whose lineage can be traced to Baumgart [22]; the algorithms embody 'Euler operators' which help to insure the topological validity of manifold solids. Mäntylä showed that any object constructed by Euler operations is a valid manifold solid *if properly embedded in* E^3, and conversely that all such manifolds are constructable by Euler operations [20]. Manifold models are a restricted class of *r*-sets that is *not closed* under regularized or conventional set operations – in our opinion, a serious deficiency. The counter argument is that non-manifold solids are physically unrealistic. Figure 11.6 presents an example – a solid joined at an edge, which is an unexceptional *r*-set but not a manifold. In the mid-1980s, Weiler proposed a generalization of manifold-oriented data structures that removed the manifold requirement, i.e. admitted 'all' *r*-sets as well as lower dimensional solids such as curves [23]. As the 1980s drew to a close, boundary representations clearly were moving toward generalizations of this sort and acceptance of *r*-sets as underlying models for solids.

In summary, it is worth noting that point-set topology and combinatorial (or algebraic) topology provide complementary perspectives and tools for solid modeling. Regularity is defined naturally in either perspective, and is

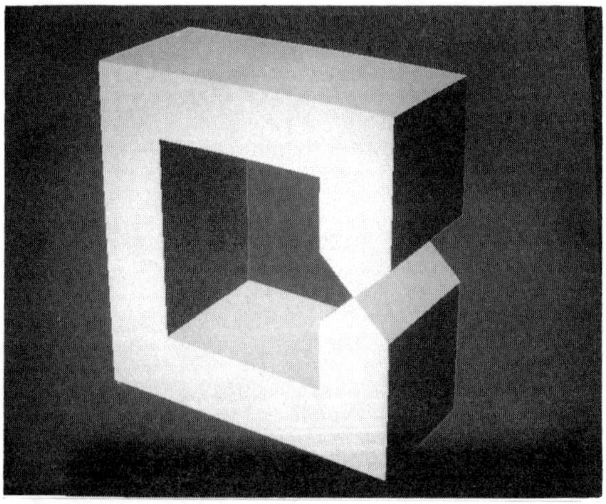

Figure 11.6. A solid that is an *r*-set but not a manifold.

[9] Arbab proposed an open-regular-set variant of the *r*-set theory in which the conventional union operator can be used to model assembly [24].

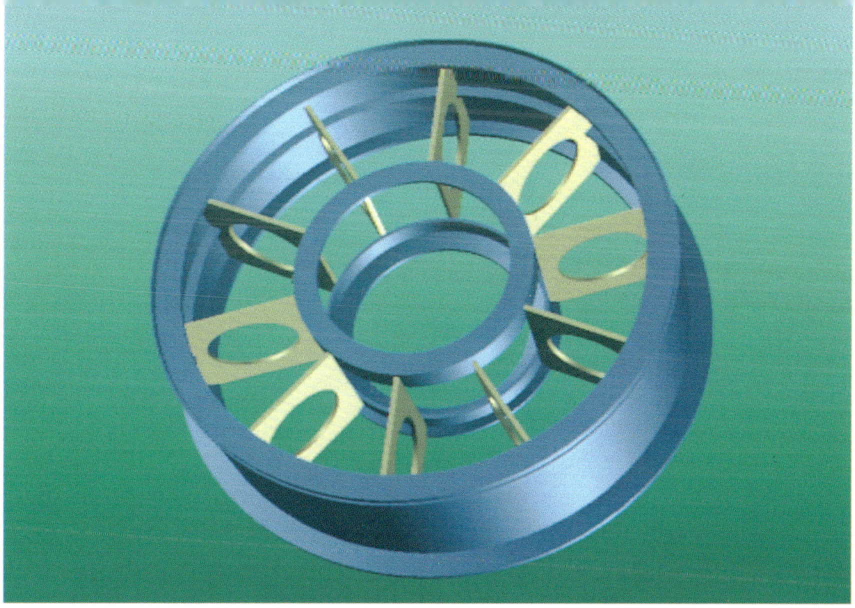

Plate 1. (Chapter 5)
Surfaces of revolution and radial structures associated with each other in a group operation.

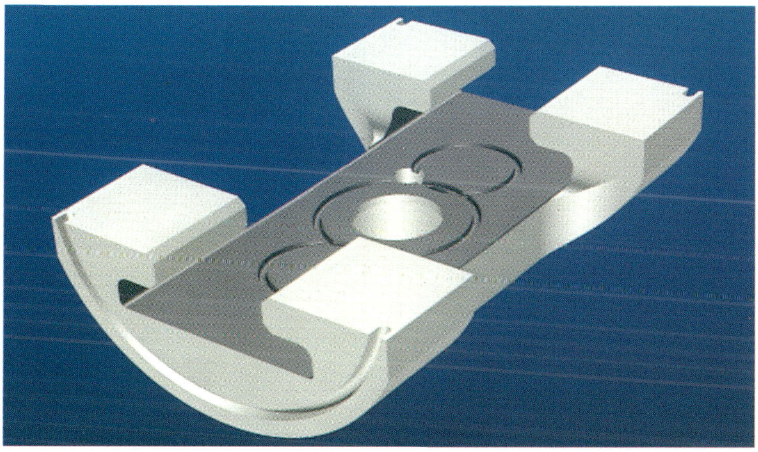

Plate 2. (Chapter 5)
Part defined by applying extrusion, revolution and Boolean operations.

Plate 3. (Chapter 5)
Scotch bottle defined by flatten and warp operations.

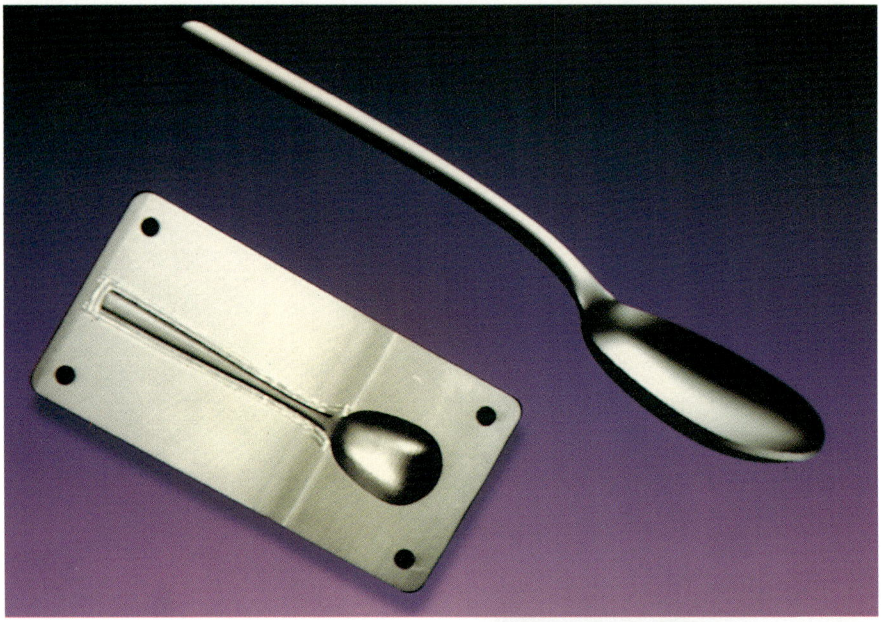

Plate 4. (Chapter 5)

Spoon defined by a variety of shape operators: warp, taper, bend and thicken.

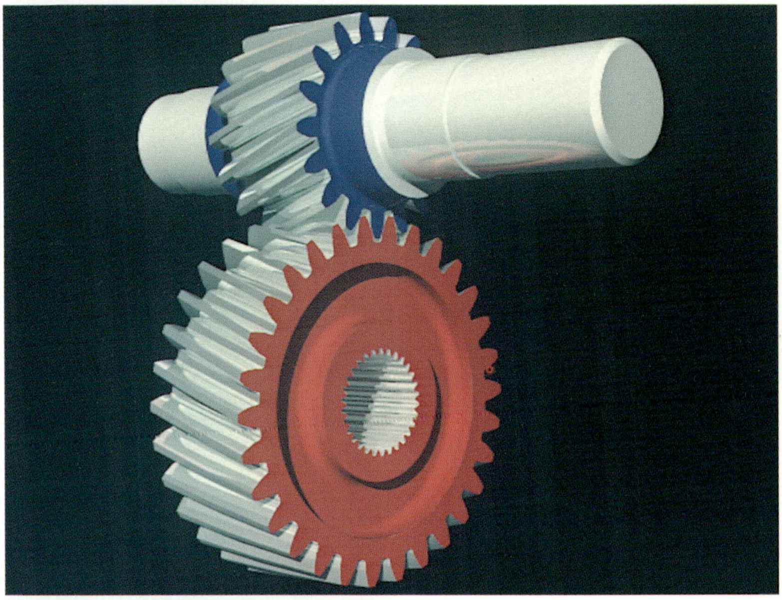

Plate 5. (Chapter 5)

Gears defined by revolution and sweep operators.

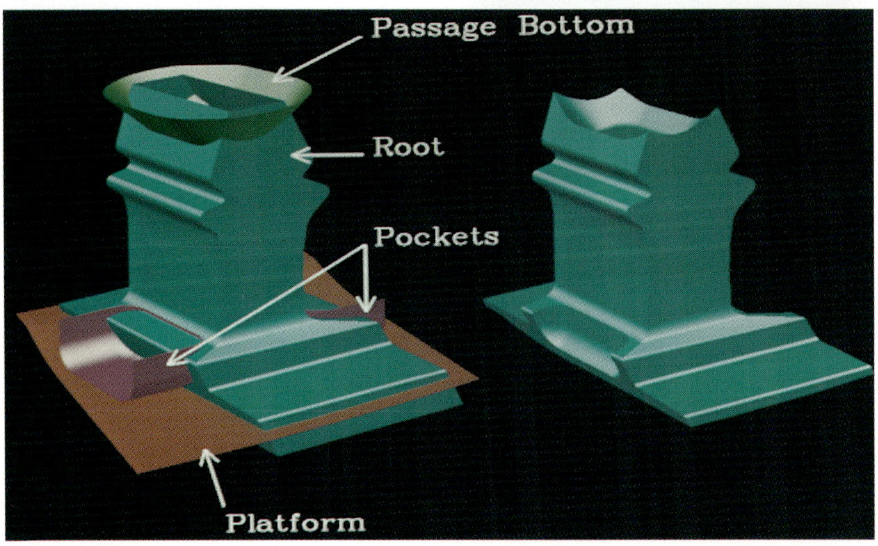

Plate 6. (Chapter 5)

Turbine blade root.

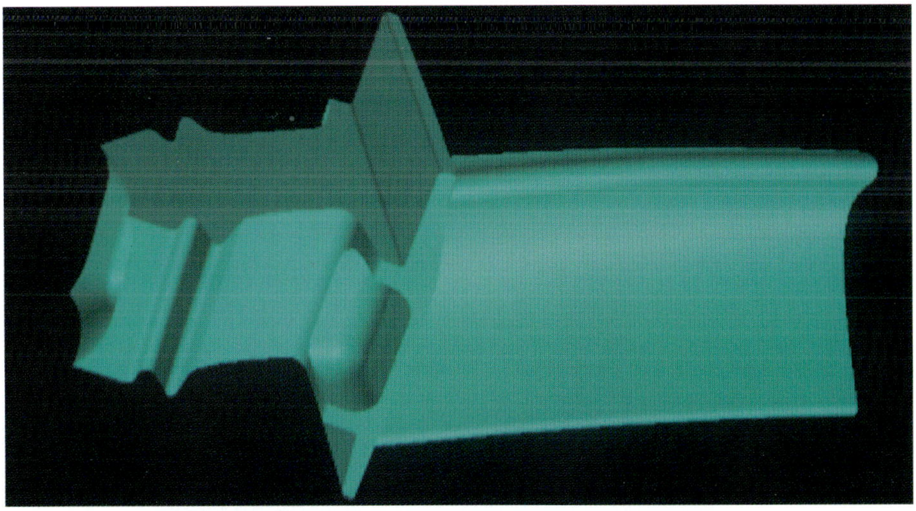

Plate 7. (Chapter 5)
Composite turbine blade structure.

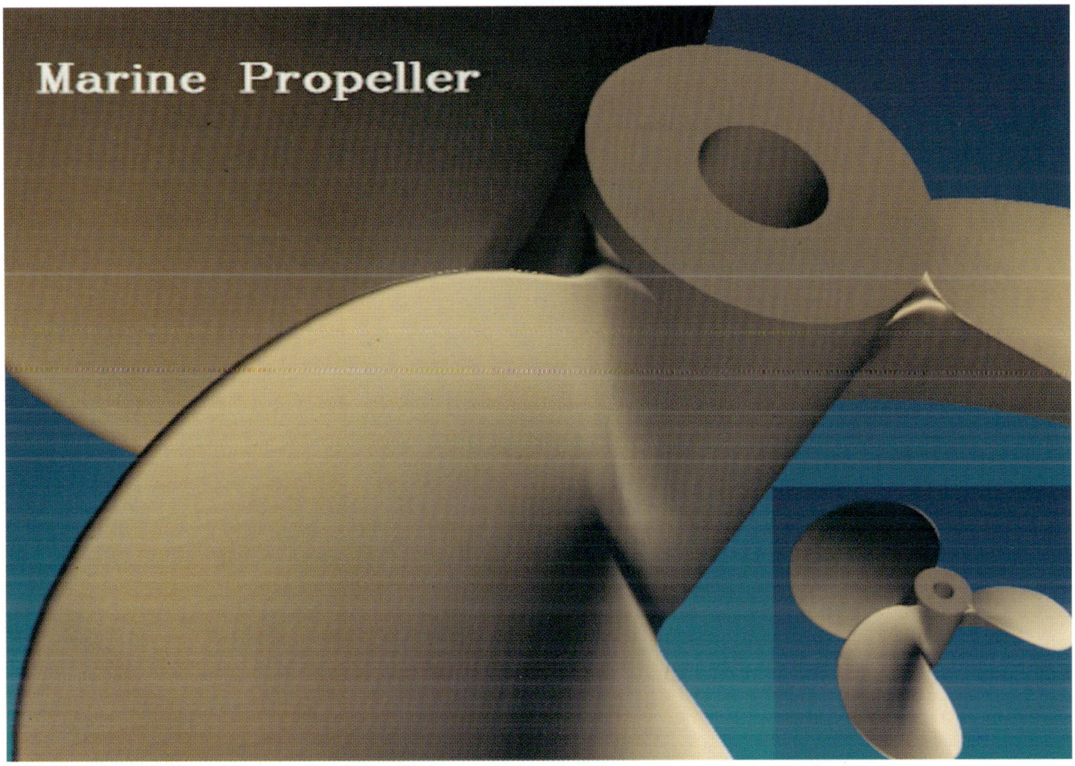

Plate 8. (Chapter 5)
Marine propeller.

Plate 9. (Chapter 7)
Concept car called 'Buick Wild Cat' (courtesy of GM Design Staff).

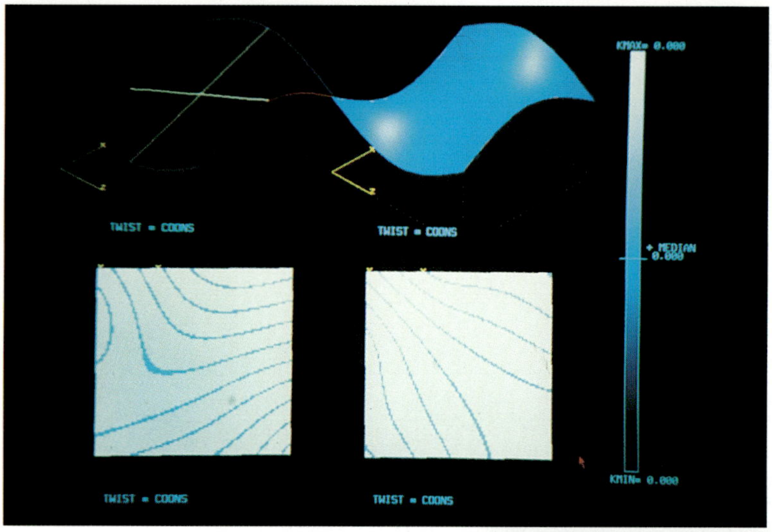

Plate 10. (Chapter 8)
A surface and its 'horizontal' and 'diagonal' reflection lines.

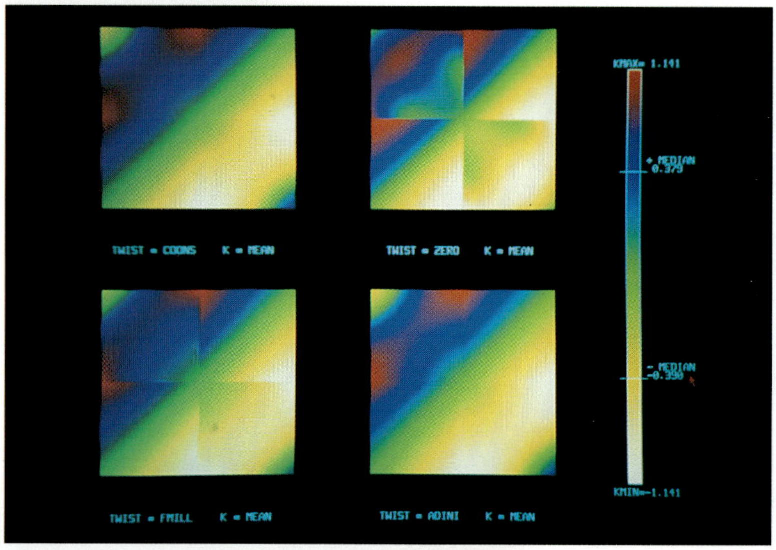

Plate 11. (Chapter 8)
Four twist approximations and their mean curvatures for the surface in Plate 10.

Plate 12. (Chapter 9)
A complex assembly made with Romulus.

Plate 13. (Chapter 11)
Rolling-sphere blending.

Plate 14. (Chapter 11)
The Ford Shader (1983).

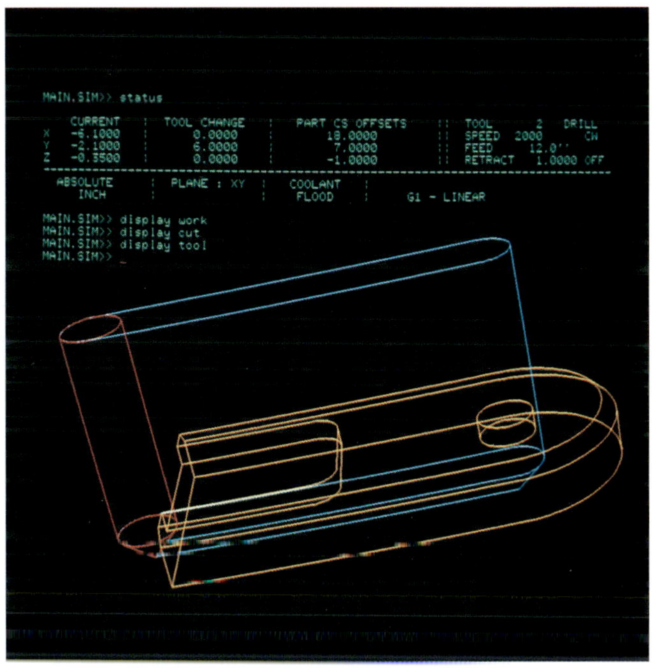

Plate 15. (Chapter 11)
Typical P2NC output (1986).

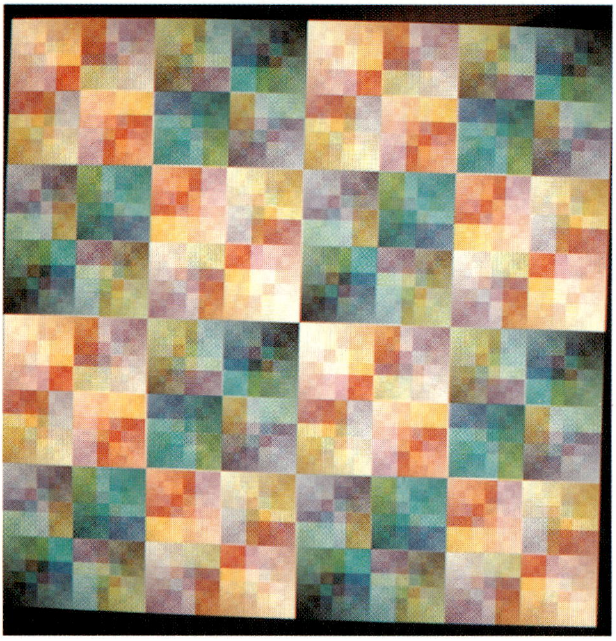

Plate 16. (Chapter 14)
Parallel Graphics Machine: dynamic test pattern in which every pixel is being computed twice per 1/25 second, with values exclusive or-ed to produce patterning. Demonstrates 450 MIPS operation.

Plate 17. (Chapter 14)
Parallel Graphics Machine: intersecting and dynamic pair of surfaces showing hidden surface removal. Top left shows result of a picture processing to discover edges, using 3 x 3 convolution in a tenth of a second.

Plate 18. (Chapter 14)
Parallel Graphics Machine: simple test scene being dynamically maintained and computed at 25 frames per second. Left half gives dynamically modified colors during recomputation, and bottom right is picture processed to discover edges.

Plate 19. (Chapter 14)
Parallel Graphics Machine: perspective view of non-flat world scene.

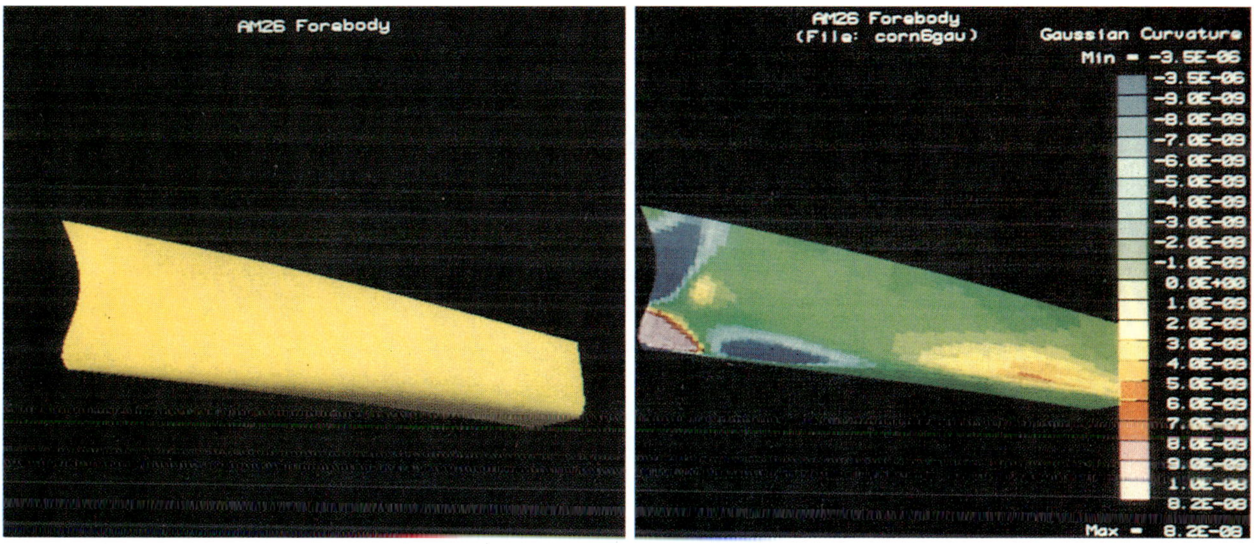

Plate 20. (Chapter 15)
Images of the ammunition ship. (a) Normally colored. (b)Color encoded Gaussian curvature.

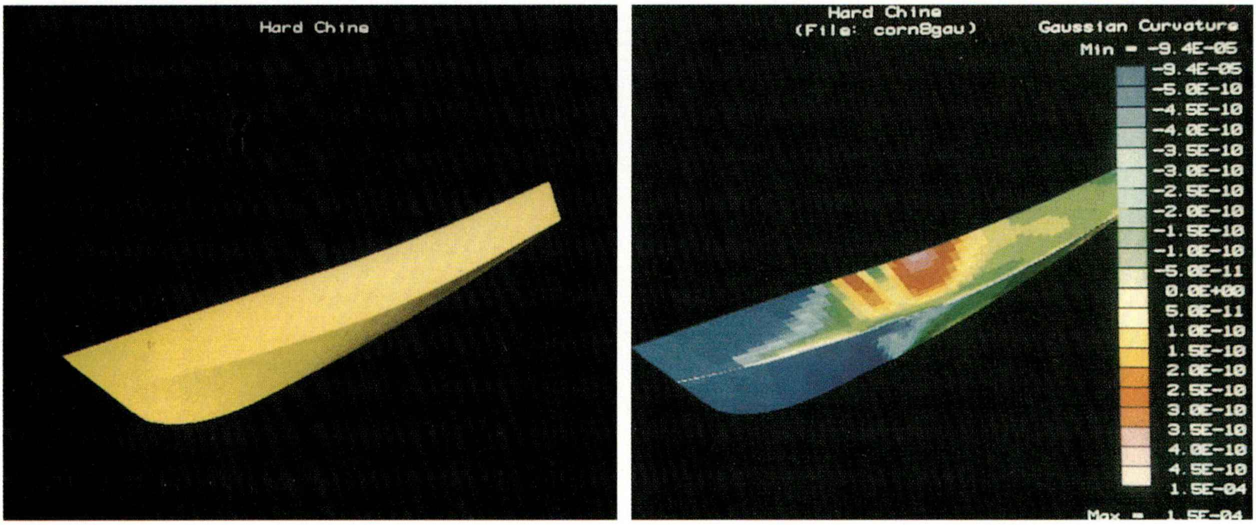

Plate 21. (Chapter 15)
Ship with hard chine. (a) Normally colored.

(b) Color encoded Gaussian curvature.

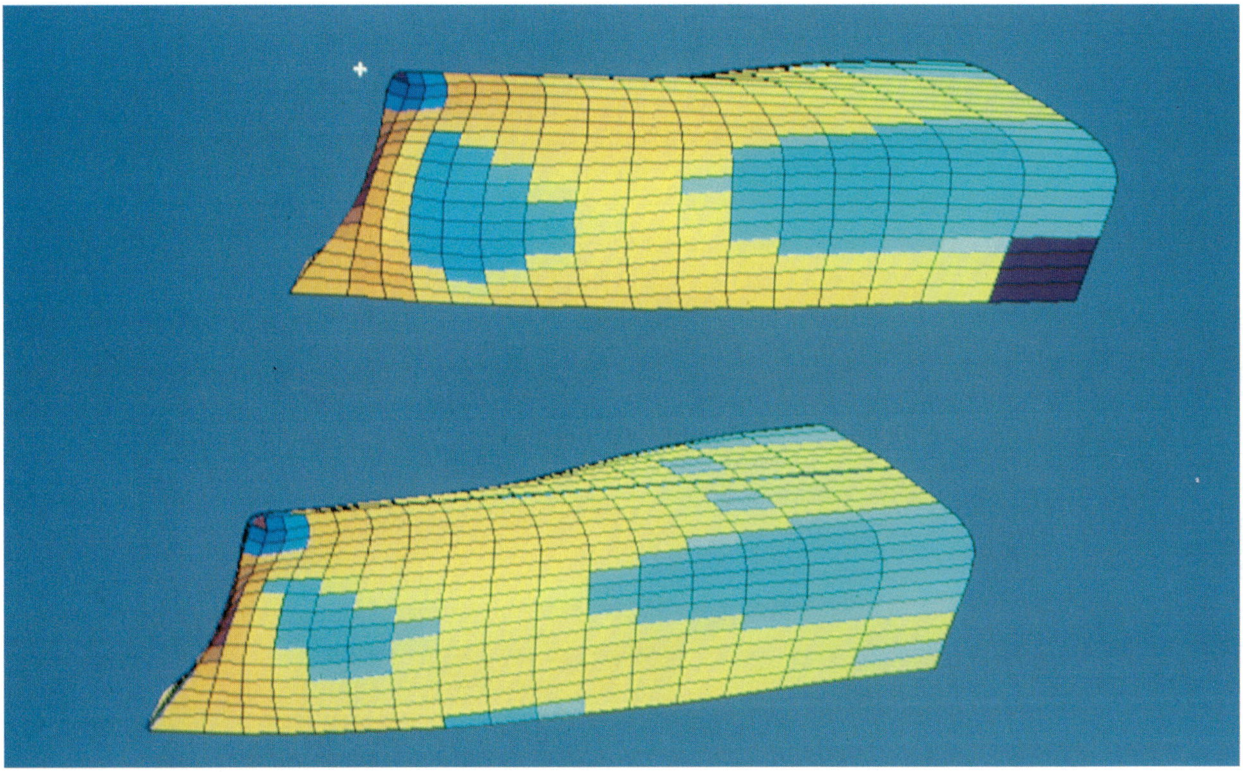

Plate 22. (Chapter 15)
Color encoded Gaussian curvature plots of constrained B-spline surface fit of the data for the ammunition ship. Lower image unmodified; upper image interactively modified.

consistent across both. In practical terms, we found combinatorial notions useful in dealing with boundary representations and intrinsically topological properties such as connectedness and number of components. The local, neighborhood-based view provided by point-set topology proved useful in establishing a host of properties for closed regular sets [14], and in designing CSG algorithms.

Complete representation schemes

The approach used to define **M** was motivated in part by the representation schemes known at the time – boundaries (hence the concern for boundary determinism), Boolean composition (issues of closure under set operations), finite-element meshes (triangulation and cell complexes in combinatorial topology), and so forth. When the modeling work was largely finished in 1977, the next obvious step was construction of an organized view of representation. Figure 11.5 depicts the relationship between modeling and representation. The refined view in Figure 11.7 leads to four formally definable properties.

Domain: **D** is the set of entities describable in scheme s.

Validity: elements $r \in \mathbf{V}$ are valid representations, i.e. images of elements of **D**; other elements, while syntactically correct, represent 'invalid' solids.

Uniqueness: a scheme s is unique if s is a function, i.e. each $m \in \mathbf{D}$ has but a single corresponding $r \in \mathbf{V}$.

Completeness: a scheme is complete or unambiguous if s^{-1} is a function, because each $r \in \mathbf{V}$ then associates with a single $m \in \mathbf{D}$.

These properties are discussed in Requicha [25, 26], together with the properties of consistency and equivalence for multiple representations and such informal properties as conciseness and efficiency. Completeness clearly was the most important property, in our opinion, and thus the issues became: of the various schemes known or proposed for solid modeling, which are complete, and do the complete schemes admit a taxonomy or grouping into families?

The six-family taxonomy of complete schemes in Requicha [25, 26] answered these questions. Summarily, the families are:

1. pure primitive instancing (or template, 'family of parts') schemes;
2. spatial enumeration schemes, both linear (voxel, hexcell, ...) and hierarchical (octree, ...);
3. cell decomposition (including triangulation and meshing) schemes;
4. constructive solid geometry (CSG: regularized Boolean composition) schemes;
5. sweep-representation schemes, both simple (translational, rotational) and general;
6. boundary-representation (B-rep) schemes.

At the time, and through the mid-1980s, B-rep and CSG schemes were thought most important and attracted the most intensive development. Figure 11.8, which appeared in many papers on solid modeling, illustrates these. Cell decompositions, in the form of finite-element meshes, also were (and are) important and under intensive development, but by a separate and somewhat older R&D community.

The two papers on representation by Requicha [25, 26] with the latter being a condensed version of the former, were the culmination of our 1970s

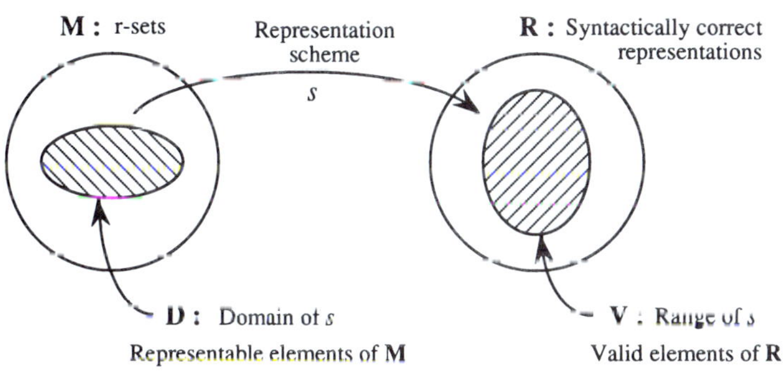

Figure 11.7. Domain and range of a representation scheme.

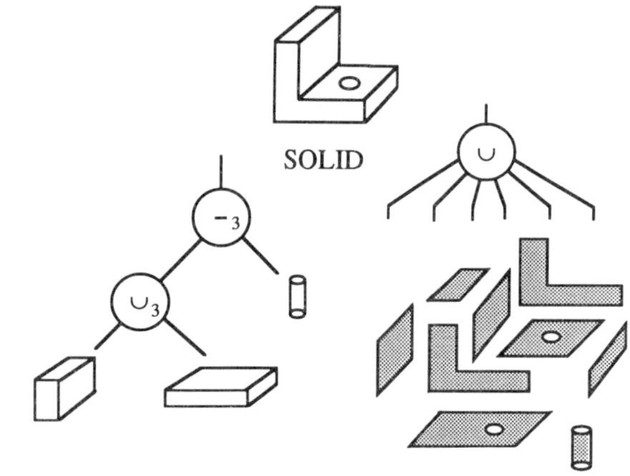

Constructive Solid Geometry   Boundary representation

Figure 11.8. CSG and B-rep schemes.

work on modeling and representing solids, and the 1970s work constitutes the bulk of our contributions to these topics. Requicha [26] is probably still the most widely cited paper in the solid modeling literature, although it is over ten years old.

Later work

In 1981, Silva [27] produced a careful mathematical characterization of boundary representations. This clarified different definitions of 'face', and was motivated by design decisions we faced in the PADL-2 Project. One of Silva's most interesting results is that B-reps can be unique (a rare property in the geometry business) if based on maximal faces ('m-faces') or *maximally connected faces* ('c-faces'). Informally, and assuming a 'constant-normal' property, for a solid S composed from halfspaces $\{H_i\}$, the m-face associated with a specific ∂H_i is

$$m\text{-}face\ (\partial H_i) = r_{n(\partial Hi)}(\partial H_i \cap \partial S) \qquad (5)$$

where regularization is in the relative topology of the boundary ∂H_i. C-faces are simply the connected components of m-faces.

In 1980/81, Tilove attacked the null-object problem in CSG and formalized the important notion of redundancy in CSG representations [28, 29]. The null-object problem arises because CSG does not have a unique representation for $\varnothing$, the empty set. (This simple fact is awkward in some important applications, e.g. collision testing in CSG machining simulators.) Tilove finessed the problem by defining the empty CSG tree as a 'canonic' representation for $\varnothing$, and providing a test that enables CSG trees representing $\varnothing$ to be reduced to empty trees. In 1984–86 Rossignac [30] generalized the notion of redundancy in a major way through 'active zones'. Informally, these are regions of space in which particular representational entities 'contribute' to a computation; non-active regions can be ignored in the computation. Active-zone theory subsumes Tilove's redundancy and is useful in null-object detection and several other applications.

The final significant contribution to modeling and representing solids was Rossignac's work on offsetting and blending [31–33]. This began with a review of Ricci's [34] provocative paper, then wound back through some nineteenth-century geometry on 'canal surfaces' and other esoterica before finally surfacing with spherical (constant-distance) offsetting and rolling-sphere blending, plus extended-CSG structures to support both. In Rossignac's method of blending, blended objects may be subjected to Boolean operations and/or further blending without raising the order of the blend surfaces, which are approximated by piecewise cylindrical and toroidal canal surfaces. Plate 13 illustrates rolling-sphere blending. Offsetting will reappear shortly, in the context of tolerancing.

Tolerances

Tolerancing information is crucial for the selection of machining processes, and for many other applications. Because automated machining was a major PAP goal from the outset, we had to face squarely the problem of representing toleranced objects. Initial research, launched in 1974, led to a language for specifying tolerances [35] that was later implemented in the PADL-1 modeler. This research showed that tolerancing practices and standards had weak mathematical foundations, and raised a host of problems, many of which are still open today. For example, Figure 11.9 shows a conventional limit ($\pm$) specification that is considered invalid (overdimensioned) in industrial practice. However, the three tolerances indicated are independent, in the sense that satisfaction of any two of the constraints does not imply satisfaction of the third. Why is overdimensioning considered unacceptable? We conjectured that the reasons lie in the 'open-loop'

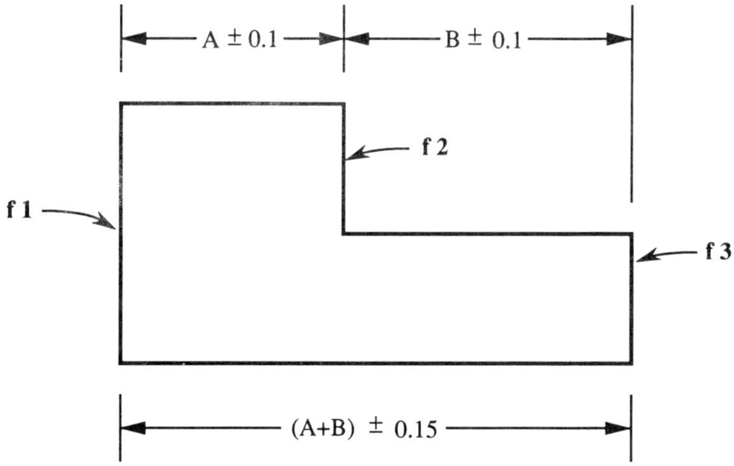

Figure 11.9. 'Overdimensioning'.

or non-adaptive nature of most manufacturing procedures. (Manufacturing plans generally do not admit *if-then-else* statements, which would enable the results of in-process measurements to be used to govern the selection of alternative parameters or processes.)

It was clear from the outset that toleranced objects should be modeled mathematically as classes, or (hyper-) sets, of *r*-sets, with each *r*-set corresponding to a solid that meets the tolerance specifications [36]. We coined the term *variational classes* to denote such hypersets, since they define precisely the allowable variations in shape and dimensions from a nominal object. But a sharp mathematical characterization of acceptable (i.e. valid) variational classes was not found, by us or anyone else, in the period covered by this essay.

A study of tolerancing practices revealed that two types of specifications were used in industry: conventional limit tolerancing, as shown in Figure 11.9, and geometric tolerancing (also called modern or true-position tolerancing) which is based on the ANSI standard [37]. Nuriel Samuel, an experienced tool engineer seconded to the PAP by the Xerox Corporation, played an important role in these early studies, and taught us much of what we know about tolerancing practices. We were unable to devise a unified theory and associated representation scheme for dealing simultaneously with conventional and geometric tolerances, and decided to focus our early work on the former. The result was a theory of conventional tolerancing which was implemented in later versions of PADL-1 [36,38]. In this theory tolerances are allowable changes in the 'dimensions' of a solid. Dimensions are defined formally as position or configuration parameters of the half spaces used to construct a solid's CSG representation. Relative position is encoded in a dimension graph, with nodes representing halfspaces and arcs representing signed distances. Dimension graphs provide elegant means for detecting under- and over-dimensioning. Specifically, overdimensioning is associated with cycles in the graph, and dimension graphs of correctly dimensioned objects must be trees (see Figure 11.10). (The representation of relative positions through structures very similar to our dimension trees was discovered independently by several groups of researchers.)

The meaning of limit-tolerance specifications is easy to define if one assumes that the surfaces of a physical object have perfect form, e.g. are perfectly planar or cylindrical. If the diameter of a cylindrical hole is specified to be 10.0 ± 0.1, a perfect cylinder with diameter 9.95 obviously is intolerance. But if one admits form variations – which are important in many applications – the semantics of limit tolerances becomes quite unclear. The parameters used to define a perfect-form surface are not sufficient to define imperfect-form versions of it. For example, an imperfect cylinder cannot be defined solely by its radius and axis.

Research on tolerance semantics led in the early 1980s to the theory described in Requicha [39]. In this theory, tolerance specifications imply that the imperfect-form surfaces of

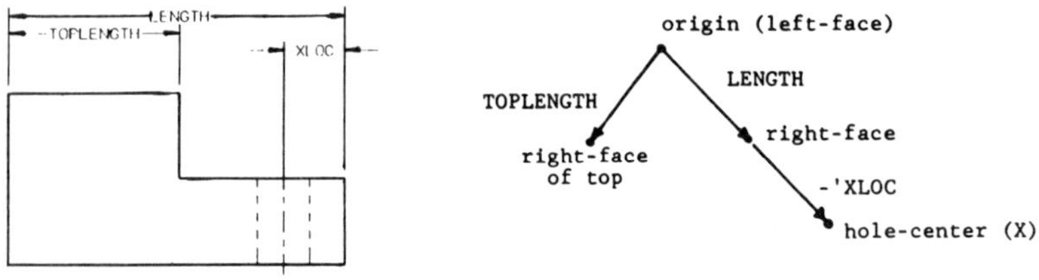

Figure 11.10. PADL-1 dimension-posting and the associated dimension tree.

physical parts must lie within regions of space called tolerance zones. A surface feature typically has several associated tolerance zones – corresponding to constraints on size, position, and so on – and these zones are not necessarily fixed in space, but may 'float' relative to other features. We proposed two approaches for building tolerance zones, one based on the parameters used to define an object, and the other 'non-parametric' [40]. Non-parametric tolerance zones are independent of specific object parameterizations, and can be defined in completely general terms and for arbitrary features. A non-parametric zone is constructed by growing and shrinking operations – called solid offsets [33]. Technically, $S \uparrow r$, the positive solid offset by a distance r for a set S, is the Minkowski sum of S with a ball of radius r, and equals the set of all points that are within a distance r of S:

$$S \uparrow r = \{x \in E^3 : \|x - y\| \leqslant r, y \in S\} \qquad (6)$$

Negative offsets are defined by growing the complement of a set, and tolerance zones are constructed by subtracting a contracted set from its expanded version.

The non-parametric theory is promising, but it has not been fully developed and it has certain subtle 'bugs'. Current research is aimed at understanding the theory's strengths and weaknesses by attempting to use it for practical applications, such as manufacturing planning and tolerance analysis.

Representations for toleranced objects were incorporated in an experimental version of PADL-2 which included a subsystem called VGraph (for Variational Graph) [41]. Briefly, tolerances are represented in VGraph as properties or attributes of surface features, which are collections of faces of a solid. VGraph representations do not require the existence of a B-rep, because object faces are represented through the primitive faces in the object's CSG definition.

Fundamental algorithms

Solid modeling algorithms can be divided loosely into three categories:

- *Application algorithms:* these generate externally useful output – graphics displays, analyses of several kinds, process simulations, and so forth.
- *Logistics algorithms:* these build, maintain, and convert representations of solids. Included here are many user-interface procedures, algorithms for converting between representations (e.g. CSG → B-rep), and a few others.
- *Utility algorithms:* these execute lower level tasks and serve as worker bees for application and logistics algorithms. Utility algorithms come in two flavors: generic, and solid-modeling specific. The former often are broadly useful across computing and include, for example, searching and sorting utilities. The latter are crucial to the effectiveness of solid modeling and include Euler-operator and set-membership-classification utilities.

In the sequel, applications are covered in a separate section. We focus below, after some introductory remarks, on set membership classification and representation conversion.

Functions and algorithms

When is a geometric algorithm functionally correct? When are two algorithms functionally equivalent? How does one compare algorithms that purport to 'do the same thing' but are quite dissimilar and operate on different representations? We began to reflect seriously on these and similar questions in 1979–80, when codifying our work for presentation in short courses and ACM tutorials. The reflections led to a conceptual framework that can be explained through Figure 11.11 and a simple example.

First, express a specific application – mass

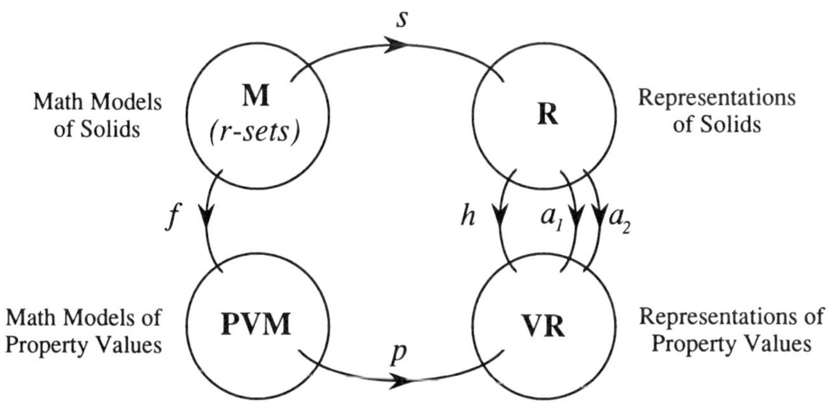

Figure 11.11. Functions (f) and algorithms (a_i).

property calculation, say – in terms of mathematical functions. Evaluation of the moment of inertia with respect to the z axis can be viewed as a mapping or function f that associates with each r-set S in the modeling domain **M** a non-negative real number in **PVM**, a property-value modeling domain. The association is defined as a volume integral

$$I_{zz} = \int_S (x^2 + y^2)\rho \, dv \tag{7}$$

where ρ is the density. Next, select representation schemes s and p for the input (r-sets) and the output (real numbers). A possible selection is the pair (CSG, floating-point numbers). A mass-property algorithm a_i operating on the selected representations is correct if it implements a symbol mapping h from CSG representations to floating point numbers that corresponds to f. Requicha [42] explains how this conception of computation can be extended to handle multiple representation schemes. Tilove [43] shows that there are open issues in the comparative analysis of 'functionally equivalent' algorithms that operate on different representations.

One might think that such matters would have been dispatched by computer scientists, but we discovered in the late 1970s that computer science was centered on what we viewed as representational issues, and offered little guidance on links to the physical semantics of computations.

Set membership classification

The notion of point membership classification

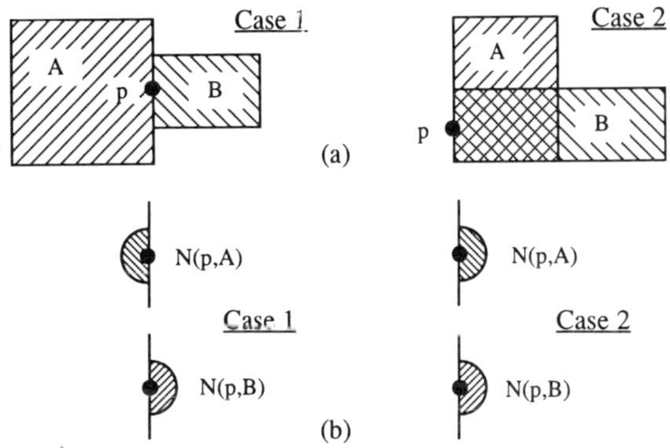

Figure 11.12. 'On/on' ambiguities for point p with respect to the solid $A \cap_2 B$, and their resolution through neighborhoods.

(PMC) emerged as an attempt to define 'geometric completeness' (Figures 11.2 and 11.4). We soon realized that a point classification algorithm would be useful as a computational utility for several applications, and assigned point classification on CSG representations as a homework problem in an experimental course we were teaching in the spring of 1975.[10] We suggested a divide-and-conquer (recursive descent) approach, which seemed natural in a CSG setting. To our embarrassment, the students concluded quickly that the divide-and-conquer paradigm could not be applied directly, because the classification of a point with respect to a regularized Boolean composition cannot be inferred solely from its classifications with

[10] This is a standard academic gambit. When unsure about how to solve a problem, assign it as homework without telling the students that it may be hard.

respect to the operands. The difficulties arise in 'on/on' cases, when the boundaries of the operand sets overlap. Figure 11.12a illustrates the problem: p is 'on' A and also 'on' B, but 'out' of $A \cap_2 B$ in Case 1 and 'on' $A \cap_2 B$ in Case 2.

A careful analysis of the problem showed that information about points in the immediate vicinity of the point being classified is necessary. This led to the definition of (regular) neighborhood of a point p with respect to a set S as

$$N(p, S; r) = B(p; r) \cap_n S \tag{8}$$

where $B(p; r)$ is an open ball of radius r centered at p (see Figure 11.12b). Although point classifications cannot be combined in a divide-and-conquer algorithm, neighborhoods can, and the classification of a point with respect to a solid may be inferred trivially from the point's neighborhood with respect to the solid.

The PMC homework problem, and its resolution through neighborhoods, led to a major algorithmic breakthrough in the summer of 1975. We discovered that the edges of a solid defined in CSG can be computed using a more general form of neighborhood-based classification, and this discovery led to the ENDRUN family of boundary evaluators described briefly below. Overall, neighborhoods and their algorithmic manipulation proved central to many of the algorithms developed by the PAP, and provided means for dealing correctly with sets whose boundaries overlap. (Boolean operations on such sets still cause incorrect behavior in some contemporary modeling systems.)

The 1975 point and edge classification concepts were generalized in 1976/77 into a formal theory of set membership classification centered on a *classification function* $M(X, S)$ defined as follows [15, 16, 44]:

$$\mathbf{M}(X, S) = \{X\text{in}S, X\text{on}S, X\text{out}S\} \tag{9a}$$

where

X = a *candidate set*, closed and regular in a topology W' of dimension m;

S = a *reference set*, closed and regular in a topology W of dimension n, where $n \geqslant m$ and $W \supset W'$;

$$X\text{in}S = r_m(X \cap iS); \tag{9b}$$

$$X\text{on}S = r_m(X \cap \partial S); \tag{9c}$$

$$X\text{out}S = r_m(X \cap cS); \tag{9d}$$

$r_m Y = k(iY)$, where the interior and closure are with respect to W'.

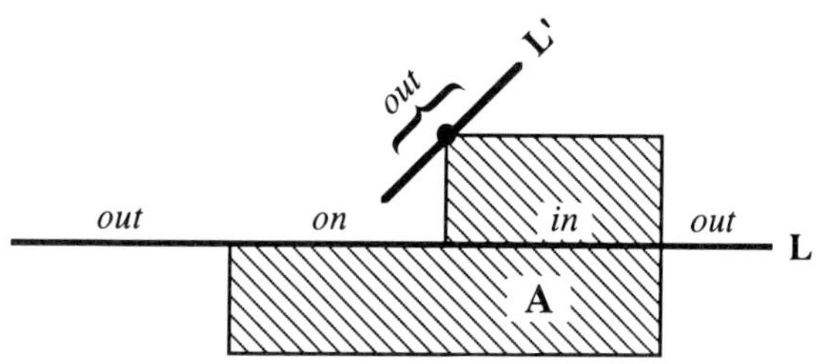

Figure 11.13. Line:polygon classification.

In effect, M partitions the candidate set into subsets that are 'in', 'on' and 'out' of the reference set. Figure 11.13 illustrates the classification of lines L and L' with respect to a 2D solid A. Note that L' classifies wholly 'out', because the single point of intersection vanishes under regularization in the 1D topology of L'.

Classifications with respect to solids represented in CSG can be computed by the following recursive (divide-and-conquer) procedure:

Procedure Classify (X, S)
    if (S is a Primitive) *then* Prim(X, S)
    else Combine (Classify(X, S-left),
    Classify(X, S-Right), S-Operator)
    end Classify

Procedure Prim(X, H)
    {Classifies candidate X against primitive solid H.}

Procedure Combine(X-Subset, Y-Subset, ⟨op⟩)
    {Performs set operation ⟨op⟩ on subsets of X.}

Figure 11.14 illustrates the 'unwound' procedure in action. The classifications on line L with respect to the 2D sets A and B are combined to produce the classification with respect to $S = A \cup_2 B$. Observe that the combination requires 1D Boolean operations on line segments.[11]. To classify a 2D set (a polygon, say) with respect to a solid defined by 3D Boolean operations, we use 2D Boolean operations. To find the edges of polygons defined through 2D Boolean operations, we can use 1D Boolean operations ... thus inducing 'dimensional

[11] If the sets A and B in Figure 11.4 had overlapping boundaries, the combining procedure would require Boolean operations on neighborhoods as well as on line segments (see Tilove [16] for details).

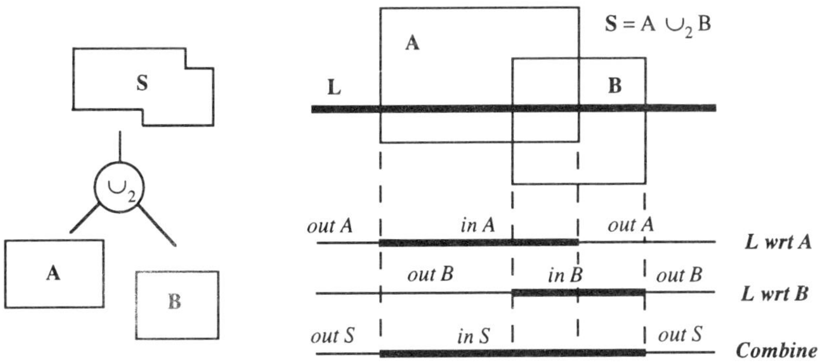

Figure 11.14. 'Bottom-up' classification of line L against solid S.

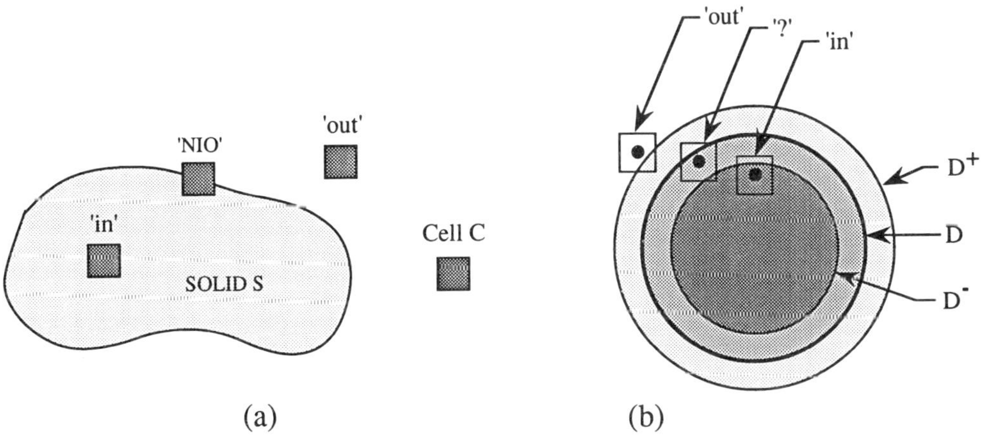

Figure 11.15. Cell:solid classification.

recursion', an algorithmic paradigm useful in many geometry computations.

The classification function M is probably the most widely used utility in all of solid modeling, although it is not always labeled as such and sometimes is implemented in hard-to-recognize guises [17]. We wish to stress that set membership classification is a representation-independent concept. While we have discussed CSG implementations, classification algorithms also are available for solids represented by their boundaries and in other schemes.

In 1979/80, Lee extended classification theory while studying CSG-to-Octree conversion by recursive spatial subdivision, which is discussed briefly below [45]. Lee wanted to do a form of (Cell, solid) classification – call it $\mathbf{M}^*(C, S)$ – wherein cell C is unambiguously 'in', or 'out', or 'NIO' (neither 'in' nor 'out') with respect to solid S, as illustrated in Figure 11.15. M* is difficult to evaluate by divide-and-conquer on CSG, but Lee defined a modified form M^{**}

$$\mathbf{M}^{**}(C, S) = \{\text{'in'} \mid \text{'out'} \mid \text{'?'}\} \tag{10}$$

where '?' usually implies 'NIO', but sometimes can be returned by 'in' or 'out' cases. **M**** values for Boolean combinations of solids can be computed by simple table lookup, and thus the essential calculation in evaluating $\mathbf{M}^{**}(C, S)$ is cell:primitive classification. $\mathbf{M}^{**}(C, \text{Primitive})$ can be implemented cheaply through two classifications of a characteristic point of C (typically the centroid).

Specifically, for a primitive solid D:

$$p \in D^- \Rightarrow \mathbf{M}^{**}(C, D) = \text{'in'} \tag{11a}$$

$$p \notin D^+ \Rightarrow \mathbf{M}^{**}(C, D) = \text{'out'} \tag{11b}$$

$$\text{neither} \Rightarrow \mathbf{M}^{**}(\mathbf{C}, \mathbf{D}) = \text{'?'} \tag{11c}$$

where D^+ and D^- are expanded and contracted versions of D, as shown in Figure 11.15b. Thus the 'in' and 'out' conditions usually can be found with binary PMC tests on scaled primitives, and in Lee's application, cells classified '?' were simply subdivided and reclassified.

Representation conversion: boundary evaluation and merging

In solid modeling parlance, phrases such as 'evaluating' or 'computing' Boolean operations usually mean one of the following [17]:

- CSG(A) → B-rep(A): compute a boundary representation of solid A from a CSG representation of A. We call this *boundary evaluation*, and algorithms that effect the conversion *boundary evaluators*. Two forms are worth distinguishing. The first operates directly on the CSG tree of A and its primitives, and is called non-incremental boundary evaluation. The second is called incremental because it uses information contained in the B-reps for subtrees of A. Incremental evaluation is more suitable than non-incremental for interactive modeling, but non-incremental evaluation is appropriate when, for example, a CSG definition is retrieved from archival memory.
- (B-rep(A), B-rep(B), $\langle op \rangle_n$) → B-rep(A $\langle op \rangle_n$ B): compute a B-rep for a regularized Boolean composition from B-reps of the operands. This is called *boundary merging*. It is a primary utility in modeling systems based on B-reps.

We shall summarize the PAP's work on boundary evaluation, because it led to some important principles and the PADL modeling systems, and also established the Project's initial credibility.

The work began in late 1973, when we were deriving boundary formulae such as (4). The first working algorithm was called a 'Star-String Evaluator' (see Figure 11.2); it processed CSG definitions in a 2D language called PDL (Part Description Language). PDL's operators were $\cap_2$, $\cup_2$, $-_2$, and rigid motions; its primitives were solid rectangles and disks defined as N-fold intersections of incrementally rotated squares. The Star-String Evaluator was organized as a compiler. It built a stack of Boolean compositions to be boundary-evaluated, and then used production rules based on the boundary formulae to construct 'object programs' of ordered calls ('star strings') to low-level geometric utilities. The utilities mainly clipped and joined lines, and did 'local closures' (the third term in (4), which can be handled neatly through neighborhoods although we did not recognize that at the time). The Evaluator began to work within the simple PDL system in early 1974 . . . slowly, and only on 2D polyhedra, but correctly.[12]

By mid-1974 we had written the specifications for the PADL family of languages [35] and started the design of a 3D boundary evaluator, called SuperClip, for PADL-1. SuperClip abandoned the compiler structure of the Star-String Evaluator and had inline geometry routines for boundary-evaluating compositions of solid blocks; it retained the boundary formulae as algorithmic controllers. The geometry routines did what we now call 'boundary merging'. They were coded on a case-by-case basis, because we lacked the insights provided a year later by set membership principles. SuperClip worked once, in the spring of 1975, but it was *really* slow (several hours of PDP-11/40 time to evaluate a three-block composition), and the prospect of extending it to handle curved primitives was daunting. We labeled SuperClip a wrong turn and cast about for alternative methods . . . with a certain degree of urgency, since we were commited to producing a working PADL-1 system in 1975/76.

We were rescued in the summer of 1975 by the classification principles that emerged from the PMC homework problem mentioned earlier. In June we discovered how to classify lines and curves with respect to CSG solids, and by August we were computing quite speedily, in a system called ENDRUN-1.0, correct wireframes for compositions of rectilinear blocks.[13] In September

---

[12] PDL and the theory and algorithms underlying the Star-String and ENDRUN boundary evaluators are documented in internal technical reports by Voelcker, Shopiro, and Fisher that can be made available to true history buffs.

[13] As explained later in the Systems section, ENDRUN was adopted as the development testbed for PADL-1 in mid-1975 and served us well into 1979, when it was honorably interred. ENDRUN lacked the I/O conveniences of its PADL counterparts but was otherwise closely similar.

ENDRUN-1.2 was computing B-reps for block compositions, and by December ENDRUN-1.4 was handling cylinders as well as blocks.

The generic, non-incremental, ENDRUN boundary algorithm may be abstracted as follows.

Procedure Boundary (CSG(S))
    $\partial S \leftarrow$ *empty*
    For every halfspace H_i in CSG(S) *do*
        $Face_i \leftarrow$ 'On'-set $(M(\partial H_i, S))$
        $\partial S \leftarrow \partial S \cup Face_i$
        end do
    end Boundary.

This is a classic embodiment of the Generate-and-Test paradigm: to build a B-rep, generate a sufficient set of candidate entities, and then test and trim each for inclusion in the result. The candidates in this case are the boundaries of all halfspaces in the CSG representation, because it is easy to prove that

$$\cup \partial H_i \supset \partial S, \quad \forall H_i \in CSG(S) \tag{12}$$

(The ENDRUN algorithms used as candidates the faces of the bounded primitive blocks and cylinders. This improves efficiency but introduces some subtleties.)

The critical calculation in the Boundary procedure is $M(\partial H_i, S)$; how does one actually *do* classification of a surface with respect to a solid? An answer can be found in the representations of the results, which are subsets of ∂H_i. Specifically,

surface-subset:(*host-surface, subset-boundary, interior indicator*)

where ∂H_i is the host surface (typically represented by an algebraic equation), the boundary may be represented by closed sets of edges, and means are provided for indicating on which side of the edge loops the subset lies. Thus the main computational problem is finding the bounding edges of the subsets, and the Generate-and-Test paradigm may again be deployed: find a sufficient set of candidate edges, then classify and trim each by curve:solid classification.

A refined version of Boundary looks like this.

Procedure Boundary (CSG(S))
    B-rep(S) $\leftarrow$ *empty*
    For every halfspace H_i in CSG(S) *do*
        face-$rep_i \leftarrow$ (∂H_i-rep, boundary $\leftarrow$ *empty*, interior-indicator)
        For each $H_j \neq H_i$ in CSG(S) *do*
            $candidate_{ij}$
                $\leftarrow$ curve-rep$(\partial H_i \cap \partial H_j)$
            $edge_{ij} \leftarrow$ On-set $(M(candidate_{ij}, CSG(S))$
            face-$rep_i \leftarrow$ face-rep_i (boundary *catenate* $edge_{ij}$)
            end do
        face-$rep_i \leftarrow Regularize_{\partial H i}$ (face-rep_i)
        B-rep(S) $\leftarrow$ B-rep(S) *catenate* face-rep_i
        end do
    end Boundary

This algorithm requires further refinement, e.g. to insure that each candidate edge is produced and classified once rather than twice, and to discard interior edges of faces via a neighborhood test. But such problems are easy to fix, and were fixed in the mid-1976 version of ENDRUN.

No particular attention is paid to maintaining 'connectivity structure' in the Boundary algorithm. Edges are not ordered, adjacency relations are not recorded, and so forth. Such information can be computed if needed, but it is not used (with minor exceptions) in the ENDRUN algorithms. Broadly, the algorithms rely on the Generate-and-Test paradigm to produce the right results without the need for extensive consistency checking of faces and edges.

In the late summer of 1976 we did the high-level design for ENDRUN-3, an incremental evaluator (see Figure 36 of Requicha [17]). The basic idea was to build B-reps for each node in a CSG tree by using known subtree faces, rather than halfspace boundaries or primitive faces, as composition-face candidates. The procedure differed from boundary-merge procedures because we could classify candidate edges against CSG subtrees using robust divide-and-conquer algorithms. We started to implement the ENDRUN-3 algorithm in 1979 for the embryonic PADL-2 system, but were deflected by some basic 'face' issues (see the earlier comments on Silva's work). A partially incremental wireframer was built in 1979/80 as a temporary substitute for an incremental boundary evaluator; it used spatial locality principles to accelerate the updating of wireframes and boundaries when CSG subtrees were edited [46]. We did not complete the incremental boundary evaluator (using 'm-faces') until 1983.

Other representation conversions

Representation conversion is the essence of many geometric computations. Other conversions we studied are summarized briefly below.

CSG → Octree conversion
Recursive subdivision coupled with cell:solid classification provides a natural mechanism for producing octrees. A skeletal algorithm follows; it is designed to be invoked with cell C enclosing solid *S*

```
Procedure Octree (C, S)
    if size(C) < ε then terminate
    else
        C ← (C1, . . . Ci, . . . C8)
                {Subdivide cell C into octants}
        For each Ci do
            class ← M**(C, S)
                        {See equation (10)}
            case of class
                'in'  : Ci ← 'full'
                'out' : Ci ← 'empty'
                '?'   : Octree(Ci, S)
            end case
        end do
    end else
end Octree
```

The algorithm, as presented, contains no code for linking the cells produced by subdivision into a tree having octal nodes, but such code is easy to install. 'Terminate' was elaborated in the PAP algorithm by inducing a classification for the terminal-size cell from the classification of a random point within the cell. (This introduced small errors, but errors are inevitable because an octree generally cannot represent a solid exactly.) Figure 11.16 provides a 2D view of the algorithm in action on an angle bracket.

The subdivision algorithm was coded for PADL-1's mass-property calculator in 1979/80 by Y. T. Lee. Alan Clark ported it, with improvements, to PADL-2 in 1982, and Ajay Kela added octree structure-building code in 1983. Figure 11.17 shows 1983 output from PADL-2's octree generator.

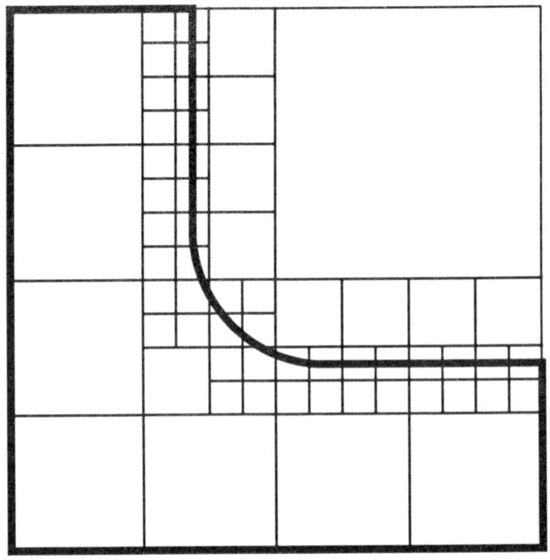

Figure 11.16. Octree generation by recursive spatial subdivision.

Figure 11.17. A six-level octree representation of the test part 'Gehause' (1983).

CSG → Cell-Complex conversion
This is discussed in the Applications section below.

B-rep → CSG conversion
In the later 1970s, as our intensive work on boundary evaluation was coming to fruition, we pondered the inverse problem, B-rep → CSG, but

found it puzzling and 'ill defined' due mainly to the strong non-uniqueness of CSG. For example, octrees and cell decompositions, which represent solids as unions of quasi-disjoint cells, can be viewed as special forms of CSG; if either representation were to be produced from a boundary representation, would that constitute B-rep → CSG? (We wanted to say, 'No.') We did not get markedly farther than this on the 3D version of the problem during the life of the PAP, but we did invest internal effort on a 2D version called 'CCSG' (Contour-to-CSG conversion).

Figure 11.18 illustrates CCSG. The goal is to produce a Boolean composition of either bounded 2D primitives (Figure 11.18b) or 2D halfspaces (Figure 11.18c) having a specified boundary composed of line segments and arcs (Figure 11.18a). The CCSG problem is intrinsically interesting, and good solutions are useful in CSG-based modeling systems to support input subsystems based on 'extruding' and 'swinging' planar contours [47]. We also hoped that good solutions to the 2D problem would provide insight into its 3D counterpart.

In the early 1980s, several papers appeared which 'solved' CCSG for linear (straight-sided) polygons by a decreasing-convex-hull technique [48–50]. Unfortunately, this technique does not accommodate curved polygons gracefully (due to termination problems, amongst others), and it produces larger than necessary CSG in some cases. The only work on curved-polygon conversion we could find was by Lim [51], using projection methods and bounded CSG primitives. We improved Lim's algorithm, and incorporated it in PADL-2 in 1983; Figure 11.19b shows one of the compositions it produced, extruded into a 3D solid. The improved algorithm was probably the best curved-polygon converter prior to Don Vossler's [47] pattern-based algorithm. In the final years we continued to work on the CCSG problem collaboratively with Vossler and Peterson [52], but the major break in B-rep → CSG was to come later.

CSG → CSG conversion
In the earlier section on models and representations for solids, we noted work by Tilove and Rossignac on detecting CSG nullity and redundancy, and more generally on 'restructuring' CSG representations. This work was really a series of exercises in CSG → CSG conversion. The properties and strategies they used are somewhat subtle, and we refer the reader to the literature rather than attempting a succinct summary here [28–30, 46]. It is worth noting that CSG restructuring makes sense only when CSG structures do not carry 'other information'. For example: parts can be defined by ordered sequences of machining or joining operations, which are representable in CSG by ordered sequences of set-difference or -union operations; such definitions cannot be restructured without doing violence to the underlying manufacturing specification. (The PAP accepted the principle set forth in Section 1.4 of Y14.5 [37], *viz.* part definitions should be independent of manufacturing processes.)

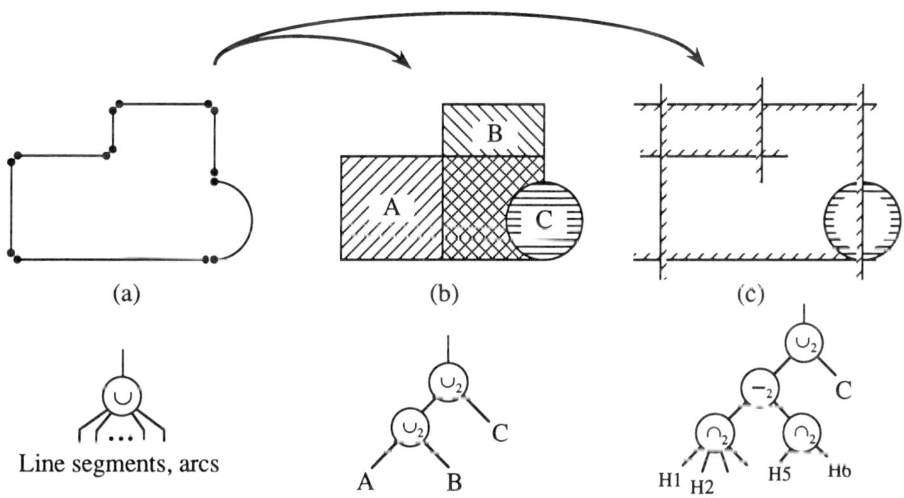

Figure 11.18. CCSG – contour-to-CSG conversion in the plane.

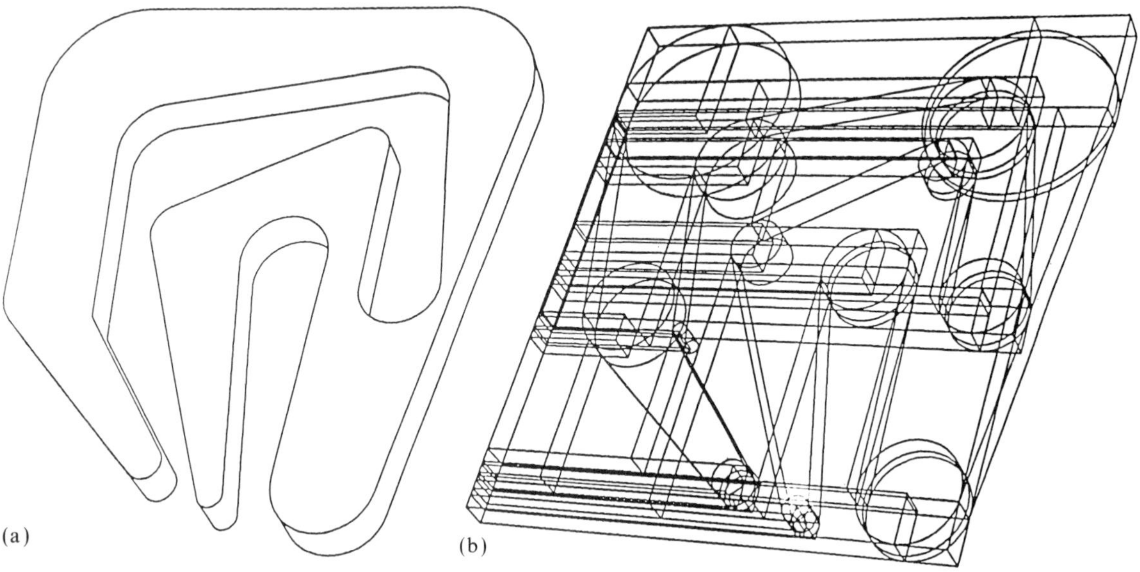

Figure 11.19. A CCSG example.

Applications

The PAP was launched in 1972 with a single broad goal: high-level programming and planning systems for NC machining. As explained earlier, we were sidetracked almost immediately into solid modeling R&D and – apart from some low key NC work in 1976–77 – did not return to serious application R&D until about 1980. Our applications work fell broadly into three areas:

- engineering graphics – mainly from CSG representations,
- engineering analysis – mass properties and finite-element meshing and analysis (FEM/A), and
- NC machining – program verification, process models, feature-level programming and planning.

The third area was the last to produce useful results. Graphics and analysis deal mainly with intrinsic properties of solids; from a geometric perspective, good solid modeling is the main tool needed to master them. By contrast, machining R&D involves solid modeling *and* a complex process *and* an 'environment'; these additional factors complicate matters considerably.

In the early days of solid modeling, many practitioners believed that useful computations could be done only on solids represented by their boundaries. We believed that CSG offered some important advantages[14] over other representations, and so in the later 1970s and early 1980s we sought to show that many applications could be supported directly from CSG – perhaps not at blazingly fast speeds, but with practically useful performances. Later, when we felt that we had made our point, we adopted a more catholic view and used whatever representations were efficacious.

Graphics

Many geometric modeling systems produce displays by first computing a B-rep, and then either displaying the object's edges, with or without hidden-line elimination, or generating a shaded image through well known visible-surface algorithms. Much of the user interaction in such systems is based on graphically 'picking' edges. But this is not the only way to produce pictures from solid models or to interact with them, as we shall explain shortly.

Our earliest work followed a boundary-based approach. From the section 'Getting Started':

[14] CSG representations are concise and exact, their validity can be established trivially, they are excellent for describing parameterized part families, and they admit relatively robust algorithms.

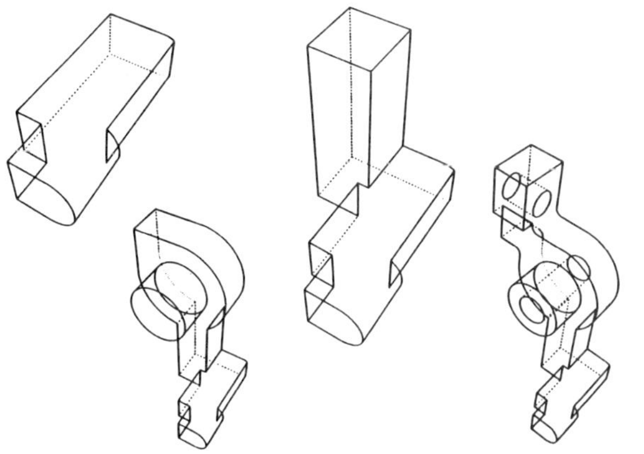

Figure 11.20. 1976 B-rep graphics from PADL-1: construction of a lever.

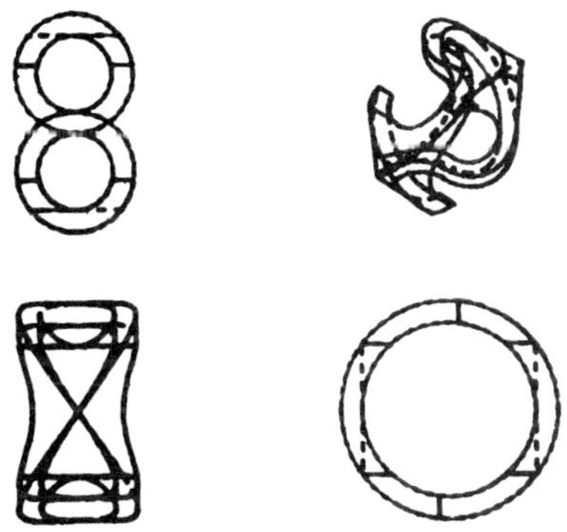

Figure 11.21. 1976 B-rep graphics from PADL-1: 'Rings'.

". . . the first hard question was, 'What and how do we compute?' We seized intuitively on boundaries of sets because, in 2D examples, these drove the vector displays of that era. (The distinctions between modeling and graphics were fuzzy . . . .)" By early 1974 the PDL system was displaying line-segment boundaries for 2D solids, and by mid-1975 ENDRUN-1.0 was displaying wireframe representations of 3D solids. In 1976 PADL-1 produced multi-view line drawings from proper B-reps using a 'poor man's visibility algorithm', wherein edge suppression or dashing was governed by tests on local surface normals; Figures 11.20 and 11.21 provide examples. About two years later PADL-1 was fitted with a proper hidden-line algorithm. After that, our graphic focus shifted to schemes for producing graphics directly from CSG, and we did almost no further work on graphics from B-reps.

Our view of graphics in the 1980s is summarized in Figure 11.22, which is an extract from the Notes for the 1982 Short Course on Solid Modeling. The top row shows major representations for solids, and the bottom row shows display options – shaded images, hidden-line and wireframe edge displays; the blobs in the middle summarize known conversions. Our work was focused mainly in the area labeled 'Ray casting algorithms/hardware'.

Our first raycaster was built in 1980 by Mark Hopkins, for PADL-2, starting from a PADL-1 mass-property raycaster built in 1979 by Y. T. Lee. Raycasting was pioneered in the 1960s by

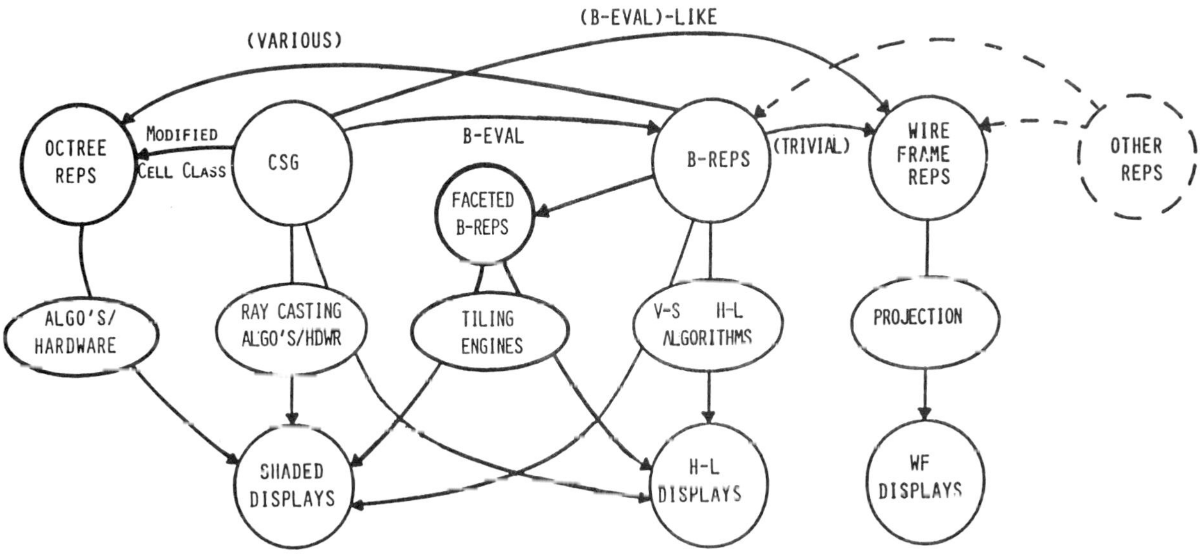

Figure 11.22. The PAP view of graphics, c. 1982.

Goldstein and co-workers for assessing the vulnerability of military targets [53]. The basic algorithm for graphics is very simple. As shown in Figure 11.23, a ray is cast (constructed) between the viewpoint and each pixel in a screen. Each ray is classified with respect to the CSG object, and the entry point extracted from the classification. The display intensity for the corresponding pixel is then calculated, using information about the surface normal and texture at the entry point and the light sources.[15]

The basic raycasting algorithm is very slow when implemented in software, but many speed-ups are available. For example, sampling may be used by subdividing the screen recursively to exploit object coherence; intersection tests may be avoided by using spatial directories; and so forth (see [54]). The PADL-2 shader uses most of the known speed-ups, plus a novel (in 1980) algorithm that seeks only the first entry point for each ray and does not attempt to produce a complete classification of the ray. Here is a simplified version of the 1980 first-entry (Hopkins) algorithm, which has not been described publicly:

For each ray *do*
- {1} Intersect the ray with each CSG primitive; label each intersection as primitive-entry or primitive-exit;
- {2} Sort the labelled points along the ray;
- {3} Traverse the sorted list conditionally, classifying each point; when an object-entry point is encountered, display it and exit.

Figure 11.24 illustrates the algorithm's operation in a simple example involving two CSG primitives, Q_1 and Q_2, and a solid S defined as their regularized intersection.

Line drawings with hidden edges removed may be generated by relatively coarse, adaptive raycasting on CSG. The basic idea may be summarized as follows:

For each scanline *do*
  Raycast coarsely across the scanline, applying the following test to the two most recent results
    If Facehit $(i-1) \neq$ Facehit (i) *then* do binary search by raycasting for an intervening edge;
  Propagate edge location(s) to the next scanline {to guide ray sampling}
  end do.

In this algorithm, the ray entry points carry pointers to the CSG halfspace boundaries or primitive faces that 'created' the points; an edge

[15] Ray*tracing* is more elaborate than raycasting, in that rays are 'traced' through multiple reflections and refractions as they 'interact' with the object and its environment. Raytracing can produce ultimate visual realism if all physical effects are modeled, whereas raycasting produces only first approximations of visual realism.

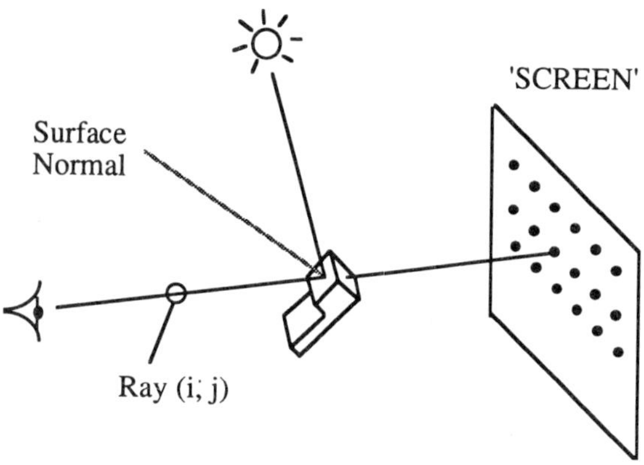

Figure 11.23. Raycasting on a regular grid for graphics.

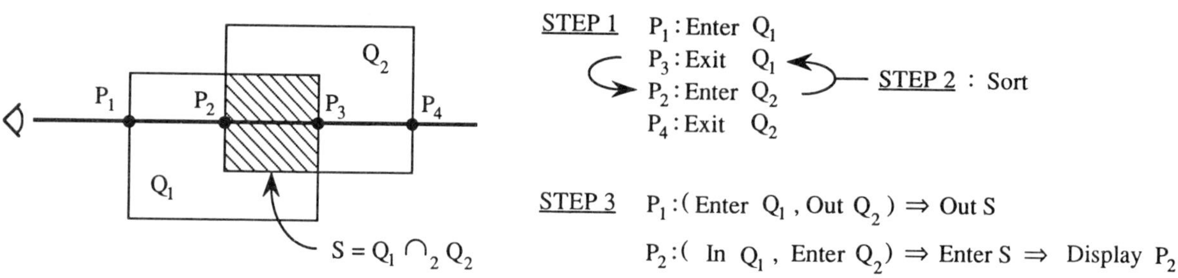

Figure 11.24. The Hopkins ray classifier for graphics (example).

is implied whenever adjacent pixel samples carry different face tags. (The algorithm as described would not display horizontal edges, but is easy to modify so that all edges visible within a specified resolution are displayed.) A well implemented version of the algorithm is included in the standard PADL-2 package.

A raycaster must be able to intersect a straight line with all of the surfaces supported by a modeler, and this requires the solution of nonlinear algebraic equations. For example: a fourth order equation must be solved to intersect a ray with a torus. An alternative CSG display algorithm that does not require intersection calculations can be based on depth- or z-buffering. The usual depth-buffering algorithm operates on B-reps: points on a solid's faces are generated and projected on the screen (deposited in a z-buffer); points that fall into the same pixel buffer are then scanned to find the point closest to the viewpoint, which is displayed. This algorithm can be modified to operate directly on CSG. CSG-primitive faces, rather than B-rep faces, are scanned to provide candidate points; however, the CSG faces generate extraneous points (because they are supersets of the B-rep faces), and the extraneous points must be discarded by point-membership classification [55].

Depth-buffering and raycasting for CSG have been shown experimentally to have roughly similar performances. Both of the basic algorithms are very slow, but many speed-ups are known. With speed-ups, the performance of CSG-based display algorithms implemented in software in sequential machines is acceptable, but still relatively slow. However, the algorithms are simple and inherently parallel, and can be implemented in special-purpose parallel and pipelined hardware . . . a topic addressed below under Systems.

On graphic interaction with CSG representations: the construction and classification of single rays through user specified 'poke points' on a screen is a fast and effective method for picking halfspace boundaries, primitive faces, or B-rep faces (if a B-rep with links to a CSG representation exists). Edges may be picked graphically through faces. PADL-2 contains the utilities needed to support these operations. For policy reasons (mentioned briefly in a later section) we did not build interactive graphic interfaces for any of our public systems, but all of our later experimental systems had them.

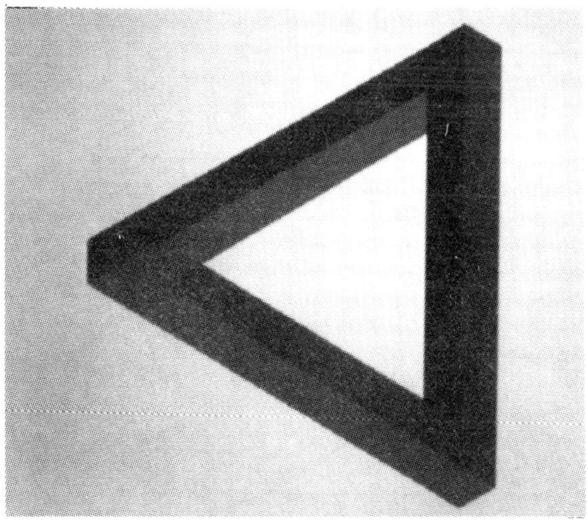

Figure 11.25. 'Devil's Triangle', a valid solid rendered from PADL-2 by a modified raycasting algorithm.

We conclude this section on graphics with Figure 11.25, an illustration produced c. 1982 to convey the message, 'Don't believe everything you see on a computer screen.'

Engineering analysis

The PAP did serious work on two analytical applications, mass property calculation and finite-element mesh generation and analysis (FEM/A), which we shall summarize in that order.

The mass properties of a solid (mass, centroid, moments and products of inertia) are important in mechanical dynamics, and may be calculated by integrating simple polynomial functions over a complex domain – the solid itself. We began to study mass property calculation in 1978, when we found that the literature on numerical analysis had little to offer in the line of solution technologies apart from Monte Carlo methods.[16] The view shown in Figure 11.26 emerged gradually [56]. In essence, each member of the six family taxonomy of complete representation schemes has a distinctive or 'natural' set of

[16] Numerical integration has been studied intensively for about a century, but in a converse form, *viz.* integration of complicated (often partial differential) functions over simple domains. Monte Carlo solutions are based on random spatial sampling. They are slow, but helpful when combined with other, deterministic techniques.

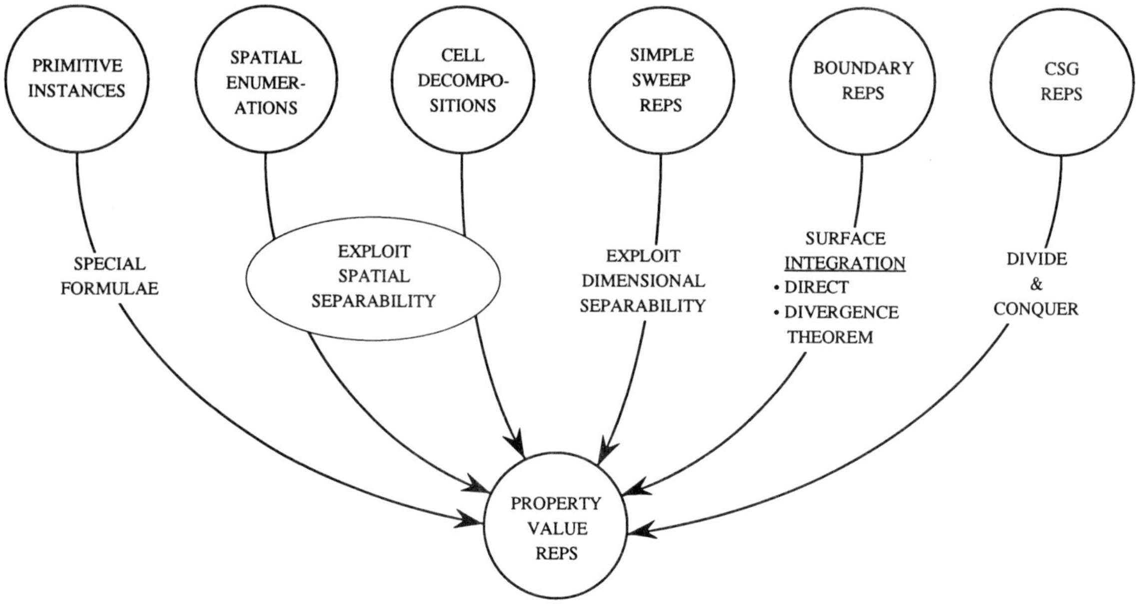

Figure 11.26. 'Natural' methods for computing integral properties.

methods that exploit a distinctive property of the scheme; thus, for example, surface integration is the natural method for boundary representations. Moreover the natural methods are reasonably efficient if one excludes divide-and-conquer on CSG, which in the form we studied can lead to exponential evaluation times.

Integration over solids represented in CSG is best approached through representation conversion. We studied three of the four approximations shown in Figure 11.27: spatial enumeration through point classification, columnar decomposition through raycasting (line classification), and the cube (octree) decomposition using the modified cell classification methods described earlier. The latter two have similar efficiencies, but octree conversion admits tighter error bounds.[17] Both techniques are implemented in the PADL-1 and PADL-2 modelers.

Low-key research on finite-element mesh generation began in 1982. We had not entered the area earlier because we lacked experience in FEM/A and expertise in solid mechanics, and in any event were fully extended on modeling-system R&D. But in early 1982, emboldened by the imminent completion of the PADL-2 Project and our success in octree generation, we decided to try to build meshes from octrees for analysis by NASTRAN. As coach we had a part-time

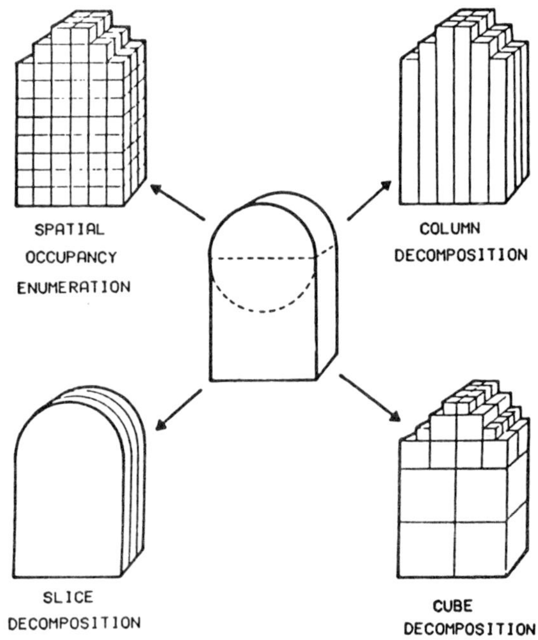

Figure 11.27. Quasi-disjoint approximations for integrating over solids represented in CSG.

[17] The Addendum to TM-35 [57] contains a useful summary of the error bound derivations not published in the archival papers [45, 56].

faculty member in mechanical engineering, Vic Genberg, who was expert in NASTRAN.

Early experiments confirmed theoretical predictions: octrees, no matter how finely graded in regions of high boundary curvature, provide unacceptable meshes for standard solvers because their 'sharp corners' induce singularities. After a study of possible alternative strategies for effecting CSG → cell-complex (i.e. FE-mesh) conversion, we settled on an octree-based hybrid strategy which we pursued from mid-1983 to the PAP's termination (and beyond). The essential ideas were to mesh the interior of the solid with an octree (Stage 1), and then to link outward to the boundary in a Stage-2 boundary traversal that used heuristics to build 'good' elements. We chose this approach because it seemed to offer, amongst other advantages, a powerful hierarchical structure for managing mesh refinement and incremental analysis [58]. John Goldak, an occasional visitor from Carleton University, helped us to formulate the approach in 1983/4 and to get a crude 2D version running.[18]

Renato Perucchio joined the Rochester faculty in mid-1984 and brought finite-element analytical expertise to research which had been focussed on mesh generation. This led to an immediate broadening of our objectives, and the decision to regard meshing and analysis as two halves of a single problem which could be coupled synergistically. By mid-1986 an experimental 2D version of the automatic analysis system (shown in Figure 11.28) was running [61, 62] and a 3D version was being built as the PAP disbanded. Figure 11.29 shows output from the 2D system. Specifically, in Figure 11.29a, a plate is meshed initially with an octree interior and custom boundary cells; Figure 11.29b shows deformations and error indicators (based on normal and tangential stress discontinuities across cell boundaries) when the plate is cantilevered on the left and traction is applied to the hole; Figure 11.29c shows *h*-type refinements to the mesh generated automatically in regions of high error.

[18] We were aware that an octree-based approach was also being pursued at RPI under Mark Shephard's leadership [59, 60]. We proceeded because we believed the PAP had better modeling and octree utilities, and because our approach, which required boundary elements to conform topologically to the octree structure, was quite different from Shephard's relaxation methods for handling boundaries and mesh 'smoothing'.

NC machining

The automation of NC machining was the PAP's *raison d'être*. Our inability to dent the problem in 1972/73 led to a decade of research on solid modeling, but during the 1970s we returned to machining whenever we could bring manpower, new ideas, and new tools to bear. Our work fell into two closely related areas: NC verification and machining-process modeling, and feature-level programming and process planning. We begin with verification.

When PADL-1 became operational in 1976, we sought a manufacturing application to show that solid modeling was more than 'automated computer graphics'. We seized on NC program verification almost immediately, because general Boolean operations were both essential and newly available (in a limited domain) through

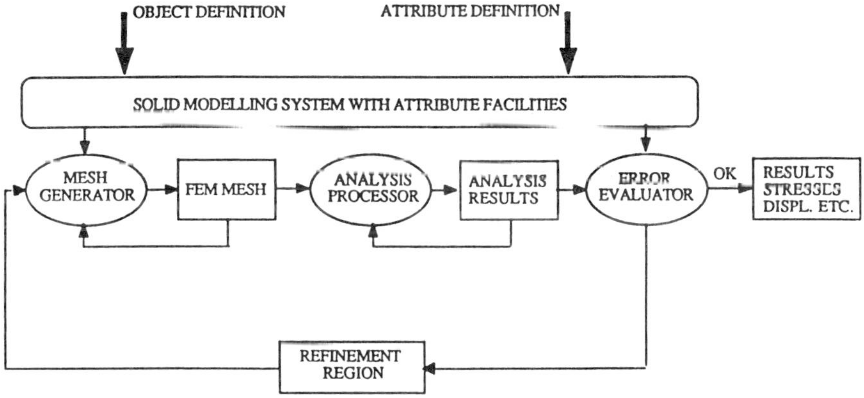

Figure 11.28. A research goal – an automatic finite element analysis system.

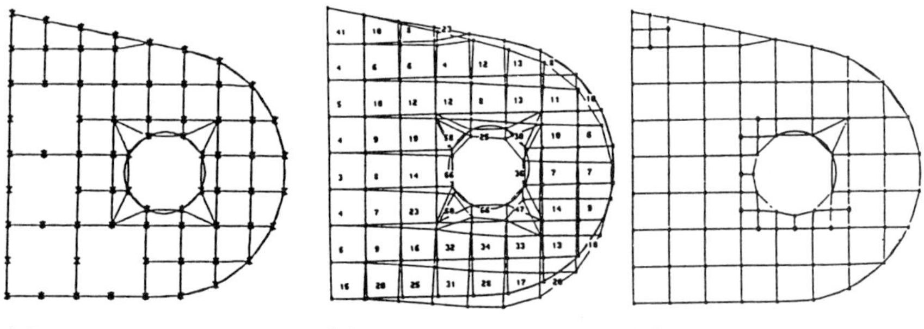

(a) (b) (c)

Figure 11.29. Results from an experimental 2D version of the automatic system.

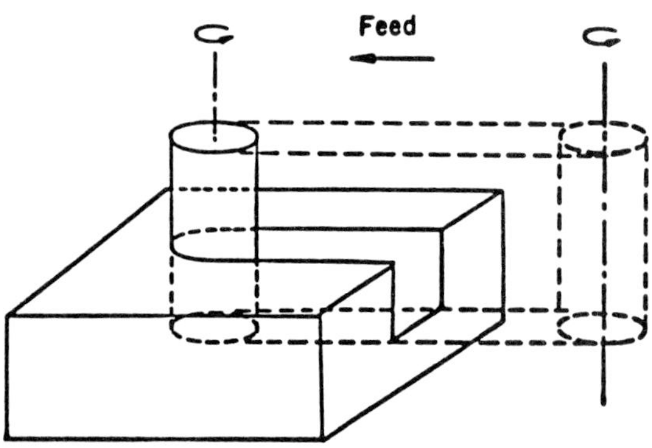

Figure 11.30. A single milling operation.

PADL-1. The initial goal, which we are still struggling to attain, was ambitious: fully automatic verification, without recourse to graphic displays for assessment by humans. The base concepts are shown in Figures 11.30 and 11.31: a rotary cutter on a feed trajectory generates a 'cutter swept (solid) region'; this may be 'subtracted' from a current workpiece to yield an updated workpiece, thereby modeling the effects of one NC motional block.

Our initial naïveté can be gauged from the first verification algorithm we implemented in 1977:

$\Delta \leftarrow \varnothing$
Until (End of NC Program)
  Read NC-command
  Generate Cutter-Swept-Region V
  $\Delta \leftarrow \Delta \cup_3 V$
  End-until
$W_f \leftarrow$ Stock $-_3 \Delta$
  If $(W_f = P)$ *then* Correct-Program *else* Error.

This verifier was a free-standing program that constructed by rule a PADL-1 composition (a text string) of blocks and cylinders to represent each swept region V induced by a linear or circular motion of a specified cutter. The finished workpiece W_f, was defined by the PADL-1 regularized difference of the Stock (predefined in PADL-1) and the accumulated V's. The sole test, goal attainment, was based on the predicate $W_f = P$, where P was the PADL-1 definition of the desired part. The predicate, called a 'same-object' predicate, was transformed into a null-object predicate through a symmetric difference, *viz.*

$$(W_f = P) \Leftrightarrow ((P -_3 W_f) \cup (W_f -_3 P)) = \varnothing \tag{13}$$

The null-object test was implemented by computing the boundary of the symmetric difference defined in PADL-1: a null boundary implied a true predicate.

The 1977 verifier computed for about an hour in a PDP-11/40 to decide that a correct 20-block, 2-axis NC program was in fact correct. But it also ruled the program correct when the NC tape was inadvertently reversed, and it ruled a second correct program erroneous! The backward-program embarrassment taught us that *every block* of an NC program must be verified as well as the program as a whole. The false error report exposed the fallacy in our goal attainment test illustrated in Figure 11.32. Observe in Figure 11.32 that Process 1 will pass the $(W_f = P)$ test while Process 2 will not, although the desired part has been produced. The 'fix' for this problem – testing P against individual *connected components* of W_f – is relatively easy to install

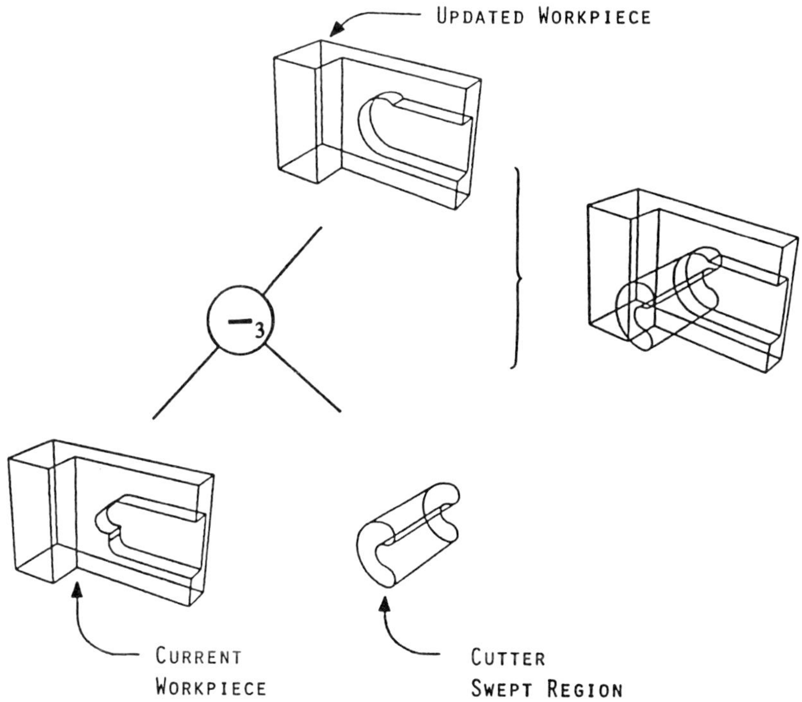

Figure 11.31. The regularized difference model for an NC motional block.

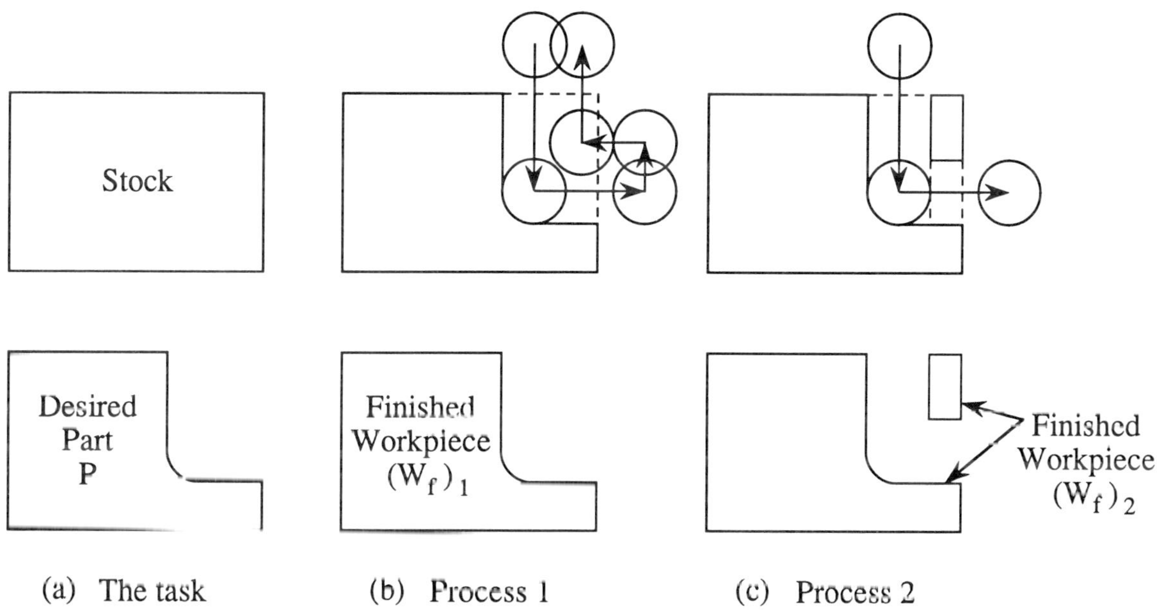

Figure 11.32. Subtleties in goal attainment testing.

in B-rep modelers, but posed a dilemma for CSG modelers until we devised a connected-component test that could be implemented through classification on CSG [63].

The 1977 experiments taught us a lot about verification and about the qualities needed in a solid modeler meant to serve as a 'geometric utility' rather than as a turnkey 'design system'. To cite but one consequence: null-object and same-object algorithms were put near the top of our priority list, and motivated the work by Tilove and Rossignac mentioned earlier. In 1978 we put the verification project into think-mode while we turned to PADL-2, and did not start another experimental verifier until 1983. In the intervening years we sharpened our models, defined tests that could be applied to each block in an NC program, and studied the computational requirements of NC verification. The results from our NC research in 1976–82 are summarized in Voelcker [64], and reported comprehensively in Hunt [63]. A few samples of this work follow.

Figure 11.33 distinguishes between a cutter's 'total swept region' TSR and its 'operative swept region' OSR. The OSR is the maximum material that a cutter can remove, and it permits sensitive tests on machining as opposed to positioning motions. Cutter-clearance testing through the predicate

$$(\mathrm{TSR}_i -_3 OSR_i) \cap_3 \mathrm{W}_{i-1} = \varnothing \qquad (14)$$

is one such test; it flags such errors as attempting to create a hole with a reamer or boring bar.

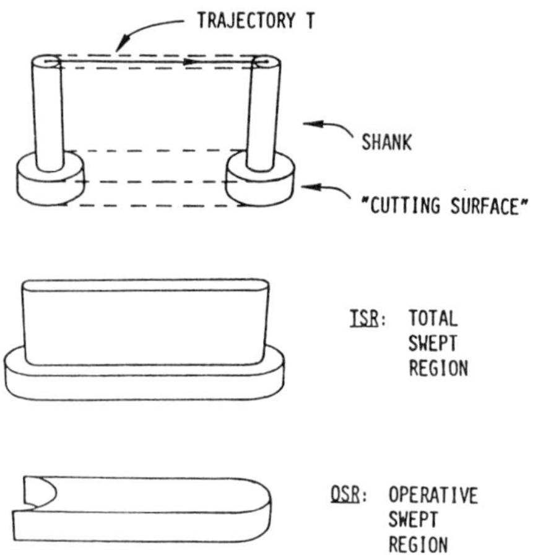

Figure 11.33. Swept-region refinements.

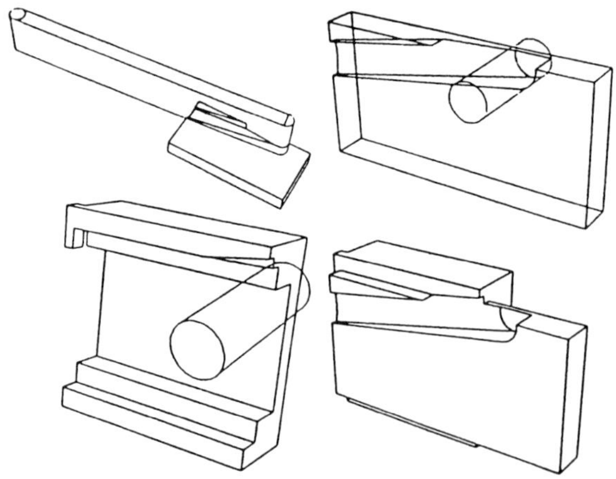

Figure 11.34. Automatic detection of cutter collisions with a workpiece and a fixture during a tool-change operation.

Figure 11.34 illustrates automatic collision detection, implemented in 1985 in the P2NC system.

In early 1983 we commissioned our first CNC machining center for teaching and experimental NC research. A few months later work started on P2NC, a PADL-2 based machining simulator (discussed later, under Systems) that evolved into a powerful, publicly available NC verifier embodying much of what we knew about the spatial and dynamical modeling of machining [65].

We turn now to process planning and feature-level programming. In our early years we tried to address NC automation as a relatively holistic problem – loosely, 'from part spec to NC code in a three-phase swoop': identification of 'removal volumes' for roughing, followed by the generation of roughing code and then, by boundary traversal, finish-machining code. Others took a similar view, as exemplified by the work of the TIPS group at Hokkaido [12], Grayer at Cambridge [66], and Armstrong at Leeds [67, 68]. The weakness in this view, as we came to understand from a study of research at Purdue [69], and from our early case studies at the Gleason Works [7] is that it acknowledges only one type of machining process – milling – and pays scant if any heed to setup requirements.

Circa 1980 we adopted the view, summarized in Figure 11.35, which captures industrial practice. Briefly, manufacturing planning occurs in two broad stages: closely linked process and setup planning, followed by operation planning (which includes NC programming). The stages are linked through prose documents called *process plans,* which specify 'unit operations' (drilling, milling, turning, grinding, painting, . . .) on 'features' (holes, slots, faces, . . .); associated setups (use of fixtures, workpiece positioning and clamping) sometimes are specified explicitly but often must be inferred in current practice. Acceptance of this view led to the establishment of parallel research streams in the PAP when talent could be redeployed from the PADL-2 Project. The 'upstream' team focused on process and setup planning, and the 'downstream' team focused on operation planning. The work was to merge at the Process Plan in Figure 11.35. This interface provides a convenient starting point for a summary of the work.

In early 1983 an old proposal [70] was resuscitated as the starting point for the MPL Project, which became the main focus of the downstream research team. MPL (Machining Process Language *or* Machining Programming Language) was envisioned as a formal language in which process plans could be written, e.g. by automatic process-planning programs, and also a relatively high-level language in which humans could program machining centers. An MPL processor (or Operation Planner, in the spirit of Figure 11.35) was to translate plans/programs into command code for machine tools and setup manipulators . . . or less futuristically, NC code and instructions for human operators. The goals of the MPL Project were to specify a version of the language (MPL-1) useful for 2.5-axis machining, and to build an experimental translator (XMPL-1). Both goals were achieved in 1983–86, although the adequacy of the language as an output medium for automatic process planners could not be tested before the PAP disbanded.

Figure 11.36 shows a segment of MPL-1 script.

Note that statements are of the form ⟨*operator*⟩⟨*operand*⟩ [/⟨*qualifier*⟩], which mimics the informal prose used in typical industrial process plans. Operators in MPL are either setup or machining commands, and operands are feature names, feature-name pairs (in some setup commands), or feature-name lists. In MPL-1 a feature is required to be a single face of the desired part, defined and named in PADL-2 (in the XMPL implementation) through the VGRAPH system (see Systems below). A basic

```
POSIT A, Vise_Face_1 /ON          }
POSIT B, Vise_Face_2 /ON          }  Setup
CLAMP Vise                        }
MILL  Slot_Bottom /ROUGH          }
DRILL  Hole_1                     }  Machining
COUNTERBORE  Hole_1               }
  ...
UNCLAMP Vise                      }
POSIT  A, Vise_Face_3 /ON         }  Setup
  ...
MILL  B_Side                      }  Machining
  ...
```

Figure 11.36. A segment of an MPL plan/program.

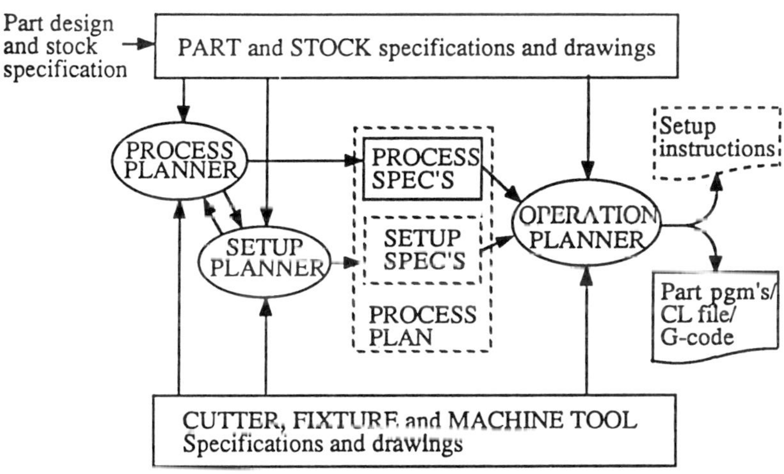

Figure 11.35. The traditional two-stage approach to manufacturing planning.

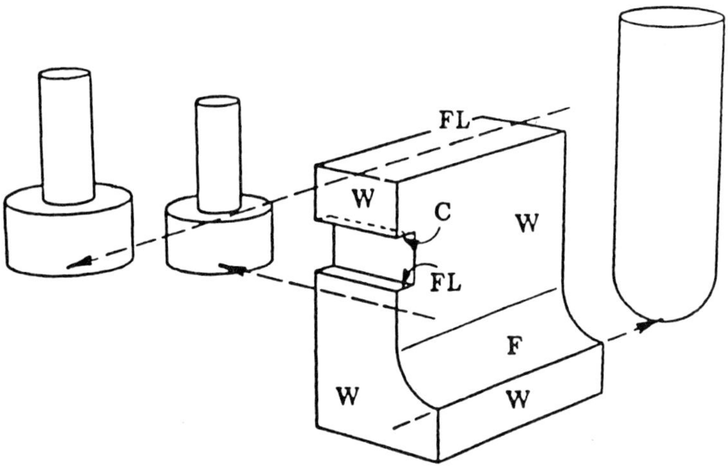

Figure 11.37. Classification of machining features in MPL-1.

design decision: machining features in MPL-1 must be single faces, or lists of faces having the same relation (parallel or perpendicular) to an implied spindle direction, because machining features are always treated as one member of a (*spindle, process, feature*) triple.[19] This enables features to be classified **in a specified setup** into one of a few generic types, as shown in Figure 11.37: floor (FL), ceiling (C), wall (W), fillet (F). Further, (process, feature) pairs in a specified setup may be associated with one of just three types of MPL internal operations: hole-making, clearing, profiling.[20] The XMPL translator had a planner for each, as indicated in Figure 11.38.

XMPL's specialized machining planners were based on a generate-trim-test paradigm discernible in Figure 11.39 which presents a simplified version of the simplest planner. Note that after a trimmed cutter path has been produced, invasive-machining and cutter-clearance tests are made using logic from the NC verification work described earlier. Note also that after the operation has been verified in isolation, it must

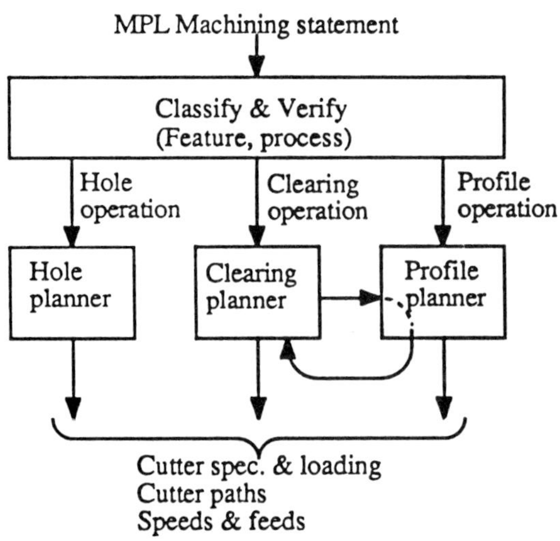

Figure 11.38. Operation planning for machining in XMPL-1.

be linked to the preceding operation by an 'InterOperation' procedure; this generates positioning and sometimes tool-change information which, after collision tests have been made, is passed to the NC code generator ('OutputNCode').

XMPL was interesting to observe in action. Usually it did what one would expect, but at times it planned in ways unlike a human NC programmer. This usually happened in clearing operations, where 'good machining practice' heuristics were used to supplement the removal information inferrable from operands. Only

[19] Thus, for example, a feature list may contain the planar and cylindrical walls of a pocket that are parallel to the spindle axis, but may not contain simultaneously the pocket's bottom, if it has one, because that would be perpendicular to the spindle axis.

[20] There are a number of subtleties, all explained in Chan [71]. For example: sometimes clearing is done as a side effect of profiling, and conversely.

```
procedure DrillSimpleHole (F)
r := radius (F)
if      Parallel (Axis(F), Spindle)
then    { generate semi-infinite swept region CSR}
        if      BlindHole (F)
        then    bottom := MinZ (BottomCone (F))
        else    bottom := MinZ (F) - DrillTip (r)
        top := MaxZ (Box (PART UN STOCK))
        CSR := CreateSweptRegion (r, bottom, top)
        { test CSR against STOCK to find OSR }
        top := MaxZ (CSR INT STOCK) + Clearance
        OSR_i := CreateSweptRegion (r, bottom, top)
        top := top + DrillShankLength (r)
        TSR_i := CreateSweptRegion (r, bottom, top)
        if      (OSR_i INT (PART UN Fixture)) <> 0
        then    Error (Invasive Machining)
        else    if      (W_{i-1} -* OSR_i) &* TSR_i <> 0
                then    Error (Insufficient Clearance)
                else    InterOperation (OSR_{i-1}, OSR_i)
                        OutputNCcode (OSR_i)
                        i := i + 1
```

Figure 11.39. XMPL's simple-hole drilling algorithm (somewhat simplified).

rarely did XMPL declare itself unable to plan an operation, and it almost never made errors because of the extensive checking it did. (XMPL contained most of the verification logic used in P2NC, as well as path and code generators and cutter, speed, and feed selectors.) In summary, MPL and XMPL brought together much of what we had learned about machining and solid modeling, and it exercised PADL-2 effectively. The highlights of the MPL work are summarized in Chan [72] and the work is described in detail in Chan [71].

Research on process planning started about a year later than operation planning research and progressed more slowly. The upstream team was introduced to AI methods in planning by Dana Nau [73] while on sabbatical leave at Rochester, and the team's first 'output' was a survey of research in process planning for machining [74]. The survey showed an almost total focus on planning from feature-based descriptions of parts, usually without precise definitions for features and with tenuous if any links to solid modeling. Integration of planning and solid modeling, which we thought essential, appeared to be an open issue.

The team decided to attack a geometry problem that seemed central to inferring features from solid models – specifically, find a (CSG) solid that corresponds in some reasonable sense to a 'localized subset' of the faces of the part to be made. The initial approach was through tree-pruning (see Voelcker [75] for an example). This work was moving toward feature recognition when the PAP disbanded, and eventually yielded the results reported in Vandenbrande's dissertation [76].

Other applications research

In 1984–87, the PAP conducted research in collaboration with the Ford Motor Company and MHP, Inc. (maker of our machining centers), aimed at investigating the role of on-line sensing in adaptive machining planning, remote command and control of machining systems, and CNC-controller modifications to support remote command and control. We were particularly interested in new research issues that might be posed by the emerging OSI and MAP communication protocols, and Neal Laurance of Ford launched a study of these while in residence at Rochester in 1985. Some results from that work are reported in Voelcker [75], but we shall not summarize them because they are largely irrelevant to solid modeling.

Vassilios Tourassis joined the Rochester faculty in mid-1985, bringing expertise in robotics to a research project that hitherto had only dabbled in the field, mainly through kinematic simulation. Tourassis' research program was acquiring a focus when the PAP disbanded.

Systems

The PAP developed three public systems – PADL-1, PADL-2 and P2NC – and several internal experimental systems. We summarize the systems work chronologically, in terms of three eras, and also cover two special projects: part surveys in the early years, and specialized computer design in the later years.

The PADL-1 era: 1972–78

The 'interaction graph' in Figure 11.2 summarizes the PADL-1 era quite fully. It opened with technology and part surveys, and closed with public dissemination of the PADL-1 system and planning for PADL-2. The highlights . . .

PDL

PDL (Part Description Language processor) was built in 1973/74 as our first experimental solid modeler. It was a modest, batch-mode, 2D regularized-CSG system that drove a Tektronix 4010 DVST vector display through the 'Star String (Boundary) Evaluator' noted earlier. It was slow, but it worked correctly and, thus emboldened, we moved into 3D.

PADL language specification
In the summer of 1974 we invested 10–12 man-months in writing a detailed specification for PADL (Part & Assembly Description Language [35]), while simultaneously starting development of a PADL processor *and* launching an ambitious survey of industrial part geometry. The language specification, TM-20, still reads well 19 years later. In 1974–78 it 'told us what we should build' with considerable precision . . . but not how to do it; it also served as a public commitment to build a PADL processor, and supplied a meta language for part surveys.

Part surveys
Our first part survey was led by Tom Nelson, a Gleason Works industrial engineer on loan to the PAP in 1972/73. About 100 Gleason parts were characterized by gross type (casting, bar, sheet, . . .), character of machining operations, and 'ease of describability with simple shapes'. This survey convinced us that (1) constructive geometry was promising as a descriptive medium for parts, (2) 'feature orthogonality' (as manifested by the predominance of 1-, 2-, and 2.5-axis machining) was important, and (3) we did not have good means to characterize part geometry.

Our second survey was more ambitious. The main goal was to assess quantitatively the distribution of 'geometric complexity' in industrial parts – as engineering research *per se*, and to guide the design of a first PADL processor (e.g. what primitive solids would be needed to make a 'convincing' system, and how many primitive instances would typical parts require?) The survey was done in 1974/75 by Nuriel Samuel, a tooling engineer on loan from Xerox Corporation. The following PADL dialects, differentiated by their sets of primitive solids, were defined as descriptive media; note that the 1.n versions do not contain a general rotational operator:

- PADL-1.0: Orthogonally positioned blocks and cylinders
- PADL-1.4: Orthogonally positioned blocks, cylinders, and wedges
- PADL-1.6: Orthogonally positioned blocks, cylinders, wedges, and cones
- PADL-2.0: Orientable blocks and cylinders (and wedges by construction)
- PADL-2.8: Orientable blocks, cylinders, spheres, cones, and tori.

Samuel generated a statistically valid sample of 128 parts from a new Xerox copier. He prepared for each, under a strict protocol, a *gedanken* definition in the lowest adequate dialect, and also studied other characteristics of the parts. The survey required about one man-year of effort and generated the wealth of data reported in TM-21 [77]. Figure 11.40 shows some of the results displayed as plots of primitive-instance counts *versus* percent of parts describable in the various dialects. For example, about 30% of the parts were describable in PADL-1.0, with the largest definition containing about 45 primitive instances. The curve labeled 'Redesigned to PADL-1.0' has interesting implications for mechanical design [77,78].

In 1975 we assessed informally about 200 parts selected from standard 'families' by Cincinnati Milacron. This survey was not done as carefully because we were seeking only a consistency check on the Xerox results. Although the Milacron and Xerox parts differed in size, material, and function, we found broad similarities in their geometric complexity distributions. In retrospect, the three part surveys yielded landmark data at the time but, insofar as we know, did not stimulate broader surveys by others using improved methodologies.[21] Is it not remarkable that hundreds of millions of dollars have been spent subsequently developing CAD/CAM systems to handle part populations whose characteristics are known only idiosyncratically?

Early PADL-1 development: SuperClip and ENDRUN
In late 1974, the accruing data from the Xerox survey led us to focus development work on PADL-1(.0), the orthogonal blocks-and-cylinders version of the language. We reasoned that a system that could handle really well about 30% of machined parts would be 'convincing', and, an important consideration at the time, would

[21] Alan dePennington's group at Leeds did a machined-part survey in 1978/79 using an improved version of our methodology, but the results have not been made public.

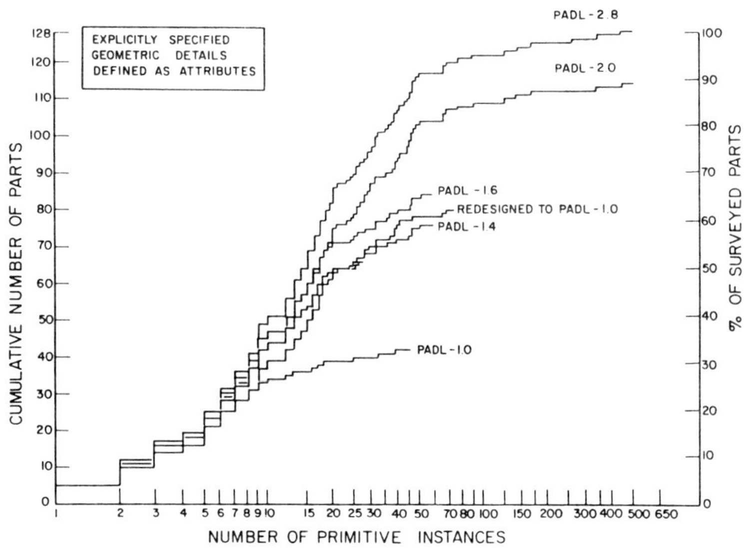

Figure 11.40. Data from the Xerox part survey.

not stand as a potential competitor to the systems that CAD/CAM vendors would develop (we thought) as soon as PADL's viability had been established. So the task for 1975 was building a PADL-1(.0) processor, and by the end of 1974 we were well launched on the design of SuperClip as the system's key boundary evaluation module. By mid-1975 we were forced to label SuperClip a 'wrong turn', for reasons given earlier in the Algorithms section. (This understates the situation. SuperClip's failure was, at the time, a frightening disaster.) Fortunately, as also noted earlier, we were rescued by the emergence of classification principles that enabled us to take an entirely different approach to boundary evaluation than the one we had been pursuing.

A crash effort was mounted in the summer of 1975 to develop the approach through a system called ENDRUN (the name comes from American football). The first version of ENDRUN – about 15,000 lines of code – was built in six weeks of intensive programming (through punched cards!). It accepted CSG trees in postfix format, and displayed correct wireframes of arbitrary collections of blocks. At that point we knew we could build an effective PADL-1 processor. By the end of 1975 cylinder code had been added to ENDRUN, as well as a crude B-rep subsystem, an automatic sectioner, and a revolutionary (for the time) static interference tester for assemblies. In late 1975 the first PADL-1 processor was designed, initially as input and output processors that provided 'sugaring' for ENDRUN.

The PADL-1 system

PADL-1 was fleshed-out in 1976 and unveiled publicly at a meeting of the CAM-I Geometric Modeling Project held at the University of Rochester in August of 1976. About 25 attendees were given one-hour user tutorials and then assigned exercises to do on the system – defining, and producing dimensioned drawings for parts from a small steam engine. This hands-on demonstration was so successful that design of a publicly disseminable version of PADL-1 was begun the following week. Some 15 months (about four man-years) were spent defining the language, designing robust B-rep and Drawing-file subsystems, tuning the boundary evaluator, writing documentation, and recoding extensively for portability and clarity. Public dissemination began in January of 1978.

Figure 11.41 shows the final, 1980 version of the system. The first (1978) public version was identical apart from the Shader and Massprop Evaluator, which were added in early 1980. The system was driven by keyboard interaction or from stored definitions. The Input Processor parsed the language statements into an interval postfix format that drove the Boundary

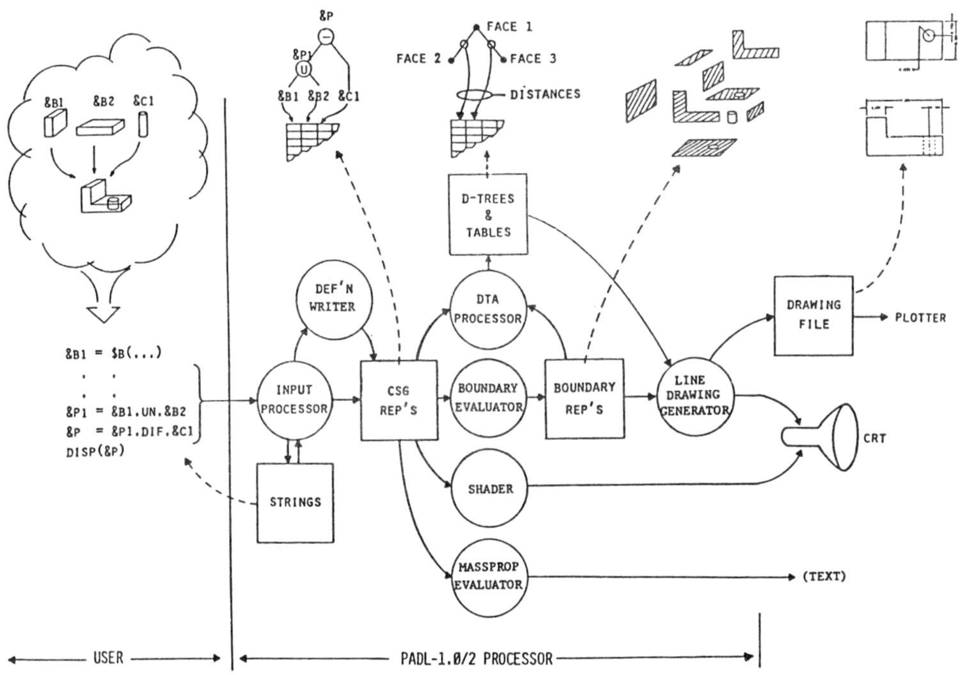

Figure 11.41. The final (1980) PAP version of PADL-1.

Evaluator. Edge displays and drawings were produced by B-reps, whereas shaded displays and mass properties were computed from CSG. The Definition Writer produced the CSG definitions appropriate for sectioning and interference analysis in assemblies (Figure 11.42), and the DTA Processor checked and posted dimensions chains and tolerances (Figure 11.43).

Public PADL-1 was designed from the outset to be a 'bullet-proof' turnkey system, with an explicitly limited geometric domain and range of functional capabilities. In many respects it was a sophisticated toy meant to demonstrate, popularize, and disseminate solid modeling concepts and a classification-based approach to their implementation. Its code was written to be read as well as executed, and by 1978, when public dissemination began, the system was well documented internally and externally. The final version contained about 80K lines of FLECS (structured FORTRAN-66 [79]) and embodied over ten man-years of effort.

PADL-1 was distributed in source format, under \$1200 and \$2200 licenses (for universities and companies, respectively) that granted essentially unlimited rights in 1981 and thereafter. Eighty licenses were issued between 1978 and 1985. The system was commissioned on Digital PDP-11s and VAXen, as well as on IBM, CDC, HP, Data General, and other computers. While designed mainly as a teaching and technology-transfer vehicle, it was also used in a surprising variety of research projects.[22]

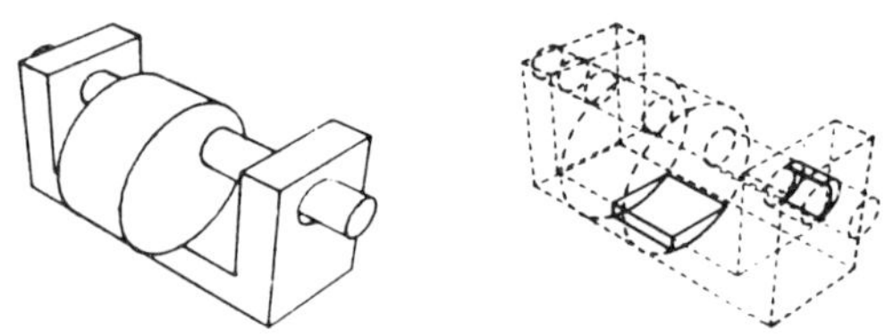

Figure 11.42. Automatic assembly interference analysis by PADL-1 (1976).

The PADL-2 era: 1979–82

We built PADL-1 carefully because we expected it to provide 'starter technology' for the CAD/CAM vendors who would develop 'industrial strength' solid modelers, and thereby release us to return to research on NC machining.

[22] PADL-1 lives on as a teaching tool because, with a simple graphic interface, students can learn to use it in fifteen minutes. It is reasonably fast on modern workstations, and offers some capabilities in its limited domain that modern systems cannot match.

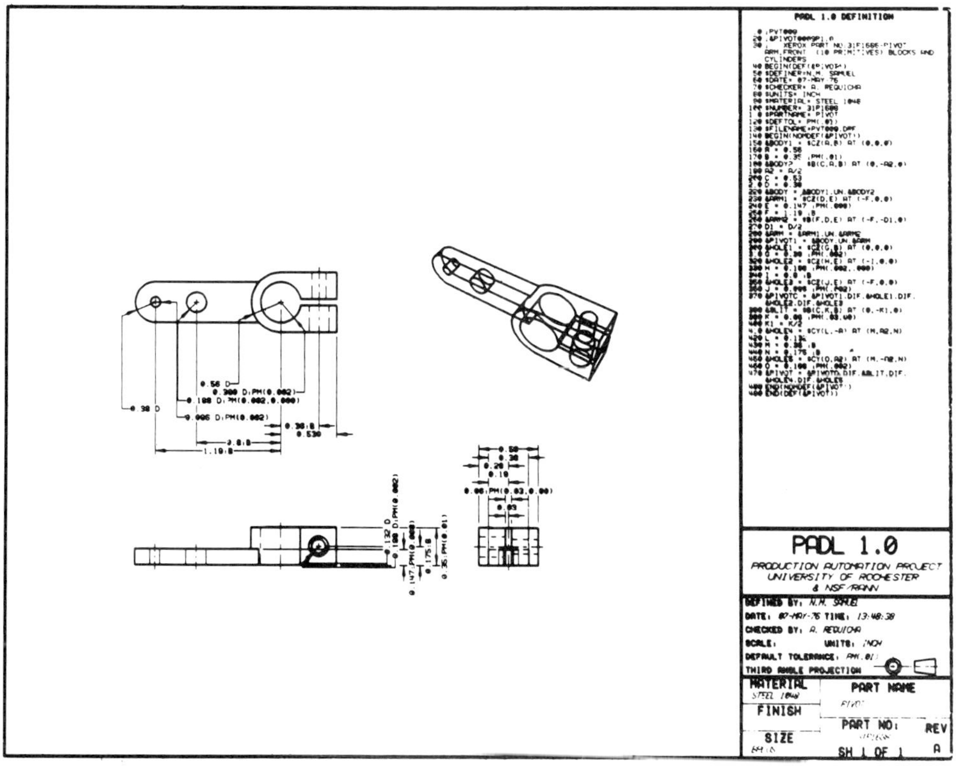

Figure 11.43. Automatic layout and dimension posting by PADL-1 (1977).

In 1977 and 1978 we learned some hard lessons about the vendor community [8]. Succinctly, some vendors saw no merit in the technology, and others were unwilling to assume the risks associated with developing a 'revolutionary' technology. The vendors' reluctance to develop PADL technology was as frustrating to some potential users – Kodak, Boeing, and Sandia, among others – as it was to us, and in 1977/78 we were urged to carry the technology to the industrial prototype stage. This we were reluctant to do because we wanted to return to research, and in any event lacked the staff and experience needed for serious system development.

The PADL-2 Project

The PADL-2 Project [80] emerged in late 1978 as a compromise: the PAP agreed to supply leadership over a two-year period for a temporary working staff of industrial residents supplemented with students. The goal was a set of 'core modules' for an industrial modeler. The constituency was defined as vendors who would develop proprietary systems based on the core, sophisticated user companies, and the research community. Several months were spent in late 1978 developing agreements that defined each party's rights – proprietary rights for the industrial sponsors, and public rights for us. (This exercise, while tedious, was time well spent because no claims or disputes of any nature arose.) We launched the Project in January 1979 with the following staff:

Director:
  Chris Brown
  (borrowed from Computer Science)
Chief Analyst:
  Bob Tilove (on leave from his PhD studies)
Senior System Engineer:
  Gene Hartquist (shared with PAP)
Computer Systems Engineer:
  Rich Marisa (shared with PAP)
Two-year Industrial Residents:
  Tom Check (Calma Corporation)
  Jeff Metzger
  (McDonnell Douglas Automation)
  Don Peterson (Sandia Laboratories)

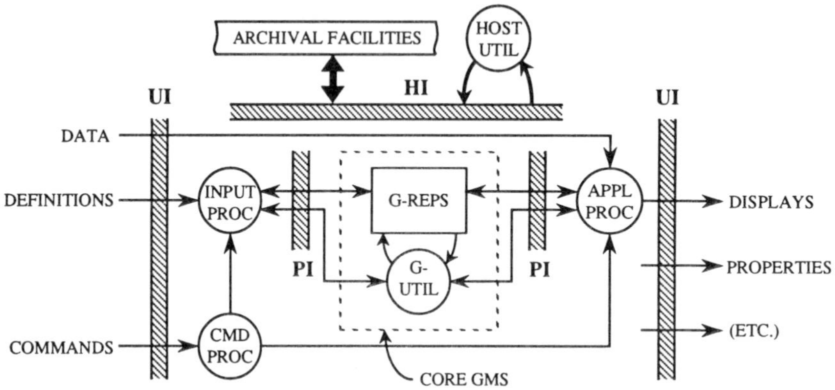

Figure 11.44. Hierarchies and interfaces in industrial geometry systems (1978/79).

and about $800,000 of funding from ten industrial sponsors and the National Science Foundation. The PADL-2 Project was designed to run as a largely autonomous activity while the PAP conducted independent research, but so many problems arose in building PADL-2 that the PAP was drawn back into solid modeling almost fully.

Design and implementation of PADL-2

We sought a core system that would cover 90–95% of unsculptured industrial parts, and this set the domain as the natural quadrics, i.e. PSDL-2.8 in the part survey taxonomy, without tori (for which partial support was provided later). We also sought multiple consistent representations (at least CSG and B-reps) in order to support a variety of applications, and we wanted a system that could be used by humans and also as a 'geometric utility' by higher level programs, e.g. NC machining planners and verifiers.

Our ideas on what constituted a 'core system' were influenced strongly by the difficulties we encountered with PADL-1 in the early NC verification research. PADL-1 was designed as a turnkey system and it did what it did nicely, but it was painful to go inside the system or to use PADL-1 as a subsystem. Thus in 1978/79 we came to the view summarized in Figure 11.44: a core geometric modeling system is a collection of representations and utilities that are procedurally accessible through programming interfaces PI. Users (humans always, automata sometimes) interact through user interfaces UI with intermediate Input, Command, and Application processors, and all processors have controllable access to host-system utilities through a host interface HI. (PADL-1 had strong user interfaces but very weak programming interfaces.)

We opted from the outset for what would now be called object-oriented programming, and designed CSG, B-rep, and other subsystems with distinct entities addressable by name and type which were operated on (only) by a well defined set of functions. But we were hampered from the outset by our choice of FLECS-77/FORTRAN-77 as programming languages; for our early work, we built a general structuring package (PACPAC) which, in effect, emulated PASCAL's Pointer and Record facilities. We chose FORTRAN (with the strong concurrence of the sponsors) only after a careful study, under a strong exportability constraint, of the alternatives we thought would be available in 1981–83. At the time only PASCAL appeared to be a potentially viable alternative, and we were concerned about the portability, capacity, and reliability of PASCAL compilers.

The range of issues that arose in PADL-2's design and implementation is too large to summarize here, but one – consistency and access control in multi-representational systems – proved to be especially challenging and is still largely open. We envisioned a Custodian module that would lie outside the core and control access to it and consistency within it. Several custodial strategies were floated and thousands of lines of test code were written, but we never settled on a satisfying strategy. (Chris Brown designed the most arcane Custodian, which he named DRACO; fragments of DRACO live on in the GRAPAK module.) PADL-2 was released without a proper Custodian, and it can be tricked into inconsistency through its programming interfaces.

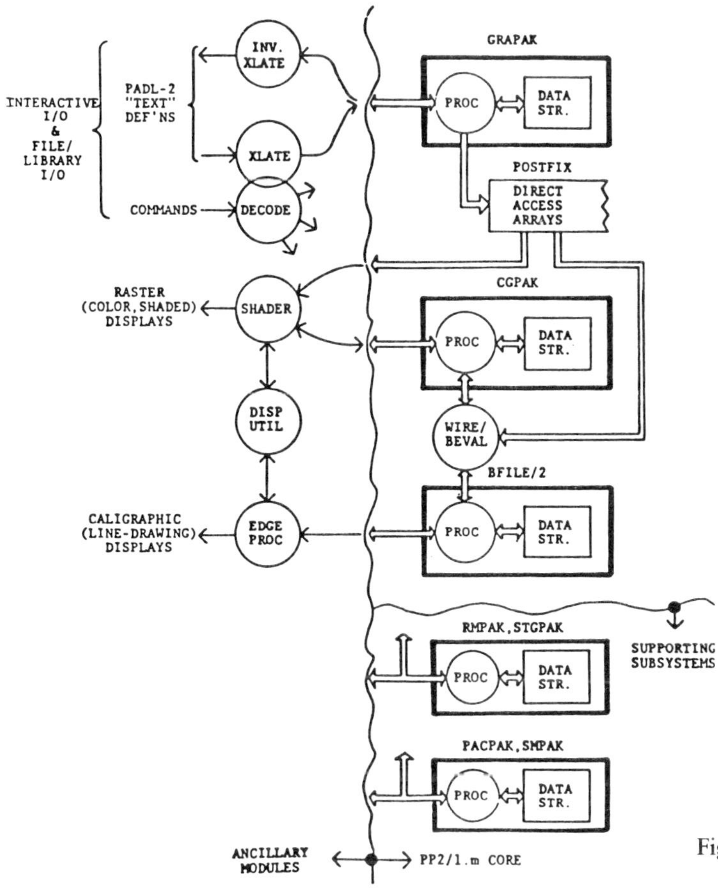

Figure 11.45. The PADL-2 system (1981).

The PADL-2 System

Figure 11.45 shows the 1981 view of the system. The core portion is on the right. Most of the Ancillary Modules on the left were regarded as 'disposable extras' that system builders would replace with fancier proprietary modules. Within the core, GRAPAK handles CSG graphs, CGPAK handles numerical geometry, and BFILE/2 provides B-reps; all are accessed through subroutine calls. The supporting systems handle rigid motions, strings, and other distinct data entities, again through subroutine calls. The exception to this encapsulation strategy is the set of direct-access arrays labeled POSTFIX: these provide CSG trees (expanded graphs, with all symbolic references resolved and algebraic expressions evaluated) in postfix format and contain the most frequently accessed data. These data were made directly accessible to get performance from the system.

Figure 11.46 shows a higher level, 1983 view which includes the octree subsystem. This illustration shows the flexibility of the architecture,

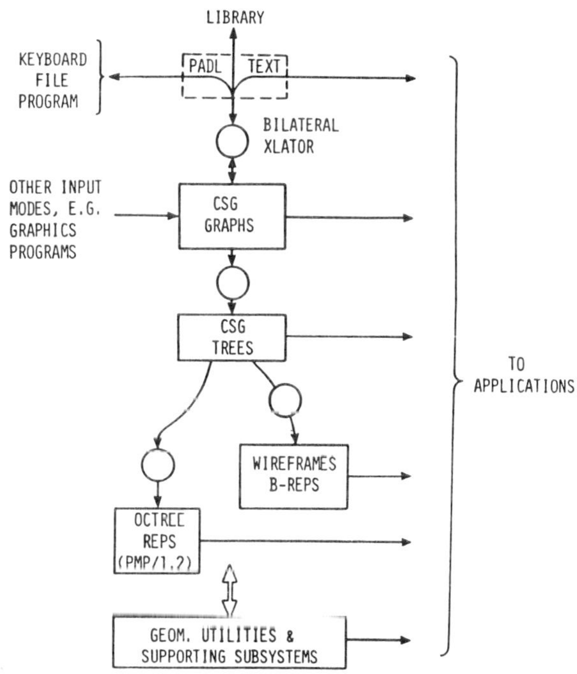

Figure 11.46. A higher level view of the system (1983).

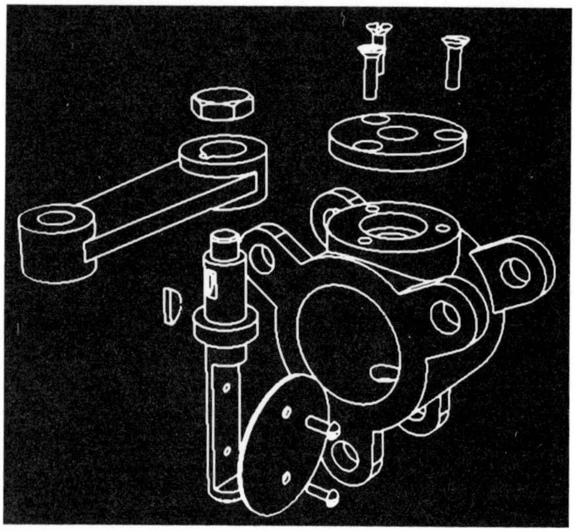

Figure 11.47. UNISOLIDS (1982).

because applications can draw on five distinct representational subsystems; we used three of the five regularly in most of our 1980s applications research.

The Industrial Residents departed in early 1981, and the PADL-2 Project ended formally in 1982. A much smaller staff focused its efforts increasingly thereafter on public dissemination, maintenance, and the packaging of Optional Modules, such as the Mass Properties Module developed by Alan Clark, of Ford, during a 1982 industrial residency. Our last significant enhancements to the core were made in 1983 (a good incremental boundary evaluator) and 1985 (implementing meta-primitives, which are user defined compositions of halfspaces).

PADL-2 dissemination

The dissemination conditions were defined in the agreements governing the PADL-2 Project. The ten industrial sponsors, who hold essentially unlimited rights to the core software, were provided with copies of the then-current system in mid-1981 and updated versions thereafter. The sponsors' subsequent use of the system varied widely. A few did nothing with it, some (notably Sandia and Kodak) used it internally for a variety of purposes, and others incorporated all or portions of PADL-2 into commercial systems. McDonnell Douglas Automation ('McAuto') charted the most aggressive course. McAuto began development of PADL-2 based UNISOLIDS in 1981, and demonstrated the system publicly at the AutoFact conference in 1982.[23] Figure 11.47 shows some typical output from the 1982 system. UNISOLIDS was developed continuously through the 1980s, and McAuto and the PAP did some collaborative R&D whose results flowed to the growing PADL-2 community at large.

Public dissemination of PADL-2 began in mid-1982, one year after the first distribution to the sponsors. The one-time fees for licenses, all subject to a non-redissemination clause, varied from $600 for educational institutions and government contractors to $20,000 for commercial concerns; redissemination licenses cost $50,000 (the Project fee paid by the original sponsors). Between 1982 and 1987, the PAP issued 143 licenses of all types, with the majority being for university research and educational use. (More licenses have been issued subsequently by Cornell, and many more copies are abroad within vendor systems.) The systems we supplied were used in a great variety of ways. At one extreme, Cadetron (later merged into Autodesk) recoded the system into C for resale on IBM-compatible PCs. At another extreme, some licensees 'disassembled' the package (which is easy to do with a modular core system) and used only one or two components – typically the GRAPAK/POSTFIX subsystems, or the numerical geometry package. Most licensees operated between these extremes, using the system for robot and machining simulation, biomedical modeling, cable routing, and many other tasks. Plate 14 illustrates one of the (literally) prettier applications: output generated at the Ford Motor Company from PADL-2 by a locally developed raycaster with advanced (and computationally expensive) illumination, texturing, and anti-aliasing models [81].

PADL-2 is now obsolescent as a system, but nearly all of it is 'good code' and portions of that code are likely to be running 20, and perhaps even 50, years hence.

The applications era: 1983–87

As the PADL-2 Project wound-down, we refocused our efforts on applications research – mainly machining, in consonance with our original goal. We summarize below one public

[23] When PADL-1 became unrestricted in 1981, McAuto distributed PADL-1 with a simple graphic interface to some potential customers in order to gather data to guide the design of UNISOLIDS' user interface.

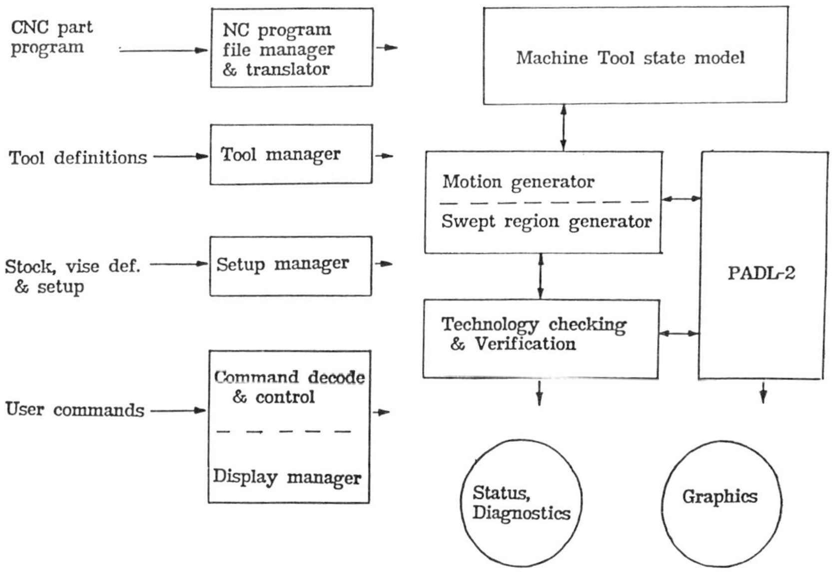

Figure 11.48. The NC simulator and verifier P2NC (1985/86).

and two experimental systems that were built in this era, plus a 'special project' on custom computer design. Our (limited) finite-element systems work was summarized earlier under Applications.

P2NC

P2NC resulted from our second foray into systems for automatic NC-program verification.[24] (Our early work in the 1970s was based on PADL-1 and did not yield a proper system.) P2NC began as a graduate student hobby project – 'fun coding' in the winter of 1982/3, to exercise PADL-2 by building a simple simulator for our new S10 machining center. In June of 1983 we taught our first NC short course without using a simulator . . . and P2NC became a serious project almost overnight, because a mere eight students trying to debug programs on the S10 brought our NC lab to its knees. (Eight months later, when a first version of P2NC was ready for use, the lab ran quite smoothly with more students.)

From mid-1983 through most of the 1980s, P2NC became the test vehicle for the machining-process modeling and program-verification research summarized under Applications. P2NC was built in a modular, evolutionary manner and had a succession of graduate student custodians. The strategy was to implement increasingly elaborate spatial-effects simulation, and then to install selectable automatic tests for collisions, invasive machining, and so forth; after that, we planned to move to dynamical effects and tolerances, but the PAP disbanded before we made much progress on the latter topics.

Figure 11.48 is drawn from Sungurtekin [65], which provides a good summary of the 1985/86 system. P2NC has a nice user interface which allows one to slip into PADL-2 to define fixtures, setups, and so forth, and then to move back to P2NC to execute the NC program, with PADL-2 called by P2NC as a 'geometric evaluator'. P2NC contains a complete state model of the machine tool (a generic vertical spindle mill, customized to the S10/S24 by values drawn from a parameter table), and thus the spindle, coolant, tool-changer, and other states may be monitored throughout. Plate 15 shows exemplary output produced at the GM Research Labs in 1986.

P2NC was designed to emulate a generic 'RS-274-D machine' [82], and Figure 11.48 does not show an elaborate preprocessor that converts programs in the extended command set of the S10/S24 controller into RS-274-D. The requirement for preprocessors to handle different brands of NC controllers is a version of 'the

[24] The NC simulator/verifier described here had various names during its evolution. In Sungurtekin [65], for example, it is 'XNCS'. 'P2NC' was adopted in 1987 and it has persisted.

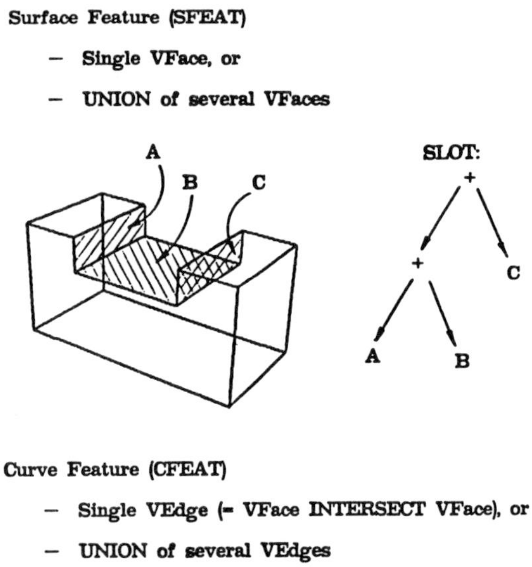

Figure 11.49. A simple surface feature in VGraph.

postprocessor problem', and was a cause of considerable concern for rather obvious reasons. Our 1985 solution was to provide a BCL interface for P2NC (see Sungurtekin [65]), but we did not really believe that BCL would be adopted as a new NC standard – and for practical purposes it has not been.

P2NC was our main vehicle for assessing and tuning PADL-2 as a program-callable 'geometric utility'. Broadly, PADL-2 fulfilled all of our expectations (within its limited domain) in respect to functionality and controllability; its major weakness was performance. For short (50-block, say) NC programs, P2NC's speed was acceptable . . . about as fast as the S10/S24 itself. For long programs, P2NC was maddeningly slow when the automatic block-by-block checking was enabled. The lesson was simple and stark: automatic verification, if done 'rigorously' by the tests summarized earlier, is computationally intensive; if snappy performance is wanted on even moderate-length NC programs, then a *lot* more flops than a VAX-11/780 could deliver must be provided.

P2NC was made publicly available in the PAP's final year, but in a low-key way. We wanted external users to evaluate the overall viability of the system, but we wanted only a few because we were apprehensive about the preprocessing problem. In the event, we got none at all.

VGraph

VGraph (from Variational Graph) was an experimental PADL-2 subsystem built by Stephen Chan in 1983 to provide a testbed for implementing tolerance facilities in CSG modelers. Variational graph concepts and their partial implementation in VGraph are covered in Requicha [41]. The basic idea is to provide facilities for defining various kinds of surface and other 'features', as for example the slot shown in Figure 11.49. This is relatively easy to do in B-rep modelers; in CSG, there are subtleties at the purely geometric level (e.g. what is a 'face'?), and even deeper subtleties associated with the inheritance and propagation of properties and attributes over CSG graphs and trees. Because there were many unknowns associated with variational graphs, VGraph was not implemented in PADL-2's GRAPAK as a proper subsystem, but effectively ran in parallel with PADL-2, with some GRAPAK facilities duplicated. VGraph was never tested in its intended role as a host for tolerance specifications, but it was used intensively by Chan as the feature-defining utility in his MPL research.

XMPL

XMPL-1 was the first (and only, to date) experimental processor for the feature-level MPL machining language [71, 72]. Figure 11.50 shows the system's organization. The MPL processor used VGraph to enable users to define machinable features through graphic pointing on a PADL-2 host definition, and it used PADL-2 as a central geometry engine to support the feature definitions, and to trim and test candidate

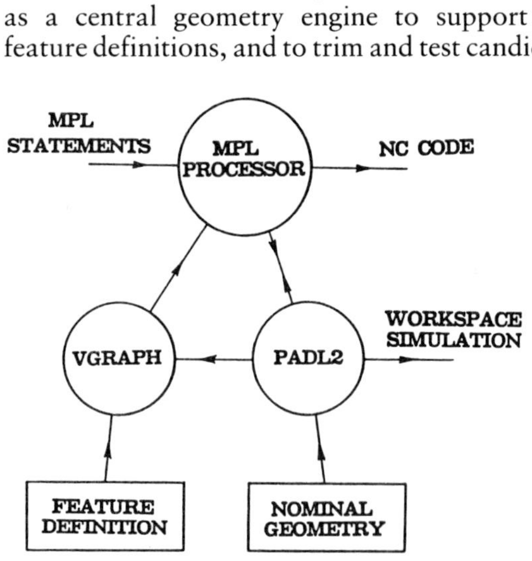

Figure 11.50. An overview of the XMPL system.

cutter paths. As noted in the earlier discussion of MPL, XMPL contained most of the checking logic used in P2NC plus a lot of additional code for feature validation and classification, candidate path generation, and so forth. It was a large, complex system, some of whose internals were almost chaotic in the later stages because the central concept that drove MPL – that machinable features are always interpreted in the context of a (spindle, process, feature) triple – did not 'jell' until relatively late in the research. (The contemporaneous versions of P2NC were much cleaner because P2NC's principles were set years earlier.) XMPL was successful in validating experimentally the ideas behind MPL, but unfortunately it began to operate properly just as the PAP was disbanding. If the PAP had continued, the system would have been thoroughly re-engineered and probably offered publicly for external evaluation, because MPL contains some nice ideas.

Specialized computer design

The recursive algorithm for evaluating the classification function M(X, S) with respect to solids represented in CSG is amenable to parallelization; to see this, one need only examine the prescribed operations when the algorithm is unwound. (We became intimately aware of the opportunities for parallelism in 1975, when FORTRAN's lack of recursion led us to implement the algorithm bottom-up, through multiple pushdown stacks.) In the later 1970s and early 1980s we often included in conference talks pleas to hardware vendors to exploit these opportunities . . . but nothing happened.

In the summer of 1982 we deciced to probe the opportunities ourselves, because we had employed an undergraduate project student, Jeffrey Peck, with a gift for TTL logic design. Jeff struggled for about six weeks, and then produced an analysis showing why an effective parallel classification machine could not be built: communication costs dissipated the savings from parallel evaluation.[25] As Peck was completing his report, we were visited by Gershon Kedem, an assistant professor of computer science who was converting himself from a numerical analyst into a designer of CAD systems for digital VLSI. Kedem needed grist for his latest CAD ideas, and he agreed to take on our problem and Peck's supervision. Within a week he realized that a classification machine must be massively parallel and quasi-systolic to avoid the problems Peck had identified, and within two weeks he had roughed-out the design of what soon came to be known as the RayCasting Machine (or RayCasting Engine, RCE, as it is now known).

Figure 11.51 shows the core of the RCE as we had come to understand it in 1983 [83–85]. The strategy is to provide a programmably configurable tree of specialized processors that can mimic the CSG tree representing the solid S. In Figure 11.51a, a CSG tree is associated with a programmable hardware tree (b) whose leaves are Primitive Classifiers (PC) and whose nodes are Classification Combiners (CC); any CSG tree may be represented as a right-heavy tree as in (c), and any N-leaf right-heavy tree may be mapped into an $N \times (N + 1) \log_2 N$ array as in (d). In the array architecture of Figure 11.51d, each PC classifies the candidate X simultaneously, and classified X-subsets are pipelined bottom-up through the tree. The result – a classification of X with respect to the CSG solid S – is 'returned' by the root processor.

In principle, X could be a solid, a 'face', a curve, or a point, but our experience led us to focus on curves represented parametrically. This decision set the character of the CCs and PCs: the former do domain-independent Boolean operations on intervals, whereas the latter do root-finding to intersect curves with primitive algebraic halfspaces. In Kedem's design, X was required to be a line, and moreover a line from a grid of parallel lines (hence 'raycasting'). These constraints were imposed to balance the computation done by the CCs and PCs and hence maximize the flow through the machine. More specifically, the CCs are intrinsically fast because their main work is sorting real numbers, whereas the PCs are potentially slow. To get hardware PCs fast enough to keep the CC pipeline full, the equations to be solved were restricted to second order (line versus quadric halfspace), and the quadratic equations were solved over the grid of lines by propagated difference techniques.

Kedem and John Ellis started to design the machine in 5μ NMOS in 1983, for fabrication through DARPA MOSIS, but the work was interrupted in 1984 when Kedem moved to Duke (Ellis followed shortly thereafter). In 1985 the

[25] Peck's analysis was a precursor to several studies done by others later in the decade on tradeoffs in implementing variously structured problems on the alternative parallel architectures that were then emerging.

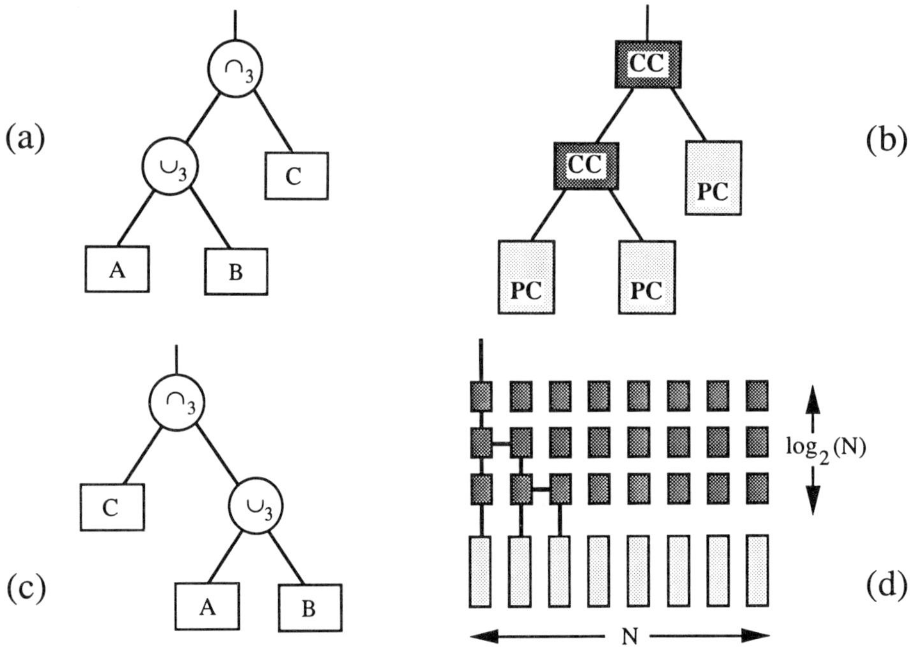

Figure 11.51. The RayCasting Engine – mapping geometry into silicon.

design was restarted at Duke in CMOS, with design tools and simulators being built in parallel. The PAP tracked Kedem's work after he left Rochester, but the early links were broken when the PAP disbanded . . . about eighteen months before the first RCE became operational.

Codification

When we began research in modeling and automation in 1972, there were no accepted scientific foundations, almost no literature, and no acknowledged community of scholars and researchers. The only areas with some semblances of organization were numerical control, computer graphics, curve and surface modeling, and robotics/AI. In the first two years we established contact with Doug Ross (APT); Bert Herzog and Bob Sproull[26] (graphics); Steven Coons and some of his students (Jeff Posdamer and Rich Riesenfeld, for curves and surfaces); Robin Forrest, Charles Lang, and Ian Braid (early geometric modeling); and Jerry Feldman, Lou Paul, and Dan Whitney (robotics). From all of these people we learned of first-class research in progress, but with a very strong focus on techniques; only the curve and surface work seemed to have some of the mathematical substance we had come to expect from our earlier work in electrical engineering and signal theory. These early explorations were unsettling, but also challenging because they led us to try to build foundations for an emerging field – foundations to be tested and refined through teaching, and ultimately recorded in the archival literature we expected to emerge.

The emergence of our ideas – distinctions between models and representations, formal properties of representations, set membership classification, and so forth – is summarized in earlier sections. Their dissemination in codified form was done initially through public reports – the PAP Technical Memorandum (TM) series – and later through journal publication. The TMs became popular documents in the late 1970s and early 1980s as new people were drawn into the field around the world. While some 40 TMs were published, only 15-or-so were 'important'; about 1500 copies of these were distributed, each in response to a specific request, from the time

[26] The Younger; the Elder, a physicist, was Provost and later President of the University of Rochester.

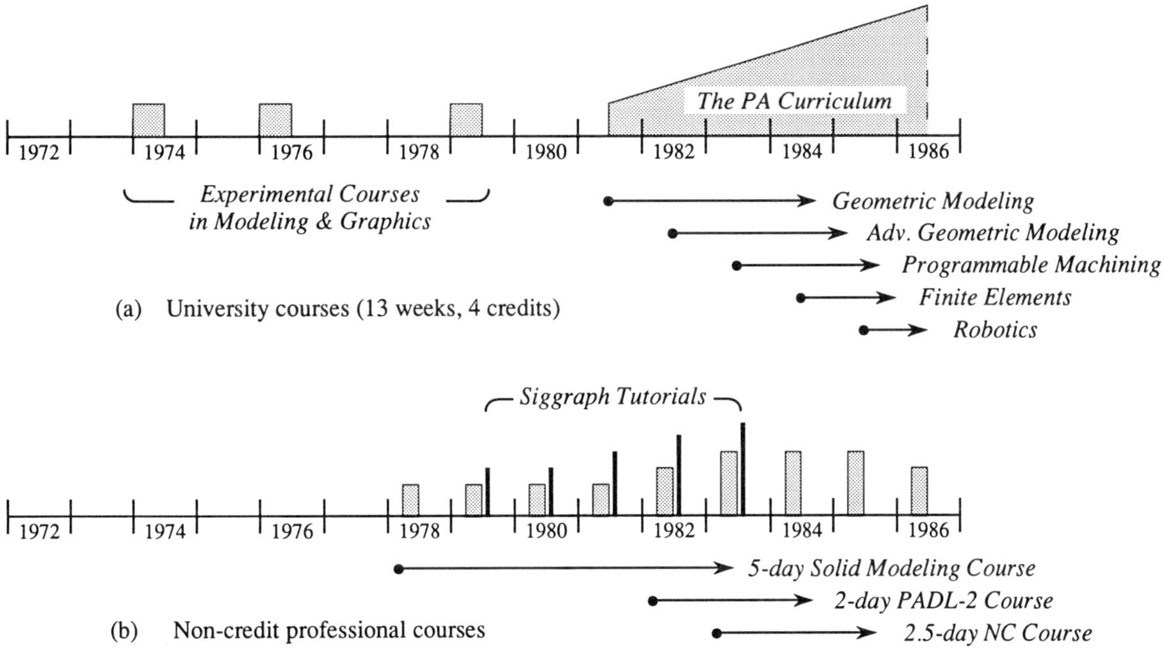

Figure 11.52. Courses developed and taught by PAP personnel.

we began keeping records in 1977 until the PAP's demise. All are available from Cornell, but most are now of only historical interest.

We relied on TMs in the early years because there were no really appropriate journals in the 1970s, the 'forcing' papers into the journals that did exist required a prolonged correspondence for each manuscript. The tide turned in 1980, when two seminal papers [16, 26] were published, and soon cited by others with increasing frequency.

Our efforts to codify through publication would have been less effective if we had not sought simultaneously to teach the material we were codifying. Figure 11.52 summarizes these efforts. Both credit and non-credit courses were offered, to serve different purposes. The credit courses were classical: about 35 lectures over 13 weeks, with assigned analytical and programming problems, examinations, and laboratory work in the machining course. The courses were aimed at graduate students and very able final year undergraduates; they drew enrollments that ranged from 6 to 30 per course. While the 1970s courses were experimental and often ragged, by 1982 we had enough material codified to launch 'The PA (Programmable Automation) Curriculum' as an area of graduate specialization spanning electrical and mechanical engineering and computer science. Geometric and solid modeling was the foundation for the curriculum. We did all of the teaching until Renato Perucchio, a new Cornell PhD, joined mechanical engineering in 1984 to handle finite elements and design, and Vassilios Tourassis, a new Carnegie Mellon PhD, joined EE in 1985 to handle robotics.

The non-credit courses were aimed mainly at industrial personnel who wanted specialized training in the new technology. About 400 people passed through these courses from 1978 to 1986. (Another 400-or-so passed through the Siggraph Solid Modeling tutorials, which grew from half-day affairs in 1979 and 1980 to two full days in 1983.) Developing and teaching these courses was hard work, but the rewards were considerable. We were able to present our approach and our technology to people who developed and bought industrial CAD/CAM, we made friends who helped us later, and we made some needed money for the PAP. The greatest benefit, however, was the discipline of having to decide what was really important, and how to communicate that effectively.

We have been asked why we did not write a textbook on geometric and solid modeling. There is no simple answer to the question. The default response has been '. . . too busy doing research'.

Technology transfer

The PAP produced three kinds of 'results' worthy of transfer to the external world:

- *codified knowledge* – mainly archival papers and organized courses, as summarized above;
- *artifacts and techniques* – prototype systems, working algorithms, heuristics and paradigms;
- *trained people* – primarily graduates who earned research degrees, but also industrial residents, undergraduate assistants, and staff members who left the university for career advancement.

We accepted responsibility for technology transfer from the outset, and sought to do it through the publishing and teaching programs described above, and through programs for software dissemination, industrial residences, and graduate placement.

Our software dissemination policy distinguished sharply between 'engineered' software and experimental software. All software was disseminated in source format, but engineered software (e.g. PADL-1 and PADL-2) was tested and internally and externally documented, and generally was as good as we could make it. It was available to anyone willing to pay relatively modest license fees, and it was meant to be *read* as well as executed because we viewed software as an important medium for transferring knowledge. By contrast, experimental software was often quirky and buggy, and was documented sketchily if at all; we released it only to parties we thought capable of handling it, and also delivering something of value to us (e.g. test data or programs) in return.

Eleven Industrial Residents worked with us for periods ranging from six months to two years (see Appendix A). By 1980 we had devised the following simple rules for industrial residences:

- No money changed hands. Companies supported their residents; we provided working space and facilities, but collected no fees.
- A problem of common interest to which the resident would devote about half of his/her effort was identified in advance; the resident's remaining time was discretionary (e.g. he/she could attend lectures).
- The university and company had unrestricted rights to whatever results the resident produced.

These rules worked well. We had no disputes of any kind, and residents made important contributions. Many continued as personal friends and occasional collaborators.

Our approach to technology transfer was conditioned strongly by three decisions made in 1977, when PADL-1 was clearly successful. We decided specifically:

- not to found a company to commercialize our technology. This decision, which was reviewed periodically but never changed, reflected a persistent 'academic mindset'; it was right for us, but could have been wrong for a different cast of characters.
- not to work with users of modeling systems (*especially* small users) except under controlled conditions, to learn specific things. We sought to use CAD/CAM vendors as our primary medium for technology dissemination, because they could leverage our efforts while also 'protecting' us.
- not to develop graphic-interactive or 'industrial strength' user interfaces for our systems, because (1) such work requires considerable effort with little research payoff, (2) we lacked the knowledge of user needs required to do such work well, and (3) we wanted to reserve an area in which vendors could add value to and customize our technology.

Finally, it is worth noting that solid modeling was a revolutionary technology in the 1970s. The traditional mechanisms for transmuting university R&D into commercial technology were not in place, and thus projects like the PAP were led to do things that would be unnecessary and probably inappropriate in 'normal times' [8].

Facilities and organization

The PAPs offices and labs were in Hopeman Hall at the University of Rochester, in space assigned to the Electrical Engineering Department.

Computing facilities

There were two distinct computing eras: the PDP-11/RT-11 era from 1972 to 1979, and the

VAX-11/VMS era from 1979 to 1985. Conversion to a workstation environment based on Micro-Vaxes was in progress when the project disbanded. Broadly, the PAP's facilities were close to the forefront of *engineering* computing but lagged well behind those available in forefront computer science research groups.

The initial system was a single-user PDP-11/40 (one of the first built) with a Tektronix 4010 DVST storage-tube terminal, a screen copier, a card reader, and a Centronix printer. All programming was done in FORTRAN-66. A magnetic tape unit and Versatec electrostatic plotter were added in 1974. In 1975 we moved to foreground/background operation under RT-11 and added a second PDP-11/40. In that year we also adopted FLECS – a preprocessor-based 'structured' dialect of Fortran from the University of Oregon [79] – as our standard programming language; FLECS raised our programming effectiveness considerably. PADL-1 was built in this environment, with boundary evaluator development done mainly through punched cards. The computing system provided only 28K words of RAM ('core', in those days) for executable code, and so considerable effort was devoted to the design of overlay structures, and to developing assembly language routines to manage 96K words of 'high core' (extended memory); in 1976 we partially automated overlay design. Documents were generated through computerized text editing from the outset, and indeed our first significant coding project in 1972/73 was building a text editor.

In the later 1970s we decided not to upgrade to PDP-11/70s and the RSX operating system, but rather to move directly to a 32-bit machine. The PADL-2 Project enabled us to do that in 1979. An early VAX-11/780 with VT-100 terminals provided, through VMS, excellent (for the era) time-sharing and wondrous conveniences. We added a Lexidata 3400 color raster-graphic frame buffer in 1980, and added in 1981 several low-cost DEC GiGi graphic terminals. While PADL-2 was programmed in FLECS-77, an increasing amount of experimental code was written in PASCAL. RUNOFF was the default document language, with TEX used for many of the later technical reports.

Machining facilities

A small CNC machining lab was built in the basement of Hopeman Hall in 1982, when we redirected some of our research effort from solid modeling to automated machining. The first machine was an MHP S-10: a 3-axis, 3 HP, vertical-spindle mill with servo-electric drives and a 10-station tool changer. The S-10 was upgraded in 1984 to an S-24 (24-station tool changer, enhanced controller) with a fourth table-mounted axis. Both machines were programmed through MHP's extended version of the RS-274-D 'M-code, G-code language' [82]. Programs were generated electronically and transferred through an RS-232 interface. UNIAPT was used in our NC courses and executed, after translation, through a locally built S-24 post-processor. The S-24 was the testbed for all of our work on NC simulation and automatic verification, and on high-level NC programming through MPL.

Organization

The PAP was loosely organized in the early years, when the roster was small and we did

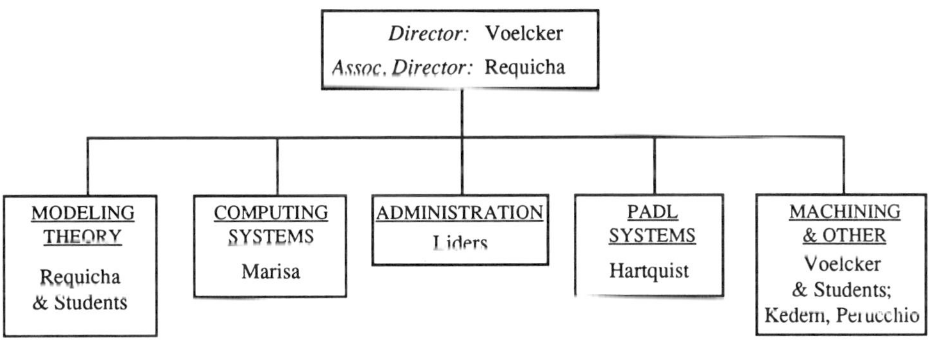

Figure 11.53. Functional organization of the PAP in 1982–86.

exploratory research. The development of PADL-1 as a disseminable system required some structure to get reliable software built, and the PADL-2 Project continued this trend. In the final five years the PAP carried 11–15 people on its roster (not all full-time), and responsibilities were distributed as shown in Figure 11.53.[27]

EPILOGUE

Lessons learned

We learned many lessons, some painfully, about building software, managing people and projects, and tactical and strategic planning. As we look back on the era, three principles stand out:

- It is important to do both theoretical research and experimental system building. They are synergistic, and the exclusive pursuit of either can lead to sterile theory or quirky, opaque systems.
- Teaching is a marvellous spur to codification, and codification imposes discipline on research. Succinctly, if you can't teach it with some degree of elegance to able students, you probably don't understand it.
- The development of 'serious' or 'industrial strength' systems by university groups hinders research, however carefully one tries to shield one activity from the other.

Strategic errors

Our main error, from a 1990 perspective, was the 1981 decision to focus our research mainly on manufacturing, rather than design, as the PADL-2 Project drew to a close. We thought that manufacturing applications would burgeon when viable solid modeling became available, but in fact CAM barely moved in the 1980s while CAD burgeoned. (The reasons for this, we see now, had little to do with technology *per se*.) Of course, there were other errors, such as not moving swiftly in the 1970s to extend the rectilinear domain of PADL-1 to the natural quadrics, and our failure in the 1980s to accept the challenge of supporting some form of free-form modeling in CSG systems.

[27] Kedem departed in 1984 and Perucchio joined it in the same year. Requicha replaced Voelcker as Director in mid-1985.

The present, past and near-term future

Some of the research themes left dangling when the PAP trifurcated are now maturing. One example, of several: the 'Contour-to-CSG' problem has been solved almost definitely and generalized to 3D, to provide the missing B-rep → CSG link needed to build symmetric dual-representation modeling systems [86–88].

In retrospect, the 1970s were a time of great conceptual ferment which produced some rigorous theory and some well identified and challenging 'implementation' problems (domain extension, user interfaces, robustness, and so forth) . . . but only a few applications were addressed satisfactorily in the 1970s. The 1980s saw little conceptual innovation but a lot of work on the problems bequeathed by the 1970s – and these proved to be considerably harder than expected. (Some new problems surfaced in the 1980s, e.g. computational intensity, which was present from the outset but was, by tacit agreement, 'swept under the table'.)

As we enter the 1990s, work continues on the hard problems and there are signs of renewed conceptual vigor. We are likely to find, for example, that new representation schemes[28] and massively parallel computers allow us to finesse some present difficulties and to provide effective solutions for problems now thought to be very difficult. We are also likely to find that some

Figure 11.54. Logotype, by PADL-1 and PADL-2.

[28] It is surprising, at least to us, that no basically new (complete) representation schemes were invented, insofar as we know, in the decade following publication of the six-family taxonomy [25, 26].

applications are simply too ill-understood to automate, no matter how good our modeling facilities. We enter the 1990s with much more 'inertia' than was present in the early 1980s – notably the large investments made by the vendor and user communities in software development, user training, and databases of part definitions. This base of in-place technology and experience is likely to limit severely the adoption of 'radical' new technology unless the new technology is either 'embeddable' in current systems or offers radical advantages.

In conclusion, we note that we hold different views on where the main challenges in solid modeling and programmable automation lie. Requicha thinks they lie mainly in deepening our understanding of information and its transformation, and thus he has cast his lot with computer science. Voelcker believes they lie mainly in understanding more fully the 'physics' of the problems to be solved computationally, and thus he has returned to his roots in mechanical engineering. We remain close friends and occasional collaborators.

ACKNOWLEDGEMENTS

The PAP was supported throughout its life by a sequence of NSF grants that provided substantially all of its funding in the 1970s, and about 50% of its funding in the 1980s. Additional funds and equipment were provided in the 1980s by 15 companies in the PAP's Industrial Associates Program, by software licensing fees, and by the University of Rochester.

We are indebted to everyone associated with the PAP (see the Appendix); each person brought something unique to the Project, and each contributed something unique to its work. Four remarkable people deserve special mention, because without them far less would have been accomplished. They are senior engineers Burns Fisher, Gene Hartquist, and Rich Marisa, and senior NSF'er Bernard Chern, who provided wise counsel at critical junctures.

APPENDIX A PAP Staff, Students, and Visitors

Rochester Faculty Members

Voelcker, Herb (EE)	1972–86
Requicha, Ari (EE: Sr Staff < 1983)	1973–86
Brown, Chris (CS)	1979–81
Genberg, Vic (ME)	1982–83
Kedem, Gershon (CS & EE)	1982–84
Perucchio, Renato (ME)	1984–87
Tourassis, Vassili (EE)	1985–87

Senior Technical Staff

Fisher, Burns	1972–79
Zuckerman, Paul	1972–74
Middleditch, Alan	1972–73
Hartquist, Gene	1974–86
Claytor, Richard	1977–78
Marisa, Rich	1979–86
Hopkins, Mark	1980–82

Junior Technical Staff

McSweeney, Jeff	1977–80
Hunt, Bill	1979–81
Armstrong, Del	1982–87
Parkins, Dave	1983–85
Maruggi, Ed	1984–85

Administrators

Pattison, Pat	1972–76
Pierce, Holly	1976–79
Hudgens, Colleen	1979–80
Rosenberg, Arlene	1980–83
Liders, Gunta	1983–86
Parker, Kathy (senior secretary)	1980–87

Resident Visitors

Nelson, Tom (Gleason Works)	1972–73
Samuel, Nuriel (Xerox)	1974–76
Armstrong, Gavin (U. Leeds)	1977
Boyse, John (GM Research)	1977
Bayer, Gerald (Kodak)	1977
Hoffmann, Peter (Hungarian Academy)	1977
Lee, Robert (Boeing Commercial Aircraft)	1977
Stark, Richard (NM State U)	1978 & 1981
Check, Tom (Calma)	1979–81
Metzger, Jeff (McDonnell Douglas)	1979–81
Peterson, Don (Sandia)	1979–81
Blezurs, Brigita (U. Leeds)	1980
Dodsworth, John (U. Leeds)	1981
Huitric, Hervé (Paris)	1981–82
Nahas, Monique (Paris)	1981–82
Carey, Graham (U. Leeds)	1982
Clark, Alan (Ford)	1982
Hornung, Christoph (T.U. Darmstadt)	1982
Kane, Peter (U. Leeds)	1982
Saia, Tony (U. Leeds)	1982
Goldak, John (Carleton U.)	1983 & 1984
Molari, Gabriele (U. Bologna)	1984
Ansaldi, Silvia (Selenia spa)	1984–85

Laurance, Neal (Ford)	1985
Nau, Dana (U. Maryland)	1985–86

Graduate Students

Shopiro, Jon (Ph.D/CS '79)	1972–74
Tilove, Bob (Ph.D/EE '81)	1975–81
Silva, Cesar (Ph.D/Math '84)	1979–81
Rossignac, Jarek (Ph.D/EE '85)	1980–85
Chan, Stephen (Ph.D/EE '87)	1981–87
Kela, Ajay (Ph.D/ME '87)	1981–86
Vandenbrande, Jan (Ph.D/EE '90)	1983–86
Ang, Marcelo (Ph.D/EE '88)	1985–87
Saxena, Mukul (Ph.D/ME '89)	1985–87

* * *

Birrell, Kirk (MS/EE '76)	1974–76
Hunt, Bill (MS/ME '81)	1976–79
Lee, Y. T. (MS/ME '80)	1978–80
McClellan, Rob (MS/EE '83)	1983
Kohli, Ugal (MS/EE '83)	1983
Sungurtekin, Ugur (MS/EE '84)	1983–86
Harris, Matt (MS/EE '86)	1983–86
Winyard, David (MS/ME '85)	1984–85
Narayana, UVL (MS/ME '85)	1984–85
Fowler, Jim (MS/EE '86)	1984–86
Parratt, Steffen (MS/EE '86)	1984–86
Beatty, Larry (MS/EE '87)	1984–86
Tang, Kit-Ming (MS/EE '87)	1986–87
Khoo, Audrey (MS/EE '87)	1986–87

Undergraduate Students

Moote, Bob	1976–77
Elkind, Steve	1975–76
Miller, Stuart	1977–79
Stevans, Mark	1979–80
Feuerstien, Steve	1979–81
Milner, Rich	1980
England, Ben	1980
Miller, Arnold	1980
Wellington, John	1980
Sawicki, Joe	1980–81
Fiederer, Jens	1980–81
Bresler, Stewart	1980–81
Kain, Brian	1982–83
Kuzawinski, Karla	1981–83
Peck, Jeffrey	1982–83
Grobe, Julie	1983
Durdan, Kathy	1983
Hersh, Jay	1983–85
Bohner, Christine	1983–84
McKinley, Phil	1983–84
Bailey, Dave	1984
Caliva, Tony	1984–85

REFERENCES[29]

1 **A. A. G. Requicha and H. B. Voelcker** Solid modeling: A historical summary and contemporary assessment, *IEEE Computer Graphics and Applications,* Vol 2 No 2, March 1982, 9–24.

2 **A. A. G. Requicha and H. B. Voelcker** Solid modeling: Current status and research directions, *IEEE Computer Graphics and Applications,* Vol 3 No 7, October 1983, 25–37.

3 **G. R. Price** How to speed up invention, *Fortune,* November 1956, 150–228.

4 **H. B. Voelcker** A professional memoir: the first twenty-five years, *IEEE Communications Society Magazine,* Vol 16 No 3, May 1978, 2–4.

5 **A. E. Middleditch** *Survey of numerical controller technology,* Appendix D of P.A.P. Technical Report TR-1, August 1973.

6 **J. E. Shopiro** *Survey of computer graphics,* Appendix A of P.A.P. Technical Report TR-1, January 1974.

7 **H. B. Voelcker *et al.*** *Discrete part manufacturing: theory & practice,* P.A.P. Technical Report TR-1-I, 1974.

8 **H. B. Voelcker** A tale of technology transfer, *IEEE Computer Graphics & Applications,* Vol 2 No 2, March 1982, 70–71.

9 **C. M. Brown** PADL-2: A technical summary, *IEEE Computer Graphics & Applications,* Vol 2 No 2, March 1982, 69–84.

10 **E. E. Hartquist** Public PADL-2, *IEEE Computer Graphics & Applications,* Vol 3 No 7, October 1983, 30–31.

11 **I. C. Braid and C. A. Lang** Computer-aided design of mechanical components with volume building bricks, in J. Hatvany, ed., *Computer Languages for Numerical Control (PROLAMAT '73),* North-Holland, Amsterdam, 1973, 173–184.

12 **N. Okino, Y. Kakazu and H. Kubo** TIPS-1: Technical information processing system for computer-aided design, drawing, and manufacturing, in J. Hatvany, ed., *Computer Languages for Numerical Control (PROLAMAT '73),* North-Holland, Amsterdam, 1973, 141–150.

13 **H. B. Voelcker and A. A. G. Requicha** Geometric modeling of mechanical parts and processes, *IEEE Computer,* Vol 10 No 12, December 1977, 48–57.

14 **A. A. G. Requicha and R. B. Tilove** *Mathematical foundations of constructive solid geometry: General topology of closed regular sets,* P.A.P. Technical Memorandom TM-27a, June 1978.

[29] Production Automation Project and CPA documents may be obtained, for moderate fees, from CPA, 195 Engineering & Theory Center, Cornell University, Ithaca, NY 14853-3801. A complete PAP document list is available. PAP documents are *not* available from the University of Rochester.

15 **R. B. Tilove** *A study of geometric set-membership classification*, MS Thesis, University of Rochester, 1977. (Available as P.A.P. Technical Memorandum TM-30, November 1977.)

16 **R. B. Tilove** Set membership classification: A unified approach to geometric intersection problems, *IEEE Trans. Computers,* Vol 29 No 10, October 1980, 874–883.

17 **A. A. G. Requicha and H. B. Voelcker** Boolean operations in solid modelling: Boundary evaluation and merging algorithms, *Proceedings IEEE,* Vol 3 No 1, January 1985, 30–44.

18 **A. A. G. Requicha** *Mathematical models of rigid solid objects,* P.A.P. Technical Memorandum TM-28, November 1977.

19 **I. C. Braid, R. C. Hillyard and I. A. Stroud** Stepwise construction of polyhedra in geometric modelling, in K. W. Brodie, ed., *Mathematical Methods in Computer Graphics and Design,* Academic Press, London, 1980, 123–141.

20 **M. Mäntylä** A note on the modeling space of Euler operators, *Computer Vision, Graphics, and Image Processing,* Vol 26, April 1984, 45–60.

21 **M. Mäntylä** Boolean operations of 2-manifolds through vertex neighborhood classification, *ACM Transactions on Graphics,* Vol 5 No 1, January 1986, 1–29.

22 **B. G. Baumgart** *Geometric modeling for computer vision,* Report STAN-CS-74-463, Stanford Artificial Intelligence Laboratory, Stanford University, CA, 1974.

23 **K. J. Weiler** *Topological structures for geometric modeling,* PhD Thesis, Rensselaer Polytechnic Institute, Troy, NY, 1986.

24 **F. Arbab** RSC: A calculus of shapes, *Proceedings CAD '84,* Brighton, UK; April 1984, 244–251.

25 **A. A. G. Requicha** Representations of rigid solid objects, in J. Encarnacao, ed., *Computer Aided Design: Lecture Notes in Computer Science No. 89,* Springer-Verlag, New York, 1980, 2–78.

26 **A. A. G. Requicha** Representation for rigid solids: Theory, methods, and systems, *ACM Computing Surveys,* Vol 12 No 4, December 1980, 437–464.

27 **C. E. Silva** *Alternative definitions of faces of boundary representations of solid objects,* P.A.P. Technical Memorandum TM-36, November 1981.

28 **R. B. Tilove** *Exploiting spatial and structural locality in geometric modeling,* PhD Thesis, University of Rochester, 1981. (Available as P.A.P. Technical Memorandum TM-38, October 1981.)

29 **R. B. Tilove** A null-object detection algorithm for constructive solid geometry, *Commun. ACM,* Vol 27 No 7, July 1984, 684–694.

30 **J. R. Rossignac and H. B. Voelcker** Active zones in CSG for accelerating boundary evaluation, redundancy elimination, interference detection, and shading algorithms, *ACM Transactions on Graphics,* Vol 8 No 1, January 1989, 51–87.

31 **J. R. Rossignac and A. A. G. Requicha** Constant-radius blending in solid modeling, *ASME Computers in Mechanical Engineering,* Vol 3 No 1, July 1984, 65–73.

32 **J. R. Rossignac** *Blending and offsetting solid models,* PhD Thesis, University of Rochester, 1985. (Available as P.A.P. Technical Memorandum TM-54, July 1985.)

33 **J. R. Rossignac and A. A. G. Requicha** Offsetting operations in solid modeling, *Computer Aided Geometric Design,* 1986, 129–148.

34 **A. Ricci** A constructive solid geometry for computer graphics, *Computer Journal,* Vol 16 No 2, May 1973, 157–160.

35 **W. B. Fisher, A. A. G. Requicha, N. M. Samuel and H. B. Voelcker** *Part and assembly description languages II: Definition facilities in PADL languages and command facilities in the PADL-1 processor,* P.A.P. Technical Memorandum TM-20a, November 1994.

36 **A. A. G. Requicha** *Part and assembly description languages I: Dimensioning and tolerancing,* P.A.P. Technical Memorandum TM-19, May 1977.

37 *Dimensioning and Tolerancing,* ANSI Y14.5-1973 (superseded by ANSI Y14.5M-1982), American Society of Mechanical Engineers, New York City; 1973 & 1982.

38 **H. B. Voelcker, A. A. G. Requicha, E. E. Hartquist, W. B. Fisher, J. Metzger, R. B. Tilove, N. K. Birrell, W. A. Hunt, G. T. Armstrong, T. F. Check, R. Moote and J. McSweeney** The PADL-1.0/2 System for Defining and Displaying Solid Objects, *Computer Graphics (Proc. SIGGRAPH '78),* Vol 12 No 3, August 1978, 257–263.

39 **A. A. G. Requicha** Toward a theory of geometric tolerancing, *International Journal of Robotics Research,* Vol 2 No 4, Winter 1983, 25–37.

40 **A. A. G. Requicha** Representation of tolerances in solid modeling: Issues and alternative approaches, in J. W. Boyse and M. S. Pickett, eds., *Solid Modeling by Computers,* Plenum Press, New York, 1984, 3–22.

41 **A. A. G. Requicha and S. C. Chan** Representation of geometric features, tolerances and attributes in solid modelers based on constructive geometry, *IEEE Journal of Robotics and Automation,* Vol 2 No 3, September 1986, 156–166.

42 **A. A. G. Requicha and H. B. Voelcker** An introduction to geometric modelling and its applications in mechanical design and production, in J. T. Tou, ed., *Advances in Information Systems Science* (Vol 8), Plenum Press, New York, 1981, 293–328.

43 **R. B. Tilove** Line/polygon classification: A study

of the complexity of geometric computation, *IEEE Computer Graphics and Applications,* Vol 1 No 2, April 1981, 75–88.

44 **A. A. G. Requicha and H. B. Voelcker** *Constructive solid geometry,* P.A.P. Technical Memorandum TM-25, November 1977.

45 **Y. T. Lee and A. A. G. Requicha** Algorithms for computing the volume and other integral properties of solids: II – A family of algorithms based on representation conversion and cellular approximation, *Commun. ACM,* Vol 25 No 9, September 1982, 642–650.

46 **R. B. Tilove, A. A. G. Requicha and M. R. Hopkins** Efficient editing of solid models by exploiting structural and spatial locality, *Computer-Aided Geometric Design,* Vol 1, 1984, 227–239.

47 **D. L. Vossler** Sweep-to-CSG conversion using pattern recognition techniques, *IEEE Computer Graphics and Applications,* Vol 5 No 8, August 1985, 61–68.

48 **T. C. Woo,** Feature extraction by volume decomposition, *Proceedings Conference on CAD/CAM Technology in Mechanical Engineering,* MIT, Cambridge, MA, March 1982.

49 **J. R. Woodward and A. L. Wallis** Graphical input to a boolean solid modeller, *Proceedings CAD '82,* Brighton, UK, 1982, 681–688.

50 **S. B. Tor and A. E. Middleditch** Convex decomposition of simple polygons, *ACM Transactions on Graphics,* Vol 3 No 4, 1984, 244–265.

51 **T. C. Lim** *Derivation of CSG definitions from 2D outline, GM/PADL: 2.5D,* Department of Mechanical Engineering, University of Leeds, UK, April 1980.

52 **D. P. Peterson** Boundary to constructive solid geometry mappings: a focus on 2D issues, *Comput. Aided Design,* Vol 18 No 1, January/February 1986, 3–14.

53 **R. Goldstein and L. Malin** 3-D modeling with the Synthavision system, *Proceedings 1st Annual Conference on Computer Graphics in CAD/CAM Systems,* MIT, Cambridge, MA, April 9–11 1979, 244–247.

54 **S. D. Roth** Ray casting for modeling solids, *Computer Graphics and Image Processing,* Vol 18 No 2, February 1982, 109–144.

55 **J. R. Rossignac and A. A. G. Requicha** Depth-buffering display techniques for constructive solid geometry, *IEEE Computer Graphics and Applications,* Vol 6, September 1986, 29–39.

56 **Y. T. Lee and A. A. G. Requicha** Algorithms for computing the volume and other integral properties of solids: I – Known methods and open issues, *Commun. ACM,* Vol 25 No 9, September 1982, 635–641.

57 **Y. T. Lee and A. A. G. Requicha** *Algorithms for computing the volume and other integral properties of solid objects,* P.A.P. Technical Memorandum TM-35, February 1980.

58 **A. Kela, H. B. Voelcker and J. Goldak** Automatic generation of hierarchical, spatially addressable finite element meshes from CSG representations of solids, *Proceedings International Conference on Accuracy Estimates and Adaptive Refinements in Finite Element Computation (ARFEC),* Lisbon, Portugal, June 1984.

59 **M. A. Yerry and M. S. Shephard** A modified quadtree approach to finite element mesh generation, *IEEE Computer Graphics and Applications,* Vol 3 No 1, January/February 1983, 39–46.

60 **M. A. Yerry and M. S. Shephard** Automatic three-dimensional mesh generation by the modified octree technique, *International Journal for Numerical Methods in Engineering,* Vol 20, 1984, 1965–1990.

61 **A. Kela, R. Perucchio and H. B. Voelcker** Toward automatic finite element analysis, *ASME Computers in Mechanical Engineering,* Vol 5 No 1, July 1986, 57–71.

62 **A. Kela** *Automatic finite element mesh generation and self-adaptive incremental analysis through geometric modeling,* PhD Thesis, University of Rochester, 1987. (Available as P.A.P. Technical Memorandum TM-64, January 1987.)

63 **W. A. Hunt and H. B. Voelcker** *An exploratory study of automatic verification of programs for numerically controlled machine tools,* P.A.P. Technical Memorandum TM-34, July 1982.

64 **H. B. Voelcker and W. A. Hunt** The role of solid modeling in machining-process modeling and NC verification, *Proceedings SAE Int. Congress and Exposition,* Detroit, MI, February, 1981, 1–8.

65 **U. A. Sungurtekin and H. B. Voelcker** Graphical simulation and automatic verification of NC machining programs, *Proceedings IEEE Int. Conference on Robotics & Automation,* San Francisco, CA, April 7–10, 1986, 156–165.

66 **A. R. Grayer** The automatic production of machine components starting from a stored geometric description, in D. McPherson, ed., *Advances in Computer-Aided Manufacture (PROLAMAT '76),* North-Holland, Amsterdam, 1977, 137–152.

67 **G. T. Armstrong,** *A study of automatic generation of non-invasive NC machine paths from geometric models,* PhD Thesis, University of Leeds, UK, April 1983.

68 **G. T. Armstrong, G. C. Carey and A. dePennington** Numerical code generation from geometry modeling system, in J. W. Boyse and M. S. Pickett, eds., *Solid Modeling by Computers,* Plenum Press, New York, 1984, 139–157.

69 **R. A. Wysk, M. M. Barash and C. L. Moodie** *Unit machining operations – Part 2 – APPAS – Automatic process selection and planning program,* Report No. 6 on NSF Grant

Apr74-15256; Purdue University, West Lafayette, IN, July 1977.

70 **H. B. Voelcker** *A preliminary proposal for a machining process language,* P.A.P. Technical Memorandum ITM-24, November 1976.

71 **S. C. Chan** *MPL: a machining process/programming language,* Technical Report CPA87-1, Sibley School of Mechanical Engineering, Cornell University, Ithaca, NY, 1987. (PhD Thesis, University of Rochester, 1987.)

72 **S. C. Chan and H. B. Voelcker** An introduction to MPL – A new machining process/programming language, *Proceedings IEEE Int. Conference on Robots and Automation,* San Francisco, CA, April 7–10, 1986, 333–344. (Re-published by request in K. Rathmill, ed., *Control and Programming in Advanced Manufacturing,* IFS Publications, UK, 1988, 155–182.)

73 **D. S. Nau and T. C. Chang** *Hierarchical representation of problem-solving knowledge in a frame-based process planning system,* P.A.P. Technical Memorandum TM-50, November 1985.

74 **A. A. G. Requicha and J. Vandenbrande** Automated systems for process planning and part programming, in A. Kusiak, ed., *Artificial Intelligence: Implications for CIM,* Springer-Verlag, New York, 1988, 299–326.

75 **H. B. Voelcker *et al.*** CNC machining: Simulation, verification, programming, planning, communication, and control, *Proceedings 13th NSF Grantees Conference on Production Research and Technology,* University of Florida, FL, November 1986.

76 **J. H. Vandenbrande** *Automatic recognition of machinable features in solid models,* PhD Thesis, University of Rochester, May 1990. (Available as IRIS Report No. 260 from the Institute for Robotics and Intelligent Systems, University of Southern California, Los Angeles, CA 90089.)

77 **N. M. Samuel, A. A. G. Requicha and S. A. Elkind** *Methodology and results of an industrial part survey,* P.A.P. Technical Memorandum TM-21, July 1976.

78 **H. B. Voelcker** Modeling in the design process, in W. D. Compton, ed., *Design and Analysis of Integrated Manufacturing Systems,* National Academy Press, Washington, DC, 1988, 167–199.

79 **T. Beyer** *Users' guide to Version 22 of the Oregon FLECS/FORTRAN programming system,* October 1975. (Reprinted as PAP CRDM-19, November 1976.)

80 **C. M. Brown and H. B. Voelcker** The PADL-2 Project, *Proceedings Seventh NSF Conf. Production Research and Technology,* Ithaca, NY, F1-F6, September 1979.

81 **A. L. Clark** Roughing it: realistic surface types and textures in solid modeling, *ASME Computers in Mechanical Engineering (CIME),* Vol 3 No 5, March 1985, 12–16.

82 **Electronic Industries Association** *Interchangeable variable block data format for positioning, contouring, contouring/positioning numerically controlled machines,* ANSI/EIA Standard RS-274-D, Washington, DC, February 1979.

83 **G. Kedem and J. L. Ellis** *Computer structures for curve-solid classification in geometric modeling,* P.A.P. Technical Memorandum TM-51, May 1984.

84 **G. Kedem and J. L. Ellis** *Computer systems for curve-solid classification and solid modelling,* Patent Application, 1984; U.S. Patent 4,649,498 awarded March 10 1987.

85 **G. Kedem and J. L. Ellis** The ray-casting machine, *Proceedings ICCD'84,* October 1984, 533–538.

86 **V. Shapiro and D. L. Vossler** Construction and optimization of CSG representations, *Comput.-Aided Design,* Vol 23 No 1, January/February 1991, 4–20.

87 **V. Shapiro and D. L. Vossler** Efficient CSG representations of planar solids, *ASME Journal of Mechanical Design,* Vol 113, September 1991, 292–305.

88 **V. Shapiro and D. L. Vossler** *B-rep → CSG conversion III: Boundary-based separation,* CPA Technical Report CPA91-6, July 1991.

ABOUT THE AUTHORS

Herbert B. Voelcker holds the Charles Lake Chair in the Sibley School of Mechanical & Aerospace Engineering at Cornell University, and directed Cornell's Manufacturing Engineering & Productivity Program (COMEPP) in 1987–90. He spent 1985–86 at the National Science Foundation, helping to organize the Design, Manufacturing, and Computer Engineering Division and the Directorate for Computer & Information Science & Engineering. Prior to that he was a member of the electrical engineering faculty at the University of Rochester for 24 years. While at Rochester he founded the Production Automation Project in 1972 and served as its Director until 1985.

Voelcker holds SB/ME and SM/EE degrees from MIT, and an engineering PhD from the Imperial College of Science and Technology. He is a Fellow of the IEEE and a member of ASME, ACM, SME, and several honorary societies. He spent seven years in the US Army, has worked or consulted for various industrial and governmental organizations, has won various research prizes, teaching prizes, and fellowships, and has served on various National Research Council and professional-society study, editorial and advisory boards.

Voelcker has worked in several technical fields. In the 1950s, he studied radio propagation, aural perception, and bandwidth compression; in the 1960s, modulation theory and digital signal processing; in the 1970s, computer science and then solid modeling; in the 1980s, machine tools and NC programming systems. He is currently studying mechanical design, dimensional tolerancing and metrology, and a few issues in geometric modeling.

Aristides A. G. Requicha was born in Monte Estoril, Portugal, in 1939. He received the *Engenheiro Electrotécnico* degree from the Instituto Superior Técnico, Lisbon, Portugal, in 1962, and a PhD in electrical engineering from the University of Rochester, Rochester, NY in 1970. He was a high school and college Valedictorian in Portugal.

Requicha is Professor of Computer Science and Electrical Engineering at the University of Southern California, where he directs the Programmable Automation Laboratory, a research unit within USC's Institute for Robotics and Intelligent Systems (IRIS). He joined USC in late 1986, after 13 years at the University of Rochester's Production Automation Project. Requicha was previously a lecturer in physics at the University of Lisbon, a lieutenant in the Portuguese Air Force, and a research scientist with NATO's SACLANTCEN in La Spezia, Italy.

Requicha is a Fellow of the IEEE, a member of AAAI, AAAS, ACM, SME and Sigma Xi. He served as an editor for the *ACM Transactions on Graphics* (1984–1990) and now edits the *IEEE Transactions on Robotics and Automation,* the Academic Press journal *CVGIP – Graphic Models and Image Processing,* Chapman and Hall's *Journal of Design and Manufacturing,* and the Springer-Verlag book series on Computer Graphics. He chaired or participated in numerous conference program committees, and is the author of about 100 scientific papers, reports and book chapters.

His current work is a blend of artificial intelligence and computational geometry, with an emphasis on spatial reasoning and its applications in intelligent systems for design and manufacturing. Specific topics include geometric uncertainty, automatic planning of manufacturing and inspection tasks, and conceptual design for mechatronic systems.

12 Numerical Master Geometry, *n*-sided Patches and Surface Interrogations

Malcolm Sabin

INTRODUCTION

I first got involved in CAD, in particular with its sculptured surfaces aspects, in May 1965. I had served my apprenticeship with what was by then the British Aircraft Corporation (BAC), and had worked for just over a year in flight test instrumentation, when I was offered a job by another company. As BAC had provided a fairly generous scholarship during my undergraduate years, I felt that I ought to look around inside the company first, and was invited to interview for a job whose mandate was to find a way of representing shapes of aircraft numerically, so that the task of lofting could be done more rapidly, and using less floor space.

Not knowing anything about lofting or aircraft design (I had regarded myself as a budding electronic engineer until then), very little about computer programming and nothing about numerical analysis, I accepted the job.

I was not the first to address this question. The particular site at which I was based (Weybridge) was still using manual lofting, in which cross sections of the aircraft in three perpendicular sets of planes, horizontal, parallel to the plane of the symmetry, and perpendicular to the plane of symmetry, were drawn out by hand and cross-faired until they were both consistent and fair. Other sites, however (Filton, Luton and Warton), had picked up the ideas of 'conic lofting', which had been developed by Liming [1] in the late 1930s and by Shelley at Gloster during the war.

CONIC LOFTING

These techniques were based on setting up two-dimensional equations for the curves which would have been drawn by hand, and the manual cross-fairing was replaced by a procedure of systematic generation which guaranteed fairness. The output from the conic lofting was not a set of drawn lines on dimensionally stable material, but a table of coordinates of points every inch or so along the 'calculated profile lines' (CPL). To cover the whole of an aircraft required a rather thick book of tables, but the need for thousands of square feet of loft space was avoided, as were the inaccuracies resulting from measurement and replotting.

The name 'conic lofting' (Figure 12.1) came from the use of second degree curves (conic sections) as the pieces from which the aircraft lines were built. These had the advantages that they had the necessary applicability, they had curvature varying smoothly with arc length, and points that could be evaluated from either coordinate, using only the facilities of desk calculators (which included square roots).

This is but one example of a class of surface definition methods which I now call *directrix-generator techniques,* with many examples differing in greater or lesser detail. The *Master Dimensions* system being developed at Lockheed at the same time used super-ellipses as fuselage cross-sections, a choice of curve for which I can identify no advantages over conics whatever. The DUCT package, developed only a few years later,

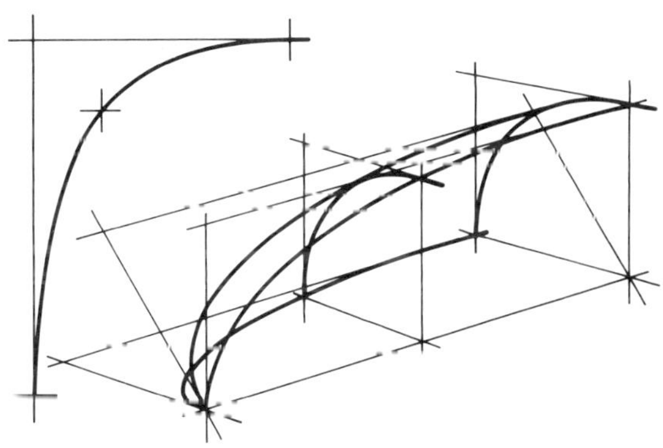

Figure 12.1 Conic lofting.

Fundamental Developments of Computer-Aided Geometric Modeling
ISBN 012-554765-X

started life very similarly, and its many variants explored the space of possible curves fairly thoroughly.

It is amusing in retrospect, though it certainly would not have been at the time, to acknowledge that the surfaces so defined do in fact have many of the properties of the then revolutionary parametric surfaces, and that my own system (see below) can also be viewed as lying in this class.

SPLINES

Also available were a number of more mathematical papers, including those of Birkhoff and Garabedian [2], Coons [3], Ferguson [4] and Schoenberg [5,6]. These introduced two other ideas, those of spline curves and of parametric surfaces.

The spline curve ideas were immediately attractive. The staff for whom I was supposed to be creating facilities were used to the idea of making a smooth curve pass through a set of data points, and this is exactly what the spline offered. Indeed, the name itself was comforting, since the strips of springy wood used for guiding the pen were also called splines. Although I did not actually conceal the fact that the mathematics we used had a few differences from a complete simulation of the shape the wood took up, I did not understand those differences sufficiently to be able to identify them precisely. What mattered was that the mathematics did the same job of passing a smooth curve through empirical points.

The actual mathematics passed through some variants before stabilizing. My first encounter with numerical instability was an attempt to evaluate the spline by working along it span by span, evaluating each slope/curvature pair at the end of a span from the values at its start. The idea was that by taking two solutions, I could discover at the end of the curve what weighted average was needed to satisfy the end condition and then either take that average everywhere, or else march again using the average to give the initial condition.

The failure of this approach was quite spectacular, hitting floating point overflow after only 20 or so spans. After a somewhat half-hearted attempt to rescue the method by backtracking a few spans when this happened, I learned what matrix arithmetic was all about, and why it was sensible. The simple elimination of the spline triband matrix works faster than the method I had invented and is totally stable.

The principle of the spline is very simple. Consider the case where a curve lies close to the x-axis, so that $dy/dx \ll 1$.

The calculus of variation leads to the conclusion that the curve which minimizes the square of the second derivative (the strain energy) consists of cubic pieces, joining with discontinuities of the third derivative at the points where the curve is constrained in y-value. Such points are called *knots*.

At each such point there is a single value of first derivative and a single value of second derivative. Each span which has four degrees of freedom, therefore gives two (linear) equations relating to ordinate, slope and curvature values at its two ends. These equations may be solved simultaneously with an additional two equations (representing end-conditions) over the entire curve to produce the values of the derivatives (Figure 12.2).

In fact, it is better to treat the slopes as independent variables, and the curvature as being defined by those equations. New equations, only one per data point instead of two, are then set up which apply the continuity of curvature condition, and these are solved to give the values of first derivative at the knots.

An alternative, used in my undergraduate text books, was to treat the curvatures as the independent variables, and solve the continuity of slope equations to get their values. We chose the first because it permitted us to constrain specific slopes if we wanted to.

The modern way of looking at these equations would be to postulate that the curve was the sum

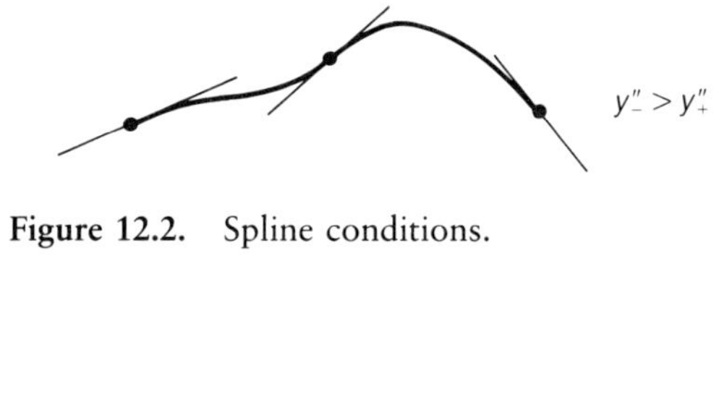

Figure 12.2. Spline conditions.

of a set of B-splines and just solve the interpolation equations to get their magnitudes, but I didn't know that at the time.

I also toyed with splines based on local chords, now known as the Fowler–Wilson spline [7], explored in more detail by Ken Butterfield, a student of Geoff Pitts at Southampton, but rejected all their complexity in favor of the parametric form, in which x, y and z are all spline functions of an auxiliary variable or *parameter*, as soon as I understood the latter.

Once the spline was well understood, I put together a simple spline/conic lofting system design, which used splines for the longitudinal curves and conics for the frames. Based on this design, Chris McDouall and I identified a development plan which covered a man-year or so of work, and made estimates of how long various manual operations would take on the computer about to be ordered. This plan got lost somewhere and the actual development (which happened without lots of hassle from the management) was based on different theory, took a lot longer, and achieved a great deal more.

My early experiments were made using Ferranti Pegasus, a valve computer capable of a few thousand instructions per second. The high level language was an interpreted autocode which ran at somewhat less than ten instructions per second. My first intersection-of-two-surfaces code generated one result point every two minutes, and took two hours one Christmas Eve to produce a set of printed coordinates which were duly plotted after the holiday by the Mould Loft staff.

The new computer was a big multi-programming machine, an ICL 1905, running Fortran (2 and a bit), with a power of a few thousand floating point operations per second and a memory of 96 Kbytes.

NUMERICAL MASTER GEOMETRY

Not long into this job I attended a lecture at NPL (the National Physical Laboratory, only a few miles from Weybridge) given by Charles Lang, and seriously embarrassed the colleague who had given me a lift there, by going up to the speaker after the lecture and monopolizing him for nearly an hour. This occasion led to an extremely fruitful collaboration with Charles' CAD group at Cambridge University, and in particular Robin Forrest, who was also looking at free-form surfaces [8].

When I first visited him, Robin had a program running on the PDP-7 which could display single patch parametric surfaces, and his research was comparing the specification by polynomial coefficients with specification by points, tangent vectors and twists.

Twist vectors were a hard matter then. Some early bicubic formulations just omitted the twist terms completely, assuming their coefficients were zero. It was a great puzzle that the first order Coons form, which interpolated only the edges of a patch, using linear blending functions, gave so much better an answer than the second order form which used cubic blending functions to interpolate both the edges and the cross-derivatives. Other formulations effectively permittted two incompatible twists to be generated by interpolation of cross slope alone each patch boundary [2], using piecewise or rational terms in their equations. I spotted the gap in the zero twist formulation and invented twist vectors, calling them 'cross-curvatures' because the coefficients were mixed second derivatives.

The current thinking was that a cubic polynomial curve segment was described Hermite fashion by its end points P_0 and P_1, and the first derivative factors T_0 and T_1 at the two ends. The actual equation was obtained by multiplying each control vector by a scalar basis function and adding the contributions together

$$P(t) = P_0 f_0(t) + P_1 f_1(t) + T_0 g_0(t) + T_1 g_1(t)$$

where

$$f_1(t) = 3t^2 - 2t^3$$
$$f_0(t) = 1 - f_1(t)$$
$$g_1(t) = t^3 - t^2$$
$$g_0(t) = -g_1(1 - t)$$

These functions were chosen so that the results of evaluating P and dP/dt at the end points ($t = 0, 1$) matched the original data (for details see Chapter 6 in this volume). The equations are repeated here because f_1 will occur again in another context below.

The early Ferguson paper we saw extended this to surfaces by associating in like fashion the products

$$f_i(u)f_j(v)$$

with the corner points, and products

$$f_i(u)g_j(v)$$

and

$$g_i(u)f_j(v)$$

with the tangents at the corner points to the patch edge curves. The entire surface of the patch was thus defined by numerical values at the corner points, and it could be proved that two patches which shared the control values at the ends of a common edge would be continuous in position and first derivative across that common edge.

I asked myself what the combinations

$$g_i(u)g_j(v)$$

should correspond to. The question once posed is easy to answer, the mixed partial derivative

$$\frac{\partial^2 P}{\partial u\,\partial v}$$

and we soon found that the inclusion of these terms in our equations gave much better surfaces in circumstances where twist was required.

Parametric splines and the full bicubic formulation formed an elegant combination, giving very smooth surfaces (continuity of second derivative) (Figure 12.3), combined with the ability to control surface first derivatives when necessary. 'Correct' slopes and twist vectors were calculated by the spline (which by then I regarded as a numerical differentiator) and did not have to be seen by the loftsman at all.

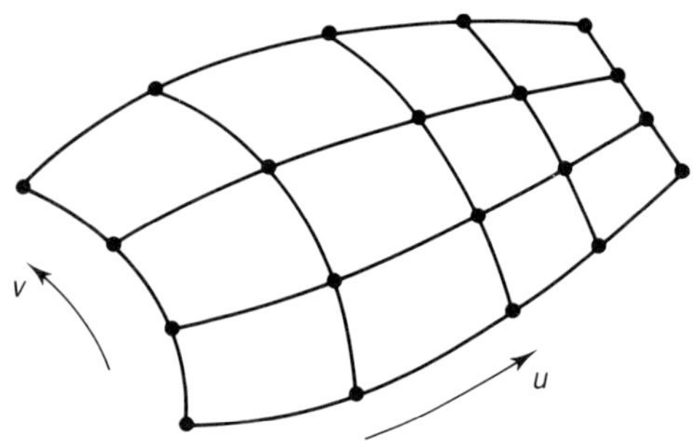

Figure 12.3. Spline carpet.

The values

$$\frac{\partial P}{\partial u}, \quad \frac{\partial P}{\partial v}, \quad \frac{\partial^2 P}{\partial u\,\partial v}$$

necessary to plug into the individual patch equations were derived by the following process:

```
for each row of points in the
increasing u direction
do
  compute values of ∂P/∂u from
  values of P
od
for each row of points inthe
increasing v direction
do
  compute values of ∂P/∂v from
  values of P
  compute values of ∂²P/∂u∂v from
  values of ∂P/∂u
od
```

Although this computation is not symmetric in u and v, the other formulation, which computes the mixed partial derivative from the values of v directional partial derivatives gives exactly the same answer.

I discovered a few years later (Robin Forrest found the reference for me) that Inaba had published a scalar-valued version of the same thing, and much later (Summer 1987) that Carl de Boor had published a vector-valued version in 1962 [9].

The Numerical Master Geometry was built and put into experimental operation using this surface form late in 1966. By mid-1967 it was being heavily used on real work, the fin of the production version of the Concorde supersonic airliner.

The system was split into three main programs (Figure 12.4), as we were constrained to run in a 36 Kbyte partition in order to allow effective multiprogramming, and these programs communicated by files on magnetic tape.

The first program took textual descriptions of surfaces to be defined, and stored the patch corner values in an array on tape. It also had facilities for copying other surfaces from an input

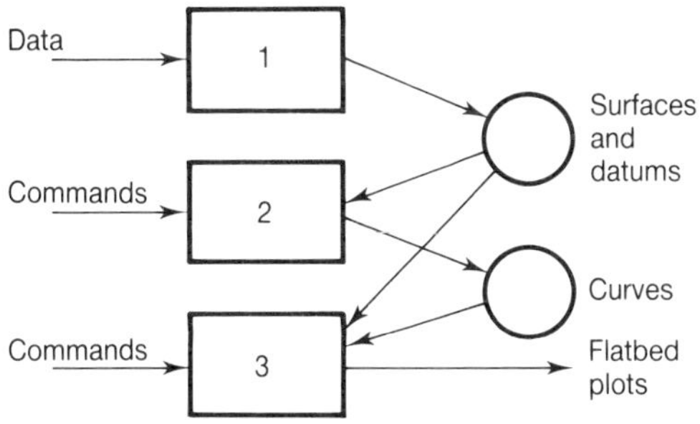

Figure 12.4. NMG structure.

tape, so that the surfaces of a particular project could be kept together. There were similar possibilities for definition and maintenance of coordinate systems, treated as first-class geometric objects from the start, called 'datums', following the current design office practice.

The second program calculated cross-sections and intersections, storing the resulting curves on its output tape. We did not hold planes as items in their own right. When a plane section was required it was assumed to be perpendicular to the x-axis of a predefined datum, and so all that was necessary was to specify the name of the datum and the distance along the x-axis from the origin to the plane.

Often the requirement was for a set of sections with many parallel planes (a 'nest' of sections), and the specification of a datum name and the initial distance, the inter-plane distance and the number of sections proved eminently practical.

Curves so calculated consisted of arrays of points, with various options for what was stored at each point. The coordinates themselves, of course, but also the curve tangent, the surface normal (possibly two) and the surface parameter values (possibly two pairs) could be stored.

A nest of sections was typically stored as a consecutive set of curves all with the same name, and we applied the convention that such a set would be the unit for the programs based on curves, rather than the individual curve.

Although iso-parametric lines were not particularly valuable for surface evaluation processes, we provided a facility for calculating any straight line in the parameter plane as an alternative curve construction.

The third program related views of stored curves, issuing them in the form of control tapes for a large 12 feet by 5 feet flatbed plotter. We had no interactive graphics at the time, and indeed it would have been of little use, since the surface flaws which were found were normally so fine that the resolution and size of the plotter were essential in finding them. For production purposes, of course, the flatbed hard-copy was what we were trying to make.

Mathematicians were often puzzled by the way our plots always plotted z against minus y. This seemingly odd combination was chosen as providing what was required. The loftsman would select a datum for calculation of his sections, and would normally want to view it square, looking along the x-axis, and this convention meant that he would simply use the same datum to position his 'camera' as he did for the calculation itself.

As well as the main programs, we also had utilities for printing the stored entities. The tape formats were designed so that the lowest level utilities, which we used for debugging rather than production work, would operate on any tape, containing any mix of the entities with which the system dealt. Every logical block contained three pieces, a control block of fixed length, which described the data to follow, and in particular the length of the other two pieces, an integer data block and a real data block. The real data block contained a number of scalars, a number of vectors, and a number of points. By knowing how many of each were present, changes of coordinate system could be applied to any entity by a standard dumb utility.

This simple software structure, with very few (but well defined) entities (surface, curve and datum) and functions invocable by the operator for constructing them, stood up well to subsequent enhancement. We added patch subdivision, offset surfaces (see below), panel development, cylindrical intersections, silhouette and reflection curves, clipping of curves by planes and surfaces, reinterpolation of curves at a nest of planes, and output to numerically controlled (NC) machine tools. Every new facility could be used in combination with all those already implemented.

If building such a system today, I would remove the artificial distinction placed upon us by memory constraints and build independent programs for each calculation. Each entity, be it surface, curve, datum, or some new object worth storing by name, would be put into a separate file in the project directory. Only when all the substantive code of the first release was fully tested would I consider wrapping a command processor round the whole facility.

N-SIDED PATCHES

One problem to which I bent my mind from the outset was the problem of what to do at wing roots (Figure 12.5). The natural parametrizations of the fuselage and of the wing did not match up if a smooth fillet was required. Drawing sample such curves left a five sided hole near the leading edge.

I had to find a way of filling it, and the search did not take long. By mid-1967 I had a five sided patch which was controlled in the same way as the Hermite bicubics, and which could meet adjacent bicubics with continuity of first derivative. Finding the pentagon within a few months of starting looking now seems all the more amazing as I have been looking for a hexagon ever since, without success.[1]

This five sided patch theory used the idea of representing the position of a point on the patch as a function of five parametric variables, each of which was zero on its own edge of the patch, and unity on the opposite edges (Figure 12.6).

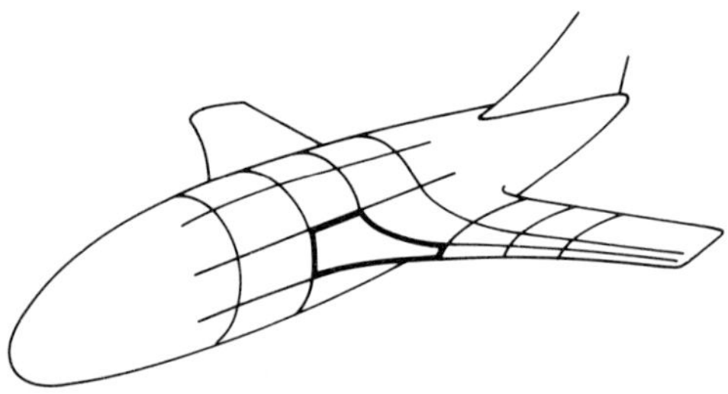

Figure 12.5. Wing root.

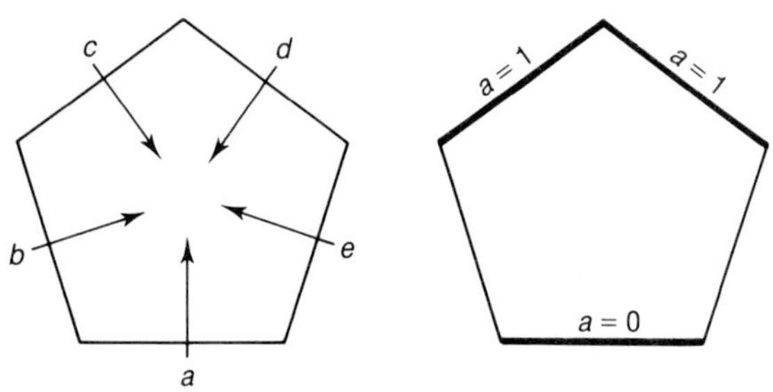

Figure 12.6. Five sided patch.

[1] The original paper includes an assertion that certain equations provide a hexagon. I re-examined these late in 1989, using a computer algebra package, and discovered that they do indeed work. This result will be published in the proceedings of the fourth IMA conference on *The Mathematics of Surfaces.*

These five values were linked together by three equations so that the patch was a mapping of a bivariate manifold in 5-space.

Suppose these five parameters are called a, b, c, d and e. The three linking equations were

$$c = \frac{1-a}{1-ab} \quad d = \frac{1-b}{1-bc} \quad e = \frac{1-c}{1-cd}$$

Because these are rational, all five parameters, and thence the coordinates of the corresponding point, can be computed without approximation or ambiguity from a and b. These equations have the surprising, but necessary property that from them can be dervied the similar looking forms which make the whole patch cyclically symmetric in formulation:

$$a = \frac{1-d}{1-de} \quad b = \frac{1-e}{1-ea}$$

and the mirror images

$$a = \frac{1-c}{1-bc} \quad b = \frac{1-d}{1-cd} \quad c = \frac{1-e}{1-de}$$

$$d = \frac{1-a}{1-ea} \quad e = \frac{1-b}{1-ab}$$

The 1968 pentagon was actually a little more complicated than this, as it used the solution of a cubic function (known to have exactly one root in (0, 1), thus not introducing any ambiguity) to achieve its slope continuity properties.

The parameters a, b, c, d and e from above were each related to another variable A, B, C, D and E with the same limit values, but related by the equations

$$a = f_1(A), \quad b = f_1(B), \quad \text{etc.}$$

where f_1 is the Hermite cubic defined earlier. This transformation was introduced so that there would be a unique cross-tangent derivative which could be matched to an adjacent four sided patch, i.e.

$$\left.\frac{\partial P}{\partial A}\right|_{E=const.} = \left.\frac{\partial P}{\partial A}\right|_{B=const.}$$

at $A = 0$. Once this had been achieved, just using the obvious association of products of f and g

functions with the corner slopes and twists gave an almost correct formulation, which it was easy to polish into shape.

Fifteen years later, in 1983 [10], I used the same linking equations to formulate a much more economical patch, which derived its cross tangent from

$$\left.\frac{\partial P}{\partial a}\right|_{e = const.}$$

and

$$\left.\frac{\partial P}{\partial a}\right|_{b = const.}$$

without forcing them into equality.

I well remember going, at Charles Lang's suggestion, to visit Malcolm Wood and Stanley Matthews at Ford at Laindon. I expected them to have done very clever things, and therefore went armed with my five sided patch pictures as the cleverest thing I could reply with. I was a little disappointed that there was no contest, but the visit started a lasting friendship with Malcolm Wood.

In fact, we never used the pentagon within NMG, because it did not fit the carpet concept which gave the values of the tangent and twist vectors. It was used in automotive systems, at Nissan and, I believe, at Fiat, whose patch oriented systems were more capable of accepting it.

OFFSET SURFACES

There were other pieces of theory which were less dramatic, but more important to operational use of the development. Foremost of these was the result that a surface defined as an offset from a parametric surface is itself a parametric surface. The theory goes like this: define the term *parametric surface* to mean a surface for which any point P, say, may be computed from the values of u and v. By differentiating the code of the evaluation routine we get the code for the evaluation of the partial derivatives. The cross product of these two vectors lies perpendicular to the surface, and may be normalized (divided by its length) to give a unit surface normal vector N. A point Q, say, on the surface offset by a constant d, say, may therefore be computed by

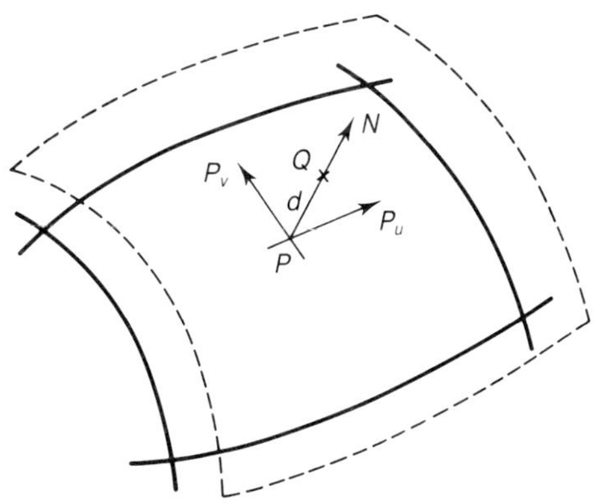

Figure 12.7. Offset construction.

the vector assignment (Figure 12.7)

$$Q = P + d \cdot N$$

Q may be computed by a chain of computation starting only from the values of u and v, and therefore lies in a parametric surface.

The same proof principle still applies when the offset amount d is not constant, but depends on the position of P, or when the direction of offset is not perpendicular but a more complicated function on the surface normal, such as is found when computing tool center points by efficient milling cutters.

By including this chain of computation in the evaluation of every point sampled, precision of offsetting is guaranteed, at the price of slower computation. This was the matter of some debate at the time. The builders of other systems, for example, Arthur Flutter, decided to implement offsetting by the construction of an explicit approximate offset surface which had the same form of equation as the base surface but different coefficients. Either way, the concept that one could drive a cutter along a path such that it just touched a surface to be machined, by driving the cutter center along an offset surface made the machining of surface intersections much simpler than attempting to construct the intersection curve first and then apply the offsets.

This result led to the use of the evaluator concept in the NMG, which isolated the surface definition from the software which carried out such interrogations as the cutting of plane sections and the calculation of surface intersections. Nowadays we would regard this is an example of an abstract data type.

This concept was used later by the VDI standard makers as providing a unifying concept for linking CAD systems with differently detailed surface definitions. The version of the NMG used at the Warton site used this very mechanism to offer Alan Ball's CONSURF surface as an alternative to the bicubic spline carpet [11–13].

INTERROGATION TECHNIQUES

Much of our effort was put into interrogation techniques. It was quite clear that the images which most other surface experiments used, of iso-parametric lines, were totally useless as a display for the loftsmen. Loftsmen had always used nests of closely spaced plane sections, and nests of plane sections we had to give them, to enable them to judge whether a surface was adequate.

I was always saddened at the reluctance of most of my academic colleagues to treat interrogations as seriously as surface definitions. The honorable exceptions were Peter Payne [14, 15], who wrote the lattice method plane section algorithm for Andrew Armit's MULTIPATCH system, and Adrian Geisow [16, 17], of whom more below.

We used the simple marching concept, previously described by South and Kelly [18], whereby from a simple starting point, found by seeking a point where the boundary of the surface crossed the intersection plane, a step in the direction of the intersection would be taken, resolved back into parametric coordinates, evaluated and iterated on to the true intersection by use of Newton's method. Because this sometimes failed to converge we wrapped a lot of recovery code round the basic inner loop, to give a system more than adequately robust for its intended use, though one would have to do better in today's solid modeling environment.

For some years we found the boundary search to be quite adequate for practical purposes, but later added the continuation of the search round the boundary so that multiple open pieces could be found, and an interior scan which could (probably) find a single interior loop if there were no open intersection pieces.

The basic method works like this. Suppose that we have a point P_0 in the curve of intersection of the parametric surface $P = P(u, v)$

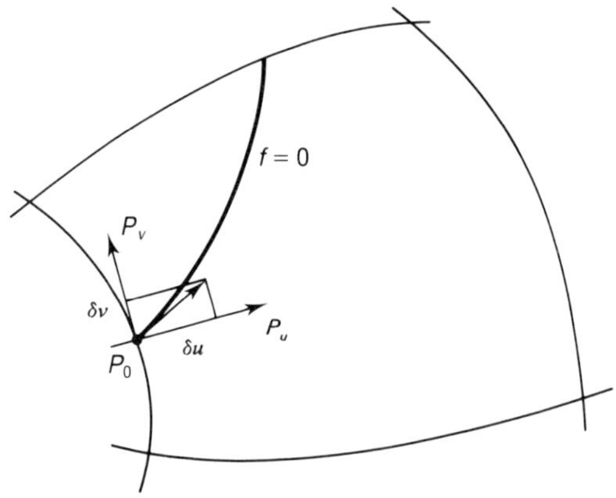

Figure 12.8. Major step.

and the potential surface $f(P) = 0$, that is

$$P_0 = P(u_0, v_0), \quad f(P_0) = 0$$

Let

$$\frac{\partial f}{\partial P} = \left(\frac{\partial f}{\partial x}, \frac{\partial f}{\partial y}, \frac{\partial f}{\partial z}\right) = N$$

Postulate a movement du, dv in the parameter plane. Then (Figure 12.8)

$$\mathrm{d}f = \frac{\partial f}{\partial P}\,\mathrm{d}P = N\left(\frac{\partial P_0}{\partial u}\,\mathrm{d}u + \frac{\partial P_0}{\partial v}\,\mathrm{d}v\right) = 0$$

for the step to lie along the intersection curve. Thus du and dv in the ratio of

$$-N\frac{\partial P}{\partial v}$$

to

$$N\frac{\partial P}{\partial u}$$

will give a step tangential to the intersection line. Knowing this ratio, we can determine the magnitude to give a step of the required size:

$$u_1 = u_0 + \mathrm{d}u, \quad v_1 = v_0 + \mathrm{d}v$$

This gives a starting point for iteration toward a point really on the intersection.

Because the non-zero higher derivatives in the surface expression bend the intersection curve, when $f(P(u_1, v_1))$ is actually evaluated, it has non-zero value f_1, say.

Near u_1, v_1 we can linearize in terms of $\partial P_1/\partial u$ and $\partial P_1/\partial v$, giving an equation which must be satisfied by the next step in u and v, to give a point lying on the intersection curve

$$f = f_1 + N_1 \frac{\partial P_1}{\partial u} du + N_1 \frac{\partial P_1}{\partial v} dv = 0$$

We also have a requirement on the step length, which maybe expressed as a step length error function e, evaluable at P_1 and giving a similar equation to that for f

$$e = e_1 + \left(\frac{\partial e}{\partial P_1}\frac{\partial P_1}{\partial u}\right) du + \left(\frac{\partial e}{\partial P_1}\frac{\partial P_1}{\partial v}\right) dv = 0$$

These two equations are solved simultaneously for du and dv to give a step to a point better approximating the $e = f = 0$ condition (Figure 12.9)

$$\begin{pmatrix} u_{i+1} \\ v_{i+1} \end{pmatrix} = \begin{pmatrix} u_i \\ v_i \end{pmatrix} - \begin{pmatrix} \partial f_i/\partial u & \partial f_i/\partial v \\ \partial e_i/\partial u & \partial e_i/\partial v \end{pmatrix}^{-1} \begin{pmatrix} f_i \\ e_i \end{pmatrix}$$

Under normal circumstances the vectors $\partial f/\partial P$ and $\partial e/\partial P$ are almost perpendicular in 3-space, and well away from parallel in uv-space, and so no problems of ill-conditioning are found in this solution. Again, if the steps being taken are small compared with the local radii of curvature of the surfaces, the initial estimate is well within the region of convergence of Newton's method, and a very small number of steps (two or three) give convergence within engineering tolerances.

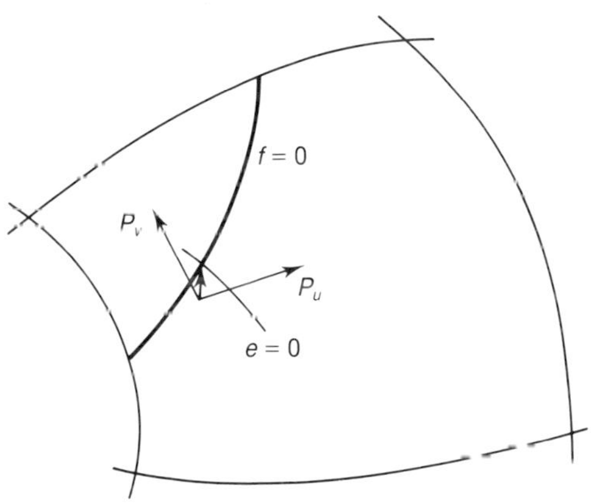

Figure 12.9. Minor step.

The above was initially implemented for plane sections with the function f being given by

$$f = (P - Q) \cdot N$$

where P is the point being tested, Q is a point on the plane and N is the vector normal to the plane. However, when a requirement was raised to calculate the intersections of surfaces with cylinders, we were able rapidly to modify the code to handle the more general implicit equation.

The carpet concept, in which an entire array of patches was treated as the unit of surface, was complemented by a desire in my mind for symmetry of treatment of the parametric variables. A major advantage of the parametric surface approach over the conic lofting methods was perceived then to be their axis-independence, and I found concepts like *calculate the intersection of each iso-u line with the plane in turn* most unesthetic. My methods must have nothing so arbitrary, and so instead we used step length criteria directly related to what the loftsman wanted to do, and to the accuracy required in the output. These related the distance between the points evaluated on the curve to the local curvature dynamically as the curve was traced out.

Experiments with a large number of draughtsmen had indicated that the ideal step length for the approximation of a real curve by straight line segments, was that which gave a chord height error of 0.002 inches. Using this as an error measure was possible but rather complicated and not totally reliable. We therefore offered a number of options for the step length function (Figure 12.10). The simplest gave steps of constant size

$$e = (P - P_0) \cdot (P - P_0) - p^2$$

where p is the required pitch. An alternative gave steps whose initial and final tangents made a constant angle, and the third used a linear combination of these two so that on very shallow curves the step length reached an upper bound. The ratio and sizes of the upper bounds were chosen to give effectively constant chord height error over the range of curvatures found in the curve being evaluated (Figure 12.10).

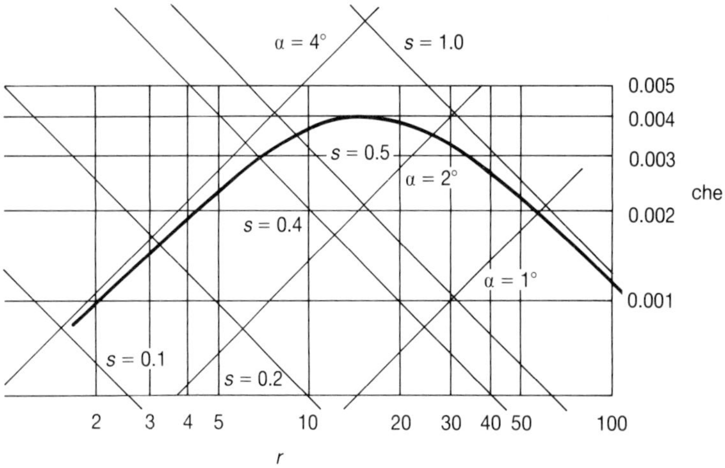

Figure 12.10. Step error functions.

DOCUMENTATION

1968 and 1969 saw a lot of documentation written of the methods we had used. Some of these documents were rather speculative, and I am not sure that everything I wrote was totally justified. However, many of the ideas which I did not test at the time have surfaced again, properly proven, since then.

SURFACE MACHINING

My responsibilities during these years included the support of the software which translated N/C part programs written in the Ferranti Profiledata language into machine tool control tapes, and also with the Wind Tunnel's use of N/C machines for the manufacture of wind tunnel models. Although the wind tunnel model makers had used the Ferranti program, PMT2, to generate their tapes, the success of the NMG in the lofting context made it natural to try to link the two, and by 1970 surfaces defined within NMG were being machined to high precision.

It was a high point when, because of the way we had implemented offsets, the first time we drove a complicated cutter along the intersection of two surfaces, it worked first time.

CONFERENCES

These achievements were declared outside the coterie of Charles Lang's group and its contacts for the first time in 1970, at the CG70 graphics conference, organized by Mike Pitteway and his colleagues at Brunel University. I was highly delighted to hear after my own talk that a delegate from McDonnell Douglas had felt himself upstaged by what I had been able to show. His talk was the day after mine. This conference gave my first opportunity to meet Steve Coons and Bert Herzog.

The following year the Royal Society organized a meeting on CAD, at which I was able to meet Even Mehlum and Pierre Bézier, and in 1972 the *CAD Journal* ran a conference at Cambridge, at which we used the NMG to modify the surface and cut a mold for the new shape within 12 hours (Figure 12.11).

BÉZIER SURFACES

I wrote a document in 1968 [18], in response to one [19] in which Robin Forrest had suggested interpolation of four interior points of a surface patch as a way of controlling the twist vectors. In my paper I described a 16-point form which allowed a bicubic patch to be specified by the positions of the 16 points which had very specific geometric relations to the shape of the patch. Robin was not convinced that this was useful, though Andrew Armit understood it very quickly. However, within a few months Robin heard and understood the same formulation from Pierre Bézier, who had invented it (and a lot

Figure 12.11. Jug photo.

more detail that I had not thought to explore) several years before.

On a subsequent visit to Japan, in 1974, I met Mamoru Hosaka, who had also come to the same ideas independently, and developed the same theory in a completely different notation.

B-SPLINES

Some time in 1971 Robin Forrest alerted me to some new theory coming out of Syracuse. Rich Riesenfeld was working on a new form of curve which had all the nice Bézier properties, but could also represent closed curves without any arbitrary sewing up of a seam [20].

When I eventually found out what this was all about, I was very cross with myself, because I had described exactly the functions which Rich was using, and had recommended them, under the name *finite width modes* for least square curve fitting, but had not spotted that they would be useful as a basis for design (in fact, Schoenberg had described them long before me, and his name *B-splines*, is the one which has stuck).

LATER DEVELOPMENT

During the early 1970s my writing turned to more speculative topics. I had been elected chairman of the CAD Specialist Group of the British Computer Society, and decided to use my term of office to run workshops where CAD professionals could exchange knowhow of how to build systems. This meant doing a bit of thinking myself about principles, and writing things down helps to sort one's ideas out.

I looked for untapped previously developed theory in what in hindsight was a rather obvious place, University Library, Cambridge, and found the algebraic geometry of about a century ago. Much to my delight, this subject has been picked up and popularized by Tom Sederberg. It is now becoming recognized as a rich lode, and Shriram Abhyankar and others at Purdue are well placed to mine its riches for the rest of us.

BÉZIER TRIANGLES

I had begun to seek more flexibility in the setting up of a shape than the regular spline carpet permitted, and I looked for ways of breaking the strict rectangular mold. One result was a theory of Bézier triangles, which I was later able to put together into a constructive theory of B-splines on a regular triangular lattice [21].

One way of looking at the univariate Bézier basis functions is to regard them as functions of homogeneous coordinates on a line. Start from the linking equation $s + t = 1$, and raise it to the nth power. The binomial terms are exactly the Bernstein basis functions, complete with numerical coefficients.

Similarly, we can start from the linking equation

$$u + v + w = 1$$

and raise it to the nth power

$$(u + v + w)^n = 1$$

The terms of the expansion of the left hand side may be laid out in a tableau (Figure 12.12).

Each term in this tableau becomes a basis function with which a vector coefficient is associated in a surface description. The position of the maximum value in the $u, v, w > 0$ triangle of each basis function corresponds exactly to the term's position in the tableau.

The value of the surface along an edge, $u = 0$ say, depends only on the coefficients of the terms of the corresponding edge of the tableau, and the value of the derivative across that edge depends only on the coefficients of those terms and the adjacent row, and so on.

Sufficient conditions for continuity across an edge can therefore be expressed in terms of the coefficients close to that edge of the combined tableau. Suppose that the following represents a part of the tableau of coefficients of two adjacent triangular patches

```
        |
        A
    B   |   C
D       E       F
    G   |   H
        J
        |
```

In the case the two triangles face each other squarely, and there is no skew, sufficient conditions for slope continuity are

...

$$B + C = A + E$$

$$G + H = E + J$$

...

$$u^n + n\,u^{n-1}v + \cdots + n\,uv^{n-1} + v^n$$

$$+ n\,u^{n-1}w + \cdots + n\,v^{n-1}w$$

$$+ \frac{n!\,u^i v^j w^k}{i!\,j!\,k!}$$

$$+ n\,uw^{n-1} + n\,v\,w^{n-1}$$

$$+ w^n$$

Figure 12.12. Triangular tableau.

and for second derivative continuity are

...

$$C + D + H = F + G + B$$

...

Sufficient conditions for more general cases have been since described in more detail by Gerald Farin [22].

These triangles may be joined together, six at each vertex, in a regular three-way grid, and I generated a family of B-spline-like basis functions of increasing continuity over these grids by applying a convolution operator (shift, subtract and integrate) in three directions [21]. Some years later, Paul Sablonnierre [23] showed me some papers by Frederickson [24] which predate my work, describing the same functions, which are now best described as Box splines, though nobody knew that in the early 1970s. Mamoru Hosaka had also invented them independently.

However, without benefit of powerful operators I also identified functions which used the additional freedom available in triangles whose boundaries are of lower order than the interior surface, to achieve slope continuity over domains in which three, four or five triangles meet at each vertex.

Subsequently I also used the modified linking equation

$$u + v + w - 2uvw = 1$$

to achieve the same kinds of effect, a quadratic suitcase-corner patch, which achieved slope continuity with adjacent triangles, rectangles or pentagons [10]. Order raising would enable higher orders to be defined, but these do not open the way to higher continuity.

I also explored the explicit curve representation (curvature is a function of distance along the curve), looking for equivalent surface forms. I tried out arc length being a function of curvature, and arc length being a function of tangent angle. What I found that worked was the pedal equation, in which the distance from the origin to a tangent plane is a function of the orientation of the normal to that plane. This representation has the interesting property that it is closed under generalized offsetting if the actual functions used are closed under addition, although this condition gives a subset of the rational polynomials. It also has the undesirable property that the surface can easily have cusps in it, and these two properties are intimately linked.

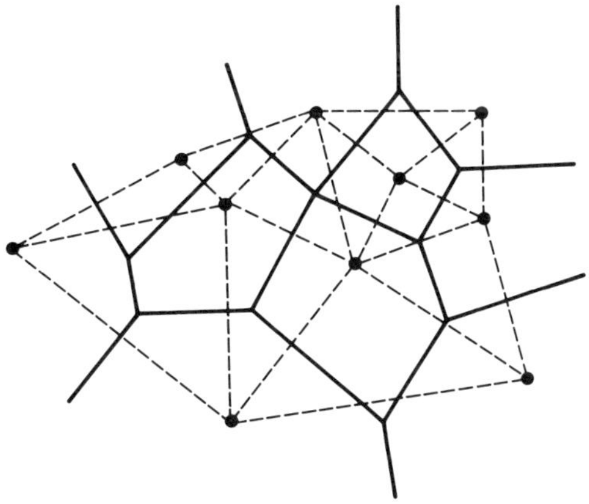

Figure 12.13. Voronoi regions and Delaunay triangulation.

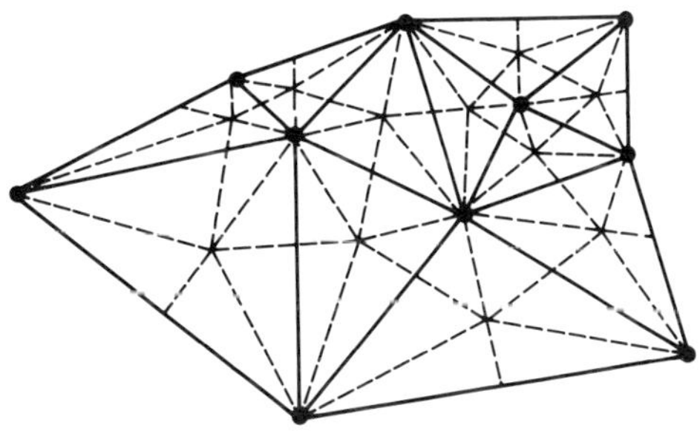

Figure 12.14. Piecewise quadratic triangles.

SCATTERED DATA INTERPOLATION

A side-interest, partly fuelled by the topology problem, was scattered data bivariate interpolation. I attended a course at Harwell, organized by Mike Powell, at which David Newton gave an excellent presentation on this topic. He took the line of explaining each method in turn very plausibly, and only then showing us the data pattern for which it behaved badly. I found that after three or four of these I could think ahead and predict what pitfalls he was going to produce. This encouraged me to set my mind to thinking up a method which would avoid them.

Over the next few days I used my spare time to build a simple contouring package based on what is now known as Delaunay triangulation. Although Robin Sibson's subsequent implementation [25] was a great deal more economical of memory than mine, my version was sold to Applied Research of Cambridge, and as far as I know still forms part of their terrain model software.

The concept on which I built was that each point should have a region nearby in which it had an influence, and that by being explicit about where the boundaries between these regions of influence lay, one might be able to avoid the problems which many methods had.

The obvious primary region of influence was the Voronoi region (Figure 12.13) of the data point (that part of the plane nearer to the data point than to any other), and simple use of such regions would give a plateau round each data point. The secondary region could be gained by building triangles from each data point to its neighbors in the Voronoi sense, which would give a piecewise linear value-continuous surface.

Rather than extending these regions to achieve higher continuity, I looked instead for slope continuous functions defined over this triangulation (Figure 12.14). This problem led to cooperation with Mike Powell, and in due course to a joint paper [26].

MORE *n*-SIDED PATCHES

The *n*-sided patch problem has never become a really fashionable topic of study, but a few others have applied themselves to it. Bob Barnhill's group [27] created a number of transfinite triangles, and this approach was continued to n-sided regions by John Gregory and Peter Charrot [28]. The rules of the transfinite game are that slope continuity is achieved by predefining the surface normal along every patch boundary and making the patch fit. It is therefore a Hermite approach, which may be contrasted with the B-spline approach of finding functions of larger support and predefined continuity.

Tamás Várady [29] has recently surveyed the n-sided patch literature, and has introduced the concept of the overlap patch. I hope and believe that this approach is capable of relating my own methods to those of others, and look forward to this comparison being made.

Tony De Rose and Charles Loop [30, 31] have constructed an algorithm stimulated by the blossoming work of Lyle Ramshaw [32], but their approach requires that adjacent patches share a common domain. It is therefore aimed more at the scattered data problem than at the topology problem.

RECURSIVE DIVISION SURFACES

One approach to the topology problem which raised considerable hopes, but which does not seem to have matured to a useful state, is that of general recursive subdivision. This was explored by Ed Catmull and Jim Clark [33], and by Daniel Doo and myself [34]. Shape Data built an experimental modeler, Remus, which did not sell at all as well as its big brother.

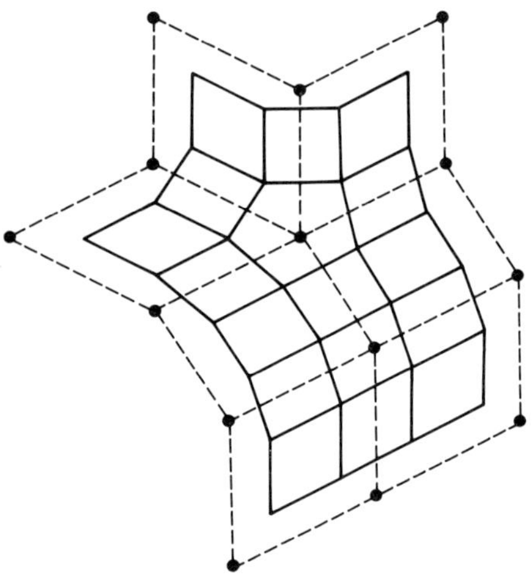

Figure 12.15. Quadratic recursive subdivision.

The principle on which this approach is founded is that a B-spline curve has a control polygon. If extra knots are introduced into the curve without changing its shape, extra vertices are added to the control polygon, which are specific positive linear combinations of the original converges. The new polygon lies between the old one and the curve, and in the limit converges to the curve itself.

In the bivariate case, repeated knot-like insertion causes the control net to converge to the surface. Each stage in, for example, a doubling of knot-line density causes new vertices to be created which are joined in a way easily described as a construction from the original control net vertices.

It is easy to generalize such a construction in such a way that it can be proved that even when the original network is not regular, and does not therefore have a B-spline control network interpretation, successive application of the construction will cause the net of control points to converge to a limit, and, if the generalization of the construction is a good one, slope continuity properties can be proved of this limit surface without having a closed algebraic form at all (Figure 12.15).

Unfortunately, no formulations have yet been published which give continuity of curvature, though many have continuity of slope.[2] Some have continuity of slope, but with unboundedly large curvature at a few specific points of a surface, so the analysis is not particularly simple.

The quadratic construction gives the only algorithm so far discovered which is totally well behaved at the singular points. Unfortunately, it gives surface shapes which are decidedly bulgy towards the control points, and is therefore unacceptable esthetically. Sad. At one time I viewed recursive subdivision surfaces as the next step in the mainstream development (Figure 12.16), but they will need to be extended to include the unequal-interval B-splines as well as the regular ones before any solution to the esthetics problem would make them a contender again.

[2] I later discovered [35] that the cubic can be made to give bounded curvature, but not quite curvature continuity.

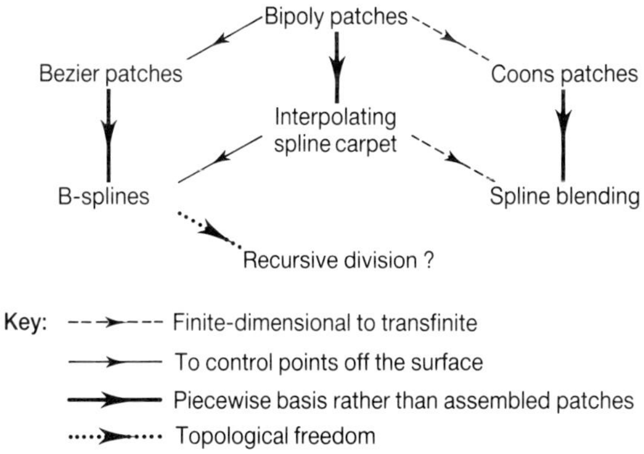

Figure 12.16. Twenty mainstream surfaces.

MORE INTERSECTION ALGORITHMS

A major advance was made in the theory of intersections when Adrian Geisow's thesis became available [16]. In it he studied a number of interrogation techniques, comparing their speed and the robustness. Robustness is far more important in the context of today's geometric modelers, than in the NMG context. There each intersection was individually requested, and in the case of occasional failure could be re-requested with different step length parameters. A geometric modeler may have to compute many intersection curves for a single interaction, and there is no way of inserting extra information to provide the algorithm with assistance.

Of the techniques Geisow described, the most promising for general purposes is recursive subdivision, in which the convex hull property is used to identify whether a pair of pieces of surface may intersect. If they cannot intersect, because the convex hull of their control nets do not overlap, the pair needs no intersection calculation. If they may intersect, either subdivision takes place, giving a bunch of simpler problems, or else the problem is simple enough already for a direct method to give a piece of the required curve (see Chapter 5 in this volume for more details).

More recently, two other approaches have been put forward. One is the computation of a parametric approximation to the intersection curve in the parameter plane. This would have been thought impractical, since the order may be very high, which should lead to the curve being of high genus. However, certain surfaces of lower order than the bicubic have intersection curves which are of low enough genus to consider parametrizing them with expressions which involve the solution of an algebraic equation of medium order for each point evaluated. The rest of the computation would just be a rote plugging in of successive parameter values and evaluating rational functions to get the coordinates. The other is the acceptance of solution procedures working in many dimensions.

Both of these are still in their early stages of development. There is still plenty of excitement left to come.

REFERENCES

1 **R. A. Liming** *Practical Analytic Geometry with Applications to Aircraft,* Macmillan, 1944.
2 **G. Birkhoff and H. L. Garabedian** Smooth surface interpolation, *J. Maths. Physics,* Vol 39, 1960, 258–268.
3 **S. A. Coons** *Surfaces for computer-aided design of space forms,* MIT Project MAC-TR-41, June 1967.
4 **J. C. Ferguson** *Multivariate curve interpolation,* Boeing Document D2-22504, July 1963.
5 **I. J. Schoenberg** Contributions to the problem of approximation of equidistant data by analytic functions, *Quarterly Applied Math.,* Vol 4, 1946, 45–99.
6 **I. J. Schoenberg** Cardinal interpolation and spline functions, *J. Approximation Theory,* Vol 2, 1969, 167–206.
7 **A. H. Fowler and C. W. Wilson** *Cubic spline, a curve fitting routine,* Report No Y-1400, Union Carbide Corporation, Oak Ridge, TN, 1966.
8 **A. R. Forrest** *Curves and surfaces for computer-aided design,* PhD thesis, University of Cambridge, 1968.
9 **C. de Boor** Bicubic spline interpolation, *J. Maths Physics,* Vol 41, September 1962, 212–218.
10 **M. A. Sabin** Non-rectangular surface patches suitable for inclusion in a B-spline surface, *Proceedings EUROGRAPHICS'83,* P. J. W. ten Hagen, ed., North-Holland, 1983, 57–69.
11 **A. A. Ball** CONSURF. Part one: introduction of the conic lofting tile, *Comput.-Aided Design,* Vol 6, 1974, 243–249.
12 **A. A. Ball** CONSURF. Part two: description of the algorithms, *Comput.-Aided Design,* Vol 7, 1975, 237–242.
13 **A. A. Ball** CONSURF. Part 3: how the program is used, *Comput.-Aided Design,* Vol 9, 1977, 9–12.

14 **P. J. Payne** *A contouring program for Coons surface patches,* CAD Group Document 16, University of Cambridge, 1968.
15 **P. J. Payne** *A contouring program for joint surface patches,* CAD Group Document 58, University of Cambridge, 1971.
16 **A. D. Geisow** *Surface interrogations,* PhD thesis, University of East Anglia, 1983.
17 **M. J. Platt and A. D. Geisow** Surface/surface intersection problems, in *The Mathematics of Surfaces,* Gregory, J. A., ed., 1986, 117–142.
18 **M. A. Sabin** *A 16-point bicubic formulation suitable for multipatch surfaces,* Rep No VTO/MS/155, BAC, 1969.
19 **A. R. Forrest** *The computation of bicubic twist terms,* CAD Group Document 20, University of Cambridge, 1969.
20 **R. F. Reisenfeld** *Applications of the B-spline approximation to geometric problems of computer-aided design,* PhD thesis, Syracuse University, 1973.
21 **M. A. Sabin** *The use of piecewise forms for the numerical representation of shape,* PhD thesis, Budapest, 1976.
22 **G. Farin** *Subsplines über Dreiecken,* PhD thesis, TU Braunschweig, 1979.
23 **P. Sablonniere** *A catalog of B-splines of degree <10 on a three direction mesh,* Report AN0-132, University of Lille, 1984.
24 **P. O. Frederikson** Quasi-interpolation, extrapolation and approximation on the plane, *Proceedings Numerical Mathematics,* Winnipeg, Canada, 1971.
25 **P. J. Green and R. Sibson** Computing Drichlet tessellation in the plane, *The Computer Journal,* Vol 21, 1978, 168–173.
26 **M. J. D. Powell and M. A. Sabin** Piecewise quadratic approximations on triangles, *ACM Transaction on Mathematical Software,* Vol 3, 1966, 316–325.
27 **R. E. Barnhill** Smooth interpolation over triangles, in *Computer Aided Geometric Design,* Barnhill, R. E. and Riesenfeld, R. F., eds., Academic Press, 1974, 45–70.
28 **J. A. Gregory** N-sided surface patches, in *Mathematics of Surfaces,* Gregory, J. A., ed., Clarendon Press, Oxford, 1986, 217–232.
29 **T. Várady** Survey and new results in N-sided patch generation, in *Mathematics of Surfaces II,* Martin, R. R., ed., 1987, 203–235.
30 **A. D. DeRose and C. T. Loop** *Generalized B-spline surfaces of arbitrary topology,* TR 90-01-10, Computer Science Department, University of Washington, 1990.
31 **C. T. Loop** *Smooth subdivision surfaces based on triangles,* MS thesis, University of Utah, 1987.
32 **L. Ramshaw** *Blossoming: a connect-the-dot approach to splines,* TR No 19, DEC Systems Research Center, Palo Alto, CA, 1987.
33 **E. Catmull and J. Clark** Recursively generated B-spline surfaces on arbitrary topological meshes, *Comput.-Aided Design,* Vol 10, 1978, 350–355.
34 **D. Doo and M. A. Sabin** Behaviour of recursive division surfaces near extraordinary points, *Comput.-Aided Design,* Vol 10, 1978, 356–360.
35 **M. A. Sabin** Cubic recursive division with bounded curvature, in *Curves and Surfaces,* P.-J. Laurent, A. Le Méhauté and L. L. Schumaker, eds., Academic Press, 1991, 411–414.

ABOUT THE AUTHOR

Malcolm A. Sabin graduated in Mechanical Sciences from Cambridge University in 1962. After completing his apprenticeship with the British Aircraft Corporation, he worked there on systems for representation of aircraft shapes and for the part-programming of NC machine tools. In 1973 he was part of the team which conducted the IIASA world survey of CAD.

After a year spent at the Hungarian Academy of Sciences, IIASA and Portsmouth Polytechnic, he joined Kongsberg Data Systems Ltd, and worked on CNC machine tool control systems.

In 1979 he moved to the CAD Centre at Cambridge to become the executive of the CAE panel, a body which, until 1981, advised the Department of Industry and the Science and Engineering Research Council on funding of CAE research. With the reorganization of DTI committees, he moved to FEGS Ltd, where he has worked on mesh generation techniques and application of massive parallelism to finite element calculations, and continued until December 1990 to act as a consultant to the ACME directorate of SERC, helping to steer academic CAE research in the UK.

During his professional career, he has been chairman of the British Computer Society CAD specialist group and member of IFIP WG5.2 and TC5. He is a member of the editorial boards of *The Computer Journal, Computer Aided Geometric Design* and *Computer-Aided Design.*

13 Geometric Methods for Computer-Aided Design

Michael J. Pratt

INTRODUCTION

Most of this volume is concerned solely with representations of curves and surfaces, mainly from the parametric point of view. This chapter attempts both to provide a balance, by presenting other types of geometric representation, and also to provide a window onto the real world of CAD by discussing the geometric interrogations which are commonly needed in an engineering context. Some of the material of the chapter is well known, and is expounded here for the sake of completeness; other topics are less well known, and are mentioned because they may become important in the future of CAD. An attempt has been made to relate the different possible representations of the same curve and surface types, and to spell out some of their relative advantages and disadvantages for practical purposes. The scheme of the chapter is as follows: after a general discussion of the different types of representation, a range of different curves and surfaces is dealt with. Important topics in elementary differential geometry are introduced, and commonly required curve and surface interrogations are discussed. A fairly extensive bibliography then follows. Since free-form parametric curves and surfaces are extensively treated elsewhere in this volume, parametric representations are only detailed in this chapter in the special cases where simple alternative types of representation exist.

Curve and surface representations

Over the nearly 30 years of existence of CAD systems there have been two distinct threads of development. The first stems from the implementation of 2D draughting systems, which gave rise successively to 3D wireframe systems and more recently solid modeling systems. The second, which arguably may be traced back in its origins to the methods used by shipbuilders in antiquity, has led to today's systems for free-form curve and surface definition. It is true that there has been some interaction between the two schools of development over the past decade or so, but their full integration has so far not been demonstrated. One of the chief reasons for this is that the types of geometry used in the two classes of systems are significantly different.

Draughting, wireframe and solid modeling systems mainly function in terms of 'traditional' geometry such as straight lines, circles and other conic curves. Where surfaces are implemented, particularly in solid modelers, they are usually planes and the other 'natural' quadrics (spheres, right circular cylinders and cones). The torus, a non-quadric surface, is also often provided; the reasons for this will be discussed later.

The free-form curve and surface systems in use for the design and manufacture of aircraft, cars, ship hulls, shoes, plastic bottles, and so forth, are on the other hand based upon rather different concepts. The underlying constructional ideas have a long history, as does the theory of parametrically defined curves and surfaces, but it was not until the advent of powerful computers in the 1950s and 1960s that a satisfactory synthesis of the two could be made. The invention of polynomial splines in the 1940s also had a strong influence on this development. Much of the rest of the present volume is devoted to these and related concepts, but this chapter will endeavour to present a broader view by discussing also a more 'traditional' approach to geometry. Interrelationships between the two frequently exist, and these will be pointed out, but for detailed expositions of parametric curve and surface representations, reference may be made to other chapters.

The statement made earlier, that full integration of traditional and parametric geometry remains to be demonstrated, perhaps requires some justification. Its truth depends of course on what is understood by 'integration' in this context, and almost certainly the present writer's criteria are more stringent than those of some vendors of commercial systems. A few examples, chosen

Fundamental Developments of Computer-Aided Geometric Modeling
ISBN 012-554765-X

rather arbitrarily from today's widely used systems, may serve to illustrate the situation at the time of writing:

1. GEOMOD (developers and vendors SDRC; also sold by IBM) is a solid modeling system implementing both natural quadric and free-form surfaces. A unified surface representation is used, in the form of non-uniform rational B-splines (NURBS), which can model either type of surface exactly. Integration is not complete, however, since for many computational purposes curved surfaces are approximated by assemblies of planar facets. The unifying factor is therefore really this approximation technique, rather than the underlying use of NURBS.
2. CATIA (developers Dassault, vendors IBM) is rather similar in nature. It also uses faceting, but the unified geometric approach is based upon polynomial rather than rational methods. This does not allow exact representation of conics and quadrics, so that for example a circle has to be approximated as a sequence of parametric polynomial curve segments. This system therefore is not fully integrated.
3. PARASOLID (developers Shape Data, vendors Electronic Data Systems and others) is one of the most highly integrated systems currently known to the author. It uses traditional exact representations for the natural quadrics and the torus, and also implements parametric polynomial curves and surfaces exactly in Bézier format. A third class of surfaces (to be discussed later) is used for the construction of blends, and it is certainly possible to model a very wide variety of objects. It is understood that certain implementational problems still remain, however. Further, the exact provision internally of all the requisite geometry does not necessarily also imply full integration from the point of view of the user, who may sometimes be forced to use inconvenient techniques to achieve desired results. Although a full discussion of the user interface aspect is not appropriate here, it should be pointed out that this is a vitally important and hitherto largely neglected area of development of integrated systems [1,2]. Essentially, what is required is that the user can operate in what is to him a totally natural way, while the system decides what is the appropriate type of underlying mathematics to use. The situation at present is that the user often has to make this decision and then use different techniques according to the chosen type of geometry.

Within each of the two major subclasses of geometry already mentioned, several different types of geometric representation are in use or have been suggested for use in CAD. These will now be listed and briefly described in preparation for the more detailed discussions of individual curve and surface types in the next section. The remainder of the chapter will be concerned with a range of important topics in differential geometry and their use in accomplishing various geometric interrogations which are frequently needed in CAD/CAM.

'Geometric' representation

The term *geometric representation* was coined by Goldman [3] for a method of describing commonly used curves and surfaces in terms of values of a few salient parameters. For example, an infinite right circular cylinder is completely specified by (1) a code denoting the nature of the surface, (2) a straight line, the axis of the cylinder, and (3) a scalar, giving the radius. The straight line may be defined in various ways, for example in terms of a point and a direction. The first item, the code, is essential to enable the proper interpretation of the ensuing data. This type of representation is conceptually very simple. It has the disadvantage that different types of curve and surface all have their own distinctive data requirements, so that no 'unified' formulation is possible. On the other hand, the information stored in a geometric representation is often in the precise form required by application programs making use of the geometry, for example for the generation of manufacturing data.

Implicit representation

Equations such as $f(x, y) = 0$ for a planar curve or $w(x, y, z) = 0$ for a surface are said to be *implicit*. In some cases these can be put into an *explicit* form, in which one of the variables is expressed as a function of the others, but often this is either inconvenient or impossible. Consider, for example, the implicit circle equation $x^2 + y^2 = 4$, which may be put into the explicit form $y = (4 - x^2)^{1/2}$. This faces us with the responsibility of deciding which sign to take for the square root when we are evaluating points on the curve. The only slightly more difficult implicit equation $2x^3 + 3xy + y^3 = 0$, on the other hand, has no explicit form; general points on the curve can only be calculated by substituting values of x and solving the resulting cubic equation in y, in practice probably by the use of some numerical technique. With representations of this kind it is therefore often not convenient to evaluate points on a curve or a surface. Conversely, it is very easy to determine whether a specified point lies on a curve or surface; it is only necessary to substitute the coordinates into the equation, and to see whether the function on the left has the value zero.

Implicit representations can often be formulated in terms either of conventional cartesian coordinates or of *homogeneous coordinates*. Examples of the latter approach will be given later, and a fuller explanation of it may be found in Riesenfeld [4].

One disadvantage of the implicit approach to geometric definition is that application of a transformation leads to a major change in representation, since the various terms of the original form combine in complex ways. For this reason, it is usual to store details of the geometry as originally defined, and to append to it the matrix representing the transformation applied [5]. This avoids the need for continual recalculation of the terms of an implicit equation.

Parametric representation

Much of this volume is concerned with parametric representations of geometry, and so few details will be given here. Conventionally, the underlying functions employed are either polynomial or rational. These have been chosen largely for computational convenience. The rationals are the most complex functions which can be evaluated exactly (in principle) in terms of the four basic arithmetic operations of addition, subtraction, multiplication and division, and the polynomials are special cases of the rationals which have constant denominators. In either case, according to the choice of basis functions we have Ferguson/Hermite, Bézier, B-spline and other more exotic formulations. The maximum degree of the polynomials used in commercial systems ranges between two and 20.

Parametric representations may however be based on different classes of functions, as witness the well-known trigonometric equation for the unit circle $\mathbf{r}(t) = \mathbf{i}\cos t + \mathbf{j}\sin t$. This has the virtue of representing a complete circle exactly, in terms of a very small amount of data. Furthermore, the parametrization is proportional to arc length, a valuable property for many applications. The circle occurs very frequently in engineering design, and commercial systems exist in which they are approximated by sequences of parametric polynomial segments. It is worth noting that this requires much more data than the trigonometric representation, particularly if a very good approximation is required, in which case either the number of segments or their degree will be large. Additionally, the parametrization is now not exactly proportional to arc length. It is known that trigonometric formulations are used in a small minority of commercial CAD systems; their virtues should be more widely considered.

Manning [6], in some hitherto unpublished work, has considered the use of bases involving powers of trigonometric functions to mimic the properties of B-spline curves, but this seems to lose the essential simplicity mentioned above without conferring any significant advantages over the polynomial and rational approaches.

Parametric representations usually take the form of a linear combination of scalar basis functions with vector-valued coefficients. They therefore have the advantage that application of a transformation leads to minimal changes in the representation of geometry; each of the defining coefficients may be transformed independently, while the basis functions remain unchanged.

Procedural representation

A brief mention will be made here of several ways of defining surfaces in terms of other geometric entities and operations performed upon them. A simple case is that of a rotational surface, defined by a curve in space, an axis, and the operation of rotating the curve about the axis. If the curve is of rational polynomial parametric form then the surface swept out by the motion can be expressed exactly in rational bipolynomial terms, but for many types of procedurally defined surfaces it is not possible to write down defining equations. This section is devoted to such cases, in which points on the surface can only be evaluated by referring to the lower-level geometric entities and the specified operations. Several examples will be given, but there is as yet no general agreement about the use of such surfaces in practical systems. They will therefore not be further analyzed in the remainder of this chapter, with the exception of blending surfaces, some of which may be considered to fall into the procedural category. The examples are as follows:

1. Surfaces generated by sweeping one parametric curve along another. Here the defining geometric entities are the two curves and the operation is a combination of translation and rotation. If the first curve is closed then a generalized duct is created. Such surfaces have been discussed by Klok [7]. It is usual for the first curve to be planar, and to be swept so that its plane is always normal to the second curve where the two intersect.
2. Chord-height blend surfaces. These are provided in the system PDGS developed by the Ford Motor Company [8]. Consider four planar parametric curve segments, not all lying in the same plane, which are regarded as the boundaries of a surface patch. Suppose also that each curve is parametrized from 0 to 1. A ruled surface may be defined by simply connecting points having corresponding parameter values on two opposite boundaries with straight lines. Now start with one of the other boundaries, and sweep it up and over the ruled surface in such a way that (1) at any stage its chord line corresponds to one of the rulings, (2) at any stage the moving curve lies in a plane normal in some mean sense to the ruled surface, and (3) as it moves the shape of the curve varies smoothly so that it matches the fourth boundary curve on completion of the sweep. The principle is illustrated in Figure 13.1. In practice, PDGS uses a generalization of the method described; in particular, the boundary curves are not necessarily planar. Ford has found this type of surface extremely useful in automotive design, but the definition is strictly procedural. No surface equation exists; for specified parameter values points on the surface must be evaluated by applying certain algorithms or procedures to points evaluated on the boundary curves.
3. Cross-product surfaces. This term was coined by Hinds and Kuan [9] (see also Farouki and Hinds [10]). Each such surface patch is defined in terms of two curves, one of which is used to define a motion of the other. The simplest case is where a reference point on one curve is swept along the other curve, the first curve remaining parallel to its original version during the motion. A second variant defines a combination of rotation and scaling of the first curve, and a third a combination of translation and scaling. There is a practical implementation of these surfaces in the CAM-I Sculptured Surfaces System [11], which was used to generate the rotational cross-product surface shown in Figure 13.2. Again, points on the surface are determined in terms of points on the defining curves and a set of numerical procedures.
4. Swept volume surfaces. These are very useful, for example, in the computer simulation of machining operations. The

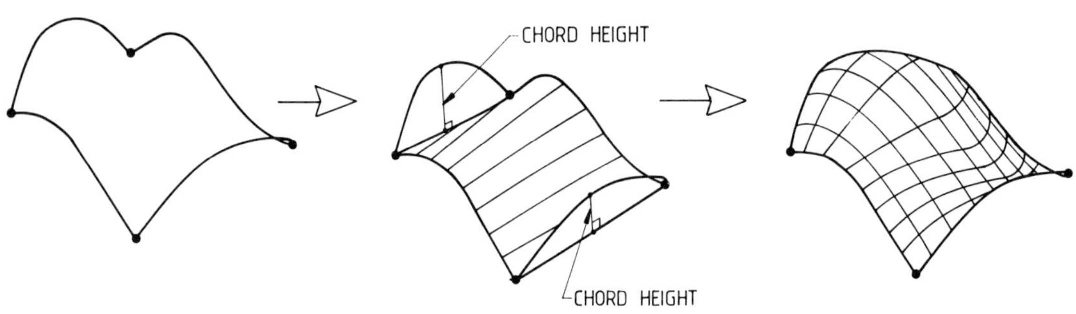

Figure 13.1. Procedure for constructing a chord-height blend surface.

cutting tool has some rotational shape, and as it moves it sweeps out a volume which corresponds to the volume of material machined from the part being manufactured. In 5-axis machining the axis of the tool may tilt continuously during the motion. If the volume swept out by the tool can be represented in a solid modeling system then it may be subtracted from a model of the part in process to show the effect of the cutter motion. This clearly requires some means of representing the swept surfaces forming the boundaries of the envelope generated by the cutting tool. Again, the definition of these surfaces is procedural, this time in terms of a volume (of the cutting tool) and a specified motion. For further details the reader is referred to Martin [12] or Wang and Wang [13].

It may have been noticed that procedural and geometric representations have much in common. For example, a torus may be defined procedurally in terms of a circle and a rotation operation (the details of how this may be done are given later), or geometrically in terms of its axis and its major and minor radii. The two are clearly equivalent; the second is perhaps more succinct, the rotation operation being implied rather than made explicit. The chief distinction seems to be as follows. Geometric representations are generally used for specialized surfaces having simple implicit formulations which can be expressed in terms of a small number of parameters. The procedural approach, on the other hand, is appropriate for surfaces of a free-form character, and gives rise to representations which for some computational purposes can be handled as though they were parametric.

Intrinsic representation

The simplest case of an *intrinsic* representation is that of a planar curve which is specified in terms of a starting point and a functional relation between arc length s and the tangent angle ψ made with the x-axis. The principle is illustrated in Figure 13.3. It may be shown (see, for example, Faux and Pratt [14]) that the derivative $d\psi/ds$ is identical with the curvature κ of the curve, and this allows the setting up of coupled linear

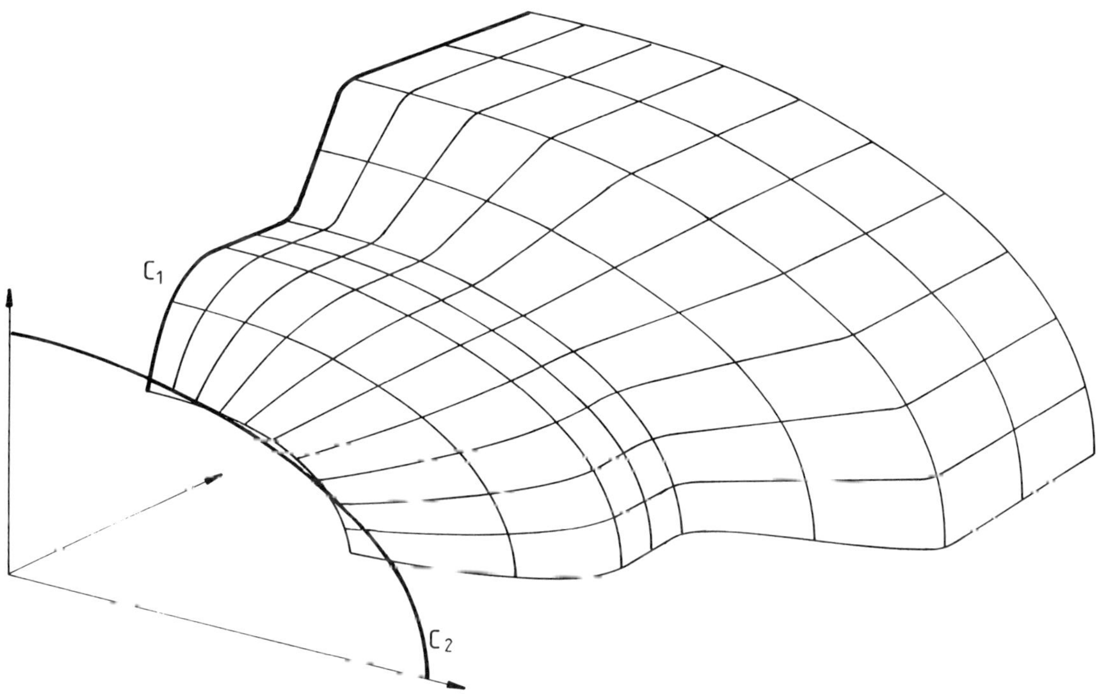

Figure 13.2. A rotational cross-product surface. Points on the curve C_2 control both the angle of rotation of C_1 about the perpendicular axis and also its scaling in terms of distance from that axis.

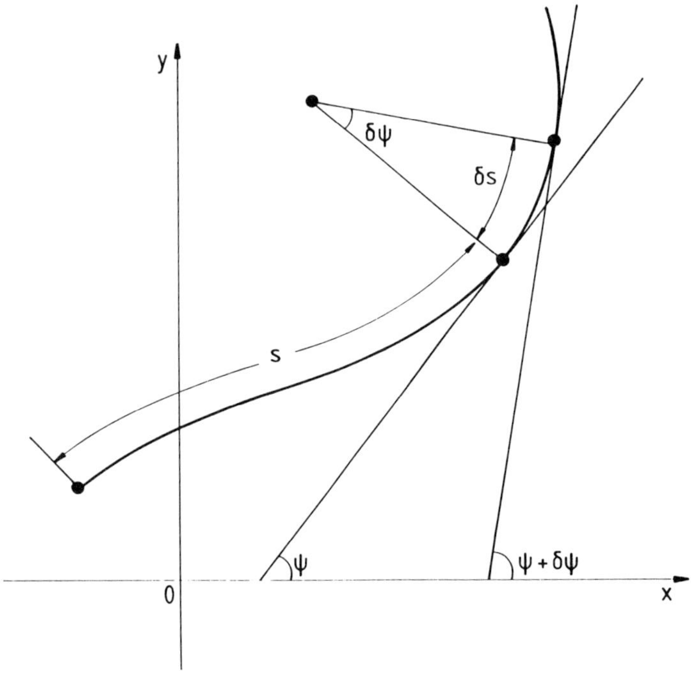

Figure 13.3. The intrinsic coordinates of a point on a curve.

second-order differential equations, in which $\kappa(s)$ appears as a coefficient, whose solutions are x and y as functions of s. Hence, an alternative intrinsic definition of the curve takes the form of an initial point, an initial tangent direction (these may be regarded as the initial conditions for the differential system) plus a specified curvature function $\kappa(s)$. Nutbourne *et al.* [15] and Adams [16] discuss methods for computing points on a curve by numerical solution of the differential equations. In Nutbourne and Martin [17] it is shown how twisted space curves have analogous intrinsic representations in terms of their curvature distribution $\kappa(s)$, and also their torsion distribution $\tau(s)$, where torsion may be thought of as a measure of deviation from planarity. They then further develop this idea to give a related definition of surface regions.

Although intrinsic representations have some interesting and attractive characteristics they are not currently in use for CAD, and accordingly this topic will only occasionally be mentioned in what follows. The interested reader may refer for more details to Nutbourne and Martin [17], which may foreshadow significant future developments in the synthesis of geometry with specified geometric properties. The concepts of curvature and torsion mentioned in the previous paragraph will, however, be discussed in more detail below.

Transformation between representations

The various representations of curves and surfaces have characteristic virtues and disadvantages in different situations. One well-known example of this, discussed below, relates to the computation of intersections between surfaces, which proves to be simpler when one surface is implicit and the other parametric rather than when both are implicit or both parametric.

It is therefore desirable to be able to transfer between one type of representation and another. In some cases this is easy; for example, geometric representations are used only in the case of 'simple' surfaces such as planes, cylinders, and so on, and in such cases the corresponding implicit and parametric forms are well known, as discussed below. In other cases, the transformation is much more difficult. It is known, for example, that not all algebraic implicit surfaces can be parametrized in terms of polynomial or rational polynomial functions, though all the most widely used such surfaces can be so expressed. On the other hand, it has only recently been realized [18, 19] that polynomial and rational parametric curves and surfaces can always in principle be expressed in algebraic implicit form, using methods developed in the Victorian era [20]. The practical difficulties of achieving this can be immense, however. Some geometric modeling systems use parametric forms exclusively, even for surfaces such as cylinders and cones which have simple implicit forms. In such cases it is highly desirable to store a geometric flag to indicate such special forms, since it is generally difficult to recognize them by examining the coefficients of the parametric equations. Without this stratagem, the determination of a corresponding geometric form requires an initial conversion to implicit form, followed by the recognition of a particularly simple surface type by examination of relations between its coefficients, and finally the subsequent determination of its salient dimensions and orientation.

Space does not permit an exhaustive general treatment of the topic of transformation between representations, but a more extended discussion of conversion specifically between the implicit and parametric forms is given later.

CURVES AND SURFACES USED IN CAD

This section discusses the most commonly used curves and surfaces of CAD, mentioning only briefly the general parametric free-form or sculptured geometry which is amply covered elsewhere in this volume. The topics covered have their greatest importance in the context of solid modeling. Much of the geometry will be very familiar, though it is hoped that some of the viewpoints will be less familiar. The material includes some little published CAD folklore and also a few excursions into less well-known geometric territory.

Curves

The straight line

Having got down to specifics, we are immediately in the midst of controversy. The straight line has many different representations, including the following:

Geometric – two obvious possibilities for defining a line are (1) for a bounded line segment, in terms of its two end points, and (2) for an infinite line, in terms of a point on the line and a direction vector. The first method is characteristic of many of the older, wireframe-based CAD systems, which model an object as a set of curve segments in space with no connectivity information. The second is found in certain more recent solid modelers. In deciding which form should be implemented in the international STEP standard for data transfer between CAD systems agreement could not be reached, and both are permitted in the current draft proposal [21]. There are several arguments in favor of the point/direction option, especially in a solid modeling context. To give one example, some such systems actually implement a 'direction' node in their datastructures, and in this case all lines which are parallel may have pointers to the same direction entity. This makes it very easy to determine which lines belong to a parallel set, a type of interrogation often required for the generation of manufacturing data. Answering such a query is clearly more difficult when the alternative formulation is used, since directions have to be computed and a search made before parallelism can be established. However, advantages are claimed for the two-point representation by certain manufacturers of large artefacts such as aircraft. If the point/direction method is used and the point is specified at the nose of the aircraft, any numerical rounding errors in the components of the direction vector can lead to significant inaccuracies in the evaluated position of points on the line at the rear of the fuselage.

Implicit – the well-known implicit equation for an infinite line in 2D space is

$$ax + by + c = 0 \tag{1}$$

and the line can therefore be economically represented in terms of values of the three coefficients a, b, c. Alternatively, if we raise the dimension to three by introducing a further variable W, and defining $Wx = X$, $Wy = Y$, then we may write

$$aX + bY + cW = 0 \tag{2}$$

in terms of the *homogeneous coordinates* X, Y, W. The name arises because the equation is now homogeneous in its variables, each term being linear in one variable. Clearly the line is still characterized by the same three coefficients, but there are certain advantages in working with the second form of the equation, notably in handling limiting cases. For example, if in equation (1) the coefficients a and b are both zero, but c is non-zero, a contradiction results. On the other hand, the use of equation (2) gives $W = 0$ for this case, so that $x = X/W$ and $y = Y/W$ are both infinite. The equation $W = 0$ therefore represents the *line at infinity*.

Other uses of homogeneous coordinates are in the unification of the matrix formulations of the commonly used transformations of translation, rotation and scaling, and in the concise representation of rational curves and surfaces [14].

A straight line in three dimensions is conventionally regarded as the intersection line of two planar surfaces, in which case it is defined by a system of two implicit equations. Alternatively, a planar system of local coordinates may be set up at some general orientation to the global system, and the line defined in this plane by the use of (1) or (2).

Parametric – there are two standard parametric representations of the straight line, in direct correspondence with the two geometric representations discussed earlier. The two-point formulation gives rise to the equation

$$\mathbf{r}(t) = (1 - t)\mathbf{p}_0 + t\mathbf{p}_1 \qquad (3)$$

where $\mathbf{p}_0$, $\mathbf{p}_1$ and $\mathbf{r}(t)$ are, respectively, the position vectors of the two end points and a general point on the line. For the parameter range $0 \leqslant t \leqslant 1$ we obtain precisely the line segment bounded by the two points. This formulation may be regarded as a Bézier curve of degree one, the two control points giving the end points of the curve segment. The point/direction formulation leads to the equation

$$\mathbf{r}(t) = \mathbf{p}_0 + t\mathbf{a} \qquad (4)$$

in which $\mathbf{p}_0$ is now a reference point on the curve and $\mathbf{a}$ a direction vector. An infinite range of the parameter t generates an infinite line. The two parametric representations are equivalent, as may be seen by noting that $(\mathbf{p}_1 - \mathbf{p}_0)$ is a direction vector for the line and substituting this for $\mathbf{a}$ in the second equation.

The conic sections

The conic sections have been used in design for many years, notably in the aircraft industry. One valuable feature which they possess is a smooth variation of curvature with arc length, a property desirable in many applications, but notably lacking in splines and other more 'modern' types of curves used in CAD.

As implied by their name, the conic sections are the curves of intersection of a cone with a plane. The generators of the cone are assumed to be of infinite length, having a common point of intersection at the vertex. If the sectioning plane is parallel to one of the generators of the cone, and is perpendicular to the plane defined by that generator and the axis, then the curve which results is a parabola. Otherwise, if the plane intersects the cone in a single closed curve an ellipse results, or if it intersects the cone on both sides of the vertex a hyperbola with two disjoint branches is generated. The latter degenerates to a pair of straight lines if the sectioning plane contains the axis of the cone. Examples are shown in Figure 13.4. The general implicit equation for a conic section in the x,y-plane is

$$ax^2 + 2hxy + by^2 + 2gx + 2fy + c = 0 \qquad (5)$$

where the nature of the curve depends upon the value of the discriminant $h^2 - ab$. Values which are negative, zero or positive give respectively an ellipse, parabola or hyperbola. Clearly, any one of the six coefficients a, b, c, f, g, h may be chosen arbitrarily, since the equation may be multiplied by any scalar without affecting the relationship between x and y. Consequently, a conic curve may be defined by specifying any five independent properties which serve to evaluate the remaining five coefficients. In a CAD context, for example, a conic curve segment may be determined in terms of its two end points, two end tangents and an interpolated point in its interior. Alternatively, it may simply be required to interpolate five points.

It should be noted that the substitutions $Wx = X$, $Wy = Y$ lead to the formulation

$$aX^2 + 2hXY + bY^2 + 2gWX + 2fWY + cW^2 = 0 \qquad (6)$$

which is homogeneous in the three variables X, Y, W, since each term is of degree two.

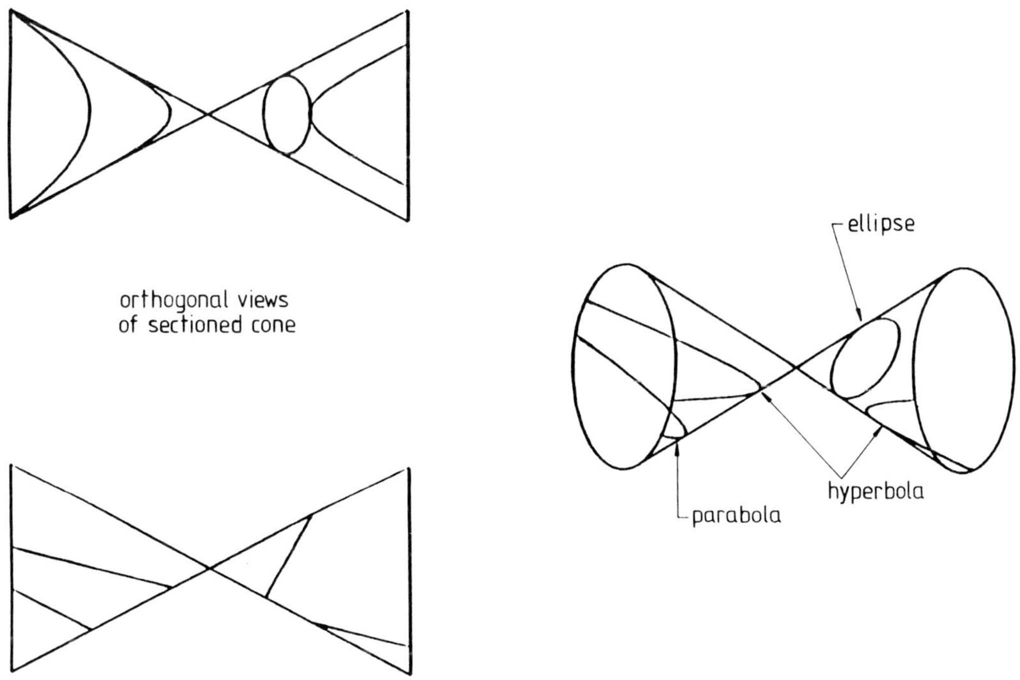

Figure 13.4. The conic curves as sections of a cone.

The circle
The circle is by far the most widely used conic curve. It is the special case of an ellipse for which the sectioning plane is perpendicular to the axis of the cone. For the circle the relationships $a = b$ and $h = 0$ must exist between the coefficients in the general conic equation. These remove two degrees of freedom, so that a circle may be determined in terms of three conditions, for example three points to be interpolated. The various representations of the circle include the following:

Geometric – in the plane, the required information is center point and radius; in three dimensions this must be supplemented by the direction of the axis.

Implicit – in the x,y-plane the equation for the general conic specializes to

$$(x - \alpha)^2 + (y - \beta)^2 = r^2 \qquad (7)$$

where (α, β) is the center and r the radius. In three dimensions it is usual to define a generally oriented circle in terms of a local 2D coordinate system.

Parametric – here we may start with the well known trigonometric formulation

$$\mathbf{r}(t) = a\{\mathbf{i}\cos t + \mathbf{j}\sin t\} \qquad (8)$$

in which a is the radius, **i**, **j** the unit vectors in the x and y directions and $\mathbf{r}(t)$ a general point on the curve. The circle lies in the x,y-plane and is centered at the origin. The parameter t corresponds to the polar angle of the radius vector, and the complete circle is therefore generated by the parameter range $0 \leqslant t < 2\pi$. As remarked earlier, the value of t is proportional to the arc length along the curve from the initial point. Transformations may be applied to give this original circle any desired position and orientation in space.

For those wishing to avoid the use of trigonometric functions a rational quadratic representation may be obtained on making the standard substitution $\tau = \tan \frac{1}{2}t$, with the result

$$\mathbf{r}(\tau) = a\{(1 - \tau^2)\mathbf{i}/(1 + \tau^2) + 2\tau\mathbf{j}/(1 + \tau^2)\} \qquad (9)$$

It is not difficult to recast this in, for example,

the rational Bézier or one of the other usual parametric forms. A problem arises in that the parameter range for the complete circle is $-\infty \leqslant \tau \leqslant \infty$, which is unsuitable for practical purposes. One remedy is to take the first quadrant, which has the parameter range $0 \leqslant \tau \leqslant 1$, and by rotating it through 90, 180 and 270 degrees generate three further segments, each parametrized over the same range. These four segments taken together provide an exact piecewise rational quadratic representation of the circle. The parametrization is not very uniform, but this may be improved by adjusting the weights in the rational form in such a way that the geometry remains unaltered [14,22].

Intrinsic – the circle has a particularly simple intrinsic representation in terms of curvature, which is, of course, constant. The analysis is given by Faux and Pratt [14, p. 41].

Biarcs
Various authors have noted that a *biarc,* or piecewise curve composed of two circular arcs joining with tangent continuity, provides similar freedom for the designer as does the widely used parametric cubic curve segment. In particular, a biarc may be constructed to interpolate two end points with associated end tangents. These four conditions do not completely determine the biarc; there remains some freedom in the choice of the junction point between the two arcs. It is possible to construct tangent-continuous sequences of biarcs, and so to generate spline-like curves made up of circular arc segments. Most of the work in this area has been concerned with planar biarcs [23–25], but an analysis of spatial biarcs is given in Nutbourne and Martin [17], based on work by Sharrock [26]. Rossignac [27] has also advocated the use of such curves. At least one commercial CAD system is known in which curves of this type are used for the approximate representation of complex surface/surface intersections.

The general conic sections
It is shown in elementary textbooks (see, for example, Towers [28]) that the conic curve of equation (5) may usually be reduced by an appropriate rotation and translation to one of the forms

$$(x/\alpha)^2 + (y/\beta)^2 = 1 \qquad (10)$$

$$(x/\alpha)^2 - (y/\beta)^2 = 1 \qquad (11)$$

$$y^2 = \gamma x \qquad (12)$$

representing respectively an ellipse, a hyperbola or a parabola. Here α, β and γ are real constants. Other, degenerate, possibilities include a pair of straight lines (possibly coincident) and cases where the curve has complex coordinates; we will not consider these further here. Equations (10)–(12) are known as *normal forms*; we may consider them to be the standard implicit representations of the three non-degenerate conics. The first two are referred to as *central conics,* since in their normal forms they are symmetrical about the two coordinate axes, and the origin is therefore a center of symmetry. Note that the parabola has only one axis of symmetry, and is not a central conic. Other possible representations of the conics are as follows:

Geometric – Geometrically, a conic is the locus of a point moving so that its distance from a fixed point (the *focus*) bears a constant ratio to its distance from a fixed line (the *directrix*). The constant ratio (the first distance divided by the second) is the *eccentricity* of the curve, usually denoted by e. The value of e determines the nature of the conic; $e < 1$ gives an ellipse, $e = 1$ a parabola and $e > 1$ a hyperbola.

The ellipse and the hyperbola, being central conics, have two-fold symmetry. If either is generated according to the definition above, the symmetry implies the existence of a second focus and a second corresponding directrix whose use would generate the same curve (see Figure 13.5). In these two cases alternative geometric characterizations exist:

- Ellipse – the locus of a point moving so that the sum of its distances from two fixed points (the foci of the ellipse) is constant.
- Hyperbola – the locus of a point moving so that the difference of its distances from two fixed points (the foci of the hyperbola) is constant.

As far as is known to the author, none of these definitions has been found particularly useful in the context of CAD; they are mentioned here for the sake of completeness.

More useful, as pointed out by Goldman [3], are the representations expressed in terms of

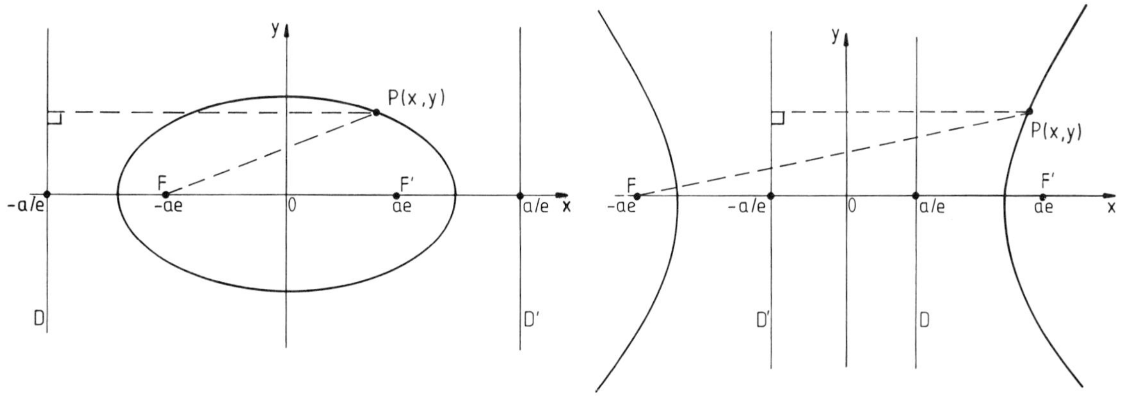

Figure 13.5. Generation of the ellipse and the hyperbola in terms of a focus and a directrix. Note that in each case the curve has two foci, F and F', and two directrices, D and D'.

local coordinate systems. Each conic in its normal form in such a system may be specified by not more than two parameters. The ellipse and hyperbola are completely specified by the lengths of their major and minor axes (α and β in equations (10) and (11)), while the parabola may be defined by the distance from the origin to its focal point (γ in equation (12)).

Parametric – It is clear from equation (10) that

$$\mathbf{r}(t) = \alpha\,\mathbf{i}\cos t + \beta\,\mathbf{j}\sin t \qquad (13)$$

is a parametrization of the ellipse in its normal form. As in the case of the circle, we may use the substitution $\tau = \tan\frac{1}{2}t$ to convert this to a rational parametrization, though the same problem concerning the infinite range of the parameter τ is encountered.

Similarly, the hyperbola may be represented by

$$\mathbf{r}(t) = \alpha\,\mathbf{i}\cosh t + \beta\,\mathbf{j}\sinh t \qquad (14)$$

where again a rational form may be obtained, this time by making the substitution $\tau = \tanh\frac{1}{2}t$.

The parabola has a simpler parametrization

$$\mathbf{r}(t) = \mathbf{i}\,t^2/\gamma + \mathbf{j}\,t \qquad (15)$$

which may of course be recast in Bézier or some other standard CAGD form; the parabola is the only conic having a polynomial rather than a rational parametrization.

It is shown in Faux and Pratt [14] that changing the weights of a rational quadratic curve segment in such a way that a certain relationship between them is maintained allows the parametrization of the segment to be altered without any modification of the actual curve geometry. The nature of the parametrization is illustrated in Figure 13.6. Here PQ is the chord joining the end-points of the segment, R is a point on the segment, and S a point on the conic lying outside the segment. If SR intersects PQ at a point which divides the chord in the ratio $(1 - t):t$, then t is the parameter value associated with the point R on the conic segment. Clearly, a different choice of the point S will lead to a different parametrization of the segment, though its geometry is, of course, unmodified. This is the geometric interpretation of the reparametrization property.

The superconics

The superconics are generalizations of the conics in which the degree of the terms differs from two. The one most frequently encountered in CAD is the *superellipse*

$$(x/a)^n + (y/b)^n = 1$$

which clearly reduces to a standard ellipse for $n = 2$. As n increases from two the curve deforms and tends towards a pair of perpendicular line segments, as shown in Figure 13.7. The first known use of this form in CAD was in the design of a roundabout or traffic circle in Stockholm, Sweden, but its use has also been reported in the cross-sectional design of aircraft fuselages [29].

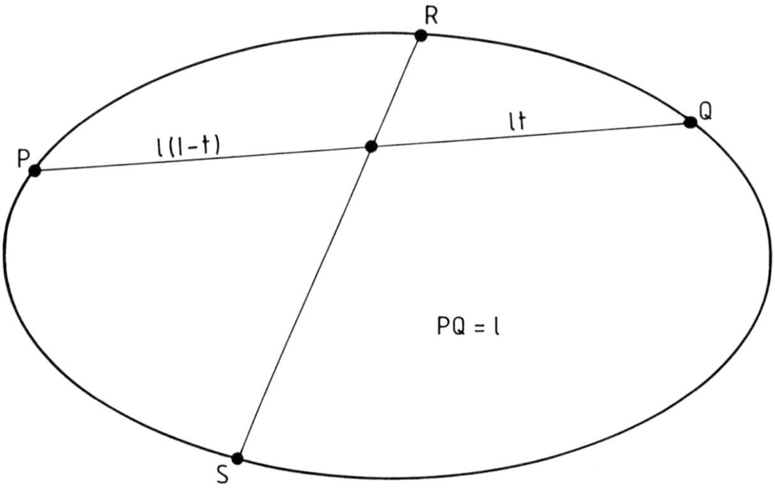

Figure 13.6. Parametrization of a conic segment PRQ in terms of the subdivision of the chord PQ by the line SR.

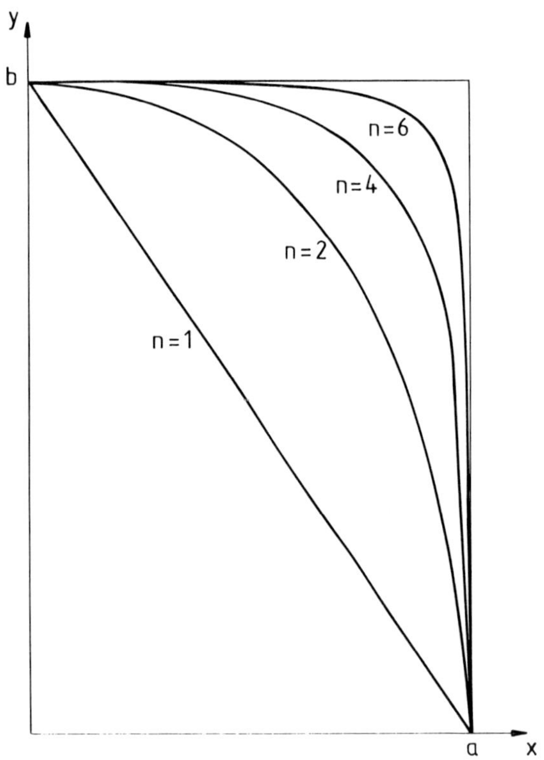

Figure 13.7. Superelliptic arcs for different values of the index *n*.

More general algebraic curves

Any equation of the form $f(x, y) = 0$, where $f(x, y)$ is a polynomial or rational function, defines a planar algebraic curve. Clearly, the conics are special cases of such curves. No very suitable methods have yet been discovered for using general algebraic curves in a CAD context, though Sederberg [30] has proposed an approach which may form the basis of a practical technique.

Free-form parametric curves

As mentioned at the beginning of this chapter, general free-form or sculptured curves will be omitted from consideration since they are extensively covered elsewhere in the book.

Intersection curves

This brief section concerns curves defined by the intersections of specified surfaces. The discussion here is confined to representational issues, the algorithmic aspects of the subject being dealt with later. As will be seen, standard methods suffice for the representation of such curves, though it is frequently necessary to resort to approximations even when exact surface geometry is used.

Certain special instances of intersections between quadric or cyclide surfaces give rise to intersection curves which are lines, circles or conic sections, but cases frequently arise in geometric modeling where more complex curves result, having no simple implicit representation. If one of the intersecting surfaces has an explicit formulation it is usually possible to obtain a parametric representation of the intersection.

Many practical systems, however, simply compute a sequence of points lying approximately on the intersection; in themselves these provide a piecewise linear approximation to the desired curve, though alternatively some form of spline may be computed to obtain a smooth fit to them. If the intersecting surfaces are represented parametrically it is possible to generate either of these types of approximation not only in Euclidian space, but also in the 2D parametric space of one or both of the surfaces. In general, these three possible representations will not correspond exactly in terms of their 3D geometry. Any one of the forms may be best for a specific application, and for this reason many practical systems store all of them.

Another strategem is to store the intersection curve exactly but indirectly, simply by reference to the two surfaces which generate it. This is a type of procedural representation, having the advantage of containing sufficient information to allow the computation of any other desired form to whatever required accuracy, at least in principle.

Surfaces

Many varieties of surfaces are used or have been proposed for use in CAD. In this section we deal with some of the most important cases, and touch upon types of surface which may be significant in the future.

The plane

The plane is fundamental to the modeling of many engineering parts. It has numerous representations, some of which are listed below:

Geometric – a plane may be defined by three points in space, by a point and a vector normal to the plane, or by its normal vector and its perpendicular distance from the origin. These require nine, six and four scalar values, respectively, if three components are stored for the vector (only the direction is important, and therefore we could use a unit vector for which one of the components may be calculated from the other two, and is therefore in principle redundant).

Implicit – the standard implicit equation of a plane is

$$ax + by + cz + d = 0 \tag{16}$$

which has the corresponding homogeneous form

$$aX + bY + cZ + dW = 0 \tag{17}$$

where $Wx = X$, $Wy = Y$ and $Wz = Z$.

Parametric – the standard parametric representation of an infinite plane is

$$\mathbf{r}(u, v) = \mathbf{r}_0 + \mathbf{a}u + \mathbf{b}v \tag{18}$$

in which $\mathbf{a}$ and $\mathbf{b}$ are two non-parallel vectors defining a family of parallel planes, and the reference point $\mathbf{r}_0$ picks out one member of that family. This is an obvious extension of equation (4) for an infinite line, but we now have two parameters u and v to provide the two degrees of freedom on a surface.

A parametric equation may also be written to represent a bounded region of a plane. Using equation (3), define two coplanar line segments by $\mathbf{r}_0 = \mathbf{L}_0(u)$, $\mathbf{r}_1 = \mathbf{L}_1(u)$, $0 \leqslant u \leqslant 1$. For any fixed value of u the equation

$$\mathbf{r}(u, v) = (1 - v)\mathbf{L}_0(u) + v\mathbf{L}_1(u) \qquad 0 \leqslant v \leqslant 1 \tag{19}$$

specifies a line segment joining corresponding points on the two original lines. But as u varies from 0 to 1, this line moves and sweeps out a planar quadrilateral region in space. Note that if the two original lines are not coplanar this *bilinear* equation still describes a bounded surface region, though obviously not a planar one. In fact, the region lies on a *hyperbolic paraboloid* (see below).

The quadric surfaces

The quadric surfaces may be regarded as the three-dimensional analogs of the conic curves. Any planar section of a quadric surface is a conic curve. The implicit form of a quadric is given by the most general equation of the second degree in three variables

$$\begin{aligned} &ax^2 + by^2 + cz^2 + 2fyz + 2gzx + 2hxy \\ &\quad + 2px + 2qy + 2rz + d = 0 \end{aligned} \tag{20}$$

By rotating and/or translating the axes, it is possible (except in certain degenerate cases) to express this equation in one or other of the two forms below, from which the *normal forms* of the quadric may be derived:

$$\pm(x/\alpha)^2 \pm (y/\beta)^2 \pm (z/\gamma)^2 = 1$$
$$\pm(x/\alpha)^2 \pm (y/\beta)^2 = (z/\gamma) \tag{21}$$

The first of these can represent only three fundamentally different surfaces, namely

$$(x/\alpha)^2 + (y/\beta)^2 + (z/\gamma)^2 = 1$$
$$(x/\alpha)^2 + (y/\beta)^2 - (z/\gamma)^2 = 1 \tag{22}$$
$$(x/\alpha)^2 - (y/\beta)^2 - (z/\gamma)^2 = 1$$

The surfaces represented by these equations are symmetrical with respect to the origin in all three variables, and for this reason they are called *central quadrics*. The second of the two previous equations has just two forms (with identical and with opposite signs), and these represent the *non-central quadrics*. The degenerate cases mentioned above represent a plane, a pair of parallel planes, a parabolic cylinder or a surface having complex coordinate values.

The non-degenerate forms of the quadric surface include the ellipsoid, hyperboloids of one and two sheets, elliptic and hyperbolic paraboloids, the cone, elliptic and hyperbolic cylinders. For further details of the classification of quadrics see, for example, Sommerville [31]. Since in CAD the use of the more general forms is unusual we omit further discussion of them, and concentrate on the more important special cases.

The natural quadrics

The *natural quadrics* are the plane, circular cylinder, circular cone and sphere. The first of these has already been discussed above. The other three are all rotationally symmetric surfaces. They have the following representations:

Geometric – **Cylinder**: Point and direction (specifying the axis), radius. **Cone**: Point (the vertex), direction (of the axis), semi-vertex angle. **Sphere**: point (the center), radius.

Implicit – The following equations are equivalent to the normal forms of the specified surfaces. The cylinder and the cone are rotationally symmetric about the z-axis. The cone has its vertex and the sphere its center at the origin:

Cylinder of radius a:

$$x^2 + y^2 = a^2 \tag{23}$$

Right circular cone of semi-vertex angle α:

$$x^2 + y^2 - \tan^2\alpha\, z^2 = 0 \tag{24}$$

Sphere of radius a:

$$x^2 + y^2 + z^2 = a^2 \tag{25}$$

Parametric – The natural quadrics are all surfaces of revolution. One way of parametrizing such a surface is as follows:

1. Define a parametric curve $\mathbf{r}(v) = \mathbf{i} f_1(v) + \mathbf{k} f_2(v)$ in the z,x-plane.
2. Replace $\mathbf{i}$ by the expression $(\mathbf{i}\cos u + \mathbf{j}\sin u)$. Then, for every fixed value of v we have a circle of radius $f_1(v)$ in the plane $z = f_2(v)$; the family of such circles forms the surface of revolution obtained by rotating the original curve about the z-axis.

Using this method we obtain the following for the three surfaces as defined above:

Cylinder of radius a:

generating curve $\mathbf{r}(v) = a\,\mathbf{i} + \mathbf{k}\,v$

surface $\mathbf{r}(u, v) =$

$$a(\mathbf{i}\cos u + \mathbf{j}\sin u) + \mathbf{k}\,v \tag{26}$$

parameter ranges $0 \leqslant u < 2\pi, \quad -\infty < v < \infty$

Cone of semi-vertex angle α:

generating curve $\mathbf{r}(v) = (\mathbf{i}\tan\alpha + \mathbf{k})v$

surface $\mathbf{r}(u, v) =$

$$\{(\mathbf{i}\cos u + \mathbf{j}\sin u)\tan\alpha + \mathbf{k}\}v \tag{27}$$

parameter ranges $0 \leqslant u < 2\pi, \quad -\infty < v < \infty$

Sphere of radius a:

generating curve $a(\mathbf{i}\cos v + \mathbf{k}\sin v)$

surface $a\{(\mathbf{i}\cos u + \mathbf{j}\sin u)\cos v + \mathbf{k}\sin v\}$ (28)

parameter ranges $0 \leqslant u < 2\pi, \quad -\pi \leqslant v \leqslant \pi$

Clearly, this approach to the parametrization of rotational surfaces is not restricted to the natural quadrics; in particular, it also leads to a simple parametric representation of the torus, and a wide variety of other non-quadrics.

The superquadrics

The superquadric surfaces were introduced by Barr [32], who also defines a generalization of the method just described which he calls the *spherical product* of two parametrically defined curves. One of these is used to define a simultaneous translation and scaling of the other in such a way that a surface is swept out.

The *superquadric* surfaces result when, in any of the standard trigonometric parametrizations of a quadric, the sines and/or cosines are raised to some power not equal to one. It is not difficult to see that these surfaces bear the same relationship to the standard quadrics as the superconics discussed earlier hold to the conics. As might be expected, then, it is possible to generate superellipsoids with varying degrees of 'squareness' (see Figure 13.8). Barr [32] gives examples of several such generalizations of the quadrics and also of the toroids; the reader is referred to his paper for further details.

Cyclides and the torus

'Traditional' solid modeling, as developed over the last 20 years, has been based to a large extent on the natural quadrics, with the addition of the torus. The latter surface was long regarded as a troublesome anomaly; while the implicit equations of the quadrics have degree two, that of the torus is four. This initially created mathematical problems in the computation of intersection curves with the torus, for example. Nevertheless, this surface was found to be essential for the modeling of certain commonly occurring situations such as 'elbows' in pipework and for the rounding of the edges where a cylinder abuts at right angles to a plane. It is only recently that the question has arisen as to whether a class of surfaces exists, including both the natural quadrics and the torus, which is reasonably easy to handle in a geometric modeling context. Such surfaces have in fact been known for some time; they are the *Dupin cyclides,* discovered by the French mathematician and naval architect Dupin

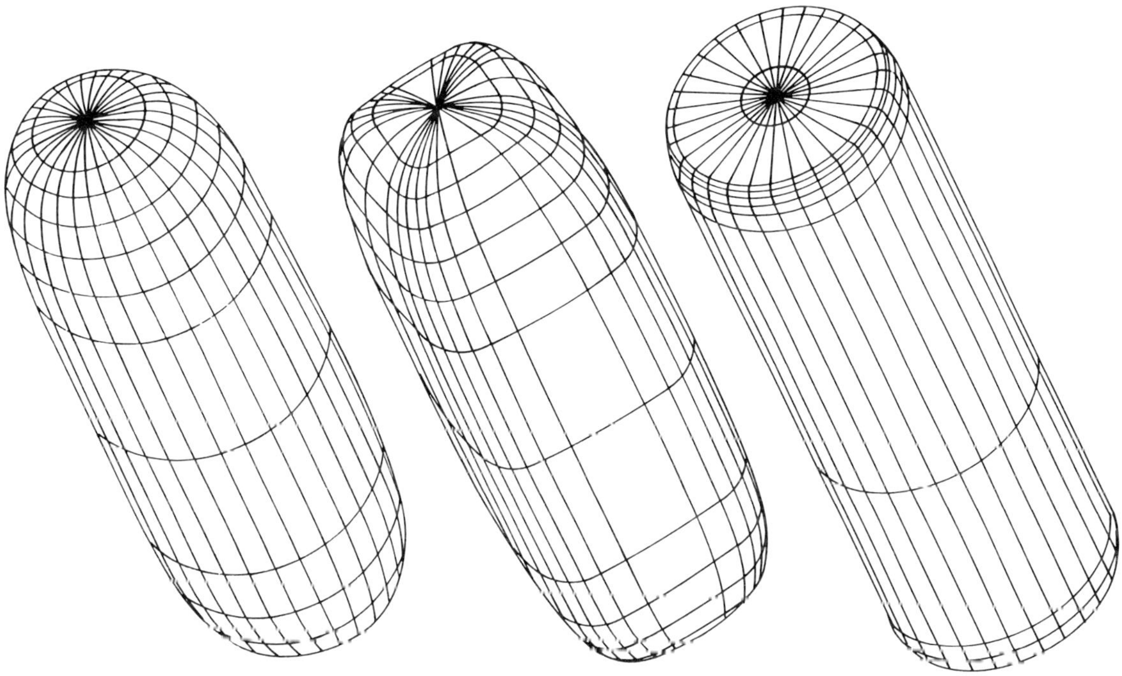

Figure 13.8. Examples of superellipsoids.

[33]. The cyclides include only the natural quadrics, and not the more general quadrics.

After the early years of the present century interest in cyclides declined but they were recently rediscovered during some research at Cambridge University into novel methods of computer aided surface design [17, 34]. This work is mainly directed towards the synthesis of patched surfaces having predetermined curvature characteristics, but subsequently cyclides have found application as blending surfaces in solid modeling [35–37].

The simplest form of the cyclide may be thought of as a deformed torus, in which the minor radius varies around the central hole. There are two planes of symmetry, conventionally taken to be the x,y-plane and the z,x-plane. The cross-sections in either of these two planes consist of two circles, as shown in Figure 13.9. This form is known as a *ring cyclide*. There are also two self-intersecting forms, one in which the circles in the x,y-plane intersect (the *horned cyclide*), and one in which the circles in the z,x-plane intersect (the *spindle cyclide* – the spindle is the closed volume with pointed ends which arises in the central region of the cyclide). The torus is a particularly simple case in which the circles in the x,y-plane are concentric and those in the z,x-plane have equal radii. The sphere is the limiting case of the spindle cyclide in which the pairs of circles in both planes coincide. If the radii of the circles in the x,y-plane tend to infinity the cyclide becomes a cylinder or a cone in the concentric and non-concentric cases, respectively (the apex of the cone corresponds to a self-intersection point of a horned cyclide). In the case of the cylinder, if the minor radius also tends to infinity then a

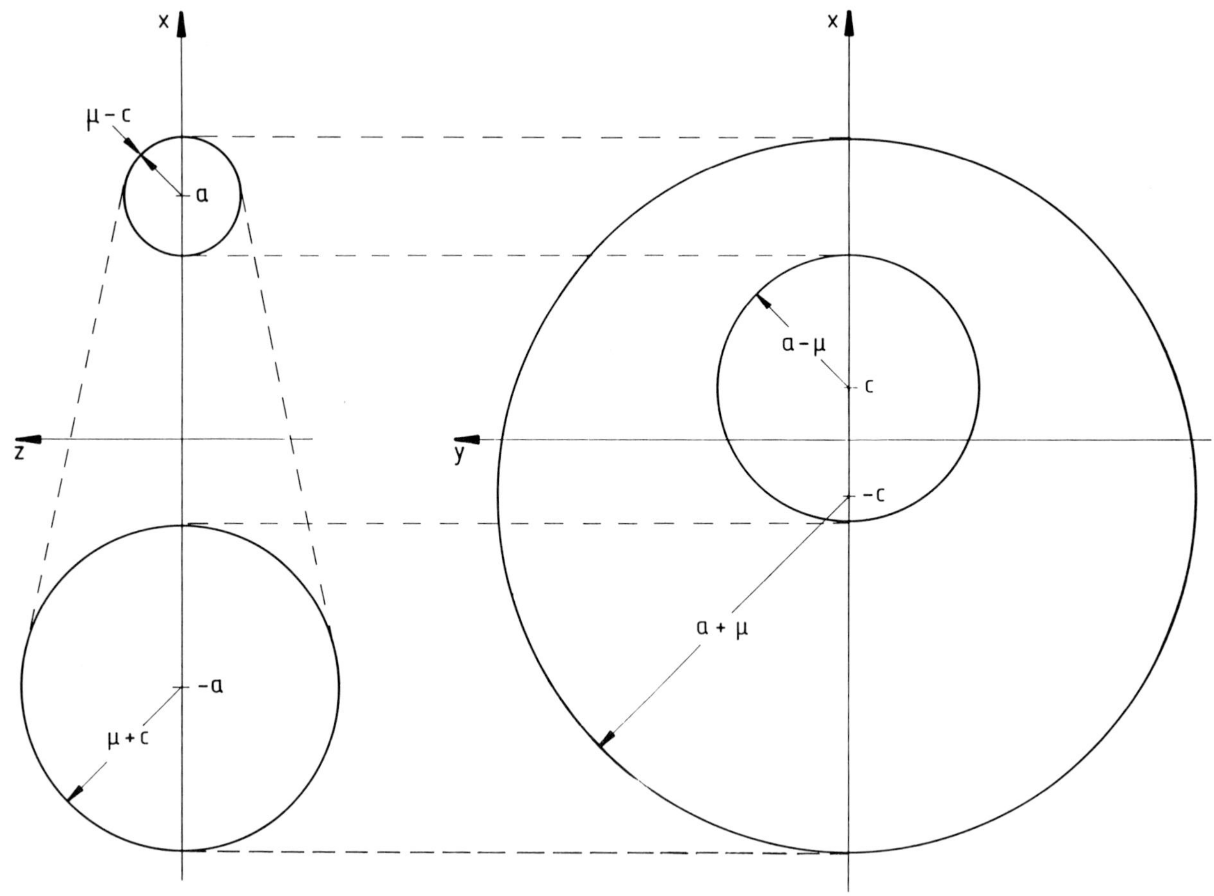

Figure 13.9. Cross-sections of a ring cyclide in its planes of symmetry.

plane results. The natural quadrics are therefore all special cases of the cyclide. Certain other limiting cases exist, but will not be discussed here.

Forsyth [38] gives two equivalent implicit equations and a parametric representation for a cyclide in a standard position and orientation (see Figure 13.9):

Implicit

$$(x^2 + y^2 + z^2 - \mu^2 + b^2)^2 = 4(ax - c\mu)^2 + 4b^2y^2 \tag{29}$$

$$(x^2 + y^2 + z^2 - \mu^2 - b^2)^2 = 4(cx - a\mu)^2 - 4b^2z^2 \tag{30}$$

Parametric (in which $0 \leqslant \theta, \phi < 2\pi$)

$$x = \frac{\{\mu(c - a\cos\theta\cos\psi) + b^2\cos\theta\}}{(a - c\cos\theta\cos\psi)} \tag{31 a}$$

$$y = \frac{b\sin\theta(a - \mu\cos\psi)}{(a - c\cos\theta\cos\psi)} \tag{31 b}$$

$$z = \frac{b\sin\psi(c\cos\theta - \mu)}{(a - c\cos\theta\cos\psi)} \tag{31 c}$$

Here x, y, z are the usual cartesian coordinates. The constants a, b, c relate to two conics involved in a geometric definition of the surface, namely the ellipse

$$(x/a)^2 + (y/b)^2 = 1, \quad z = 0$$

and the hyperbola

$$(x/c)^2 - (z/b)^2 = 1, \quad y = 0$$

lying in perpendicular planes. Each conic is required to pass through the foci of the other, which imposes the additional constraint

$$c^2 = a^2 - b^2 \tag{32}$$

According to Cayley [39], the cyclide is generated as the envelope of a sphere of variable radius moving so that its center lies on one of the conics whilst it remains in contact with a fixed sphere centered at a given point on the other conic. The fourth constant μ in Forsyth's equations is determined by the position and radius of Cayley's second sphere, and hence relates the radius of the moving sphere to the position of its center. The four parameters a, b, c and μ in fact determine the centers and radii of the circles in the planes of symmetry, as shown in Figure 13.9. Note that putting $c = 0$ gives rise to a torus with major radius a and minor radius μ.

It appears that the cyclides may have a useful role to play in future geometric modelers, since they provide simple solutions to blending problems in a range of cases of frequent occurrence in engineering design. Although the trigonometric parametrization given above is actually quite convenient for CAD purposes, a more conventional rational quadratic Bézier form has been derived by Pratt [35], following a suggestion by Sabin.

More general algebraic surfaces

An algebraic surface has an implicit equation of the form $f(x, y, z) = 0$, where f is a polynomial or rational function. The quadrics are therefore algebraic surfaces of degree two, whilst the cyclides are special cases of the quartic algebraic surfaces. The chief problem with using yet more general algebraic surfaces in CAD is that of providing the user with suitable 'handles' on the geometric freedom available, to enable him to achieve his desired results in a reasonably intuitive way. Sederberg [30, 40] has proposed techniques based on a variation of the Bézier approach to curve and surface definition. Whilst ingenious, these methods may not prove appropriate for mathematically unsophisticated users of practical design systems.

Research into the use of general algebraic surfaces for geometric modeling is currently at an early stage. Most of the initial work has been concerned with their use for blending in solid modeling (see below), but wider applications are now being studied. A good review of some of the associated problems which need to be solved is given by Bajaj [41].

Free-form parametric surfaces

Surfaces of this type are extensively treated in other chapters of this book, and no further discussion of them will be given here.

Blend surfaces

The need for blend surfaces is well illustrated by what has become known as the 'Cranfield Object', in reality part of an oil rig (Figure 13.10). With conventional solid modeling systems the gross geometry of the object is very easy to model; the conical body, flanges, etc., can all be generated by defining a 2D profile and rotating it through 360°, while the cylindrical and toroidal projections can be added by the use of Boolean operations. The remaining problem is to add the blends between the projections and the main body; the cone/torus blend is highlighted in the figure. From the designer's point of view, these last operations should be quick and largely automatic, since most of the design is complete and the addition of the blends is merely putting the finishing touches. However, in the current state of the art, many commercially available systems are not capable of constructing these blends. Of those that can model them, the majority use approximate rather than exact blending surfaces; even so, the procedures required for constructing the blends are often lengthy and prone to failure.

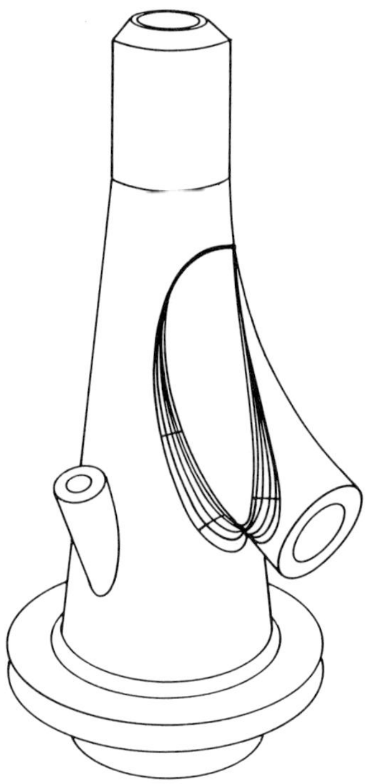

Figure 13.10. The 'Cranfield Object', exhibiting a cone/torus blend.

One popular approach to the generation of exact blends is based on mathematical combinations of the equations of the two surfaces being blended. In some cases, these result in implicit algebraic surfaces having the required properties, while in others the blend surface representations are procedural in nature. Often some freedom is available to the designer, who may for example have some control over the extent and 'fullness' of the blend. Some authors trace the basic idea back to a technique developed by Liming [42], also explained in Faux and Pratt [14], for the generation of smooth fuselage shapes in aircraft design. Variations on the general theme create blends involving trigonometric or other non-algebraic functions. Currently, these methods are best suited to blending between two surfaces which are themselves algebraic in nature. Several authors have been working independently but simultaneously on the blending problem in recent years; some relevant papers are Middleditch and Sears [43], Hoffmann and Hopcroft [44,45] and Holmström [46]. Only one commercial system is currently known to have a working implementation of exact blends [47]. Several other contributions are cited in a survey by Woodwark [48].

An alternative technique for exact blends has been suggested by Pratt [49]; this requires a change in the representation of faces lying on simple algebraic surfaces to Bézier or B-spline form, followed by local surface modifications to give smooth transitions between those faces. Chiyokura [50] has demonstrated yet another method, in which a geometric model is initially constructed in wireframe form. Edges to be blended are replaced by sets of edges forming appropriate boundaries to blending faces. As a final step, parametrically defined faces are interpolated to all the wireframe data in such a way that smooth transitions result wherever they are desired.

Several methods have been tried for the approximate generation of blends. The first is by the use of faceted approximations, but these are clearly bound to run into trouble along the boundaries of the blend surfaces where, if the geometry were exact, the blending and blended surfaces would be tangential. In the faceted case this gives problems of ill-conditioning or (frequently) non-existence of segments of the blend boundary, since the system is attempting to compute intersections between facets which may well be mutually parallel.

A second approach uses segments of implicit surfaces to form approximations to complex blends. Rossignac and Requicha [51] use piecewise cylinders and toruses to approximate constant radius blends of the 'rolling ball' type while Chandru *et al.* [37] use piecewise cyclides for variable radius blends.

A third method for approximate blending employs parametrically defined surfaces for the blends. This involves determining points on the blend boundary curves, which are then approximated by an appropriate type of parametric curve. The variation of surface tangent across these curves must be similarly approximated. A 'lofting' technique based on cubic Hermite or some similar interpolation method is then used to generate the desired blend surface. A recent paper on this approach is by Várady *et al.* [52].

The object of this section has been to show that, whilst a wide variety of techniques has been employed for the generation of blending surfaces, all of them fall into the categories of surface representations discussed earlier. The last-mentioned paper by Várady *et al.* contains a useful summary of work in this area; a slightly less recent but more detailed review is given by Woodwark [48].

DIFFERENTIAL GEOMETRY OF CURVES AND SURFACES

This section is concerned with those geometric properties of curves and surfaces which are important for CAD purposes. The emphasis is on implicit and parametric representations, since these are the most widely used in practical systems. Space does not permit the derivation of most of the results quoted, which are listed primarily for reference purposes. Further information may be found in, for example, Faux and Pratt [14], or Nutbourne and Martin [17].

Geometry of space curves

It is convenient to start by considering a curve $\mathbf{r}(s)$ parametrized in terms of its arc length s. At a point on the curve corresponding to a particular value of s three mutually orthogonal planes may be defined (see Figure 13.11). The *normal plane* is perpendicular to the tangent direction to the curve. The *osculating plane* (Latin *osculum*, kiss) is the plane in which an infinitesimal neighborhood of the point is considered to lie; specifically, it is the plane containing the circle passing through the three points $\mathbf{r}(s-\delta s)$, $\mathbf{r}(s)$ and $\mathbf{r}(s+\delta s)$ in the limit as δs tends to zero. Since it clearly contains the tangent direction to the curve, the osculating plane is perpendicular to the normal plane. The limiting circle is known as the *osculating circle*. Finally, the *rectifying plane* is the plane through the point $\mathbf{r}(s)$ which is perpendicular to both the others. Its name arises from the fact that certain curves are *rectifiable*, which is to say that their length can be computed exactly. Geometrically, such a curve lies on a surface which is the envelope of all its rectifying planes; this surface is a developable surface which can be rolled out flat without distortion onto a plane. A rectifiable curve becomes a straight line on this flattened surface, and its length can be measured.

In terms of the vector calculus, the following equations hold, where a dot denotes differentiation with respect to the indicated parameter:

$$\dot{\mathbf{r}}(s) = \mathbf{T}(s) \tag{33}$$

$$\dot{\mathbf{T}}(s) = \kappa(s)\mathbf{N}(s) \tag{34}$$

$$\dot{\mathbf{N}}(s) = \tau(s)\mathbf{B}(s) - \kappa(s)\mathbf{T}(s) \tag{35}$$

$$\dot{\mathbf{B}}(s) = -\tau(s)\mathbf{N}(s) \tag{36}$$

Here **T**, **N** and **B** are respectively the unit *tangent*, *principal normal* and *binormal* vectors of the curve at the point $\mathbf{r}(s)$. The significance of the tangent vector is obvious, and this vector lies in the intersection of the osculating and rectifying planes. The principal normal vector points towards the *center of curvature* of the curve at the point $\mathbf{r}(s)$, which is the center of the osculating circle referred to above. It lies in the intersection of the osculating and normal planes. The binormal vector is perpendicular to the other two, and therefore lies in the intersection of the rectifying and normal planes. The scalar κ is the *curvature* of the curve at $\mathbf{r}(s)$; it is the reciprocal of the *radius of curvature*, which is in fact the radius of the osculating circle. The second scalar τ is the *torsion* of the curve at $\mathbf{r}(s)$; it may be regarded as a measure of deviation of the curve from planarity, i.e. of how fast it is twisting out of the osculating plane (note that a planar curve has a constant osculating plane and zero torsion).

If for simplicity we drop explicit mention of the parameter s, the last three of the above equations may be expressed in matrix form as

$$\frac{d}{ds}\begin{pmatrix}\mathbf{T}\\\mathbf{N}\\\mathbf{B}\end{pmatrix} = \begin{pmatrix}0 & \kappa & 0\\-\kappa & 0 & \tau\\0 & -\tau & 0\end{pmatrix}\begin{pmatrix}\mathbf{T}\\\mathbf{N}\\\mathbf{B}\end{pmatrix} \tag{37}$$

a system of three coupled linear differential equations in which the matrix of coefficients is skew-symmetric. These are known as the *Frenet–Serret equations*. As s varies the point $\mathbf{r}(s)$ will move along the curve, its associated triad of unit vectors $\mathbf{T}$, $\mathbf{N}$ and $\mathbf{B}$ (sometimes referred to as the *Frenet frame*) moving with it. In general, the motion of the Frenet frame will be compounded of a simultaneous translation and rotation. The disposition of the Frenet frame at a particular point on a curve is illustrated in Figure 13.11.

Arc-length parametrization of curves has the advantage of giving simple formulae for the geometric properties, as shown above. Unfortunately, however, the methods of curve synthesis generally used in CAD lead to parametrizations which, although usually monotone with respect to arc length, are only in very

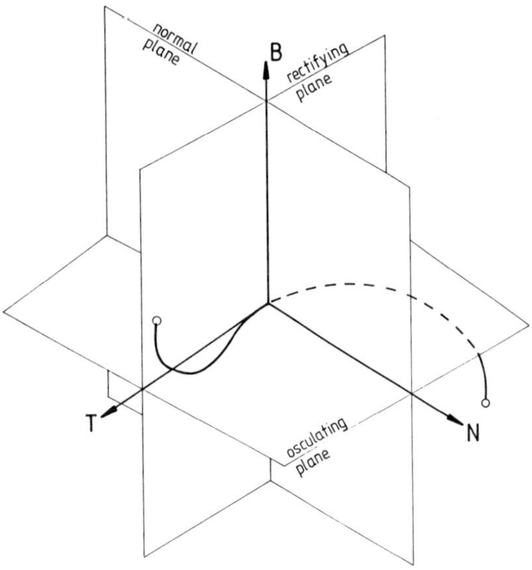

Figure 13.11. The Frenet frame at a point on a curve.

special cases proportional to it. The corresponding formulae then become more complicated. For a parametric curve $\mathbf{r}(t)$, a convenient set of formulae for practical purposes is

$$\dot{\mathbf{r}}(t) = \dot{s}(t)\mathbf{T}(t) \tag{38}$$

$$\kappa(t)\mathbf{B}(t) = \{\dot{\mathbf{r}}(t) \times \ddot{\mathbf{r}}(t)\}/\{\dot{s}(t)\}^3 \tag{39}$$

$$\mathbf{N}(t) = \mathbf{B}(t) \times \mathbf{T}(t) \tag{40}$$

and

$$\tau(t) = \dot{\mathbf{r}}\cdot\{\ddot{\mathbf{r}}(t) \times \dddot{\mathbf{r}}(t)\}/\{\dot{s}(t)\}^6\kappa^2(t) \tag{41}$$

Since $\mathbf{T}(t)$ is by definition a unit vector, it is clear from the first of these equations that

$$\dot{s}(t) = |\dot{\mathbf{r}}(t)| \tag{42}$$

These five new equations permit the determination of $\mathbf{T}, \mathbf{N}, \mathbf{B}, \kappa$ and τ at any point of the curve.

To illustrate a typical use of these relations we consider calculation of the curvature at the end points of a Bézier curve segment of degree n parametrized from 0 to 1. Evaluation of the first and second derivatives and use of equation (39) gives the results

$$\kappa(0) = \frac{(n-1)|(\mathbf{r}_1 - \mathbf{r}_0) \times (\mathbf{r}_2 - \mathbf{r}_1)|}{n|\mathbf{r}_1 - \mathbf{r}_0|^3} \tag{43}$$

$$\kappa(1) = \frac{(n-1)|(\mathbf{r}_{n-1} - \mathbf{r}_{n-2}) \times (\mathbf{r}_n - \mathbf{r}_{n-1})|}{n|\mathbf{r}_n - \mathbf{r}_{n-1}|^3} \tag{44}$$

These are comparatively simple formulae, expressing the desired results in terms of the sides of the control polygon. Note that the cross-product in the numerator of equation (43) gives twice the area of the triangle whose vertices are at $\mathbf{r}_0$, $\mathbf{r}_1$ and $\mathbf{r}_2$, and that there is a similar interpretation for equation (44). The triangles concerned are illustrated in Figure 13.12. The computation of end-point torsions has a related geometric interpretation [53].

The corresponding equation for the curvature at a general point on a Bézier curve segment is considerably more complex. However, it is easily possible to subdivide a curve segment at an arbitrary interior point into two smaller segments with [0, 1] parametrization; if this is done, equations (43) or (44) may be used to determine the curvature at the chosen point.

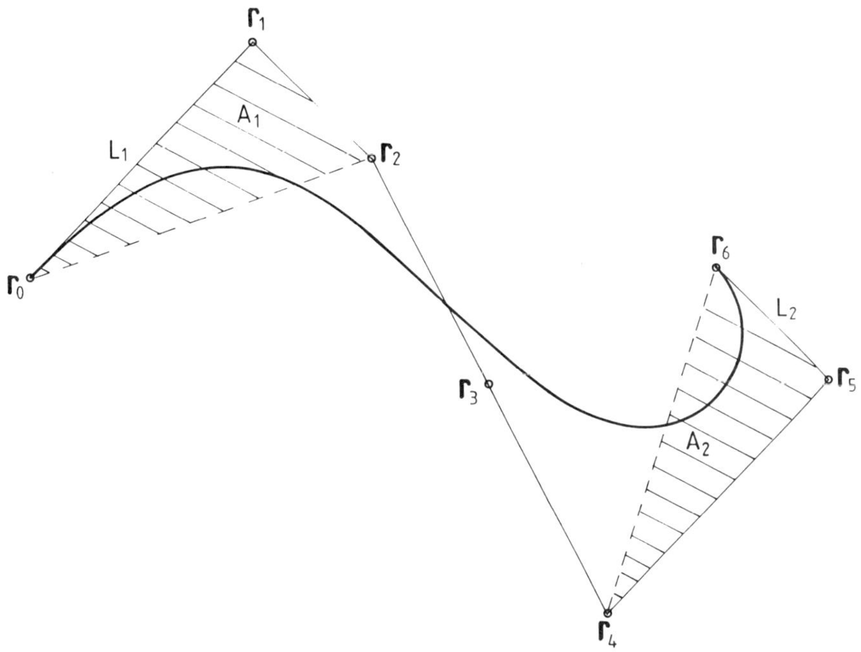

Figure 13.12. The areas A_1, A_2 and the lengths L_1, L_2 determining the curvatures at the end points of a Bézier curve segment.

It is much easier to compute the geometric properties of a 3D space curve from its parametric representation than from any other. For planar curves expressed in the explicit form $y = f(x)$ or the implicit form $g(x, y) = 0$ the method of calculating tangents is well known, while for the curvature we have respectively

$$\kappa = y''/(1 + y'^2)^{3/2} \tag{45}$$

$$\kappa = -(g_{xx}g_y^2 - 2g_{xy}g_xg_y + g_{yy}g_x^2)/(g_x^2 + g_y^2)^{3/2} \tag{46}$$

Here subscripts denote partial differentiation with respect to the indicated variable. Since planar curves have zero torsion there are no corresponding formulae for τ.

Geometry of surfaces

On a parametric surface $\mathbf{r}(u, v)$ there exist two families of isoparametric curves, $u =$ constant and $v =$ constant. A curve of constant u has v as its variable parameter, and *vice versa*. At any point on the surface having parameter values u, v we can form two first partial derivatives which will be denoted by $\mathbf{r}_u(u, v)$ and $\mathbf{r}_v(u, v)$. These represent respectively the tangent vectors to the isoparametric curves of constant v and of constant u through the point $\mathbf{r}(u, v)$. Provided these two vectors are non-null and not parallel they define the *tangent plane* of the surface at the point in question, and their cross product gives a vector perpendicular to this plane. If we denote the unit normal vector to the surface by $\mathbf{n}(u, v)$, then

$$\mathbf{n} = \mathbf{r}_u \times \mathbf{r}_v / |\mathbf{r}_u \times \mathbf{r}_v| \tag{47}$$

where all quantities are understood to be evaluated at (u, v).

Curves on surfaces

Any curve lying on a surface has a unit tangent vector $\mathbf{T}$ at all nonsingular points. By contrast with the space curve, it now seems more natural to think of a local coordinate frame based on the surface normal $\mathbf{n}$ rather than the curve principal normal $\mathbf{N}$. The frame must be completed by a third unit vector perpendicular to the first two, which must lie in the intersection

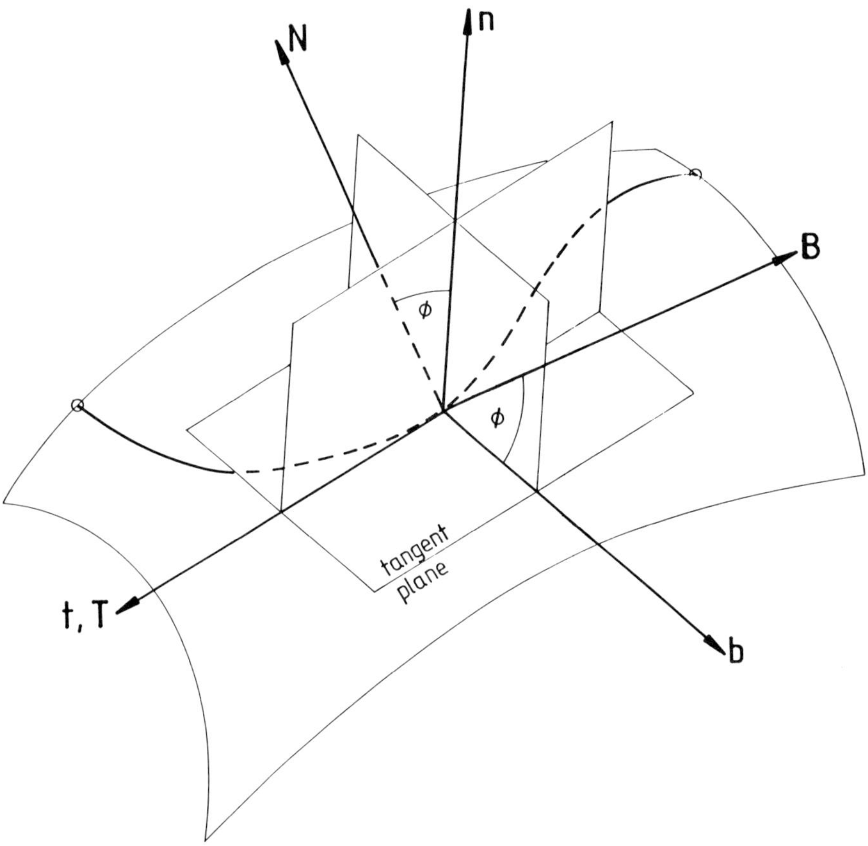

Figure 13.13. Relation between the Frenet and surface frames at a point on a curve defined on a surface.

of the normal plane to the curve and the tangent plane to the surface. This vector does not seem to have an accepted name; Nutbourne and Martin [17] refer to it as the *bitangent vector,* by analogy with the binormal vector to a space curve. The notation **b** is then a logical choice for it. Then **T**, **n** and **b** constitute a local system of coordinates at any point on the curve. It should be noted that this frame shares the vector **T** with the Frenet frame of the curve regarded as a space curve. Thus the two frames are related by a rotation about **T**, as shown in Figure 13.13. The angle of rotation is generally denoted by ϕ, which may be defined as the angle between the curve principal normal **N** and the surface normal **n**, having the positive sense if the rotation from the first to the second is clockwise when viewed in the direction of the tangent vector **T** (it should be noted that some writers define ϕ in slightly different but related ways). If the curve frame is denoted by **F** and the surface frame by **f**, then the relation between the two is

$$\mathbf{f} = \mathbf{\Phi F} \tag{48}$$

where $\mathbf{\Phi}$ is a rotation matrix which in general is a function of position along the curve, here assumed to be specified in terms of arc length s.

Now the curve frame is given by the Frenet–Serret equations above, and this last relation may be used to obtain the corresponding result for the surface frame, known as the Bonnet–Kovalevsky equations

$$\frac{d}{ds}\begin{pmatrix}\mathbf{T}\\ \mathbf{b}\\ \mathbf{n}\end{pmatrix} = \begin{pmatrix}0 & \kappa_g & \kappa_n\\ -\kappa_g & 0 & \tau_g\\ -\kappa_n & -\tau_g & 0\end{pmatrix}\begin{pmatrix}\mathbf{T}\\ \mathbf{b}\\ \mathbf{n}\end{pmatrix} \tag{49}$$

Here the symbols in the matrix of coefficients have the following significance (they are all in general functions of s):

$$\kappa_g = \kappa \sin \phi \text{ (geodesic curvature)} \quad (50)$$

$$\kappa_n = \kappa \cos \phi \text{ (normal curvature)} \quad (51)$$

$$\tau_g = \phi' + \tau \text{ (geodesic torsion)} \quad (52)$$

The geometric interpretation of these quantities is as follows:

The geodesic curvature is the curvature of the normal projection of the curve onto the tangent plane of the surface; it is therefore a property of the curve itself.

The normal curvature is the curvature of the normal projection of the curve onto the plane containing the tangent and the surface normal; it is therefore the *normal curvature* of the surface itself (see below) in the direction of the curve. It is the same for all curves passing through a given point in a given direction, and therefore depends only on the surface.

Note that these two quantities are components of the curvature κ of the curve regarded as a space curve:

The geodesic torsion is compounded of the space curve torsion τ and the rate ϕ' at which the surface frame **f** is rotating with respect to the curve frame **F**. The first quantity depends only on the curve; the second measures the rate of rotation of the surface normal **n** about the tangent vector **T**, which is the same for all curves which go through a given point in a given direction, and is therefore a property of the surface itself.

In terms of the three quantities just introduced, three specialized types of curves may be defined on a surface:

- *Geodesic curves* – these have $\kappa_g = 0$ throughout their length, which implies that either the curve is a straight line or that it has a constant value of $\phi = \pm 90°$. In this case $\phi' = 0$ and hence $\tau_g = \tau$, which explains why τ_g is called the *geodesic torsion*; it is the torsion of a geodesic curve. There is a geodesic curve through any point on a surface in any specified direction tangent to the surface. It may be shown that the geodesic curve segment joining two points on a surface has extremal (minimal or maximal) length. This fact is obviously of potential importance in applications such as, for example, the machining of surfaces.
- *Asymptotic curves* – these have $\kappa_n = 0$ throughout their length. Such curves are less common than geodesic curves (for example, they do not occur at all on surfaces which are purely convex or concave), and do not at present appear to have any practical applications.
- *Lines of curvature* – these might more correctly be called 'curves of curvature', but the customary terminology is used here. Such curves have $\tau_g = 0$ along their length. They are discussed more fully in the next section.

Properties of surfaces

Any continuous relation between u and v defines a curve on the surface $\mathbf{r} = \mathbf{r}(u, v)$. In what follows it is convenient to specify such a relation in terms of a parameter t, writing

$$u = u(t), \quad v = v(t)$$

or in vector form

$$\mathbf{u}(t) = (u(t), v(t))^{\mathrm{T}} \quad (53)$$

Here the superscript T denotes a transpose. The above notation allows the use of $\mathbf{r}(u, v)$ and $\mathbf{r}(t)$ to denote general points on the surface and on the curve respectively.

Surface tangents and the first fundamental form
A tangent vector to a curve on the surface $\mathbf{r} = \mathbf{r}(u, v)$, in which u and v in turn depend upon the single parameter t, may be written using the chain rule as

$$\dot{\mathbf{r}} = \mathbf{r}_u \dot{u} + \mathbf{r}_v \dot{v} = \mathbf{A}\dot{\mathbf{u}} \quad (54)$$

where

$$\mathbf{A} = (\mathbf{r}_u, \mathbf{r}_v) \quad (55)$$

and a dot denotes differentiation with respect to the parameter t. The length of the tangent vector is given by

$$\dot{s}^2 = |\dot{\mathbf{r}}|^2 = \dot{\mathbf{r}}^{\mathrm{T}}\dot{\mathbf{r}} = \dot{\mathbf{u}}^{\mathrm{T}}\mathbf{A}^{\mathrm{T}}\mathbf{A}\dot{\mathbf{u}} = \dot{\mathbf{u}}^{\mathrm{T}}\mathbf{G}\dot{\mathbf{u}} \quad (56)$$

where

$$G = A^T A = \begin{pmatrix} \mathbf{r}_u \cdot \mathbf{r}_u & \mathbf{r}_u \cdot \mathbf{r}_v \\ \mathbf{r}_v \cdot \mathbf{r}_u & \mathbf{r}_v \cdot \mathbf{r}_v \end{pmatrix} \tag{57}$$

The symmetric matrix **G** is called the *first fundamental matrix* or *first fundamental form* of the surface. It is required for the calculation of several important metrical properties of the surface; for example

1. The unit tangent vector along the curve $\mathbf{u}(t)$ is

$$\mathbf{T} = \mathbf{A\dot{u}}/(\mathbf{\dot{u}}^T\mathbf{G\dot{u}})^{1/2} \tag{58}$$

2. The length of the curve segment with $t_0 \leqslant t \leqslant t_1$ is

$$s = \int_{t_0}^{t_1} (\mathbf{\dot{u}}^T\mathbf{G\dot{u}})^{1/2}\, dt \tag{59}$$

3. The angle of intersection between two curves $\mathbf{u}_1(t)$ and $\mathbf{u}_2(t)$ is given by

$$\cos\theta = \mathbf{\dot{u}}_1^T\mathbf{G\dot{u}}_2/\{(\mathbf{\dot{u}}_1^T\mathbf{G\dot{u}}_1)^{1/2}(\mathbf{\dot{u}}_2^T\mathbf{G\dot{u}}_2)^{1/2}\} \tag{60}$$

4. The area of a bounded region of a surface corresponding to a region R in the u,v-plane is

$$S = \iint_R (|\mathbf{G}|)^{1/2}\, du\, dv \tag{61}$$

Further details of the derivation of these formulae may be found, for instance, in Faux and Pratt [14].

Before moving on to discuss the curvature properties of surfaces, certain pathological cases should be mentioned for which the unit tangent vector **T** to a curve on a surface is not well-defined. From equation (58) we see that $\mathbf{\dot{u}}^T\mathbf{G\dot{u}}$ must be greater than zero for **T** to exist. Clearly, $\mathbf{\dot{u}} \neq 0$ is a necessary condition; **T** will then be defined for any curve through any point on the surface provided **G** is positive definite. This last condition requires $g_{11} > 0$ for the leading element of **G**, and also $\det(\mathbf{G}) \neq 0$. These are equivalent, respectively, to $\mathbf{r}_u \neq 0$ and $\mathbf{r}_u \times \mathbf{r}_v \neq 0$ (since $|\mathbf{G}|$ may be shown to be equal to the scalar product of the left-hand side of this last equation with itself). It follows that the tangent to the curve specified by $\mathbf{u}(t)$ is well-defined at all surface points for which (i) $\mathbf{\dot{u}} \neq 0$ and (ii) $\mathbf{r}_u \times \mathbf{r}_v \neq 0$. If the second condition holds at a particular point the surface has a well-defined tangent plane there; failure of the first condition then implies that the curve on the surface has a discontinuity in its tangent direction. The second condition may fail to be satisfied if either one of the derivatives is zero (which implies a tangent discontinuity in one of the isoparametric lines, and hence the possible existence of a cusp or ridge in the surface) or because the u and v isoparametric lines happen to be parallel at the point under consideration. Finally, then, we see that the non-existence of **T** at any point of a curve on a surface may stem (1) from a kink in the curve itself, even though the surface is smooth, (2) from a cusp or ridge in the underlying surface, or (3) from a badly chosen parametrization of the surface.

Surface curvatures and the second fundamental form

A further differentiation of equation (54) with respect to t, followed by the formation of the scalar product of both sides of the resulting equation with **n**, leads to the result

$$\dot{s}^2\kappa\mathbf{N}\cdot\mathbf{n} = \mathbf{\dot{u}}^T\mathbf{D\dot{u}} \tag{62}$$

where

$$\mathbf{D} = \begin{pmatrix} \mathbf{n}\cdot\mathbf{r}_{uu} & \mathbf{n}\cdot\mathbf{r}_{uv} \\ \mathbf{n}\cdot\mathbf{r}_{vu} & \mathbf{n}\cdot\mathbf{r}_{vv} \end{pmatrix} \tag{63}$$

The matrix **D** is the *second fundamental form* of the surface. For well-behaved surfaces having the property that $\mathbf{r}_{uv} = \mathbf{r}_{vu}$ it should be noted that **D**, like **G**, is symmetric.

Now consider an arbitrary point on a general smooth surface. The unit surface normal there will be denoted as usual by **n**, and a unit surface tangent vector **T** may be freely chosen in any direction orthogonal to **n**. The vectors **n** and **T** define a plane which intersects with the surface in a curve. The curvature of this curve at the point in question is known as the *normal curvature* κ_n in the chosen direction. A curve so defined has its principal normal **N** parallel to the surface normal **n**, so that equations (56) and (62) give

$$\kappa_n = \mathbf{\dot{u}}^T\mathbf{D\dot{u}}/\dot{s}^2 = \mathbf{\dot{u}}^T\mathbf{D\dot{u}}/\mathbf{\dot{u}}^T\mathbf{G\dot{u}} \tag{64}$$

Thus defined, the normal curvature is positive

when the curve is concave in the direction of the surface normal. If the tangent direction **T** is rotated through 360° about **n** it is found in general that κ_n has two maxima (given by oppositely directed values of **T**) and, similarly, two minima. The directions of maximum and minimum curvature are perpendicular to each other. They are referred to as the *principal directions* of normal curvature at the point under consideration, and the corresponding curvature values as *principal curvatures*. Two scalar equations for their values may be obtained from equation (64); elimination of $\dot{u}$ and $\dot{v}$ from these equations gives a single quadratic whose roots are the desired principal curvatures. Alternatively, elimination of the principal curvature values allows determination of the ratio $\dot{u}:\dot{v}$, and hence the principal directions.

Let the principal curvatures at a point on a surface be denoted by κ_1, κ_2. Then the normal curvature there in a direction making an angle Θ in the tangent plane with the direction of normal curvature κ_1 is given by a formula due to Euler

$$\kappa_n(\Theta) = \kappa_1 \cos^2 \Theta + \kappa_2 \sin^2 \Theta \qquad (65)$$

A related result due to Germain gives the geodesic torsion of a curve passing through the point in question in the specified direction

$$\tau_g(\Theta) = (\kappa_2 - \kappa_1) \sin \Theta \cos \Theta \qquad (66)$$

Note that this is consistent with the previous statement that $\tau_g = 0$ in the principal directions of curvature.

Two commonly used diagnostics of surface curvature are the *mean curvature* M (the sum of the principal curvatures) and the *Gaussian curvature* K (their product). The second is more interesting from a practical point of view, since its value enables certain conclusions to be drawn about the local geometry of a surface. If K is positive at a point, both principal curvatures are in the same sense there, and the surface is either convex or concave about the normal direction **n**. If K is negative, on the other hand, the local nature of the surface is that of a saddle-point about the direction of the normal.

The local behavior of a surface at a point may be quantified more precisely by means of the *Dupin indicatrix*. This is the curve obtained in the tangent plane by taking the point as origin and the two principal directions as axes, and plotting $|\kappa_n(\Theta)|^{-1/2}$ as a function of Θ as defined in the last paragraph but one. It is easy to show from equation (65) that when $K > 0$ this curve is an ellipse, when $K < 0$ it is a hyperbola, and when $K = 0$ it is a straight line. For this reason, points with positive and negative Gaussian curvature are said to be *elliptic* and *hyperbolic*, respectively. As might be expected, the limiting case between the two, for which $K = 0$, is referred to as a *parabolic* point, though the geometric analogy in this instance is less clear. The Dupin indicatrix has the following characteristics in the three possible cases:

- *Elliptic* – The major and minor axes of the ellipse have lengths κ_1, κ_2, and are in the directions of principal curvature at the point in question.
- *Hyperbolic* – The directions of principal curvature bisect the angles between the asymptotic directions of the hyperbola. In fact, the indicatrix comprises not one but *two* hyperbolae, one for positive and one for negative values of κ_n. They share the same asymptotes but occupy alternate sectors between them.
- *Parabolic* – For this case one of the principal curvatures is zero. The local nature of the surface is that of a cylinder, and the generator passing through the point in question coincides with the Dupin indicatrix. This, being a line of minimum curvature, lies along one of the principal directions; the direction of maximum curvature is therefore perpendicular to it.

As previously mentioned, there are usually two orthogonal directions of principal curvature through a point on a surface (exceptional cases will be described shortly). It is therefore possible to define a curve on a surface having the property that its tangent at any point is always in one of the principal directions at that point. Such a curve is a *line of curvature*, a concept introduced earlier as a surface curve with zero geodesic torsion. The equivalence between the two definitions would take too long to prove here, but is demonstrated in any book on differential geometry. Note that most surfaces are covered by two mutually orthogonal families of lines of curvature, tracking respectively the directions of minimum and of maximum curvature.

Consider now the special case of planar lines of curvature. Since these have zero curve torsion, equation (52) implies that such curves have $\phi' = 0$ and hence ϕ = constant. Along a planar line of curvature, then, the angle between the surface normal and the principal normal to the curve remains constant. The lines of curvature on a Dupin cyclide have this property, since, as mentioned earlier, they are all circles and hence planar.

There are exceptions to the general rule that only two principal directions exist at any point on a surface. One such exception occurs when the two principal curvatures are equal, in which case the Dupin indicatrix is a circle and the surface curvature is locally spherical. There are infinitely many lines of curvature through such a point, since any direction may be regarded as a principal direction there. Points at which more than two principal directions exist are called *umbilics*. Note that every point on a plane or a sphere is umbilical in nature. Other types of umbilics also exist; for example, a surface with 3-fold symmetry will have an umbilic where three ridges meet, as shown in Figure 13.14. Clearly, there are precisely three principal directions of both minimum and maximum curvature at this type of umbilic. Such situations may obviously be generalized to cases which locally have n-fold symmetry. One way of regarding umbilics is as the points on a surface at which lines of curvature begin and terminate. Some classes of surfaces, notably the cyclide and its special case the torus, do not exhibit any umbilical points.

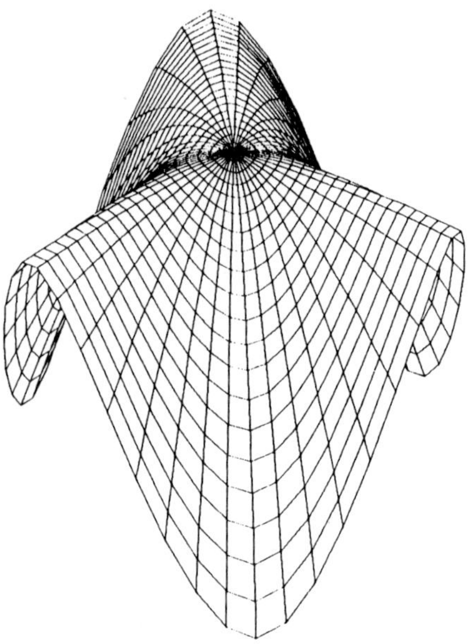

Figure 13.14. Surface with three-fold symmetry exhibiting an umbilical point.

The methods of computing principal curvatures, lines of curvature and the location of umbilical points for parametrically defined surfaces are straightforward but too lengthy to explain here. The reader is referred to Nutbourne and Martin [17], where the salient formulae are derived and worked examples given.

Ruled surfaces

A ruled surface is a surface generated by a family of straight lines. If $\mathbf{r} = \mathbf{r}_0(u)$ is a parametrically defined space curve, then we may associate a direction vector $\mathbf{d}(u)$ with every point on the curve. This allows the definition of the ruled surface

$$\mathbf{r} = \mathbf{r}(u, v) = \mathbf{r}_0(u) + v\mathbf{d}(u) \tag{67}$$

Alternatively, a ruled surface may be defined by constructing linear rulings between corresponding points on two space curves $\mathbf{r} = \mathbf{r}_1(u)$ and $\mathbf{r} = \mathbf{r}_2(u)$:

$$\mathbf{r} = \mathbf{r}(u, v) = (1 - v)\mathbf{r}_1(u) + v\mathbf{r}_2(u) \tag{68}$$

In either case, the constructional curves are called *directrices* and the rulings *generators*. Note that in the second case the shape of the resulting surface depends strongly on the parametrization of the directrices. By way of example, consider two uniformly parametrized coaxial circles of equal radius, lying in parallel planes. If, looking along the common axis, the parametrizations of the two circles are in phase, a cylinder results. If they are 180° out of phase the surface is a cone; the rulings all intersect in a common point. For other cases the surface is a hyperboloid of one sheet.

A further point to note is that both the above representations for ruled surfaces exhibit a typical characteristic of such surfaces; they possess a parametrization that is linear in one variable.

Developable surfaces

A developable surface is a special case of a ruled surface, having the property that it can be flattened or *developed* onto a plane without stretching or tearing. Conversely, a planar surface may be distorted into a developable surface by pure bending, with no other form of distortion. For this reason, developable surfaces are very important in engineering applications. Certain particular cases are well known; for example, if a circular cylinder or cone is cut along one of its generators it can be unrolled onto a flat surface.

On any ruled surface, the surface normal curvature along a ruling is clearly zero. It is therefore not possible for a ruled surface to be elliptic in nature, as defined above, since at any elliptic point the maximum and minimum (principal) normal curvatures are either both positive or both negative, the value zero lying outside their range. Points on a ruled surface must, therefore, be parabolic or hyperbolic. In the hyperbolic case the two principal curvatures at a point differ in sign; the zero curvature value lies within the range, and it occurs in the directions of the rulings, though these are not principal directions. The parabolic case is the most important for present purposes; now one of the principal curvatures is zero, and the corresponding principal directions are clearly those of the rulings. Such surfaces have Gaussian curvature $K = 0$ everywhere, and it proves to be this condition which characterizes the developable surface from amongst the more general ruled surfaces.

Another property which distinguishes the developable surface from other ruled surfaces is that all tangent planes of a developable are tangent along entire rulings, not just at single points. We shall return to this characteristic shortly.

One approach to the analysis of developable surfaces, given in Faux and Pratt [14], starts with an approximation to such a surface in the form of an initially planar sheet with a sequence of folds imposed upon it. Since the 'facets' of the 3D configuration are planar, it is clear that pairs of successive fold lines are coplanar. Hence, they cannot be skew to each other, and must either be parallel or have a point of intersection. If the number of folds is increased indefinitely, a smooth developable surface is obtained in the limit, the finite number of folds becoming replaced by an infinite number of rulings or generators.

We may think of this surface as being the envelope of a family of planes, the tangent planes to the surface. Since each such plane is tangent along an entire generator, this is a *one-parameter* family. Matters are different for non-developable ruled surfaces, which have the characteristic of surfaces in general that their tangent planes belong to two-parameter families. If the generators of a developable surface are all parallel the result is obviously a generalized cylinder, but the case when this is not so is more interesting. In either case, each ruling may be regarded, in the limit, as the line of intersection of two successive planes, but when the surface is not a generalized cylinder, three successive planes will intersect in a point. The locus of all such points is a space curve known as the *edge of regression* or the *cuspidal edge* of the developable surface. In fact, the generators prove to be the tangent lines to this curve, while the enveloping planes mentioned earlier in this paragraph are the osculating planes of the curve. Conversely, we may start by defining a space curve $\mathbf{r} = \mathbf{r}_0(u)$ and then generate the surface from it, in the form

$$\mathbf{r}(u, v) = \mathbf{r}_0(u) + v\mathbf{T}(u) \tag{69}$$

A developable surface of this type is called a *tangent developable*. Apart from tangent developables and generalized cylinders, other cases of developable surfaces are the plane (which may be regarded as the tangent developable of any planar curve) and the cone (whose edge of regression has degenerated to a point).

CURVE AND SURFACE INTERROGATIONS FOR CAD

In the previous section, methods were described for the computation of geometric properties of curves and surfaces. Such properties are often needed in practice, usually in the context of some more complex interrogation carried out for a practical purpose. Examples include the generation of a graphical display or of a set of instructions to a computer-controlled machine tool for machining a surface. In this section, some of these more complex curve and surface interrogations are examined. Methods similar to those described are implemented in many practical CAD/CAM systems, and there is a constant search for new algorithms which are faster, more efficient or more robust than those currently available.

Attention here is largely confined to interrogations of parametric curve and surface representations. Before proceeding further, it is worth mentioning that there are advantages in terms of accuracy and numerical stability in the exclusive use of Bézier representations, and specially configured algorithms operating in terms of Bernstein bases. This has recently been shown by Farouki and Rajan [54,55]. As yet their findings have not had much effect on the internal functionality of established systems, however; at the time of writing there seems still to be a strong movement towards increased implementation of the fashionable non-uniform rational B-splines (NURBS) and algorithms tailored for this type of representation [56].

Intersections

Geometric modelling frequently requires the calculation of intersections. The intersecting entities are usually two curves, two surfaces or a curve and a surface. In solid modeling, for example, the edge shared by two sculptured faces lies on the curve of intersection of the two surfaces on which they lie. Note, moreover, that the vertices bounding that edge lie at points corresponding to the intersections of *three* surfaces. There are three highly desirable qualities in an intersection algorithm, namely accuracy, robustness and speed. These are discussed in turn below.

Accuracy is necessary because geometric modelers are often used for engineering purposes where precision is important. A further reason concerns the relation between geometry and topology in solid modeling systems. Consider Figure 13.15, which depicts the intersection between two cylinders of different diameters. Note that very small changes can lead to totally different topologies for the intersection curve. Numerical inaccuracy leading to deduction of the wrong topology can lead to erroneous results in subsequent modeling operations because the system is trying to determine topological changes from an incorrect starting point. This problem is a hazard frequently encountered in modeling systems using faceted geometry.

Robustness of an intersection algorithm implies that it is not subject to failure, even in limiting cases. It also requires the algorithm to find all the points of intersection of a curve with another curve or a surface, and all the portions of the intersection curve between two surfaces, which may have many disjoint branches. Unfortunately for implementers of geometric modeling systems, designers seem to like working with limiting cases involving, for example, surfaces which are mutually tangent. This can lead to ill-conditioning in intersection computations, and extreme care is therefore necessary in designing and coding the algorithms.

Speed of operation is of primary importance in an interactive system, particularly if it is to be used for productive purposes in engineering. However, there is a conflict here with the first two requirements of accuracy and robustness. As a result, most geometric modeling systems are not as fast as their users would wish in

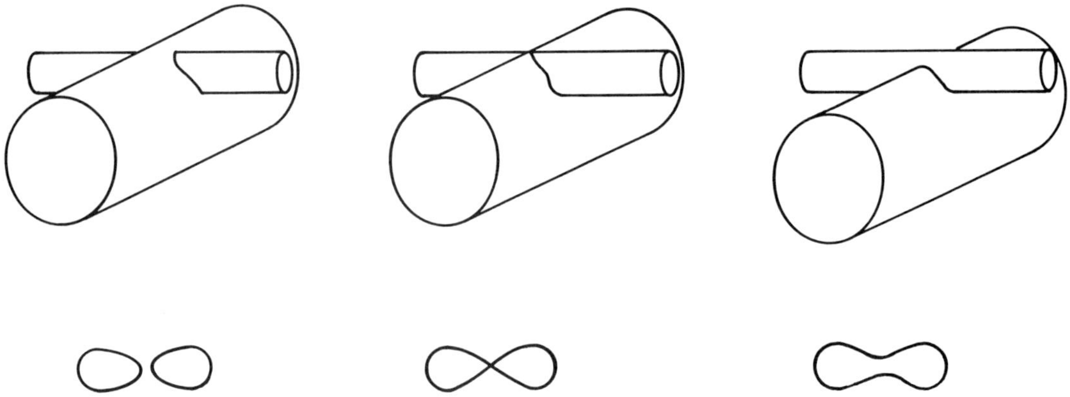

Figure 13.15. Three possibilities for the topology of an intersection curve between two cylinders having different diameters.

performing intersection calculations. Matters are improving however as new algorithms are developed and the speed of computing hardware increases.

Although intersections between such apparently simple entities as lines and circles are even now presenting computational problems [119]. Attention will here be concentrated on surface-surface intersections. These are the most difficult of the intersection calculations performed by most systems, and this type of problem is also the most widely discussed in the literature. A review of the methods used for parametric surfaces is given by Pratt and Geisow [57], and other recent references to specific techniques will be cited as appropriate.

There are two types of complexity which occur in the surface intersection problem, referred to by Pratt and Geisow as 'global' and 'local'. *Global* complexity is concerned with the number of disjoint branches of the intersection curve, which may be many. Clearly all these branches should be detected (the missing of small closed loops is a common problem), and each of them should be dealt with once and once only. *Local* complexity is concerned with the behavior of an intersection curve in the neighborhood of its singular points such as cusps, self-intersections and self-tangencies. It is easy for errors to arise in the computed connectivity of segments of the curve in such regions. The two types of complexity are illustrated in Figure 13.16. Farouki [58] has suggested a way of overcoming local complexity problems, at least for the intersection of parametric with algebraic surfaces, by identifying all the singular points first, and by then breaking the curve up into segments such that all singular points are segment end-points. This permits the reliable computation of all the segments, since none of them have singularities in their interior, and also assists in the determination of the correct connectivity at points where the curve intersects or is tangent to itself.

Intersections between implicitly defined surfaces

Developers of solid modeling systems have given much attention to this problem, especially as regards intersections between quadric surfaces. An algorithm originally due to Levin [59] forms the basis for a number of practical methods (see, for example, Sarraga [60]). Levin uses the fact that if two quadric surfaces $f = 0$ and $g = 0$ intersect in a curve C, then the family of quadric surfaces $f + \lambda g = 0$, where λ is any real scalar, also contains C. It is always possible to choose λ so that the linear combination is a ruled surface.

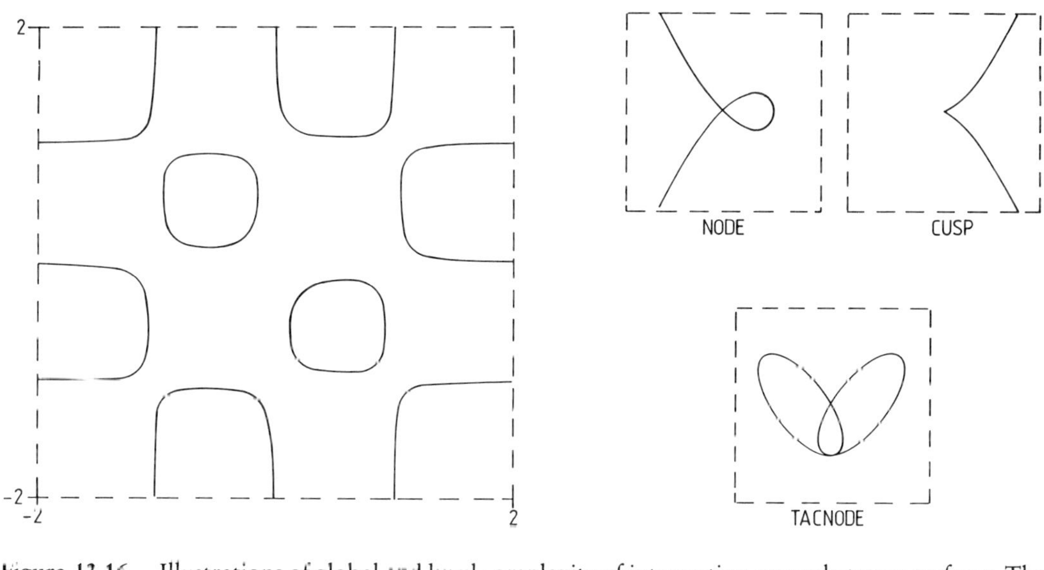

Figure 13.16. Illustrations of global and local complexity of intersection curves between surfaces. The left-hand figure shows the global complexity of the $z = 0$ contour of the bicubic surface $z = (x - 1)x(x + 1)(y - 1)y(y + 1) + 0.05$ on the square $[-2, 2] \times [-2, 2]$. The example is taken from Geisow [75].

The intersection of either of the original surfaces with this new surface gives the desired intersection curve, which, owing to the fact that a ruled surface has a simple linear parametrization in one direction, can be expressed in a comparatively simple quadratic parametric form.

Surface intersection computation in 'traditional' solid modelers is complicated by the need to implement the torus, which, of course, is not a quadric. Methods suitable for use with the torus and more general cyclides are discussed by de Pont [61]. A suitable technique is to express one of the surfaces concerned in implicit and the other in parametric form; the resulting 'hybrid' problem has a solution which, in the case of two cyclides, is expressible as a quartic relation between u and v in the parameter space of the parametric surface.

Recently, there has developed an interest in the use of general algebraic surfaces for geometric modeling, and several authors have published algorithms for computing their intersection curves, including Bajaj *et al.* [62] and Garrity and Warren [63]. A practical intersection method for completely general implicit surfaces has been described by Owen and Rockwood [64].

An interesting approach to the computation of individual points on intersection curves has been investigated by Morgan [65]. A point is specified as the intersection of three implicitly defined surfaces, and its coordinates are found by solving the three surface equations simultaneously. Morgan's method uses *homotopy continuation*, starting from a simplified set of equations whose solution can be easily found. The equations are then progressively modified by the gradual reintroduction of the neglected terms, and the corresponding changes in their solution are followed; this requires certain differential equations to be solved numerically. Eventually, the modification leads back to the original set of surface equations and hence to the desired intersection point.

Intersections between geometrically defined surfaces

The use of geometric surface definitions for the commonly used quadric surfaces was advocated by Goldman [3], and subsequently by Miller [2, 66]. A more recent and detailed paper by Piegl [67] discusses the use of this approach for computing intersections between natural quadrics, and concludes that it has significant advantages in speed, accuracy and robustness over other available methods, including that of Levin outlined in the previous section.

Intersections between parametrically defined surfaces

This is the type of surface intersection problem which has attracted the most attention in the literature, and it is also the one which presents the greatest combination of difficulties. It is shown by Pratt and Geisow [57] that even low-degree polynomial surfaces can have very complex intersection curves with many disjoint branches (see Figure 13.16). Four main classes of approach are identified, as listed below; further details may be found in the original paper.

1. Algebra-based methods – clearly, explicit solutions exist in certain geometrically simple cases, such as plane/quadric intersections which always give rise to conic sections, and such curves may be determined by algebraic means. However, algebra may also sometimes be used to simplify a problem which has no explicit solution by putting it into a form where a computed solution is easier to obtain. As will be explained below, the simplest form of intersection calculation occurs when one of the surfaces is implicitly and the other parametrically defined. Sederberg *et al.* [19] have recently shown how one type of representation may be converted into the other to obtain this desirable situation, using algebraic techniques developed more than a century ago. Some further discussion of representation conversion is given below.
2. Lattice-based methods – little will be said about these, since they are generally used for contouring algorithms, i.e. for computing intersections of general surfaces with horizontal planes. Such methods function by relating discrete data extracted from the general surface with a regular grid defined upon the plane concerned and then calculating solutions to simpler sub-problems in each of the grid cells.
3. Marching methods – these are methods which require an initial point on the

intersection curve to be determined by some means, and which build up the curve from this point using computations dependent upon local surface geometry. A step-length must be specified for the determination of each new point, and this is usually done adaptively, often using a local estimate of the curvature of the intersection. Marching methods are very sensitive to the step-length mechanism, since if an inappropriate choice is made it is easy for the procedure to step from one branch of the curve onto another nearby branch, and so to give a topologically incorrect result.

4. Recursive subdivision methods – this class of techniques is appropriate when Bézier or B-spline surface representations are used. They depend upon the convex hull and variation-diminishing properties of these representations, and are essentially stratagems of the 'divide and conquer' type. At each subdivision a pair of similar subproblems is generated, and the process proceeds until all the resulting subproblems have solutions which are in effect trivial.

A further novel method has recently been suggested by Cheng [68] based on ideas from two-dimensional vector field theory.

Hybrid intersection calculations

The intersection curve between two implicit surfaces with equations $f(x, y, z) = 0$ and $g(x, y, z) = 0$ may be computed by the simultaneous solution of these two equations in three variables. The degree of freedom which remains represents the freedom for a point on the intersection to move along the curve; in practice, it is effectively removed by the imposition of a step-length constraint which is equivalent to a third surface intersecting the first two to determine a discrete point on the curve. The surface equations are often nonlinear, and the Newton–Raphson method or some equivalent must then be used for their solution.

Now consider the case of two parametric surfaces, $\mathbf{r} = \mathbf{r}_1(u_1, v_1)$ and $\mathbf{r} = \mathbf{r}_2(u_2, v_2)$. Their intersection is given by the single vector equation $\mathbf{r}_1 - \mathbf{r}_2 = 0$. This is equivalent to three scalar equations in the four variables u_1, v_1, u_2, v_2, whose simultaneous solution once again leaves one degree of freedom. Computationally, this appears to be a more difficult problem to solve than the one in the last paragraph.

Finally, consider an implicit surface $f(x, y, z) = 0$ and a parametric surface $\mathbf{r} = \mathbf{r}(u, v)$. This proves to be computationally the simplest case of the three, since the three components of $\mathbf{r}$ may simply be substituted into the first equation to give a single equation in two variables

$$f(x\{u, v\}, y\{u, v\}, z\{u, v\}) = F(u, v) = 0$$

This result explains the desirability of conversion between implicit and parametric form, and *vice versa,* since either of the first two problems may be reduced to the apparently simpler third problem by this means. It should be noted however that Hoffmann [69] argues that the simplification is illusory for several reasons, including problems of numerical stability, and that it is better to work with a larger number of variables and consequently more, but simpler, equations. This matter still leaves much room for debate.

General versus specialized intersection algorithms

Some modeling systems employ specialized intersection routines for each pair of surface types. This strategy implies that whenever a new surface type is implemented in the system, new routines must be provided for the computation of intersections of this surface type with each of the other types previously implemented. The task clearly increases in magnitude as the geometric capabilities of the system are increased. On the other hand, the proponents of the use of non-uniform rational B-splines (NURBS) point out that surfaces of this type subsume the plane, all the quadric surfaces and the torus, and also provide a fully free-form capability. It is therefore possible to write a single routine for intersecting a pair of NURBS surfaces which obviates the need for separate procedures for intersecting a cone with a sphere, a cylinder with a torus, etc. This possibility certainly exists, but unfortunately such a general routine turns out to be rather inefficient when used for certain simple but frequently occurring special cases, in particular plane–plane intersections. It therefore proves expedient to provide specialized code for dealing with these cases, which implies a move back towards the philosophy of using tailored routines for each pair of surface types. At present, it must be said that the optimal strategy is still a matter for debate which will only be resolved by further experience.

Some intersection methods are designed for use with particular types of surface representation; in the Bézier and B-spline cases, for example, methods based on recursive subdivision have been devised. On the other hand, there is some virtue in the use of methods which will function with, say, any pair of parametrically defined surfaces. Techniques of this kind are based on what is known as the *parametric evaluator* approach. They rely solely on the existence for each surface type of a routine which, when passed values of the two surface parameters, will return the position of the corresponding space point and the values of the first and second derivatives at that point. Algorithms of this kind have been demonstrated [70], but at present they tend to be slow. To implement a new surface type in a system employing this approach it is only necessary to write an appropriate new evaluator routine; the actual algorithms used remain unchanged. The same philosophy may obviously also be applied to other types of curve and surface interrogations.

Representation of intersection curves

It has been found that intersection curves between either algebraically or parametrically defined surfaces of the polynomial or rational type possess algebraic representations (see below). These are, however, generally of such high degree that their computation is not a practical proposition. For this reason such curves are usually calculated in the first instance as sequences of points using methods such as those described above. This leads to various alternative possibilities for the representation of the curve, as has been discussed previously.

Trimmed surfaces

A *trimmed surface* is a bounded region defined upon either a larger surface or an unbounded surface. In a solid modeling context, the faces of an object are trimmed surfaces, at least from the geometric point of view. Such regions may be quite complex with multiple non-intersecting boundaries. A comparatively simple example is shown in Figure 13.17. Some modeling systems seek to avoid such complications by subdividing regions of this kind in such a way that each subregion has a single boundary. To illustrate,

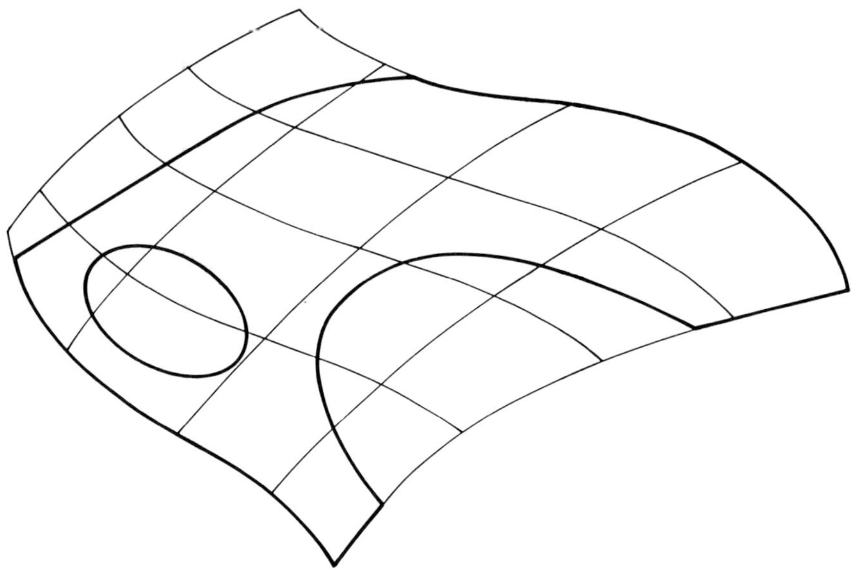

Figure 13.17. A trimmed surface region defined upon a patched parametric surface.

this may be achieved for a square face containing a central hole by the interpolation of two new dummy edges linking the square face boundary to the hole periphery. In complex cases there is no obviously optimal strategy for such a decomposition. There are two major approaches to the modeling of trimmed surfaces, as described in the following paragraphs.

Some modeling systems bound trimmed surfaces by topological means. This is to say that the representation of the original larger surface is left unmodified, the trimmed region being defined by pointers to that surface, and also to one or more loops of bounding edges lying upon it. These edges will be specified, in turn, by the curves on which they lie and the vertices which bound them. This technique in fact uses the standard hierarchical face/edge/vertex data structure typical of solid modelers of the boundary representation type [71]. If such facilities are available, the major computational problems are those of computing the edge curves (which are often initially specified as intersection curves) and of representing them in some economical manner. These matters were discussed above.

Where the topological bounding method cannot be used it is necessary to represent a trimmed surface as a separate surface in its own right, possibly defined in a piecewise manner. Consider, for example, the trimmed region created by projecting a square onto a parametrically defined sculptured surface. First the boundaries of the region must be computed; these will become the boundaries of the new surface. It is easy to create a surface fitting those boundaries, for example by the use of Coons blending. Obviously, the result will only be an approximation to the desired result, and there are various means of improving this approximation, for example by taking into account interior data computed from the original surface. Some appropriate techniques will be briefly reviewed below. In more complex cases, especially those involving interior loops of edges bounding holes in the trimmed surface, some automatic means will be needed of subdividing the region into an assembly of three- or four-sided patches, depending upon the type of mathematical representation used in the underlying system. This problem is easier if triangular patches are used, but this is not a satisfactory option with most commercially available CAD systems.

Farouki [72] is a useful source of further information on the 'topological' approach to the representation of trimmed surfaces, and also on their algorithmic requirements. Casale [73] outlines a distinctive method for defining trimmed surfaces, based on the use of 2D Boolean operations in parameter space.

Representation conversion

Problems of *representation conversion* occur in several fields of geometric modeling. One such problem concerns the generation of one type of solid modeling representation from another; for example, methods for the transformation of a constructive solid geometry (CSG) model into a boundary representation model are well known, whereas no satisfactory algorithm has yet been published for the reverse process [74]. This type of conversion has both geometrical and topological aspects, and since topology falls outside the scope of this chapter attention will here be concentrated on conversions of a purely geometric nature.

Implicitation and parametrization

In this section the conversion of curve and surface representations from parametric to implicit form, and *vice versa,* will be briefly discussed. Each type of representation has certain advantages, and as pointed out above, intersection calculations are easiest if, given a pair of intersecting entities in either implicit or parametric form, a hybrid situation can be created by converting one of them into the other form.

The idea behind conversion from parametric to implicit form is a simple one. For example, a parametric surface may be expressed in component form as

$$x = x(u, v)$$
$$y = y(u, v)$$
$$z = z(u, v)$$

In principle, the process of implicitization involves the elimination of the parameters u and v between these three equations to leave a relation between x, y and z. In practice, straightforward techniques of algebraic elimination lead to enormous complexity even for simple parametric forms such as bicubics, to such an extent that until comparatively recently implicitization was thought to be a practical impossibility. However, it was then discovered that the concept of *resultants,* developed in the Victorian era but subsequently neglected, provided a practical way forward. This approach was first published by Sederberg *et al.* [18], though similar ideas were being pursued independently by Geisow [75]. Still more recently, alternative methods for implicitization have been developed using *Gröbner bases* [76].

As regards the outcome of the conversion, a planar parametric curve of degree n, either polynomial or rational in its parameter, converts to an implicit algebraic curve of the same degree in x and y. A four-sided parametric surface patch of degree $n \times n$, either polynomial or rational, converts to an implicit algebraic surface of degree $2n^2$ in x, y and z, though it is perhaps worth noting that this reduces to n^2 for triangular patches. The commonly used four-sided bicubic polynomial patch therefore has an implicit representation of degree 18, containing no less than 1330 terms. This alarming complexity is one reason for the growing interest in modeling with algebraic rather than parametric surfaces. All polynomial and rational parametric surfaces have corresponding algebraic forms, but the reverse is not true; the algebraic surfaces are therefore a superset of these parametric surfaces, and offer more freedom for the designer whilst presenting less algebraic complexity for the system implementer. One problem, as previously mentioned, is that of presenting the geometric freedom to the user in an intuitively understandable way; more progress is needed on this topic.

The process of implicitization also applies, in principle, to the generation of implicit algebraic representations of intersection curves. However, the algebraic degree of the intersection curve between two parametric bicubic surfaces is 324 and its exact computation, even using symbolic methods, is not currently practicable.

The reverse process, parametrization, has also received some attention [77–81]. Work has so far been concentrated on curves and surfaces of low degree. As pointed out by Sederberg and Snively [77], the cubic algebraic surfaces are of particular interest since they are the surfaces of *lowest* degree which can be used in a piecewise manner to construct composite surfaces with tangent plane continuity, whilst being also the *highest* degree surfaces which generally permit a rational parametrization. Note that when parametric surface patches are used to construct smooth composite surfaces the degree employed is usually bicubic or greater. The bicubic case, as mentioned above, has algebraic degree 18, compared with the cubic degree required to obtain a similar result by the direct use of algebraic surfaces. Algebraic surface patches, however, have certain disadvantages. In particular, their boundaries do not in general correspond to isoparametric lines of the equivalent parametric forms, and this leads to difficulties which do not arise with conventional parametric patches.

Exact conversion between polynomial bases

Conversions between different types of parametric polynomial curve and surface representations are required for the transfer of geometric data between different CAD systems. The usual means for achieving this is by the use of a standard *neutral format,* the best-known current example of which is IGES [82], although a new draft international standard format called STEP [21] has recently been published. The idea is that each CAD system is furnished with two translators. These are a *preprocessor,* which converts the native internal geometry representations of the system into the neutral format for transmission to another system, and a *postprocessor* which performs the reverse process for data received from another system. This philosophy avoids the vast proliferation of translators which would occur if specialized software were written to accomplish transfer between all possible pairs of different CAD systems, whose numbers must now run into hundreds.

Conversions are necessary even for very simple geometry. As mentioned above, there are rival representations for straight lines, and the same is certainly true for circular arcs. The advantages and disadvantages of different representations of conic curves have been discussed by Wilson [83]. However, the main concern here is translation between free-form curve and surface representations, many of which are mathematically equivalent. For example, it may be desired to transfer geometric data from a Bézier-based system into a polynomial B-spline system. This may actually require more than one conversion, depending upon the type of representation specified by the neutral file format; there is in fact a German standard called VDAFS which represents free-form polynomial surface data in terms of explicit polynomial coefficients rather than Bézier or B-spline coefficients, and the use of this would necessitate two conversions. Exact conversion is, of course, not possible, even in principle, if the degree of the polynomials transmitted by the sending system exceeds the maximum degree which can be handled by the receiving system. The latter ranges from 2 to 20 over the spectrum of commercially available systems. Clearly, exact conversion is frequently impossible, and approximation must then be used; this topic is dealt with below.

In principle, conversion between different polynomial bases of the same degree is simple. Consider two surface patches of degree 3 × 3, for example

$$\mathbf{r}_1(u, v) = \mathbf{UAMA}^{\mathrm{T}}\mathbf{V}^{\mathrm{T}}$$

$$\mathbf{r}_2(u, v) = \mathbf{UBNB}^{\mathrm{T}}\mathbf{V}^{\mathrm{T}}$$

where $\mathbf{U} = (1, u, u^2, u^3)$, $\mathbf{V} = (1, v, v^2, v^3)$, **A** and **B** are constant 3 × 3 matrices and **M**, **N** are 3 × 3 matrices containing the defining vector-valued coefficients of the patches. Different nonsingular matrices **A**, **B** generate different sets of basis functions of cubic degree, for example Hermite, Bézier or B-spline bases [84]. If the two patches are required to be geometrically identical, even though they are expressed in different terms, equating coefficients in the two bicubic functions gives

$$\mathbf{AMA}^{\mathrm{T}} = \mathbf{BNB}^{\mathrm{T}}$$

and hence, since **A** and **B** are nonsingular,

$$\mathbf{M} = \mathbf{A}^{-1}\mathbf{BNB}^{\mathrm{T}}\mathbf{A}^{-\mathrm{T}}$$

This expresses the defining coefficients of the first patch in terms of those of the second; obviously, the inverse relation may be similarly obtained. Unfortunately, although these exact algebraic relations exist, the conversion as carried out in practice using a computer working to finite precision is subject to arithmetic rounding errors. It is therefore important, particularly for engineering purposes where precision of geometric information matters a great deal, that numerically stable algorithms are used to keep these introduced errors to a minimum.

References containing details of conversions between different polynomial bases include de Boor [85], Sablonnière [86], Cohen *et al.* [87], Boehm [88, 89], Cohen and Riesenfeld [90] and Lyche and Mørken [91].

Curve and surface approximations

The necessity to transfer geometric information between different CAD systems has already been mentioned above. In practice the sending and receiving systems are frequently not mathematically compatible with each other, so that exact conversion of the data is not possible. Polynomial-based systems known to the present writer range in maximum degree implemented from 2 to 20; the range of variation amongst rational systems is narrower, but nonetheless significant.

The problem addressed here arises, for example, when it is desired to transfer geometry from a polynomial system with surfaces of degree 20 × 20 into a system limited to bicubics. Clearly, this cannot be done exactly. Similarly, transfer from a rational system into a polynomial system must involve approximation, since the former handles a wider class of functions. The reverse process can, however, be carried out exactly in principle, subject to obvious provisos on the degrees of the functions involved, since the polynomials are a subset of the rationals.

Clearly, when any approximation of engineering data is performed the user must be able to specify a tolerance on the accuracy of the results. In practice, this means that a high-degree curve segment will have to be split into several segments if it is transferred into a system limited to handling a lower degree. Similarly, a single high-degree surface patch may be split into an assembly of lower-degree patches. Subdivision is necessary to gain the added flexibility needed in the receiving system for accurate representation of the transmitted data.

Degree reduction for pure polynomial systems

It is possible to base approximation algorithms for parametric curves and surfaces on classical methods, though certain modifications are necessary. A given scalar-valued polynomial $p_n(x)$ of degree n has a best uniform approximation $p_{n-1}(x)$ of degree $(n - 1)$ on the interval $[-1, 1]$ which may be found by *Chebyshev economization* [92]. The term *best uniform approximation* means that the maximum value of the quantity

$$\varepsilon(x) = |p_n(x) - p_{n-1}(x)|$$

on the interval is minimized. The method is based on the properties of *Chebyshev polynomials*, and the approximation is derived by subtracting from $p_n(x)$ an appropriate multiple of the Chebyshev polynomial of degree n so that the terms of degree n cancel. The effect is to distribute the error over the interval $[-1, 1]$ in such a way that it oscillates between maximum and minimum values having the same magnitude; the number of oscillations depends on the degree.

Since a parametric polynomial curve has two or three components which are themselves scalar polynomial functions of the parameter, the method of Chebyshev economization may be applied directly to each component to achieve a degree reduction. A minor problem is that classically the Chebyshev polynomials have their crucial properties on $[-1, 1]$, whereas parametric curves are often defined on the interval $[0, 1]$ in their parameter. A simple linear transformation of variable generates the *shifted* Chebyshev polynomials appropriate for this or any other interval. A more serious problem stems from the fact that the Chebyshev polynomials are non-zero at the ends of the interval concerned. In geometric terms this implies that the end-points of the original and the economized curves do not coincide. Then if it is desired to approximate a *piecewise* polynomial curve the economized segments resulting from this procedure will, in general, not even be connected to each other. In practice, the original curve may well have continuity up to first or second derivatives at its junctions, and a satisfactory approximation algorithm should be able to reproduce these conditions if desired as well as mere positional continuity.

Lachance [93] has proposed a method for achieving the aims described above, based on the use of *constrained* Chebyshev polynomials. These share the property known as equioscillation with the standard and shifted Chebyshev polynomials, but they are constrained so that their values and those of their first m derivatives are zero at the end points of the parameter range of interest, usually $[0, 1]$. A different set of constrained polynomials results for each choice of m, though for practical purposes it is not usually necessary to go beyond $m = 2$. This method gives a realistic upper bound on the error of the approximation for each curve segment; if the error estimate is greater than some predefined tolerance then the segment being approximated may be subdivided at its parametric mid-point and the procedure applied separately to the two resulting smaller segments. This subdivision may be repeated recursively until the desired tolerance is achieved over the entire length of the segment. The method has the advantage that it approximates not only the geometry of the curve, but also its parametrization, which may be important for some applications.

Lachance's paper also describes the extension of his method to the economization of polynomial surface patches, which may be of different degrees in their two parameters. In this case the degree is reduced by one in each parametric direction.

Multiple applications of the Chebyshev economization method are needed to achieve degree reductions of more than one. In this case, it can no longer be guaranteed that the resulting approximation is best possible in the uniform sense. Nevertheless, the procedure has been found to work well in practice; the use of subdivision as described above permits the reliable approximation of the original geometry to any desired lower degree and to within any specified tolerance.

A more general approximation algorithm

The Chebyshev-based method described above is restricted to degree reduction of polynomial curves and surfaces. However, it is frequently desired to obtain a polynomial approximation of some specified degree to an entity originally represented in non-polynomial terms, usually rational or procedural. For this purpose Goult [94] has devised an approach based on least-squares rather than uniform approximation. What is now minimized is the integral of the square of the distance between corresponding points on the original and the approximating curve or surface. Classically, this is achieved most efficiently in terms of Legendre polynomials. In a CAD context, for reasons mentioned in the last section, it proves necessary to define and work in terms of *constrained* Legendre polynomials. Goult's method requires the evaluation of integrals, which is conveniently

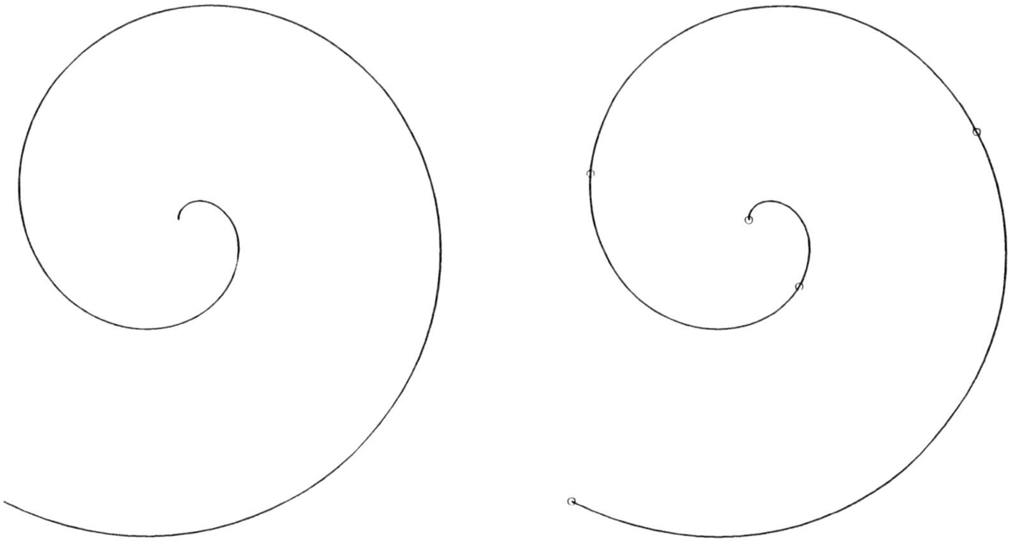

Figure 13.18. An equiangular spiral (exact trigonometric representation) and a piecewise quartic approximation having four segments.

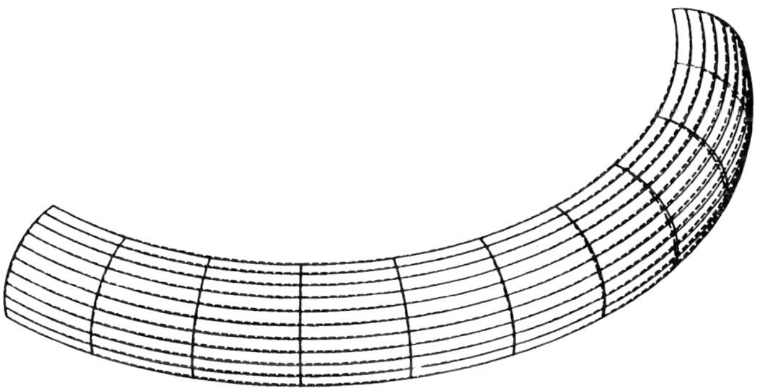

Figure 13.19. A toroidal patch in an exact trigonometric representation (solid lines) and a single patch bicubic polynomial approximation (dashed lines).

done numerically using Gaussian quadrature. Then the curve or surface being approximated need only be sampled at a set of discrete points. Provided this is possible, the actual nature of the curve or surface being approximated is immaterial. It should be noted that whereas Lachance's algorithm reduces the degree by one on each application, a single application of Goult's method will generate a polynomial approximation of any specified degree. In common with the former method, the parametrization as well as the geometry of the original curve or surface is approximated. Examples of curve and surface approximation using this approach are shown in Figures 13.18 and 13.19.

Since the functionality of Goult's method does not depend upon the nature of the curve or surface being approximated, this technique may be used for the approximation of procedurally defined entities. It is also applicable to *degree elevation*, i.e. the approximation of composite curves made up of low degree segments in terms of a smaller number of higher-degree segments.

Other curve and surface approximation methods

Some other techniques for parametric curve and surface approximation have also been proposed. Watkins and Worsey [95] achieve degree reduction (for Bézier curves only) using unconstrained Chebyshev polynomials. This has the effect of moving the end points as pointed out above; they overcome this by simply moving the end points back to their original positions after the degree reduction. An error analysis indicates that the results are acceptable for single segments, but it is not possible to obtain continuity of first or higher derivatives when piecewise curves are dealt with in this manner. Dannenberg and Nowacki [96] describe a more general method for curve and surface approximation based on a subdivision strategy and a knot placement algorithm due to Hölzle [97]. Hoschek [98] and Hoschek and Wissel [99] describe a technique for lowering or raising the degree of spline curves whilst retaining derivative continuity, using modification of the curve parametrization to achieve geometrically optimal results. It has already been mentioned, however, that significant changes in the parametrization may not be acceptable for certain applications. Patrikalakis [100], and Bardis and Patrikalakis [101] describe techniques for approximate curve and surface conversion from rational to non-rational B-spline form, based on iterative adjustment of knot vectors and control points. Lyche and Mørken [102] use an approach based on an initial dense sampling of the entity to be approximated. In the curve case, this gives a piecewise linear approximation which is regarded as a B-spline curve of degree 1. Then the degree of the B-spline curve is progressively raised and data points discarded in such a way that the curve always remains within some specified tolerance of the data. The process can be stopped at whatever degree is required for the final approximation.

Application of approximation to the trimmed surface problem

As mentioned above, some CAD systems use topological methods for the representation of trimmed surfaces, while others either have no trimmed surface capability at all or use reparametrization. Consider the transfer of data from a system of the first type into a polynomial-based system of the second or third type. Two operations must be performed: firstly, the trimmed surface must be subdivided into an assembly of (conventionally) four-sided patches, and secondly, each patch must be approximated in terms of whatever degree of polynomials is appropriate for the receiving system. As regards the first step, note that any patch structure existing on the surface which has been trimmed must almost invariably be modified for the second step because the trim curves usually give rise to three or five-sided regions around the boundary (see Figure 13.17). Once the new patch structure has been determined, any of the approximation methods mentioned above can be used to achieve the reparametrization of the second stage.

Graphical interrogations

In this section two types of interrogation required for the graphical rendering of geometric models will be briefly reviewed. The first is needed for the computer generation of realistic line drawings of objects with curved surfaces. Consider, for example, a cone. Visually, this appears to have a boundary consisting of a pair of intersecting straight lines. The lines are not actual edges of the cone, however, they are view-dependent horizon lines separating the region of the cone's surface which can be seen by the viewer from the region which is hidden from him. Such lines (or more generally curves in the case of doubly curved surfaces) are usually referred to as *profile lines* or *silhouette lines*. They are important for the generation of convincing drawings; for example, a perspective view of a cylinder of finite length will appear simply as a pair of disconnected ellipses if the edges alone are drawn. The addition of the profile lines gives a much better impression of the shape depicted.

The silhouette line of a smooth curved surface is the locus of points at which straight lines through the viewpoint are tangential to the surface. Equivalently, at these points the lines of sight are perpendicular to the surface normal. For a parametric surface $\mathbf{r} = \mathbf{r}(u, v)$ viewed under parallel projection such points satisfy

$$\mathbf{s} \cdot (\mathbf{r}_u(u, v) \times \mathbf{r}_v(u, v)) = 0$$

where $\mathbf{s}$ is the direction of sight. The corresponding equation for a perspective projection is

$$(\mathbf{r}(u, v) - \mathbf{e}) \cdot (\mathbf{r}_u(u, v) \times \mathbf{r}_v(u, v)) = 0$$

where $\mathbf{e}$ is the position vector of the eyepoint.

The equivalent results for implicit surfaces are given by Geisow [35], who also shows that any silhouette of a quadric surface is a conic curve, a fact known to projective geometers for many years. For a specified surface function $\mathbf{r}(u, v)$ both the above equations reduce to the form $h(u, v) = 0$, where h is a scalar-valued function. This is the equation for the silhouette line in terms of the parametric coordinates of the surface. It is easy to work out the degree of the equation when the surfaces are of parametric polynomial type; this is biquadratic and biquintic respectively for the two above examples when the surface itself is bicubic. Some appropriate strategy must be used for stepping along the curve to compute points upon it. This is equivalent to introducing another relation between u and v, and so points are effectively calculated by solving two simultaneous nonlinear equations in those two variables. A simpler strategy (but one which does not work well in practice) is simply to assign a sequence of equally spaced values to either u or v, in which case it is only necessary at each step to solve a polynomial equation in the remaining variable.

Whereas line drawings are appropriate for many engineering purposes, today's graphical workstations can now generate shaded surface renderings. These too have their engineering uses, allowing realistic 3D visualization at the design stage and providing illustrations for technical and sales documentation. However, it is undoubtedly in the leisure industry that this type of graphical technique has its greatest value. The basic principles of the *ray-firing* method of shaded surface rendering are simple. The existence of a surfaced model is assumed. An eyepoint is chosen, and the positions of one or more sources of illumination are specified. Now consider a line of sight through the eyepoint which passes through the model. This line (or *ray*) in general intersects the surface of the model at least twice, once on entry and once on exit. Intersection calculations can be used to compute the intersections, the nearest to the eyepoint giving a point on the surface of the model which can be seen by the viewer. The surface normal can now be computed at this point. The surface illumination there is governed by the strengths and distances of the light sources and the angles made by the rays from these sources with the surface normal. What the viewer observes is light reflected from the surface. There are two types of reflectivity, specular (mirror-like, with equal angles of incidence and reflection) and diffuse (where the light may be considered to be absorbed by the surface and then re-emitted in some radially symmetric distribution about the normal direction). It is easy to model specular reflection, and simple heuristic formulae such as Lambert's Cosine Law give good agreement with observed patterns of diffuse reflection. The two types of model may be combined in varying proportions to give, for example, a metallic appearance to the modeled surface.

With these preliminaries, the rendering process is as follows. The plane of the screen is assumed to lie between the model and the observer. A ray is fired from the eyepoint through one pixel of the screen to intersect the model. The method described earlier is used to determine the illumination of the model surface at the appropriate point of intersection, and the chosen reflection model then determines the amount of light reflected back to the observer along the initial ray. This process is then repeated for all other pixels through which rays from the eyepoint intersect the model. For a high-resolution screen there is likely to be a large number of these, and so the generation of shaded surface images is highly computation-intensive. Further details may be found in any textbook on computer graphics (for example Foley and Van Dam [103]). The point of providing a brief description here is to show that the procedures used depend only upon simple geometric calculations of the types described earlier.

There are cruder and faster methods than the one described, based on the use of faceted approximations to curved surfaces. In the simplest case, the facets may be regarded as flat and uniformly illuminated, but various techniques have been devised for smoothing the apparent illumination to blur the resulting sharp changes in brightness between facets. The best known are Gouraud and Phong shading; both require little more than the computation of surface normals. Foley and Van Dam [103] again give details.

Ray-firing and faceting approaches to the generation of shaded surface images have been briefly outlined above. Other methods for creating hidden-line and hidden-surface pictures from geometric models will not be discussed here. Many of the algorithms used require only simple geometric calculations and are dominated by sorting procedures for determining which faces obscure which other faces and so forth.

Again, these techniques are covered in many textbooks on computer graphics.

Approximate development of surfaces

Little has been published on approximate development of non-developable surfaces, though several CAD systems provide some capability in this respect. Consider first a ruled surface which *is* developable, and recall that this requires all points on any generator to share the same tangent plane. Suppose that a curve C_1 is drawn on the original surface; there will be a corresponding planar curve C_2 on the developed surface. The process of development uses the fact [104] that the curvature at any point of C_2 is equal to the curvature at the corresponding point of C_1 of the projection of C_1 onto the tangent plane at that point. The latter is the *geodesic curvatuve* κ_g. If the parametric equations of the boundary of the developable surface region are known the geodesic curvature along these curves may be computed from

$$\kappa_g = \mathbf{n}\cdot(\dot{\mathbf{r}} \times \ddot{\mathbf{r}})/(\dot{s})^3$$

the projection of the curvature vector as given by equation (39) in the surface normal direction, i.e. onto the tangent plane. With κ_g known at any point on the boundary, the curvature at the corresponding point of the equivalent boundary of the developed surface is known. The boundary curve may then be computed by integration of the *intrinsic equations* mentioned earlier, which are coupled linear differential equations expressing the curve coordinates in terms of the curvature along the curve. Faux and Pratt [14] give further details.

For a non-developable ruled surface approximate methods must be used; such surfaces cannot be flattened without some stretching or tearing. One crude but often reasonably effective method is simply to polygonalize the two generating curves, using the same number of linear segments in approximating each of them. Joining corresponding vertices now gives an approximation to the surface in terms of a series of strips which are in general, however, not planar but twisted. If diagonals are next constructed in each strip the surface is approximated by an assembly of planar triangular facets. The approximate development now proceeds by holding the first facet fixed and rotating the second about their shared diagonal until it becomes coplanar with the first. The third facet is now similarly rotated about its shared boundary with the second, until finally the whole approximate surface has been unrolled onto a plane.

For non-ruled surfaces no such simple techniques exist, and, as already mentioned, little has been published on this topic. One relevant reference is Manning [105], in which it is proposed to use classical map projection techniques to map a tree-like graph of feature lines on the surface onto a plane. This is accomplished for each feature line by constructing an auxiliary developable surface containing it and then unrolling this surface. Thus the lengths of the original and the mapped feature lines are the same. It is then possible to compute the amount of distortion in the regions of surface between these lines.

Offset curves and surfaces

The *offsetting* problem frequently occurs in applications of CAD, particularly in manufacturing. One common manifestation is in peripheral machining around a horizontal contour, when the center of the cutting tool moves along a curve which is offset from the contour curve by a constant distance, the cutter radius. Another is the machining of a parametric surface with a cutting tool having a spherical end; the tool reference point is the center of the sphere, which moves on a surface offset from the original by the radius of the sphere. Other applications include collision avoidance in robot motion planning [106], analysis of rocket fuel consumption [107], determination of tolerance zones in mechanical enginering design [108], and various approaches to the definition of blends and fillets in solid modeling [47, 109].

Offset curves

Initially, parallel offsets to planar curves will be discussed, where the offset lies in the same plane as the original. Such an offset curve is frequently more complicated geometrically than the curve from which it is generated, in that it may possess cusps or self-intersections even though the original exhibits neither of these. Self-intersections occur, for example, when the offset distance exceeds the radius of curvature of the

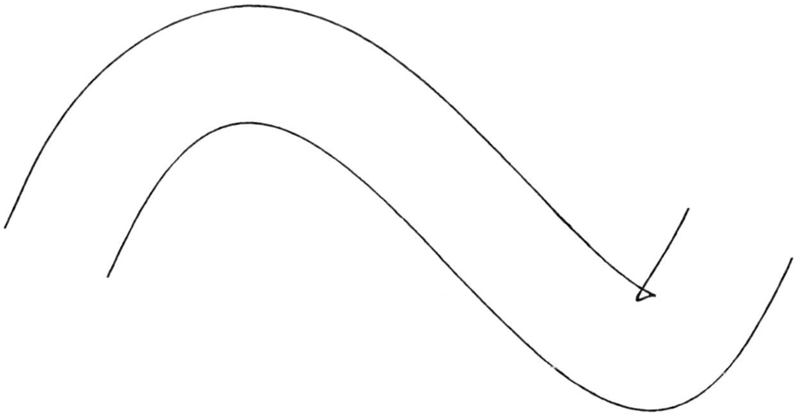

Figure 13.20. A parametric curve with an offset curve exhibiting a self-intersection.

original curve (see Figure 13.20). The usual procedure for computing an offset to a planar curve is straightforward, requiring nothing more complicated than the calculation of the unit principal normal to the curve. To every point on the original curve there corresponds a point on the offset curve, which is found simply by moving in the direction of the normal by the desired offset distance. If a sequence of points on the offset curve is calculated in this manner, then some form of spline curve may be used to interpolate them to give a continuous offset.

Certain practical problems remain to be solved. The first is that at a point of inflection on the original curve the direction of the principal normal flips from one side of the curve to the other, whereas what is usually required in practice is an offset which remains consistently on the same side of the original. A simple modification detects inflection points and reverses the direction of the principal normal as appropriate to ensure this. A second problem concerns the accuracy of the computed offset curve, which is usually required to satisfy some specified tolerance. It is clearly necessary to check the offset distance at points intermediate to those used in setting up the curve; if the tolerance is not met then new offset points must be computed at these intermediate points, a new spline curve calculated and further checks made. This process must be repeated until a satisfactory result is obtained. A third problem concerns offset curves with self-intersections, which often occur in the computation of cutter paths for NC machining. These arise in various situations, but are inevitable when the offset distance is greater than the radius of curvature of the original curve and the offset is on the same side as the center of curvature. In these cases it is often necessary to determine the points of self-intersection and to trim away the unwanted loops to obtain the desired trajectories, as shown in Figure 13.21.

Published papers on the approximation of planar offset curves include Klass [110], Tiller and Hanson [107], Hoschek [111] and Hoschek and Wissel [99]. Farouki and Sakkalis [112] have identified certain restricted classes of planar polynomial Bézier curves possessing offsets which are expressible exactly as rational curves of higher degree.

The computation of an approximate normal offset to a space curve is complicated by the fact that the geometry is not now constrained to lie in a plane. Consequently, the curve normal is not uniquely defined, and some additional constraint must be imposed. One possible choice is to take the offset to be in the direction of the *principal* normal at any point on the curve. Another is for the user to specify a fixed direction for the offset. Coquillart [113] describes a method for offsetting 3D rational B-spline curves using the concept of a *quasinormal*. This is defined so as to avoid certain problems arising when the principal normal is used, notably the 180° change of direction at an inflection point and the lack of uniqueness along a straight line segment.

A curve on a surface again has no uniquely defined offset. In algorithms for machining surfaces it is usual to take the offset direction at

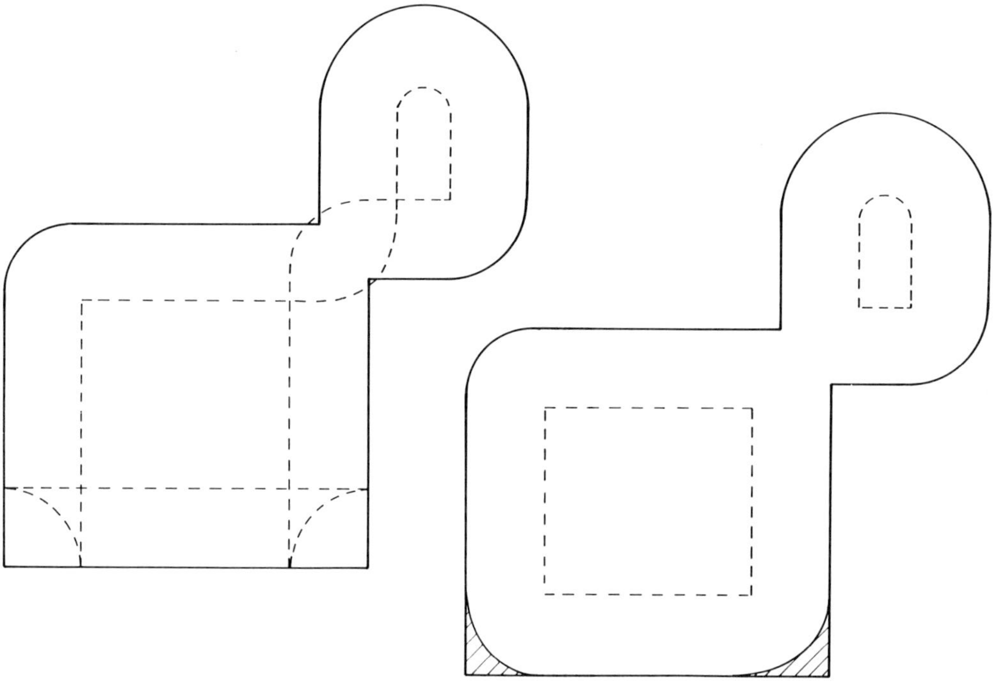

Figure 13.21. A pocket profile with initially computed and trimmed offset profiles. The latter will be used as the path of the cutter center in machining the walls of the pocket. Note the excess material which will remain in the lower corners of the pocket after machining.

any point to be that of the surface normal (see below). However, for design purposes it is sometimes useful to be able to define an offset curve which lies on the same surface as the original. Patrikalakis and Bardis [114] give a method for computing offsets of this kind, in which the offset distance is measured along geodesics on the surface, each geodesic passing through a specified point on the original curve in the 'bitangent' direction normal to the curve but tangent to the surface.

Offset surfaces

The computation of a constant distance offset surface is very simple in principle. To calculate a point on such a surface, it is only necessary to offset from any point on the original surface by the required distance in the direction of the unit surface normal. This may be done along isoparametric curves on the original surface to create a grid of points which are then interpolated to give an approximate offset surface. As in the case of offset curves, it is often necessary for practical purposes to impose some tolerance on this approximation, which requires sampling the offset distance at intermediate points and if necessary recomputing with more closely spaced data. There are also similar problems with cusps, ridges and self-intersections which can occur in the offset surface if the offset distance is chosen larger than the minimum principal radius of curvature of the original surface. Farouki [115] gives a good discussion of these problems.

Three further points are worth mentioning. The first is that some commercial CAD systems provide *tapered offsets,* for both curves and surfaces. These are simple to implement; the methods are the same as those described above, except that the offset distance is made to vary, usually linearly in the curve parameter or one of the surface parameters. The second point is that it is possible to create piecewise cyclide free-form surfaces having exact offsets which may be determined algebraically [34, 35]. Finally, it should be pointed out that the definitions of both offset curves and surfaces are essentially procedural in nature. Certain general approximation algorithms, such as that of Goult [94] are therefore well suited to the computation of approximate offsets.

Recommended further reading on the topic of offset surfaces are two papers by Farouki [115, 116]. The first of these gives a detailed theoretical introduction to the subject and discusses the generation of exact offsets to simple solid shapes. The second paper is concerned with approximate offsets of parametric surfaces. Two other papers on the theory of offsetting, with a solid modeling orientation, are Rossignac and Requicha [109] and Ghosh [117].

Interrogation for NC machining

Just one application will be considered here, the generation of cutter contact paths for the machining of surfaces. Depending to some extent on the type of surface representations used, there are two strategies which are widely implemented. These are dealt with briefly in the following paragraphs.

Conventional APT machining

APT stands for Automatically Programmed Tools; it is essentially a standardized language for defining certain types of geometry and for specifying cutting tool motions with respect to this geometry. If a program is written in this language then the computer can carry out low-level calculations to determine all the necessary detailed tool motions in terms of the reference coordinates of the machine tool to be used. This was one of the first applications of computers in manufacturing, dating in origin from the mid-1950s.

The surfaces conventionally provided in an APT system include the plane, the quadrics, and ruled surfaces including tabulated cylinders. These last are surfaces generated by linear translation of Wilson–Fowler splines, planar curvature-continuous piecewise cubics, in which each segment is defined in its own local coordinate system with the chord between two successive data points as its x-axis; Fritsch [118] provides an accessible reference. Modern APT systems often provide various types of parametric curve and surface definitions. The motion of a cutting tool is specified in terms of three surfaces. The first is the *part surface*, the surface being machined. The second is the *drive surface*, which may be thought of as guiding the motion of the tool along the part surface. The third surface is the *check surface*, which terminates the tool motion in any particular cutter pass. The tool moves so that it maintains simultaneous contact with the part and drive surfaces until it contacts the check surface, when it stops. An area of a surface may be cleared by defining a sequence of drive surfaces which might be, for example equally spaced parallel planes.

Isoparametric machining

A very common strategy for the machining of parametrically defined surfaces is to use isoparametric lines as the cutter contact paths. Area clearance is then achieved by employing a sequence of such lines as cutter paths, stepping by uniform increments in the transverse surface parameter between passes. This works reasonably well when the parametrization of the surface is fairly uniform, since then the scale of the irregularities remaining on the surface after machining is completed are predictable and controllable. For surfaces with highly non-uniform parametrization it is often preferable to machine along curves of intersection of the surface with a sequence of equally spaced parallel planes. This appears to be related back to the APT approach, but the computational details are usually rather different when parametric surfaces are involved; in particular, the intersection curve is used to define the cutter contact path explicitly, whereas in the APT approach the contact path is only implicitly defined in terms of the part and drive surfaces and the geometry of the cutting tool.

Computational aspects of NC machining

Initially, two common machining strategies will be described which have simple computational requirements. The necessary generalizations to more complex situations will then be discussed. First, contour milling is a means of rough machining, designed to remove most of the material to be machined away in preparation for the finer finishing cuts. Contours, in the usual geographical sense, are obtained by intersecting the surface to be machined with a series of parallel horizontal planes. For roughing it is usual to use a cylindrical cutter, with cutting edges on the sides of the cylinder. The motion of the tool reference point, the center of the base of the cutter, is on a profile which is simply the contour itself offset outwards by the cutter radius; this may be calculated using the methods introduced above. For finishing, machining along isoparametric lines is often used, with a cutter having a ball or spherical end. Here the center of the sphere moves on an offset surface; points on its path are easily computed using the method described earlier. In both cases, the major computational problem relates to the fact that machine tool controllers usually provide either linear or circular arc interpolation between calculated tool reference points. Thus the actual cutter path is inevitably an approximation to the ideal path, and the spacing of computed points needs to be carefully controlled to ensure that the machined surface remains within any specified tolerance of the nominal surface.

The finish machining of parametric surfaces can be very much more complicated than indicated in the last paragraph. First, there are certain advantages in the use of cutters with more complex geometry than the ball-end. Radiused end-mill or toroidal cutters are frequently used in practice; these have cylindrical sides and flat ends with a blending radius between the two. Optimal cutting conditions result when the tool contact point is on the toroidal blending region of the tool. It is clear that given the tool contact path on the surface, the computation of points on the path of the tool reference point (the center of the flat end) is now more difficult than for the ball-end cutter. It is even worse for other cutter geometries which are sometimes used, for example barrel-shaped or conical (in both these examples the sides of the tool are the cutting surfaces). The more complex tool geometries are used to best advantage in the context of what is known as 4- or 5-axis machining. In 3-axis machining as described above the tool can be driven simultaneously in the x, y and z senses, but the direction of its axis remains constant. In 4- and 5-axis machining the direction of the tool axis can also in effect be controlled, though this is sometimes achieved indirectly by tilting or rotating the table holding the part being machined. The advantage of these more complex strategies is that a much finer finish can be obtained on the machined surface, which obviates the need for expensive manual finishing with hand-held grinders. The disadvantage, it will readily be appreciated, is the much greater mathematical complexity of the necessary cutter path calculations. Another disadvantage of the use of more general cutter geometries is that while the tool may contact the surface at the desired point on the designated cutter path, it may also interfere with the surface in an undesired manner at some other point causing *gouging*. Some commercial CAD/CAD systems provide means for the automatic detection of such situations, but there is scope for the development of algorithms much more efficient than those so far demonstrated. One point worth making here is that if these complex NC machining algorithms can be implemented in terms of a standard set of input quantities, they will function for any type of parametrically defined surface, thus avoiding the need for multiple special case approaches. The *parametric evaluator* approach mentioned earlier also has application in the NC machining context. It requires algorithms based solely on the use of positional, first and second derivative information for any point on a surface specified in terms of two parameter values.

So far little attention has been given to the way in which the surface region to be machined is bounded. It may, in fact, be surrounded by other surface regions, each of which also has to be machined, and it is clearly important that in machining the first surface the cutter should not gouge any of the others. Accordingly, there is a requirement for computing the locus of tool contact points on one surface such that the tool is simultaneously in contact with (but not gouging) a second surface. Note that a similar situation arises in the APT machining strategy when a tool is driven in simultaneous contact with a part and a drive surface. The computations are simplest when a ball-end cutter is used, but the following paragraphs outline a procedure which works for more general tool geometries.

Consider first the curve of intersection between two surfaces. At any point on the curve the normal direction to both surfaces may be computed as previously described. The cross-product of these normals gives a vector which, since it is perpendicular to both of them, lies in the tangent plane of both surfaces. This vector is therefore in the direction of the line of intersection of the two tangent planes, and in marching methods for computing intersections

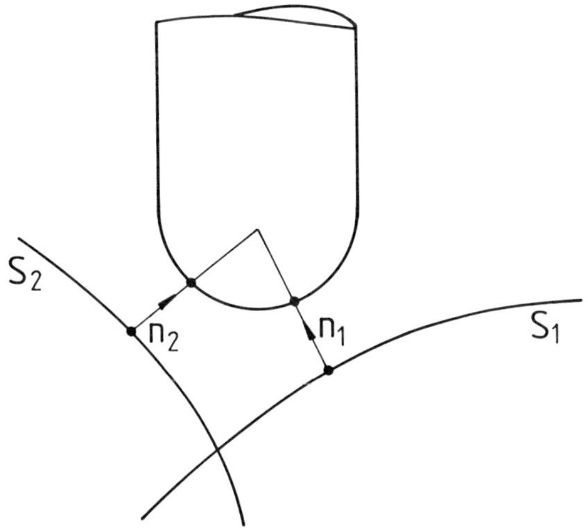

Figure 13.22. Common normals used in computing tool positions for machining with the tool in contact with two surfaces.

may be used as a first approximation to the direction of a computed step along the curve.

The foregoing procedure may be generalized for the calculation of tool paths. In practice, the aim of the computation is to determine a sequence of tool positions where the tool is not actually in contact with both surfaces, but rather is within some predefined tolerance of both of them. In place of the surface normals at a point on the intersection curve, the algorithm computes what are known as *common normals*. These are two vectors, each normal to one of the surfaces but also to the surface of the cutting tool, as shown in Figure 13.22. The determination of these vectors requires a complex iterative procedure. Once they are known, their cross-product gives a first approximation to the direction of tool motion, and a notional step is taken in this direction, the step length being determined by some estimate of the curvature of the tool path. The resulting position of the tool will, in general, leave it out of tolerance with both surfaces, but a further iterative process may be used to bring it back within tolerance. The cycle is now repeated for the determination of the next step in the tool path. This is necessarily a very abridged description of the calculation process; further details are given in Faux and Pratt [14].

ACKNOWLEDGEMENTS

The author is grateful to his former colleagues Kamal Bechkoum, Ray Goult and Peter Sherar for generating some of the illustrations used in this chapter.

REFERENCES

1 **T. Várady and M. J. Pratt** Design Techniques for the Definition of Solid Objects with Free-form Geometry, *Comput.-Aided Geometric Design,* Vol 1 No 2, 1984, 207–225.

2 **J. R. Miller** Sculptured Surfaces in Solid Models: Issues and Alternative Approaches, *IEEE Computer Graphics & Applications,* Vol 6 No 12, 1986, 33–43.

3 **R. N. Goldman** Two Approaches to a Computer Model for Quadric Surfaces, *IEEE Computer Graphics & Applications,* Vol 3 No 6, 1983, 21–24.

4 **R. F. Riesenfeld** Homogeneous Coordinates and Projective Planes in Computer Graphics, *IEEE Computer Graphics & Applications,* Vol 1 No 1, 1981, 50–55.

5 **I. C. Braid** *Notes on a Geometric Modeller,* CAD Group Document 101, University of Cambridge, England (revised 1980).

6 **J. R. Manning** *Feature Lines on Curved Surfaces,* SATRA Information Publication IP 146, Shoe and Allied Trades Research Association, Kettering, UK, 1974.

7 **F. Klok** Two Moving Coordinate Frames for Sweeping along a 3D Trajectory, *Comput.-Aided Geometric Design,* Vol 3 No 3, 1986, 217–229.

8 **M. A. Lachance** Data Proliferation in the Exchange of Surfaces, in R. R. Martin, ed., *The Mathematics of Surfaces II,* Oxford University Press, 1987.

9 **J. K. Hinds and L. P. Kuan** *Sculptured Surface Technology as a Unified Approach to Geometric Definition,* Technical Paper MS78-146, Computer & Automated Systems Association of SME, 1978.

10 **R. T. Farouki and J. K. Hinds** A Hierarchy of Geometric Forms, *IEEE Computer Graphics & Applications,* Vol 5 No 5, 1985, 51–78.

11 **CAM-I** *APT IV/Sculptured Surfaces Part Programmer's Manual (Revised Version), Vol 2,* Document No. TM-84-SS-01.1, Computer Aided Manufacturing International, Inc., Arlington, TX, 1985.

12 **R. R. Martin** Aspects of the Interference Problem, in G. Renner and M. J. Pratt, eds., *Proceedings 4th Anglo/Hungarian Seminar on CAD,* Budapest, October 1985 (Computer &

Automatic Institute of the Hungarian Academy of Sciences.)

13 **W. P. Wang and K. K. Wang** Geometric Modeling for Swept Volume of Moving Solids, *IEEE Computer Graphics & Applications,* Vol 6 No 12, 1986, 8–17.

14 **I. D. Faux and M. J. Pratt** *Computational Geometry for Design and Manufacture,* Ellis Horwood, 1979.

15 **A. W. Nutbourne, P. M. McLellan and R. M. L. Kensit** Curvature Profiles for Plane Curves, *Comput.-Aided Design,* Vol 4 No 4, 1972, 176–184.

16 **J. A. Adams** The Intrinsic Method for Curve Definition, *Comput.-Aided Design,* Vol 7 No 4, 1975, 243–249.

17 **A. W. Nutbourne and R. R. Martin** *Differential Geometry Applied to Curve and Surface Design,* Ellis Horwood, 1988.

18 **T. W. Sederberg, D. C. Anderson and R. N. Goldman** Implicit Representation of Parametric Curves and Surfaces, *Computer Vision, Graphics & Image Processing,* Vol 28, 1984, 72–84.

19 **T. W. Sederberg, D. C. Anderson and R. N. Goldman** Implicitization, Inversion and Intersection of Planar Rational Cubic Curves, *Computer Vision, Graphics & Image Processing,* Vol 31, 1985, 89–102.

20 **G. Salmon** *Modern Higher Algebra,* Hodges, Smith & Co., Dublin, 1866.

21 **ISO** *First Draft Proposal for the STEP Version 1.0 CAD/CAM Product Data Exchange Standard,* NIST, Gaithersburg, MD, 1989.

22 **E. T. Y. Lee** The Rational Bézier Representation for Conics, in G. Farin, ed., *Geometric Modeling: Algorithms and New Trends,* SIAM, Philadelphia, 1987.

23 **K. M. Bolton** Biarc Curves, *Comput.-Aided Design,* Vol 7 No 2, 1975, 89–92.

24 **M. A. Sabin** *The Use of Piecewise Forms for the Numerical Representation of Shape,* Tanulmanyok 60/1977, Computer & Automation Institute of the Hungarian Academy of Sciences, 1977.

25 **J. R. Manning** *Circular Arc Interpolation,* SATRA Report SR 30, Shoe and Allied Trades Research Association, Kettering, UK, 1982.

26 **T. J. Sharrock** *Surface Design with Cyclide Patches,* PhD thesis, Cambridge University Engineering Department, 1985.

27 **J. R. Rossignac** *Blending and Offsetting Solid Models,* Technical Memo. TM-54 (PhD dissertation), Production Automation Project, University of Rochester, New York, 1985.

28 **D. A. Towers** *Guide to Linear Algebra,* Macmillan, 1988.

29 **D. L. Flanagan and O. V. Hefner** Surface Molding – New Tool for the Engineer, *Aeronautics and Astronautics,* April 1967, 58–62.

30 **T. W. Sederberg** Planar Piecewise Algebraic Curves, *Comput.-Aided Geometric Design,* Vol 1 No 3, 1984, 241–255.

31 **D. M. Y. Sommerville** *Analytical Geometry of Three Dimensions,* Cambridge University Press, 1934.

32 **A. H. Barr** Superquadrics and Angle-preserving Transformations, *IEEE Computer Graphics & Applications,* Vol 1 No 1, 1981, 11–23.

33 **C. Dupin** *Applications de géometrie et de méchanique,* Bachelier, Paris, 1822.

34 **R. R. Martin, J. de Pont and T. J. Sharrock** Cyclide Surfaces in Computer Aided Design, in J. A. Gregory, ed., *The Mathematics of Surfaces,* Oxford University Press, 1986.

35 **M. J. Pratt** Applications of Cyclide Surfaces in Geometric Modelling, in D. C. Handscomb, ed., *The Mathematics of Surfaces III,* Oxford University Press, 1989.

36 **M. J. Pratt** Cyclides in Computer-Aided Design, *Comput.-Aided Geometric Design,* Vol. 7, 1990, 221–242.

37 **V. Chandru, D. Dutta and C. M. Hoffmann** Variable Radius Blending using Dupin Cyclides, in M. J. Wozny, J. Turner and K. Preiss, eds., *Geometric Modelling for Product Engineering: Proc. IFIP/NSF Workshop on Geometric Modelling,* Rensselaerville, NY, September 1988.

38 **A. R. Forsyth** *Lectures on Differential Geometry of Curves and Surfaces,* Cambridge University Press, 1912.

39 **A. Cayley** On the Cyclide, *Quarterly J. Pure & Applied Maths,* Vol XII, 1873, 148–165.

40 **T. W. Sederberg** Piecewise Algebraic Surface Patches, *Comput.-Aided Geometric Design,* Vol 2 No 1-3, 1985, 53–59.

41 **C. L. Bajaj** Geometric Modelling with Algebraic Surfaces, in D. C. Handscomb, eds., *The Mathematics of Surfaces III,* Oxford University Press, 1989.

42 **R. Liming** *Mathematics for Computer Graphics,* Aero Publishers, Fallbrook, CA, 1979.

43 **A. E. Middleditch and K. H. Sears** Blend Surfaces for Set-Theoretic Volume Modelling Systems, *ACM Computer Graphics,* Vol 19 No 3 (*Proceedings SIGGRAPH Conf. 1985,* San Francisco, CA), 1985, 161–170.

44 **C. M. Hoffmann and J. E. Hopcroft** Quadratic Blending Surfaces, *Comput.-Aided Design,* Vol 18 No 6, 1986, 301–306.

45 **C. M. Hoffmann and J. E. Hopcroft** The Potential Method for Blending Surfaces and Corners, in G. Farin, ed., *Geometric Modeling: Algorithms and New Trends,* SIAM, Philadelphia, 1987.

46 **L. Holmström** Piecewise Quadratic Blending of Implicitly Defined Surfaces, *Comput.-Aided Geometric Design,* Vol 4 No 3, 1987, 171–189.

47 **A. P. Rockwood and J. C. Owen** Blending

Surfaces in Solid Modelling, in G. Farin, ed., *Geometric Modeling: Algorithms and New Trends,* SIAM, Philadelphia, 1987.

48 **J. R. Woodwark** Blends in Geometric Modelling, in R. R. Martin, ed., *The Mathematics of Surfaces II,* Oxford University Press, 1987.

49 **M. J. Pratt** Bézier Surface Blending in Solid Modelling, in R. B. Fisher, ed., The Mathematics of Surfaces V, Oxford University Press (in preparation).

50 **H. Chiyokura** *Solid Modelling with DESIGN-BASE,* Addison-Wesley, 1988.

51 **J. R. Rossignac and A. A. G. Requicha** Constant Radius Blending in Solid Modeling, *Computers in Mechanical Engineering,* July 1984, 65–73.

52 **T. Várady, J. Vida and R. R. Martin** Parametric Blending in a Boundary Representation Solid Modeller, in D. C. Handscomb, ed., *The Mathematics of Surfaces III,* Oxford University Press, 1989.

53 **G. Farin** *Curves and Surfaces for Computer Aided Geometric Design,* Academic Press, 1988.

54 **R. T. Farouki and V. T. Rajan** On the Numerical Condition of Polynomials in Bernstein Form, *Comput.-Aided Geometric Design,* Vol 4 No 3, 1987, 191–216.

55 **R. T. Farouki and V. T. Rajan** Algorithms for Polynomials in Bernstein Form, *Comput.-Aided Geometric Design,* Vol 5 No 1, 1988, 1–26.

56 **L. Piegl and W. Tiller** Curve and Surface Constructions using Rational B-splines, *Comput.-Aided Design,* Vol 19 No 9, 1987, 485–498.

57 **M. J. Pratt and A. D. Geisow** Surface/Surface Intersection Problems, in J. A. Gregory, ed., *The Mathematics of Surfaces,* Oxford University Press, 1986.

58 **R. T. Farouki** The Characterization of Parametric Surface Sections, *Computer Vision, Graphics & Image Processing,* Vol 33, 1986, 209–236.

59 **J. Z. Levin** A Parametric Algorithm for Drawing Pictures of Solid Objects Composed of Quadric Surfaces, *Commun. ACM,* Vol 19, 1976, 555–563.

60 **R. F. Sarraga** Algebraic Methods for Intersections of Quadric Surfaces in GMSOLID, *Computer Vision, Graphics & Image Processing,* Vol 22, 1983, 222–238.

61 **J. J. de Pont** *Essays on the Cyclide Patch,* PhD thesis, Cambridge University Engineering Department, 1984.

62 **C. L. Bajaj, C. M. Hoffmann, J. E. Hopcroft and R. E. Lynch** Tracing Surface Intersections, *Comput.-Aided Geometric Design,* Vol 5 No 4, 1988, 285–307.

63 **T. Garrity and J. Warren** On Computing the Intersection of a Pair of Algebraic Surfaces, *Comput.-Aided Geometric Design,* Vol 6 No 2, 1989, 137–153.

64 **J. C. Owen and A. P. Rockwood** Intersection of General Implicit Surfaces, in G. Farin, ed., *Geometric Modeling: Algorithms and New Trends,* SIAM, Philadelphia, 1987.

65 **A. P. Morgan** A Method for Computing All Solutions to Systems of Polynomial Equations, *ACM Trans. Mathematical Software,* Vol 9 No 1, 1983, 1–17.

66 **J. R. Miller** Geometric Approaches to Nonplanar Quadric Surface Intersection Curves, *ACM Trans. on Graphics,* Vol 6 No 4, 1987, 274–307.

67 **L. Piegl** Geometric Method of Intersecting Natural Quadrics Represented in Trimmed Surface Form, *Comput.-Aided Design,* Vol 21 No 4, 1989, 201–212

68 **K.-P. Cheng** Using Plane Vector Fields to Obtain All the Intersection Curves of Two General Surfaces, in W. Strasser and H.-P. Seidel, eds., *Theory and Practice of Geometric Modeling,* Springer-Verlag, 1989.

69 **C. M. Hoffmann** *A Dimensionality Paradigm for Surface Interrogations,* Technical Report CSD-TR-837 (CAPO Report CER-87-45), Dept. of Computer Science, Purdue University, 1988.

70 **E. G. Houghton, R. F. Emnett, J. D. Factor and L. Sabharwal** Implementation of a Divide-and-Conquer Method for Intersection of Parametric Surfaces, *Comput.-Aided Geometric Design,* Vol 2 Nos 1–3, 1985, 173–183.

71 **J. Rooney and P. Steadman** *Principles of Computer Aided Design,* Pitman/Open University, 1987.

72 **R. T. Farouki** Trimmed-surface Algorithms for the Evaluation and Interrogation of Solid Boundary Representations, *IBM J. Research & Development,* Vol 31 No 3, 1987, 314–334.

73 **M. S. Casale** Free-form Solid Modeling with Trimmed Surface Patches, *IEEE Computer Graphics & Applications,* Vol 7 No 1, 1987, 33–43.

74 **A. A. G. Requicha and H. B. Voelcker** Solid Modeling: Current Status and Research Directions, *IEEE Computer Graphics & Applications,* Vol 3 No 7, 1983, 25–37.

75 **A. D. Geisow** *Surface Interrogations,* PhD thesis, School of Computing Studies & Accountancy, University of East Anglia, 1983.

76 **C. A. Neff** Decomposing Algebraic Sets using Gröbner Bases, *Comput.-Aided Geometric Design,* Vol 6 No 3, 1989, 249–264.

77 **T. W. Sederberg and J. P. Snively** Parametrization of Cubic Algebraic Surfaces, in R. R. Martin, ed., *The Mathematics of Surfaces II,* Oxford University Press, 1987.

78 **S. S. Abhyankar and C. L. Bajaj** Automatic Rational Parameterization of Curves and Surfaces I: Conics and Conicoids, *Comput.-Aided Design,* Vol 19 No 1, 1987, 11–14.

79 **S. S. Abhyankar and C. L. Bajaj** Automatic

Rational Parameterization of Curves and Surfaces II: Cubics and Cubicoids, *Comput.-Aided Design,* Vol 19 No 9, 1987, 499–502.

80 **S. S. Abhyankar and C. L. Bajaj** Automatic Rational Parameterization of Curves and Surfaces III: Algebraic Plane Curves, *Comput.-Aided Geometric Design,* Vol 5 No 4, 1988, 309–321.

81 **S. S. Abhyankar and C. L. Bajaj** *Automatic Rational Parameterization of Curves and Surfaces IV: Algebraic Space Curves,* Technical Report CSD-TR-703 (CAPO Report CER-87-5), Dept. of Computer Science, Purdue University, 1988.

82 **IGES** *Initial Graphics Exchange Specification, Version 5.1,* NCGA Technical Services & Standards, Fairfax, VA, 1991.

83 **P. R. Wilson** Conic Representations for Shape Description, *IEEE Computer Graphics & Applications,* Vol 7 No 4, 1987, 23–30.

84 **M. J. Pratt** Parametric Curves and Surfaces as used in Computer Aided Design, in J. A. Gregory, ed., *The Mathematics of Surfaces,* Oxford University Press, 1986.

85 **C. de Boor** *A Practical Guide to Splines,* Springer-Verlag, 1978.

86 **P. Sablonnière** Spline and Bézier Polygons associated with a Polynomial Spline Curve, *Comput.-Aided Design,* Vol 10 No 4, 1978, 257–261.

87 **E. Cohen, T. Lyche and R. F. Riesenfeld** Discrete B-splines and Subdivision Techniques in Computer Aided Geometric Design and Computer Graphics, *Comput. Graphics & Image Processing,* Vol 14, 1989, 87–111.

88 **W. Boehm** Inserting New Knots into B-spline Curves, *Comput.-Aided Design,* Vol 12 No 4, 1980, 199–201.

89 **W. Boehm** Generating the Bézier Points of B-spline Curves and Surfaces, *Comput.-Aided Design,* Vol 13 No 6, 1981, 365–366.

90 **E. Cohen and R. F. Riesenfeld** General Matrix Representations for Bézier and B-spline Curves, *Computers in Industry,* Vol 3 Nos 1–2, 1982, 9–15.

91 **T. Lyche and K. Mørken** Making the Oslo Algorithm More Efficient, *SIAM J. Numerical Analysis,* Vol 23 No 3, 1986, 663–675.

92 **P. J. Davis** *Interpolation and Approximation,* Dover, New York, 1963.

93 **M. A. Lachance** Chebyshev Economisation for Parametric Surfaces, *Comput.-Aided Geometric Design,* Vol 5 No 3, 1988, 195–208.

94 **R. J. Goult** Parametric Curve and Surface Approximation, in D. C. Handscomb, ed., *The Mathematics of Surface III,* Oxford University Press, 1989.

95 **M. A. Watkins and A. J. Worsey** Degree Reduction of Bézier Curves, *Comput.-Aided Design,* Vol 20 No 7, 1988, 398–405.

96 **L. Dannenberg and H. Nowacki** Approximate Conversion of Surface Representations with Polynomial Bases, *Comput.-Aided Geometric Design,* Vol 2 Nos 1–3, 1985, 123–131.

97 **G. E. Hölzle** Knot Placement for Piecewise Polynomial Approximation of Curves, *Comput.-Aided Design,* Vol 15 No 5, 1983, 295–296.

98 **J. Hoschek** Approximate Conversion of Spline Curves, *Comput.-Aided Geometric Design,* Vol 4 No 1, 1987, 59–66.

99 **J. Hoschek and N. Wissel** Optimal Approximate Conversion of Spline Curves and Spline Approximation of Offset Curves, *Comput.-Aided Design,* Vol 20 No 8, 1988, 475–483.

100 **N. M. Patrikalakis** Approximate Conversion of Rational Splines, *Comput.-Aided Geometric Design,* Vol 6 No 2, 1989, 155–165.

101 **L. Bardis and N. M. Patrikalakis** Approximate Conversion of Rational B-spline Patches, *Comput.-Aided Geometric Design,* Vol 6 No 3, 1989, 189–204.

102 **T. Lyche and K. Mørken** Knot Removal for Parametric B-spline Curves and Surfaces, *Comput.-Aided Geometric Design,* Vol 4 No 3, 1987, 217–230.

103 **J. D. Foley and A. Vam Dam** *Fundamentals of Interactive Computer Graphics,* Addison-Wesley, 1982.

104 **T. J. Willmore** *Differential Geometry,* Oxford University Press, 1958.

105 **J. R. Manning** *Computerised Pattern Cutting 1: Methods Based on an Isometric Tree,* SATRA Research Report RR 287, Shoe and Allied Trades Research Association, Kettering, UK, 1978.

106 **T. Lozano-Perez** Spatial Planning: A Configuration Space Approach, *IEEE Trans. on Computers,* Vol 32 No 2, 1983, 108–120.

107 **W. Tiller and E. G. Hanson** Offsets of Two-dimensional Profiles, *IEEE Computer Graphics & Applications,* Vol 4 No 9, 1984, 36–46.

108 **A. A. G. Requicha** Representation of Tolerances in Solid Modeling: Issues and Alternative Approaches, in M. S. Pickett and J. W. Boyse, eds., *Solid Modeling by Computers: Proceedings General Motors Solid Modeling Symposium,* Warren, MI, September 1983, Plenum Press, New York, 1984.

109 **J. R. Rossignac and A. A. G. Requicha** Offsetting Operations in Solid Modeling, *Comput.-Aided Geometric Design,* Vol 3 No 2, 1986, 129–148.

110 **R. Klass** An Offset Spline Approximation for Plane Cubic Splines, *Comput.-Aided Design,* Vol 15 No 5, 1983, 297–299.

111 **J. Hoschek** Offset Curves in the Plane, *Comput.-Aided Design,* Vol 17 No 2, 1985, 77–82.

112 **R. T. Farouki and T. Sakkalis** *Pythagorean*

Hodographs, Research Report RC 15223 (#67894), IBM Research Division, TJ Watson Research Center, Yorktown Heights, NY, 1989.

113 **S. Coquillart** Computing Offsets of B-spline Curves, *Comput.-Aided Design,* Vol 19 No 6, 1987, 305–309.

114 **N. M. Patrikalakis and L. Bardis** Offsets of Curves on Rational B-Spline Surfaces, *Engineering with Computers,* Vol 5 No 1, 1989, 39–46.

115 **R. T. Farouki** The Approximation of Non-Degenerate Offset Surfaces, *Comput.-Aided Geometric Design,* Vol 3 No 1, 1986, 15–43.

116 **R. T. Farouki** Exact Offset Procedures for Simple Solids, *Comput.-Aided Geometric Design,* Vol 2 No 4, 1985, 257–279.

117 **P. K. Ghosh** A Mathematical Model for Shape Descriptions using Minkowski Operators, *Computer Vision, Graphics & Image Processing,* Vol 44, 1988, 239–269.

118 **F. N. Fritsch** The Wilson-Fowler Spline is a *v*-spline, *Comput.-Aided Geometric Design,* Vol 3 No 2, 1986, 155–162.

119 **A. E. Middleditch, T. W. Stacey and S. B. Tor** Intersection Algorithms for Lines and Circles, *ACM Trans. on Graphics*, Vol 6, No 1, 1989, 25–40.

ABOUT THE AUTHOR

Michael J. Pratt holds an MA in physics from Oxford University and an MSc in Aeronautical Science and a PhD in Mechanical Engineering from Cranfield Institute of Technology. He joined the research staff at Cranfield in the 1960s, pursuing theoretical and experimental studies in hypersonic gasdynamics. Subsequently he transferred to the academic staff, teaching mathematics. From the early 1970s, together with his former colleague Ivor Faux, he developed a teaching and research activity in applications of geometry, particularly in computer aided design and manufacture. He was for some years Professor of Computer Aided Engineering and Head of the Department of Applied Computing and Mathematics at Cranfield, but has recently taken up a senior position in the Design Research Center of Rensselaer Polytechnic Institute in the USA.

Professor Pratt's current research interests include the computer modeling of surfaces and solids for CAD, the automation of downstream applications, particularly in finite element analysis and manufacturing, and standard data communication interfaces for CAD/CAM systems. His most recent papers have been concerned with form features and their applications and with the use of cyclide surfaces for blending in solid modeling. He is part author of several books, notably (with Ivor Faux) the standard text *Computational Geometry for Design and Manufacture* (Ellis Horwood/Simon & Schuster, 1979). He is a frequent invited speaker at international conferences, and is on the editorial board of the journals *Computer-Aided Drafting, Design and Manufacture, Computer Aided Geometric Design* and *Engineering Applications of Artificial Intelligence.* British professional organizations of which he is a member include the Royal Aeronautical Society, the British Computer Society and the Institute of Mathematics and its Applications. He also belongs to the IEEE Computer Society and the American Society of Mechanical Engineers. He is registered as a practising engineer both in Britain (CEng) and in Europe (EurEng).

During recent years he has been involved in a variety of activities on an international scale. These include working with the European ESPRIT and BRITE research programs and with the US-based research organization Computer Aided Manufacturing International Inc. (CAM-I). He has been responsible for a collaborative research agreement between CIT and the Computer and Automation Institute of the Hungarian Academy of Sciences and for the setting up of a Master's degree program in Computer Aided Engineering run specially for the organization Institut du Logiciel et Systemes in Bayonne, France. Recent visits abroad have included consultancy visits to the People's Republic of China and to Korea under the auspices of the World Bank and UNIDO respectively, and a 6-month Visiting Fellowship to the General Electric Corporate Research and Development Center, Schenectady, NY.

14 Computer Systems for Interactive Design: The Description of Multipatch, Multiobject and Other Existing Systems

Andrew P. Armit

INTRODUCTION

Twenty years ago I was constructing and writing about Multipatch and Multiobject. From those beginnings in Computer Aided Design I must admit to remaining fascinated with the creation of interactive systems – and mostly concerned with 3D graphics. A further admission is that the fascination stems more from wishing to explore my personal capabilities and to make useful products than from a need to influence others – though it seems this has occurred as a side-effect.

Many authors will be capable of presenting a 'balanced view' with their own work set neatly in historical context (and probably in a position of importance), but for me it is ongoing, and I can but describe various aspects of system design as they have grown and mutated under my pen over the years. Some years ago now, after I had given an invited paper upon the interactive languages of Multipatch and Multiobject, an associate of mine enquired of William Newman how it had gone – to be told that "Andrew just talked about the systems he has made". I remain open to the same criticism.

Of course, in the last 20 years computers have advanced so much that some of the problems in the construction of those early systems are mere history, and anyway nothing looks difficult in retrospect. Attempt is made to avoid dwelling upon the parochial difficulties of the past, and to concentrate instead upon the 'nuts and bolts' that seem to have proven their relevance in a variety of systems over the years. Do not believe that I see anything here as unassailably right or even totally desirable: rather, the information and thoughts and methods are offered in the knowledge that they did at least allow some complicated systems to be built. Others could well be successful by following a completely different approach.

To set the scene for my opinions, a description of some of the software and hardware developments is given – a sort of 'what I did next' list. I have tried to keep this brief, leaving the detail to the following sections, which trace the development by topic for data-structures, languages, geometric representations, graphical forms and representation of interactions, etc., in the face of ever increasing requirements. Along with all this goes the advancement in graphics hardware – and this too is discussed.

OUTLINE OF SYSTEMS

Multipatch (1968)

This was constructed during my first six months of PhD work at Cambridge University Computer Laboratory (so it can't have been very difficult to have happened so quickly). At that time, others (including Robin Forrest) had arranged to show a single Coons' patch on screen and to manipulate its shape by the supply of values to any of the 48 defining parameters [1] (see also this volume). It was quite clear that complex shapes needed to use lots of patches, and indeed, that their control and visualization was of importance if a useable system was to result.

The computer was a DEC 8K PDP7 (18-bit word) with DEC 340 Display with peripherals of paper tape input/output, a Teletype and a joystick and lightpen. This early display was of the 'steered beam' variety, in which the internal record of the picture to appear on screen gives vector position data for lines, together with character codes for text.

Multipatch dealt only with Coons' Bicubic patches, and allowed the user to create and manipulate patches by command with real-time computations and presentations of their wire-frame representations on-screen. Multipatch

The colour plates for this chapter appear between pp. 210 and 211.

Fundamental Developments of Computer-Aided Geometric Modeling
ISBN 012-554765-X

allowed virtually instant rotation of the view – and joystick control of this remained 'live' during any command input.

The surface computation for the patches was the easy bit of the system – the major problem was how to arrange for the user to effectively modify a particular patch in each of its 48 values, and to be able to indicate that patches should be 'joined', and possibly made continuous in slope across their joined edges. What should happen if a joined patch is to be moved or even deleted? How many views of the assembly should be shown, and how shall views be dynamically controlled and presented to aid visualization of the 3D shape? I remember that while I was wrestling with these problems, Robin Forrest observed to me that he was worried that I was spending too much time programming and not enough time thinking – with the latter more important for the PhD.

Now most systems of the time used lightpen and single key hits (i.e. using the Teletype as a button box) – but there were too many things to be expressed for this to be an adequate method. The problem was solved by the creation of an 'Interactive Language' which was no doubt simple in itself, but allowed the ready definition of typed commands together with the connection path to routines for execution of effects. Certainly, the user had to hit more than one key to give a command, but the implementation and language structure allowed the differences between the previous command and the required command to be sufficient input – so the repeat of a command (e.g. an incremental change) could be achieved in a single key hit – or its effect reversed by two key hits.

At a time when Crispin Gray and Charles Lang were giving papers upon ring structures, the data structures of Multipatch must have been deeply offensive in their crude stark simplicity – with absolutely no possibility of a 'space jam', since no garbage ever got left to be collected (or be impossible to collect within the environment). Multipatch used fixed size blocks for patches with reference by name rather than pointer, and thus all was inherently relocatable. 'Duplication' of data at patch joins was made invisible by programs which maintained any continuity conditions set up by the user by 'appropriate' reflections into other patches.

Multipatch allowed multiple views on-screen and allowed these to be rotated under joystick control. All the picture computation (and even vector clipping) was by software with all sorts of measures taken to yield new pictures in a fraction of a second. No doubt the robustness of the system (i.e. it worked) together with the dynamic pictures and evident usefulness even at that time were responsible for the droves of enthusiastic visitors from all over the world. Figures 14.1 and 14.2 give example screen pictures, while [2–11] give further information.

Multiobject (1970)

As implied in the name, Multiobject allowed various 'objects' to be included in a design: parallelepipeds, various order Bézier curves ([12], and see also this volume), various order Bézier surfaces, points lines, etc. This much larger system was also built at the Cambridge Computer Laboratory, and employed the Titan

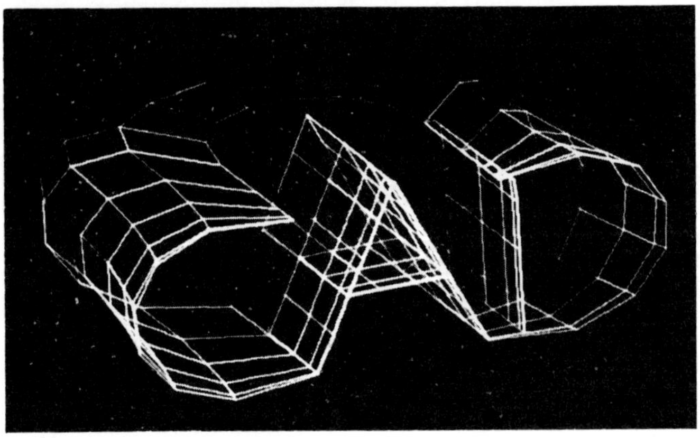

Figure 14.1. Multipatch 'Insignia' (1969).

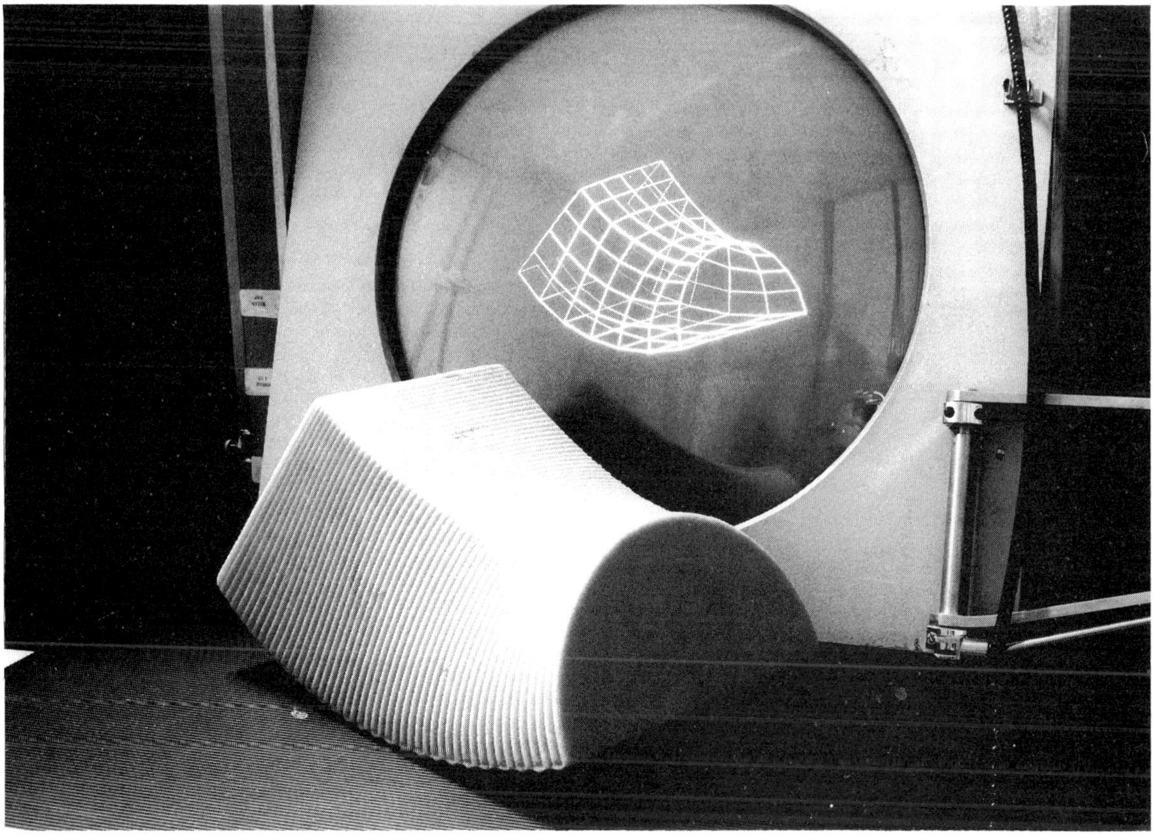

Figure 14.2. Multipatch showing screen of DEC 340 display. The design was sent on paper tape to Professor P. Bézier of Renault to allow it to be cut in polystyrene on the UNISURF system – with the result shown.

(Prototype Atlas II) computer linked to the DEC PDP7 and DEC 340 Display. The link allowed about 10K words per second transfer to or from the PDP7, and the 48 bit word length and 24K words in Titan allowed great expansion of facilities compared with Multipatch.

Whereas Multipatch allowed surface design by a collection of patches, Multiobject offered the construction of a framework of curves to be designed first, followed by the creation of the surfaces depending upon these curves for their edges. The surface design method was thus improved, needing fewer overall patches, as higher order curves and surfaces could be used for the regions of higher curvature. Figures 14.3–14.5 are from Multiobject.

The simple block data structures of Multipatch were retained in Multiobject, preserving inherent relocatability by recording inter-object references by name rather than address. Thus

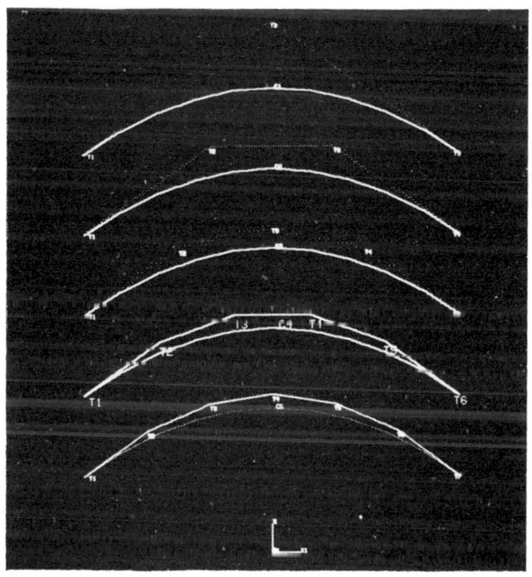

Figure 14.3. Multiobject: Bézier curve types (1970).

Figure 14.4. Multiobject: Parallelepipeds (1970).

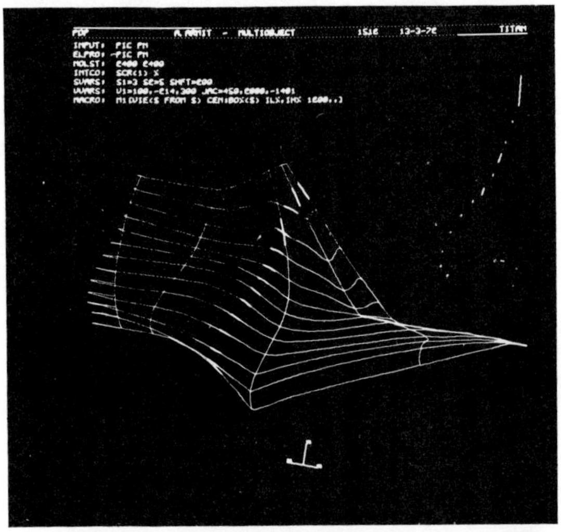

Figure 14.5. Multiobject: Shoe Last design showing brightened pseudo section upon the Last, and also in true view (1972).

two data structures could be added by simply adjusting names wherever any clash existed.

The interactive language was much extended in Multiobject to allow any value to be an expression, including scalar, vector and matrix arithmetic, together with the possibility of extracting any parameter from any object or view data structure by name (e.g. for use in a command). The language form thus allowed 'right hand side processing' to extract values from a data structure, while its 'left hand side processing' continued to express the command to be executed – often leading to the supply of values to some data structure.

Apart from these extensive improvements, Multiobject occasioned an even greater concern for computational efficiency than Multipatch, because Titan was running the time-sharing system for online users: with a resulting limit of three minutes maximum computation in the hour being available. The user of the link to the PDP7 was awarded a few seconds priority execution time at each 'new' request upon Titan: and Multiobject set out to achieve command execution within these few seconds by a policy of 'minimum recomputations' (and hence also link communications). Thus according to the command (or set of commands by macro invocation, etc.), Multiobject worked out the view(s) and object(s) needing recomputation and then performed just these. Thus the command execution 'phase' operated by making data structure changes and marking objects or views needing recomputation for display. Typically, command sequences might re-mark the same objects several times. Thereafter, the picture computation 'phase' operated according to the marks, with Titan transferring the necessary pieces of display file to the PDP7, which had the job of 'editing' the collection of display files for objects in views. The PDP7 kept a compacted collection of object display files and arranged efficient use of the 340 Display upon these. The input command was buffered at the PDP7 and passed to Titan for execution.

Multiobject surpassed the *Z*-brightness modulation and surface normal presentations of Multipatch by also giving a 'boxing' process in its view computations. Three pairs of planes (mutually orthogonal) could be positioned in object space and at any orientation. The lines and curves of screen representations could be clipped against any or all of these planes before being offset and rotated to the viewing angles. In addition, a Box direction could be used for either progressive brightness modulation or switched brightness – a square wave – yielding sections of the lines etc. with their tips at true (but presentably skew) sections (Figure 14.5).

The inclusion of perspective showed its necessity for some applications (e.g. architectural) and irrelevance to others (Figures 14.4, 14.5) [3–11].

ICON (Interactive Creation of NASTRAN Data, 1975)

The opportunity came to work on the latest and greatest equipment: DEC PDP11/45 with Vector General Series 3 Display. The system was for Lloyd's Register of Shipping, and had to allow screen design of Finite Element meshes for Ships' Hulls, and to create exact NASTRAN data format in accord. The FE analysis was not performed upon the PDP11/45 (!).

A memory management system (not my design) allowed the separation of Data and Program space and thus stretched the PDP11/45 to offer 32K words (16-bit) for program and 32K for data. This latter was split up to allow a 'disc window' through which the large text associated with the NASTRAN data format could be wound, 4K at a time.

After the claims of system stacks, etc., just 16K words remained for the geometry structure: simple sums showed that this was too little for the number of points and lines implied by the quantity of 'nodes' needed for the FE. Unlike Multipatch or Multiobject, the ICON data structure had to deal with lots of small geometric items (points and lines) rather than curves and surfaces.

It was noticeable that many FE meshes included 'long lines' which could form the edge of several elements, and that subdivision of such lines into segments was unhelpful for both system and user – giving the system a larger data with less structure and the user the problem of keeping a set of points collinear when making design changes. Thus points were divided into three subtypes:

Subtype	*Method of definition*
Explicit	*x*, *y*, *z* components
Along line	Ref line and parameter
Intersection of lines	Ref line, ref line

and lines could freely refer to any kind of point. Obviously, this led to the need for recursive evaluation, since the change to a point could change lines and points, etc.

Certainly, the pointer-oriented data structure gave some new and interesting problems, such as what to do when a geometric change caused an intersection definition point to no longer exist? Clearly it, and all its 'dependents' become uncomputable. Such a situation has indeed to be tolerated, since it may arise as the user applies some perfectly reasonable change piecemeal. The solution is simply to inform the user what has happened, and to make uncomputable geometry invisible. When the situation is rectified all the dependents turn up again in accord.

The designs within ICON consisted of the geometry, together with a large body of text giving the NASTRAN data as required for the Finite Element representations. There was no system requirement (until output of the exact NASTRAN data form for analysis) that the geometry and text should exist in exact sympathy – allowing ready creation of geometry followed by automated creation of the FE text embodying references to the geometry wherever relevant. There were consequences.

Changes to geometry structure such as deletes or 'upgrades' of one kind of point to another could occasion movement of items in the geometry structure (which was kept compacted). All interactive performance would have been lost if the disc-based text, including the geometry references, had been updated directly, so rather than this a 'history of changes' was maintained. A sequence of these took but little space, and allowed the winding of the disc-based text to be delayed until necessary with simple speedy correction of references by application of the history.

It was necessary to be able to find the geometry references within the text, and this was arranged by the construction of a language definition of the NASTRAN card format. This data description was viewed as instructions to a 'pseudo machine'. ICON included two pseudo machines, one for the representation of interactive sequences (i.e. choices on-screen) and another for the definition of the syntax of the text information. This latter pseudo machine was moded to allow all sorts of effects, such as

1. Generation of screen text of NASTRAN data card with geometry references.
2. Access of text of card to find and correct geometry references against history of changes.
3. Automatic coercion to produce syntactically correct NASTRAN data cards by directing the choice mechanism.

There were about 11 modes of operation of the 'syntax machine', and of course, there is no difficulty in placing instructions in the pseudo code that have exactly no effect in specific modes.

ICON remained in use for some years, and a further system with auto-generation facilities for repetitive structures for geometry and related NASTRAN data was supplied to the Ocean Engineering Department of Lloyd's Register of Shipping for work upon oil rigs. Armit and Filby [13] give the first use of the pseudo machine approach, with Armit, Elliott and Lemke [14–16] giving more information upon ICON.

TAG: High performance graphics system (1981)

This multi-user high performance graphics system is running today upon the DEC PDP11/70 and Vector General VG3400 at Wellworthy, Lymington. TAG is a large 3D interactive design system, built to surpass all the previous systems of my design: taking more than three years upon the software, together with a year to put the VG3400 to a satisfactory working state.

From the outset it was intended to have a large program and data space, and this was arranged by the design of a memory management system 'MEGA' to give demand paging from disc to physical memory and demand mapping of the segments in memory. This is entirely automatic and invisible to the programs of TAG, which simply compose long addresses, and make access via them, to provoke disc transfer and remaps in accord. As configured, a 64 million word program and data space is achieved. Provided a program loop does not visit more than the stack and own data segment, together with four other data segments, then it will run at full machine speed (as soon as the segments are in memory and mapped) until some other graphics user operates.

MEGA is no doubt unusual in offering a 'pure-code' environment in which just one code copy exists. MEGA is used to manage TAG, which is itself 100K of assembler code as one program. Some special routines allow TAG to be built and linked piecemeal: so a 2K segment can be assembled, linked and inserted in about four minutes: an important consideration once the system gets large.

TAG draws upon Multiobject in offering a variety of geometric entities for design, including point, line, infinite line, arc, plane, curve, bicubic patch, etc. It also includes transforms of offset, scale, rotate and mirror. By extension of the notion of pseudo machine it proved possible to include a syntax definition of the various items creatable, and to ensure that its operation directs system behavior. In this way, new objects can be modeled by merely writing the pseudo machine instructions embodying their syntax definition, and indicating their picture computation process (which can call freely upon any available computational processes already embodied).

The scheme thus supports the application of transforms to the definitions of geometric objects – which is typically vastly superior to applying transforms to the line segments of which their screen representations may be composed (Figures 14.6, 14.7).

TAG data structures (and programs) come in 2K chunks, and thus garbage collection within segments, segment splitting, segment merging, etc., all feature in the scheme. Indeed, the data structures are full of references, some implicitly within the segment and others to any segment. The relocation processes required in any situation make use of the pseudo machine and syntax definition of objects so that no special action is required to cope with the introduction of new objects.

The interactive language of TAG far surpasses that of Multiobject, and uses the same pseudo machine as employed for data-structure representation. TAG allows keyed-in commands or menu picking using a 'puck' upon the tablet. More than this, picks upon the geometry can be made and, according to the context implied by the current part-input command, further command text will be generated. In any event, TAG builds an unambiguous text form of every interactive command, and journalizes those executed together with the system response.

The synthesis of command from key and pick inputs allows the user, for example, to communicate with features of the design not currently represented on screen. By keying or picking the menu item 'PNT' the system can be made to 'expect' the identifying number of a point (rather than any other object). If the user then makes a pick upon any geometric item, the system will determine the identifying number of the point referenced by or referencing the picked object nearest to the position of the pick (in projected screen space). This is useful.

Indeed, geometry positions can be supplied by picks even upon perspective views of skewed objects, with TAG performing the necessary arithmetic to resolve the position in screen space and back transform to object space. Thus, in creating a geometry item, the syntax of the item type controls the menu of choices and the context within which picks will be interpreted as supplying references or explicit component or vector positions – all with text synthesis and coordinate transforms in accord.

The role of the Vector General 3400 needs to be mentioned. It proved necessary to obtain the Writeable Control Store to obtain the performance characteristics originally sought. The 'firmware' was thus purpose built for TAG, and allows multiple views and multiple users and updates to objects or views without the need to recompute everything. The VG3400 accesses the PDP11/70 by Direct Memory Access, and the special Peripheral Mapping features of the PDP11/70 are used to enable the VG3400 to steer itself over memory to obtain its data. Indeed, the VG3400 is made to maintain 'greased lighting forms' in the PDP11/70 memory to give near real-time (fraction of a second for worst case) rotations and transforms as it re-absorbs these.

It would be wrong to suggest that the VG3400 offered easy solution to problems. As delivered the manufacturer's firmware showed many troubles, and once the Writeable Control Store was obtained it was possible to see that there were many hardware problems too. It took a year to fix these to allow a near crash-free situation while using the hardware to its limit and operating all day. That situation has been maintained since 1981. Figures 14.6 and 14.7 show some typical pictures from one of the screens ([17] gives but brief description of TAG).

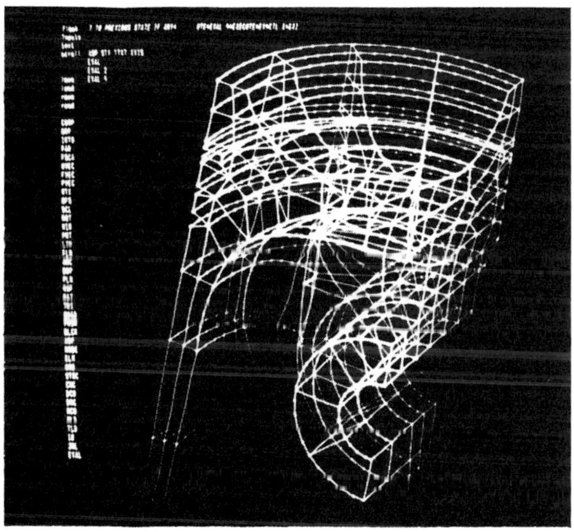

Figure 14.6. TAG: Design of a quarter piston (1982).

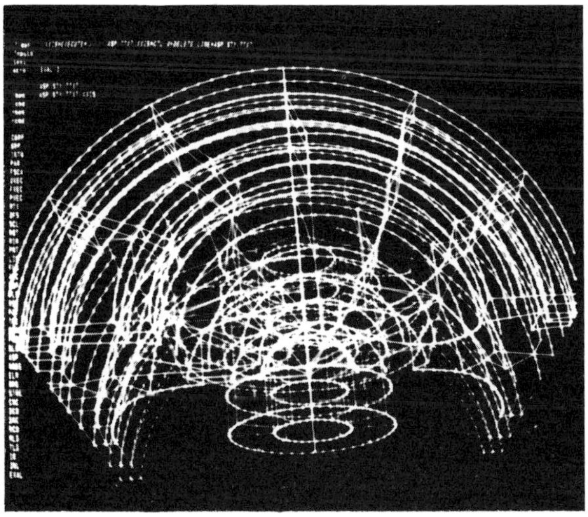

Figure 14.7. TAG: Design of a half piston (1982).

Custom firmware on the Vector General 3400 (1982, 1984)

The programmability of the VG3400 (though only available with the Writeable Control Store) is (was) its strong point. The hardware of the machine needed some significant effort to find and cure various bus and operational problems, none of which were chip malfunction. Before cure such problems manifested themselves as system crashes or crazy pictures or inability to pick properly, etc. These experiences did not improve the machine's impact, and the achieved performance of the manufacturer's firmware offered justification for the belief that the GPU was too slow to give real-time performance.

With the problems cured it became possible to fit a VG3400 at the Royal Aircraft Establishment with 'World Scene Firmware', with the result that real-time perspective views of a 'world' with objects is obtained. To give a good 'flying time', the scheme uses 32-bit space, and is yet able to support the perspective computations and present moving and rotating objects as observed from a moving and rotating position. Figure 14.8 shows the sort of picture achieved, with perspective world lines and false horizon.

Back in the realm of computer aided design

Figure 14.8. 'World Scene Firmware': perspective view of grid and objects (1983).

(rather than simulation), the VG3400 equipments at British Aerospace were fitted with firmware offering a performance increase of times 10 in speed and times 3 in quantity of visual information within the stored binary representation compared with the manufacturer's firmware.

Regrettably, most VG3400 equipments never realized their potential.

Parallel Graphics Machine (1987)

The 'dally' into simulation work with the VG3400 led to the opportunity to design a high performance machine capable of computing real-time color shaded pictures with hidden surfaces removed. The work has led to the construction of a prototype Parallel Graphics Machine (PGM) for the Royal Aircraft Establishment, Farnborough. The PGM is more fully described in the closing section.

PROBLEM SOLVING

All the former section upon the historical (and/or hysterical) detail should serve to inform upon the background responsible for producing my opinions. Inevitably, we are widened by our experience, but is some sense narrowed also, since tendency to cling to outdated yet comfortable ideas is readily observed (in others).

Anyway, upon the chance that there is yet some relevance to others upon the development of my schemes and ideas, the following sections trace these in some specific technical areas.

Before launching into this – with potential total loss of readership – let me take the chance to observe that really all the work and problems and solutions are merely representational changes. All the perceived hard work is really just gaining an adequate understanding of a problem to determine how to transform to the solution. Within this process, many routes are possible, and inside software edifices the opportunity to invent methods of representation to aid thought, and hence solution is indeed rich. Beware of 'total generality', since it will take all space and solve no problem: rather, arrange the architecture of representation to be a little more than adequate to cope with the problem in hand, to allow for the true requirement as it becomes revealed.

I have been surprised by the possibility of coping with apparently very complex situations by the invention of a very small set of intellectual tools with which and within which to express the problem. Of course, the magnificent advantage of the small set is that it is concisely representable in a machine: the smaller the set of expressible possibilities, the smaller the possible representation of each. However, there is no requirement that the meaning of each possibility be trivial, or invariant upon context. Thus, in general, the approach of finding an adequate model of the problem and a form within which its expression can be concise has been the familiar starting point.

A recent, albeit atypical, example arose in the implementation of the logic for the control of a bit-slice processor. The equations had too many terms to fit into ordinary programmable logic chips. By simply interchanging the binary coding of some instructions, the logic equations could be reduced to a form that did fit: thus by re-expressing the problem it became solvable.

DATA STRUCTURES

The design of the data structure is of paramount importance, since it holds the machine's model of the problem or situation under investigation or design: what can't be represented can't be questioned or designed.

At a level of indirection, it is entirely reasonable to use a data-structure to define the content and structure of a separate data structure: and we may according to our preference view such a definition as either a 'syntax definition structure' (with thoughts of 'syntax directed compilers') or as the instructions to a pseudo machine. This latter view allows us ready comprehension that the exact meaning of the 'instructions' is determined by the pseudo machine – which can readily choose to interpret differently according to the job in hand – be it questioning, creation, deletion, output, picture generation, relocation, etc., of the 'target' data structure.

It proves possible to represent the content of a data structure with only a few basic pseudo instructions which (just like ordinary program) are usefully arranged with 'substructure' call, return and jumps, etc. As any machine designer knows, conditionals and status are necessary to make a machine useful – and the pseudo machines are no exception.

Thus, in a few sentences we move from the problem of wishing to be able to describe the content of a 'target' data structure into the construction of a description – which is viewable as either data or instruction, and can indeed include algorithm. This is all of no matter: they are merely the same thing in a different form or arrangement.

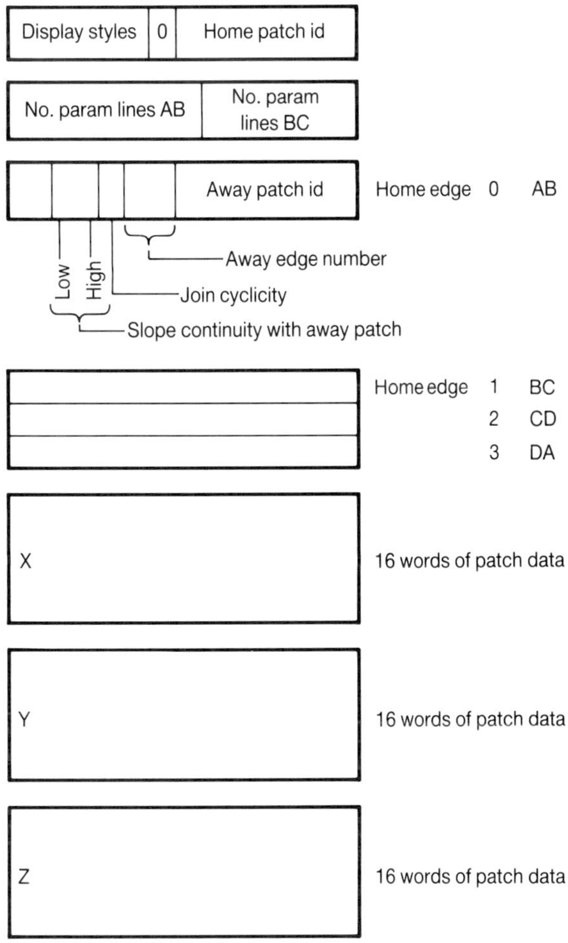

Figure 14.9. Patch data structure of Multipatch.

Data structures of Multipatch

Looking back at Multipatch, there were some very good reasons for the simple and simplistic data structure consisting of fixed sized blocks for each patch, as shown in Figure 14.9.

A sequence of such blocks was terminated by a −1. As shown, the join information for each of the four patch edges consists of a single word indicating

- id of the patch to which it joined
- edge number (2 bits) of the 'away' patch
- join cyclicity (1 bit) of the join
- whether slope continuity existed across the edge (2 bits, 1 for each end of the edge)

It will be noticed that in such a scheme all reference to other patches was by name: no pointers. Of consequence, the data structure was inherently relocatable, and indeed, the addition of data structures need only allow for name clashes between existing patches and incoming patches, and make the necessary name changes to incoming patches.

In a small scheme the necessity to search the entire data structure (or half on average) to locate a joined patch was of no computational significance – except that it is much less work than the correction of pointers!

As implemented the scheme did not allow multiway joins (i.e. three patches joined at a common edge) because the join was recorded in both the participating patches, filling out the space in each for the relevant edge and 'cyclicity' (i.e. whether movement along the joined edge corresponded to parametric movement in the same or opposing directions in the pair of patches).

In this example, after patches become joined there is no reduction in the 48 numbers per patch body: rather the 'joined' patch has its content modified to achieve the position for the joined edge, and the join information is filled out in both 'home' and 'away' patches. Thereafter, movement of either patch (at that edge) is automatically reflected into the joined patch by programs sensitive to the joins recorded in the data structure.

We may argue that the data structure is at fault whenever data are recorded multiply: in every instance we will have taken upon work to record how to keep the multiple versions 'in step'. Unfortunately, it is a little more complex, because the slope continuity information for Coons' Bicubic patches does not simply duplicate slope data – it undergoes sign reversal according to which edge joins which edge and with which cyclicity. If the patches are re-written in Bézier form, then slope information (noted as a 'control point') needs reflection via the common corner point.

So to avoid holding data multiply we would need to allow for a way of using shared slope information and point information, *and* be able to continue to allow for the possibility of detaching previously joined patches, or even deleting them – and *without* consequential loss of their previous neighbors.

As can be seen, in an arbitrary connected array of patches, changes to one patch can give rise to the necessity for changes in adjacent patches, and so on, to maintain the applied continuity conditions. Multipatch solved this by program applied to the join definitions recorded in the data structure, rather than modifying the data *structure* to itself reflect the geometric structure's current interpretation.

Data structures of Multiobject

The various geometric items of Multiobject were again within fixed size blocks and kept compacted, and joins were similarly recorded by name rather than by address. However, the design method ensured that patches were placed upon a mesh of curves: so that patches did not exist without curves upon which to depend. This encouraged the user to get the curves right before worrying about the surface, but of course, after surfaces were created the opportunity to modify curves was not lost! Their deletion was made to inform the user of the consequential loss of patches (before doing it). So, in Multiobject the reference (albeit by name) had expanded to the possibility of naming objects of a different type. Armit [5] gives a complete description of the data structures of Multipatch and Multiobject.

Data structures of ICON

Within ICON the references were by address rather than by name, and indeed the line and points, etc., had no unique identifying number. Of consequence, they could only be referenced by lightpen picks upon geometry, or via NASTRAN data containing reference to such points. This made for compactness, and the adoption of three kinds of point allowed the geometry structure to be directly reflected within the internal model:

1. *P*-point 7 words: type, *X*, *Y*, *Z* (all components in 2 word floating point)
2. *Q*-point 4 words: type, ref to LINE, fractional distance along line
3. *R*-point 3 words: type, ref to LINE, ref to LINE

with

LINE 3 words: type, ref to Point, ref to Point

Not only did *Q* and *R*-Points reduce the data structure size (a vital need to fit within 16K words), but their existence allowed the creation of points to match the geometric conditions. Since LINE could refer to any kind of point, this little scheme immediately allowed recursive dependency of line upon point upon line upon point, etc. Thus, the movement of any point could cause any number of other lines or points to move: exactly in accord with the topology of creation adopted by the user. This was (and is) a highly desirable feature, corresponding to the flexibility to move the drains after building the house. While all the systems of my construction have offered this, ICON was the first to need recursive evaluation of the geometry – i.e. to compute the geometric position of some point it could be necessary to first compute the positions of other points or lines.

This was solved by a 'geometry control stack' which was setup with the need to compute some item by holding a reference to it as the only entry. A 'data stack' was initialized as empty. The normal recursive scheme was to examine the control stack for a 'job' and then either extend the control stack by further job and/or the data stack by values until the control stack became empty and the data stack held the result. Thus the computation of the Q-Point would cause the data stack to be extended by a fraction obtained from the data structure definition of the Q-Point, and the control stack to be popped of its Q-Point reference but then extended by *both*

- indicator of the need to compute distance along line
- ref LINE

where each of these is a job. The ref LINE would be picked up, and eventually lead to a situation with two position vectors upon the data stack, and then the final computation would be upon a control stack

indicator of the need to compute distance along line

with data stack holding

- fraction obtained from original Q-Point = f
- position vector of start of line = START
- position vector of end of line = END

and the result of the job would be the popping of the control stack (to empty, and hence done) after the computation, yielding the position vector of

$$(1 - f) \times \text{START} + f \times \text{END}$$

upon the data stack.

Data structures of TAG

Within TAG the data structure received an almighty upgrade from the previous systems, mostly because of my increased awareness of the advantage of using the pseudo machine approach to allow the definition and management of a more complex situation. This confidence had come from the successful use of the pseudo machine in ICON for the specification of the NASTRAN data card format within the system.

Whereas ICON had to employ the pseudo machine approach because of the syntactic 'nasties' of the language whose syntax needed description, TAG was under my control, and the pseudo machine was not put in to solve any mess. Rather, it allowed a formal definition of the content of the data structure, together with a formal description of how to compute the geometric position of any item in the data structure. These naturally interleave: as content is described it is classified into syntactic type, and thus its relevance to 'naming' an item or controlling the position of an item is fully discernible. Add to this just the indication of basic computational processes such as transforms, intersection routines, etc., and the full control of the job and data stacks (of the form described above in regard to ICON) becomes a mere side-effect of one operational mode of the pseudo machine running upon geometry definition structures (or 'instructions').

Object form within data segments of TAG

TAG organizes its data structure into 2K word segments, and no object needs to span any segment. Each segment commences at offset 0 with a segment header to indicate the identifying number of the data segment, together with the offset to the start of the free space chain and the offset to the first entry in the object chain. These are simply 0 if none such exist. The free space chain within a segment uses entries of just two words:

- offset of next entry in chain from segment beginning
- number of words of free space here

The chain is ordered to give the smallest first, so that the least fragmentation of space occurs within the segment as space is taken.

The object chain is in no particular order, and includes objects in the form

header:	offset to next in chain
	length of this object
	type of object
	subtype of object
	offset to lower level form, 0 if none
	offset to in-ref chain, 0 if none
	identifying number
	control of line style of presentation
	control of text style of presentation
body:	mixture of integer and floating point values and references to other objects according to type and subtype
tail:	optional long form reference to a transform object

The header of any object type is in exactly the same form (because this was found to be adequate). Although all data structure references between objects are by *address*, every object is given a unique identifying number within its *type*. So, TAG obtained the best of the earlier systems by allowing the internal address reference to avoid data structure search, but also allowed the user to quote specific objects by their identifying numbers. This is of particular importance to the overall system, since TAG automatically synthesizes an unambiguous textual form of every command the user gives – even if given solely by picks upon geometry and button pushes, and journalizes this. Thus, with saved versions of data structure together with the journal, the user can always go back upon what he has done: kind of useful when the design task gets really difficult.

TAG keeps the association of type and is quoted by the user to segment in which the object resides (for the most recently quoted 666 such). Outside this a data-structure search is required.

The header of each item includes two offsets (or 'short-form references') which are guaranteed to be within the segment holding the object – and thus only take one word for reference. The reference to 'lower level form' is non-zero to indicate the presence and position of the computed 'explicit form' of the object. This information is exactly as produced by provoking geometric computation of an item: the resulting numeric (not necessarily all floating point) information is recorded in the segment, topped off by a word holding its length. The purpose of this is explained below.

The other offset in the header is to the 'in-ref chain'. If non-zero, this offset leads to a list of 'full-form' (i.e. two words of offset, segment) references to objects which directly refer in their bodies to the object in hand. Thus the in-ref list indicates all objects dependent for their geometric computation upon this object.

Within the body of the object is a mix of values, components and references to other objects exactly in accord with the object type and subtype. For example, there are indeed many ways of defining a point – from the 'explicit' in which the x, y, z components are specified, to the referencing examples, say at the intersection of three planes.

Now when we use our recursive evaluation process to compute the intersection of the three planes we arrive at the so-called *explicit form* of the point: which is the same for any subtype. Having done the work to find the position, it is recorded as the lower level form of the object in the segment, as mentioned above. This serves two purposes: first, it avoids the recursive evaluation process from having to go to the greatest depth under all circumstances: if the lower level form exists then it is simply re-captured to data-stack rather than re-computed; second, and in strict corollary, its non-existence indicates the need that it should exist for potential display upon the screen. Hence the modification of geometry processes arrange to remove the lower level forms to any object they change, together with lower level forms to the recursive in-reffing set of objects to the objects changed.

After a set of design changes the data structure will have lost lower level forms and, for each object changed (rather than consequently changed), a reference (two word address) is pushed to a stack of such. When design changes are through, TAG proceeds to perform picture computation, taking each reference from the stack together with any (recursively) referenced object for which a lower level form is not already present. All such objects are for re-output to display in each of the views created by the user. TAG thus proceeds to output only the necessary recomputed picture parts – like Multiobject – but it also recognizes that beyond some hundreds of picture changes it becomes more efficient overall to recompute the complete view (because of display considerations in maintaining and shuffling the picture on-screen as held in the hardware 'refresh buffer').

In TAG each object (including transforms) may quote an optional transform. So apart from the many ways of object definition the position of the object may undergo further (and possibly chained) transforms. It is typically useful to allow several objects to reference the same transform so that changes to the latter can provoke gross design changes in some desired manner. This can be considered to be 'grouping by common reference', and indeed, TAG also supports arbitrary groups (consisting of two-word references to objects) through which commands can be directed as appropriate (i.e. according to object type) to the members of the group. The provision of the data structure flexibility demands that the user can communicate flexibly through the structure he creates – the interactive language allows commands to be given via any object to its referenced or referencing (or either of these recursively) objects.

Garbage collection and segment splitting within TAG

Garbage collection within TAG operates within the segment environment. Given the availability of 64M word demand paged demand mapped space, it is entirely trivial to obtain more 2K word segments. Thus, data structure expansion and contraction comes down to the problems of segment splitting and merging (of the content). In either case, the address (of offset, segment words) of the objects are likely to change (though merge can typically avoid changing them all). So when it is desired to deposit into a specific segment (e.g. to record the lower level form to an object), it may first be necessary to split the segment, modifying the position of the objects. This will be determined as necessary when the segment's free space chain fails to yield sufficient space.

Normal space take-and-release within a segment simply operates upon the free space chain, always keeping a size ordered chain to allow the smallest available space to be reused at any request. It will be noted that the free space chain cannot record that a single word is available, because each free space chain entry takes two words. However, no item in the data structure can be as short as one word either, so at return of space to the free chain these one word losses are also recoverable, with consequent merge of returned space with one or two other free space entries, and free chain re-ordering in accord.

Very typically, the 'garbage collection problem' arises at an inconvenient moment. In TAG we may pick a piece of geometry (using puck upon tablet surface) which causes the hardware to interrupt and the software to identify the object – such as a line – upon which the pick is made. Given the type and id of the object the software can consult the index to obtain the data structure two word reference address at which the object resides. According to the *context* (as setup by the interactive command being synthesized by TAG from the user's key hits and picks), the specific pick may be interpreted as the need to access an object referencing the picked object. For example, a point upon a line can be picked by selecting 'PNT' from the 'menu' (or keying this) and then picking the line itself, leaving TAG to work out the nearest point to the pick position directly related to the line (in projected screen space). It is further possible that such a point is currently invisible, and also that its lower level form is non-existent (following some geometry change). It is also possible that the segment containing the relevant point is full: so the pick upon a piece of geometry can provoke the need to split a segment.

Such an event is no problem to TAG even though the system is so deep into data structure access. While it is obviously necessary that the system can determine how to correct the pointers within its data structure under segment split or merge, most systems attempt to arrange that computational processes are not in progress at the same moment. As described above, TAG holds explicit indication of both referenced and referencing objects for each object, and this avoids all data structure search as objects are moved from one segment to another: the immediate in-referencing set need their pointers corrected. In addition, and most specially, TAG is organized so that every external reference to the data structure can be located. This tiny statement is of large consequence, for it allows the segment splitting and merging to occur any time, and within any process whatever. A space jam is impossible.

Operationally in our example, the pick will have set up the general address of the item picked, and a consequent data structure segment split can have caused the written address to have changed.

The scheme is organized by the formualizing of the access of data structures into stacks, each containing 0 or more levels, and with each level containing a pointer to the data structure. Each such stack has a fixed format description of a few words to indicate the position of the stack and the size of levels, and the top, current and off-end pointers of the content. On top of this is a separate description (at simplest a list), indicating all the stack descriptors. Thus, all the long form addresses used for data structure access are in memory, and can be found for invisible relocation during the operation of any process.

It would be misleading to suggest that such a scheme is trivial, but equally, its evident elegance seems to deserve its description. In our example, the pseudo machine will run upon a geometry description to control the menu presentation and context (for the meaning of picks) during geometry creation, and *within* this can be required to control the recursive evaluation of some item referencing the picked item (again running from the geometry definition structure), and *within* this can be required to control the comprehension of geometry objects as their contained references are relocated as part of the segment splitting activity. The pseudo machine stacks its incarnations and gets used within itself.

INTERACTIVE LANGUAGES

Interactive language of Multipatch

In the early interactive graphics systems, researchers were much taken by lightpens and styli with tablets. When constructing Multipatch in 1968, the system design itself was as much how to communicate a design and its change as how to represent the result in the machine, and to cope with the incoming changes. The need for various commands with clear and reversible effects was evident. The 'steered beam' displays available had only limited drawing capacity if flicker (and large display files) were to be avoided. The clumsiness of the lightpen (the 'one-bit input device') and the impracticality of single key stroke 'commands' encouraged the construction of the interactive language of Multipatch.

The language was based around hierarchical name-address structures, and simply associated the characters of a 'name' with an address – which had a 1-bit typifier to distinguish between the address of a 'lower level' name-address structure and the address of the routine to be entered to perform the effect of the command. Thus, the language processor simply needed to know the fixed starting address of the top level of the name-address structure, together with the input command line to 'present' the 'names' given against the structure, working down to the address of the execution routine. Simplicity indeed:

```
DEFN:  'PAT'      ; top level mnemonic
                    for patch controls
       1$         ; address of next level
                    structure
       'VIE'      ; top level mnemonic
                    for view controls
       2$
       0          ; terminator

1$:    'A'        ; patch controls:
                    names of corners
       3$
       'B'
       3$
       'C'
       3$
       'D'
       3$
       0          ; terminator

3$:    'IX'       ; incremental
                    commands for patch
                    corners
       XCT RT1    ; XCT for typifying as
                    routine address
       'IY'
       XCT RT1    ; use same execution
                    routine
       'IZ'
       XCT RT1
       0          ; terminator

2$:               ; view commands
                    similarly

RT1:   subroutine to execute incremental
       changes to patch corner positions
```

Thus our structure embodies the syntax:

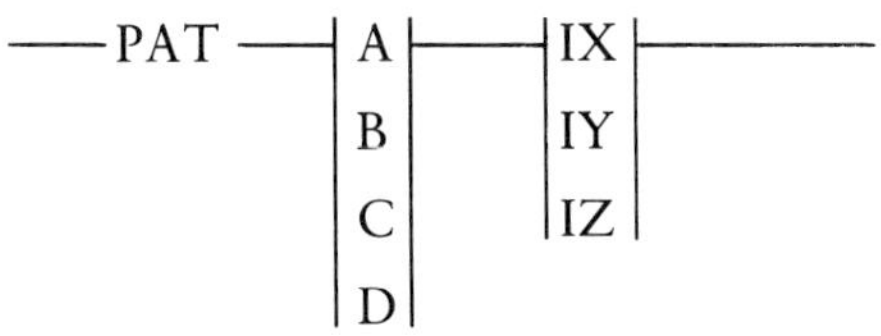

and indicates that the routine at RT1 is to execute the effect for any of these. The diagram shows that following PAT is to come any of *A, B, C, D* and following this is to come any of *IX, IY, IZ*. Given that the patch corners are labeled *A, B, C, D* this set of commands allows incremental movement to any corner.

This example shows the organization and the ready sharing of definition structures and execution routines. To be effective, the execution routine needs to know how it got entered – solved by the creation of

LEV0:	word to hold top level 0 mnemonic
	word to hold level 1 mnemonic
	word to hold level 2 mnemonic

By allowing the execution routine to test LEV0 record it can readily become efficient in achieving incremental or absolute setting to *X* or *Y* or *Z* components for corners *A, B, C* or *D* of the patch. The scheme also allowed input of a single number with the necessary status of

NUM0:	binary of last input number
NUM1:	set if NUM0 input in last command, else clear

Typically, the hierarchy can be limited to a maximum of three levels to avoid clumsiness. Even this represents too much input (whether typing or not). Multipatch took on a design restriction in the interactive language, that 'no name may exist upon any path leading to that name' to allow the user to supply just the differences between the command required and the last given. This was coupled with the special recognition of a minus sign without following digits to mean a reversal in the value recorded in NUM0. Thus, interactive adjustment of patch position could proceed:

```
PAT 2 SEL   ; select specific patch
A IX 20     ; increment X component of
              corner A of patch 2
-           ; made a mistake: put it back
              again
-10         ; go the other way by half as
              much
```

after each command line the on-screen picture would update virtually at once.

At the time of design of all this, there was a gap between those who studied languages and those who made interactive systems (perhaps there still is?). Anyway, it turned out that this little language did not properly classify according to Chomsky, and attempts by others to show it to be simply produceable by compiler techniques led to the ridiculous situation of feeding in a language description to produce a large execution version of the language processor unable even to handle the minimum input features: the whole point of it. In any case, the whole thing took only a weekend to devise and program.

The ease of use of a system is much like a work of art – you know if you don't like it, and can always see something wrong with it. For Multipatch, some were amazed by the range of effects so readily achieved, whereas others could only see that typing ability was an advantage.

Interactive language of Multiobject

In Multiobject there was much more to say, and the interactive language was extended to allow macros, expressions wherever numbers, arguments to any 'name', scalars, vectors and matrices, complete with operators. On top of all of this came 'right hand side' processing, with a language definition structure exactly as 'left hand side' but for value extraction, rather than value supply. Hence it was possible to give

LIN (1) *A XYZ* = LIN (2) *A XYZ*

where the right hand side of the equation was used against a name-address language definition structure, just like that for Multipatch with the setting up of the internal path leading to a 'fetch' rather than an execution routine. In the example, the *XYZ* vector position of end *A* of line 2 is extracted for supply to the command: with the result that end *A* of line 1 receives the extracted value, to become coincident on screen *but not joined*. Thus, geometry placements are readily expressed in the language by arranging that every value that can be set (by name) can be questioned (also by name in the same form).

The minimizing of input by supply of differences was implemented (separately, of course) for left and right hand processing. The scheme really allowed any number to be an expression including these fetches, possibly within each other – which typically has no application in a geometry design system.

The supply of the calculator facilities and trig functions and macros did allow some quite complex commands to be constructed and stored as macros for ready invocation. For example, the evolute to a curve was obtained, complete with its graphical representation on-screen ([5,7,9] give further information).

Interactive language of ICON

Within ICON, the geometry did not include unique identifying numbers for each object within each type – and hence reference by address rather than name was required. In consequence, the interactive language allowed for choice picking upon menu items and picks upon the geometry and keying of 'string arguments' for file names, etc. Thus, the language definition allowed the legality and meaning of these to be given, so that a pick input would be accepted at relevant times within the command input. Indeed, at each pick or key input, the language processor would re-assess the input so far to determine the relevant set of screen choices and prompts. At command execution the specific routine was entered similarly to Multipatch and Multiobject, and it had to determine whether input pick data should be used to supply positional or reference information according to context. ICON bore the shift from 'language definition with processor' to 'instructions and pseudo machine'.

Interaction language of TAG

TAG's use of picked and keyed input

In TAG the approaches of the former systems were combined in a most advantageous manner. Like Multiobject, every number could be an expression, with scalar vector, matrix arithmetic and trig functions. Also, every value in every data structure could be questioned by name, since unique type and id were assigned by user (or system by default). In TAG, all picks upon menu choices or the geometry of the design are used directly to synthesize the text command – which can be attacked by the delete key just as though it had been typed. Thus, like Multiobject, every command has an unambiguous text form, even though the user may operate the system entirely by picks, and even picks upon geometry items rather than text choices of a menu.

In TAG the context of the interactions is determined by the pseudo machine, so that not only are the choices and messages generated according to the part-input command, but the meaning of geometry picks for the supply of item type, id, position in X, Y, Z or vector XYZ, etc., is all determined. It is this that allows TAG to accept a screen space pick upon a perspective view, and to resolve the position in the rotated screen space, back-transform to object space, and then extract relevant component(s) and synthesize the text of these with automatic insertion into the command being input – the values appear in text on-screen just as though they had been typed.

Thus it is possible to pick invisible geometry by first picking from a choice menu to indicate the item type – as line or arc, etc., and then to pick the geometry – to have the system trace the nearest placed such item related to the picked item. In this manner, TAG synthesizes not the floating point position vector, but the unique identifying number – i.e. the argument to specify the relevant line or arc.

Thus, the pseudo instructions include definition of how to treat picked geometry in each instance, but unlike ICON, this is kept within the language processor: the execution routines are kept unaware.

The option to supply an expression, including fetch upon geometry, is always available: the user merely keys an arithmetic operator (e.g. plus) to have the right hand side choice menu and prompts offered. These too can be supplied by picks.

TAG's pseudo machine instructions and modes for interactions control

The control instructions of the pseudo machine are

- ordinary UP DWN BRN SKP controls

- conditional UPE DWNE BRNE SKPE for effect if 'Equal'
- conditional UPD DWND BRND SKPD for effect if 'Different'

The syntax of the control instructions is simply that UP or SKP are complete, whereas DWN and BRN quote the address of other pseudo instructions. Thus the DWN, UP and BRN are strictly analogous to call, return and jump in an ordinary computing context.

There are instructions specific to interactive language comprehension and execution:

instruction	defines and controls
NOARG	possibility of absence of argument
EXEC	routine to perform effect of command
EXES	structure to perform effect of command
SCA	scalar argument
VEC	vector argument (x, y, z components)
MAT	matrix argument (3×3 rotation)
ARB	arbitrary length list argument
STRN	string argument of any text
LSTRN	literal string argument of specific text
SMBL	symbol of the hierarchical language
TSMBL	lowest level symbol in the hierarchical language
EMS	'enter major state' to allow whole new language within language
CMT	comment: defines non-pickable text
MSG	message: defined non-pickable text

The meaning of the content definition instructions depends upon the execution mode of the pseudo machine – of which there are twelve, including

choice:	in which the menus of choices for screen presentation are determined and output.
presentation:	input text is 'presented' against the language definition and left or right hand memories of last input command or fetch in order to understand the input.
pick:	to determine the legality of the pick input from the user and to synthesize text in accord.
execution:	to consider the user's request for execution by first ensuring that the command is complete with supply from memory of last command if necessary and legal, followed by entry to specific routine or routines to perform the action.

Thus to specify choices on-screen we may construct a little pseudo code such as

```
MSG ⟨any text we like to appear on screen⟩
MSG ⟨more text to appear on next screen
      line⟩
SMBL PNT TLA1 FPNT1 PNT PK1
CMT ⟨any text we like to appear in lower
     case to explain the live choice of PNT
     just given⟩
. . more of such
UP   ;indicate conclusion of the definitions
```

The messages and comments appearing on-screen are not pickable, and simply inform the user. Whereas messages commence upon consecutive lines, comments follow a specific choice (or 'menu' item) which is pickable, and are upon the same line. TAG arranges that comments and messages are suppressed or enabled for screen presentation at any receipt of double space (which is always syntactically equivalent to single space, but is given this side-effect).

The entry of interest is the SMBL, which defines a symbol or 'name' within the interactive language. The form of this pseudo instruction is

```
SMBL ⟨exact ASCII chars of name with
        leading alpha and no space⟩
     ⟨address of argument definition
        structure with NA if none⟩
     ⟨address of next level name
        definition structure⟩
     ⟨internal value to be associated with
        the symbol⟩
     ⟨coding to indicate legality and
        method of text synthesis from a
        geometry pick⟩
```

Hence, in the example, the SMBL pseudo instruction is defining a screen choice with text

PNT, with an argument definition structure at TLA1, with next level's symbols, etc., definition at FPNT1, with an internal code value from the symbol PNT (distinct from the text), and a PK1 internal coding for text synthesis. The top level argument definition structure is

```
TLA1:  MSG ⟨give no arg or non-existing id
          if 'create'⟩
       LSTRN ALLX 1$ ZIC PK0
1$:    SCA 1$ ZIC PK1
       NOARG
```

Here we have defined the 'literal string' ALLX as a screen choice. If picked it will be added to the command under construction enclosed by square brackets. Syntactically, this merely allows the specific string to be a legal argument to the top level symbol. Operationally, TAG recognizes the ALLX forms to apply the given command to ALL the objects of the given type eXcept any quoted exceptions. In addition to allowing ALLX the top level argument can be a SCAlar – the integer identifying number of (in this case) a point. In addition a NOARG is legal – i.e. an absence of argument. The form of the SCA pseudo instruction is

```
SCA ⟨address of next argument definition
       structure, NA if none⟩
    ⟨associated internal code: fill by ZIC if
       none relevant⟩
    ⟨coding to indicate legality and
       method of pick text synthesis from a
       geometry pick⟩
```

The form of the LSTRN pseudo instruction is the same except it commences with the quoted literal string.

The overall effect of the three pseudo instructions is to allow the written forms

[ALLX] followed by none or more scalars
or none or more scalars

The language allows the application of the command to be determined as ALL (with exceptions) or the quoted list of object ids. If no id is given, then the 'currently selected' object of the type is affected. The language structure and implementation scheme arrange that the 'demultiplexing' of which objects get modified is entirely outside the modification routine – and thus the demultiplexing is written once for the whole system!

In choice presentation mode the fact that the arguments are optionally absent (in this example) causes the further use of the pseudo machine upon the next symbol level choices – to be found at address FPNT1 in the example.

At some lower level (and this is kept within a total of three within TAG for user convenience) there is no lower level symbol because the command is complete – except for the supply of values to the overall command. It has been no snag to allow that arguments may, in general, occur following any level's SYM, but that in practice they occur after the top level SYM to determine the 'point of application' of the command, and very rarely (and beyond the scope of this explanation) to the next level. It has been beneficial to avoid arguments to the lowest symbol level (indicated syntactically by TSYM) in each command, but to allow instead the supply of floating point values to the total command, This comes to much the same thing, except that when the user provokes the ability to supply just the differences between a command and the last, the supply of values not preceded by a SYM will indeed allow repeat of the last command with new values.

To wrap up the example, by running the pseudo machine in presentation (rather than choice) mode, the input text from any source is considered in succession against the defined legal possibilities – with the construction of an internal fixed form of the part input command from both the input text and the memory of the last command. When a pick input is found in the device input buffer (rather than a bunch of one or more characters) the pseudo machine is run in pick mode, with the specific function of comparing the pick input with each syntactic possibility. If a particular symbol encodes that its supply cannot be by picking then consideration passes to the next syntactic possibility (by continuing the pseudo machine). Pick text synthesis is also sensitive to the language 'level' being supplied. Hence it becomes possible to synthesize identifying numbers at the top level (e.g. for picking of a point) or the *X*, *Y*, *Z* component or vector at the lowest symbol level (as indicated by TSYM pseudo instruction).

Execution mode is also sensitive to TSYM and in contrast with SYM pseudo instruction finds the execution routine address (rather than address of lower level SYM definitions) – just like Multipatch.

GEOMETRIC REPRESENTATION

Geometric representations in Multipatch, Multiobject and ICON

Now, clearly, the section upon data structure is relevant in defining where and how definitions of geometry get recorded. The question here is, what to record and how to understand the recorded content?

At simplest we build the detail into the programs which write and read the content of the data structure. This is exactly what Multipatch, Multiobject and ICON did. However, ICON maintained a large text definition of finite element definitions, which referenced the geometry, and in this ICON used the pseudo machine approach to define the recorded content and to control its access and change.

Geometric representations in TAG

Again, TAG hit it big, and used the pseudo machine approach to define the data structure content, its access, relocation, etc. The formalism allows new objects to be included in the data structure by merely constructing their syntax definition: this is sufficient to know that at segment splitting all the necessary corrections to incoming and outgoing pointers will be properly maintained by the system. Indeed, the object creation process itself is driven from the syntax definitions of objects that can exist in the data structure – so this too is available for direct use!

Example geometry syntax in TAG

To aid comprehension a brief description of the relevant pseudo instructions is given:

pseudo instsr	defining and controlling
HDR ⟨arg⟩	current word of 'header'. The arg gives the specific content, as detailed below.
DTR ⟨arg⟩	current word of body is data, with specific type according to arg.
REF2 ⟨arg⟩	current and next word of body is a two word reference to an object or group as specified by the arg.
RESD ⟨arg⟩	indicates a 'resolve data' routine according to arg. It is used in geometry computation.
SPLT ⟨args⟩	provides a de-multiplex capability by getting a data-structure value and adding it to the pseudo machine's 'program counter'. Typically, such a split is performed upon object type or sub-type.

Thus the construction 'HDR TYPE' gives definition that the current header word yields that object's unique type – distinguishing it from all other object and group types. A 'DTR PVEC' denotes that the body contains a data entry of a position vector – held as three floating point numbers (for x, y, z). A 'REF2 GFLN' indicates that the body contains a two word address onto an object within the group GFLN (group of finite linears). This group includes finite line, arc, B-spline curve – and the referenced object is one of these.

The syntactic definition of objects may proceed:

```
; definition of 'header'
; only the first three words of every object
  have an invariable format
SHDR1: HDR NXT     ; offset form seg top
                     to 'next' entry
       HDR LNG     ; length of object
       HDR TYPE    ; indicate syntactic
                     type
; geometry objects have a further six fixed
  format words
       HDR STYPE   ; sub-type within type
       HDR LLF     ; offset from seg top
                     to 'lower level form'
       HDR IRF     ; offset from seg top
                     to first entry on
                     'in-ref chain'
       HDR GID     ; identifying number
       HDR TLIN    ; control of line styles
                     for display output
       HDR TTXT    ; control of text styles
                     for display output
       UP
```

The above allows the header content to be described using the HDR pseudo instruction and thus to define say a point we use

```
SPNT:  DWN SHDR1 ; use the ordinary
                   header
       SPLT DLR IND ORIG STYPE
       ; split upon object subtype
       1$  ; explicit
       2$  ; on refd GFLN at explicit PAR
       3$  ; GPLN, GPLN, GPLN
             intersection

; explicit
1$:    DTR PVEC
       BRN SEO2

; on refd GFLN at explicit PAR
2$:    REF2 GFLN ; reference to group of
                   finite linears
       DTR PAR
       RESD SI15 ; resolve point
       BRN SEO2

; on GPLN, GPLN, GPLN intersection
3$:    REF2 GPLN ; reference to group of
                   objects defining planes
       REF2 GPLN
       RESL SI7

       REF2 GPLN
       RESL SI22 ; truncate arc to plane if
                   reqd
       RESD SI2
       BRN SEO2

SEO2:  code to allow optional transform
       and if not resolve mode then done
       else resolve mode and lower level
       form to be transferred to data
       structure and then done
```

Now just as a 'split' upon object type is used to give entry to SPNT definitions for point, so a split upon object subtype gives entry to the definitions in accord. Only three of TAG's 12 ways of point definition are shown – the explicit, on finite linear and at intersection of three planes. The explicit is easy: the DTR pseudo instruction defines data content – in this case a position vector. Points defined on a finite line, arc or B-spline curve are all treated by the 'finite linear' group reference under REF2 (meaning a two word reference by offset, segment), and this is followed by the data definition of a PARameter (a scalar indicating the fractional distance along the finite linear). While the REF2 and DTR pseudo instructions thus define the syntactic content, they do not define the evaluation process.

Resolution of geometric position in TAG

At the start of the geometry resolution process, we have an empty data stack and a single pointer level in the data access control stack to the item to be resolved. This level contains more than just a two word pointer, and records the offset to the start of the object in the segment and the type of object accessed. It is arranged that in resolve mode each DTR causes extraction of data from the object data structure to the 'data stack'. In contrast, at the pseudo machine execution of each REF2 from the body of an object, it pushes the data structure reference from the object onto the data access control stack and does an auto-DWN in pseudo code execution to re-enter the decoding of objects.

So, by the time the DTR PAR is met the pseudo machine has recursed to ensure that the data stack holds the explicit form of the referenced finite line, arc or B-spline curve. The RESD instruction is for 'resolve data', and operates only in resolve mode to indicate the program to be entered to crunch the data stack content, and in this case to determine the position of a point along the finite linear given its parametric position.

Each geometry syntax definition structure thus includes sufficient information to allow its computation process (resolves) and also makes a common definition 'exit' via the optional transform reference (at SEO2 in the example). A non-null reference in an object body is taken as a reference to a transform, which is similarly resolved and applied to the object's definition as held in the data stack. It will perhaps be clear that the data stack gets to hold the 'explicit form' of an object as a result of the pseudo machine recursions – i.e. there are never any references but only data in the data stack. Thus, we say, transforms are readily applied. But careful, we won't want to transform every number the same way, e.g. if an arc is defined as a fraction of a revolution (and other skew definition data) we better not change the length of the arc! So, we need to know the data type within the data stack to be able to apply transforms to any object without needing to know the object type.

This is achieved by the fact that TAG's data definitions under DTR pseudo instruction descend to any of

name of data	meaning
INTG	integer in word
STRG	string of characters
PAR	parameter – not subject to scale changes
FSCA	scalar – is subject to scale changes
UVEC	unit vector, three components
FVEC	free vector, three components
PVEC	position vector, three components
MTX	matrix (3 × 3 for rotation)

TAG includes transforms of

OFS SCL ROT MIR

The above names are the syntactic types of objects within TAG, and it will be noticed that there are unit vectors, free vectors and position vectors. The FSCA is unlike the PAR in that its value is subject to scale change (e.g. radius of a circle is FSCA and thus transformed, whereas parametric distance along a curve is PAR, and thus not). The data definitions within the pseudo machine and under DTR instruction reduce all data to the definitions listed – and in accord the execution of DTR instruction in resolve puts both data type and data value information into the data stack to allow proper evaluation and transform processes upon it.

Looking at the last example, of a point existing at the intersection of three planes, we again see that the syntactic reference is to a GPLN in which both arc and plane are allowed (since both give definition of a plane). The data stack thus becomes extended by the explicit forms of a plane or arc and another plane or arc. Then the resolve routine code SI7 uses these to replace them by the explicit form of an infinite line (their intersection). The data stack is then extended by the third plane or arc and resolve routine code SI22 does nothing if plane, or truncates arc definition to that of plane. Finally, the resolve routine code SI2 intersects infinite line and infinite plane to yield the point.

It may come as a surprise that in a system with 38 or so geometry types, each with many subtypes and various groups, that only 39 resolve routines are needed. This occurs because of the flexibility with which they can be concatenated under the pseudo machine instructions.

The resolve routines are the geometry computation of the system, and a few simple aspect calculations of conversions to integers, etc., for output to display are the only other such arithmetic. I consider their positioning to be most advantageous, since their operation is upon the latest entries on a stack – giving full recursive and re-entrant possibilities.

GRAPHICS AND HARDWARE

Picture computation in Multipatch and Multiobject

Much of the fascination comes from the screen presentations and the possible speed and effects achievable. When writing Multipatch it was simply a matter of reading the book to find how the hardware interpreted binary patterns to control the steered beam display to draw vectors or characters. There were options upon line style in that early DEC 340 display by spacing the dots, there were eight levels of brightness, and flashing could be obtained by intermittent inclusion of pieces of display file.

The complete set of transforms was for software to do, i.e. to offset and rotate the object into screen space, then clip to screen limits, decompose into vector instructions, and then construct the binary form required by the display hardware and deposit into memory. Given the (lack of) power of the machine, together with the fact that curves were being displayed by using lots of small straight line segments, exact clipping to screen edges was not done in Multipatch – rather, the small segment going off the screen was simply omitted. Even the process of vector computation was tricky, in that a line across a large portion of screen needed to be represented within several instructions to the display, and it was important to arrange these to be of equal length to avoid ready discernment of slight change in angle (or short closing vector), etc.

The visualization of 3D shapes from a 2D screen was investigated in Multipatch – e.g. short surface normals were drawn upon surfaces (and immediately named 'hedge hogs'). These give an interesting understanding of the way in which a surface shape is changing. Unfortunately, they use up lots of display file space in requiring discontiguous lines – leading to a full machine and flickery pictures. Brightness modulation had been overlooked by others, especially as the machine had so few distinct levels. However, by auto-ranging the brightness modulation it was made to be most successful. Specifically, the picture computation process kept track of Z_{min} and Z_{max} from all picture on-screen. Then, upon the next computed frame, this allowed the available brightness levels to be applied to maximum effect, i.e. spread over this range of positions in Z. At a time when hidden surface algorithms were hot news (though impossibly expensive in equipment and time), Multipatch achieved a most effective scheme in real-time, with only about 15 instructions per line drawn added.

Multiobject used the same 340 display, but calculated the screen edge intersections fully – which is more expensive in programming space than in time, because most vectors don't intersect the screen edge.

Picture computation in ICON

With ICON came the Vector General Series 3 display, offering hardware rotation and Z-axis giving brightness control. The software had a notionally easier time, needing to convert from floating point and to obtain vectors within a number space larger than the screen, leaving the hardware to perform the final offset and rotation and screen edge clean-up. Though the screen was much higher resolution (12 bits rather than 10), the scaling and offsetting could be driven to situations where it was essential to re-extract from the floating point positions to have more significance reaching the screen after scaling by hardware. Thus, the splitting of the offsets and transforms between software and hardware gave every bit as much program as doing it all by software, but there was the significant time advantage in the hardware! Perspective views were not possible. Interestingly, the Series 3 was unable to return the position of a pick with the lightpen, but only the end-point of the line picked. Further software and use of the hardware of the display allowed the determination of the beginning position of the line, and from this a process of binary division by drawing the original line from ends to center allowed the field of view of the lightpen to be determined with consequent position determination. Once written it was just as though the hardware had the facility.

Picture computation in TAG

TAG employs the Vector General 3400 display, which includes a Graphics Processor Unit (GPU) clocking at 4 MIPS to perform the rotations, etc. It proved that the manufacturer's firmware would simply not provide the facilities sought, but in a state of desperation their Writable Control Store was obtained, and firmware was purpose built. This showed the machine to be indeed capable of impressive performance, and allows full rotation, offset and perspective facilities with multiple views and, indeed, multiple users. The operation of the hardware is to extract from the host machine (DEC PDP11/70) by direct memory access and to compute pictures within the GPU, sending the binary representations over the internal GPBUS to the display system consisting of Refresh Buffer Unit (RBU), Display Control Unit (DCU), Vector Generator Unit (VGU) and Font Control Unit (FGU). Thus, while the GPU is computing and depositing into the RBU, the DCU is concurrently extracting the last computed frame from the RBU and sending out control and data to the VGU and FGU to maintain the picture on-screen. Interestingly, the 11/70 includes peripheral mapping hardware allowing the 18-bit Unibus addresses to be mapped anywhere in physical memory of the PDP11/70. Even more interesting is that these registers can be WRITTEN by a DMA peripheral – allowing the VG3400 to steer itself through segments of picture definition data all by itself. (The MEGA operating system allows its 2K word managed segments to be locked into physical memory.) The RBU is thus the memory which holds the binary of the computed pictures, giving the beam steering and character drawing information for maintenance of picture on-screen at 25 frames per second. Typically, the GPU can cause RBU changes within a fraction of a second to give new screen representation or positional change for an object, or indeed, to rotate a particular view of the assembly of objects.

Thus from Multipatch we come full circle to the VG3400, with all the picture computation process being performed by program again – albeit rather faster in the latter. Thus the facilities of perspective auto-ranging Z-brightness, etc., all become matters of programming – and are provided within the firmware. The overall picture computation process involves the host in performing the floating point computations for the resolution of object positions, and then for offset and scale and integer conversion. Thereafter, the display GPU performs offset, rotation, scale, perspective, clip, screen offset processes together with vector generation to yield the binary for the RBU. The coarse offsetting and scaling operation of the host is invoked automatically as the user pans or zooms beyond preset limits to 're-center' the 16-bit integer space handled by the GPU upon the center of interest in the object space. This involves the host in some computation of a few seconds. Within the local ranges the GPU will recompute complex views in a fraction of a second.

From high performance steered beam to raster displays

Now 'modern' displays no longer use the steered beam technology of the VG3400 (which was available in 1975). We have raster displays which can advantage us with more colors, but also give jagged lines and many other visual 'nasties' which become very noticeable if we have enough performance to compute real-time rotations.

Certainly, very few raster displays can compute and present line drawings as fast as the VG3400, and even fewer will show a crisper picture. Raster display technology is cheaper, and gives the possibility of area fills – quite impossible with the steered beam.

The problem in the computation of pictures for a raster display is in two parts:

- arithmetic for transformation of objects into screen space and for deciding how to represent the faces of the objects;
- filling out pixels in the memory with the required color encodings.

Now the arithmetic of transforms is exactly akin to that job within the VG3400, and a fast programmable processor remains a good approach. The quantity of necessary arithmetic is much the same as before, though we also consider the topic of hidden surface removal. With our steered beam display we had no opportunity to use the display hardware to remove hidden lines because the hardware had no notion of 'surface' upon which to proceed. In a 3D design system it is often convenient to be able to design the middle or back of some item, and the line drawings showing features readily support this provided that brightness or perspective is used to aid front-back determination. Probably the worst pictures on a raster display are line drawings, but shaded pictures demand hidden surface removal in 3D design situations. There are many ways to proceed, and some of these involve ordering or otherwise imply extra computational effort *prior* to pixel depositing.

The major bottleneck in production of a raster picture is simply computing and writing down in memory the color required for each pixel. There is no problem getting out to the display because the very ordered sequence of memory addressing allows several pixels to be sent upon a wide data bus for last instant serializing and output to video. The computation for each pixel will involve determining color and position, and there are lots of pixels even for modest resolution. The end point positioning upon a steered beam display such as the VG3400 was to 12 bits – today a 4096 × 4096 raster display in color would be very very expensive.

Let us consider the resolution of 1024 × 768 for which we need to compute 1024 × 768 × 25 pixels per second to achieve a real-time update of the picture every frame. Even this is short measure if we are to cope with hidden surface removal using a 'Z-buffer'. In such a scheme the pixel computation gives color and x, y, z position for a pixel, where the z value refers to the distance out of the screen. For each pixel the hardware includes not just memory for the computed color(s), but for the associated z value. Then, deposit of pixel (color and z) information can be made conditional upon the computed z value being nearer to the observer than that currently held in memory. If we allow that any pixel might get filled three times in the process of computing a hidden surface removed picture, then we need 58 million pixels to be recomputed per second for our example resolution.

While dependent upon exact hardware design, we will need to compute X, Y, Z, R, G, B for these 58 million pixels/s, as well as make the Z comparison. Per frame we will need to initialize the picture memory. Worse still is the potential need to compute yet more detail of 'sub-pixels' to arrive at a color for a pixel based upon its actual size rather than yielding a color from an infinitessimally small sampling (at the pixel center) and then spreading this to the whole area of the pixel. It is this sampling process that yields the 'aliasing' and 'jaggies' problems.

Parallel Graphics Machine design

This was briefly mentioned above, as it represents a current and ongoing graphics involvement, with recent design and construction of a prototype high performance graphics machine.

Apart from the knowledge of the process of picture computations, the main input to realizing the design was manufacturers' data sheets for their (available) chips. Study showed that standard microprocessors or signal processing chips were

- too slow at ordinary instructions
- too slow at multiplies
- not interlinkable for concurrent synchronous operation.

In this situation, the design was created by employing the simplest arithmetic unit, the fastest multiplying accumulator and fast registers. The whole thing was made good by writing down many logic equations and then putting these into programmable logic chips (without which the design could not have been achieved). In all there are about 100 different programmings of chips (with 60 different just upon an Arithmetic Machine). The prototype equipment consists of 24 cards with about 200 chips per card, with no custom VLSI to shrink things.

The machine architecture of Figure 14.10 shows that the PGM is connected to the VME Bus via an interface, and that on the PGM side are the 'Arithmetic Machines' (AMC) which are generally concerned with the geometric transforms (rotation, etc.) of the data extracted from the host, together with the setup of surface drawing requests upon the Picture Machines.

The DMA upon the host is organized by the Interface on behalf of a specific AMC, so AMCs compete with each other for the right to set this process up.

An entirely separate competition is for AMC ownership of the Result Bus, through which the AMCs arrange to set up jobs for specific Pictures Machines (PMC). These processors are not the same as the AMCs, and are specifically concerned with pixel filling from perhaps surface or line requests.

Each PMC consists of an SIMD (single instruction multiple data) processor, with a total of eight processors. Associated with each PMC is a memory system for a screen area of 256 × 256 pixels, and for each pixel there 8-bits for each of Red, Green and Blue memory. Further, this memory is double buffered, i.e. there are two sets, to allow the video to extract the last computed 'frame' while the PMCs operate to compute the next. There is also a 16-bit Z memory for each pixel to allow hidden surface removal during picture computation. Hence there is a grand total of 64-bits per pixel.

There is one Video Machine (VMC) which is responsible for setting up an addressing sequence upon the pixel memory within the various PMCs, and which thus collects pixel data over the video bus. The video card includes multiplexers and look-up tables for the video, and is responsible for the serialization of the data and output to the monitor. The parameters for various video rates are output over the Result Bus to the video card.

Figure 14.10 shows a connection path from Interface to Result Bus, and this intends to portray that the PMCs and VMC can be accessed from the Interface without the need to route a request via an AMC.

There is a design limit of 16 maximum AMCs, and a current equipment sizing limit of 4. The AMCs are identical processors offering 10 MIPS each. The quantity of PMCs is an independent decision – each covers a screen area of 256 × 256 (as standard). Thus 4 PMCs give 512 × 512 on screen.

The performance of the component parts is 10 MIPS per AMC and 80 MIPS per PMC, with an additional 40 MIPS per PMC for the concatenation of computed R, G, B values with those currently in memory together with the Z comparison for hidden surface removal. Plates 16–19 show some simple PMC computations, though the dynamic nature of the test pattern and scene need to be observed to be believed.

The design aim is to allow the re-computation of every pixel three times over within 1/25 second, while supporting the shading and hidden surface computations at every pixel (i.e. not just block fill by some color at some depth). Each PMC can compute next pixel upon some line in X, Y, Z, R, G, B in a two-instruction loop, allowing 5 million pixels per second computation

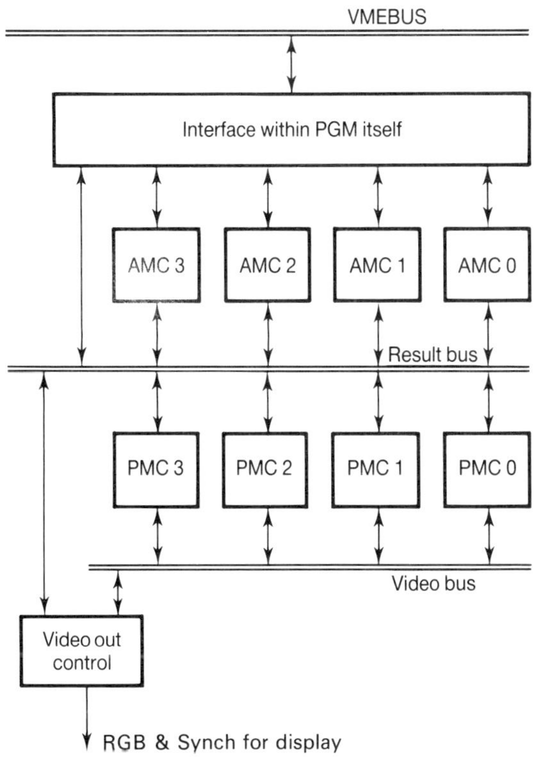

Figure 14.10. Parallel Graphics Machine.

rate for each. The memory system will absorb at the same rate, complete with the hidden surface removal.

For some sort of historical completeness, it seemed necessary to interface the VG3400 steered beam display to the PGM by interfacing it to the Result Bus, leaving the parallel AMCs to do all the computation formerly done by the GPU and associated logic of the VG3400. Thus, the PGM can offer raster and steered beam display outputs concurrently.

In 'honoring my past' the PGM includes no Read-only-Memory, and all the processors are programmable and down-loadable from the host. Thus the AMCs of the PGM have a historical link with the VG3400 GPU, and are the current home of the programmable transforms that appeared in Multipatch.

To bring the software 'full circle' the assembler and debugging aid for the PGM depends heavily upon a pseudo machine approach to record syntax definitions of the assembler languages of the PGM processors, together with the syntax definition of the 'sequence language', which is both the interactive language and the test and control language for the PGM. It has been possible to provide single-pass assemblers to make the debugging aid the same as the assembler: all facilities are fully available, allowing the user to key in a modification with full forward referencing, symbolic definitions, globals, etc. Symbolic reverse assembly is fully available.

It was Richard Parkins at the Computer Laboratory in 1971 who mentioned he felt that assemblers really should be single-pass – and in 1981 I had the pseudo machine approach to achieve this with full assembly, disassembly and debugging facilities for a Z80 based machine (still in use). The 1988 construction for the PGM, while much grander in offering the possibilities of concurrent assembly to each of three types of machine, with distinct code assemblies for different machines of the same type if required, was nevertheless built while glancing at the flow charts of the earlier construction.

As a final anecdote, I remember Peter Payne at the Computer Laboratory quipping that every Cambridge PhD student at the Laboratory came and invented a new language and then spent the next three years trying to write a compiler for it. At the time I saw myself as outside his generalization, though now I'm not so sure, except perhaps in timescale!

REFERENCES

1 **S. A. Coons** *Surfaces for Computer-Aided Design of Space Forms,* MIT Project MAC, MAC-TR-41, June 1967.

2 **A. P. Armit** A Multipatch Design System for Coons' Patches, *IEE Conference Publication 51,* April 1969.

3 **A. P. Armit** Example of an existing system in university research. Multipatch and Multiobject design systems, *Proc. Roy. Soc. Lond. A,* Vol 321, 1971, 235–242.

4 **A. P. Armit and A. R. Forrest** Interactive Surface Design, *Computer Graphics 1970,* Brunel University, UK, April 1970.

5 **A. P. Armit** *Computer systems for interactive design of three dimensional shapes,* PhD Thesis, Cambridge University, UK, 1970.

6 **A. P. Armit** Curve and Surface design using Multipatch and Multiobject design systems, *Comput.-Aided Design,* Summer 1971, 3–12.

7 **A. P. Armit** The Interactive Languages of Multipatch and Multiobject Design Systems, *Comput. Aided Design,* Vol 4 No 1, Autumn 1971, 10–15.

8 **A. P. Armit** Multiobject – The Organization of this Interactive Design System, *IEE Conference Publication Number 86 on Computer Aided Design,* April 1972.
9 **A. P. Armit** A Language for Interactive Design, *Proceedings IFIP Working Conference on Graphic Languages,* Vancouver, Canada, May 1972.
10 **A. P. Armit** Interactive 3-D Shape Design – Multipatch and Multiobject, *Curved Surfaces in Engineering,* IPC Science and Technology Press, 1972, 26–38.
11 **A. P. Armit** Organization and Control of Interactive Design Systems, *Conference Proceedings of Journees Graphiques 1973,* AFCET/IRIA, Rocquencourt, France, December 1973.
12 **P. Bézier** *How Renault uses Numerical Control for Car Body Design and Tooling,* Society of Automative Engineering Paper SAE 680010, 1968.
13 **A. P. Armit and F. S. Filby** Production of Syntactically Correct ATLAS, *Automatic Testing Conference Proceedings* (Session 4, 'Software'), November 1973.
14 **A. P. Armit and H. U. Lemke** ICON: the interactive creation of NASTRAN data. A system description, *Comput.-Aided Design,* Vol 7 No 3, July 1975.
15 **W. S. Elliott and A. P. Armit** An interactive three-dimensional graphics system for preparation and checking of structural idealisation, in *Computer Applications in the Automation of Shipyard Operation and Ship Designs II,* Jacobsson *et al.*, eds., North-Holland, Amsterdam, 1976.
16 **A. P. Armit and W. S. Elliott** The design of interactive systems for CAD, *CAD Conference Proceedings,* 1976.
17 **A. P. Armit** TAG – A High Performance Interactive 3D Graphics System, *Computers in Industry,* Vol 3 Nos 1/2, March/June 1982, 117–123.

FURTHER INFORMATION

A. P. Armit A Tracking Cross Algorithm, *SIGGRAPH Newsletter,* No 5, January 1969.
J. D. Foley and A. Vam Dam *Fundamentals of Interactive Computer Graphics,* Addison-Wesley, 1982.
D. F. Rogers *Procedural Elements for Computer Graphics,* McGraw Hill, 1985.
B. J. Schachter *Computer Image Generation,* John Wiley, 1983.

ABOUT THE AUTHOR

Andrew P. Armit graduated in Mechanical Sciences in 1966, and later joined British Aerospace to work on design of analog and digital circuits. In 1967 he returned to Cambridge to read for the PhD in computer systems for interactive design of three dimensional shapes – during this time *Multipatch* and *Multiobject* were constructed. In 1968 he published a textbook entitled *Advanced Level Vectors.* From 1967 he has designed and constructed substantial interactive systems, including ICON for Lloyds Register of Shipping and TAG for Wellworthy Ltd, Lymington. An increasing involvement in hardware started in 1981 with the building of graphics custom firmware for the Royal Aircraft Establishment and then British Aerospace. In 1984 design commenced upon a high performance Parallel Graphics Machine. Since then, the machine has been built for the Royal Aircraft Establishment by Andrew Armit and his wife Sheila, within the environment of their company. Current activity centers upon the software and application of the PGM.

15 Interactive Graphics and Numerical Control for Computer-Aided Design

David F. Rogers

This paper discusses the development of techniques for using B-spline curves and surfaces for real-time interactive surface design and techniques for numerical control production of the resulting surfaces. The discussion is presented from a personal viewpoint and in the context of ship hull design. However, it is applicable to general sculptured surface design and manufacturing.

As with most work on B-splines in Computer-Aided Design, the roots reside with Pierre Bézier. Although my long time colleague Al Adams and I had been involved with computer graphics for three to four years, it was his decision to spend the 1971/72 academic year on sabbatical at Cambridge University in England that crystallized our efforts. During December 1971 and January 1972, Bézier presented a series of lectures at Cambridge. These lectures, along with access to the internal documents of the Computer Aided Design Group, specifically Robin Forrest's papers [1–3] and an advanced copy of Rich Riesenfeld's thesis [4], introduced Al Adams to Bézier and B-spline curves and surfaces. Discussions during an extended visit to Cambridge in the early Spring of 1972 convinced me of the importance of these techniques.

Events during the next academic year were crucial. First Al Adams completed and published a report of his sabbitical studies [5].[1] Louie Knapp, then a graduate student at Syracuse University, provided a FORTRAN program for generating B-spline curves,[2] the Naval Academy decided to build a new Engineering Studies Complex and I returned to Europe for a visit to Bézier's laboratory in Billancourt outside Paris.

As an aeronautical engineer, rather than a pure mathematician, two aspects of Bézier's work impressed me. The first was his ability to generate actual parts from the computer description using numerical control (NC) techniques. The five-axis NC machining center used in Bézier's laboratory is shown in Figure 15.1. Two parts that had been designed using Bézier's system and cut with the mill, are shown in Figure 15.2. Figure 15.2a shows an aircraft wing for a wind tunnel model. Note that additional 'hand' finishing is required to 'complete' the model. The underlying philosophy here is that when generating sculptured surfaces there is an 'optimal' mix, based on economic/time considerations, between 'machine' work and 'hand' work, i.e. an optimal man-machine symbiosis. Also note that a ball-nosed cutter was used to machine the wing.

Figure 15.2b shows a trial die punch for a seat for the French National Railway. The punch is machined from plywood, and is used without further finishing. The plywood die is sufficiently durable to generate the three or four test pieces needed to check the stretch and springback characteristics of the metal seat. Note that here a metal machining center is used to cut wood.

The second aspect of Bézier's work that was of interest was real time interaction. Real time interaction, like surface fairness, is 'in the eye of the beholder'. Bézier's interactive setup was rather unique. It consisted of a single user computer, a teletype like 'console' and a Königberg flat bed plotter approximately 10 meters long by 2 meters wide[3] mounted nearly vertically. The interactive design technique consisted of typing the location or movement of a defining polygon vertex on the console, watching the plotter draw the curve, walking over to the plotter to examine the result, returning to the console to modify the result etc. The designer was definitely 'in the loop'. My experience as an engineer, a designer and a pilot, as well as knowledge of the rather disastrous

The colour plates for this chapter appear between pp. 210 and 211.

[1] Revised and expanded, this report became the basis for the first edition of *Mathematical Elements for Computer Graphics* (MECG1) [6].

[2] Rewritten, this program was the basis of the B-spline curve algorithm in MECG1.

[3] At least two full sized views of an automobile would fit on the plotter.

Fundamental Developments of Computer-Aided Geometric Modeling
ISBN 012-554765-X

Figure 15.1. The five axis NC machining center in Bézier's laboratory.

results of 'automatic computer design optimization' then occurring in the aerospace industry, made Bézier's approach appealing.

At about the same time, the decision was made to build a new Engineering Studies Complex at the Naval Academy. As a consequence, funds became available to significantly increase available computer facilities. Further, several key faculty members were intimately involved in the design of the building's laboratory spaces and of its communication facilities. As a result, a computer jack (twisted pair) was placed in every room with multiple jacks in laboratory spaces. A large contiguous area (5800 square feet) was designed with raised floors and additional air conditioning to accommodate the computer aided design facilities. Other areas scattered throughout the complex were designated as terminal rooms. A large model shop and a large area for numerical control machines (1500 square feet) in the machine shop were also provided.

In the 1973/74 time frame, design and selection of computer equipment began in earnest. Two basic facilities were decided upon, one to support real time largely interactive computer aided design/computer aided manufacturing (CAD/CAM) and a teaching facility with emphasis on computer graphics, data acquisition and analysis.

During late 1973 and early 1974, the author heard rumors that a small company in Salt Lake City, Evans & Sutherland, was designing and marketing a highly interactive refresh line drawing display. So the author went off to Salt Lake City for an inspection visit. Dave Evans and Ivan Sutherland were most cordial. The device, the Evans and Sutherland PS1, was most impressive. The procurement process was initiated and serial number 9, complete with tablet, control dials, function switches and alphanumeric keyboard, was delivered in the Spring of 1975. Its support computer, a DEC PDP 11/45, was procured separately and delivered shortly prior to the PS1.[4] The combination created a single user real time engineering workstation circa 1975. The PS1 is shown in Figure 15.3.

Since completely designing something like a 1000 foot long ship on a 10 × 10 inch display didn't seem reasonable, a medium size (5 × 8 feet) Xynetics flat bed plotter was also acquired. Large scale drawings of ship's lines along with subsequent modification would insure fairness. The plotter is shown in Figure 15.4.

At about the same time, Tektronix announced a new storage tube graphics 'personal work-

[4] Government procurement is rather 'fun'. The two machines were obtained as separate 'split sole source' procurements. Justification required several rather high level discussions in Washington. Frankly, they all thought the author was crazy but humored him anyway.

(a)

(b)

Figure 15.2. Numerically controlled cut parts from Bézier's laboratory. (a) Aircraft wind tunnel model wing, (b) die punch for a railway seat.

station' (using today's terminology), the Tektronix 4051. It had an integral storage tube graphics display, microprocessor (Motorola 6500), mass storage (tape drive), 32K of memory, communications interface (RS232C), IEEE 488 external data bus and a fast and powerful language in ROM (BASIC). It could be used either stand alone for programming, data acquisition and control, or as a terminal hooked to the Academy's new time sharing system. Forty units (of the first 100 off the Tektronix assembly line) complete with data tablets along with 10 hard copy units were acquired and delivered in the Summer of 1975. Twenty units were placed in a single 'classroom' for student use (see Figure 15.5) The remainder were scattered about in laboratories, faculty offices and the machine shop. Availability of the Tektronix 4051 would subsequently be crucial to the NC effort.

The CAD/CAM hardware system was com-

Figure 15.3. Evans & Sutherland PS1 real-time graphics system.

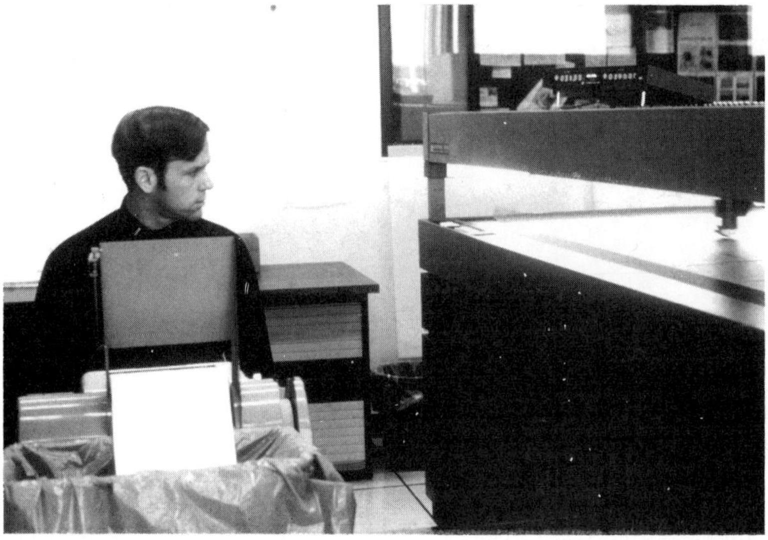

Figure 15.4. Xynetics flatbed plotter.

pleted by acquisition of a medium size 3-axis machining center, a Pratt & Whitney Trimac XV, shown in Figure 15.6.

With the opening of the new Engineering Studies Complex buildings and acquisition of equipment, the Computer Aided Design/Interactive Graphic Group (CADIG) expanded. Francisco 'Paco' Rodriguez joined the group in early 1975 and Steve Satterfield joined in the Summer of 1975, Satterfield did most of the interface programming and Rodriguez most of the numerical control programming, while the author concentrated on program functionality, interaction and development and programming of the fundamental algorithms.

Once the CAD/CAM system, E&S PS-1, PDP 11/45, Xynetrics plotter and P&W Trimac XV were in place, development of a three-dimensional manipulator, Threed, began. The basic design philosophy behind Threed was to

Figure 15.5. Computer graphics classroom (c. 1975).

Figure 15.6. Pratt & Whitney Trimac XV machining center.

read data from a simple ASCII formatted file and allow the user to interactively manipulate, i.e. rotate, translate, zoom, scale, slice, etc. a wireframe representation in real-time. This program became the backbone of a number of subsequent efforts including the CAD/CAM programs. Even today Threed is typically the first graphics program brought up on any new machine.

The new Engineering Studies Complex also included a new 380 foot modern high-speed towing tank with digital data acquisition system [7, 8], a refurbished 120 foot towing tank as well as an array of subsonic, transonic and supersonic wind tunnels. Discussions with faculty indicated that significant numbers of wooden towing tank models would be required to support research and development efforts. A modest number of wind tunnel models would also be required. Computer-Aided Design of the ship hulls and numerical control production of towing tank models offered the potential for cost reduction, ensured the similarity of models of different scales and offered the potential for regularizing test results by eliminating model variations caused by traditional hand manufacturing techniques. Finally, the US Coast Guard was willing to provide modest but long term support for the effort.[5]

CAMILL

Although we preferred to immediately develop a surface design program, there was resistance from Naval Architects who, when designing and fairing a ship hull, normally work with a surface described by three sets of lines in orthogonal planes: station or section lines in planes perpendicular to the longitudinal axis of the ship and to the water surface, waterlines in planes parallel to the water surface and buttock lines perpendicular to the water surface and parallel to the longitudinal axis of the ship (see Figure 15.7). To gain user acceptance and valuable feedback, the decision was taken to develop a lines fairing program. The program was called CAMILL (Computer Aided MILLing). In developing CAMILL several design requirements were of primary importance. Some of these were:

- Generation of lines could occur either from an external source or *ab initio* internal to the system.
- Various lines fairing techniques would be available.
- Full interactive fairing capability would be available.
- Three-dimensional perspective and orthographic views would be available.
- Hardcopy of anything displayed would be available.
- Production of the NC milling data base would occur without part programming.
- Proofing of the NC milling data base would be possible.
- The machine operator should have precise graphical and/or machining information on the shop floor without complex drawings or lists.
- The end goal was to produce wooden towing tank models.

In all cases as simple an approach to meeting these goals as was possible would be adopted.

[5] The US Coast Guard's continuing support to this day is gratefully acknowledged.

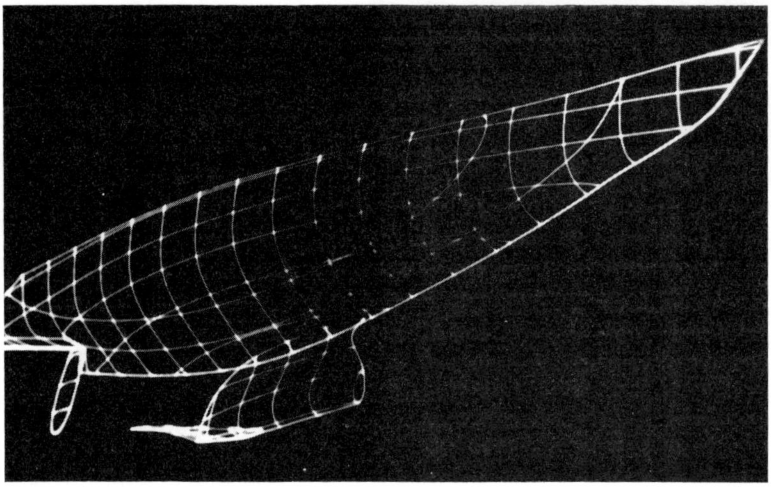

Figure 15.7. A ship surface described as a net of lines (Courtesy George Hazen).

General description

The CAMILL program allows ship's lines to be generated by either digitizing an existing lines plan and interactively fairing the digitized data or by interactively creating a body plan *ab initio* at the graphics device.

The body plan along with the profile line of the ship is considered fundamental. Waterlines, buttock lines and diagonals[6] are automatically generated from the body plan and the profile line. In addition, a three-dimensional perspective or orthographic view of the ship can be generated and interactively manipulated in real time. This proved a useful technique for discovering small 'unfair' regions or problem spots which do not become obvious on a 2D lines drawing.

Facilities for creating additional lines, deleting lines or modifying a line including adding, deleting, or moving points on individual lines, are provided. Interactive selection of the type of fairing technique is also provided. Individual lines may use different fairing algorithms. Among these are cubic splines, parabolically blended curves and B-spline curves of arbitrary order. Cross fairing from one view to another is accomplished interactively.

Once the ship lines are considered fair and correct, a numerical control milling file is generated and stored on disk. The milling file consists of 3D data points which describe waterlines. The waterlines are sorted and stored from the keel to the deck. Waterlines may contain as many points as desired. As many waterlines as desired may be stored. Once stored, the mill file is then transferred to magnetic tape in a stand alone microprocessor based graphics system.

Operating environment

The original version of CAMILL was written in FORTRAN and implemented on the PDP 11/45. As configured, the PDP 11/45 had 64K core memory, a 1K Cache memory, floating point processor, a 256K fixed head disk, two 2.5 Mbyte cartridge disks, DEC tape, 9 track 800 bpi industrial compatible magnetic tape, paper tape reader/punch and various communication interfaces. The operating system was RT-11 version 3B configured as a single user system. This implementation of RT-11 allowed up to 28K of user program and operating system in the lower memory, plus data arrays assigned locations in the upper 36K of the available 64K of memory. Because of this program size restriction, CAMILL was overlaided in five segments. In addition, several functions were written as separate modules and accessed via a program chaining facility.

CAMILL was implemented for the E&S PS1. The PS1 was a refresh display internally equipped with 16K of 36 bit refresh buffer operated in doubled buffered mode. Hardware segmentation allowed the refresh buffer to be divided into a dynamic region and several static regions. The regions could be individually accessed for picture update. An internal hardware matrix multiplier provided for fast, efficient real-time rotation, translation, perspective and scaling in three dimensions. An internal three-dimensional hardware clipper was utilized. These features, in combination with separate update and refresh rates, made screen flicker nearly an unknown problem. The PS1 was connected to the PDP 11/45 via the UNIBUS utilizing DMA transfers for communication.

The flat bed plotter was a Xynetics Model 1200 with a drawing area of 59 × 89 inches. Maximum plotter speed was 40 ips. Ballpoint, ink or felt tip pens were available for plotting. Local scaling, reflection and rotation of drawings were available through plotter software. Drawing data was normally transferred from the PS1/PDP 11/45 to the plotter by creating a magnetic tape. Plotting was thus accomplished off-line and did not 'tie-up' the design system.

Input techiques

Two methods of creating an initial database for a ship were available. An existing set of lines, either preliminary or final, could be digitized. The digitizing program allowed arbitrary placement of the lines plan on the digitizer, scaled the resulting digitized points to ship scale and provided a graphical preview of the digitized data. Alternatively, the lines and hence the data base for the ship could be interactively created *ab initio* at the Picture System.

[6] Diagonals are formed by cutting planes through the longitudinal axis of the ship and at an angle to the water surface.

Fairing techniques

Experience [9] showed that no single fairing technique was a panacea. Different types of lines are best faired using different fairing techniques. Three techniques for interactively fairing a set of ship's lines were available. These were cubic spline curves, parabolically blended curves and B-spline curves. Bézier curves are a subset of the B-spline curves. Details of the mathematical basis for each curve type are given elsewhere in this volume, and in Rogers [9] or Rogers and Adams [10]. Of the available fairing algorithms the B-spline curve proved most generally used. Automatic fitting of B-spline curves to a set of digitized points proved unsatisfactory. To overcome this difficulty an interactive real-time fairing technique was generally used.

When using the interactive real-time fairing technique, the digitized points from the lines plan were displayed on the Picture System. Using lines creation and modification techniques, a B-spline curve of appropriate order was interactively created using the tablet and pen. This curve was a first approximation to the digitized line. The curve modification facilities described later were then used to interactively fair the created line to the digitized points representing the original line. A zoom feature and the excellent local control characteristics of B-spline curves allowed matching of critical points to better than one part in 65536. This represents an accuracy of better than 3/16 inch on a 1000 foot ship and better than 0.0025 inch on a typical towing tank model. Using this technique, the time required to interactively fair a typical line is 2–5 minutes. This compares favorably with automatic techniques which require subsequent interactive adjustment. This process is shown in Figures 15.8–15.10. Once the created lines are satisfactory, the original digitized data are removed.

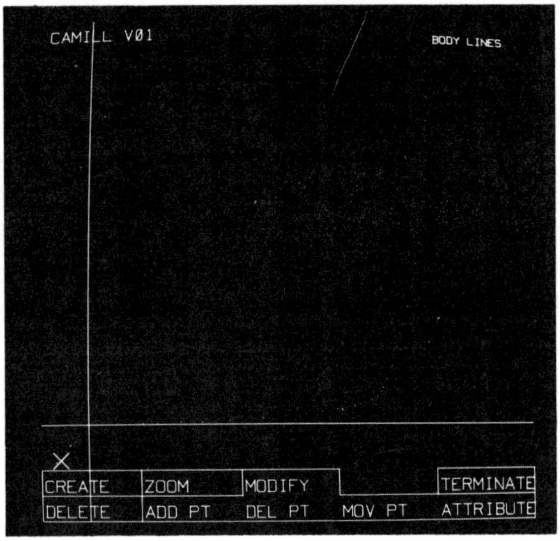

Figure 15.8. Digitized data points on the Picture system.

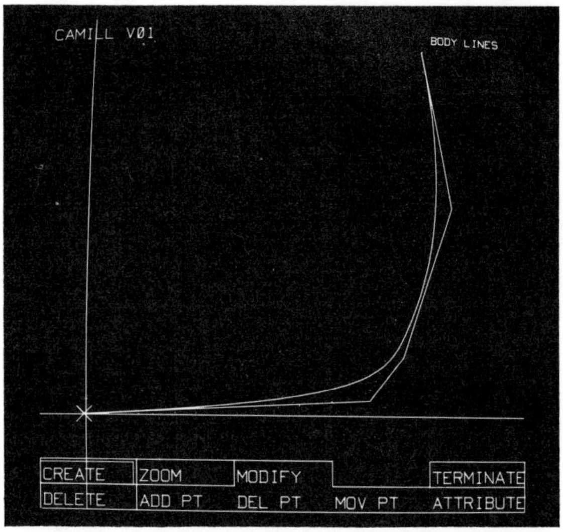

Figure 15.9. Creating the approximate faired curve.

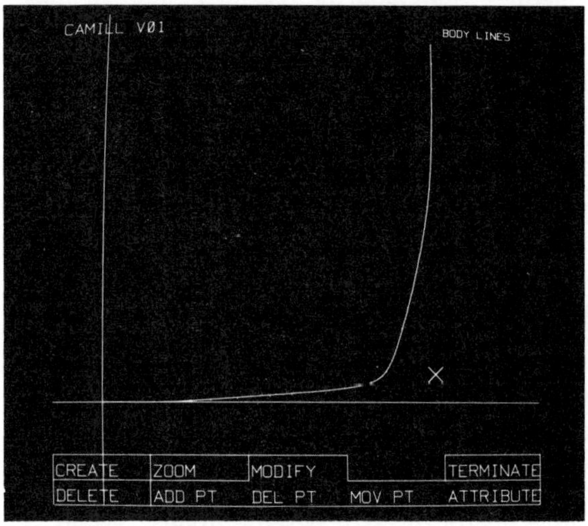

Figure 15.10. The final faired curve.

Lines facilities

CAMILL provided facilities for creating any of the four traditional two-dimensional views of the ship surface, i.e. body plan, waterlines, buttock

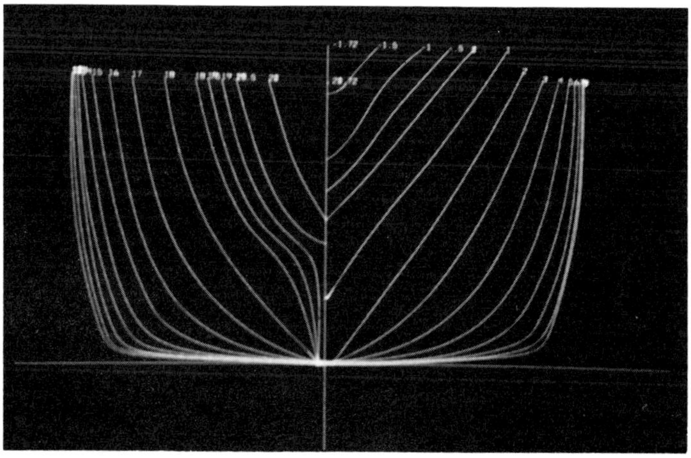

Figure 15.11. A typical body plan faired with B-spline curves.

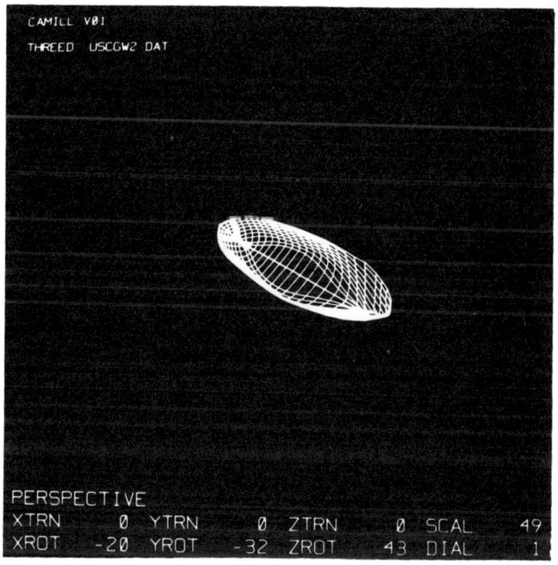

Figure 15.12. A typical 3D perspective view.

lines or diagonals. However, the body plan and the inboard profile line are normally considered fundamental. Waterlines, buttock lines and diagonals are calculated from the body plan and the inboard profile lines. These lines are then checked for fairness and interactively modified. To avoid slowing down the interactive speed, modifications to one set of lines were not automatically incorporated into the other sets. Once the modifications to a set of lines were complete, the other lines were generated from this modified set. For example, a new body plan, new buttock lines and diagonals would be generated from the modified waterlines. These modified lines were then displayed as points along with the unmodified lines in that view. In this way, the effect of modifications in one view on the other views was evaluated. If these efforts were acceptable, the interactive fairing technique described above was used to refair the affected lines. In this way a designer could work back and forth between the various views until the entire surface was fair. Figure 15.11 illustrates a typical display incorporating a body plan.

Three-dimensional display

In checking and examining a 3D surface for shape and fairness a 3D presentation is most advantageous. Three-dimensional perspective and orthographic displays were provided in CAMILL. A typical perspective view is shown in Figure 15.12. Selection of perspective or orthographic views was accomplished in real-time using a single function switch. Three-dimensional rotation about each axis was implemented using control dials, as was three-dimensional translation and overall scaling. The eighth control dial governed the sensitivity of the other dials. This allowed very fine tuning of the control dial functions. The hardware matrix multiplier implementation of three-dimensional transformations in the PS1 allowed very rapid interaction, e.g. simultaneous rotations about all three axes at rates in excess of 360°/s for pictures with in excess of 2500 lines was possible. The display provided a continuous update of the current rotation angles, translation distances, scale factor and the dial scalar setting (see Figure 15.12).

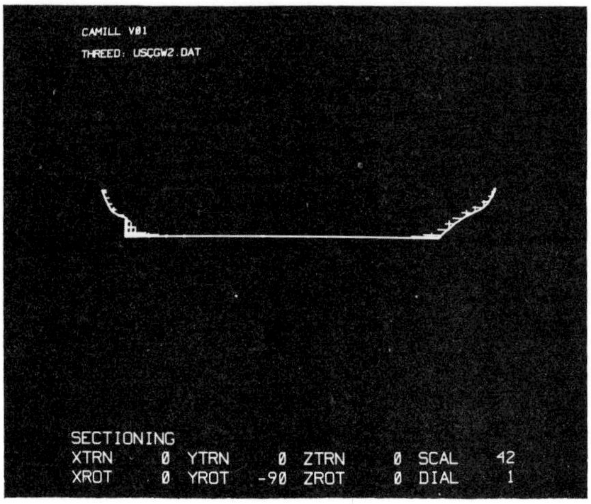

Figure 15.13. A sectional slice of a ship.

A sectioning capability was implemented by operating a single function switch to activate the 3D hardware clipper in the Picture System. Control dials were used to rotate the ship to a suitable orientation and then the *z*-translation control dial used to 'walk' the section through the ship. A typical display is shown in Figure 15.13.

Interaction facilities

CAMILL provided real-time interactive facilities when creating, deleting or modifying lines. The menu for these facilities is shown in Figures 15.8–15.10. Items from the menu were selected using the tablet and pen. When a menu item was activated it blinked. Selecting a particular menu item might cause additional requests for information to appear on the screen. A simulated key pad proved faster for numeric input than using the physical alphanumeric keyboard for input of secondary information. This eliminated the need to lay the tablet pen down, type on the keyboard and then pick up the pen to continue.

The first step when creating a line was to select the type of curve. The tablet and pen were used to interactively create the desired line. Lines could be modified by moving points, adding points or deleting points. Positive confirmation was provided by visual feedback. For example, when moving a point on a line, rubber banding was used to provide visual feedback to the user. Again, when deleting a point, positive confirmation was required prior to the final act of deletion. The attributes (types of curves) of individual lines could be interactively changed. For example, the order of a B-spline curve could easily be changed.

Generation of NC mill file

Once the design was finalized, a numerical control (NC) milling file was created and stored on disk. The NC milling file consisted of waterlines sorted from the keel upward, since the model was cut upside down. Any number of points on each waterline could be generated. The actual NC mill file was plotted on the Xynetics plotter for proofing purposes. Once the milling file was considered satisfactory, it was transferred from the PDP 11/45 disk file to the magnetic tape unit in the Tektronix 4051 graphics system using a standard RS232C serial interface. It was this data, stored on tape as waterline data points, which represented the direct interface between the design facility and the manufacturing facility. No separate part programming occurred.

NUMERICAL CONTROL MILLING PROGRAM

General description

The NC milling program was written in Tektronix 4051 BASIC and run on the Tektronix 4051. The program provided the NC machine operator with assistance in setting up the mill. The NC program provided the capability to graphically preview the milling file. This could be done either before or during the cutting operation. During the cutting operation the NC program provided, on the shop floor, in real-time, a graphical display of the shape of the waterline to be cut, the location of the cutter and all cutter motions.

The numerical control machining center is a Pratt and Whitney Trimac XV. The machining center is capable of 54 inches of travel in the *x* direction, 28 inches in the *y* direction and 24 inches in the *z* direction. The Trimac XV is a full three axis contouring machine. Control of the machining center is accomplished using a minicomputer supplied as standard equipment. Normal input to the CNC (minicomputer) controller is paper tape.

The hardware interface between the CNC controller and the 4051 was accomplished by using the GPIB of the 4051 and a simple in-house designed interface to simulate the normal paper tape input to the CNC controller. Using this technique required no modification of the standard NC machine software. Unplugging the cable from the 4051 to the CNC controller restored the NC machine to normal operation with paper tape input.

NC program details

The NC program provided previewing or checking facilities. A graphical display of any waterline and the center line of the tool with or without small circles drawn to represent the tool was available. Any area of the model could be expanded to allow examination of the details of the cutter path. Figures 15.14–15.16 illustrate a typical preview sequence.

The machine operator controlled the operation of the mill from the Tektronix 4051 keyboard. The program allowed him to start milling at any particular waterline, set or change the spindle speed, feed rate and cutter diameter interactively at the keyboard. The operator could use the 4051 function keys to stop the program and to start or restart the program at any point. The function keys were also used to invoke various options such as only drawing the waterlines and the cutter path or both drawing and actual cutting on the mill, or to draw or not draw the cutter circle. However, the NC mill's standard control console functions could be used at any time to override the 4051 program.

Control of the machining center was accomplished on line with all calculations accomplished in real time. Data for each individual waterline were read from the internal magnetic tape. For each point, cutter offsets were generated, slope checking accomplished, if required, and the necessary codes to drive the mill generated. These

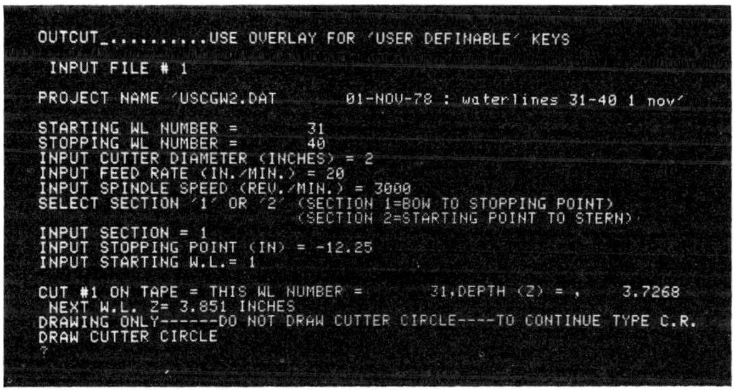

Figure 15.14. NC program interactive setup commands.

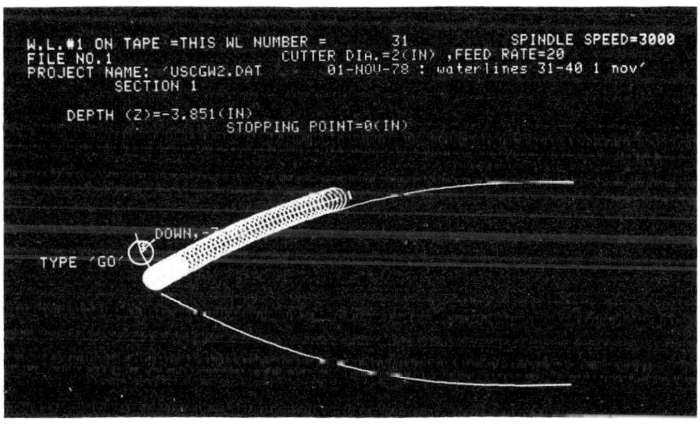

Figure 15.15. Typical preview display.

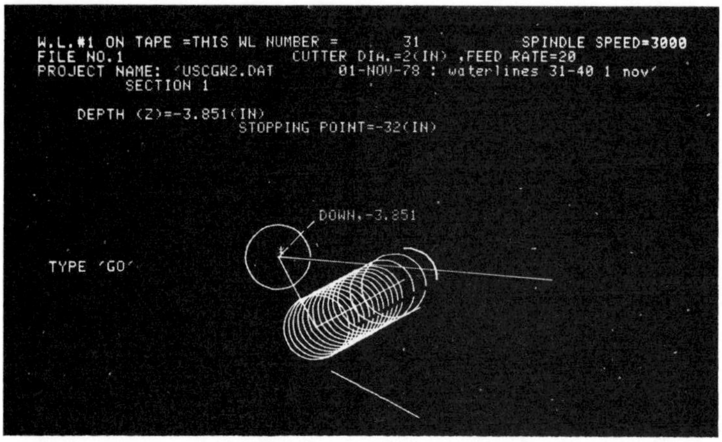

Figure 15.16. Expanded preview display.

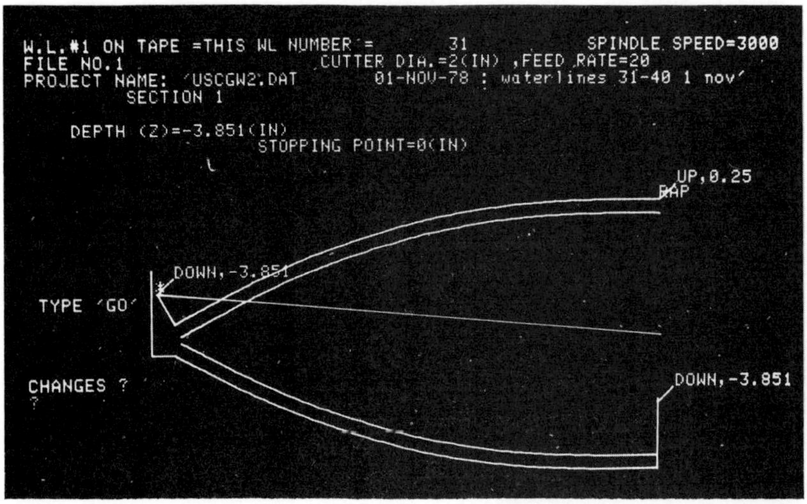

Figure 15.17. NC program shop floor display.

codes were then transferred to the machining center controller using the Tektronix 4051 GPIB bus and the special hardware interface.

Waterlines for only one side of the model are stored on the magnetic tape. If the model is symmetrical, then the waterline is reflected and the other side of the model is cut. In this case the program also calculated a retraction and movement path which transported the cutter from one side of the model to the other without interference. The program actually calculated ahead of the mill. This graphical preview of the location of the cutter allowed the operator to continuously monitor possible interferences. A typical display as seen by the operator is shown in Figure 15.17.

EXPERIENCES

Cutter characteristics for wood

Cutting softwood (sugar pine) as normally used in towing tank models at the low rotational speed available with a metal cutting NC milling machine imposes severe requirements on cutter design. The principal problems are associated with peripheral speed, rake and clearance angles,

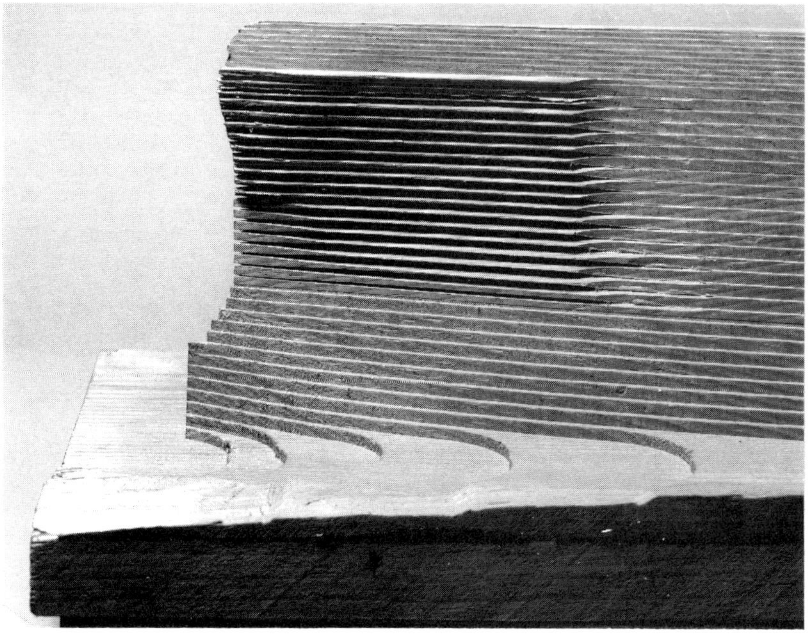

Figure 15.18. Model showing under cut.

as well as the number of cutting teeth. Several experimental cutters were designed and used. These included standard metal cutting end mills, standard metal cutting shell mills, carbide tipped circular saw blades and specially reground shell mills. Currently, high speed carbide router bits as used in the furniture industry are proving most satisfactory and considerably less expensive.

Optimum number of waterlines

The choice of the number of waterlines is a trade-off between machining time and hand finishing time. Experience indicated that between 8 and 10 waterlines per inch of model surface gave an acceptable result.

Undercuts

In cutting the ship upside down, i.e. from the keel to the deck, it is possible that the next waterline to be cut is 'narrower than' or 'inside' one or more of the previously cut waterlines. In this case a cylindrical cutter on a 3-axis machining center could not be used. The simple solution was to use a carbide circular saw blade with a V cutting profile. In this case, the surface exists at the bottom of the V. A model cut using this technique is shown in Figure 15.18.

Lifts

Towing tank models are not made from solid blocks of wood, they are made from glued up hollow laminated shells stacked from the keel to the deck. Each lamination is called a lift (see Figure 15.19). Generally, a lift is from $1\frac{1}{2}$ to 2 inches thick. The basic problem is to determine both the inside and outside shape of each lift so that the final hull surface is contained within the rough shape. Traditionally this is done by the model maker performing a rough 'fairing' of the hull at model size at appropriate waterlines. This is a time consuming process that is somewhat subject to errors.

CAMILL significantly simplified and automated the task. Waterlines at the appropriate model scale and locations were created for each lift. The bottom and top waterlines for each lift were then drawn full sized on the flatbed plotter. Each drawing constituted a pattern for the lift. The model maker then glued up the lift over this pattern.

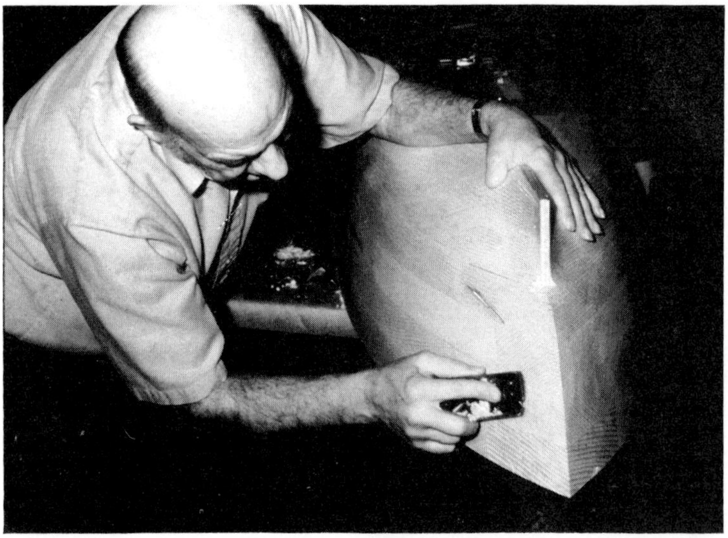

Figure 15.19. Hand finishing a model.

Generation of larger models

The Trimac XV NC machining center is capable of 54 inches of travel in the x direction, which allows cutting of up to 48 inch long models in one continuous pass. Longer models are cut using an in-house designed auxiliary sliding table. These models are machined in sections using multiple passes. Models up to 11 feet can be cut in one piece. Above 11 feet the models must be fabricated in sections and the sections assembled after cutting. A section to be cut can be specified interactively on the shop floor without any change in the database.

Model finishing

Upon completion of machining the model surface is defined by the interior corner of the machined cut. During machining approximately 0.004 inch is added to the model for hand finishing. In the model shop the model maker runs a sharp pencil along each one of these corners. No templates are used. The model is finished by cutting away the excess wood to the pencil lines which precisely define the surface (see Figure 15.19).

3D digitizer

After the model is finished it is measured using the Trimac XV connected, as before, to the Tektronix 4051. The Trimac XV/Tektronix 4051 combination forms a three-dimensional automatic digitizer system.

A sensor, consisting of a modified mechanical distance indicator, is mounted in the tool holder of the Trimac XV. Using the Tektronix 4051 as a controller, the sensor is moved toward the surface. When the sensor hits the surface, an interrupt to the 4051 is generated on the GPIB. The 3D location is then recorded on magnetic tape, the sensor is moved away from the surface and positioned to seek a new point. Where to digitize and at what resolution is selected interactively from the Tektronix 4051 on the shop floor. Resolution is 0.002 inches.

Tank testing

The finished model is tested in the towing tank. An example is shown in Figure 15.20.

ANALYSIS

CAMILL provided an integrated CAD/CAM system for the design, production and testing of towing tank models of ships. Developing and using CAMILL provided several insights useful in designing and developing future systems.

Experience in using CAMILL for surface design showed that designers preferred B-spline

Figure 15.20. Model being tank tested.

curves and had no difficulty in manipulating them using the defining polygons. Generally, less than an hour of instruction plus answering occasional detailed questions was all that was necessary before productive work was being generated.

As an experiment, a model maker (Tom Price) from the woodshop was introduced to the system and asked to fair a ship hull using a previously digitized body plan. He had essentially no computer background and only secondary school mathematics. He was given about two hours of instruction on the computer system and CAMILL. He was quite successful. His only problems were basic computer system problems, e.g. how do I load/unload the disk drive? He also provided valuable feedback on the program interface and suggested new and interesting ways to use the program for more efficient production of the models.

Although CAMILL was quite successful, with several ships being faired by several different designers, a number of problems surfaced with use. The principal problem was that it was a lines fairing program. Typically fairing of some 80 waterlines was required to manufacture a towing tank model. That's a lot of lines. This is especially true when one considers that a Naval Architect normally provides about 20 waterlines. The model maker implicitly generated the rest of them because he treated the surface as a surface rather than as a net of orthogonal lines when fairing the model. Thus, the program imposed a greater burden on the designer – specifically, the surface had to be specified in much greater detail.

A B-spline curve fitting program (see Rogers [9]) was implemented to somewhat automate the problem. Unfortunately, for several reasons this was less successful than anticipated. First, because of memory limitations it had to be implemented as a separate module. This made it somewhat inconvenient to use. Second, because the digitized data were almost never fair, the resulting lines were almost never fair. Third, because only an unconstrained fit was possible (see below and [11]) the defining polygons for adjacent lines frequently had no relationship one to another nor to the ship's standard stations. This made their interactive manipulation much less intuitive.

Most designers quickly developed design techniques that allowed them to efficiently generate and fair the large numbers of required waterlines. The most successful of these techniques used a successive subdivision (generation) method. Specifically, about 10 waterlines would initially be developed and cross-faired (see below) with the body and buttock lines. Once satisfied with this fairing, the designer generated and faired the waterline between each of these initial waterlines. The process was repeated until the requisite number of waterlines was generated.

Designers rapidly learned that a minimum number of defining polygon vertices should be used, that where possible, the same number of defining polygons should be used for adjacent waterlines, that if the defining polygons for a series of adjacent waterlines were 'fair', then the resulting surface was more likely to be fair.

Cross fairing, the process of ensuring that the surface was fair in all directions and that body lines, waterlines and buttock lines all passed through the same point in three space, was nearly as difficult, but a bit faster, using CAMILL as doing the job manually.

These difficulties convinced all involved that a program that treated a surface as a 'real' surface rather than as a net of lines was required. Consequently, work on BSSD (B-spline Surfaces for Ship Design) was started in the Spring of 1979 [12].

BSSD

Based on experience with CAMILL the decision was taken to use B-spline surfaces in implementing the surface design program. Some of the reasons for this decision were: B-spline surface descriptions are powerful and flexible, both smooth surfaces and surfaces with discontinuities (knuckles or hard chines) could be described, a fast incremental algorithm for real time rubber sheeting of moderately complex B-spline surfaces was possible as was a technique for automatically fitting B-spline surfaces to known data. The BSSD program was implemented in the same environment as CAMILL. It also used the same environment for NC production of the towing tank models.

In the following discussion it is important to realize that two different concepts have been developed with respect to use of a polygonal net to generate and control the B-spline surface – local and global. Local B-spline nets consider that the surface is composed of several (many) quadrilateral patches, each defined by component vectors at each corner, e.g. for bicubic surfaces these are the 16 three component position, tangent and twist vectors. Continuity of some arbitrary degree is maintained across the patch boundaries. Except for continuity requirements each patch can be separately manipulated. The global approach assumes that a single large polygonal net controls the entire B-spline surface. Although the resulting surface is constructed of piecewise patches, individual quadrilateral patches cannot be separately manipulated. The present author has used and recommends this latter approach [12].

Fast incremental algorithm

A Cartesian product B-spline surface is given by

$$Q(u, w) = \sum_{i=1}^{n+1} \sum_{j=1}^{m+1} B_{i,j} N_{i,k}(u) M_{j,\ell}(w) \quad (1)$$

where $N_{i,k}(u)$ and $M_{j,\ell}(w)$ are the B-spline basis functions in the biparametric u and w directions, respectively, given by

$$N_{i,k}(u) = \begin{cases} 1 & \text{if } x_i \leqslant u < x_{i+1} \\ 0 & \text{otherwise} \end{cases} \quad (2\text{a})$$

$$N_{i,k}(u) = \frac{(u - x_i) N_{i,k-1}(u)}{x_{i+k-1} - x_i} + \frac{(x_{i+k} - u) N_{i+1,k-1}(u)}{x_{i+k} - x_{i+1}} \quad (2\text{b})$$

and

$$M_{j,\ell}(w) = \begin{cases} 1 & \text{if } y_j \leqslant w < y_{j+1} \\ 0 & \text{otherwise} \end{cases} \quad (3\text{a})$$

$$M_{j,\ell}(w) = \frac{(w - y_j) M_{j,\ell-1}(w)}{y_{j+\ell-1} - y_j} + \frac{(y_{j+\ell} - w) M_{j+1,\ell-1}(w)}{y_{j+\ell} - y_{j+1}} \quad (3\text{b})$$

where the x_i and y_j are elements of knot vectors in the u and w parametric directions. The $B_{i,j}$'s are the vertices of a defining polygon net. For quadrilateral surface 'patches' the defining polygon net must be topologically rectangular. The indices n and m are one less than the number of defining polygon vertices in the u and w parametric directions, respectively. Additional details are given in Rogers and Adams [10].

To develop a dynamic B-spline surface algorithm, consider the Cartesian product B-spline surface governed by equation (1), see

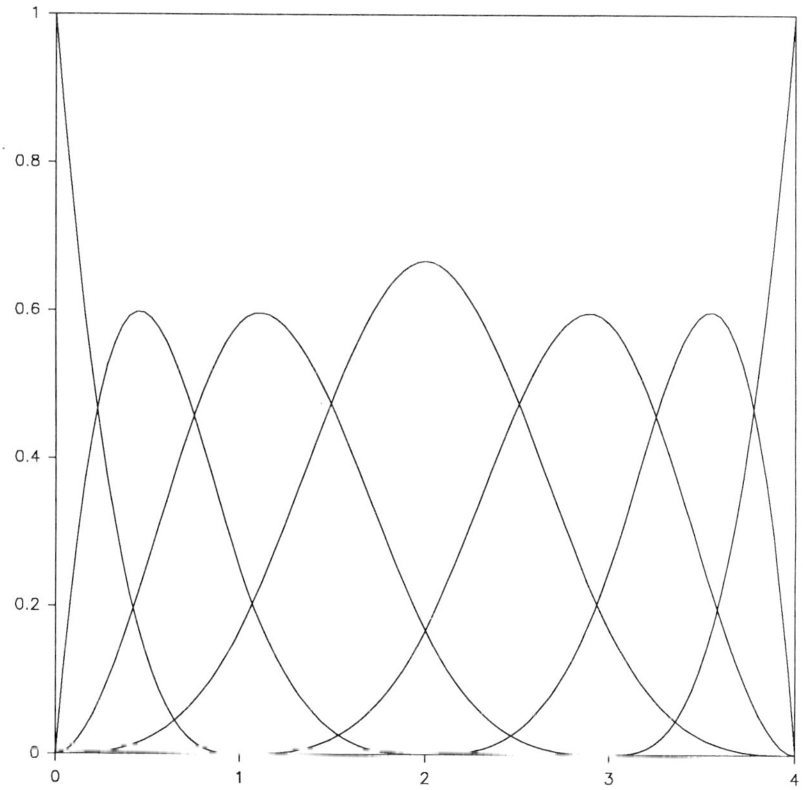

Figure 15.21. Fourth order B-spline basis functions for the open knot vector [0 0 0 0 1 2 3 4 4 4 4].

also [15] and [30]. The blending functions in the u, w parametric space are the B-spline basis functions N and M, respectively. The basis functions depend upon the number of polygon net points and the character of the knot vector (see [10]). As shown by the recursion relations given in equation (1), the B-spline basis function provides only local support for the surface. That is, the extent of the surface influenced by a given net point is limited. For example, the basis function $N_{i,k}$ for a fourth-order B-spline with seven polygon points (grid or net lines) is shown in Figure 15.21. The open knot vector is [0 0 0 0 1 2 3 4 4 4 4]. The maximum parameter value is 4. The individual basis functions represent the extent of the influence of any given polygon point. Note that in parameter space the maximum extent of the influence of any polygon point is $\pm$ half the order of the basis function. A knowledge of the extent of this influence is useful in increasing the speed of calculation of the surface algorithm.

Further, note that if the number of polygon net points in the u and w directions does not change the knot vector does not change, and that if the parameter values used to calculate parametric lines on the surface do not change, then the basis functions N and M do not change and may be precalculated for any arbitrary location of the polygon net points, i.e. the $B_{i,j}$'s.

Examination of equation (1) then leads to an incremental method for calculating a new or updated surface from a known B-spline surface when a change in a single polygon net point occurs. In particular,

$$Q_{\text{new}}(u, w) - Q_{\text{old}}(u, w) + \sum_{i=1}^{n+1} \sum_{j=1}^{m+1} (B_{(s,t)_{\text{new}}} - B_{(s,t)_{\text{old}}})[N_{i,k}(u) M_{j,\ell}(w)] \quad (4)$$

where $B_{s,t}$ is the specific polygon net point that has been changed. Using equation (4) for each parametric value on the surface, a fast algorithm for the dynamic update of a B-spline surface is

```
Start:
    If surface order or number of polygon net points or number of
    required parametric lines for either u or w has not changed then goto 4
Comment: calculate complete surface
    calculate knot vector in the u direction - [X]
    calculate knot vector in the w direction - [Y]
    for u = min to max for increment Xmax / (number of u parametric lines - 1)
        calculate N(u)
    next u
    for w = min to max for increment Ymax / (number of w parametric lines - 1)
        calculate M(w)
    next w
    Q = 0
    for i = 1 to number of parametric lines in the u direction
        if N(i) = 0 then 3
        for j = 1 to number of parametric lines in the w direction
            if M(j) = 0 then 2
            Q(i,j) = Q(i,j) + B(i,j) * N(i) * M(j)
        2 next j
  3 next i
    QOLD = Q
    BOLD = B
    Done
Comment: calculate incremental change in surface
  4 BCHANG = (new polygon net point) - (old polygon net point)
    for i = 1 to number of parametric lines in u direction
        if N(i) = 0 then 6
        for j = 1 to number of parametric lines in w direction
            if M(j) - 0 then 5
            Q(i,j) = QOLD(i,j) + BCHANG * N(i) * M(j)
        5 next j
  6 next i
    Done
```

This algorithm, written in FORTRAN and implemented on a PDP-11/45 with floating point processor and a 2K byte cache memory driving the Evans & Sutherland PS1, allowed real time interactive surface dragging or rubber sheeting at 4–20 updates per second, depending on the complexity of the surface. The algorithm was incorporated into the BSSD pilot program [12]. Examples of surfaces constructed using the dynamic B-spline surface algorithm are shown in Figures 15.22–15.24. Figure 15.22 shows a US Navy ammunition ship. Figure 15.23 shows a modified bulb on the ammunition ship of Figure 15.22. Modification of the surface required approximately three minutes of user time sitting at the display. The defining polygon is 5×8 and the defining surface has 15×24 parametric lines. The algorithm allowed eight updates per second. The algorithm can be further optimized by premultiplying $N(u)$ and $M(w)$ together and storing the values. This was not done.

Automatic surface fitting

Ab initio surfaces designed with B-splines are most efficiently generated from an initially flat defining polygon. Although surfaces can be interactively fit to a set of displayed surface points starting from an initially flat defining polygen net, when fitting to an existing surface it is somewhat more efficient to obtain a preliminary surface by automatic fitting.

Here it is necessary to determine the defining polygonal net from an existing network of three-dimensional surface point data. This problem has been studied by Barsky and Greenberg [13], who exploited the known characteristics of bicubic B-spline surfaces to obtain a computationally efficient algorithm. The approach described here is conceptually simpler but computationally less efficient. It is also applicable to any order B-spline surface. Since in the present context the algorithm is only to be used to obtain an initial approximate

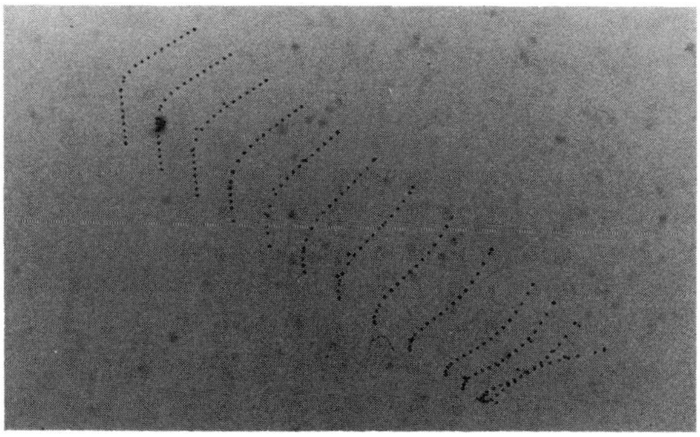

(a)

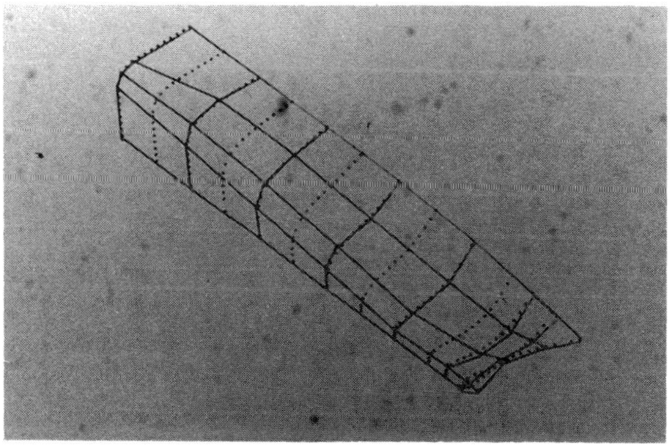

(b)

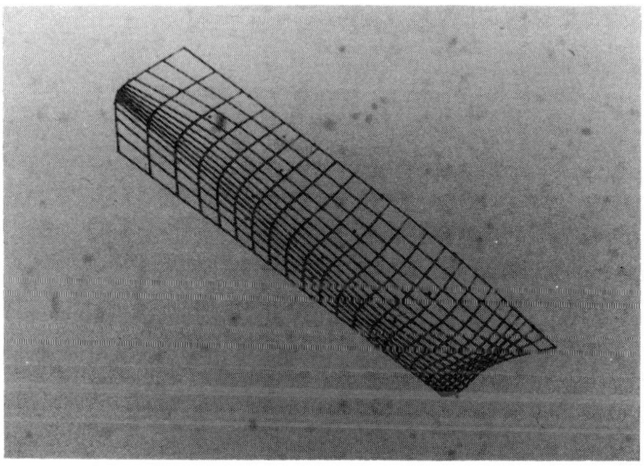

(c)

Figure 15.22. Forebody for a US Navy Ammunition ship. (a) Digitized points, (b) interactively fit B-spline polygon net, (c) bicubic B-spline surface.

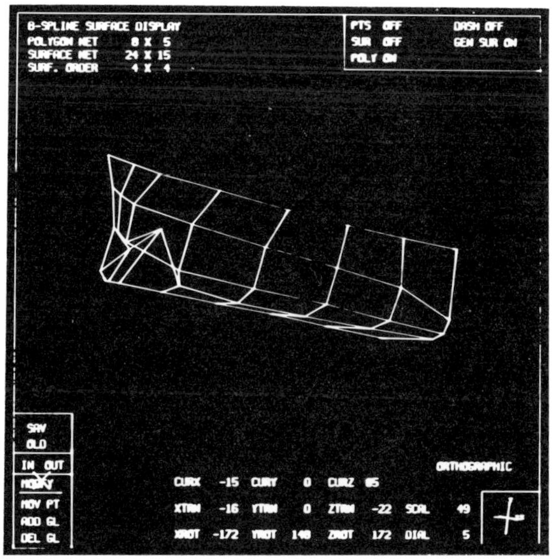

(a)

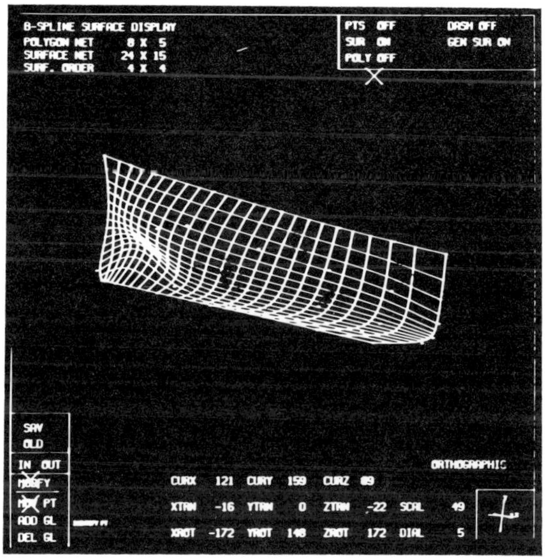

(b)

Figure 15.23. Modified bulb on the ammunition ship. (a) Polygon net, (b) bicubic B-spline surface.

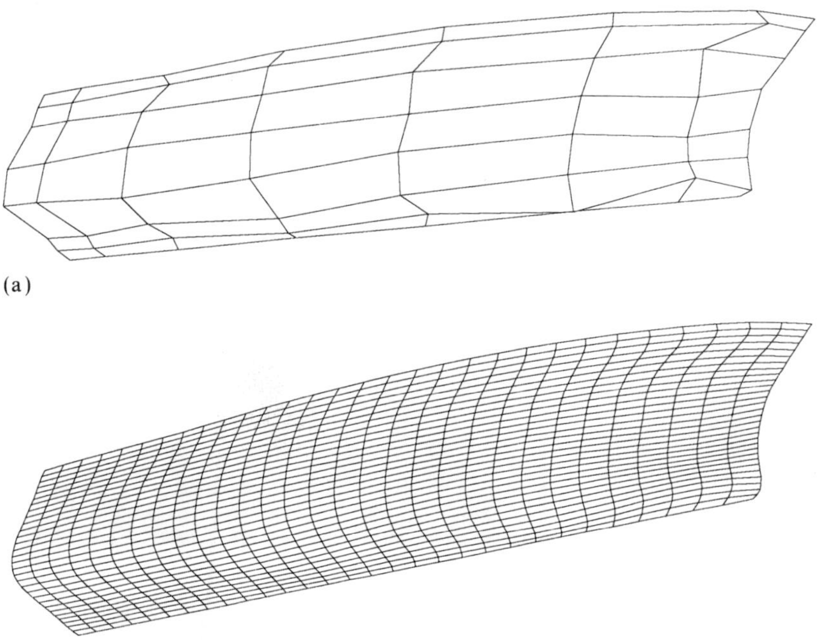

(a)

(b)

Figure 15.24. Fitted B-spline surface for the forebody of the US Navy ammunition ship of Figure 15.23. (a) Automatically fit defining polygon net, (b) surface derived from the net.

surface for any given design, the computational inefficiencies are acceptable.

Referring to equation (1), we note that here the surface data points are known. The N and M basis functions can be determined from the knot vector and the parameter values u and w. The $B_{i,j}$'s are the unknown required polygon net points. For each known surface data point, $D(u, w)$, equation (1) provides a linear equation in the unknown $B_{i,j}$'s. Writing this out for a single surface data point, $D_{1,1}$, yields

$$\begin{aligned} D_{1,1}(u_1, w_1) &= N_{1,k}(u_1)[M_{1,\ell}(w_1)B_{1,1} + M_{2,\ell}(w_1)B_{1,2} \\ &\quad + \cdots + M_{m+1,\ell}(w_1)B_{1,m+1}] + \\ &\quad \vdots \\ &\quad N_{n+1,k}(u_1)[M_{1,\ell}(w_1)B_{n+1,1} \\ &\quad + M_{2,\ell}(w_1)B_{n+1,2} + \cdots \\ &\quad + M_{m+1,\ell}(w_1)B_{n+1,m+1}] \end{aligned}$$

where for an $r \times s$ topologically rectangular set of data $2 \leqslant k \leqslant n+1 \leqslant r$ and $2 \leqslant \ell \leqslant m+1 \leqslant s$. Writing an equation of this form for each data point yields a system of simultaneous equations. In matrix form the equations are

$$[D] = [C][B] \tag{5}$$

where $C_{i,j} = N_{i,k}M_{j,\ell}$. For an $r \times s$ topologically rectangular surface point data, $[D]$ is an $r*s \times 3$ matrix containing the three-dimensional coordinates of the surface point data, $[C]$ is an $r*s \times n*m$ matrix of the products of the B-spline basis functions and $[B]$ is an $n*m \times 3$ matrix of the three-dimensional coordinates of the required polygon net points.

If $[C]$ is square, the defining polygon net is obtained directly by matrix inversion, i.e.

$$[B] = [C]^{-1}[D] \tag{6}$$

In this case the resulting surface passes through each data point. Although the resulting surface is everywhere $C^{k-2,\ell-2}$ continuous, it may not be fair. Experience has shown that, in general, the fewer the defining polygon net points the fairer the surface.

If $[C]$ is not square, the problem is overspecified and a solution can only be obtained in some mean sense. In particular, the solution is given by (see [10])

$$[B] = [[C]^T[C]]^{-1}[C]^T[D] \tag{7}$$

The u and w parametric values for each surface data point are obtained using a chord length approximation. Specifically, for r data points the parameter value at the ℓth data point in the u parametric direction is

$$u_1 = 0; \quad \frac{u_g}{u_{max}} = \frac{\sum_{g=2}^{\ell} |D_{g,r} - D_{g-1,r}|}{\sum_{g=2}^{r} |D_{g,r} - D_{g-1,r}|}$$

Similarly, for s data points in the w parametric direction

$$w_1 = 0; \quad \frac{w_g}{w_{max}} = \frac{\sum_{g=2}^{\ell} |D_{s,g} - D_{s,g-1}|}{\sum_{g=2}^{s} |D_{s,g} - D_{s,g-1}|}$$

where u_{max} and w_{max} are the maximum values of the appropriate knot vector. Recalling the surface data for the US Navy Ammunition ship shown in Figure 15.22a, the defining polygon net generated using this technique is shown in Figure 15.24a. Note that the vertices of the defining polygon net lie anywhere in three space. Figure 15.24b shows the B-spline surface generated from this net.

Although the initial surface fit obtained using equation (7) is generally quite acceptable, it can be improved by iteration on the parameter values (see Rogers [11] for a discussion of this technique for curves and surfaces). Using a two parameter first-order Taylor series expansion, the component errors between the point on the B-spline surface calculated at u_r, w_s using equation (1) with the $B_{i,j}$'s determined using equation (7), and the data point corresponding to u_r, w_s can be written as

$$\delta_x = (Q_{x_c} - Q_x) + Q_{x_u} \Delta u + Q_{x_w} \Delta w \quad (8a)$$

$$\delta_y = (Q_{y_c} - Q_y) + Q_{y_u} \Delta u + Q_{y_w} \Delta w \quad (8b)$$

$$\delta_z = (Q_{z_c} - Q_z) + Q_{z_u} \Delta u + Q_{z_w} \Delta w \quad (8c)$$

The error between the calculated values and the required values at a given point is defined as the sum of the squares of the component errors, i.e. by

$$E = \delta_x^2 + \delta_y^2 + \delta_z^2 \quad (9)$$

Differentiating equation (9) with respect to both u and w and setting the result equal to zero minimizes the error. Thus,

$$\frac{\partial E}{\partial u} = 2\delta_x \frac{\partial \delta_x}{\partial u} + 2\delta_y \frac{\partial \delta_y}{\partial u} + 2\delta_z \frac{\partial \delta_z}{\partial u} = 0 \quad (10)$$

$$\frac{\partial E}{\partial w} = 2\delta_x \frac{\partial \delta_x}{\partial w} + 2\delta_y \frac{\partial \delta_y}{\partial w} + 2\delta_z \frac{\partial \delta_z}{\partial w} = 0 \quad (11)$$

with

$$\frac{\partial \delta_x}{\partial u} = Q_{x_u}; \quad \frac{\partial \delta_y}{\partial u} = Q_{y_u}; \quad \frac{\partial \delta_z}{\partial u} = Q_{z_u} \quad (12)$$

$$\frac{\partial \delta_x}{\partial w} = Q_{x_w}; \quad \frac{\partial \delta_y}{\partial w} = Q_{y_w}; \quad \frac{\partial \delta_z}{\partial w} = Q_{z_w} \quad (13)$$

Substituting equations (9), (12) and (13) into equations (10) and (11) yields

$$\begin{aligned} &[Q_{x_u}^2 + Q_{y_u}^2 + Q_{z_u}^2]\Delta u \\ &+ [Q_{x_u}Q_{x_w} + Q_{y_u}Q_{y_w} + Q_{z_u}Q_{z_w}]\Delta w \\ &= (Q_x - Q_{x_c})Q_{x_u} + (Q_y - Q_{y_c})Q_{y_u} \\ &+ (Q_z - Q_{z_c})Q_{z_u} \end{aligned} \quad (14)$$

$$\begin{aligned} &[Q_{x_u}Q_{x_w} + Q_{y_u}Q_{y_w} + Q_{z_u}Q_{z_w}]\Delta u \\ &+ [Q_{x_w}^2 + Q_{y_w}^2 + Q_{z_w}^2]\Delta w \\ &= (Q_x - Q_{x_c})Q_{x_w} + (Q_y - Q_{y_c})Q_{y_w} \\ &+ (Q_z - Q_{z_c})Q_{z_w} \end{aligned} \quad (15)$$

Rewriting in a more compact form gives

$$A_{uu}\Delta u + A_{uw}\Delta w = A_{cu} \quad (16)$$

$$A_{uw}\Delta u + A_{ww}\Delta w = A_{cw} \quad (17)$$

Solving for Δu and Δw, the corrections to the parametric values at each surface data point, yields

$$\Delta u = \frac{A_{ww}A_{cu} - A_{cw}A_{uw}}{A_{ww}A_{uu} - A_{uw}^2} \quad (18)$$

$$\Delta w = \frac{A_{uu}A_{cw} - A_{cu}A_{uw}}{A_{ww}A_{uu} - A_{uw}^2} \quad (19)$$

The derivatives in equations (18) and (19) can be obtained by formally differentiating equations (1) to (3) (see Rogers and Adams [10]) or by successive approximation as the iteration proceeds.

As pointed out in Rogers [9] for B-spline curves, this technique cannot yield hard points, knuckles or hard chines (discontinuities in first and second derivative) in the resulting B-spline curve or surface.[7] The Barsky and Greenberg [13] technique also exhibits this characteristic. To illustrate this, consider the ship hull shown in Figure 15.25. The surface contains a hard chine. Digitized data points for this surface were used to generate an initial polygon net using the technique described above. The resulting polygon net is shown along with the resulting surface in Figure 15.25b and c. Note that the new polygon net does not have any coincident lines and that the resulting surface does not exhibit a hard chine. Stroobant and Mars [14] present a technique for automatically introducing hard chines or knuckles into the surface by modification of the knot vectors, i.e. by introducing duplicate knot values. In BSSD hard chines or knuckles are introduced interactively by using multiple polygon net points with a uniform knot vector. Figures 15.25d and 15.25e show the defining polygon net and surface after interactively adding the hard chine. The surface is defined by a 9×7 polygon net with 18×14 parametric lines display.

Additional difficulties encountered when automatically fitting a surface to known three-dimensional data points are illustrated in Figure 15.26. Figure 15.26 illustrates three-dimensional digitized data representing the afterbody station lines for a US Navy ammunition ship. The ship hull is smooth, i.e. does not exhibit any hard chines or knuckles. The resulting automatically generated polygon net and the resulting bicubic B-spline surface is also shown. Of particular interest is the shape of the polygon net and surface in this region. The midbody of the ship is essentially wall sided (flat). Thus, the digitized data should lie on a straight line and the surface should be flat. Looking carefully at Figure 15.26 shows that the digitized data for the midship section from the deck edge downward do not line on a straight line but rather exhibit a small oscillation as a result of the digitization process. Examination of the digitized numerical values confirm this [15]. The automatically generated polygon net (see Figure 15.26) and resulting surface exhibit the same oscillation and hence are not considered fair. Nonanalytic, sculptured, digitized surface data is almost never fair. Hence the resulting automatic fit to this data is almost never fair. Thus, either very careful and laborious preprocessing of the numerical data is required or interactive manipulation techniques must be used.

[7] However, a technique for accomplishing this using rational B-spline homogeneous weighting factors is under development.

Constrained automatic surface fitting

Generation of an initial defining polygon or polygon net for known curve or surface point data potentially simplifies the task of interactively fairing a mathematical curve or surface to the data. The technique discussed above generates polygon net points arbitrarily located anywhere in three space. Manipulation of these arbitrarily located net points for further surface refinement or modification is not always intuitive. It is much more desirable to constrain the polygon net points to lie in planes or along lines, e.g. station planes, water planes or buttock planes.

The polygon net points are constrained by solving equation (7) for $[B]$ and then replacing

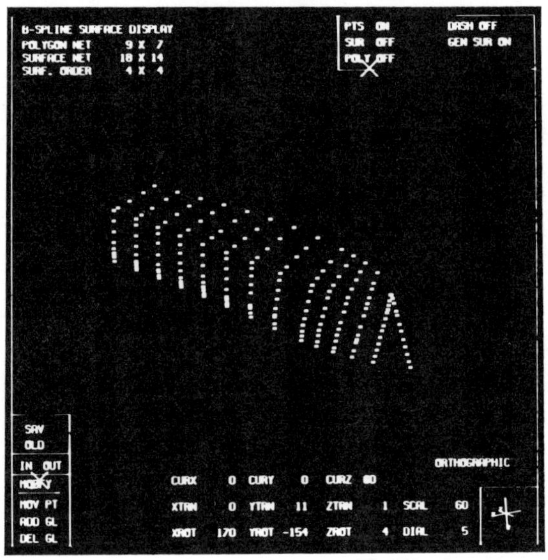

(a)

Figure 15.25. Ship hull with a hard chine. (a) Digitized points, (b) automatically fit polygon net, (c) automatically fit surface, (d) interactively modified polygon net, (e) interactively modified surface.

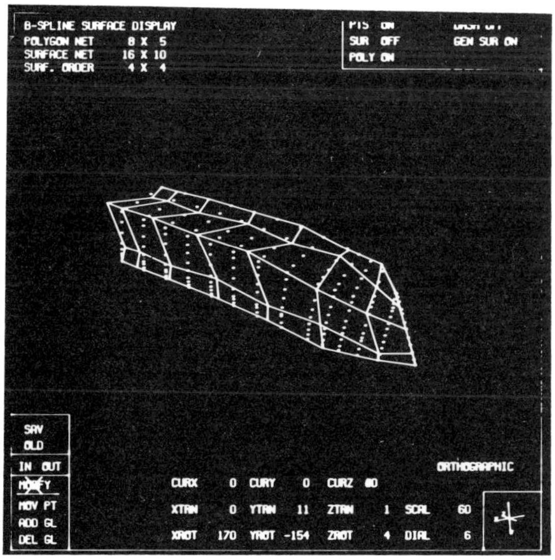

(b)

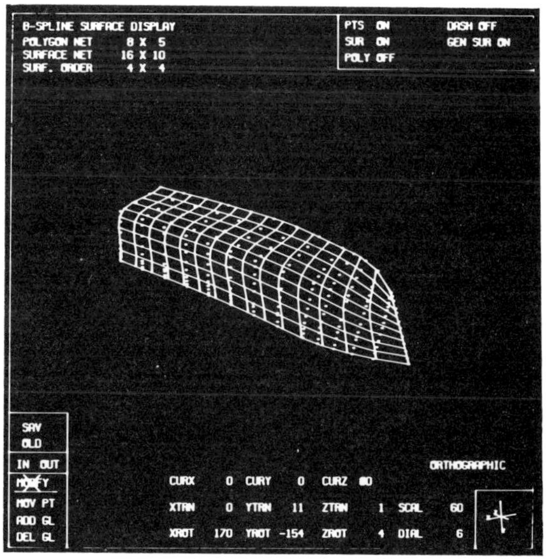

(c)

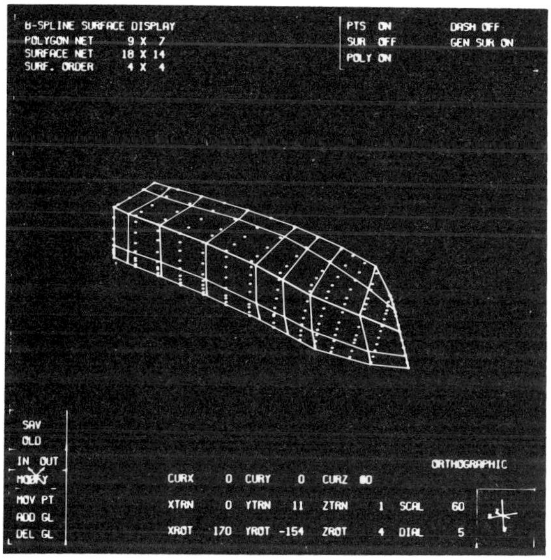

(d)

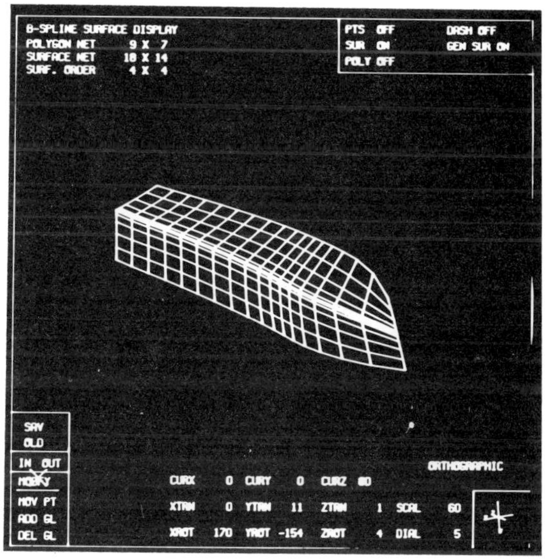

(e)

the appropriate component values of the $B_{i,j}$'s with the constrained component values [11]. The new approximate fitted surface is then generated using these $B_{i,j}$ values to yield Q_{x_c}, Q_{y_c}, Q_{z_c}. Equations (18) and (19) give the corrections necessary to generate new estimates for u_r and w_s at the data points. Iteration proceeds until convergence to some acceptable error level is achieved.

The data for the forward half of the ammunition ship shown above was used to test the constrained B-spline surface fitting technique. The data shown in Figure 15.22a consisted of 12 section curves and the stem profile, each defined with 17 data points. Experience has shown that using the fewest number of defining polygon points yields the fairest curve or surface. With this in mind an 8 × 8 defining polygon net was

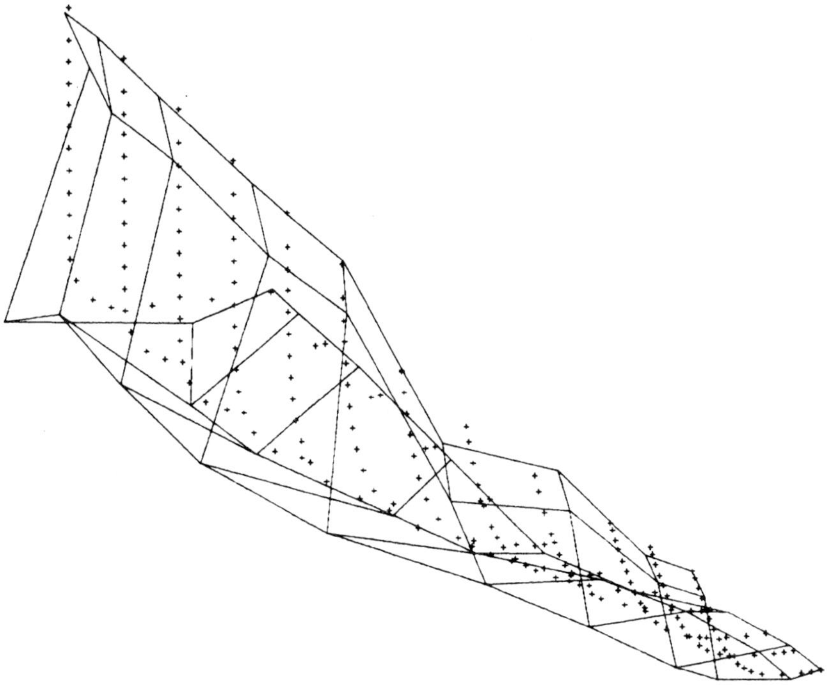

Figure 15.26. Digitized data and automatically fit polygon net for the aft half of the ammunition ship.

chosen for the constrained tests. Uniform open knot vectors were used and the fitted surface is fourth-order in both u and w.

The defining polygon net was constrained to lie in the centerline plane for the stem, the first two section planes aft of the stem and every other section plane thereafter. These locations are marked by arrows in Figure 15.27b. After 10 iterations the constrained B-spline surface fit error was 0.5. This represents a maximum deviation from the data of 0.71 in any single coordinate, or an approximate 1.75% deviation from the surface data.

Figure 15.27b clearly shows that the polygon net points lie in the constraint planes. Figure 15.28 shows a perspective view of the constrained defining polygon net. Figure 15.29 shows a perspective view of the resulting B-spline surface as an orthogonal net of parametric lines along with the defining data points. Figure 15.30 shows a perspective view of the surface as a dense net of parametric lines.

Examining Figures 15.27b and 15.28 shows that the surface is not fair. Here the area near the stem is particularly notable. Additional interactive modification is required to create a fair surface. Clearly, manipulating the current net by planar movement of the defining polygon net points is more intuitive than that of the unconstrained net shown in Figure 15.24a.

Surface fairness

Determining and visualizing the fairness of a surface is of considerable importance in design. Although a B-spline surface is everywhere $C^{k-2,\ell-2}$ continuous, it is not necessarily fair. Dill [16] and Dill and Rogers [17] used color encoded Gaussian curvature displays to determine the fairness of surfaces. Dill [16] originally applied the technique to Gordon surfaces. Dill and Rogers [17] extended the method to bicubic B-spline surfaces. Munchmeyer [18, 19] used both Eulerian nets of minimum and maximum surface curvature as well as Gaussian curvature plots to determine surface fairness of biquintic B-spline surfaces. Mason [20] uses both the Gaussian curvature technique developed by Dill and Rogers [17] and transverse and longitudinal curvatures to determine fairness of B-spline surfaces.

The curvature of a biparametric surface at any point P is defined by the curvature, κ, of the

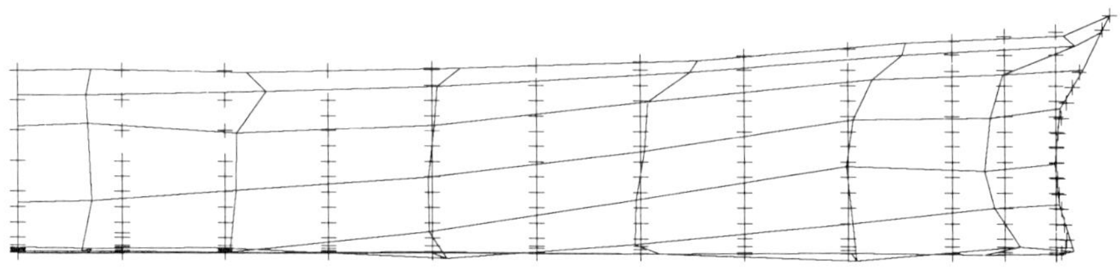

(a)

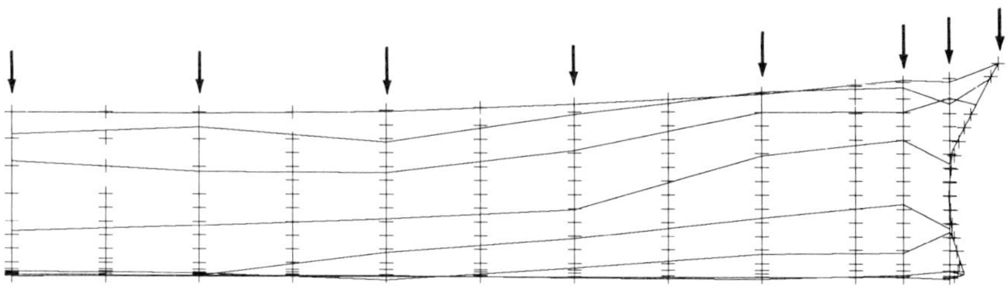

(b)

Figure 15.27. Projection of the 8 × 8 polygon nets for the ammunition ship. (a) Unconstrained, (b) constrained.

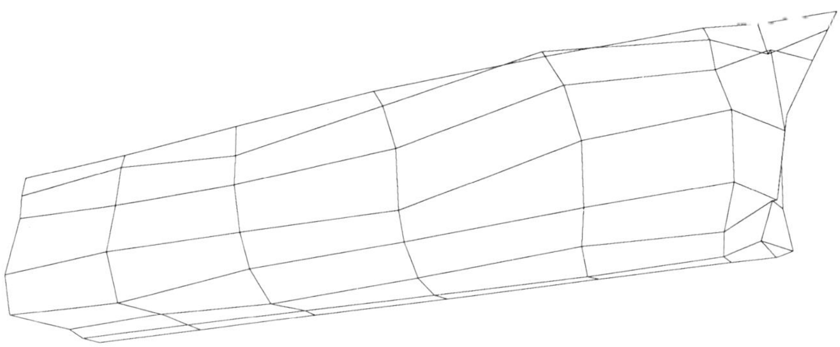

Figure 15.28. Perspective view of the 8 × 8 constrained polygon net.

curve of intersection of a plane through the normal at P and the surface as shown in Figure 15.31.

Euler, the great Swiss mathematician, showed that unique directions for which the curvature is a minimum and a maximum exist. The curvatures in these directions are called the principal curvatures, κ_{min} and κ_{max}. Further, the principal curvature directions are orthogonal. Two combinations of the principal curvatures are of particular interest, the average and the Gaussian curvatures:

$$H = \frac{\kappa_{min} + \kappa_{max}}{2} \tag{20}$$

$$K = \kappa_{min}\kappa_{max} \tag{21}$$

For a biparametric surface the average and Gaussian curvatures are given by

$$H = \frac{A|Q_w|^2 - 2BQ_u \cdot Q_w + C|Q_u|^2}{2|Q_u \times Q_w|^3} \tag{22}$$

$$K = \frac{AC - B^2}{|Q_u \times Q_w|^4} \tag{23}$$

where

$$(A\ B\ C) = [Q_u \times Q_w] \cdot [Q_{uu} \quad Q_{uw} \quad Q_{ww}]$$

The sign of the Gaussian curvature serves to characterize the local shape of the surface: positive–elliptic; negative–hyperbolic; zero–cylindrical or conical.

To color encode the curvature of the surface the Gaussian curvature was determined at each intersection of a pair of parametric net lines. The Gaussian curvature values at the four corners of each surface polygon formed by pairs of orthogonal parametric net lines were averaged to obtain a single value for each of the displayed polygons. The maximum and minimum of the values of all the displayed polygons were determined and the resulting range divided into a number of intervals. Each interval was assigned a different color. In general, colors ranging from yellow-green through yellow and red to magenta representing positive values and colors from yellow-green through turquoise to blue representing negative values yield the best results.

Plates 20a,b show a normally colored and a Gaussian curvature encoded image for the forebody of the US Navy Ammunition ship shown previously in Figure 15.24b. In general, the color encoded Gaussian curvature images show few surprises that close examination of the surface itself does not reveal. However, the color encoded images make any unfairness obvious. Looking at Plate 20b shows that there is an unfair spot just aft of the bow of the ship midway up the side as shown by the small patch of relatively high Gaussian curvature (yellow spot) in the middle of a large area of small, negative Gaussian curvature (green). Also shown is an unfair area

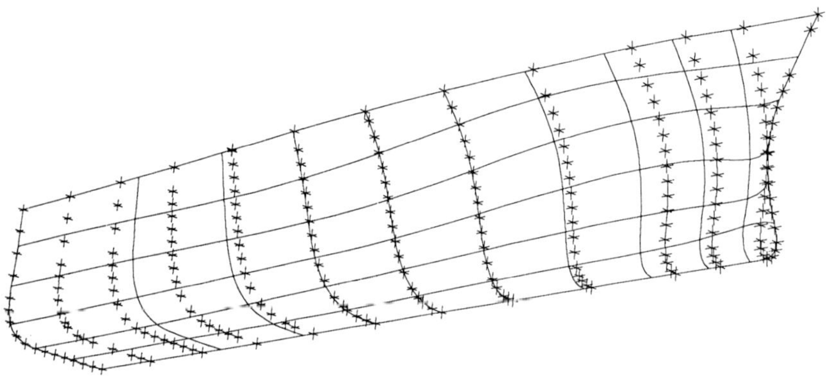

Figure 15.29. Parametric ship surface from the 8 × 8 constrained polygon net.

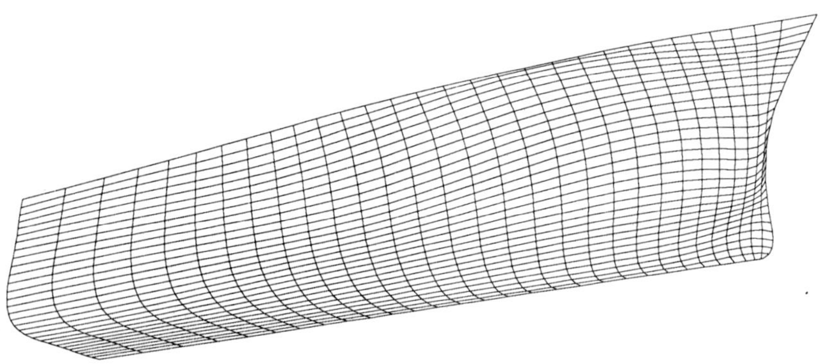

Figure 15.30. Dense parametric ship surface from the 8 × 8 constrained polygon net.

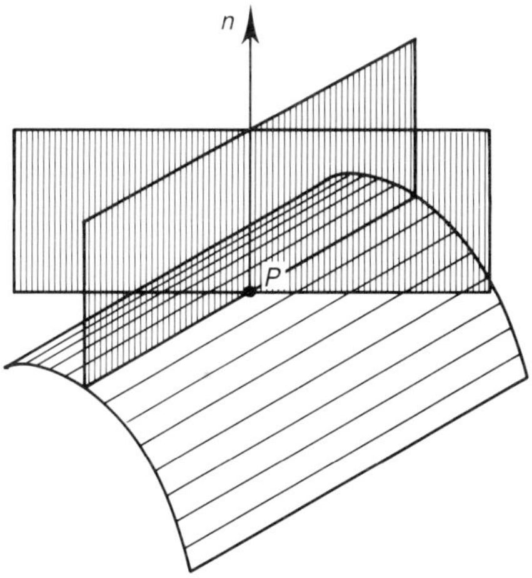

Figure 15.31. Curvature of a biparametric surface.

just at the bottom forward end of the bow. Otherwise, the ship is reasonably fair. Note, in particular, the smooth transition along the side of the ship and around the turn of the bilge. Finally, notice that the unfair spot just aft of the bow is not evident in the normally colored image shown in Plate 20a. The color encoded image makes it obvious.

Plates 21a and b show the ship hull with a hard chine. Looking at Plate 21a, notice that this surface is definitely not fair. In particular, note the ripple in the surface along the side of the ship near the sheer line (top) as evidenced by the region of negative curvature (green) between two positive regions (red). Also notice the unfair bump slightly aft of the middle of the side. The bottom of the ship also exhibits an unfair ripple. Of particular interest is the fact that none of these effects are evident in the normally colored image shown in Plate 21a. Again, the color encoded image makes it obvious. However, note that the unfairness along the side of ship is discernible in the wireframe drawing shown in Figure 15.25e.

Generating contour lines

Once the surface is fair it is necessary to generate waterlines to interface with the numerically controlled machining center to obtain a model. This is equivalent to generating contours of the surface. Because of the flexibility of B-spline surfaces the need for closed disconnected contours was anticipated.

The problem of generating contour lines would seem to be the same as intersecting the B-spline surface with a plane. However, generating true contour lines requires more information than finding a group of planar points. The points must be connected in a logical order that 'follows' the desired contour. Accuracy of the resulting contour lines is of considerable interest. After all, if the designer has spent considerable energy obtaining a fair surface, he does not want to throw that effort away by using an inaccurate contouring algorithm. Consequently, contouring algorithms that simply polygonalize the surface and then find the contour for the polygonalized surface are not sufficient.

Satterfield and Rogers [21] developed an accurate technique for contour determination for B-spline surfaces which builds on that due to Heap [22] for contour lines of triangularized surfaces. The original Heap algorithm intersected each triangular element of the surface with the contour plane. Following the intersection from adjacent triangle to adjacent triangle yields the desired contour. Figure 15.32 shows a single triangular element where P_1, P_2 and P_3 are points lying on the surface. The solid curved lines indicate the true surface. The dashed lines indicate a single triangle on the approximate polygonalized surface. The contour-line segment produced by the Heap algorithm is the bold line segment joining two edges of the approximating triangle.

The Satterfield and Rogers method modifies the Heap algorithm to produce contour bands of triangles, as shown in Figure 15.33. The vertices of each triangle, which are points on the actual B-spline surface, are remembered. The triangular contour band follows the desired contour and spans the actual contour plane; i.e. two vertices of each triangle lie on opposite sides of the contour plane. The triangular contours produced by the modified Heap algorithm are now used as the basis for a piecewise B-spline calculation of the surface. The parametric values for the vertices of each triangle of the triangular contour band are used to calculate a small subregion of the full B spline surface. As shown in Figure 15.32, a number of parametric lines from P_2 to positions on the parametric line between P_1 and P_3 are generated. Each of these parametric lines crosses the contour plane. A

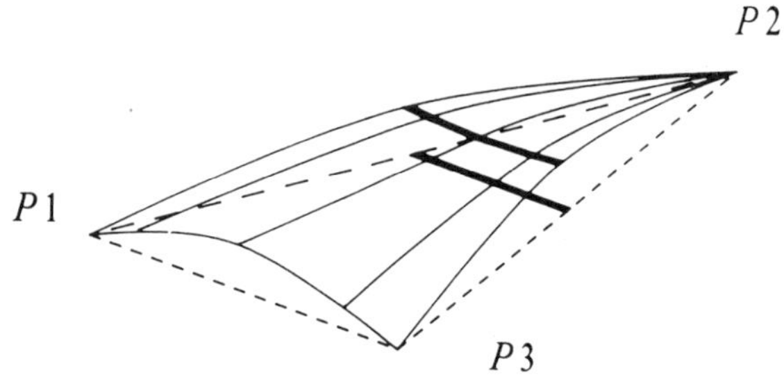

Figure 15.32. Comparison of the Heap and Satterfield and Rogers contour intersections.

search along each parametric line is performed until the contour plane is found. In practice, this search is performed by rotating and translating the contour plane until it is coincident with one of the coordinate planes. The search then consists of following the parametric line until a change in sign of the coordinate normal to the transformed contour plane is found. The bold solid line shown connecting the parametric lines in Figure 15.32 illustrates the difference between the Heap and the Satterfield and Rogers algorithms. The contour line on the B-spline surface can be obtained to arbitrary accuracy using this technique with considerably less computational effort. Tests indicate that the Satterfield and Rogers algorithm is at least twice as fast as the Heap algorithm and requires considerably less storage. Figure 15.34 shows the traditional section, waterlines and buttock lines for a ship generated with this technique.

EXPERIENCES

Generating the surface

BSSD has proven quite successful. A number of different designers from a number of different backgrounds, including academics, students and Coast Guard Naval Arctic tests, have successfully used the system. Initially, the system was used by digitizing a set of body lines and displaying these digitized points along with an initially flat defining polygon net of the appropriate size. The polygon net was then interactively modified until the generated B-spline surface and the digitized points agreed. Designers had no difficulty in adapting to this mode of operation. An initial fair surface was easily obtained within a span of 1–2 hours at the display.

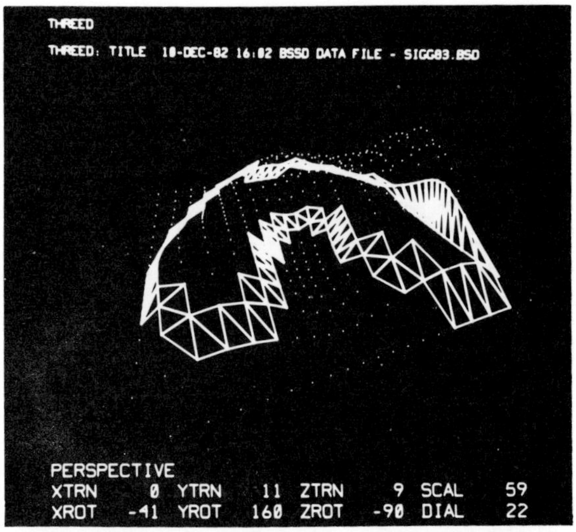

Figure 15.33. A single contour band of triangles.

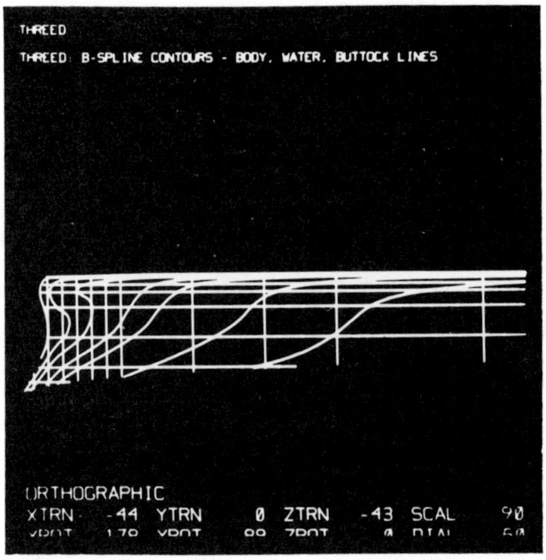

Figure 15.34. Section, water and buttock lines from the contouring algorithm.

When an automatically fit initial surface with the defining polygons arbitrarily located anywhere in three space (see above) was generated and displayed along with the digitized body points, designers had considerable difficulty in further manipulating the surface to obtain a fair surface. In fact, the time spent fairing this initially automatically fit surface generally exceeded that for the initially flat surface. This is attributed to the poor relationship between the arbitrary location of a polygon point in three space and its effect on the B-spline surface.

Using a constrained B-spline surface (see above) significantly improved the designer's performance. As an example, the unfair constrained surface fit for the ammunition surface shown in Figure 15.27b and Figures 15.28–15.30 was refaired in approximately 15 minutes of interactive design time. Color Gaussian curvature plots of the ship before and after refairing are shown in Plate 22. The original constrained fit is the lower image. The refaired ship is the upper image. Notice the smooth blending of colors on the refaired ship.

Cutting the model

Using the contouring algorithm discussed above, sufficient waterlines are generated to cut the model using the technique presented above in the discussion of the CAMILL program. A number of models have been manufactured.

ANALYSIS

BSSD confirmed that real-time interactive design of complex surfaces using B-splines is a viable may in fact be less efficient as evidenced by experiences with automatic surface fitting.

Even though successful, several areas of further research are needed. One is developing a technique for designing developable surfaces. Of even more importance is providing the designer with adequate tools for designing developable surfaces. Note this is *not* the same problem as identifying a developable surface. Further, the technique must *not* limit the surfaces to second degree but allow developable portions of more complex surfaces. This is a particularly critical problem for ship hull design since many ships are wall (flat) sided. Recent reports indicate that the constraint technique discussed above is useful in designing flat of side and flat of bottom hulls [20].

Along with the requirement for developable surfaces is the need to fair more complex surfaces into typical quadratic surfaces, e.g. cylinders or cones. A typical problem is to fair the bow of a ship into a cylindrical or conical stem. The obvious solution to this particular problem is to abandon nonrational B-splines and use rational B-splines. In fact, this is the direction that a new implementation of BSSD on the Silicon Graphics Iris workstation by the present author and Linda Adlum has taken. The new program is called technique. Designers were successful in adapting to the interface provided. However, it is extremely important to pay attention to feedback provided by the designers. Even though an automatic technique may at first appear more advantageous than an interactive technique, it

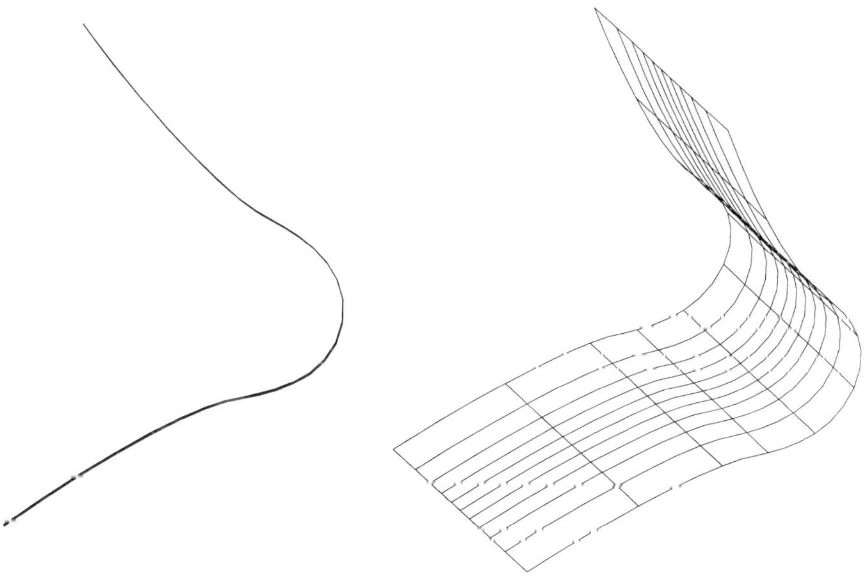

Figure 15.35. Rational B-spline surface faired into a cylindrical stem.

RBSSD (Rational B-spline Surfaces for Ship Design). An example of a rather complex bow section faired into a warped cylindrical stem taken from Rogers and Adams [10] is shown in Figure 15.35. However, rational B-splines introduce the homogeneous coordinate weighting factor (see below) as an additional parameter that the designer must be aware of and be provided tools to manipulate. An excellent introduction to appropriate manipula tion techniques is given in Piegl [23, 24]. Rogers and Adams [10], Piegl and Tiller [25], Tiller [26] and Rogers and Adlum [30] provide additional discussion on rational B-splines.

COMMERCIAL DEVELOPMENT

One of the fundamental tests of academic research and development is whether it is transferred to the commercial environment. The basic concepts and techniques developed in BSSD became the basis for two commercial yacht design systems, FAST YACHT by George Hazen [27] of Design Systems Inc. located in Annapolis, Maryland, USA and MACSURF by Andrew Mason [20] of Graphic Magic Ltd. located in Perth, Australia.[8] FAST YACHT is implemented on Hewlett-Packard equipment. MACSURF, as the name implies, is implemented on MacIntosh computers. Both systems use the basic interactive design and modification concepts developed in BSSD. Both systems provide good interfaces to typical Naval Architecture analysis systems, including standard intact stability and an interface to computational fluid dynamics programs such as VSAERO [28]. Both systems are capable of generating appropriate numerical control manufacturing data. Hundreds of copies of these systems are in place.

[8] Although the current author has acted as an unpaid consultant to both Mason and Hazen, he has no financial interest in either program or company.

Table 15.1. Formal B-spline surface file format description

Surface identification string	for example: hull number
number of surfaces, n	the object consists of n surfaces
nonrational, rational	rational or nonrational in u, w, respectively
open, periodic	open, periodic, nonuniform in u, w, respectively
order of basis, order of basis	order in u, order in w
no. net pts., no. net pts.	no. net pts. in u dir., no. nets pts. in w dir.
knot vector in u	sequence of numbers 1, 2, 3, . . .
knot vector in w	sequence of numbers 1, 2, 3, . . .
4D net points list	$B(i, j)$ a ((no. net pts. in u) × (no. net pts. in w), 4) array. For $B(i, j)$, j varies fastest with fixed i. Each row contains x, y, z, h components of $B(i, j)$ where h is the homogeneous (weighting) factor for each net point. Note: For nonrational B-splines, all $h = 1$.

Example file

Example surface	
1	A single surface is specified
nonrational, nonrational	The surface is nonrational in both directions
open, open	Open basis functions are used in both directions
4, 4	Fourth-order in both directions
8, 8	An 8 × 8 defining polygon net is used
0, 0, 0, 0, 1, 2, 3, 4, 5, 5, 5, 5	Open uniform knot vector for u direction.
0, 0, 0, 0, 1, 2, 3, 4, 5, 5, 5, 5	Open uniform knot vector for w direction.

0 ,	62.7896 ,	−14.2345 ,	1	Net points $B(i, j)$, j varies fastest with fixed i.
0 ,	58.3546 ,	−11.7746 ,	1	
0 ,	50.7321 ,	−7.7246 ,	1	
0 ,	37.2752 ,	−0.7445 ,	1	
0 ,	24.6114 ,	0.2266 ,	1	
0 ,	12.2841 ,	−0.7321 ,	1	
0 ,	0.9638 ,	−4.2096 ,	1	
0 ,	−0.02351 ,	−0.06207 ,	1	

Table 15.1. *continued*

13.2994	,	57.2518	,	0	,	1
8.6079	,	50.7918	,	0	,	1
7.2922	,	52.4931	,	0	,	1
−0.6518	,	33.3121	,	0	,	1
1.5411	,	14.3331	,	0	,	1
3.7183	,	13.4063	,	0	,	1
7.87419	,	1.8224	,	0	,	1
0.9025	,	−0.4589	,	0	,	1
17.0482	,	57.6166	,	13.5	,	1
14.0925	,	55.5365	,	13.5	,	1
11.067	,	48.282	,	13.5	,	1
3.0433	,	40.1302	,	13.5	,	1
0.2289	,	21.5281	,	13.5	,	1
3.5539	,	7.0775	,	13.5	,	1
3.0443	,	0.7293	,	13.5	,	1
−0.2092	,	0.08977	,	13.5	,	1
29.2903	,	53.3467	,	54	,	1
30.7639	,	54.1187	,	54	,	1
21.1234	,	48.1403	,	54	,	1
12.0645	,	35.461	,	54	,	1
9.0546	,	17.3787	,	54	,	1
8.6407	,	6.64932	,	54	,	1
5.1127	,	−0.7182	,	54	,	1
0.1572	,	−0.03143	,	54	,	1
37.8884	,	50.5464	,	108	,	1
33.861	,	47.4467	,	108	,	1
27.1525	,	36.9685	,	108	,	1
23.2976	,	19.6017	,	108	,	1
19.832	,	11.186	,	108	,	1
12.7107	,	2.5924	,	108	,	1
5.5457	,	−0.4150	,	108	,	1
−0.01007	,	−0.04313	,	108	,	1
41.0213	,	48.1632	,	162	,	1
39.6319	,	39.2661	,	162	,	1
38.6344	,	30.4192	,	162	,	1
33.8422	,	16.8768	,	162	,	1
28.2826	,	6.96425	,	162	,	1
18.2607	,	−1.0973	,	162	,	1
4.4469	,	0.2564	,	162	,	1
0.1344	,	0.01421	,	162	,	1
40.4748	,	47.8315	,	216	,	1
41.1403	,	43.5986	,	216	,	1
40.9666	,	30.6183	,	216	,	1
41.4111	,	14.9213	,	216	,	1
34.093	,	0.2635	,	216	,	1
16.9322	,	0.9311	,	216	,	1
5.55319	,	0.167384	,	216	,	1
0.06208	,	−0.07885	,	216	,	1
40.536	,	47.9986	,	270	,	1
40.8065	,	41.4721	,	270	,	1
39.8169	,	33.6058	,	270	,	1
41.8009	,	13.1977	,	270	,	1
37.0837	,	0.3062	,	270	,	1
13.1546	,	1.2999	,	270	,	1
7.0730	,	−0.2726	,	270	,	1
0.07475	,	−0.04795	,	270	,	1

Well over 100 yacht designs have been completed by more than three dozen different designers using these two systems. The designs range from a sailboard and a 16 foot canoe, through a typical IOR One Ton sailing yacht, a 37 foot steel tug, a 45 foot developable lobster boat to a 110 foot luxury motor yacht. The hull for the America's Cup challenger *New Zealand* was totally designed using FAST YACHT *in 10 days*! The hulls of the defending catamaran were also partially designed using MACSURF and FAST YACHT.

SURFACE TRANSFER

One of the problems in B-spline surface design is the transfer of data between design systems. Although IGES [29] and its successors could be used, support of a standard as complex as IGES is beyond the resources of many small developers. Consequently, a simpler surface file transfer format, shown in Table 15.1, has been proposed within the Naval Architecture community. One of the advantages of this proposal is that Bézier, nonrational and rational B-spline surfaces are contained within the single format.

CONCLUSIONS

The feasibility of interactive real-time surface design using B-spline surfaces has been demonstrated in both a research and a commercial environment. Additional work needs to be done on providing the necessary tools for designers to allow them to take full advantage of the increased flexibility offered by rational B-splines.

REFERENCES

1 **A. R. Forrest** *Interpolation and approximation by Bézier polynomials,* CAD Group Doc. No. 45, UML, Cambridge University, October 1970.

2 **A. R. Forrest** *Shape classification of the nonrational twisted cubic curve in terms of Bézier polynomials,* CAD Group Doc. No. 52, UML, Cambridge University, December 1970.

3 **A. R. Forrest** *A new curve form for computer aided design,* CAD Group Doc. No. 66, UML, Cambridge University, June 1972.

4 **R. F. Riesenfeld** *Application of B-spline approximation to geometric problems of computer aided design,* PhD Thesis, Syracuse University (advanced copy 1972).

5 **J. A. Adams** *Geometric concepts for computer graphics,* Engr. & Weap. Rep. No. EW-72-4, U.S. Naval Academy, Annapolis, MD, September 1972.

6 **D. F. Rogers and J. A. Adams** *Mathematical Elements for Computer Graphics, 1st ed.,* McGraw-Hill, New York, 1976.

7 **F. J. Schroeder and B. Johnson** Design Summary of the USNA 360 Foot High Performance Towing Tank, *Proceedings of the 16th American Towing Tank Conference,* São Paulo, Brazil, 9–13 August 1971. (Also US Naval Academy Engr. & Weap. Report E-71-2.)

8 **J. C. Gebhardt, S. Williams III and R. H. Compton** *The USNA Hydromechanics Laboratory Data Acquisition, Control and Analysis System, Proc. of the 18th General Meeting of the American Towing Tank Conf.,* Vol 3, 23–25 August 1977.

9 **D. F. Rogers** B-spline curves and surfaces for ship hull definition, *Proceedings of SNAME, SCAHD77, First Inter. Symp. on Computer Aided Hull Surface Definition,* Annapolis, MD, 26–27 September 1977.

10 **D. F. Rogers and J. A. Adams** *Mathematical Elements for Computer Graphics, 2nd ed.,* McGraw-Hill, New York, 1990.

11 **D. F. Rogers** Constrained B-spline curve and surface fitting, *CAD J.,* Vol 22 No 9, 1990, 609–616.

12 **D. F. Rogers and S. G. Satterfield** B-spline surfaces for ship hull design, *Comp. Graphics,* Vol 4, 1980, 211–217.

13 **B. A. Barsky and D. P. Greenberg** Determining a set of B-spline control vertices to generate an interpolating surface, *Comp. Graphics and Image Proc.,* Vol 14, 1980, 203–226.

14 **G. Stroobant and D. Mars** Ship Hull Form Fairing, in *Comp. Appl. in the Auto. of Shipyard Operation and Ship Design* IV, D. F. Rogers, Nehrling, B. C. and Kuo, C., eds., North-Holland, Amsterdam, 1982.

15 **D. F. Rogers and S. G. Satterfield** Dynamic B-spline surfaces, in *Comp. Appl. in the Auto. of Shipyard Operation and Ship Design IV,* D. F. Rogers, Nehrling, B. C. and Kuo, C., eds., North-Holland, Amsterdam, 1982.

16 **J. C. Dill** An application of color graphics to the display of surface curvature. *Computer Graphics,* Vol 15, 1981, 153–161.

17 **J. C. Dill and D. F. Rogers** Color graphics and ship hull surface curvature, in *Comp. Appl. in the Auto. of Shipyard Operation and Ship Design IV,* D. F. Rogers, Nehrling, B. C. and Kuo, C., eds., North-Holland, Amsterdam, 1982, 197–205.

18 **F. C. Munchmeyer, C. Schubert and H. Nowacki** Interactive design of fair hull surfaces using B-splines, in *Comp. Appl. in the Auto. of Shipyard Operation and Ship Design III*, Kuo, C., MacCallum, K. and Williams, T. J., eds., North-Holland, Amsterdam, 1979, 67–76.

19 **F. C. Munchmeyer** The Gaussian curvature of Coons biquintic patches, *Proceedings of ASME Century 2 Inter. Comp. Tech. Conf.*, 12–15 August 1980.

20 **A. Mason** *MacSurf*, private communication, April and November 1989.

21 **S. G. Satterfield and D. F. Rogers** A procedure for generating contour lines from a B-spline surface. *IEEE Comp. Graph & Appl.*, Vol 5, April 1985, 71–75.

22 **B. R. Heap** *Algorithms for the production of contour maps over an irregular triangular mesh*, Nat. Phy. Lab. report NAC 10, February 1972.

23 **L. Piegl** Modifying the shape of rational B-splines. Part 1: curves, *CAD J.*, Vol 21, 1989, 509–518.

24 **L. Piegl** Modifying the shape of rational B-splines. Part 2: surfaces, *CAD J.*, Vol 21, 1989, 538–546.

25 **L. Piegl and W. Tiller** Curve and surface constructions using Rational B-splines, *CAD J.*, Vol 19, 1987, 485–498.

26 **W. Tiller** Rational B-splines for curve and surface representation, *IEEE Comp. Graph. & Appl.*, Vol 3 No 6, September 1983, 61–69.

27 **G. Hazen** *Fast Yacht*, private communication, April 1989.

28 **B. Maskew** *Program VSAERO: Theory Document*, Analytical Methods Inc., 2047 152nd Avenue NE, Redmond, WA 98052, NASA Contract NAS2-11945, December 1984.

29 **Initial Graphics Exchange Specification** (IGES), Version 3.0, National Bureau of Standards Report NBSIR 86-3359, April 1986.

30 **D. F. Rogers and L. Adlum** Dynamic Rational and Nonrational B-spline Surface for Display and Manipulation (invited paper in the commemorative issue honoring Pierre Bezier on his 80th birthday), *Comput. Aided Design Journal*, Vol 22, 1990, 609–616.

ABOUT THE AUTHOR

David F. Rogers is Professor of Aerospace Engineering at the US Naval Academy. In 1959, he earned a Bachelor of Aeronautical Engineering degree from Rensselaer Polytechnic Institute and subsequently was awarded the MSAE and PhD degrees from the same institute. Professor Rogers is the author of three textbooks on computer graphics, including *Mathematical Elements for Computer Graphics* and *Procedural Elements for Computer Graphics*. He is a member of SIGGRAPH, ACM, the Society of Naval Architects and Marine Engineers, and an Associate Fellow of the American Institute of Aeronautics and Astronautics. Dr Rogers is the founder and former Director of the Computer-Aided Design/Interactive Graphics Group at the US Naval Academy. He is editor of the Springer-Verlag Series *Monographs in Visual Communication* and is a founding editor of *Computers & Education*. He is also a member of the editorial board of *The Visual Computer* and of the *Computer-Aided Design Journal*.

Professor Rogers was the co-chair of the BCS/ACM International Summer Institute on *State of the Art in Computer Graphics* held in Scotland in 1986, in Exeter in 1988, in Edinburgh in 1990 and in Reading in 1992. He was also co-chairman of ICCAS'82, The International Conference on Computer Applications in the Automation of Shipyard Operation and Ship Design. He is member of the International Program Committee for ICCAS. He was also co-chairman of the International Program Committee for Computer Graphics Tokyo'85.

Professor Rogers was the Fujitsu Research Fellow at the Royal Melbourne Institute of Technology in Melbourne, Australia in 1987 and a Visiting Professor at the University of New South Wales, Sydney, Australia in 1982. He was an Honorary Research Fellow at University College London in England during 1977–78 where he studied Naval Architecture.

Professor Rogers was one of the original faculty who established the Aerospace Engineering Department at the US Naval Academy in 1964. He has both an experimental and a theoretical research background. He has research interests in the areas of highly interactive graphics, computer-aided design and manufacturing, numerical control, computer-aided education, hypersonic viscous flow, boundary layer theory, and computational fluid mechanics.

16 Computer-Aided Design – 1959 Through 1965 – in the Design and Graphics Division of MIT's Mechanical Engineering Department

Robert W. Mann

The Computer-Aided Design project at MIT was inaugurated by the intersection of two groups who approached the idea from different directions, and who represented different talents and experiences. One group was staffed and led by computer software specialists in a research laboratory, while the other was comprised of experienced design engineers who, as faculty in an academic department, were also responsible for graphics and design education. But first, a sketch of the historical background.

Post-World War II, the United States Air Force addressed the increasing complexity of the machining of metal parts for evermore sophisticated aircraft by funding at the MIT Servomechanisms Laboratory the development of a numerically-controlled 3-axis milling machine, and then the development of a computer programming language which facilitated the transfer of the graphic and numerical information on mechanical drawings into computer code to control the path of the milling machine cutter blade [1]. By the time of inception of the CAD project in 1959 the Servomechanisms Laboratory had been renamed the Electronics Systems Laboratory, and its predominantly professional staff was directed by Professor J. Francis Reintjes, Professor of Electrical Engineering. The APT (Automatic Programming of Tools) language was developed under the supervision of Douglas T. Ross, Head of the ESL Computer Applications Group. The ESL staff pursued a 'bottom-up' approach to addressing design, i.e. the numerically-controlled machine came first and then the APT language, after which they began to consider the process which culminated in the mechanical drawing of the machined part. From a companion article in this compendium authored by Ross [2]:

> "It is hoped that the result of this effort [e.g. computer-aided design] can be blended with the APT system . . . to provide an efficient mechanism for going almost directly from the requirement for a machined part to the finished product."

By contrast, because they practised and taught engineering design, the faculty group in the Engineering Design and Graphics Division of the Mechanical Engineering Department took a 'top down' view of CAD. The Division head, then Associate Professor Robert W. Mann, had been a draftsman at the Bell Telephone Laboratories before and after his military service in the US Army Signal Corps where he installed and maintained teletype and cryptographic apparatus. A vocational high school graduate, the G.I. Bill made possible his belated entry to MIT, where he completed the Bachelor's and Master's degrees in mechanical engineering in four years. He then joined an offspring of the original Servomechanisms Laboratory, the Dynamic Analysis and Control Laboratory at MIT, where he became engaged in research and development of components for air-to-air missiles, and the then world's largest and most complex electro-mechanical analog computer. Used to simulate the flight of future missiles, an electrohydraulic multi-axis flight table, actuated by the computer, subjected prototype autopilots to the dynamic environment of the future airframe, with the analog computer calculating the aerodynamic responses and the missile trajectory. Shortly after assuming responsibility for the Design Division in the DACL in 1952, Mann was asked to join the faculty to also teach design. By the time of the inception of the CAD project in 1959, as a result of the departure of the more senior faculty, he had become head of the Engineering Design and Graphics Division of the Mechanical Engineering Department.

As an undergraduate, Mann became aware of digital computation and graphic displays through his fiancée and later wife, Margaret I. Florencourt, a physicist and charter member on

Fundamental Developments of Computer-Aided Geometric Modeling
ISBN 012-554765-X

the staff of the Whirlwind computer [3]. His undergraduate classmate and roommate, Kenneth H. Olsen, joined Whirlwind for his Master's degree program, and Mann and Olsen continued their friendship through the transition of Whirlwind into the Lincoln Laboratory, where Olsen designed an early solid-state computer, the TX-0.

In 1953, MIT had restructured the undergraduate curriculum, eliminating a requirement that all freshmen take drawing and descriptive geometry; the prior free-standing Division of Graphics and its faculty were merged into the Mechanical Engineering Department's Design Division. Steven Anson Coons, who taught graphics subjects, thus became a member of the Design Division in Mechanical Engineering. Although his forte was graphics, he was an experienced and accomplished design engineer. A member of the MIT class of 1932, between 1940 and 1947 he was a mathematician and designer at Chance-Vought Aircraft, where he developed the mathematical methods for defining and computing aerodynamic shapes. His 'algorithms', implemented on banks of mechanical adding machines operated by clerks, became the forerunner of computer programs subsequently used throughout the aircraft industry [4]. After a brief period as chief designer in a small electronics firm specializing in high-quality recording and playback equipment, Coons joined the MIT faculty as an Instructor in the Division of Graphics, and was an Assistant Professor at the time of the merger of Graphics into Mechanical Engineering. Beginning in 1936 and throughout this period, he was also a photographic illustrator and color consultant to a major book publisher.

Professor Coons' and his wife Jane's sense of style was exemplified by their primary residence in Chichester, New Hampshire. Situated on a high rise with a magnificent view of the White Mountains to the North, the classic brick structure, built originally as an inn, had been completely refurbished by Steve and Jane and decorated and furnished with period pieces appropriate to its 19th century origins. The long dirt driveway from the nearest public road was not always passable in snow and/or mud season when the Coons were 'imprisoned' in a small apartment near MIT.

The demise of required graphics instruction at MIT was in part driven by the need for curricular space to accommodate the new engineering science subjects. But it also reflected a certain stodginess of approach and message of the graphics teaching faculty. Not so for Steve Coons! Extraordinarily fluent and eloquent in

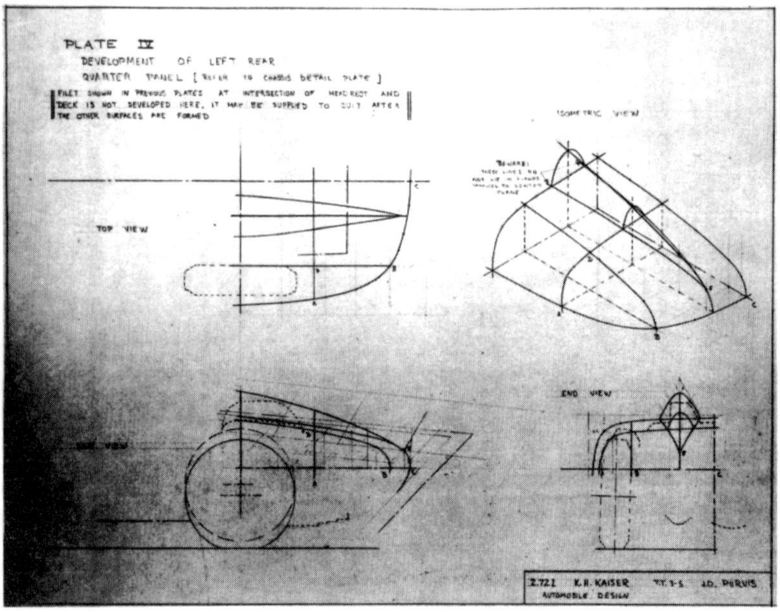

(a)

Figure 16.1. Undergraduate student solutions to graphics assignments of Professor Coons.

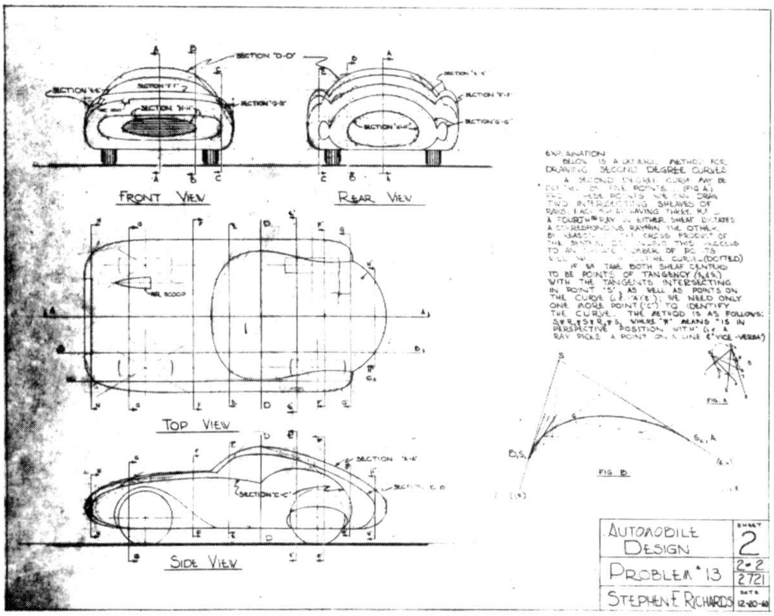

(b)

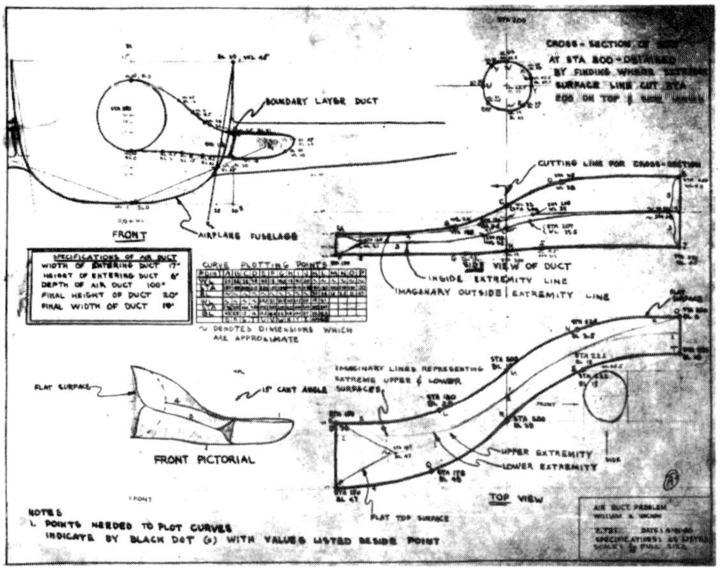

(c)

Figure 16.1 *continued*

presenting and using graphical techniques – ranging from sketching and orthographic projection on to projective geometrical transformations, non-linear differential equations, curvelinear networks (for other topics, see his text [5] and articles [6–12]) – he exuded that 'personal magnetism and inquisitive enthusiasm that makes for excellent teaching. Naturally creative, he was regarded by his colleagues as brilliant'. These quotations are excerpted from his personnel record at MIT as of 1949. Examples of student work (Figures 16.1a–c) are evidence of how he challenged undergraduate students in his projective geometry subject; this same ability to inspire and challenge graduate students will be apparent in his Master's and Doctoral thesis supervision in the CAD project.

The third academic collaborator in the ME CAD project was Dwight M. B. Baumann, a North Dakota native who joined the DACL for his Master's and then PhD degree, and was an Instructor in the Design Division in 1959. Raised on a farm, he was conversant with a broad range of machinery, and became equally competent in the then emerging computer world – his 1960 doctoral thesis was a technique for automatic character recognition.

As mentioned earlier, the 1950s were a period of transition in engineering education. The weak pre-World War II science- and mathematics-base and the over emphasis on 'copy-cat' design was giving way to the new engineering science approach – displacing drawing and graphics and challenging the utility of engineering design education. Much discussion ensued locally and nationally on how to characterize the design process *per se*; how it was practised in industry was explored and the content and validity of academic design education was argued. Examples of this scrutiny would include a study conducted at MIT during the summer of 1959, sponsored by the Carnegie Corporation of New York [13], and a Committee on Authentic Involvement in Engineering Design of the National Commission on Engineering Education, organized and chaired by Professor Mann, which held national conferences on engineering design education in 1960, 1962, 1964 and 1966 [14]. Thus, concurrent with the CAD research reported herein, the design faculty at MIT were organizing and documenting their understanding of the design process, and exploring how its pursuit might be enhanced in education and practice, both locally and nationally.

The role that the digital computer might play in the design process, especially to those who were familiar with the graphics capability of Whirlwind, was reinforced by the transfer in 1958 of the TX-0 computer and its graphics display from Lincoln Laboratory to the MIT campus, where access was offered to all interested parties. In January of 1959, the 'bottom-up' computer programming staff of ESL and the 'top-down' engineering design faculty of the ME Department joined in the planning of a seminar series, which mapped out what became the Computer-Aided Design Project.

The respective philosophies of the two groups are documented in concurrent Electronics Systems Laboratory's reports. The one authored by Ross [15] reflects his computer programming wisdom and discusses 'outside-in problem statements', 'flexible language structures', 'problem-statement languages', etc., for 20 pages. Only then are input-output issues raised: 'in addition to typewriter and push-button languages, the system will employ picture languages, for both input and output, as well as physical manipulation of pictures.' In the companion Coons and Mann document [16], the engineering design process is discussed at length and diagrammed, as reproduced in Figure 16.2. The essential role of graphical input and output in the development of the initial concept for a design and in its subsequent analysis, improvement and documentation permeate the report which concludes with a list of research tasks, the 'Primary Areas of Current Investigations'.

Staffing the CAD Project strained the diminished Design and Graphics Division faculty. Instructor Baumann competed with his own PhD dissertation to organize and participate in the seminar series. Professor Mann and a few other engineering design faculty were assuming the responsibility for subjects previously taught by departed colleagues. The Dynamic Analysis and Control Laboratory was transitioning into what became the Engineering Projects Laboratory under Professor Mann's direction. Steven Anson Coons was the logical CAD Project Director for the ME group. Coons' unique knowledge and skills in graphics, as well as in design, assured the vision and supervisory ability required. And the slackened demand on the graphics faculty meant he had the time needed for the job. To staff the 'Primary Areas of Current Investigations' [16], the academic faculty adopted the tried and true recruitment of graduate students who, in the conduct of research, would satisfy their Master's or Doctoral degree requirements, and thereby complete their graduate education. Much of the rest of this report will identify these research theses, and relate their contributions to the overall effort.

The first two priorities set by Coons and Mann in their 1960 document [16] were:

1. "A graphical input device and associated programming so as to enable the designer to introduce a geometrical shape description into computer memory and process this data for ultimate use in manufacture."

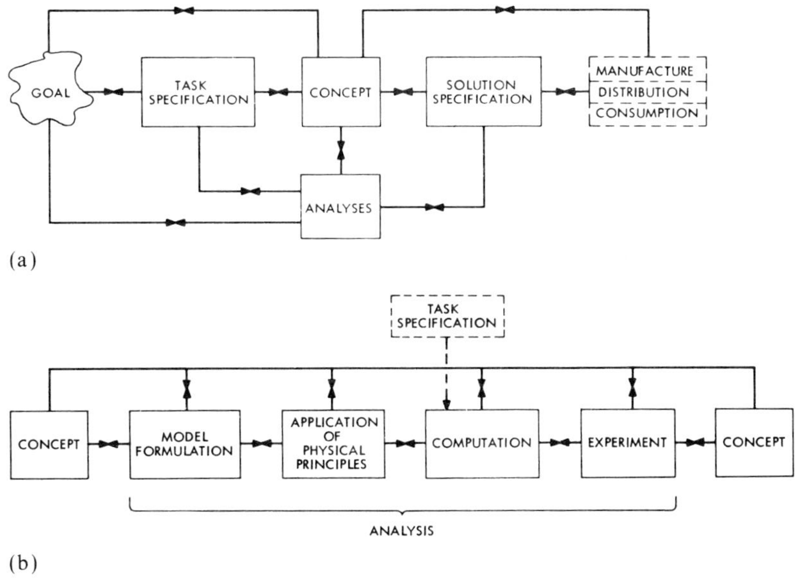

Figure 16.2. The Engineering Design Process and the Role of Analysis [16].

2. "A graphical output (oscilloscope, pantograph, etc.) to permit the designer to query the computer about the state of graphical information content."

In Coons', Mann's and Baumann's view, a future CAD system, to fully capitalize on the extant skills of the engineer/designer, would have to support drawing and subsequent alteration and/or augmentation in a manner akin to the contemporary practices employed in sketching and mechanical drawing. Not surprisingly, the two Master's degree candidates who elected to address the human/machine computer interface for drawing were disposed towards electromechanical input/output devices. Laird Johnston's thesis [17], June 1961, supervised by now Assistant Professor Baumann, describes a mechanized drafting arrangement supplemented with push-buttons, while Joseph Verderber's [18] concurrent effort completed in May 1961 and supervised by Mann, adopted sketching as the input mode using a tracing apparatus or joystick, with the computer transforming the features of the sketch into a three-dimensional orthographic projection.

The definitive interface was described in the January 1963 PhD thesis of Ivan E. Sutherland, 'Sketchpad' [19]. Sutherland, an Electrical Engineering graduate student at MIT, had independent support from a National Science Foundation fellowship. His dissertation committee chairman was Claude Shannon of information theory fame. Sutherland became familiar with light pen input by experimenting with the TX-0 computer now at MIT. Through various academic contacts, he became aware of its successor at Lincoln Laboratory, the TX-2, probably the most powerful machine of its kind at that time. The TX-2 was equipped with a variety of input/output devices including light-pen, push-buttons and oscilloscopes, as illustrated in the upper portion of Figure 16.3, the frontispiece of an article by Mann and Coons in the *1965 Yearbook of the McGraw Hill Encyclopedia of Science and Technology* [20]. Sutherland began to develop the software which became Sketchpad, and recruited Coons onto his doctoral committee, an alliance which was to persevere to their mutual benefit for several years beyond the completion of Sutherland's thesis.

The assignment of the respective contributions of the graduate student *versus* that of dissertation committee members is always a difficult task. In this case, suffice it to say that Coons was the only member of Sutherland's committee with graphics and mechanical design experience, some of which shines through Sutherland's document. As an example, Figure 16.4 illustrates Sutherland's concept of constraints – in this case making initially sketched lines parallel; this mimics the features of lines produced by mechanical drafting

Figure 16.3. Ivan Sutherland at the TX-2 computer console using Sketchpad [20].

devices. An even more explicit example of Coons' influence would be Sutherland's experimentation with kinematic linkages, quite outside of his experience and that of his electrical engineering education. The bottom part of Figure 16.3 demonstrates Sutherland's animating of an initially sketched four-bar linkage. The several 'tear-drop' figures below Sutherland's left hand are Sketchpad's generation of the complex trajectories of a point moving with, but displaced from the linkage connecting rod. Figure 16.4 was sketched and animated by Sutherland in the presence of both Coons and Mann. Coons then suggested testing the veracity of a linkage mechanism purported to generate a 'straight-line', and attributed to James Watt of steam engine fame. The linkage was sketched, rectified and then automatically 'driven' to several different positions, with Sketchpad reporting the distance from the same point on the linkage to a reference line (see Figure 16.5). The display's iteration of '1087' demonstrated that the 19th century linkage design could, in fact, generate a straight line. Another example of the application of Sutherland's constraint approach was his demonstration of how tension and compression forces in members of a truss structure changed when the structure was 'loaded' by imposing inconsistent constraints. Figure 16.6 illustrates the differences between a cantilever bridge and an arch bridge; the numbers on individual members depict their tension and compression when 'loaded' in the center by the vertical line (with the large number).

An illustration of an ex-draftsman's joy with Sketchpad manipulation is shown in Figure 16.7 [22]. The angle bracket is sketched, then rectified orthogonally by applying constraints. A rivet is designed and sized to fit a hole in the bracket (an obvious extension would have the computer present the designer with a range of manufacturing tolerances for the hole and the rivet).

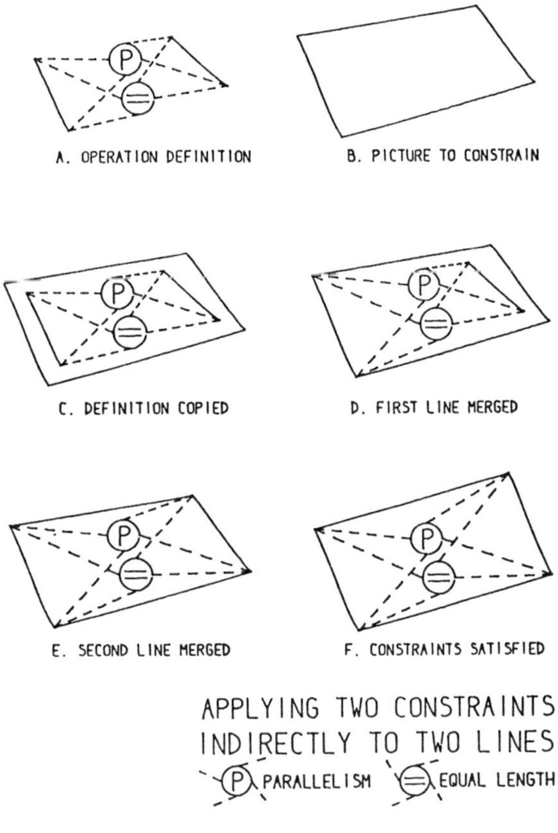

Figure 16.4. Constraint application as displayed on the TX-2 display [19].

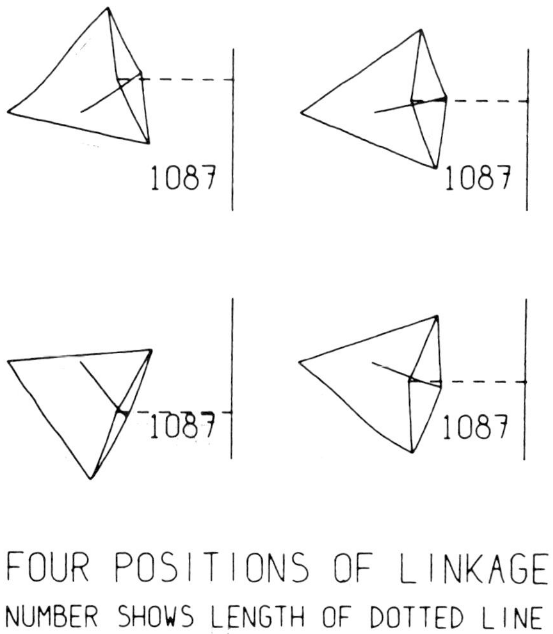

Figure 16.5. Testing the James Watt 'straight-line' linkage on Sketchpad [21].

As drawn the rivet is too short and therefore functionally useless, however, the principle is clear. Once defined geometrically the *generic* rivet can subsequently be distorted to any desired diameter, length, head size, etc. And after many applications of the 'rivet', if one wishes to make a change, for example, removing the diagonal lines used initially to define the arc of the head, retrieving the generic rivet and removing the diagonals eliminates those diagonals in all prior copies in the computer database. To a draftsman/designer the prospect of such automatic 'change orders' rippling through the various part and assembly drawings of a system are profound and liberating. Finally, since in a sense all new drawing is but the selection from, and the assembly of, prior drawing, with Sketchpad, copying is invoked effortlessly, without error, and virtually instantaneously.

During the latter stages of Sutherland's thesis, the third graphically-oriented mechanical engineering Master's candidate, Timothy Johnson, supervised by Coons, extended the two-dimensional input/output pictorial competence of Sketchpad I to Sketchpad III, realizing

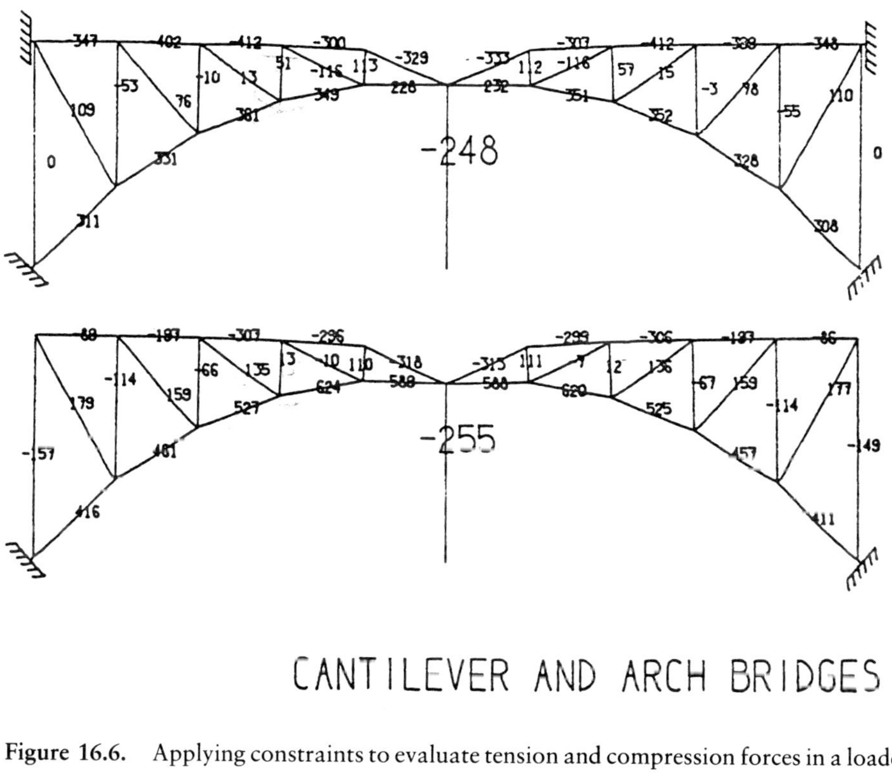

Figure 16.6. Applying constraints to evaluate tension and compression forces in a loaded truss structure [19].

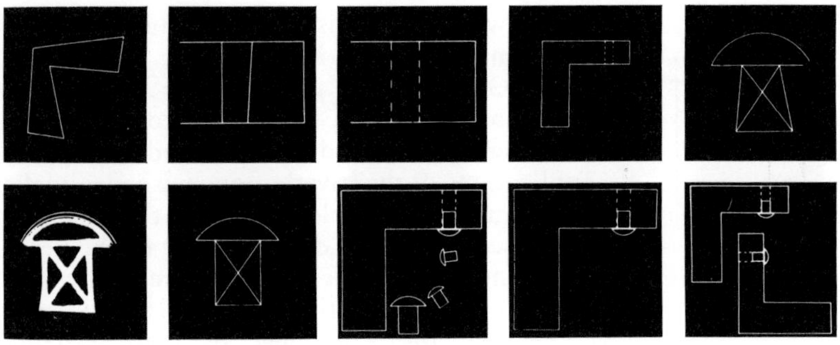

Figure 16.7. A demonstration of Sketchpad part drawing, assembly and copying [22].

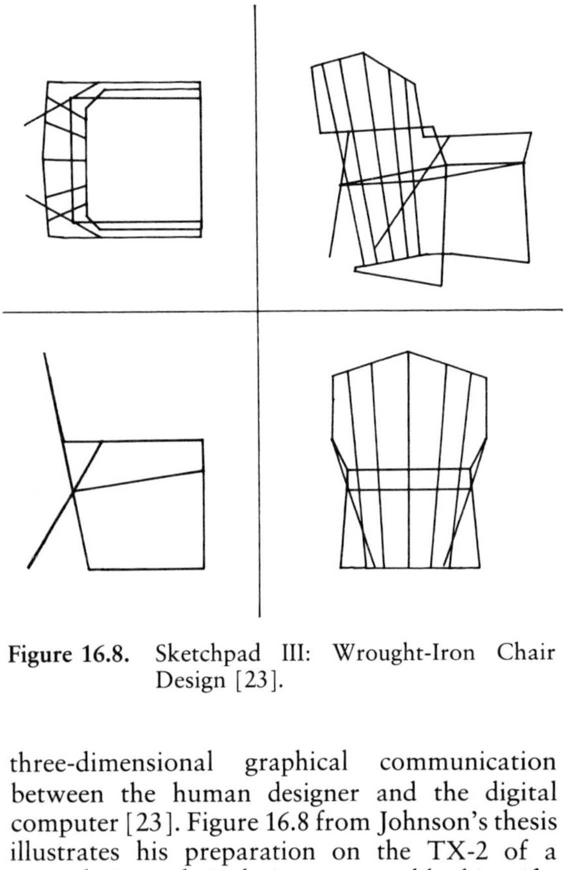

Figure 16.8. Sketchpad III: Wrought-Iron Chair Design [23].

three-dimensional graphical communication between the human designer and the digital computer [23]. Figure 16.8 from Johnson's thesis illustrates his preparation on the TX-2 of a wrought-iron chair design suggested by his wife. The usual front, side and top projections are complemented with a perspective view in the upper right hand quadrant, with designer control of viewing distance and perspective vantage points.

Although for decades following termination of the MIT project CAD applications were exclusively restricted to computer *drawing*, from the outset the mechanical engineering group envisioned the computer as ally during the entire design process as diagrammed in Figure 16.2. Thus, analyses to establish whether the part under design meets specifications must include consideration of required strength and/or stiffness. Among the Primary Areas of Current Investigations in Coons and Mann 1960 [16] was:

> "Programs for strength of material calculations, to be selected and initiated by the designer, with parameters introduced both externally and by interconnections with the shape memory when needed."

The Mechanical Engineering graduate student who proceeded to convert this 1960 CAD aspiration into practical capability was Richard P. Parmelee [24]. His May 1961 SM thesis, supervised by Coons, *A Study of a Stress Analysis Facility for Computer-Aided Design,* developed computer methods for automating stress calculations for the simplified shapes, such as beams and columns, which designers traditionally abstract from a developing design to make manual computation practical. Figure 16.9 illustrates how a 'sculptured part' might be modeled as a cantilever beam. But real machined parts more likely look like the left illustration of Figure 16.10, taken from Parmelee's Doctoral thesis [25]; here simplified modeling is not reliable in evaluating stress and strain. In his PhD thesis, Parmelee developed a finite-difference approach to modeling and analyzing complex shapes and demonstrated its feasibility by applying his numerical technique to classical problems where analytical solutions existed. Figure 16.11 is an error analysis comparing his finite difference solutions with the analytical

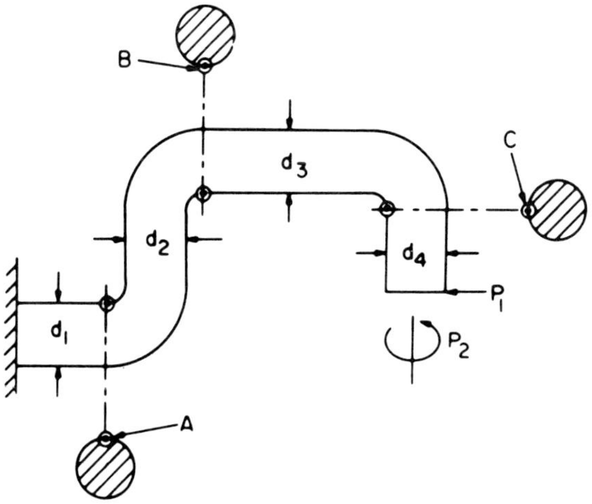

Figure 16.9. 'Model' of a hypothetical part for stress analysis [24].

many standard parts, and that frequently the features of the standard part dictated geometrical and material decisions. Accordingly as the third of their Primary Areas of Current Investigation in the 1960 Report [16], they specified the need for:

> "A catalog of standard parts, including descriptive specifications, strength data and duty cycle data; this memory system should be accessible to the symbolic input system."

In 1959, as Johnston and Verderber were exploring graphical input/output, and Parmelee stress analysis, Joe D. Purvis, Jr. addressed what became his August 1961 Master's thesis, *An Investigation of a Standard Parts Selection Facility for Computer Aided Design* [27], stress distributions around a hole in a plate, documenting the increasing numerical burden of reducing error.

Another Master's thesis supervised by Professor Coons, that of Peter T. Bennett, completed in January of 1963 [26] extended the beam analysis approach of Parmelee's Master's thesis. Bennett developed computer-aided routines for analyzing beam stresses in more detail, considering both normal and bending stresses and transverse and torsional shear. He also included computational means for modifying the beam cross-section and therefore its moment of inertia. Material failure criteria were also integrated into the routines.

Beyond first conceiving and recording the geometric form of a design and then analyzing the consequences of load and deformation, as experienced designers Coons, Mann and Baumann knew that any real design incorporated

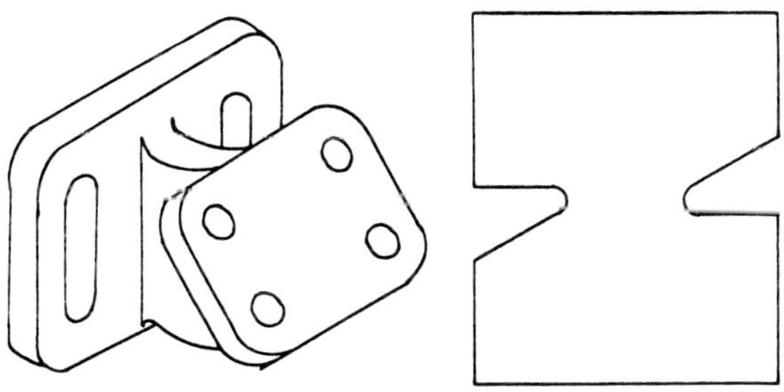

Figure 16.10. Examples of realistic parts for stress analysis [25].

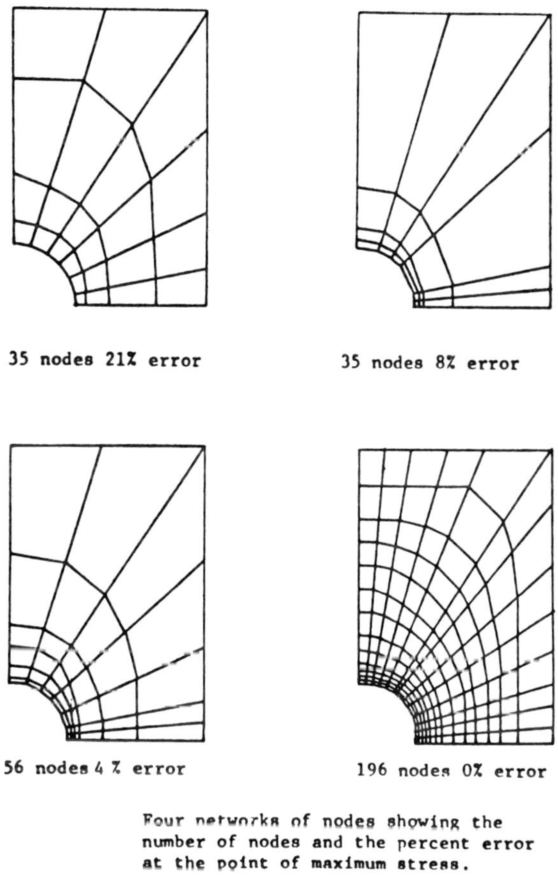

Figure 16.11. Finite difference calculation of stress field in a plate with a hole [25].

supervised by Mann. Of the vast range of catalog parts the designer draws upon in the prosecution of a design – ball, roller and needle bearings, connectors, pins, snap rings, seals, O-rings, etc. – Purvis focussed on threaded fasteners, exploring the criteria the designer must satisfy – static and fluctuating loads, factors of safety, appearance, etc. He proposed a computer mediated selection process, including fastener and material identification and special design requirements, leading to the development of a database which included these and other relevant parameters. Subsequently Coons supervised the Master's thesis of Robert E. Starzec, completed in June of 1966 [28], to further elaborate a scheme for information storage and retrieval that could be used as a 'standard part-selection' component of a computer-aided design system.

Concurrent with his supervision of these Master's and Doctoral students, Coons was developing what has become his CAD signature – the Coons' Patch [29]. Since that topic is described in more detail elsewhere in this compendium, this chapter is limited to discussion of three applications of Coons' patches, early demonstrations of computer-aided design as it existed at MIT in the mid-1960s.

A thesis which applies Coons' Patches in naval architecture was completed in 1964 by two graduate student officers, M. L. Hamilton and A. D. Weiss [30], jointly supervised by Professor Coons and a naval professor of naval architecture at MIT. The goal was to use Coons' surface description techniques in the design and evaluation of a ship's hull. The lines of a US destroyer were simulated, including calculation of the midship and prismatic coefficients, the displacement, the wetted surface area, and the center of buoyancy (see Figure 16.12). The computer-aided design was shown to closely resemble that of the actual ship. A brief economic analysis demonstrated a great savings in time was possible with a CAD system, providing the opportunity to explore alternate designs.

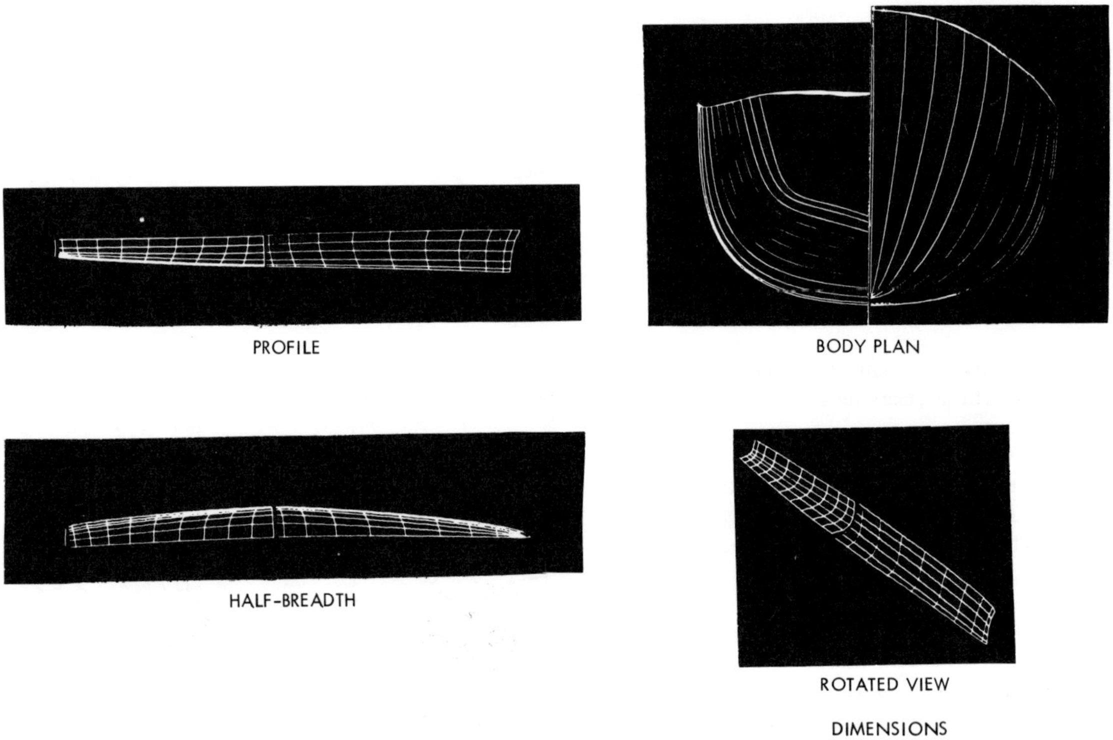

These pictures were taken from the face of the display scope. The hull shape is modeled by two surface patches photographed separately and pieced together above.

DIMENSIONS

L	369'
B	39'
H	12.5'
C_p	.625
Δ	2750

Figure 16.12. CAD design and evaluation of a destroyer hull [30].

In 1963, with the CAD project just over three years old, knowledge of the MIT effort was spreading, with the Spring Joint Computer Conference that year a particular impetus [31–33]. Coons became much sought-after for presentations, including participating as a principal in many design conferences. A specific illustration would be what became the annual week-long 'Application of Computers to Automated Design' conference, sponsored by the University of Michigan in cooperation with the University of Detroit [34]. Paul M. Reinhard of the latter was the conference director. Coons collaborated with S. H. Chasen, who then directed the computer-aided project at the Lockheed Corporation in Marietta, GA, in the application of CAD techniques to the design of components of the C5-A transport aircraft. A (for then) huge CDC mainframe was the computer; Mann recalls demonstrations of CAD aiding the design of linkages for the landing gear, first devising and then refining the ability to extend and retract into the fuselage. Coons also consulted with the Ford Motor Company, together with Professor Bertram Hertzog of the University of Michigan. One artifact of that collaboration, produced in 1966, is a short 16 mm silent movie of an oscilloscope screen on which Coons' Patches assemble body panels into what becomes an early vintage Mustang-like automobile. The movie sequence continues with demonstrations of rotations to observe various views of the just-designed body. The driver's door swings open to check clearances, and then is abruptly slammed shut; were audio imposed one might expect the solid 'chunk' that used to be the test of the solidity of an automobile design! The Mustang is then placed on a roadway and drives off, back and forth at ever reduced scale, 'into the sunset'. Figure 16.13 is a composite of photographs made from several frames of this delightful, avant-garde CAD movie.

In December 1965, support by the Electronics Systems Laboratory of the Mechanical Engineering effort in the CAD project ended. The ME faculty's desire that students conduct 'real-world' theses on sponsored research projects was limited in the CAD effort by the difficult access to the TX-2 computer, the only machine on which real-time graphics was then possible. Lincoln Laboratory was a high-security facility which prohibited access to some students. It was 15 miles from the MIT campus, and the TX-2 was so heavily scheduled that the only available student time was in the wee hours. Mann and Baumann had turned their attention to what is now called rehabilitation engineering. They, together with their students, were developing English-to-braille translation software and braille-embossing typewriters [35], ultrasonic devices to aid the mobility of blind persons [36], and human nervous system-controlled, servo-mechanism-driven, artificial limbs for amputees [37].

Travel and conferences, combined with his teaching obligations kept Professor Coons very busy. In 1963, the interdepartmental MIT Project MAC (Machine Aided Cognition or Multiple Access Computer) became a co-sponsor of the CAD project, with Coons conducting research at that facility. In 1966/67, he took a sabbatical, writing the Department Head: "I think it is important that I get time to do some thinking and generate a new head of steam. I want to get away from the telephone and from academic duties and from the constant stream of visitors that intrude upon my time and energy." He spent part of that academic year at MAC with Professor Ted Glaser, and part of the time with Dr Sutherland at the Harvard Aitken Computer Center, where among other things they worked on an early form of virtual environment. The computer graphics were displayed on two small CRO tubes, headmounted with appropriate visual disparity, giving the impression of 3D spatial viewing. Coons continued at Harvard during 1967/68, now as a Research Associate in Computer Graphics, was back on the MIT campus in 1968/69, and announced in late Fall of 1968 that he had been offered a much-sought-after (and deserved) promotion to full Professor at the University of Syracuse in their Systems and Information Sciences, and that he would leave MIT effective June of 1969. During his Syracuse tenure, Professor Coons also conducted research in computer science in Budapest, Hungary, under the sponsorship of the Hungarian Academy of Sciences. Subsequently, he joined the faculty at the University of Colorado in Boulder, where he died on August 19, 1979, at the age of 67. In his obituary in *Technology Review* [38], Steven Anson Coons was characterized by Professor Mann as "an endearing, modest gentleman, of great imagination and inquisitiveness who was loved by all who know him, but especially by his students, faculty colleagues and industrial collaborators."

The summary description of the Mechanical

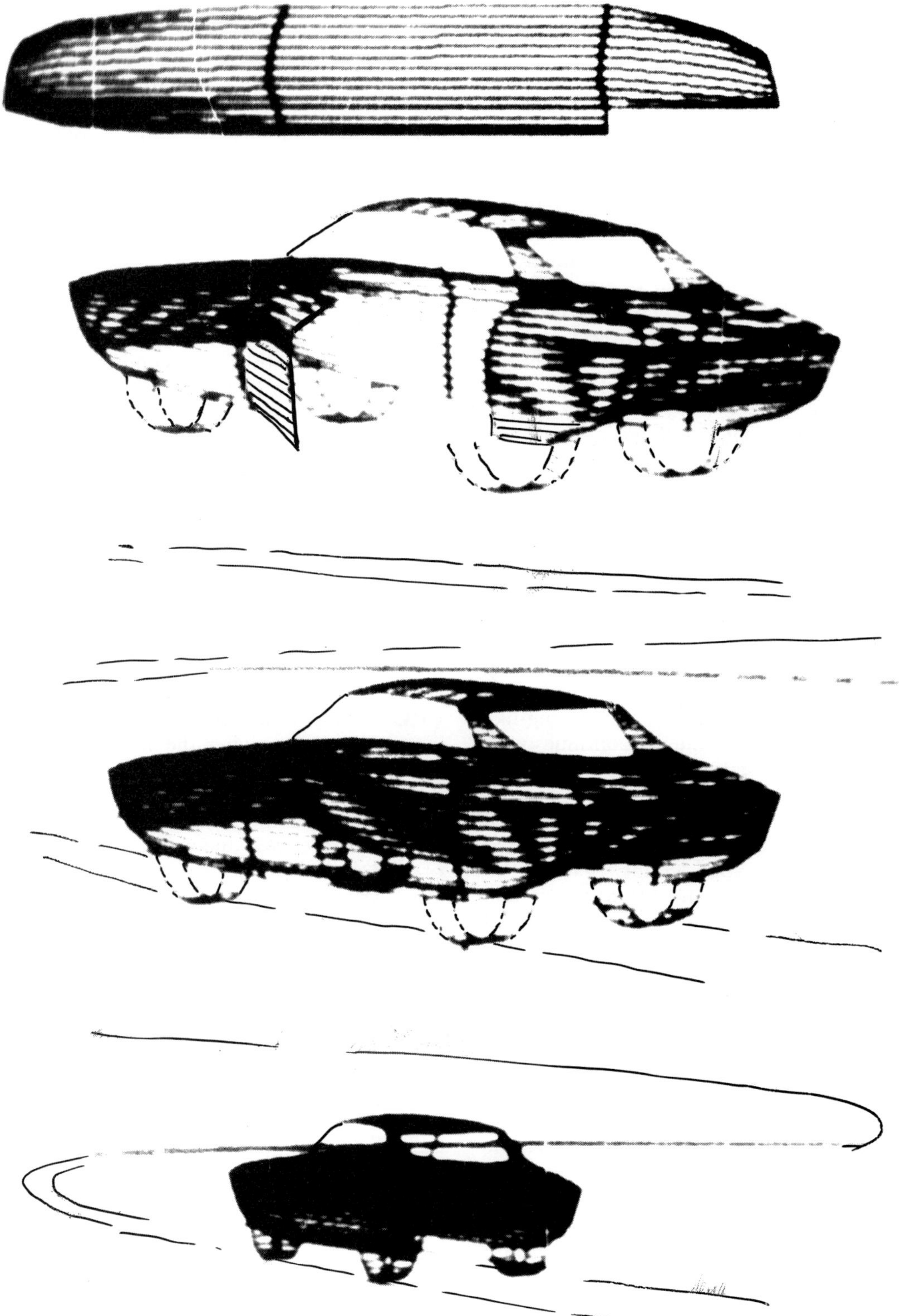

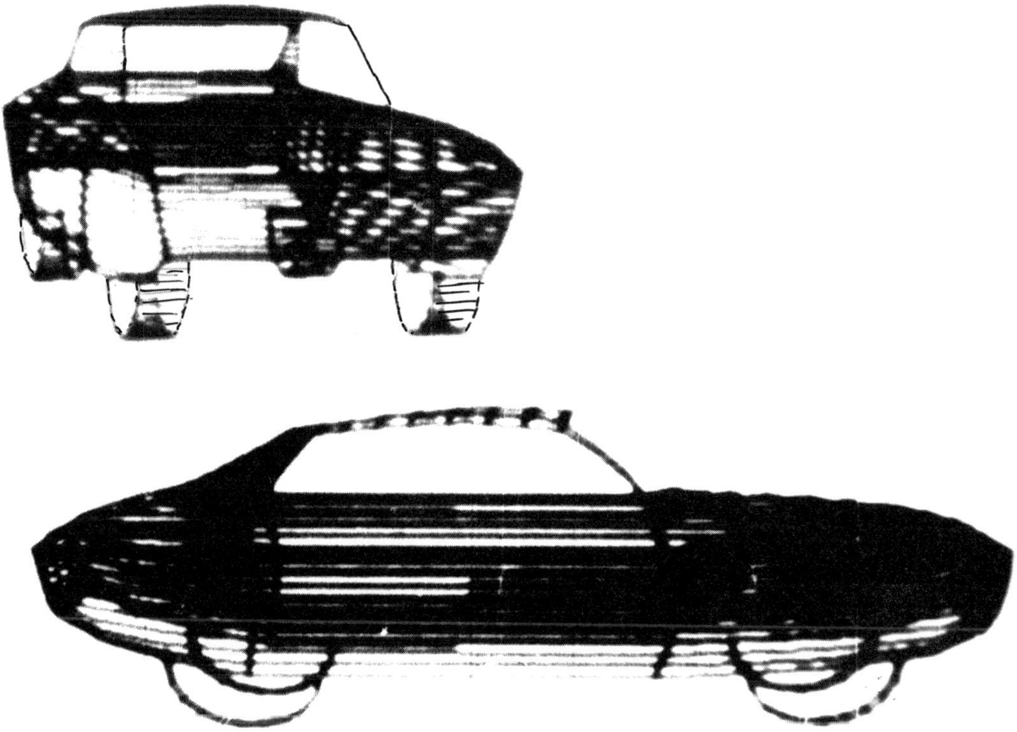

Figure 16.13. Composite of movie frames from film showing CAD design and visualization of automobile body (courtesy Ford Motor Company).

Engineering effort in computer-aided design is captured in an article published in the *1965 McGraw-Hill Yearbook of Science and Technology* [20]. The rationale for CAD is presented in the context of a description of the overall engineering design process, including how resource commitments are made, over time, in design. The chapter stresses the ME group's commitment to an interactive (computer *hyphen* aided) process with the machine's unsurpassable memory, speed and accuracy complementing the human's unique abilities in goal setting and evaluation, with both the human and machine exploiting different kinds of associative memory. To describe this relationship in Steve Coons' own words, from the transcript of a talk he gave on computer-aided design at MIT in December 1965,

> "It is natural to expect that if computers can behave themselves in a quasi-intelligent way they ought to be able to act as assistants to people who are engaged in creative tasks (like engineering) where a great deal of hard, tedious, complicated detail, must be kept track of and dealt with; where elaborate calculations must be performed with speed and precision. Such an assistant could perform all of the non-creative tasks, leaving the engineer free to perform the creative tasks that people do so well."

Like most innovators enamored of their creations, in the 1965 article [20], the authors speculated on the impact of CAD on design across a spectrum from engineering through the arts, to education, with due regard for socio-economic implications. Given the progress they had experienced in half a decade, they predicted that in 'no more than one more decade' computer-aided design, in its broader sense, including the capability for analyses of all kinds, standard parts selection, etc., would be fully operational. That expectation was premature, delayed by limited computer power and inadequate graphic displays. Even when these became available, for an extended period computer-aided design was synonymous with computer-aided drafting. Only now, over a quarter century after the termination of the engineering design group's computer-aided design research, can it be said that full-fledged computer-aided design is reaching the capability they envisioned.

REFERENCES

1 **F. J. Reintjes** *Numerical Control,* Oxford University Press, 1991.

2 **D. T. Ross** A Philosophy of Computer-Aided Design. This volume, pp 397–409.

3 **K. C. Redmond and T. M. Smith,** *Project Whirlwind: A Case History in Contemporary Technology,* Boston: The MITRE Corporation, November 1975.

4 **S. A. Coons** Graphical and Analytic Methods as Applied to Aircraft Design, *J. of Eng. Education,* Vol 37 No 10, June 1947.

5 **J. T. Rule and S. A. Coons** *GRAPHICS,* McGraw-Hill, NY, 1961.

6 **S. A. Coons** Cross Ratio in Elementary Engineering Drawing, *J. of Eng. Drawing,* February 1950.

7 **S. A. Coons** Conic Constructions from the Projective Viewpoint, *J. of Eng. Drawing,* February 1953.

8 **S. A. Coons** N-Dimensional Descriptive Geometry, *J. of Eng. Drawing,* Vol 18 No 2, May 1954.

9 **S. A. Coons** Operational Symbolism in Graphical Processes, *J. of Eng. Drawing,* Vol 21 No 3, 1957, 27–39.

10 **S. A. Coons** Research in Graphics, *J. of Eng. Education,* Vol 48 No 2, 1957, 157–160.

11 **S. A. Coons** Graphics in Research, *J. of Eng. Education,* Vol 50, January 1960, 330–334.

12 **S. A. Coons** The Future of Engineering Graphics – from the Faculty Viewpoint, *The Journal of Graphics,* Vol 26, May 1962, 22–25.

13 Report on Engineering Design, *J. of Engineering Education,* Vol 51 No 8, April 1961.

14 **R. W. Mann** Engineering Design Education: US Retrospective and Contemporary, *ASME Journal of Mechanical Engineering,* Vol 103, October 1981, 696–701.

15 **D. T. Ross** *Computer-Aided Design – a Statement of Objectives,* Cambridge, MA, MIT Electronics Systems Laboratory Report, # 8436-TM-2, 1960.

16 **S. A. Coons and R. W. Mann** *Computer-Aided Design Related to the Engineering Design Process,* Cambridge, MA, MIT Electronics Systems Laboratory Report, #8436-TM-5, 1960.

17 **L. E. Johnston** *Graphical Communication with a Digital Computer,* SM Thesis, Dept. of Mechanical Engineering, MIT, May 1961.

18 **J. A. Verderber** *Graphical Input-Output Devices for Computer Aided Design,* SB/SM Thesis, Dept. of Mechanical Engineering, MIT, June 1961.

19 **I. E. Sutherland** *Sketchpad, A Man-Machine Graphical Communication System,* PhD Thesis, Dept. of Electrical Engineering, MIT, January 1963.

20 **R. W. Mann and S. A. Coons** Computer-Aided Design, *McGraw-Hill Yearbook of Science and Technology,* McGraw-Hill, NY, 1965.

21 **R. W. Mann** Computer-Aided Mechanism Design, *ASME Mechanisms Conf.,* Lafayette, ID, October 1974, 1–4.

22 **W. Strubble** Computer-Aided Design System, *Technology Review,* March 1963, 27–28.

23 **T. E. Johnson** *Sketchpad III, Three Dimensional Graphical Communication with a Digital Computer,* SM Thesis, Dept. of Mechanical Engineering, MIT, May 1963.

24 **R. P. Parmelee** *A Study of a Stress Analysis Facility for Computer-Aided Design,* SM Thesis, Dept. of Mechanical Engineering, MIT, May 1961.

25 **R. P. Parmelee** *Three-Dimensional Stress Analysis for Computer-Aided Design,* PhD Thesis, Dept. of Mechanical Engineering, MIT, April 1966.

26 **P. T. Bennett** *Automatic Strength of Materials Routines for Computer-Aided Design,* SM Thesis, Dept. of Mechanical Engineering, MIT, January 1963.

27 **J. D. Purvis Jr.** *An Investigation of a Standard Parts Selection Facility for Computer-Aided Design,* SM Thesis, Dept. of Mechanical Engineering, MIT, August 1961.

28 **R. E. Starzec** *A Scheme for Information Storage and Retrieval,* SM Thesis, Dept. of Mechanical Engineering, MIT, June 1966.

29 **S. A. Coons,** *Surfaces for Computer-Aided Design of Space Forms,* Project MAC, Cambridge, MA: Massachusetts Institute of Technology, June 1967.

30 **M. L. Hamilton and A. D. Weiss** *An Approach to Computer-Aided Preliminary Design,* SM Thesis, Dept. of Mechanical Engineering, MIT, September 1964.

31 **S. A. Coons** An Outline of the Requirements for a Computer-Aided Design System, *AFIPS Conf. Proc., Vol 23, Spring Joint Computer Conf.,* Detroit, MI, 1963.

32 **I. E. Sutherland** Sketchpad: A Man-Machine Graphical Communication System, *AFIPS Conf.*

Proc., Vol 23, Spring Joint Computer Conf., Detroit, MI, 1963.
33 **T. E. Johnson** Sketchpad III: A Computer Program for Drawing in Three Dimensions, *AFIPS Conf. Proc., Vol 23, Spring Joint Computer Conf.*, Detroit, MI, 1963.
34 **R. W. Mann** Engineering Specifications for a Man-Computer System for Design, *University of Michigan – University of Detroit Engineering Summer Conf., Application of Computers to Automated Design,* August 1964.
35 **R. W. Mann** Enhancing the Availability of Braille, *Proceedings of the International Congress on Technology and Blindness*, New York, Vol 1, 1963.
36 **R. W. Mann** The Evaluation and Simulation of Mobility Aids for the Blind, *The Research Bulletin,* The American Foundation for the Blind, October 1965, 93–98.
37 **R. W. Mann and R. A. Rothchild** An EMG Controlled, Force Sensing, Proportional Rate Elbow Prosthesis, *Symposium on Biomedical Engineering,* Milwaukee, WI: Marquette University, Vol 1, 1966.
38 *Technology Review,* December/January 1980, A27.

ABOUT THE AUTHOR

Robert W. Mann joined the MIT Mechanical Engineering faculty in 1953, was appointed Germeshausen Professor in 1970 and then Whitaker Professor of Biomedical Engineering in 1974, an endowed chair he continues to occupy. A Brooklyn, New York, vocational high school graduate, he was employed as a draftsman at the Bell Telephone Laboratories in New York City before and after World War II military service in the Signal Corps in the Southwest Pacific arena. He matriculated at MIT as a freshman on the G.I. Bill in February of 1947, received his Bachelor's degree in Mechanical Engineering in 1950, his Master's in 1951 and his Doctor of Science in 1957.

At MIT through the 1950s, he conducted research and development on air-to-air missile components with an emphasis on internal power systems, including the power supplies for Sparrow I and III, subsequently applied in the Hawk ground-to-air missile. He became conversant with early digital computers and graphics displays through his wife Margaret Florencourt Mann, a charter member of Project Whirlwind, and his undergraduate roommate, Kenneth H. Olsen. Combining his draftsman and design experiences with this computer awareness, Professor Mann inaugurated the Mechanical Engineering Department contribution to the Computer-Aided Design project at MIT as described in this volume.

Throughout his youth experiences in crafts, his high school education, his military service, his preprofessional employment and his MIT education and research, Mann's *raison d'être* has been and is engineering design. Thus, as a pedagogue, he has been committed to involving engineering students in the design process. In the missile period, over 60 theses, mostly graduate, contributed to the success of the hardware developed. As the article in this volume describes, the ME CAD project was performed by graduate students supervised by faculty. When restricted computer access and limited graphics capability precipitated a search for a new research area satisfying his dual criteria of design opportunity and student involvement, technology to ameliorate the disabilities resulting from physical handicaps looked promising. Through the 1960s, Mann organized and led a group addressing the communication, mobility, recreational and vocational needs of blind persons. Outcomes included English-to-braille computer translation systems, now routinely applied both in the mass production of braille and on personal computers, electronic travel aids for blind travellers, and a number of recreational and vocational devices.

His rehabilitation technology expanded to embrace the musculoskeletal system with research that demonstrated the feasibility of Norbert Wiener's cybernetic notion to use brain signals to control a prosthesis; Mann's research became manifest in the 'Boston Arm', a servo-mechanism joint for the above-elbow amputee controlled by minute electrical signals originating in the amputee's brain and expressed in his otherwise useless stump muscles. Over the intervening years, other musculoskeletal disabilities have been and are being addressed in the Newman Laboratory for Biomechanics and Human Rehabilitation which Mann founded in 1975, and which he directs along with the Harvard-MIT Rehabilitation Engineering Center. Mann and five faculty colleagues supervise annually 35–40 Master's and Doctoral candidates with a cumulative thesis production as of 1990 in this lab of 47 Doctoral degrees, 136 Master's degrees, and 197 Bachelor's degrees.

Professor Mann is currently revisiting CAD in a project called Computer-Aided Surgical Simulation. His research explaining the remarkable tribology of human joints, which includes *in vitro* and *in vivo* instrumentation, made him aware of conservative orthopaedic surgical procedures, which because of difficulty in planning, produce mixed outcomes. CASS sits the surgeon before a workstation presenting a three-dimensional reconstruction from CT or MRI scans of the relevant anatomy, and permits the interactive simulation of bone cutting, realignment, reattachment, and test of efficacy of alternative surgical procedures, with the surgeon entering the operating room with a well-conceived and validated plan.

Professor Mann's contributions have been acknowledged at MIT through the award of two endowed chairs and his designation as the James R. Killian, Jr. Faculty Achievement Awardee in 1983–84, established to 'recognize extraordinary professional accomplishments of full-time members of the MIT faculty'. That same year, Mann also served as President of the MIT Alumni/ae Association, the first faculty member to so serve this century.

Externally, he has been recognized by election to the National Academy of Sciences, the National Academy of Engineering and the Institute of Medicine, and is a Fellow of the ASME, IEEE, the American Academy of Arts and Sciences and the American Association for the Advancement of Science. He has received the Gold Medal of the ASME, was the inaugural recipient of the ASME Lissner Award in Biomedical Engineering, and has been honored by many other awards and recognitions. He is the author and co-author of 213 publications, and inventor on four patents.

17 A Philosophy of Computer-Aided Design

Douglas T. Ross

Computer-Aided Design. What does it mean? By that I don't mean what does that *term* mean, as a *name* – but what does its *meaning* mean? What is Computer-Aided Design *as an actuality* – regardless of how its name may be defined with reference to other named things? What is it that *changed* about the world, when Computer-Aided Design became a reality? – for we didn't *always* have Computer-Aided Design, but we do now. What *is* it really? And how did it get here? – and when?

Either I or Dwight Baumann probably originated the term, but I can't be certain of that. The term appears in both of our write-ups from the 1959 discussions between the MIT Electronic Systems Laboratory and the Mechanical Engineering Department preceding the formation of the MIT Computer-Aided Design Project. It was brand new then, for both of our write-ups give more prominence to the 'computer aids to design' term, which I later fought hard to suppress because it relegated the computer to being only a *tool* for, rather than a *partner* in the design process. I *am* sure, however, that I was the first to use it with most of today's *modern* meaning, almost 30 years ago. But you can be the judge of that, for the body of this article reprints my **Statement of Objectives** for the MIT Computer-Aided Design Project, written in the summer of 1960. That decade-long project was a direct outgrowth of the preceding APT Project (1956–1959), which introduced the Automatically Programmed Tool (APT) Language and System, still the international standard for numerical control programming – now with greatly expanded capabilities. APT, itself, was preceded by the Numerically Controlled Milling Machine (NCMM) Project (1949–1955), which originated the field, itself, and overlapped with my earliest work at MIT on man-machine systems, as part of which I carried out the *first graphic input* to a computer (1954) by handwriting my name on a Whirlwind computer output scope, well before the days of the lightpen. I already have published detailed accounts of those beginnings [1,2], but on this occasion I prefer to let my original words speak for themselves. Except for the omission of some Summary sections, and minor reformatting of headings, what follows is my 1960 expression [3], verbatim. Its opening chapter bore the same title as this paper and began:

Introduction

On December 1st, 1959 under sponsorship of the Manufacturing Methods Division, Air Material Command of the United States Air Force, the Computer Applications Group of the Electronics Systems Laboratory, MIT, in cooperation with a group from the Mechanical Engineering Department, MIT, officially began an investigation of the possibilities for computer-aided design. This memorandum is an attempt to set down, in more or less explicit form, the type of man-machine system which is the goal of the project, and to indicate some of the fundamental reasons for the establishment of a goal of such magnitude. Descriptions of progress and achievements will be found in Interim Reports issued every six months by the project and in Technical Memoranda and Reports published as major phases of the work are completed. Undoubtedly the philosophy which is espoused here will be subject to many modifications as the work proceeds, so that this memorandum should not be considered to be a definitive specification for the work to be done, but rather an indication of the approach which will be followed.

What is not *Computer-Aided Design*

It is very difficult to define what is meant by computer-aided design, since the complete definition is, in fact, the sum and substance of the total project effort which has only begun. It is much easier to describe what is *not* computer-aided design as we mean it. So we begin by describing several ways in which computers have already been employed in design – work which we do not intend to duplicate or mimic. In the final analysis, after a complete definition of what we do in fact mean by

Fundamental Developments of Computer-Aided Geometric Modeling
ISBN 012-554765-X

computer-aided design has been shown and demonstrated through an actual working system of some sort, almost all of these ways of using computers in the design process, which we now disown, will then be incorporated and used in one form or another. These are all useful techniques and ones well deserving of further study, but all seem to be subsidiary to the more elaborate problems to which we wish to address our efforts. What, then, are these things that computer-aided design is *not*?

Computer-aided design is *not* automatic design, although it must include many automatic design features. By automatic design we mean design procedures which are capable of being completely specified in a form which a computer can execute without human intervention. There is a growing number of examples of automatic design programs of this type for designing electrical motors, transformers, electronic circuit layouts, cut-and-fill procedures in civil engineering, and many others. In all these systems there is an essentially fixed design or class of designs which are already completely worked out. The problem which is solved by the computer program is that of varying the parameters of the generalized design to match the constraints imposed by a particular application. Programs of this type can have a very important economic advantage over manual solution of the same problem, and in some cases can be used to yield optimum designs rather than ones which are merely satisfactory or adequate. Automatic design procedures of this type, however, offer little opportunity for modifying the basic features of a design and in general are applicable only to very restricted design areas.

Computer-aided design also is not automatic programming, although automatic programming techniques must necessarily play an important role in computer-aided design. Automatic programming systems such as APT have the features of flexibility and adaptability in problem specification by the human which are found wanting in the automatic design area. However, automatic programming systems usually represent only an aid in machine coding, and in general contain very little of a programming, problem-solving, or design nature. In an automatic programming system the human is still left with the burden of solution, which is why automatic programming cannot be considered computer-aided design.

Thus we see that automatic design has the computer do too much and the human do too little, whereas automatic programming has the human do too much and the computer do too little. Both techniques are important, but are not representative of what we wish to mean by computer-aided design.

Before launching into a description of what we think computer-aided design ought to mean, let us mention one other thing which it should not be. Computer-aided design should *not* just be a mirror of present design practice, although it should, or course, be related to it. Present design practice has many forms, and the art of design is practiced differently in all fields and in all industries. Many features of present design practice are undoubtedly of immortal worth and are good whether or not the computer is employed. Others, however, can quite probably be replaced by more efficient, more natural, more expressive techniques which are made possible only through the combining of man and machine into a design system. Therefore, computer-aided design, whatever its ultimate meaning may be, is not the same as our present understanding of design as it is practiced today.

Design versus problem-solving

Is design a special case of problem-solving or is problem-solving a special case of design? This is a moot question; for it is possible to argue that one only works out a design as a solution to a problem, or one can argue equally strongly that the problem-solving portion of the solution to a problem has all of the characteristics of creativity, esthetics, and adjustment of relevant factors which are normally associated with design. In this discussion we shall use the two terms almost interchangeably. In one instance the terminology of problem-solving may be more conducive to agreement, while in another, the terminology of design may yield a more lucid exposition. It is quite important to recognize that these terms are purposefully being confused. The appropriate quote from Webster's is "to plan mentally; to outline, to scheme; – distinguished from execute". In Webster's this definition is listed under *design*. It should be borne in mind that we wish to apply it also to problem-solving. As soon as the primary object of a procedure ceases to be scheming, and changes to execution or carrying out of schemata, then the problem-solving or design terminology may no longer properly be applied.

At this point, it does not seem advantageous to belabor further the definition of design and problem-solving as individual words. Any attempt to be more explicit would be ill-suited to our purpose. Extended discussion of specifics would be patently specious, since our primary purpose here is not to propound proposed dogma, but rather to express a philosophy in a general but compelling way. Bearing these thoughts in mind – the nebulous difference between scheming and executing schemata, and that the discussion is of necessity philosophical in nature – we may now resume our pursuit of the elusive meaning of computer-aided design.

To solve or to state?

We begin with a paradoxical twist. We have just finished pointing out the essential equivalence of design and problem-solving, which would seem to indicate that we were going to turn our attention to the solution of problems. Instead, however, we now declaim that our main objective is not to *solve* problems, but to *state* problems.

In actuality, this apparent exercise in mental gymnastics is well founded. At the grand level in which we want to consider problems, we don't really know *what* problem to solve in the first place. Take, for example, any cogent design problem such as designing an airplane. Admittedly, there are many very difficult problems associated with the design of an airplane, many of which can be solved with the aid of a computer. But if all we say is 'I want to design an airplane', there is no possible way of proceeding. Such a situation is reminiscent of certain instances of childhood in which a youngster will express vehemently the demand that some want be satisfied, without adequately describing the difficulty. Whether the designer is computer-aided or people-aided, he must certainly be more explicit than this.

This contention that the primary objective is not to solve problems but to state problems is further supported by the fact that we know many problems that we don't know how to solve. We also know many problems we can 'solve' in many ways with or without computer programs. In a particular case, we don't know which solution method to use. Furthermore, we don't want to be restricted to solutions we already have even though they at first seem adequate. A new problem environment may stimulate better solutions. In all of these instances the basic inadequacy – the first problem to be solved – is to achieve an adequate statement of our problem.

Outside-in problem statement

The *manner* of stating problems is also of primary importance. We must be able to state problems from *outside-in,* not *inside-out*. The normal practice of being explicit about a problem is to build a description of the solution procedure in small steps, i.e. to code the problem. This is particularly evident if a computer is being used, but is also true in activities among people when those activities are organized into a well-structured system. In other words, whether a 'machine' to assist in solving problems is composed of mechanical and electronic hardware or of people with specialized training and talents, the method of employing this machine in the solution of the problem is essentially the same. The solution procedure must be built up step-by-step to achieve the desired result. The most complete statement of the problem is, in fact, its solution procedure.

Stating a problem step-by-step from the inside-out in this manner is satisfactory if the problem is well understood to begin with. But there is an alternate procedure which is frequently used among people, but is virtually nonexistent as regards computers. In this alternate procedure. which we call stating a problem from the outside-in, we state the problem originally in general terms, and then explain further and in more and more detail what we mean. In the case of designing an airplane, for example, we might add a general statement such as 'It must be a high-performance commercial aircraft'. As the specification of the design proceeds the general terms 'high-performance' and 'commercial', and possibly even 'aircraft' will be elaborated further.

It is quite apparent that this procedure of stating a problem from the outside-in by means of further and further refinements is the only feasible way to attack grand problems. The inside-out method of stating problems, which is normal practice with computers, cannot possibly be carried out until a problem is stated in sufficiently detailed form, and this statement itself can come only from an outside-in consideration of the problem. Normally, this preliminary outside-in study of the problem must be carried out entirely by people. A principal aim of computer-aided design as we mean it is to permit the computer to play a part in this scheming portion of problem-solving by providing a mechanism for performing outside-in problem statement.

The necessity for idealization

What is a problem? If we are to design a procedure for stating problems from the outside-in (or even from the inside-out) we should first have a firm idea of what a problem consists of, so that our efforts may have a solid foundation. What, then, *really* is a problem? A seemingly irrefutable argument may be made that if we really knew what a problem is, in a profound way, we would have no real problems. The argument returns again and again to the proposition with which we are already laboring – that the statement of the problem is of paramount importance. Therefore, we will not attempt to invent a rigorous definition of 'problem', since any such definition which could be stated precisely would of necessity be so restrictive as to leave us playing games of little consequence.

If we are unwilling to define precisely what we mean by problem, it might seem that a reasonable substitute would be to be specific about the kinds of problems we wish to consider. This, too, is not an appropriate path to follow. Initially, we don't care about detailed solutions to problems, nor even tightly circumscribed problem areas. We can only get to setting up explicit sub-problems at the very end of the problem statement sequence. As far as the outside-in problem statement sequence is concerned, an abstract or idealized problem is treated in the same way as a detailed practical problem, and in fact is more pertinent to the study. We only arrive at explicit problems after a long series of hazy, incomplete, ambiguous refinements of the original goal statement, which taken together acquire precision, so that our main emphasis must be on the consideration of problems which are in a sense abstractions or idealizations of practical problems.

Neither of the terms 'abstraction' or 'idealization' really conveys the concept which is important for our study. We do not wish pertinaciously to insist upon the abstract or the ideal due to any whimsical misconception of our role, or of the magnitude or importance of our task. The plain fact of the matter is that problems being what they are, the outside-in form of problem statement having inherently the characteristics it possesses, we have no alternative but to concentrate our attention on things which are poorly, incompletely, and for the most part actually incorrectly defined, if they are considered relative to a 'complete and definitive' problem statement. In the sense that these things do not accurately mirror the real world of our problem, they are abstract or ideal, but it must clearly be recognized that these traits are not arrived at by choice, but are forced upon us by circumstances. The very reality of our problems forces them to be idealized. We will later see how concomitant idealizations can yield solutions to very real, practical, and difficult problems.

Reprise

We now appear to have worked ourselves into a position which is even worse than that from which we started. Originally we were concerned with a description of what we meant by computer-aided design, a topic which admittedly needed some discussion, but one which appeared to be reasonably concrete and tangible. Now, however, by insisting on outside-in problem statement and by being painfully honest about how little we know about problems in general, we find ourselves concerned about things which are not really abstract, not really ideal, but are nebulous and incomplete, ambiguous and ill-defined.

This very lack of form, however, this very chaos, which seems to limit and almost preclude any chance of success is, in fact, our principle strength. By discarding all pretense concerning our ability to force tractability on basically intractable problems, we must consider the world of problems in its barest essentials, in terms of what it really is. We admit freely that we cannot define succinctly what a problem is, but we admit to no such incapacity with regard to the elements out of which problems are composed.

Representing the structure of problems

Our goal at this point in the discussion is to devise a scheme for representing the elements out of which problems are composed. All substantive problems are internally structured. We recognize that they are made up of sub-problems, which in turn have internal structures of their own. The sequence of sub-structures is terminated finally in some elemental quantities which are intimately related to the particular aspect of 'reality' with which the overall problem is concerned. We wish to devise a scheme, (a mathematical model, if you will), which we will consider in terms of a computer structure, for manipulating problem elements. Problem elements may be of arbitrary form, and our primary objective is to have a computer *structure* capable of expressing relationships between general objects in a natural, non-artificial way.

The initial object for any particular problem will be the general goal statement. Successive objects will be the constituents of subsequent refinement or explanation statements concerning the problem. Since the problem elements with which we must deal are so varied, we prefer not to attempt to convey what we mean by attaching very specialized meanings to existing words such as *element* and *object,* and instead we borrow the French word *objet,* meaning thing or object, and assign specialized meaning to this distinctive but mnemonic word.

Objets are merely 'things' with properties we will define. Some properties are measurable and quantitative, some properties are qualitative, some properties are expressions of relationships with other objets. Objets may range in meaning from very concrete, physical things, to the utmost in abstraction. Objets may, themselves, be properties of other objets and conversely any specific property of an objet may itself be considered an objet.

In general, an objet will be some entity of *meaning,* the meaning being made implicitly explicit through the list of properties which relate the objet to measuring scales or other objets. Thus an objet is literally anything you wish it to be. There is no standard or stock procedure for defining the properties of an objet. One merely describes how an objet is related to already established things such as units of measure, colors, qualities such as big and small, good or bad, or in terms of other objets, whether they have previously been defined or will be defined subsequently. The total combination of statements about an objet constitute the definition of the objet. The computer structure we seek must be capable of representing all of these complex interrelationships in an efficient and workable fashion.

A flexible language structure

Equally important as a primary objective is a natural *language* for expressing statements concerning objets. The way in which the computer structure will be set up and adapted to a particular problem will be the result of a sequence of statements made by the human concerning the problem. As a matter of fact, the language will represent the computer structure of the human and will be parallel, or in a sense isomorphic, with this structure. Thus each objet has a corresponding word, phrase or other means of referencing the objet from the language, and this word or phrase obtains its definition in turn from the property list of the objet, i.e. the word and the objet are equivalent representations of the same 'thing'. The human understands the thing in terms of the language representation; the computer understands the same thing in terms of the property lists stored in computer structure.

The language, itself, and the rules for its meaning and translation will be structured and defined by means of objets. In other words, the language is self-defining. It can express general relationships about general things, and in particular it can express its own rules and structure. The human's understanding of the meaning of the language is mirrored by a corresponding understanding on the part of the computer in the form of interrelated property lists among the objets of the language stored in the computer structure. Just as new objets are introduced as the various aspects of a problem are unfolded, so also the objet of the languages itself may be modified and altered through use. As long as the human and the computer maintain at all times a mutual understanding of the meanings of everything, nothing is lost and a great deal is gained by permitting the basic defintions of the language itself to change with the problem.

The combination of a language of this type with outside-in problem statement is equivalent to a language which can be continuously modified and molded to suit a particular problem. Successive refinement statements elaborate and particularize meanings of words already used, by amplifying the interrelationships between technical words representing the objects of the problem and the standard words represented by measuring systems. As words change meaning, translation procedures also change, since a translation procedure for understanding one meaning of a word may no longer be valid when a new meaning for that word is obtained by additional modifying relationships with other words.

AN OUTLINE OF SYSTEM FEATURES

The previous section has described a broad philosophy for combining the talents of a human and a computer into a man-machine system for the effective solution of complex problems. Such a philosophical discussion is necessarily concerned with intangibles. The type of man-machine system which we propose should actually be quite tangible, however, so that in this section we attempt to describe some of its features in more concrete terms, and show how they are related to those of present day automatic programming techniques. Once again, these remarks should be interpreted only as an attempt to indicate the type of system being proposed. The actual formulation of the system and the precise features which it will have will be the result of further project activity.

The problem-statement language

Initially, the system will look little different from any other modern automatic programming or compiling system. A basic set of routines will permit coding directly in machine language with free use of symbolic names for quantities and locations. A macro instruction capability will also be present so that arbitrary collections of statements in the language can be used to define new functions (possibly recursive) which can be used in the same way as basic machine instructions. The system will also permit ordinary algebraic notation to be employed, as in FORTRAN and other automatic coding systems. In general, all of the widely used techniques of modern programming will be incorporated into the system.

The principle difference between the new system and others will be internal, and the effects of this internal difference will not show externally until the system is fairly well developed. The entire system will be organized on the basis of objets with property lists, and therefore, all of the features described in the previous chapter will be available for modifying and expanding the basic language itself. By using the features such as algebraic notation and macro instructions, as described in the previous paragraph, which will be provided by the basic objets, it will be possible to define new objets which, when taken in conjunction with themselves and the basic objets, will permit essentially arbitrary expression of the language.

In particular, it will not be necessary to define an objet completely before it is used within a statement. This feature is already widely available with respect to symbolic addresses, in almost all programming systems, but in the new system the same feature will apply to operators, functions, and entire programs. If the computer is called upon to actually perform an undefined function, then it will stop and ask for a suitable definition before proceeding. Depending on circumstances, this definition may or may not be the final definition of the function. It will be perfectly possible to put in a quick approximation for testing purposes, and replace the definition later by one that is more complete.

In human languages, most words have many possible meanings all of which are listed in a dictionary, and the appropriate meaning is selected by considering the context in which the word is used. In the system it will also be possible to assign multiple meanings to individual words. For example, the word ADD may be used to add fixed-point numbers, floating-point numbers, vectors, or matrices, since the context within which ADD is used will automatically determine the appropriate corresponding computer program.

There are many other features which are planned and could be mentioned at this time. Each one taken by itself seems inconsequential, and many of them already have exact or at least very similar counterparts in other systems. The difference in internal organization of the new systems, however, will combine all of these small effects into an overall performance capability which should be a significant improvement over present day systems.

The language of control and experimentation

Since the new system is considered as a man-machine system and not just an automatic programming system, all of its features will be capable of being executed on-line from typewriter or pushbutton language control, as well as automatically from punched card or magnetic tape control. This philosophy has important consequences in terms of the point of application of statements made in the language. The same language which is used for stating and solving problems may be used for controlling the operation of the system itself. Thus, the full range of algebraic notation, machine coding, and arbitrary macro instructions as are used for creation of an object program may be applied also to the setting up of individual arguments for functions or operations. Thus, instead of laboriously figuring out the precise symbolic or numerical address of a quantity which is to be used as the argument of a function, it will be possible to merely describe a procedure by which the system itself can determine the appropriate location of the argument, since simplified special-purpose techniques are used.

An important aspect of problems of the degree of difficulty in which we are interested, is that it is extremely unlikely that the user knows all about his problem even at a very general level. He must learn about the problem as he attempts to state it and solve it. Much of this learning comes as a result of testing hypotheses. To assist this process materially, the entire sweep of the scientific method must be built into the system, in a sense. The user must be able to establish a controlled environment, set up an experiment, try test cases, analyze results, and modify whatever is appropriate – all by simulation on the computer. All of these various facets of the actual solution of the problem must be carried out efficiently and conveniently by statements back and forth in the language, in a conversation or discussion about the problem, between the man and the computer. Therefore, the system must be an operator system. Here again the operator features must be capable of either automatic or manual control without the use of restrictive or unnatural control techniques. In other words, the problem of setting up, executing, and analyzing results must be handled with the same facility as is available for programming of the object problem itself.

A COORDINATED PROGRAM OF RESEARCH

A philosophy in a vacuum is of little value. To be effective a philosophy must not only indicate the establishment of appropriate goals, but must also materially assist in the achievement of these goals. We, therefore, close with a few remarks showing how the espoused philosophy provides concrete guidance in the selection of appropriate project activities.

The efficacy of a cosmopolitan approach

It might at first appear that direct application of the philosophy to the selection of appropriate project activity would dictate that the project should concentrate exclusively upon the design and construction of the compiler and programming system so essential to its realization. At some appropriate future date, after the system was more or less fully developed and functioning, it could be applied to various problems of design, and computer-aided design would then be a reality. Although such an approach might lead to a functioning system of some sort fairly quickly, the resulting system in all probability would not be representative of what computer-aided design *ought* to be. Such a premature advance to a supposedly finished system would ignore completely the fact that despite the high-sounding phrases of the overall philosophy, there is much more still to be discovered about computer-aided design than we already know at this time. It would indeed be possible to commence immediately on an all-out effort to construct a complete system along the lines which we presently envision, but a much more sensible approach is to consider the actual construction of the computer system as only a part of an overall program of research aimed at a more complete understanding of computer-aided design.

Perhaps the best illustration of the validity of the contention that a system should be developed in conjunction with other project activities is that the philosophy as it stands today is in fact the end product of just such a situation. Both the broader aspects of the philosophy, as well as the detailed computer programming considerations which are now actively being pursued, are the result of innumerable influences from past activities of the Computer Applications Group.

Ideas, concepts, and new approaches cannot be generated spontaneously. They must result from the interplay of research goals with actual concrete problems of significance. By working with problems which are so complex and difficult that they cannot be solved acceptably with existing techniques, the generation of new approaches, or more frequently the modification of old approaches and then re-application with a slight twist, is fostered.

The importance of problem environment

There is little challenge and virtually no prospect of significant achievement in solving problems which are known immediately to lie well within the scope of existing techniques, and the same also applies to solving pseudo problems which are merely dreamed up to lend support to some otherwise untenable hypothesis.

Most of the problems which have been attacked by the Computer Applications Group have been of the type which fosters sound research. They have been difficult problems from the real world sufficiently beyond the capabilities of existing techniques to cry out for new techniques, and yet capable of being formulated in such a way that bridging from the old to the new was possible of achievement. The net effect of these past experiences has led almost directly to the present outline of a philosophy for computer-aided design. Our new task is to elaborate on this philosophy by maintaining a fertile problem environment in which to work and think.

What are the attributes which a suitable problem environment must exhibit? The problem environment must be varied but cohesive; it must be sufficiently difficult to be challenging, and yet not so difficult as to be unrealistic; it must be sufficiently detailed to stimulate concrete advancement, but broad enough to avoid fruitless endeavors to resolve parochial oddities. Above all the problem environment must be dynamic and vital; capable of being molded to suit various purposes; pliable enough to be compressed for sweeping generalized attacks, and yet able to be stretched taut for penetrating attention to detail without coming apart. Design and problem-solving are areas of sweeping scope, so that care must be exercised in selecting from them those aspects which will establish the proper environment conducive to worthwhile research.

An outline of project activities

The mandate of the project is to carry out research leading toward the formulation of a procedure for computer-aided design, with emphasis on mechanical design. It is hoped that the result of this effort can be blended with the APT System for automatic programming of numerically controlled machine tools to provide an efficient mechanism for going almost directly from the requirement for a machined part to the finished product. Although investigations will concentrate on design problems, in this rather restricted area, if other types of design appear more appropriate for the generation of ideas for handling certain types of problems, they will not be excluded.

On the basis of this mandate and the remarks of the preceding sections, project activity may be broken into five general areas:

1. Computer programming techniques.
2. Mathematical and analytical techniques.
3. Man-machine communication devices and language design.
4. Analysis of the design process itself.
5. Extension and expansion of the APT System.

We will now elaborate briefly on each of these general areas.

Under the heading of computer programming techniques would come the basic work of designing and programming the computer system itself. This topic has received sufficient discussion previously to require no further elaboration here. Also under this heading, however, would come investigations into computer application techniques which are peculiar to artificial intelligence and symbol manipulation programming. The problem of computer-aided design, at least when it is approached on the grand scale which is contemplated here, requires the elaboration and extension of these rather esoteric techniques to achieve satisfactory solutions. Many of the features of the compiler and computer system will be strongly influenced by considerations from these areas.

Investigations of mathematical and analytical techniques will be concerned primarily with procedures for solving some of the extremely complex problems of mechanical design in ways which are appropriate for computer solution. Special emphasis must be placed upon efficiency of computation and extensive use of approximations, since we are considering using these computational techniques in an essentially real-time application. Many of the computer programs which are now in use for stress analysis and other problems of design are far too cumbersome and expensive to operate to be considered for use on the incomplete or faulty designs which will be an inherent feature of a design system. It therefore seems probable that we will concentrate on techniques for getting good approximate results efficiently, and then incorporate more elaborate and exact techniques to be applied as a check to a final analysis of the resultant design. There is a remote possibility that an approach to problems of this type can be found which can be applied either coarsely at high speed or finely at low speed, so that the necessity for alternate techniques would not be required. Also under the heading of mathematical and analytical techniques will come more general problems of the efficient handling of computations of functions of many variables of differing dimensions, as well as development of generalized techniques for speeding the convergence of iterative solutions etc.

Since the computer-aided design system which is our goal is a man-machine system, the problem of obtaining a good transfer of ideas and information between the human and the computer is of paramount importance. In addition to typewriter and pushbutton languages, the system will employ picture languages, for both input and output, as well as physical manipulations of pictures. Initial developments will concentrate on output oscilloscope and lightpen input techniques because of their availability and versatility, but later investigations may branch out to consider other devices as appropriate. A great deal remains to be learned concerning the language form represented by picture devices and the correspondence of this language form with other language forms. All of these language considerations will, of course, have an important interplay with the developments of the computer programming system itself.

Although it is anticipated that development of a system for computer-aided design will introduce many techniques and approaches to design problems, it is nonetheless likely that the majority of the techniques used in the application of the system will be very closely related to, if not identical with, those that are used now in manual design. For this reason, the procedures used by mechanical designers, and problem-solvers in general, will be investigated to determine how the burden can be shared between man and machine, and also to act as a check that chosen techniques for computer-aided design are indeed suitable for actual design problems. Also, under this heading will come studies of the organization of design information as it is presently available. Many parts of design are determined not so much by mathematical and analytical considerations as the availability of standard items or the results of empirically determined test results contained in a designer's handbook. Consideration of these various forms in which information about design exists may very well influence in a fundamental way the organization of the overall computer-aided design system.

The extension and expansion of the APT System for the automatic programming of numerically controlled machine tools may at first appear to be a peripheral item which is considered in the overall project activities purely for historical reasons. The actual purpose of continuing these studies, however, is that the problems involved in expanding the APT System are a good instance of difficult problems which lie just beyond the reach of existing techniques, and yet which are well enough understood to yield many hints and suggestions for improving the state-of-the-art. Already there are indications that many of the computational techniques which have been pioneered in the APT System will be very useful in design work, and there is every reason to anticipate that this valuable cross-fertilization will continue. An additional important reason for continuing work on the APT System is that it provides an ideal outlet for results of the project. It is expected that many of the techniques which are worked out on the basis of computer-aided design will not be of practical use in design for several years at least, since to apply them to actual design problems may require as yet uncompleted portions of the system. These same techniques, however, may have immediate application in improving the performance of the APT System. Therefore, the ideas can be put to work and given extensive testing in the APT framework, even though in terms of design they may have been completed prematurely.

A final project activity which is planned and which has not been mentioned previously is the carrying out of a simple pilot study of a computer-aided design system. The precise area of application for this pilot study has not yet been decided upon, but it is hoped that a class of problem can be selected which will bring into focus all of the various aspects of the project in a simple way, so that experience can be gained with the different parts working together in a system. It is not intended that this pilot study be in any definitive way representative of what we will ultimately mean by computer-aided design, but in addition to providing a vital connecting link between the various project activities as they proceed, it will also serve a useful function in exposing our nascent ideas to public criticism.

The complexion of the project will undoubtedly change as the work proceeds, but it is hoped that this brief summary of planned project activities indicates how the broad philosophy of computer-aided design, which will continue to develop as physical realizations of its various aspects are completed, serves as a goal toward which the project can work. It is hoped that initial efforts in all these areas can begin to exert influences on each other at an early date, so that the evolution of both the philosophy and the actual man-machine system can proceed in an orderly fashion toward the desired result.

BACKGROUND

At the invitation of the Aircraft Industries Association (AIA), in 1957 the MIT APT Project, funded by the Air Force, had formed and led a massive cooperative development project for the first APT System, with most major companies participating, and had turned over further leadership responsibility to AIA at a major press conference in February 1959. (The present-day Computer Aided Manufacturing – International (CAM–I) organization ultimately evolved from that beginning.) I had written in the first APT Interim Report [4]:

> "... Conceivably, it [APT] can grow until all design and analysis for machined parts is included in the automatic programming system, and the APT Computer is programmed simply by describing the function which a machined part is to perform." (p. 10)

and we and the Air Force planned to refocus the MIT efforts toward that end, while continuing to support APT where we were needed. (Our last contribution to the AIA APT effort (1960/61) was a complete generalized Arithmetic Element (ARELEM) program for efficiently computing three-dimensional cut vectors of guaranteed tolerance for arbitrary tool paths defined by arbitrary surfaces machined by rotating cutting tools of arbitrary profile. To do that we first used the idea of a *modeling plex*, described below. We had no special language for this yet, using assembly language for our own work, and grotesque contortions with COMMON statements in Fortran for AIA use in programming APT, but the power for CAD was obvious from the beginning.)

The MIT Servomechanisms Laboratory (renamed Electronic Systems Laboratory (ESL), and now the Laboratory for Information and Decision Systems (LIDS)) in which the NCMM and APT Projects took place is an integral part of the Electrical Engineering Department at MIT. To include mechanical design, as such, into the picture, a new AF contract to ESL, with myself as Project Engineer, would subcontract to the MIT Mechanical Engineering Department Design and Graphics Division (led by Robert W. Mann, with S. A. Coons) to form the MIT Computer-Aided Design Project. Since the beginning of APT, my group, which worked on several other projects, as well, during those years, was called the Computer Applications Group, in ESL. So at its beginning, the MIT CAD Project had two groups approaching the subject from the programming and from the design perspectives.

Relevance to the present

Except that the practicing CAD, software technology, and computer science communities have settled on different terminology in many places, the philosophy espoused is fully up to date. 'Automatic programming' has fallen into disuse in favor of 'compiling', but the distinction that 'design' must embody *scheming* still distinguishes it from *processing* or *calculation* as subcategories of problem-solving. The focus on problem *statement* rather than *solution* is central to today's interest in *declarative* languages and requirements definition methodologies. Today's terminology of 'top-down', although easier to pronounce, seems to me to be *weaker* than my original 'outside-in' view in properly matching the *object-oriented* thrust, conceptually, in that the penetration to successively-deeper layers of detail leaves the object and its description *together,* whereas 'top-down' applies only to the description *separately,* and sometimes engenders confusion about which is the 'topmost top', whereas 'outside-in' *naturally* joins the *multiple viewpoints* appropriate to *any* interesting object *subjectively* viewed to highlight various interests.

The 'procedure of stating a problem from the outside in by means of further and further refinements' became the touchstone of 'stepwise refinement' when its power and generality were rediscovered in later years. But only recently has this been matched to 'object-oriented' methods and 'abstract data type' encapsulations of what I then called 'objet' – and even so, only the most recent efforts to endow *icons,* themselves, with behavioral properties matches the isomorphism between (graphical) word and object I had in mind. My *Algorithmic Theory of Language* (ATL) [5] used later in the Project, carried out the idea that "The language, itself, and the rules for its meaning and translation will be structured and defined by means of objets ... [It] is self-defining ... It can express its own rules and structure," as far as available hardware and software technology allowed. It should be revisited today, linking *knowledge modeling,* object *message passing,* and *expert reasoning* systems methodologies, with today's increased computer power.

The *Outline of System Features* presented was not mere wishful thinking nor totally visionary. Throughout the APT Project I worked with my staff on many research topics besides the AIA APT effort, and these fed later APT and CAD developments. Specifically, *Future Plans: Automated Design* [13], for the last half of 1958 outlines what later became the CAD Project, stating the need for design studies and "handwritten sketches and motions [with] lightpen input equipment", citing my 1954 graphical input method. It also cites the emerging *list processing* of IPL and LISP, then under development, "as a starting place for the work of this project. Detailed design is also already underway on a unique form of list structure which includes not only lists but also tabulated or ordered information for use in both the development of the action-verbal language as well as representing design problems themselves." Within the year I had coined the name *modeling plex* for the new form, matched by the elegant programming technique of *reversed index register* use (so that $A(I)$ references the *arbitrary-typed* value of the A component of an n-component element *pointed to* by I, as opposed to the standard array A of *fixed-type* values *indexed* by I) – what now is called *records* with *fields* [6].

The next Interim Report [7] describes the 'point-line diagram' study we did using LISP, which "As was expected ... proved to be both inflexible and inefficient for ... numerical calculation," but demonstrated the important concepts of *generic/specific elements* and *constraint satisfaction* critical to true CAD modeling. When he was introduced to them, some years later, these and other plex processing techniques stimulated Sutherland into a major reworking of his TX-2 programming (which, like most early CAD systems, in its first completed version reproduced the actions of a draftsman at a drafting table) to yield the final *Sketchpad.* The advanced interrupt handling and computing power of the TX-2 brought these important concepts to life, launching CAD to reality through the peripatetic proselytizing of Steve Coons in his many presentations.

The language developments of the project took on a life of their own. The first language to permit plex processing directly was that of the *Basic Bootstrap Compiler* (1960/61) [8], which began as a seed of 75 machine instructions plus some 300 words of stored tables, including a built-in symbol table. Generated code could augment the compiler dynamically, and by programming in the expanding language, a complete one-pass system with free format input and output, as well as conditional infix macro notation was built and was used to write the initial version of AED-0 – the first *software engineering language* and system [9]. Standing either for Automated Engineering Design or for Algol Extended for Design, AED-0 was so named because it was intended to be used only to build AED-1 (never begun because evolution was so rapid and useful). The early syntax was frozen and all development proceeded through semantic extensions of the language by *integrated packages* of routines – what now are called *abstract data types* – and by system-building systems known as *The AED Approach* (the first *compiler-compiler* technology) [10]. Lexical, parsing, macro, debugging, and other table-driven generators, with extensive libraries of modular building-blocks, yielded an integrated *software technology environment.* Engineered for machine-independent *portability* from the beginning, AED was used for dozens of major projects at MIT and in industry and government [11].

As to the MIT CAD Project, itself, it proceeded according to plan, except that other than AED development, no specific pilot study resulted from the project. The *Coons' Patch* for generalized 3D shape description is famous, and it was through Professor Coons that Ivan E. Sutherland (who had independent NSF Fellowship support, complemented by Lincoln Laboratory TX-2 Computer support) entered the picture with his notable *Sketchpad* thesis. In 1963 the CAD Project linked up with the ARPA-sponsored Project MAC, for time-sharing, and the ESL Display Group under John E. Ward was formed. The ELS Display Console ('The Kludge'), the first *real time 3D workstation* was the work of Robert M. Stotz. Charles A. Lang and Richard U. Bayles (who left to lead IBM's CP/CMS effort) handled the *time-shared display software* developments. CAD Project support of the ME Department work ceased in 1965, as all efforts were focused on the AED Project component, which in 1964 had begun another five-year, industry-wide AED Cooperative Program, with visiting staff working at MIT, stimulating the growth of many major company CAD organizations. The MIT CAD work ceased when we left to found SofTech, Inc. in the summer of 1969 [12]. Further related developments through SofTech were my origination of *structured analysis* (1973) in SoftTech's Structured Analysis and Design Technique (SADT^{TM}), and its use (as the IDEF-0 standard for *system modeling*) in the ten-year, DoD-wide ICAM (Integrated Computer Aided Manufacturing) program, again through an industry-wide coalition, initially led by SofTech.

Coda

As you can see, the 1960 philosophy was fruitful and has stood the test of time well. Many new ideas and techniques from many sources have, with today's massive memories and astounding speeds, expanded the theme to dazzling heights. Quite outside those exciting developments I have continued to deepen the philosophy, itself, which I now simply call ***Plex*** – a rigorous scientific philosophy enabling the understanding of reality, itself. Above, I said "we will not attempt to invent a rigorous definition of 'problem', since any such definition which could be stated precisely would of necessity be so restrictive as to leave us playing games of little consequence." Precisely the opposite is the case. The Plex definition (rigorously discovered, not invented) is:

A *problem* is a state of affairs
that violates some set of *requirements*.

–and a *solution* is one that does not. It turns out that in Plex *every* problem has a solution. Nature *has* no problems; it just *runs*. The meaning of meaning of meaning cannot be Nothing, which engenders the impossibility of impossibility. (The First Definition of Plex is *Nothing doesn't exist*. All of reality follows from that.) A truly grand problem, one befitting nature, itself, cannot properly be *posed* until its solution *also* is at hand. Until then the alleged 'problem' structuring of nature actually is Nothing – not real, and not a problem at all. But when all present possibility appears to be exhausted in the structuring of a *real* problem, it is the nature of nature to present a *new* possibility, and nature rushes on to the *next* such problem. That is the sequence of consequence, the only game that plays. That is the grand design. *Our* problems are man-made, not nature's. We need computer-aided design to state them well enough to solve them.

REFERENCES

1 **D. T. Ross** Origins of the APT Language for Automatically Programmed Tools, in *History of Programming Languages,* R. L. Wexelblat, ed., Academic Press, New York, 1981, 279–367.

2 **D. T. Ross** A personal view of the Personal Work Station – Some firsts in the Fifties, in *A History of Personal Workstations,* A. Goldberg, ed., ACM Press, New York, 1988, 51–114.

3 **D. T. Ross** *Computer-Aided Design: A Statement of Objectives,* Cambridge, MA: MIT Electronic Systems Lab. Rep. No. 8436-TM-4, September 1960.

4 **D. T. Ross and H. E. Pople Jr.** *Automatic Programming of Numerically Controlled Machine Tools,* Cambridge, MA: MIT Servo Lab. Rep. Nos 6863-IR-1 and 6873-IR-2, 1956.

5 **D. T. Ross** *An Algorithm Theory of Language,* Cambridge, MA: MIT Electronic Systems Lab.

Rep. No. ESL-TM-156, DDC No. AD296998, November 1962.

6 **D. T. Ross** A Generalized Technique for Symbol Manipulation and Numerical Calculation, *Commun. ACM*, Vol 4 No 3, 1961, 147–150.

7 **D. T. Ross** ***et al.*** *Automatic Programming of Numerically Controlled Machine Tools,* Cambridge, MA: MIT Electronic Systems Lab. Rep. No. 6873-IR-10 and 11, 1959.

8 **D. T. Ross, S. A. Coons** ***et al.*** *Investigations in Computer-Aided Design for Numerically Controlled Production,* Cambridge, MA: MIT Electronic Systems Lab. Rep. No. ESL-IR-138, ASD-TR-820 (IR 3 and 4), 1961.

9 **D. T. Ross** *Introduction to Software Engineering with the AED-0 Language,* Cambridge, MA: MIT Electronic Systems Lab. Rep. No. ESL-R-405, October 1969. (Also (1970) Waltham, MA: SofTech, Inc.)

10 **D. T. Ross,** The AED Approach to Generalized Computer-Aided Design, in *Proceedings ACM National Conference,* ACM Press, New York, August 1967, 367–385.

11 **D. T. Ross and J. E. Ward** *Investigations in Computer Aided Design for Numerically Controlled Production – Final Report,* Cambridge, MA: MIT Electronic Systems Lab. Rep. No. ESL-FR-351; AFML-TR-68-206, 1968.

12 **J. E. Ward** *Computer-Aided Design for Numerically Controlled Production – Final Report,* Cambridge, MA: MIT Electronic Systems Lab. Rep. No. ESL-FR-420; AFML-TR-70-78, 1967.

13 **D. T. Ross** ***et al.*** *Automatic Programming of Numerically Controlled Machine Tools,* Cambridge, MA: MIT Electronic Systems Lab. Rep. No. 6873-IR-8- and 9, 1958.

ABOUT THE AUTHOR

Douglas T. Ross is the founder and Chairman of the Board of SofTech, Inc., and since 1983 a Lecturer in the Electrical Engineering and Computer Science Department of the Massachusetts Institute of Technology (MIT). He served as the SofTech's President from its establishment until 1975, and remains its Principal Investigator. He started his career in 1951 as a Teaching Assistant in the Mathematics Department at MIT, and became Head of the Computer Applications Group in the Electronic Systems Laboratory in 1952. He was also a lecturer in the Electrical Engineering Department between 1960 and 1969. He received an AB degree in mathematics from Oberlin College in 1951, and an MS degree in Electrical Engineering from MIT in 1954.

Ross is the creator of the APT System for automatic programming of numerically controlled machine tools; the AED Language and Programming System, the first software engineering environment; Structured Analysis, a new technique bringing the disciplined approach of software engineering to the system analysis area; and the Plex Theory of problem modeling on which these achievements are based. He also worked in the field of on-line computer graphics and interactive languages. He is the author of one book and co-author of a second; he has written over 30 publications, and more than 200 reports, papers and presentations. His papers have received prizes at AFIPS and ACM conferences, and he won the Joseph Marie Jacquard Award from the Numerical Control Society in 1975.

Ross is a member of numerous societies including the Association for Computing Machinery (ACM), Sigma XI, the American Association for Advancement of Science and the Computer Society of the IEEE, and he has served on various committees of ACM, IFIP and NATO. He is on the Editorial Board of *Software: Practice and Experience,* and *Computers in Industry,* and his international recognitions include listings in nine Marquis *Who's Who* publications, *Who's Who in America,* as well as in *Men of Achievement.*

Index